CELL BIOLOGY

A Comprehensive Treatise

Volume 3

Gene Expression: The Production of RNA's

CELL BIOLOGY
A Comprehensive Treatise

Volume 3
Gene Expression: The Production
of RNA's

Edited by

LESTER GOLDSTEIN

DAVID M. PRESCOTT

Department of Molecular, Cellular and Developmental Biology
University of Colorado
Boulder, Colorado

1980

Academic Press
A Subsidiary of Harcourt Brace Jovanovich, Publishers
New York   London   Toronto   Sydney   San Francisco

ACADEMIC PRESS, INC.
111 Fifth Avenue, New York, New York 10003

United Kingdom Edition published by
ACADEMIC PRESS, INC. (LONDON) LTD.
24/28 Oval Road, London NW1 7DX

Library of Congress Cataloging in Publication Data
Main entry under title:

Gene expression, the production of RNA's.

   (Cell biology, a comprehensive treatise ; v. 3)
   Includes bibliographies and index.
   1. Gene expression.   2.   Ribonucleic acid synthesis.
3.   Genetic transcription.   I.   Goldstein, Lester.
II.   Prescott, David M. , Date   III.   Series.
[DNLM: 1.   Genetics, Biochemical.   2.   RNA––Bio–
synthesis. 3.   DNA––Biosynthesis.   QH574 C393 1977
v. 3]
QH574.C43   vol. 3   [QH450.2]   574.87s   80–10715
ISBN   0–12–289503–7   [574.87'322]

PRINTED IN THE UNITED STATES OF AMERICA

80 81 82 83   9 8 7 6 5 4 3 2 1

Contents

1 The Organization of DNA Sequences in Chromosomes

Christopher J. Bostock

2 Gene Reiteration and Gene Amplification

Adrian P. Bird

3 Basic Enzymology of Transcription in Prokaryotes and Eukaryotes

Samson T. Jacob and Kathleen M. Rose

4 Regulation of Transcription in Prokaryotes, Their Plasmids, and Viruses

Geoffrey Zubay

5 Structural Manifestation of Nonribosomal Gene Activity

J.-E. Edström

6 The Expression of Animal Virus Genes

Raymond L. Erikson

7 Aspects of Cytoplasmic and Environmental Influences on Gene Expression

Peter M. M. Rae

8 Molecular Aspects of the Regulation of Eukaryotic Transcription: Nucleosomal Proteins and Their Postsynthetic Modifications in the Control of DNA Conformation and Template Function

Vincent G. Allfrey

9 Maturation Events Leading to Transfer RNA and Ribosomal RNA

Gail P. Mazzara, Guy Plunkett, III, and William H. McClain

10   The Processing of hnRNA and Its Relation to mRNA

Robert Williamson

11   Recombinant DNA Procedures in the Study of Eukaryotic Genes

Tom Maniatis

12   Basic Characteristics of Different Classes of Cellular RNA's: A Directory

Yong C. Choi and Tae-Suk Ro-Choi

List of Contributors

Numbers in parentheses indicate the pages on which the authors' contributions begin.

Vincent G. Allfrey (347), The Rockefeller University, New York, New York 10021

Adrian P. Bird (61), Mammalian Genome Unit, Medical Research Council, Edinburgh EH9 3JN, Scotland

Christopher J. Bostock (1), MRC Mammalian Genome Unit, Department of Zoology, University of Edinburgh, Edinburgh EH9 3JN, Scotland

Yong C. Choi (609), Department of Pharmacology, Baylor College of Medicine, Houston, Texas 77030

J.-E. Edström* (215), Department of Histology, Karolinska Institutet, S-104 01 Stockholm, Sweden

Raymond L. Erikson (265), Department of Pathology, University of Colorado Medical Center, Denver, Colorado 80262

Samson T. Jacob (113), Department of Pharmacology, The Milton S. Hershey Medical Center, The Pennsylvania State University College of Medicine, Hershey, Pennsylvania 17033

William H. McClain (439), Department of Bacteriology, University of Wisconsin, Madison, Wisconsin 53706

Tom Maniatis (563), Division of Biology, California Institute of Technology, Pasadena, California 91125

Gail P. Mazzara (439), Department of Bacteriology, University of Wisconsin, Madison, Wisconsin 53706

Guy Plunkett, III (439), Department of Bacteriology, University of Wisconsin, Madison, Wisconsin 53706

Peter M. M. Rae (301), Department of Biology, Yale University, New Haven, Connecticut 06520

Tae-Suk Ro-Choi (609), Department of Pharmacology, Baylor College of Medicine, Houston, Texas 77030

Kathleen M. Rose (113), Department of Pharmacology, The Milton S.

* Present address: European Molecular Biology Laboratory, Postfach 10.2209, 6900 Heidelberg, Germany.

Hershey Medical Center, The Pennsylvania State University College of Medicine, Hershey, Pennsylvania 17033

Robert Williamson (547), Department of Biochemistry, St. Mary's Hospital Medical School, University of London, London W2 1PG, England

Geoffrey Zubay (153), Department of Biological Sciences, Columbia University, New York, New York 10027

Preface

The four volumes of this treatise deal with cell genetics. Volumes 1 and 2 were devoted to cell inheritance and the molecular basis of genetics. This volume extends the subject into the molecular and cytological basis of gene expression and begins with discussions of the organization of DNA and gene sequences in chromosomes as a prelude to the detailed coverage of gene expression. The prelude is followed by the dominant theme of the volume, the synthesis and processing of RNA molecules and the regulation of these events. Volume 4 will be devoted to the translation and behavior of proteins and will complete consideration of cell inheritance and gene expression.

Volume 3 continues the general objectives stated in the Preface to Volume 1. Each volume is designed to serve as a comprehensive source of primary knowledge at a level suitable for graduate students and researchers in need of information on some particular aspect of cell biology. Thus we have asked contributors to avoid emphasizing up-to-the-minute reviews with the latest experiments, but instead to concentrate on reasonably well-established facts and concepts in cell biology.

David M. Prescott
Lester Goldstein

Contents of Other Volumes

Volume 4   Genetic Expression: Translation and the Behavior of Proteins

1

The Organization of DNA Sequences in Chromosomes

Christopher J. Bostock

1

I. INTRODUCTION

A century of research has seen the development of the concept of the chromosome from "a colorable substance" of the nucleus to our present day view as the carrier of genetic information. Soon after the discovery that DNA could transfer genetically determined characters (Avery *et al.,* 1944), a number of biochemical and cytological studies (e.g., Swift, 1950a,b) pointed to the fact that, within a eukaryotic species, all the nuclei of all tissues contained the same amount of DNA, or simple multiples of it. The amount of DNA contained in a haploid nucleus is called the C value. Over the animal and plant kingdoms as a whole there is a general increase in C value as the evolutionary complexity increases. These two properties—DNA constancy and increasing C values—are what would be expected for a molecule that is the carrier of genetic information.

The simple notion that all the DNA encodes genetic information falters upon closer examination of C values, for there is considerable variation in the C values of organisms even within closely related groups, and occasionally evolutionarily complex organisms contain less DNA than their simpler counterparts. For example, within the single genus *Anenome* there is a fivefold variation in C value, and in the phylum Vertebrata C values range over two orders of magnitude (Rees and Jones, 1972). With DNA being the genetic material it is difficult to see why closely related organisms, with seemingly similar requirements for genetic information, should possess such widely differing amounts of DNA. This apparent contradiction has been called "the C value paradox" and lays open the possibility that not all DNA, in eukaryotes at least, is informational in the sense that it is transcribed and eventually translated into proteins.

In considering the organization of the sequences contained in DNA one must first identify what kinds of sequences are present. We know that there must be structural genes, both those that code for proteins and those that code for the various nontranslated RNA's. There must also be a fraction of the DNA turned over to gene regulatory functions, since we suspect that all the cells of an organism contain the same complement of DNA and yet display a great variation of differentiated types. As we shall see, it turns out that quite large proportions of the genome exist to which it is difficult to ascribe either structural or regulatory gene functions. The purpose of this chapter is first to examine some of the ways in which DNA can be analyzed, and then to consider the properties of the different fractions that can be identified, their occurrence in the chromosomes of different organisms, and finally their spatial and functional relationships in the linear order of the DNA molecule that constitutes the chromosome. The major emphasis will relate to eukaryotic chromosomes, although reference will be made to prokaryotic and viral chromosomes where appropriate.

II. METHODS OF ANALYZING DNA SEQUENCES

A. DNA Reassociation

The evidence which demonstrates the existence of DNA sequences that are repeated within the genomes of many organisms is based on the ability of complementary nucleotide sequences in single-stranded nucleic acids to anneal to form duplex structures. Such reassociation (in the case of homologous DNA–DNA reactions) or hybridization (in the case of heterologous DNA–DNA reactions) is essentially the reverse of DNA denaturation. When native DNA in aqueous solution is heated slowly, or subjected to alkaline titration, it will denature. Denaturation occurs when the hydrogen bonds, which hold the complementary single strands together, are broken, thus allowing the single strands to dissociate from each other. The process can be followed optically by measuring the increase in absorption at 260 nm (hyperchromicity) or biochemically using hydroxylapatite (HAP) binding to distinguish between single- and double-stranded DNA. The increase in absorption at 260 nm upon denaturation results from the fact that the component bases absorb less light when stacked in the DNA double helix than they do when in the single-stranded random coil configuration. Hydroxylapatite has the property of binding both single- and double-stranded nucleic acids at low molarities of phosphate buffer. Single-stranded nucleic acids can then be eluted from HAP in $0.12\,M$ phosphate buffer, the double-stranded nucleic acids being eluted at a concentration of $0.27\,M$ phosphate (Bernardi, 1965; Miyazawa and Thomas, 1965).

By plotting either increase in absorption at 260 nm or elution of single-stranded DNA from HAP, as a function of temperature, one obtains a denaturation or melting curve. The point at which the helix is half denatured is referred to as T_m. Measured by hyperchromicity, T_m is a few degrees below that measured by HAP chromatography. This is because hyperchromicity monitors the breaking of base pairs throughout the denaturation process, whereas HAP chromatography measures the release of single strands only after all base pairing with their complements has been destroyed.

The T_m is characteristic for a particular DNA and, if everything else is held constant, is dependent on the base composition of the DNA. Since $G \equiv C$ base pairs are more stable than $A \equiv T$ base pairs, DNA rich in

G + C will have a higher T_m than DNA rich in A + T (Marmur and Doty, 1962). T_m is also greatly affected by the experimental conditions, for example, increase in either the ionic strength or the molecular weight of the DNA will increase the T_m, but these can be measured and corrected for.

Knowledge of the thermal melting profile of a DNA is important for a correct understanding of reassociation experiments because it defines the optimal conditions under which reassociation should take place (generally 25°C below T_m). It can also help to establish the physical state of the reassociated duplex structure. A decrease in T_m over that of the native duplex (called ΔT_m) indicates some level of mismatching between the bases of the reassociated duplex. That is to say that the base sequences that participate in the formation of the reassociated duplex are not perfectly complementary. It is estimated that, for each °C increase in ΔT_m, there is between 1% (Bonner et al., 1973) and 1.5% (Laird et al., 1969) mismatching in the reassociated molecule.

The ability of the complementary single strands of denatured duplex DNA to reassociate was first shown for bacterial DNA (Marmur and Lane, 1960; Doty et al., 1960). This was followed by the demonstration that the single strands of DNA's from different bacterial species could anneal to produce hybrid duplex structures (Schildkraut et al., 1961). Reassociation of DNA (or hybridization of DNA and RNA) is totally dependent on the reacting single-stranded nucleic acids having stretches of bases with complementary sequences. The initial stage of reassociation is one of recognition or *nucleation,* when hydrogen bonds form between the bases of short lengths of complementary sequence. If the bases, which are adjacent to the nucleation site, are also complementary, the region of duplex will grow and tend to stabilize the association of the two strands as it goes. This second phase is called the *zippering* phase. Both nucleation and zippering are reversible, but with increasing length of duplex structure the association between two single strands becomes more stable (Wetmur and Davidson, 1968). Experimental conditions can markedly affect whether two strands will form a stable duplex. An increase in the salt concentration tends to stabilize duplex molecules, since cations neutralize the mutually repelling negatively charged phosphate groups of DNA and RNA. A decrease in temperature also stabilizes duplex molecules, since hydrogen bonds formed between bases are less likely to be broken at temperatures well below the T_m.

The ability to detect reassociation is based on the fact that single- and double-stranded nucleic acids can be distinguished by a number of methods. This can be done optically by measuring the decrease in absorption at 260 nm that accompanies the reformation of base-paired structures. This method has the limitation that it can only be used over certain ranges of DNA concentration and that reassociation measured in this way does not always follow second-order kinetics (see the following paragraph; see also Roberts *et al.*, 1976). Reassociation can also be measured in terms of the resistance of duplex structures to the action of the single strand-specific endonuclease S1 (Ando, 1966). By using this enzyme in conjunction with radioactive DNA, the limitations to the range of DNA concentrations accessible by optical methods can be overcome, but those due to imperfect second-order kinetics still apply. The third and most widely used method involves HAP, which, as discussed above, can be used to separate single- from double-stranded nucleic acids. Since partially duplex molecules bind to HAP as though they were double stranded, this method measures the reassociation between fragments of DNA. As such, the reaction tends to follow second-order kinetics (Smith *et al.*, 1975), but does tend to overestimate the extent of truly base-paired duplex.

Since two strands with complementary sequences must interact to form a nucleation site before reassociation can occur, the rate at which reassociation takes place will be dependent on the concentration of both reacting strands. The reaction is second order and can be described in terms of a rate constant k_2, which is in units of liters/mole/second. Britten and Kohne (1966) showed that the rate of reassociation could be usefully expressed as the product of the initial molar concentration of denatured DNA (C_0) and the time of incubation (t) in seconds, to give a term C_0t. C_0t is related to k_2 as shown by

$$\frac{C_t}{C_0} = \frac{1}{1 + k_2(C_0t)} \tag{1}$$

where C_t is the concentration of single-stranded DNA remaining non-reassociated after time t. Thus C_t/C_0 is a measure of the amount of reassociation, so that when half the DNA has reassociated

$$C_t = \frac{1}{2} C_0 \quad \text{or} \quad \frac{C_t}{C_0} = \frac{1}{2}$$

Substituting this in Eq. (1) it can be seen that

$$\frac{1}{2} = \frac{1}{1 + k_2(C_0t)}$$

which simplifies to $k_2(C_0t) = 1$. This shows that when half the DNA has reassociated, C_0t is inversely proportional to the second order rate constant k_2. The value of C_0t when one-half the DNA has reassociated is referred to as $C_0t_{1/2}$.

B. The Concept of Sequence Complexity

When considering any nucleic acid reassociation experiment it is important to remember that it is the nucleotide sequences that are reacting. While the data may be expressed as concentration of single-stranded nucleic acid, it is the effective concentrations of the sequences that determine the rate of reassociation. The distinction between total nucleic acid concentration and sequence concentration is made clearer by considering a simple example. The small animal virus SV40 has a single circular DNA chromosome containing about 5375 nucleotide pairs, whereas the bacterium *Escherichia coli* has a single circular chromosome containing about 3×10^6 nucleotide pairs. The chromosome of *E. coli* is therefore approximately 600 times as large as the SV40 chromosome. DNA from SV40 could be dissolved at the same concentration as DNA from *E. coli,* but in doing so the solution of SV40 DNA would contain about 600 times as many copies of its entire chromosomal sequence as would the solution of *E. coli* DNA. The effective concentration of SV40 sequences would therefore be 600 times greater, and consequently it would reassociate 600 times faster than the sequences of *E. coli* at the same concentration of DNA.

This illustrates the concept of sequence complexity, which is simply the length of a sequence expressed in nucleotide pairs. Provided that there are no sequences that are repeated within a single genome, sequence complexity will be equal to the size of the genome. This is found to be essentially true for all viruses and prokaryotes (Britten and Kohne, 1966; Gillis *et al.,* 1970; Kingsbury, 1969). Among these organisms $C_0t_{1/2}$ is proportional to genome size, and their DNA's reassociate as single kinetic components to produce S-shaped curves covering two orders of magnitude of C_0t (see Fig. 1). With the exception of the genes for ribosomal RNA (rRNA), some transfer RNA's (tRNA) and some specifically selected structural genes, prokaryotes do not contain sequences that are repeated within the genome.

C. DNA Fractions of Differing Complexity in Eukaryotes

While prokaryotes do not have sequences which are repeated within a single genome, the genomes of eukaryotes do. DNA reassociation experiments show that eukaryotic DNA's contain at least two kinetically distinct components. Invariably there is one component—the unique or single-copy fraction—that reassociates with a rate characteristic of DNA sequences with a complexity equal to that of the genome size. However, there are also fractions which reassociate with rates characteristic of DNA sequences with complexities far below that expected on the basis of genome size. These are the repeated DNA fractions.

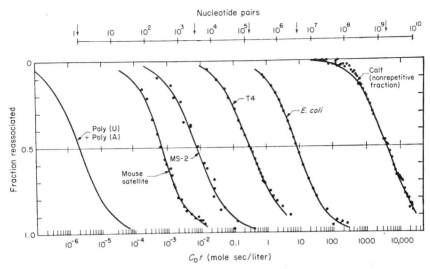

Fig. 1. Reassociation curves of various double-stranded nucleic acids, which show the relationship between genome size, sequence complexity, and the rate of reassociation measured at C_0t. Poly(U) and poly(A) are complementary homopolymers and thus have a sequence complexity equal to one, their rate of reassociation being maximal. MS-2 is a small RNA bacteriophage and T4 is a DNA bacteriophage. The reassociation curve for calf DNA is the nonrepetitive fraction only. The sizes of the genomes are indicated by the arrows above the C_0t curves and are expressed in nucleotide pairs. Notice that reassociation occurs over two orders of magnitude of C_0t and when plotted in this way the curve is S shaped. (From Britten and Kohne, 1968, *Science* **161**, 529–540. Reproduced by permission of the authors; copyright 1968 by the American Association for the Advancement of Science.)

Figures 2a and 2b show examples of reassociation curves of two eukaryotic DNA's. At one extreme of evolutionary development there is human DNA, corresponding to a C value of 3.8×10^{-12} gm DNA, and at the other extreme there is DNA from *Panagrellus* (a nematode worm), which has a C value of 0.09×10^{-12} gm DNA. Considering the nematode DNA first (Fig. 2a), it can be seen that reassociation occurs in two distinct phases. During the first phase of reassociation about 25% of the DNA reassociates with a $C_0t_{1/2}$ value of about 5.3×10^{-2}, whereas during the second phase of reassociation 60% of the DNA reassociates with a $C_0t_{1/2}$ value of 57. The latter is characteristic of single-copy DNA, when the kinetic complexity (4.1×10^7 nucleotide pairs) is equal to the genome size. The $C_0t_{1/2}$ value for the first phase of reassociation is approximately 1000 times smaller than that for the single-copy DNA, indicating the presence of a fraction of DNA composed of a sequence with an *apparent* complexity of about 4×10^4 nucleotide pairs. To a first approximation this

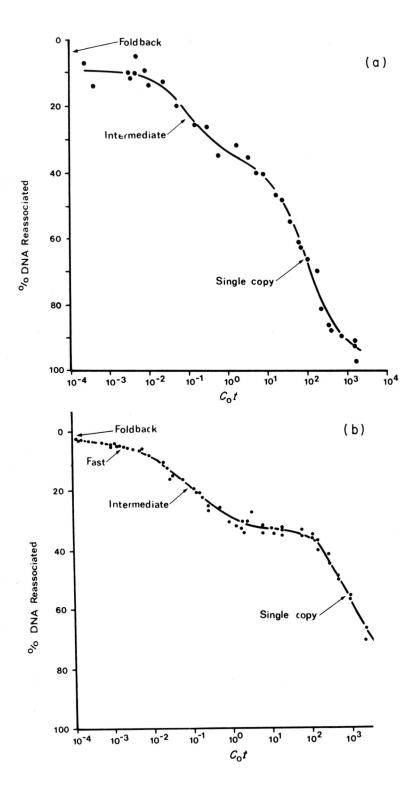

would be repeated about 1000 times per haploid genome. [It is an *apparent* complexity since the presence of mismatching during reassociation reduces the rate of reassociation and thus affects the estimate of sequence complexity (see Section II,A,1).]

The reassociation curve of total human DNA (Fig. 2b) is more complex, but has at least three distinguishable components to it. The slowest reassociating component comprises about 70% of the total DNA and has a $C_0t_{1/2}$ value of about 10^3. Using *E. coli* DNA as a standard, this value is characteristic for a DNA sequence with a complexity of about 6×10^9 nucleotide pairs—the size of the human genome—and is therefore identified as the single-copy fraction. The *intermediate* component comprises about 20% of the total and reassociates over C_0t values ranging from 10^{-2} to 10^2. This broad range of rate of reassociation indicates that this intermediate fraction contains a mixture of sequences with varying complexities and repetition frequencies. The third, *fast* reassociating fraction is only 5% of the total and has a $C_0t_{1/2}$ value of about 10^{-2}. The extremely low $C_0t_{1/2}$ value as compared to that expected for single-copy sequences, together with the limited proportion of the genome that it represents, indicates that the DNA of this fraction is composed of sequences of very low complexity repeated many times in each haploid genome. A large number of eukaryotic DNA's have these rapidly reassociating fractions, and in many cases they can be identified and purified as buoyant density satellite components. These will be discussed in Section II,D.

The reassociation curves of both *Panagrellus* and human DNA's in Fig. 2 show the presence of another kinetic component, usually called "foldback" or "zero-time binding" DNA. Careful examination of the curves shows that neither of them begin with all the DNA being single stranded. Even at the lowest values of C_0t and well below those at which reassociation of separated single strands should occur, some of the single strands appear to form duplex structures. Such a phenomenon could be due either to cross-linking of the complementary strands so that they never fully separate upon denaturation, or to the presence of inverted complementary sequences on the same polynucleotide chain. It is now known that much of the zero-time binding DNA reflects the presence of inverted repeated sequences, and these will be discussed in Section III,D.

Fig. 2. Reassociation of total genomic eukaryotic DNA's. (a) *Panagrellus silusiae* (a nematode worm). (b) Human. Reassociation is expressed as the percentage of DNA that has reassociated at given values of C_0t. Both DNA's have "foldback," "intermediate," and "single-copy" sequences. Human DNA has additional "fast" reassociating sequences. (a) Unpublished data of A. R. Mitchell. (b) Unpublished data of R. S. Beauchamp.

D. Cesium Salt Density Centrifugation and Satellite DNA's

Centrifugation of DNA to equilibrium in neutral solutions of cesium chloride results in the formation of a gradient of density, the DNA banding at a position of equal density to itself—its buoyant density position (Vinograd and Hearst, 1962). The buoyant density (ρ) of duplex DNA in neutral CsCl is largely determined by the base composition of the DNA, over the range of 20 to 80% G + C. The relationship between G + C content and buoyant density is given by the empirically derived formula

$$\rho = 1.660 + 0.098 \ (GC)$$

where GC is the mole fraction of guanine plus cytosine (Schildkraut et al., 1962).

Provided DNA is of uniform base composition, it will form at equilibrium a band in which the concentration of DNA changes in gaussian fashion about the peak position. Such homogeneity is the general rule for most prokaryotic DNA's, but many eukaryotic DNA's show asymmetric major bands and the presence of additional satellite DNA bands. The presence of a minor band was first demonstrated for mouse DNA (Kit, 1961). Mouse satellite DNA comprises 10% of the total DNA and has a buoyant density of 1.691 gm/cm^3 as compared to a density of 1.700 gm/cm^3 for mouse main band DNA. Since this initial discovery, a large number of eukaryotic DNA's have been shown to contain one or more density satellite components by a variety of cesium salt density gradient techniques. These include CsCl gradients, with or without added dye molecules (such as Hoechst 33258: Manuelidis, 1977) or antibiotics (such as actinomycin D, Netropsin, or distamycin: Kersten et al., 1966; Zimmer and Luck, 1972), or in Cs$_2$SO$_4$ gradients containing heavy metal ions such as Ag$^+$ or Hg^{2+} (Nandi et al., 1965). The addition of extra factors, which bind to DNA, is designed to accentuate differences in buoyant density. For example, actinomycin and Ag$^+$ bind preferentially to G + C-rich DNA, whereas Hoechst 33258 and Hg^{2+} bind preferentially to A + T-rich DNA.

The presence of satellite density bands in high molecular weight nuclear DNA was initially a surprise, since it demonstrated for the first time the existence of subfractions of eukaryotic DNA which were sufficiently homogeneous to band as a single component and yet were sufficiently different from the main portion of nuclear DNA to separate as a distinct band. As more DNA's were examined it became clear that both the proportions and the buoyant density values of satellite DNA's differed markedly in different organisms (reviewed in Walker, 1971; Bostock, 1971). The properties of some of the well-characterized satellite DNA's are shown here in Table I.

Apart from their common ability to separate as distinguishable components from the main band of DNA, another peculiar property of many satellite DNA's is the nature of the densities of their single strands in alkaline CsCl gradients. Centrifugation of DNA under denaturing conditions in alkaline CsCl gradients results in the separation of the single strands, which can then band at buoyant density positions according to their content of T + G (Wells and Blair, 1967). Many satellite DNA's form two bands which differ widely in buoyant density, reflecting a strong inequality in the distribution of T + G between the two complementary strands. Such a strand bias is confirmed by direct base ratio analysis, as can be seen in Table I.

The buoyant density properties of satellite DNA's allow their purification with relative ease, which is a prerequisite for any detailed studies on their sequences and structural organisation. Considerable evidence has accumulated about the DNA sequences of satellite DNA's and this is discussed in the following section.

III. CHARACTERISTICS OF THE DIFFERENT KINETIC CLASSES OF SEQUENCES

A. Satellite DNA's

1. Reassociation Studies

It is clear from the data listed in Table I and discussed in Section III,A,2 that the satellite DNA's have the property of containing very short sequences, which must therefore be repeated many times in each genome. However, well before the sequence data became available, reassociation experiments had shown that the satellite DNA's contain highly repeated sequences. Waring and Britten (1966) first showed this for mouse satellite DNA by comparing its rate of reassociation with that of SV40 DNA. The mouse satellite DNA reassociated some fifteen times faster than the DNA of SV40, indicating that it had an apparent sequence complexity of between 300 and 400 nucleotide pairs. The total genomic content of mouse satellite DNA is about 3×10^8 nucleotide pairs (10% of the haploid genome of 3×10^9 nucleotide pairs), so it can be calculated that there would be around 10^6 copies of the 300 nucleotide pair sequence of mouse satellite DNA per haploid genome.

Denaturation of the reassociated duplexes of mouse satellite results in a ΔT_m of 5°C (Flamm et al., 1967), indicating that in the reassociated molecules about 7% of the bases are mismatched. Thus, although the sequence of mouse satellite DNA is highly repeated, the individual copies of the

TABLE I

Characteristics of Some Satellite DNA's[a]

Organism	Satellite	Percentage of total	Density of duplex in neutral CsCl	Density of single strands in alkaline CsCl	G + C content of duplex (%)	Base composition				Basic nucleotide sequence
						A	T	G	C	5'——3' 3'——5'
Human (*Homo sapiens*)	I	0.17	1.687		26.4	45 / 31	32 / 42	9 / 15	14 / 12	
	II	0.7	1.693		37.1	30 / 27	26 / 30	20 / 21	24 / 22	
Cow (*Bos taurus*)		4	1.705	1.752	48.9					
		1.5	1.710	1.755	54.2					
		7	1.714	1.756 / 1.773	58.6					
		1.5	1.723	1.770	67.1					
Mouse (*Mus musculus*)		10	1.691	1.725 / 1.752	34.5	44 / 21	22 / 45	20 / 13	14 / 22	GAAAAATGA CTTTTTACT

Guinea pig (*Cavia porcellus*)	α or I	5.5	1.705	1.692 / 1.778	39.0	37 / 25	24 / 37	3 / 35	36 / 3	CCCTAA / GGGATT
	β or II	2.5	1.704	1.738 / 1.769	44.0					
	γ or III	2.5	1.704	1.740 / 1.762						
Kangaroo rat (*Dipodomys ordii*)	MS	22	1.707	1.737 / 1.759	39.5	49.6 / 8.5	11.7 / 51.2·	25.7 / 13.8	12.9 / 26.7	GTCTTCTTC / CAGAAGAAG
	HS-α	19	1.713	1.709 / 1.796	43.5	33.3 / 21.8	22.1 / 34.8	5.0 / 39.2	38.6 / 4.0	TAACCC / ATTGGG
	HS-β	11	1.713	1.761 / 1.771	66.1	17.4	16.5	33.6	33.4	ACACAGGCGGG / TGTGTCCGCCC
Fruitfly (*Drosophila melanogaster*)	I		1.692	1.687 / 1.798						ACAAACT / TGTTTGA
	II		1.688	1.697 / 1.768						ATAAACT / TATTTGA
	III		1.671	1.697 / 1.756						ACAAATT / TGTTTAA

a Compiled with data from Biro *et al.* (1975); Corneo *et al.* (1968a,b, 1970a,b, 1971, 1972); Filipski *et al.* (1973); Flamm *et al.* (1967); Fry *et al.* (1973); Gall and Atherton (1974); Gall *et al.* (1973); Hatch and Mazrimas (1974); Moar *et al.* (1975); A. R. Mitchell (unpublished results); Schildkraut and Maio (1969).

sequence have diverged from one another. Mismatching of sequences has been shown to affect the rate at which partially homologous sequences will reassociate (Southern, 1971). Southern showed that the rate of reassociation (R_m) of a mismatched duplex can be related to the rate for the same sequence if there was no mismatching (R_0) by the equation

$$\log R_m/R_0 = n \log (1 - p) \tag{2}$$

In this equation n is the minimum number of base pairs required to form a stable nucleation site and p is the fraction of mismatched bases in the final duplex. This relationship was verified for mouse satellite DNA by melting reassociated DNA and collecting fractions with different thermal stabilities. The single strands eluted at each temperature were reassociated a second time, the rate of reassociation of each thermal fraction being measured. It was found that reassociated duplex molecules with low thermal stabilities (and therefore containing more mismatched sequences) subsequently reassociate at a slower rate than do those with high thermal stabilities (and therefore containing more well matched sequences; Sutton and McCallum, 1971). Applying the correction for the effect of mismatching on the rate of reassociation reduces the complexity of the repeated sequence of mouse satellite DNA to approximately 140 nucleotide pairs.

An extreme example of the effect of mismatching on the rate of reassociation of satellite DNA is afforded by the satellite DNA of *Apodemus agrarius* (Allan, 1974). Reassociation of this satellite gives an apparent sequence complexity of 17,500 nucleotide pairs, which is hardly a short sequence! On the other hand, denaturation of the reformed duplexes reveals a ΔT_m of 19°C, which indicates about 27% mismatching in the sequences. Making allowance for this mismatching, the corrected complexity of the sequence reduces to about 330 nucleotide pairs, although considerable divergence exists between the multiple copies.

Considering the reassociation data we find that satellite DNA's from a variety of sources have the common property of consisting of sequences of low complexity repeated to varying degrees, but up to a million times per genome. Since satellite DNA's can be isolated as pure high molecular weight components (e.g., Flamm *et al.*, 1966, 1967; Filipski *et al.*, 1973), the multiple copies must be arranged in blocks. The investigation of the precise arrangement of the sequences in these blocks has been greatly aided by the use of restriction enzymes. This evidence is considered in Section III,A,4.

2. Sequencing Studies

The buoyant density properties of satellite DNA's allows their purification with relative ease, which is a prerequisite for any further detailed studies on their sequences and structural organization. A number of satellites have now been sequenced by a variety of methods. These include the guinea pig α-satellite by pyrimidine tract analysis (Southern, 1970), the HS-α and HS-β satellites of the kangaroo rat by a ribose substitution method (Fry *et al.*, 1973; Fry and Salser, 1977), and *Drosophila virilis* satellite DNA's by sequencing a complementary RNA made *in vitro* from templates of isolated *D. virilis* satellite DNA's (Gall *et al.*, 1973).

The basic sequences of some of the satellite DNA's are listed in Table I. They have the common property of being extremely short, and consequently being present in millions of copies per cell. The term basic sequence has been used because the sequence data shows a degree of heterogeneity, with quite high proportions of the satellites having variants of the predominant sequence. The presence of these variants has been interpreted as being due to divergence from the original basic sequence, perhaps with subsequent amplification of some of the variants (e.g., Southern, 1970; Biro *et al.*, 1975), or resulting from the differential amplification of different members of a library of potential satellite sequences (Fry and Salser, 1977).

3. Chromosomal Location of Satellite DNA in Condensed Regions

Clustering of the many repeats of satellite sequences is evident from cytological studies, which link satellite DNA with a limited number of sites of constitutive heterochromatin. Hybridization *in situ* of ^{3}H-labeled mouse satellite DNA (Jones, 1970) or a [^{3}H]cRNA copy of the sequence (Pardue and Gall, 1970) to cytological preparations of mouse metaphase chromosomes shows that the satellite sequences are localized in the centromeric heterochromatin of all chromosomes except the Y chromosome. In interphase nuclei the satellite sequences hybridize *in situ* to blocks of condensed chromatin, which are arranged around the periphery of the nucleus (Rae and Franke, 1972). These observations on the mouse have been followed by a host of similar findings on a variety of satellites in a number of different species. These differ in some respects from those for the mouse; for example, not all chromosomes necessarily contain satellite DNA and more than one satellite sequence can be located on the same chromosome, but they all reinforce the notion that satellite DNA's are localized in constitutive heterochromatin (see reviews in Hennig, 1973; Rae, 1972).

The same conclusion is reached from biochemical studies on the composition of isolated condensed chromatin fractions from interphase nuclei, and studies which relate the time of replication of satellite DNA's during S phase with the S-phase labeling patterns of metaphase chromosomes (reviewed in Franke, 1974; Bostock and Sumner, 1978). Although the experiments are less precise than those of *in situ* hybridization, they do suggest that the satellite DNA sequences are located in condensed chromatin. However, none of the experiments excludes the possibility that other sequences are contained within the constitutive heterochromatin, or that satellite sequences are localized at other sites in the chromosome.

A location in an inactive, condensed chromosome fraction is consistent with the inability to detect convincingly any *in vivo* transcription products of satellite DNA sequences. Flamm *et al.* (1969) incubated separately the light and heavy strands of mouse satellite DNA with a large excess of mouse total RNA, but could not find any detectable hybridization. Similarly, using total cellular RNA from human HeLa cells, Melli *et al.* (1975) could not find any sequences that were homologous to various human satellite DNA's. Very low levels of hybridization can be detected between heterogeneous nuclear RNA and satellite DNA preparations in both mouse (Harel *et al.*, 1968) and human (Melli *et al.*, 1975). In the latter case, the levels are so low that the hybridization that was detected could have been due to contamination of the satellite DNA's by other sequences. A further argument against satellite DNA's being transcribed, at least into an RNA with a coding function, is the fact that in most cases their sequences are either nontranslatable owing to a high proportion of nonsense codons, or that they encode information which would result in proteins of extremely peculiar primary structure. Although it has not yet been proved that satellite sequences are never transcribed *in vivo,* it is clear that they are not transcribed to any significant extent. In this respect, as well as others, they are similar to the nontranscribed spacer regions of repeated genes.

4. Restriction Endonuclease Studies

Restriction endonucleases are enzymes which recognize short but specific nucleotide sequences in duplex DNA and cleave both strands of the duplex within the recognition sequence, although the points of cleavage in the complementary strands may not be immediately opposite each other. Returning to satellite DNA's one can picture (Fig. 3) the repeating sequence being arranged in tandem along the length of the linear duplex molecule. Should the repeating sequence of the satellite contain a single sequence, which is recognized and cleaved by a restriction enzyme, it can be seen that digestion by that enzyme will result in a number of fragments which are equal in size to that of the repeating unit. If some base sequence divergence has occurred in the satellite, some of the restriction enzyme recognition sites may have been lost. Digestion of this DNA would, therefore, yield a collection of different-sized fragments, the smallest being the

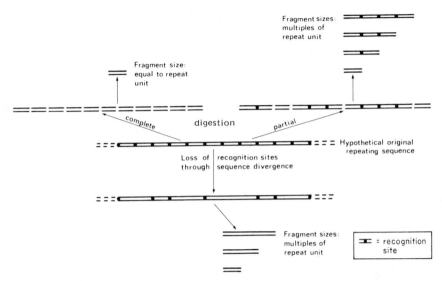

Fig. 3. Diagrammatic representation of restriction endonuclease analysis of satellite DNA sequences. In the center is a hypothetical arrangement of the basic sequence, which contains the enzyme recognition site and which is tandemly repeated. Complete digestion with the endonuclease cleaves the DNA at each recognition site and yields a homogeneous population of DNA fragments equal in size to the length of the repeating sequence (top left). Incomplete digestion results in a heterogeneous collection of fragments whose sizes are multiples of the repeated DNA sequence (top right). Similar sizes of fragments are produced by complete digestion of the DNA if some of the recognition sites have been lost by sequence divergence (below). However, the fragments produced by incomplete digestion differ from those resulting from sequence divergence in that the former still contain endonuclease recognition sites and can be digested further to yield fragments that are the size of the monomer.

size of the repeating unit and the others being simple multiples of it. Partial digestion will also yield fragments whose sizes are simple multiples of the basic repeating unit. The products of restriction enzyme digestions are usually analyzed and sized by polyacrylamide or agarose gel electrophoresis.

Digestion of mouse satellite DNA to completion with the restriction endonucleases *Eco*RII or *Ava*II produces a major set of fragments of size 240 nucleotide pairs (Fig. 4, track A), a less abundant set of fragments of size 120 nucleotide pairs, and a few minor fragments which are of sizes that are simple multiples of these two smallest fragments (Southern, 1975). This shows that the bulk of mouse satellite DNA contains the *Eco*RII and *Ava*II recognition sites distributed at regular intervals of 240 nucleotide pairs; a smaller fraction contains the sites at an interval of 120 nucleotide pairs, and still smaller fractions have the sites at intervals

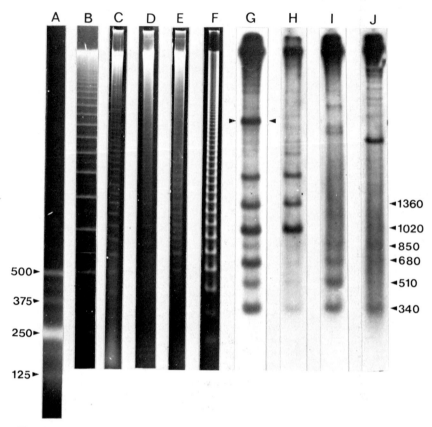

Fig. 4. Analysis by gel electrophoresis of the DNA fragments produced by restriction endonuclease digestion of mouse and human satellite DNA's. Tracks A–F are digests of mouse satellite DNA purified from mouse liver. The gels were stained with the fluorescent dye ethidium bromide and photographed after excitation at 260 nm. (A) Complete digest with *Ava*II. (B) Complete digest with *Taq*I. (C) Complete digest with *Hae*III. (D) Complete digest with *Eco*RI. (E) Complete digest with *Pst*. (F) Partial digest with *Ava*II. Tracks G–J are complete digests with *Hae*III of total cellular DNA which had been separated by gel electrophoresis, denatured *in situ,* and transferred to a nitrocellulose filter. Sequences homologous to human satellite III were identified after hybridization with ³²P nick-translated human satellite III by autoradiography. (G) Male placental DNA. (H) Mouse PG19 × human cell hybrid carrying essentially only human chromosome 22. (I) Mouse L cell × human hybrid cell containing essentially only human chromosome 15. (J) Mouse macrophage × SV40-transformed human fibroblast containing only human chromosome 7. The pattern of bands in track G is similar to that observed by ethidium bromide staining of gels containing *Hae*III digested purified human satellite III. The presence of only some of the different sized fragments in tracks H, I, and J suggests that different human chromosomes contain specific subsets of human satellite III DNA sequences. A–F kindly provided by N. Ellis; G–J unpublished data of A. R. Mitchell, R. S. Beauchamp, R. A. Buckland, and C. J. Bostock.

which are simple multiples of these two spacings. Partial digestion (Fig. 4, track F) gives two "ladders" of fragments; the major ladder is based on multiples of the 240 nucleotide pair spacing, and the minor "ladder" is based on the 120 nucleotide pair spacing. Thus the repeating units that contain the 240 nucleotide pair spacing must be juxtaposed, as are the 120 nucleotide pair units, demonstrating that the repeating units are arranged in tandem. Since virtually all the cellular satellite DNA contains this arrangement, it must be common to the different blocks of satellite sequences present on the different chromosomes.

Digestion of mouse satellite DNA with different restriction enzymes (for example, HaeIII, HindIII, or TaqI) produces a minor series of fragments based on the same size intervals as those for EcoRII (Hörz and Zachau, 1977) but leaves the majority of the satellite intact (Fig. 4, tracks B–E). The recognition sites for these different restriction enzymes appear to be clustered together in separate blocks, perhaps representing small differences in the satellite DNA sequences located on different chromosomes (Hörz and Zachau, 1977).

Sequencing data for mouse satellite (Biro et al., 1975; see also Table I) suggest that it is composed of a basic repeating sequence of about 18 nucleotide pairs. This, together with the reassociation and restriction enzyme data, suggests the following arrangement of sequences. At the lowest level there are the remnants of the original sequence, some 18 nucleotide pairs long, which are arranged in tandem, but which are sufficiently diverged to be unable to reassociate with kinetics characteristic of an 18 nucleotide sequence. Perhaps eight adjacent copies of these diverged basic sequence units form the repeated sequence detected by reassociation experiments that has a complexity of 120 nucleotide pairs. Two "reassociation" units, therefore, form the major 240 nucleotide pair fragment identified after EcoRII or AvaII restriction endonuclease digestion. These repeating units are themselves arranged in tandem in large blocks. Similar hierarchies of tandemly repeating units of sequences have also been demonstrated for other satellite DNA's (e.g., bovine satellite DNA I: Botchan, 1974; Mowbray et al., 1975), so a structure based on tandem repetition would seem to be a general feature of satellite DNA's. Perhaps the best examples of highly conserved tandemly repeated satellite sequences are those found in Drosophila melanogaster. Satellite IV of D. melanogaster (density of 1.705 gm/cm^3) has a basic sequence of —AGAAG—. Digestion of this satellite with MboII restriction endonuclease produces fragments of this size and sequence, and over 95% of all the fragments are less than 25 nucleotides in length (Endow, 1977). Similar, highly conserved, short repeating sequences are present in satellites I and II of D. melanogaster (Brutlag and Peacock, 1975). On the other hand, not all satellite DNA's of apparently homogeneous density in cesium salt gradients are homogeneous from the point of view of their sequences or the fragments produced by restriction endonuclease digestion.

An example of a satellite DNA which is composed of a mixture of different sequences is satellite III of humans. Satellite III is isolated by a combination of Cs_2SO_4 + Ag^+ and CsCl density gradients (Corneo et al., 1971) and may comprise only about 0.75% of the human genome (Moar et al., 1975). Reassociation experiments show that it reassociates as a simple sequence DNA with a $C_0t_{1/2}$ value of about 10^{-3}. Complete digestion of this satellite DNA with the restriction enzyme Hae III produces a complex set of fragments (Fig. 4, track G). Many of the fragments have sizes that are simple multiples of the smallest (174 nucleotide pairs long), but they are not present in stoichiometric amounts expected if the loss of the Hae III recognition site has been solely by sequence divergence. Cross-hybridization of the DNA sequences contained in the differently sized fragments shows that there are at least three distinct sequences present in human male satellite DNA (Bostock et al., 1978). Another example of a satellite DNA having a complex sequence arrangement revealed by restriction endonuclease digestion is the 1.688 gm/cm³ density component of Drosophila melanogaster DNA (Carlson and Brutlag, 1977).

Examination by restriction endonuclease digestion of the human satellite III sequences present in DNA isolated from hybrid cells carrying essentially single human chromosomes reveals an extremely complex organization of this apparently simple sequence DNA (Fig. 4, tracks H–J). The clearest example of a specific satellite III DNA fragment being localized on a specific chromosome is given by the association of the 3400 nucleotide pair fragment with the Y chromosome (Cooke, 1976). Other individual chromosomes contain subsets of the major spacings of restriction enzyme recognition sites identified in total satellite III, but, in addition, at this level a large number of extra minor bands can be seen. Thus, although at the level of buoyant density satellite III appears to be a homogeneous sequence, digestion of total satellite III with the Hae III restriction endonuclease reveals some heterogeneity in the organization of sequences. At the single chromosome level further heterogeneity is revealed so that no longer can we consider this satellite DNA to be a homogeneous component.

B. Intermediate Kinetic Fraction

Of the four kinetic classes of reassociating sequences discussed in Section II,C, the so-called intermediate class is generally considered to be composed of sequences that are repeated between 10^2 and 10^5 times per genome. There are large variations both in the proportions of genomes that are occupied by these sequences and in the patterns of occurrence of sequences of particular repetition frequencies and complexities. In Fig. 2 it was seen that about 25% of human DNA reassociates with kinetics that indicate apparent frequencies of repetition of between 10^2 and 10^4. In contrast, 80% of salmon DNA reassociates over similar ranges of C_0t (Britten and Kohne, 1968). In bovine DNA the repeated sequences constitute about half the DNA and reassociate over a more restricted range of C_0t, characteristic of sequences having a complexity of 17,000 nucleotide pairs, each being represented 66,000 times per genome (Britten and Smith, 1969).

Sequences that reassociate in the intermediate class form duplex molecules with greatly varying degrees of mismatching ranging from little or no mismatching up to 30% (Britten and Smith, 1969). The slower reassociating intermediate sequences contain more mismatched bases than do those that reassociate at lower C_0t values (Reed, 1969), which is consistent with the effects of mismatching on the rate of reassociation (discussed in Section III,A,1).

These characteristics—varying degrees of mismatching, heterogeneity of rate of reassociation, and large differences in the amounts of this kinetic component—suggest that the intermediate class is in fact a collection of *families* of repeated sequences, each family having a different degree of internal homology (Britten and Kohne, 1968). Those families that reassociate at low C_0t values have properties similar to diverged satellite DNA's, and may well have had their origins there (Southern, 1974). At the high C_0t end of the scale the families may have become so divergent that the distinction between "repeated" and "unique" is largely dependent on the stringency under which the reassociation is permitted to occur (e.g., McCarthy and Farquhar, 1972; McCarthy and Duerksen, 1971; Britten *et al.*, 1974).

C. Single-Copy Sequences

A single-copy sequence is, by definition, one that occurs only once in each haploid genome. From an experimental point of view a single-copy sequence is one that reassociates with a $C_0t_{1/2}$ value expected for a sequence complexity equal to that of the size of the genome. In practice, it is difficult to estimate exactly the proportion of the genome that is unique, since it is highly dependent on the conditions of reassociation. The reassociation of two single strands depends on the formation and stabilization of a nucleation site, and this depends on a combination of cation concentration and temperature. Thus, a particular copy of a diverged repeated sequence could form highly mismatched duplex structures with the other members of the family under very relaxed conditions or a well-matched duplex with its exact (and unique) complement under stringent conditions. Reassociation of corn, mouse, and sea urchin DNA's under conditions of low stringency suggests that 44, 63 and 70%, respectively, are single-copy sequences. Under more stringent conditions, these same DNA's are estimated to have 65, 88, and 85% single-copy sequences (McCarthy and Farquhar, 1972; McCarthy and Duerksen, 1971).

While there is ambiguity about exactly what proportion of the genome should be considered unique, there is no doubt that a portion of this fraction contains the sequences of many structural genes. With the exception of histone messenger RNA's, all purified specific mRNA's have been shown to be transcribed from structural genes that are present once, or at the most, a few times per haploid genome. Although total polysomal mRNA, or the polyadenylated fraction of polysomal mRNA, contains sequences that are complementary to both repeated and nonrepeated DNA, the larger proportion of sequences are complementary to the non-repeated fraction (see review in Lewin, 1975). The repeated and non-repeated sequences present in the population of mRNA molecules are contained in separate molecules (Campo and Bishop, 1974), so the repeated sequences that are transcribed into mRNA may be structural genes that are present in multiple copies within the genome (see section V). Alternatively, subpopulations of mRNA may contain leader sequences in common, as a result of posttranscriptional events which link the same leader sequence to different coding sequences (Section VI,C,1; see also Williamson, chapter 10, this volume).

Although the single-copy sequence fraction contains structural genes, it is not composed entirely of structural genes, in higher organisms at least. Comparison of the proportion of the genome that is single-copy with that which is complementary to total polysomal mRNA shows that the sequences that are complementary to mRNA represent a minority of the single-copy DNA (Bishop *et al.*, 1974; Rosbash *et al.*, 1974; Hahn and Laird, 1971; Brown and Church, 1972). It thus appears that there are both coding and noncoding single-copy sequences. The same conclusion is reached on the basis of the rate of divergence of single-copy sequences between related organisms. This was first suggested by a comparison of the rate of base substitution in total single-copy sequences and the rate of amino acid substitution in proteins, the ultimate products of the encoded information in genes (Laird *et al.*, 1969). The rate of base substitution exceeds that of amino acid substitution, suggesting that a major portion of single-copy sequences are free to undergo base substitution without the constraints imposed by having a coding function. Subsequently, Rosbash *et al.* (1975) showed that single-copy sequences that are transcribed and processed into the total polysomal mRNA population are more conserved between the rat and the mouse than are the single-copy sequences that are not transcribed. This suggests that there are single-copy sequences whose function is not directly related to a coding function, and which are presumably not subject to such strong selection pressures.

D. Inverted Repeated Sequences

Figure 5 shows a diagram of how a sequence, whose complement is repeated in reverse order on the same strand of DNA, can give rise to a folded duplex structure upon denaturation and reassociation. The length of the repeated sequence will determine the length of the duplex region (or stem) upon reassociation. The presence and size of any intervening nucleotides will determine whether a loop is formed and how large the loop is. Single-strand tails, projecting from the stem, show the presence of noncomplementary sequences at distal ends of the inverted repeated sequence. Analysis of inverted repeated sequences is usually carried out by a combination of techniques. They can be isolated by binding to hydroxylapatite (HAP) after reassociation to very low C_0t values, and then analyzed by visualization in the electron microscope or in terms of their resistance or sensitivity to S1 endonuclease.

Wilson and Thomas (1974) showed that these inverted repeated sequences are found in a variety of organisms. In general they comprise between 2 and 6% of the genome (e.g., Perlman *et al.,* 1976; Dott *et al.,*

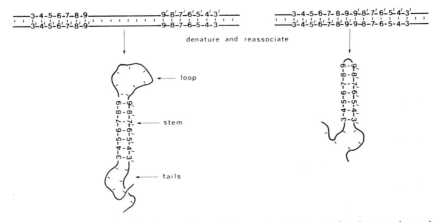

Fig. 5. Diagram to show how inverted repeated sequences fold back after denaturation and renaturation to form loops and hairpin structures.

1976; Schmid *et al.*, 1975; Cech *et al.*, 1973). The length of the inverted sequence appears to vary depending on the organism. In vertebrate and plant DNA's the sequences that anneal to form the stem structure are estimated to be between 150 and 200 nucleotides long (Dott *et al.*, 1976; Walbot and Dure, 1976), whereas in *Drosophila* (Schmid *et al.*, 1975) and the mouse (Cech and Hearst, 1975) the average lengths of the stem structures are 1100 and 1350 nucleotides, respectively.

The lengths of sequences that separate the complements of the inverted repeats appear to be variable. Wilson and Thomas (1974) could not detect looped structures either by electron microscopy or by S1 endonuclease digestion. Subsequently, loops up to 2700 nucleotides have been found in human DNA (Dott *et al.*, 1976) and up to 4000 nucleotides in mouse DNA (Cech and Hearst, 1975).

Investigation of the types of sequences that form the stem structure of foldback DNA must be preceded by the physical separation of the inverted complementary sequences located on the same DNA strand. This is because they must be free to reassociate with the other sequences present in solution rather than reassociate with themselves in the formation of a foldback structure. While it is possible to prepare a probe which is representative of such sequences (by digestion of foldback structures with S1 endonuclease to destroy the single-stranded interstitial nucleotides), it is not possible to do the same for total DNA without losing some DNA sequences. One has to rely on the probability that random shearing to very low fragment sizes will separate the majority of the complements of inverted repeated sequences. Reassociation of foldback stem sequences in the presence of an excess of highly sheared total DNA indicates that the inverted sequences are derived from all the kinetic classes (Schmid and Deininger, 1975; Schmid *et al.*, 1975) including satellite DNA's. Since there must be a minimum of two copies per genome, they cannot, strictly speaking, ever be unique sequences, although reassociation methods used in these studies were not sensitive enough to distinguish between one and two copies. All the kinetic classes also appear to be represented in the sequences that flank the inverted repeats at either end (Cech and Hearst, 1975; Perlman *et al.*, 1976).

Knowledge of the size of the genome, the proportion of foldback DNA in the genome, and the length of inverted sequences allows an estimate of the number of such structures in each genome. Such estimates tend to be approximate on account of a number of experimental factors, but they suggest that the fraction of inverted sequences in the genome is in proportion to the size of the genome. Larger genomes, such as those of humans (Dott *et al.*, 1976) and *Xenopus* (Perlman *et al.*, 1976) have about 2×10^6 and 10^5 inverted repeats per genome, respectively, whereas smaller genomes such as *Drosophila* (Schmid *et al.*, 1975) and *Panagrellus* (Beauchamp, 1977) have 3×10^4 and 2×10^4 inverted repeats per genome, respectively.

Although the general properties of inverted repeated sequences found in eukaryotic DNA's have been described, we have little definite information about what their functions may be. The fact that they are transcribed into heterogeneous nuclear RNA (Jelinek *et al.*, 1974; Molloy *et al.*, 1974), while being absent from cytoplasmic messenger RNA (Ryskov *et al.*, 1973), suggests some role in the recognition of which sequences are to be transported to the cytoplasm. Inverted repetitions, potentially capable of forming hairpin structures, are found in a wide variety of situations in which some aspect of a recognition function is suggested. Probably the most familiar example is found in the structure of tRNA, but other examples include the promotor region of the *lac* operon in *E. coli* (Gilbert and Maxam, 1973), the ends of the genome of the adenovirus (Garon *et al.*, 1972; Wolfson and Dressler, 1972; Roberts *et al.*, 1974), the ends of the translocatable elements found in prokaryotes (reviewed in Kleckner, 1977), and the ends of the "gene-sized" pieces of DNA found in the macronucleus of hypotrichous ciliates (e.g., Wesley, 1975; Lawn, 1977).

IV. INTERSPERSION OF SEQUENCES OF DIFFERENT KINDS

A. Introduction

While discussing the properties of the different kinds of sequence classes, it was convenient to draw operational distinctions between them. These distinctions are probably more a reflection of the experimental approaches used to analyze the DNA, than of any inherent properties of the fractions that are identified. The various kinetic fractions may well be different parts of a continuum representing various stages in the divergence of once highly reiterated satellitelike sequences through to the highly mismatched sequences that are eventually lost to the single-copy fraction. Even if such a view is correct, it does not preclude the possibility that each of the various types of sequences has a specific role to play in the structure or function of the chromosome. There are a number of theories that ascribe a variety of specific functions to the different fractions of the DNA. While there is not space to discuss these hypotheses in detail here, it should be noted that some give sequence-specific regulatory functions to the intermediate classes of sequences (e.g., Britten and Davidson, 1969; Georgiev, 1969), whereas others suggest a non-sequence-specific role for these intermediate sequences in the control of gene transcription (e.g., Paul, 1972; Crick, 1971). Various ideas have been put forward as to the function of satellite DNA's based largely on structural (see reviews in Walker, 1971; Bostock, 1971) or evolutionary (e.g., Hatch *et al.,* 1976) roles. Tests for these hypotheses lie not only in the properties of the sequences themselves but also in their linear arrangement and juxtaposition along the DNA molecule. In the following section we will consider in general terms how the various sequence classes are arranged.

B. Analysis of DNA Fragments of Varying Size

Given that there are two classes of sequences—repeated and unique— in the genomes of eukaryotes, there are two extreme hypothetical ways in which these can be arranged. At one extreme all the repeated sequences would occur in one block and the unique sequences in another block. At the other extreme each individual sequence of the repeated class would be separated from another member by a segment of the unique sequence fraction.

These two arrangements can be distinguished experimentally by measuring the proportion of DNA fragments of different sizes that reasso- ciate with kinetics characteristic of repeated sequences and which thus contain one or more repeated sequence. If it were possible to isolate DNA of a size equal to the whole genome, the two extremes would be indistin- guishable. All molecules with either of the arrangements would contain a repeated sequence, and thus part of their denatured single strands would form a duplex structure that would bind to HAP as a reassociated duplex. However, if the DNA was sheared to produce DNA of size equal to half the genome, the two extremes could be distinguished. In the former case, half of the fragments would contain repeated sequences, whereas the other half would be composed entirely of unique sequences. In the case of the interspersed sequences, all fragments would still contain repeated sequences. Not until the DNA fragment size approached that of the length of the unique sequence sections would the absence of repeated sequences be detected.

Analysis of the proportion of single-stranded DNA fragments that will reassociate with kinetics characteristic of repeated sequences, as a func- tion of the size of fragments, has formed the basis of a wide range of experiments designed to investigate sequence interspersion in eukaryotic DNA. A generalized plot of the proportion of duplex molecules formed at low C_0t values as a function of the size of the reacting single strands is shown in Fig. 6. At very small fragment sizes the proportion of reassociat- ing strands reflects the proportion of repeated sequences in total DNA (P). As the fragment size (L) increases so does the proportion of DNA that

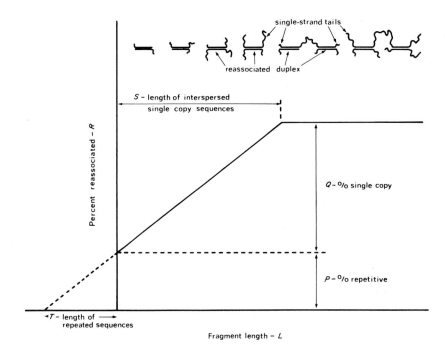

Fig. 6. Interspersion of intermediate repeated and single-copy sequences. Idealized plot of the percentage of DNA fragments that reassociate with kinetics characteristic of repeated sequences as a function of the fragment size. The characteristics of the curve are used to describe the parameters of interspersed sequences. For a full explanation, see text. Above the curve are shown diagrammatic examples of what forms the reassociated duplex molecules are assumed to be. (Based on Crain *et al.*, 1976a; Graham *et al.*, 1974.)

reassociates, until a plateau is reached and any further increase in fragment size does not result in more single strands reassociating. The difference (Q) between the plateau value for the fraction reassociated and that (P) found at small fragment sizes gives a measure of the total amount of single-copy sequences *interspersed* with repetitive sequences. The increase in single-stranded fragment length (S) over this part of the curve gives the average length of the interspersed single-copy sequences. Finally, extrapolation of the rising part of the curve to the ordinate gives the average length of the interspersed repetitive elements. Apart from the straightforward measurement of the proportion of reassociating single strands, certain refinements can be added. For example, hyperchromicity measurements can show the size of single-stranded tails attached to duplex regions, and electron microscopy or digestion with S1 endonuclease can give more direct measurements of the size of the duplex (repeated) regions (e.g., Chamberlain *et al.*, 1975; Goldberg al., 1975).

From experiments of this kind it appears that unique and repeated sequences of an organism are interspersed in either or both of two generalized patterns. These are patterns with short periodicities and patterns with long periodicities. Although the lengths of both repeated and single-copy sequences vary widely within each pattern, the short period interspersion is considered to be characterized by having short lengths (300 nucleotides) of intermediate repeated sequences interspersed between unique sequences of about 1000 nucleotides in length. This pattern of organization accounts for the major part, about 70%, of the genome in *Xenopus* (Davidson *et al.,* 1973a), sea urchin (Graham *et al.,* 1974), the mollusk *Aplysia californica* (Angerer *et al.,* 1975), and human (Schmid and Deininger, 1975; see also Davidson *et al.,* 1975).

As the name implies, long period interspersion is characterized by having longer stretches of repeated sequences interspersed with relatively long single-copy sequence elements. For example, *Drosophila melanogaster* DNA contains intermediate repetitive sequences of average length 5600 nucleotides interspersed with single-copy DNA at intervals of more than 1.3×10^4 nucleotides (Manning *et al.,* 1975; Crain *et al.,* 1976a). Most metazoan organisms studied to date have a small part of their genome organized in this way, usually in addition to the short period interspersion (see, e.g., Britten *et al.,* 1976). However, both long and short period interspersions are not always found within the same genome, as some organisms with small C values apparently lack the short period interspersion pattern. Several Diptera possess only the long periodicity pattern (*Chironomus tentans:* Wells *et al.,* 1976; *Apis mellifera:* Crain *et al.,* 1976b; *Drosophila melanogaster:* Manning *et al.,* 1975), as is the case for the nematode *Panagrellus silusiae* (Beauchamp, 1977).

While the arrangement and size of the repeated and single-copy sequence elements have been used to distinguish the long and short period interspersion patterns, the nature of the repeated sequences contained in these different patterns are also different. The difference lies in the relatedness of the different copies of the repeated sequences within each class. Denaturation of the duplex structures formed by reassociation of short period interspersed repeated sequences shows large ΔT_m values, when compared to native DNA, whereas the reassociation of long period interspersed sequences forms duplexes with thermal stabilities close to that of native DNA (Davidson *et al.,* 1973b). Thus, the long period repeats show little divergence and would appear to be of recent origin or are maintained in a homogeneous state in some way. The short period repeats must have accumulated many base substitutions, resulting in poorly base-paired reassociated duplexes, and would appear to be of older origin.

These different rates of divergence suggest that the repetitive sequences that are present in the two types of interspersion are derived from different families of sequences. This question has been explored by isolating separately the long and short repeated segments, and testing for sequence homology by cross-reassociating their sequences. In sea urchin DNA, fewer than 10% of the long repetitive DNA sequences are able to reassociate with short repetitive sequences, suggesting that these two sequence classes do indeed contain independent sequences (Eden *et al.*, 1977). On the other hand, very similar experiments on rat DNA show that the long and short repetitive DNA's share common sequences (Wu *et al.*, 1977). Thus, in the rat genome, not only are the single-copy sequences interspersed with repetitive sequences in the two basic patterns but the repetitive sequences themselves would appear to be interspersed with one another.

It should be stressed that all the experiments described above reveal only the predominant arrangement of sequences in chromosomes. Since only a fraction of the single-copy sequences appear to represent structural genes, we do not know whether the single-copy sequences detected in the interspersion experiments are in fact structural genes. The gene sequences might be arranged in some other fashion which could go undetected by these experiments.

C. Analysis of Tandem Repetition by Circle Formation

Before the advent of restriction enzymes for studying the tandem arrangement of repeated sequences, Thomas and his colleagues (Thomas *et al.*, 1970, 1973) devised two methods for revealing the presence of tandemly repeated sequences. These methods are based on the fact that, if the various copies of a repeated sequence are arranged in tandem, DNA fragments of a size larger than the size of the repeat will contain more than one direct repeat of the sequence. In the first method denaturation, followed by reassociation, of such fragments could result in the complementary single strands reassociating in a different register to that present in the original native DNA. Such a partially duplex structure would have single-stranded tails at each end, the sequences of which would be complementary and could reassociate to form a circular molecule—a "slipped" circle (see Fig. 7).

The second method, also shown in Fig. 7, involves the removal of nucleotides from the 3'-chains at each end of the native DNA fragment by exonuclease III digestion. This leaves the 5'-ended chains exposed. These ends would be free to reassociate with each other to form a "folded"

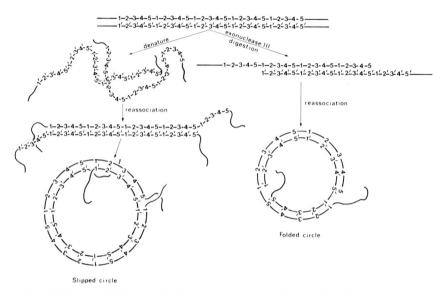

Fig. 7. Slipped and folded circle formation as a means for revealing the tandem arrangement of repeated sequences. A sequence, designated as ¹₋²₋³₋⁴₋⁵, is tandemly repeated in duplex DNA. On the left, denaturation followed by reassociation could result in the two complementary strands coming together out of their original registers. This would result in complementary sequences protruding as single strands from either end of the duplex region. These complementary sequences could undergo further reassociation to form circular structures. Folded circles are produced by digesting away part of the opposite single strands from both ends of duplex DNA fragments with exonuclease III. As with the slipped circles, incubation under reassociation conditions will allow the complementary sequences to anneal to produce circular structures. The formation of both these circular structures is dependent on there being tandem repetition of sequences.

circular structure since their sequences would be complementary. With both these methods, estimation of the proportion of fragments of total DNA that will form circular structures gives a measure of the proportion of the genome that is organized as tandem repeats.

Variations on the basic methods can allow further characterization of the sequences involved in the tandem arrangement. For example, the degree of mismatching in the regions that hold the fragment in a circular configuration can be assessed from their melting profiles. This will indicate the degree of homology between the adjacent repeats. Estimates of the length of the repeating sequences, and the sizes of the blocks in which they occur in tandem, can be obtained by scoring the frequency of circle formation both as a function of the size of the fragments and the length of the single-stranded tails exposed by nuclease digestion.

Experiments of this kind (reviewed in Thomas *et al.*, 1973) suggest that between 15 and 30% of DNA from a wide variety of eukaryotic organisms can form folded circular structures. The lengths of the repeated sequences involved are estimated to be between 600 and 6000 nucleotide pairs, their tandem arrangement forming blocks of between 3000 and 150,000 nucleotide pairs in length. It is not clear at present whether all repeated sequence classes can be organized in this way. We now know from restriction enzyme studies that satellite DNA's are composed of tandemly arranged repeated sequences, and, as expected, a high proportion of fragments of satellite DNA will form circles (e.g., 60% of mouse satellite DNA; see Pyeritz *et al.*, 1971). Whether some of the nonsatellite repeated sequences are also arranged in a tandem manner has not yet been fully resolved.

Thomas *et al.* (1970) found that circles could be produced in main band DNA of the mouse, and that the frequency of formation of these circles did not vary significantly between main band fractions from various positions in a CsCl density gradient. This constant proportion across the gradient would argue against the possibility that circles are formed solely by the presence of contaminating mouse satellite DNA. Another observation which indicates the tandem arrangement of nonsatellite repeated sequences is the detection of circle formation by DNA isolated from polytene chromosomes of *D. melanogaster* or *Drosophila virilis* (Lee and Thomas, 1973). Since satellite DNA's are drastically underreplicated during polytenization (Gall *et al.*, 1971), this result suggests that some nonsatellite DNA's have a sequence organization that permits the formation of circles. On the other hand, there is the observation that the vast majority of circles formed by *Drosophila* DNA extracted from nonpolytenized tissue are derived from highly repeated satellitelike sequences (Schachat and Hogness, 1973; Peacock *et al.*, 1973). Although these conflicting results remain unexplained, it seems unlikely that a significant proportion of tandemly arranged repeated sequences, which do not originate from blocks of satellite DNA's, exist in the genome of *Drosophila*. However, the *Drosophila* genome may not be fully representative of the majority of higher organisms, for it has been shown to lack the short period interspersion pattern found in many other higher eukaryotes (see Section IV,B). While it is probable that many circles originate from satellitelike sequences, it is nevertheless the case that the measured proportions of fragments which form circles is often higher than the proportion of their genomes that are satellites. It is, therefore, still an open question as to whether repeated sequences, other than satellite DNA's and the few specific structural genes (see Section V), are arranged in tandem, and, if they are, what their functional role might be.

V. ARRANGEMENT OF SPECIFIC STRUCTURAL GENES

A. Introduction

In the previous sections the existence, nature, and distribution of various types of eukaryotic sequence classes have been described in general terms. In the remaining part of this chapter the information that is available about certain specific structural genes will be considered. A prerequisite for studying the molecular biology of a gene is the possession of a pure "probe" of the gene sequence, the probe usually being in the form of an RNA transcript or a DNA copy (cDNA) of the sequence. Until recently the only RNA's that could be purified to a nearly homogeneous state were 28 S and 18 S rRNA's, 7 S rRNA, 5 S rRNA and the tRNA's. Early hybridization experiments (reviewed in Bostock, 1971; Tartof, 1975) showed that all these gene sequences are present in multiple copies per haploid genome, but the fact that none of them is translated into a protein sequence suggested that their use as models for the organization of the sequences associated with protein coding genes might be unsatisfactory.

In 1971 it was shown that the mRNA's for histones could be partially purified from sea urchin embryos in the early stages of development. Using the preparation of total histone mRNA as a hybridization probe, Birnstiel and co-workers (Kedes and Birnstiel, 1971; Weinberg *et al.,* 1972) showed that there are several hundred copies of the genes for histones per haploid genome of the sea urchin. More recently, other transcribed repeated structural gene sequences have been identified in poly(A)-containing mRNA preparations from *Drosophila* cells (see Section V,F), although it is not known for which specific proteins they code. In addition, mRNA's for a number of specific proteins (other than histones) have been purified from differentiating and virally infected cells, and, without exception, it appears that the genes that code for these proteins are not repeated. We will start by considering individually the known repeated genes, and then go on to discuss what is known about the structure of nonrepeated genes. Both these areas of study have been greatly aided by two new technologies, the use of sequence-specific restriction endonucleases and the ability to clone specific DNA sequences after recombination with bacterial plasmid DNA's (see Maniatis, Chapter 11, this volume).

B. The Genes for Ribosomal RNA

1. Eukaryotes

The genes for rRNA are localized in the nucleolus (Wallace and Birnstiel, 1966) and can be purified from high molecular weight DNA preparations by density gradient centrifugation (Brown and Weber, 1968; Birnstiel et al., 1968). Such properties suggest that the multiple copies of rRNA genes must be clustered. It turns out not only that they are clustered, but also that they are arranged in tandem arrays.

In most eukaryotes, rRNA's are transcribed as part of a large precursor molecule which is processed in the nucleolus (see, e.g., Weinberg and Penman, 1970). The transcribed region contains the gene sequences and also ''spacer'' sequences that are discarded from the precursor during cleavage and processing. In addition to the transcribed gene and spacer regions there are also nontranscribed spacer regions; the three types of sequences together forming a repeating unit (Brown and Weber, 1968; Birnstiel et al., 1968; Dawid et al., 1970). The transcribed region is richer in A + T than the nontranscribed spacer region in *Xenopus,* so the two can be distinguished by partial denaturation with subsequent observation in the electron microscope. The A + T-rich transcribed regions appear as eye forms (owing to the denaturation of the single strands) regularly spaced along the molecule, and separated by regions of duplex DNA of the nontranscribed spacer region (Wensink and Brown, 1971). This observation confirms the tandem arrangement of the repeated genes. By a combination of denaturation mapping of this kind and nucleic acid hybridization techniques, it was shown that the length and base sequence of the transcribed spacer region of ribosomal DNA in two species of *Xenopus* remain fairly constant (Wensink and Brown, 1971; Brown et al., 1972). In contrast, the nontranscribed spacer region is found to be heterogeneous in terms of its length and its base sequence, both within and between the two species (Brown et al., 1972; Forsheit et al., 1974; Wellauer et al., 1974).

Heterogeneity of length of the nontranscribed spacer regions has been investigated more precisely by the use of restriction endonucleases, in an analogous way to that described for satellite DNA's (Section III.A.4) and is based on examination of fragments containing transcribed regions. The ribosomal DNA of *Xenopus* contains two recognition sites for the restriction endonuclease *Eco*RI; one of these is located in the 18 S rRNA sequence and the other is within the 28 S rRNA sequence (Wellauer et al., 1974; see also Fig. 8). If the repeating unit of ribosomal DNA is homogeneous in size, only two homogeneously sized fragments should be produced by *Eco*RI digestion: a small fragment containing only transcribed sequences and including most of the 28 S rRNA and part of the 18 S rRNA sequences, and a large fragment containing all the nontranscribed spacer sequences and the major part of the 18 S rRNA sequences. In practice several different-sized fragments can be identified. The small fragment is produced and it is homogeneous in size, but the larger fragment is heterogeneous because of the variability in the length of the nontranscribed spacer region (Wellauer et al., 1974).

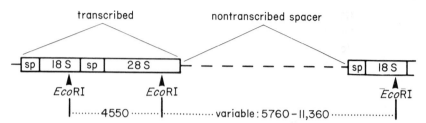

Fig. 8. The structure of the ribosomal DNA unit of *Xenopus laevis*. The unit contains both a transcribed region and a nontranscribed region. The former is of constant length and contains the 18 S and 28 S rRNA genes and transcribed spacer. The nontranscribed spacer is of variable length. *Eco*RI sites are indicated by the arrows. The numbers refer to lengths in nucleotide pairs of different segments. For fuller explanation and the sources from which this was compiled, see text.

Analysis of ribosomal DNA fragments from *Xenopus* which had been cloned in bacterial plasmids shows that the nontranscribed spacer region is itself composed of a repeating 50 nucleotide pair sequence. The observed variation in the length of the nontranscribed spacer region, between 5760 and 11,360 nucleotide pairs, occurs in increments of 50 nucleotide pairs, and appears to result from a variation in the number of copies of the internal repeat of 50 nucleotide pairs (Wellauer *et al.,* 1976a,b; Reeder *et al.,* 1976).

A different type of heterogeneity has been demonstrated for the ribosomal genes of *Drosophila melanogaster*. This organism contains two nucleolus organizers, one on the X chromosome and one on the Y chromosome (Ritossa *et al.,* 1966). Each nucleolus contains a variable, but multiple, number of genes arranged in tandemly repeating structures which are similar to that described for *Xenopus* (Tartof, 1973; Hamkalo *et al.,* 1973; Meyer and Hennig, 1974). The X and the Y chromosomes carry repeating units in common with each other, at least with respect to size. However, the X chromosome also carries a specific and larger-sized repeating unit (Tartof and Dawid, 1976).

Examination by restriction enzyme mapping and heteroduplex mapping, of fragments of *Drosophila* ribosomal DNA which had been cloned in bacterial plasmids shows that the smaller (X and Y common) 11,500 nucleotide pair unit contains an intact arrangement of the gene, transcribed spacer, and nontranscribed spacer sequences. A novel finding is that the larger, X-specific unit, 17,000 nucleotide pairs in length, contains an additional segment of DNA, which is 5400 nucleotide pairs in length, inserted *within* the coding region of the 28 S rRNA gene. The 28 S rRNA coding sequence is therefore split into two, each part being separated by a large length of DNA of unknown function (Glover *et al.,* 1975; Glover and

Hogness, 1977; White and Hogness, 1977). It is not known whether the copies of the ribosomal genes that carry this inserted sequence are transcribed (cf. Section VI,C,2). If they are, this arrangement would require a novel form of rRNA transcription and processing, similar to that discovered for structural genes that contain leader and insertion sequences. Either the inserted sequence would have to be bypassed by the polymerase during transcription, or the two halves of the 28 S rRNA sequence would have to be joined during processing with the exclusion of the inserted sequence. While both these possibilities have seemed unlikely until recently, the demonstration that leader sequences are joined to coding sequences in mRNA of eukaryotic viruses and that inserted sequences exist in structural genes known to be transcribed, suggests that such mechanisms may be quite widespread in eukaryotes.

In considering the rRNA genes we have so far talked in terms of tandemly repeated units transmitted with the haploid genome of germ cells. Such an arrangement appears to be almost ubiquitous in eukaryotes, but the presence of a single copy of the gene unit in the germinal nucleus has been shown in at least one primitive unicellular eukaryotic organism, *Tetrahymena pyriformis* (see Bird, Chapter 2, this volume).

2. Prokaryotes

In bacteria the organization of the ribosomal genes is quite different from those discussed above. Nucleic acid hybridization experiments and gene mapping studies show that in *E. coli* there are between 5 and 10 copies of the genes for 16 S, 23 S, and 5 S rRNA's per chromosome. These studies also suggest that the various rRNA genes are organized as clustered but separate operons with the structure: promotor–16 S rRNA gene–23 S rRNA gene–5 S rRNA gene (see reviews in Jaskunas *et al.,* 1974; Pace, 1973; Davies and Nomura, 1972; Schlessinger, 1974). It was, however, the ability to isolate individual rRNA operons, initially in transducing phages (Doenier *et al.,* 1974; Jorgensen, 1976) and subsequently by *in vitro* recombination with Col E1 plasmid DNA (Kenerley *et al.,* 1977; Clarke and Carbon, 1976), that allowed a detailed analysis of their fine structure.

Dissection of the transducing phages carrying the rRNA operons with restriction endonucleases, followed by hybridization of different RNA's to the fragments and electron microscope heteroduplex mapping, confirmed the order of the rRNA genes relative to the promotor. However, spacer regions were also detected between the 16 S rRNA and 23 S rRNA genes in the different plasmids. The spacer sequences in the different rRNA operons were found to be nonhomologous (Doenier *et al.,* 1971; Ohtsubo *et al.,* 1974). The spacer regions of these and other rRNA operons have been shown to be tRNA genes (Morgan *et al.*, 1977). Transfer RNA genes are also present in some rRNA operons at a point distal to (at the 3′-end of) the 5 S rRNA gene (Ikemura and Nomura, 1977). The tRNA genes are transcribed along with the rRNA's which flank them, and the control of their transcription appears to be determined by the promotor of the ribosomal genes (Morgan *et al.*, 1977; Ikemura and Nomura, 1977).

Similarities between the rRNA genes of *E. coli* and those of eukaryotes revolve around their multiplicity, albeit limited, and the fact that one copy of each of 16 S and 23 S rRNA gene is present in the transcription unit. The possession of the 5 S rRNA gene and "spacer" tRNA genes in the transcription unit, and the absence of tandem repetition, contrast with the eukaryotic rRNA gene organization.

C. The Genes for 5 S RNA

Whereas the genes for rRNA are repeated about 450 times in the haploid genome of *Xenopus* (Birnstiel *et al.*, 1968), the genes for 5 S RNA (5 S DNA) are present in about 25,000 copies per haploid genome in a nonnucleolar location (Brown and Weber, 1968). Nevertheless, the arrangement of the multiple gene sequences for 5 S RNA (see Fig. 9) show characteristics similar to those of rRNA's. 5 S DNA's from both *Xenopus laevis* and *Xenopus borealis* (formerly *Xenopus mulleri*) contain a transcribed region of 120 nucleotide pairs, and a nontranscribed spacer region which is some 600 nucleotide pairs long in *X. laevis* and 1000 nucleotide pairs long in *X.*

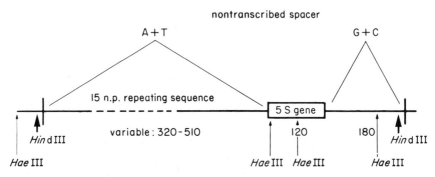

Fig. 9. The structure of the repeating unit of DNA containing the genes for oocyte 5 S RNA in *Xenopus laevis*. The 5 S gene and G + C-rich spacer sequences are of constant length. The A + T-rich nontranscribed spacer region is of variable length and is itself composed of a repeating sequence 15 nucleotide pairs long. The positions of *Hae*III and *Hind*III restriction endonuclease recognition sites are shown by the arrows. The numbers refer to lengths in nucleotide pairs of various segments. For a fuller explanation and the sources from which this figure was compiled, see the text.

borealis (Brown and Sugimoto, 1973a,b). The transcribed gene region and nontranscribed spacer region combine to form the repeating unit which is arranged in tandem arrays. The spacer region of 5 S DNA of *X. laevis* is made up of two subregions with distinguishable overall base composition: an A + T-rich region which forms the major part of the spacer and a G + C-rich region which is proximal to the 3'-end of the transcribed sequence.

Digestion of purified 5 S DNA with the restriction endonuclease *Hae* III produces fragments of three basic sizes. There are two small fragments that contain the gene sequence and the G + C-rich nontranscribed spacer, both of which are homogeneous in size. The largest fragment contains the entire A + T-rich spacer region and is heterogeneous in size. The heterogeneity of sizes is based on increments of 15 nucleotide pairs (Carroll and Brown, 1976a), which is equal to the length of an internally repeated sequence of the A + T-rich spacer region (Brownlee *et al.*, 1974). Analysis of cloned fragments of 5 S DNA from *X. laevis,* in which several tandem copies are retained intact, shows that the heterogeneity of length of the spacer sequence can occur even between adjacent copies of the repeating unit within a single genome (Carroll and Brown, 1976b).

D. The Genes for Transfer RNA

The genes for transfer RNA's are present in multiple copies in eukaryotes (see review in Bostock, 1971) and in prokaryotes (see below), but the analysis of their structure is not as simple as for the ribosomal and 5 S RNA's. This is because there are about 56 distinct tRNA species, reflecting the 20 different amino acids and the different isoaccepting tRNA's for given amino acids (Gallo and Pestka, 1970). Individual tRNA sequences can be analyzed either by prior purification of the molecules or by specific binding of a labeled amino acid. The kinetics of hybridization of a total tRNA preparation from *X. laevis* to its DNA shows that the complexity of the tRNA preparation is consistent with there being at least 43 different tRNA sequences. That is to say that the nucleotide sequence of each tRNA type is sufficiently different from the others not to cross-react during hybridization (Clarkson *et al.*, 1973).

Like the other RNA genes we have considered, the tRNA genes of *X. laevis* are clustered in some form of tandem array which also involves a nontranscribed spacer region comprising about 90% of the repeating unit (Clarkson *et al.*, 1973). On average, each tRNA gene, 74 nucleotides long, is associated with a nontranscribed spacer region 740 nucleotide pairs long. Digestion of DNA, which has been enriched for sequences homologous to $tRNA_1^{Met}$, with either *Eco*RI or *Hpa*I restriction endonucleases

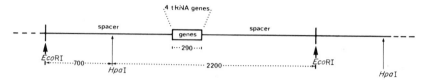

Fig. 10. The structure of the repeating unit that contains the gene for tRNA$_1^{Met}$ in *X. laevis*. Four tRNA genes are clustered close together and are separated from the next cluster of four genes by a nontranscribed spacer region. In this case the spacer appears to be of uniform length. The *Eco*RI and *Hpa*I restriction endonuclease recognition sites are shown by the arrows. The numbers refer to lengths in nucleotide pairs of segments of DNA. (From data of Clarkson *et al.*, 1973; Clarkson and Kurer, 1976.)

produces a homogeneously sized fragment, 3100 nucleotide pairs long, that will hybridize total tRNA. Partial digestion with these enzymes yields a series of fragments whose sizes are simple multiples of 3100 nucleotide pairs, indicating the tandem association of the repeating units (Clarkson and Kurer, 1976). The size of 3100 nucleotide pairs suggests that the repeating unit contains four tRNA genes with associated spacer regions—4 × (74 + 740). These four genes appear to be clustered within the unit, since tRNA$_1^{Met}$ will only hybridize to the larger of two fragments produced by the combined digestion with *Eco*RI and *Hpa*I (Clarkson and Kurer, 1976; see also Fig. 10). From the evidence as it stands at present, the arrangement of the tRNA genes in *Xenopus* has similarities to the arrangements of the genes for other structural RNAs, except that there does not seem to be heterogeneity in the nontranscribed spacer region.

Bacterial genomes have also been shown to contain multiple copies of transfer RNA genes, although the degree of repetition is much lower than for the eukaryotes. Saturation hybridization experiments suggested that the total number of tRNA genes in a single chromosome of *E. coli* is about 60 and that most of them are arranged contiguously in a block (see, e.g., Brenner *et al.*, 1970). Previously, transduction experiments had shown that various tRNA loci mapped to the same region of the chromosome (Cutler and Evans, 1967). Limited multiplicity of specific tRNA genes was first suggested by evidence from suppressor mutations (see, e.g., Russell *et al.*, 1970). Study of a temperature-sensitive mutant of *E. coli* in which the processing of the tRNA becomes defective at the restrictive temperature (Altman and Smith, 1971) has shown that tRNA genes can be transcribed as part of a large precursor molecule. The precursor contains multiple copies of the same tRNA species in tandem array (Sakano *et al.*, 1974; Ilgren *et al.*, 1976). Processing of the precursor molecules involves endonucleolytic cleavage as well as degradation of parts of the molecule—transcribed spacer regions (Altman, 1975).

Thus, in bacteria, multiple copies of tRNA genes are found in tandem arrays in which the gene sequences are separated from their neighbors by a short length of transcribed "spacer" sequence. Nontranscribed spacers would not appear to be present, at least between tRNA genes of the same type. Not all tRNA genes occur next to other tRNA genes, as some are found inserted between the genes for the rRNA's (see Section V,B).

E. The Genes for Histones

Most of the detailed structural analysis of the histone genes has been done on sea urchin DNA. The sea urchin contains several hundred copies of the genes for histones (Kedes and Birnstiel, 1971; Weinberg *et al.*, 1972), although other eukaryotes such as man, mouse or *Drosophila*, contain between 10 and 50 copies only.

In keeping with the properties of other repeated eukaryotic genes, the histone genes of sea urchins are associated with nontranscribed spacer DNA, which is rich in A + T (Birnstiel *et al.*, 1974). The presence of a tandemly repeating unit in histone DNA has been revealed by the use of restriction endonucleases. The enzyme *Hind*III cleaves histone DNA from either *Strongylocentrotus purpuratus* or *Psammechinus miliaris* into fragments of a single homogeneous size of about 6000 nucleotide pairs (Weinberg *et al.*, 1975; Gross *et al.*, 1976b; Schaffner *et al.*, 1976). As Fig. 11 shows, the 6000 nucleotide pair unit contains one copy each of all five histones. This was shown for *P. miliaris* histone DNA by subdividing the unit into five fragments by the combined use of *Eco*RI, *Hind*III, *Hind*II,

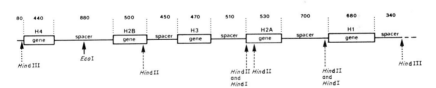

Fig. 11. The structure of the repeating unit that contains the histone genes in *Psammechinus miliaris*. The repeating unit contains one copy of the gene sequence for each of the five histones, H1, H2A, H2B, H3, and H4. Each histone gene sequence is separated from its neighboring gene sequences by a length of spacer DNA. The restriction endonuclease *Hind*III cuts histone DNA at only one site in the repeating unit in the spacer region between the H1 and H4 gene sequences. Whole unit-sized fragments produced by *Hind*III digestion, as well as subfragments of the repeating unit produced by *Eco*RI, *Hind*II, and *Hpa*I digestion, are homogeneous in size. This shows that there is no heterogeneity in the length of the spacer sequences in histone DNA. The numbers refer to the lengths in nucleotide pairs of the various segments of the unit. The arrows indicate the positions of recognition sites for the various restriction endonucleases. For a fuller explanation and the sources from which this figure was compiled, see text.

and *Hpa*I restriction endonucleases. Using mRNA's for each of the five individual histones, Gross *et al.* (1976a) found that each histone mRNA would hybridize to only one of the five fragments produced by restriction enzyme digestion (Schaffner *et al.*, 1976). Analysis of the fragments produced by digestion with different combinations of the four enzymes enabled the positioning of the histone genes with respect to each other and to the ends of the repeating unit (see Fig. 11).

Electron microscope visualization, following partial denaturation, of the 6000 nucleotide pair fragment confirms the "map" and shows clearly the separation of adjacent histone gene sequences by the A + T-rich non-transcribed spacer DNA (Portmann *et al.*, 1976). The coding sequences of the genes are all arranged on the same strand of DNA, transcription proceeding from the histone H4 gene through to the histone H1 gene (Gross *et al.*, 1976b). For sea urchins there is no information to date about whether the repeating unit is transcribed as a whole to produce a precursor RNA that is subsequently cleaved to individual histone mRNA's or whether each histone gene is a unit of transcription. In HeLa cells, however, the histone genes are transcribed as part of a high molecular weight nucleus-restricted RNA molecule that is later processed to yield the 9 S mRNA's identified in the cytoplasm (Melli *et al.*, 1977a,b). It would seem that the spacer sequences associated with histone genes are transcribed and that the five genes together form a transcription unit.

F. Repeated Gene Sequences That Are Not Arranged in Tandem

The repeated structural genes that have been discussed above all have the common property of being arranged in tandem. However, that the tandem arrangement is not universal has become evident from studies on cloned DNA fragments of *D. melanogaster,* which contain gene sequences of the *copia* and *412* gene families (reviewed in Finnegan *et al.*, 1978). Cloned DNA fragments, containing one of the sequences of a repeated gene set, were isolated in the following manner. Randomly sheared DNA from *D. melanogaster* was inserted into the Col E1 plasmid after adding poly(dA) and poly(dT) tails, respectively, to the two DNA's. After infection of *E. coli* with the recombinant plasmid, different clones were selected for their ability to hybridize the sequences present in the abundant class of cytoplasmic poly(A)-containing mRNA's. This procedure will identify any cloned DNA sequence that is homologous to a sequence present in the population of abundant mRNA's. Thus, if a gene sequence is present in the genome in multiple copies, any one of the copies could be selected by this procedure, even though perhaps only one of the copies was actually being transcribed to produce the abundant mRNA.

Using this approach, two groups of clones, *copia* and *412*, were identified which carried DNA fragments that were homologous to the abundant mRNA's. The frequency of occurrence of these clones was far higher than that expected for DNA sequences present only once per haploid genome, suggesting that all or part of the cloned *Drosophila* sequences were present in multiple copies in the total genome (Young and Hogness, 1978). The cloned *copia* and *412* fragments can be subdivided by digestion with restriction enzymes and mapped with respect to the sites of cleavage by these enzymes. Hybridization of the subfractions with the abundant mRNA preparation shows that the transcribed regions occupy only a portion of the cloned *Drosophila* DNA. The cloned DNA segments, therefore, consist of a region of transcribed repeated gene which is flanked on either side by lengths of other *Drosophila* DNA sequences.

In situ hybridization of all or part of either *copia* or *412* sequences to polytene chromosomes shows that they are present at multiple and distinctly *separate* loci in the chromosomes. [This contrasts markedly with the *in situ* hybridization of 5 S RNA, a tandemly repeated gene, to a single site in polytene chromosomes of *Drosophila* (Wimber and Steffensen, 1970).] Following dissection of the DNA fragments with restriction endonucleases, the different internal subfractions, containing the sequences that are homologous to mRNA, all hybridize *in situ* to the same locations on polytene chromosomes. The sites of *in situ* hybridization of the flanking sequences vary between the different cloned fragments. This variation is thought to reflect the other chromosomal DNA sequences present at either end of the different sites in which these repeated gene families are located.

Copia and *412* gene families have one other intriguing property in common, namely, that the sequences at either end of the transcribed gene region are direct repeats. Direct terminal repetition is difficult to analyze in complex eukaryotic DNA's but is a feature of the ends of some phages and viruses. The significance of this finding in *Drosophila* DNA is not clear at present, although it has been suggested that it may be related to the possibility that the repeated gene families could be inserted at specific chromosomal locations in each organismal generation (Finnegan *et al.*, 1978).

VI. SINGLE-COPY STRUCTURAL GENES

A. Arrangement in Prokaryotes

Current ideas about the organization of the various elements of single-copy structural genes are largely based on bacterial gene systems. This is because bacterial genes are usually present in single copies in a haploid chromosome of an organism that divides rapidly, and are thus readily amenable to genetic analysis. More recently, DNA sequencing studies of particular regions of structural gene elements have complemented the genetic analysis and added to our understanding of the structure and organization of control regions of structural genes. These are discussed in detail by Zubay (Chapter 4, this volume). Below is a very brief summary of some of the features of prokaryotic structural genes to facilitate comparison with the features discussed in the following two sections.

Structural genes in prokaryotes have the following properties. The origin of transcription is at the promotor site, which contains a polymerase binding site and can contain a positive regulatory site (for example, the CAP binding site). In the direction of transcription the promotor is followed by the operator (the repressor binding site), which can be followed by a further regulatory sequence, called the leader sequence. Distal to the operator or leader sequences are the structural genes or gene. Single gene operons may exist, but if more than one gene is present, they can be transcribed to form a polygenic mRNA. The initiation signals and reading frames for transcription and translation of all the structural genes of an operon are fixed near the 5′-end of the mRNA. This is shown by analyzing frame-shift mutations at points after the initiation signal. In general a single frame-shift mutation near the 5′-end of the mRNA affects all the genes distal to the mutation. There are, therefore, no regions within or between the structural genes of a prokaryotic operon that can "absorb" the effects of a mutation and restore the correct reading frame in the coding regions of distal genes (cf. the "insert" sequences of eukaryotic genes discussed in Section VI,C,2).

B. Structural Genes That Overlap

Bacteriophage ϕX174 contains a circular genome which consists of a single strand of DNA some 5375 nucleotides long (Barrell *et al.,* 1976; Sanger *et al.,* 1977). The genome codes for nine known proteins, the genes for these proteins being designated *A* to *H* (see Fig. 12; see also Benbow *et al.,* 1974). The combined molecular weight of the nine proteins is in excess of the coding capacity of a piece of DNA the size of the ϕX174 genome, assuming that all the coding sequences are arranged in linear succession along the DNA (see, e.g., Barrell *et al.,* 1976).

This paradox has been resolved following the determination of the complete nucleotide sequence of ϕX174 DNA (Sanger *et al.,* 1977; Smith *et al.,* 1977) and relating it to the known genetic map, the restriction enzyme cleavage site maps, and the known amino acid sequences of the proteins for which the genes code. As shown in Fig. 12, there are two

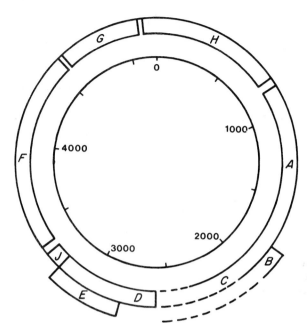

Fig. 12. The map of the circular genome of bacteriophage ϕX174. The blocks designated A to H represent the sizes and locations of the known genes in the ϕX174 chromosome. Notice that gene E is contained within gene D, and that gene B starts within gene A, although the termination points of genes A and B are not fully established. The numbers refer to distances, in nucleotide pairs, around the circular chromosome. (Based on data frѻm Barrell *et al.*, 1976; Smith *et al.*, 1977; Sanger *et al.*, 1977.)

areas of the genome in which coding sequences overlap. Barrell *et al.* (1976) first showed that the nucleotide sequence of gene E is completely contained within the nucleotide sequence of gene D, but is read into a message in a different reading frame, being displaced one nucleotide to the "right" or in the 5'- to 3'-direction. The termination signal (TGA) for the E gene is five nucleotides before the termination signal for the D gene (TAA), which itself overlaps with the start signal for the next gene, the J gene. The second region of overlapping genes is found where gene B is wholly contained within the sequence of gene A, again in a different reading frame (Sanger *et al.*, 1977; Smith *et al.*, 1977).

Overlapping structural genes have also been identified in the eukaryotic virus SV40. The genome of this virus, like ϕX174, is circular and about 5375 nucleotide pairs long, but it is composed of double-stranded DNA (reviewed in Tooze, 1973). Two distinct sets of genes can be identified according to the time during lytic infection at which they are expressed. Soon after infection an "early" viral-specific 19 S mRNA appears in the cytoplasm. This is coded for by about one-half of one of the genomic strands. Later on in the lytic cycle, after viral DNA replication has been initiated, three "late" viral-specific mRNA's appear. These are transcribed in the opposite direction from about 60% of the other strand of the genome, and code for the viral structural proteins VP1, VP2, and VP3 (Levine, 1976). The 5'-end of the mRNA that codes for the VP2 protein has been localized to a point at about 0.76 of the fractional gene map, at a point close to the junction between two DNA fragments produced by a combination of HindII and HindIII restriction enzyme digestion (May et al., 1977; Dahr et al., 1978). A few nucleotides further on, in the direction of transcription, there is an initiation triplet AUG which would permit the "in-phase" reading of the subsequent 340 triplets. This is the only "readable" sequence that would be long enough to code for the entire VP2 gene product. One hundred and seven codons further on, in the same reading frame, there is another methionine codon which probably serves as the initiator for the translation of the smaller VP3 protein (Reddy et al., 1978). Thus, the gene for the VP3 protein appears to be contained within the gene for the VP2 protein, but unlike the overlapping genes of ϕX174, the two genes are translated in the same register. There is, however, another short region of overlap between the VP1 and the VP2 and VP3 genes in which the codons read in different registers (Reddy et al., 1978).

These are two examples, one prokaryotic and the other eukaryotic, in which genes coding for different proteins overlap and use the same basic nucleotide sequence. In the case of ϕX174, the overlap makes use of the different information obtained by reading the basic sequence in different registers. In SV40 both "in-phase" and "out-of-phase" overlaps occur. At present it seems reasonable to assume that the overlapping of genes is a mechanism for maximizing the information content of small genomes, although it is not known whether similar arrangements exist in larger genomes where size would not be a limiting factor.

C. Noncontinuity of Structural Gene Sequences

1. Leader Sequences

Until very recently it has been assumed that the initial process of information transfer from gene to the final gene product involved the transcription of a message (in the form of mRNA with or without subsequent processing) from a continuous stretch of DNA. Recent experiments on animal virus genes and selected single-copy structural genes of higher animals suggest that the functional unit of a structural gene may span noncontiguous segments of DNA.

Comparison of the size and nucleotide sequence of purified individual mRNAs with the size and amino acid sequence of the proteins for which they code shows that the mRNA contains additional sequences that are not involved in the specification of the amino acid sequence (Barelle, 1977; Proudfoot, 1977; Efstratiadis *et al.*, 1977; see also the review in Lewin, 1975). These extra sequences occur at both the 3'- and 5'-ends of the message, in addition to the polyadenylation that is also found at the 3'-end of many mRNA's. Thus, transcription of a gene produces an mRNA which includes sequences other than just the coding sequences.

Adenovirus, a linear double-stranded DNA virus of about 60,000 nucleotide pairs, synthesizes both "early" and "late" mRNAs during infection (see review in Tooze, 1973). Like the mRNA's of the eukaryotic host cells, these appear as polyadenylated molecules in the cytoplasm (Philipson *et al.*, 1971) with the 5'-ends capped (Sommer *et al.*, 1976; see also Williamson, Chapter 10, this volume). Hybridization of the late mRNA's to adenovirus DNA produces hybrid molecules with the common property of sensitivity of the 5'-end of the mRNA's to ribonuclease digestion (Gelinas and Roberts, 1977). Sensitivity to ribonuclease shows that the 5'-end of the mRNA is not involved in a hybrid duplex structure with DNA and suggests that its complementary sequence is not juxtaposed to the complementary DNA sequence that hybridizes the major part of the mRNA molecule. Assuming that they are composed of sequences that are coded for by the DNA, the sequences present at the 5'-ends of late mRNA's must, therefore, be transcribed from DNA sequences elsewhere in the adenovirus genome.

Chow *et al.* (1977) hybridized late mRNA's to double-stranded DNA fragments which had been produced by restriction endonuclease cleavage under conditions in which the RNA–DNA hybrids were more stable that the DNA–DNA duplexes (R looping; see Thomas *et al.*, 1976). Visualization in the electron microscope of the structures produced showed that short stretches of nucleotides, which are complementary to three separate locations in the chromosome, are joined in some way in the final mRNA molecule to form a "leader" sequence of about 150 to 200 nucleotides located at the 5'-end of many late mRNA's (Chow *et al.*, 1977). Further studies (Klessig, 1977; Berget *et al.*, 1977) examined the ability of DNA restriction fragments from various parts of the genome to protect regions of two specific late mRNA's from ribonuclease digestion. It was found that both mRNA's contain a long sequence in common, which is complementary to at least two different sites on the adenovirus genome, both of which are remote from the coding portion of these two genes. A similar finding has been shown for the mRNA transcribed from the VP1 gene of SV40 (Aloni *et al.*, 1977; Celma *et al.*, 1977).

This raises the question of how precisely we can define a structural gene. Is it simply the sequence that codes for a protein, or does it also include the additional elements that are found in the mRNA and/or the insertion sequences in the DNA? By analogy with the trytophan operon of *E. coli,* the leader sequence at the 5'-end of the mRNA should be distinguished from the structural gene sequence, but in eukaryotes we do not know what function the leader sequences have. If we include the leader sequences with the structural gene in eukaryotes, the results on adenovirus and SV40 imply that structural genes are not necessarily arranged as a continuous unit, but that various parts of the gene can be distributed at different chromosomal locations. These parts can than be subsequently joined up to form a single unit either by some selective transcription process or by processing and ligation of selected regions of a large RNA transcript (see Williamson, Chapter 10, this volume).

2. Structural Gene Coding Sequences That Are Interrupted

We do not know at present what functions leader sequences may have in the mRNAs that contain them, although it is clear that they do not have a coding function. However, from the information that is beginning to emerge for the ovalbumin gene of the chicken and the β-globin gene of the rabbit, it looks as though even the coding sequences of structural genes can be separated into different regions of the chromosome. The experimental procedures used for the study of these two genes are similar. The mRNAs, which code for these proteins, are isolated and purified and a complementary DNA is synthesized on the mRNA template. This cDNA can then be cloned in a bacterial plasmid to purify the sequence further or can be used directly after conversion to a full length double-stranded form. The cDNA fragment is first mapped with respect to sites of cleavage by certain restriction endonucleases. If the RNA is a copy of a continuous genomic sequence, the sites of restriction endonuclease cleavage should be the same in the genomic DNA as they are in the cDNA. Thus, the sizes of fragments produced by restriction endonuclease digestion of genomic DNA should correspond to those produced from the cDNA, with the exception of those fragments that correspond to the ends of the mRNA, where the sequences would be expected to be joined to additional flanking sequences in the chromosomes. On the other hand, if the sequences present in the mRNA (and therefore those present in the cDNA) are transcribed from noncontiguous parts of the chromosome, the distribution of sizes of fragments of genomic DNA, containing sequences complementary to the mRNA, will not necessarily be the same as those of the cDNA. This is because the insertion of an additional length of DNA would change the distance relationships between restriction endonuclease cleavage sites (see Fig. 13).

The ovalbumin gene of the chicken has been studied in this way using both double-stranded cDNA directly (Doel *et al.*, 1977), or using cDNA cloned in a bacterial plasmid (Breathnach *et al.*, 1977). The cDNA prepared from ovalbumin mRNA does not contain a recognition site for either *Hin*dIII or *Eco*RI restriction endonucleases. Thus, digestion of ovalbumin cDNA with either of these enzymes would yield intact cDNA molecules. If the ovalbumin mRNA is transcribed from a single continuous sequence in genomic DNA, its sequence should be contained in a single fragment of genomic DNA whose size is equal to, or larger than, the size of intact cDNA after *Hin*dIII or *Eco*RI digestion. In practice, *Hin*dIII or *Eco*RI digestion of chromosomal DNA results in the production of three or four fragments, respectively, which contain sequences that are complementary to ovalbumin cDNA. This means that in the chromosome the ovalbumin gene contains sequences other than those contained in the mRNA inserted at various internal locations (see Fig. 13). At least one of the "inserted" sequences must be in the coding region and cannot be simply due to the separation of leaderlike sequences from coding sequences. The pattern of interspersion of complementary and noncomplementary sequences to the mRNA is the same in oviduct cells, which actively synthesize ovalbumin mRNA, as it is in erythrocyte cells

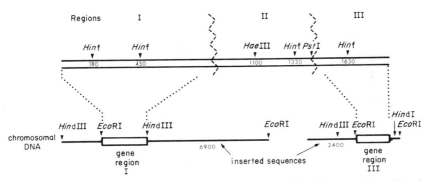

Fig. 13. The structure of ovalbumin cDNA and the regions of the chromosomal ovalbumin gene that contain inserted sequences. At the top is shown the map of some restriction sites in the cDNA made on a template of ovalbumin mRNA. Notice that there are no *Eco*RI or *Hin*dIII recognition sites present. In chromosomal DNA the cDNA sequence is "split up" into at least three different regions, I, II, and III. These regions are separated from one another by additional noncoding sequences, which can be identified by the presence of *Eco*RI or *Hin*dIII recognition sites within them. In the bottom half of the figure these inserted sequences are shown as single lines on either side of the blocks that represent the gene regions. (Redrawn and modified from Breathnach *et al.*, 1977.)

which do not make this mRNA. The presence of the intervening sequences in the coding regions therefore does not appear to be a result of transcriptional activity (Doel *et al.*, 1977; Breathnach *et al.*, 1977).

Jeffreys and Flavell (1977a,b) have shown that the same phenomenon exists for the β-globin gene of rabbit, using a similar experimental approach (and undoubtedly other examples are being found monthly). A plasmid carrying cDNA to β-globin mRNA of rabbit was first mapped with respect to the distribution of various restriction endonuclease cleavage sites. The cDNA map was compared to the map obtained if total genomic DNA was digested with these enzymes, and consistent anomalies were found. For example, the coding sequence portion of the cDNA of the plasmid contains a *Bam*H1 cleavage site 67 nucleotides away from an *Eco*RI cleavage site. In genomic DNA these two sites are separated by about 700 nucleotide pairs. Other inconsistencies between the two maps suggest that in chromosomal DNA there must be a noncoding sequence, 600 nucleotide pairs in length, inserted within the coding sequence of the β-globin gene (Jeffreys and Flavell, 1977b).

VII. CONCLUDING REMARKS

It is now clear that at the whole genome level the DNA sequences of eukaryotic chromosomes can be classified into a number of different categories, although the categories themselves do not have well-defined limits and can overlap. Four main classes can be identified using rate of reassociation as the basis for classification. These are foldback, fast reassociating, intermediate reassociating, and single-copy sequences. The fast reassociating sequences often include those sequences that can be identified as buoyant density satellites. Many of the properties of intermediate reassociating sequences are those that would be expected for sequences that were once fast reassociating or satellitelike sequences, but which have diverged through base substitutions. In a similar way, many of the sequences that are identified kinetically as being single-copy sequences may be highly diverged intermediate sequences. Such an interpretation does not prove a common mechanism for the origin of these sequences and says little about their present-day functional roles in the cell.

A second way of classifying sequences is on the basis of their patterns of interspersion with other sequences. The intermediate reassociating sequences would appear to fall into two main categories with respect to their interspersion with single-copy sequences—the long and short periodicity interspersion patterns—although the functional significance of such arrangements is not clear at present. The functional units of single-copy sequences themselves appear to be interspersed with sequences of unknown type, as is evidenced by the splitting of functional coding sequences by noncoding sequences.

The repeated structural genes show a third general pattern of sequence arrangement. This is the interspersion of transcribed coding sequences and spacer sequences with nontranscribed spacer regions, the three regions together forming a unit which is tandemly repeated. The nontranscribed spacer regions of repeated structural genes have many properties in common with satellitelike sequences; both consist of short basic sequences repeated many times within the genome and organized in hierarchies of tandemly arranged units. On the other hand, not all repeated structural genes are arranged in tandem, since in *Drosophila* two repeated gene sequences that are transcribed into mRNA have been shown to be dispersed throughout the genome.

The study of the different types of sequences present in eukaryotic chromosomes, and their organization, is of significance to the whole area of gene expression and transcription. An understanding of the way different sequences, transcribed or not, are juxtaposed or separated sheds new light on the way control of gene expression may be effected and gives some insight into the relationship between heterogeneous nuclear RNA and cytoplasmic mRNA. The existence of large proportions of nontranscribed sequences in the genome goes some way toward an explanation of the C value paradox, although it is still a major question about what function the vast majority of repeated DNA sequences serve in the cell.

REFERENCES

Allan, J. (1974). Comparative studies on the satellite DNA of related rodent species. Ph.D. Thesis, Univ. of Edinburgh, Edinburgh.

Aloni, Y., Dhar, R., Laub, O., Horowitz, M., and Khoury, G. (1977). Novel mechanism for mRNA maturation: The leader sequences of simian virus 40 mRNA are not transcribed adjacent to the coding sequences. *Proc. Natl. Acad. Sci. U.S.A.* **74**, 3686–3690.

Altman, S. (1975). Biosynthesis of transfer RNA in *Escherichia coli. Cell* **4**, 21–29.

Altman, S., and Smith, J. D. (1971). Tyrosine tRNA precursor molecule polynucleotide sequence. *Nature (London), New Biol.* **233**, 35–39.

Ando, T. (1966). A nuclease specific for heat-denatured DNA isolated from a product of *Aspergillus oryzae. Biochim. Biophys. Acta* **114**, 158–168.

Angerer, R. C., Davidson, E. H., and Britten, R. J. (1975). DNA sequence organisation in the mollusc *Aplysia californica. Cell* **6**, 29–34.

Avery, O. T., Macleod, C. M., and McCarty, M. (1944). Studies on the chemical nature of the substance inducing transformation of pneumococcal types. Induction of transformation by a desoxyribonucleic acid fraction isolated from *Pneumococcus* type III. *J. Exp. Med.* **79**, 137–158.

Barelle, F. E. (1977). Complete nucleotide sequence of the 5' non-coding region of rabbit β-globin mRNA. *Cell* **10,** 549–558.

Barrell, B. G., Air, G. M., and Hutchison, C. A., III (1976). Overlapping genes in bacteriophage φX174. *Nature (London)* **264,** 34–41.

Beauchamp, R. S. (1977). Sequence complexity and organisation of the DNA in the free-living nematode *Panagrellus silusiae:* Absence of short period interspersion. Ph.D. Thesis, Univ. of Waterloo, Waterloo, Ontario.

Benbow, R. M., Zuccarelli, A. J., Davis, G. C., and Sinshiemer, R. L. (1974). Genetic recombination in a bacteriophage φX174. *J. Virol.* **13,** 898–907.

Berget, S. M., Moore, C., and Sharp, P. A. (1977). Spliced segments at the 5' terminus of adenovirus late mRNA. *Proc. Natl. Acad. Sci. U.S.A.* **74,** 3171–3175.

Bernardi, G. (1965). Chromatography of nucleic acids on hydroxyapatite *Nature (London)* **206,** 779–783.

Birnstiel, M. L., Spiers, J., Purdom, I., Jones, K., and Loening, U. E. (1968). Properties and composition of the isolated ribosomal DNA satellite of *Xenopus laevis. Nature (London)* **219,** 454–463.

Birnstiel, M. L., Telford, J., Weinberg, E., and Stafford, D. (1974). Isolation and some properties of the genes coding for histone proteins. *Proc. Natl. Acad. Sci. U.S.A.* **71,** 2900–2904.

Biro, P. A., Carr-Brown, A., Southern, E. M., and Walker, P. M. B. (1975). Partial sequence analysis of mouse satellite DNA: Evidence for short-range periodicities. *J. Mol. Biol.* **94,** 71–86.

Bishop, J. O., Morton, J. G., Rosbash, M., and Richardson, M. (1974). Three abundance classes in HeLa cell messenger RNA. *Nature (London)* **250,** 199–204.

Bonner, T. I., Brenner, D. J., Neufeld, B. R., and Britten, R. J. (1973). Reduction in the rate of DNA reassociation by sequence divergence. *J. Mol. Biol.* **81,** 123–135.

Bostock, C. J. (1971). Repetitious DNA. *Adv. Cell Biol.* **2,** 153–224.

Bostock, C. J., and Sumner, A. T. (1978). "The Eukaryotic Chromosome." North-Holland Publ., Amsterdam.

Bostock, C. J., Gosden, J. R., and Mitchell, A. R. (1978). Localisation of a male-specific DNA fragment to a sub-region of the human Y-chromosome. *Nature (London)* **272,** 324–328.

Botchan, M. R. (1974). Bovine satellite I DNA consists of repetitive units 1,400 base pairs in length. *Nature (London)* **251,** 288–292.

Breathnach, R., Mandel, J. L., and Chambon, P. (1977). Ovalbumin gene is split in chicken DNA. *Nature (London)* **270,** 314–319.

Brenner, D. J., Fournier, M. J., and Doctor, B. P. (1970). Isolation and partial characterisation of the transfer ribonucleic acid cistrons from *Escherichia coli. Nature (London)* **227,** 448–451.

Britten, R. J., and Davidson, E. H. (1969). Gene regulation for higher cells: A theory. *Science* **165,** 349–357.

Britten, R. J., and Kohne, D. E. (1966). Nucleotide sequence repetition in DNA. *Carnegie Inst. Wash., Yearb.* **65,** 78–106.

Britten, R. J., and Kohne, D. E. (1968). Repeated sequences in DNA. *Science* **161,** 529–540.

Britten, R. J., and Smith, J. (1969). A bovine genome. *Carnegie Inst. Wash., Yearb.* **68,** 378–386.

Britten, R. J., Graham, D. E., and Neufeld, B. R. (1974). Analysis of repeating DNA sequences by reassociation. *In* "Nucleic Acids and Proteins Synthesis," Part E (L. Grossman and K. Moldave, eds.), Methods in Enzymology, Vol. 29, pp. 363–418. Academic Press, New York.

Britten, R. J., Graham, D. E., Eden, F. G., Painchaud, D. M., and Davidson, E. H. (1976). Evolutionary divergence and length of repetitive sequences in sea urchin DNA. *J. Mol. Evol.* **9**, 111–138.

Brown, D. D., and Sugimoto, K. (1973a). The structure and evolution of ribosomal and 5 S DNAs in *Xenopus laevis* and *Xenopus mulleri. Cold Spring Harbor Symp. Quant. Biol.* **38**, 501–505.

Brown, D. D., and Sugimoto, K. (1973b). 5 S DNAs of *Xenopus laevis* and *Xenopus mulleri:* Evolution of a gene family. *J. Mol. Biol.* **78**, 397–415.

Brown, D. D., and Weber, C. S. (1968). Gene linkage by RNA–DNA hybridisation. I. Unique DNA sequences homologous to 4 S RNA, 5 S RNA and ribosomal RNA. *J. Mol. Biol.* **34**, 661–680.

Brown, D. D., Wensink, P. C., and Jordan, E. (1972). A comparison of the ribosomal DNAs of *Xenopus laevis* and *Xenopus mulleri:* The evolution of tandem genes. *J. Mol. Biol.* **63**, 57–73.

Brown, I. R., and Church, R. B. (1972). Transcription of non-repeated DNA during mouse and rat development. *Dev. Biol.* **29**, 73–84.

Brownlee, G. G., Cartwright, E. M., and Brown, D. D. (1974). Sequence studies of the 5 S DNA of *Xenopus laevis. J. Mol. Biol.* **89**, 703–718.

Brutlag, D. L., and Peacock, W. J. (1975). Sequences of highly repeated DNA in Drosophila melanogaster. *In* "The Eukaryote Chromosome" (W. J. Peacock and R. D. Brock, eds.), pp. 35–45. Aust. Natl. Univ. Press, Canberra.

Campo, M. S., and Bishop, J. O. (1974). Two classes of messenger RNA in cultured rat cells: Repetitive sequence transcripts and unique sequence transcripts. *J. Mol. Biol.* **90**, 649–663.

Carlson, M., and Brutlag, D. (1977). Cloning and characterisation of a complex satellite DNA from *Drosophila melanogaster. Cell* **11**, 371–381.

Carroll, D., and Brown, D. D. (1976a). Repeating units of *Xenopus laevis* oocyte-type 5 S DNA are hetergeneous in length. *Cell* **7**, 467–475.

Carroll, D., and Brown, D. D. (1976b). Adjacent repeating units of *Xenopus laevis* 5 S DNA can be heterogeneous in length. *Cell* **7**, 477–486.

Cech, T. R., and Hearst, J. E. (1975). An electron microscope study of mouse foldback DNA. *Cell* **5**, 429–446.

Cech, T. R., Rosenfeld, A., and Hearst, J. E. (1973). Characterisation of the most rapidly renaturing sequences in mouse main-band DNA. *J. Mol. Biol.* **81**, 299–325.

Celma, M. L., Dhar, R., Pan, J., and Weissman, S. M. (1977). Comparison of the nucleotide sequence of the messenger RNA for the major structural protein of SV40 with the DNA sequence encoding the amino acids of the protein. *Nucleic Acids Res.* **4**, 2549–2559.

Chamberlain, M. E., Britten, R. J., and Davidson, E. H. (1975). Sequence organisation in *Xenopus* DNA studied by the electron microscope. *J. Mol. Biol.* **96**, 317–333.

Chow, L. T., Gelinas, R. E., Broker, T. R., and Roberts, R. J. (1977). An amazing sequence arrangement at the 5′ ends of adenovirus 2 messenger RNA. *Cell* **12**, 1–8.

Clarke, L., and Carbon, J. (1976). A colony bank containing synthetic Col E1 hybrid plasmids representative of the entire *E. coli* genome. *Cell* **9**, 91–99.

Clarkson, S. G., and Kurer, V. (1976). Isolation and some properties of DNA coding for tRNA$_1^{Met}$ from *Xenopus laevis. Cell* **8**, 183–195.

Clarkson, S. G., Birnstiel, M. L., and Serra, V. (1973). Reiterated transfer RNA genes of *Xenopus laevis. J. Mol. Biol.* **79**, 391–410.

Cooke, H. J. (1976). Repeated sequence specific to human males. *Nature (London)* **262**, 182–186.

Corneo, G., Ginelli, E., Soave, C., and Bernardi, G. (1968a). Isolation and characterisation of mouse and guinea pig satellite deoxyribonucleic acids. *Biochemistry* **7**, 4373–4379.

Corneo, G., Ginelli, E., and Polli, E. (1968b). Isolation of the complementary strands of a human satellite DNA. *J. Mol. Biol.* **33**, 331–335.

Corneo, G., Ginelli, E. and Polli, E. (1970a). Different satellite deoxyribonucleic acids of guinea pig and ox. *Biochemistry* **9**, 1565–1571.

Corneo, G., Ginelli, E., and Polli, E. (1970b). Repeated sequences in human DNA. *J. Mol. Biol.* **48**, 319–327.

Corneo, G., Ginelli, E., and Polli, E. (1971). Renaturation properties and localisation in heterochromatin of human satellite DNAs. *Biochim. Biophys. Acta* **247**, 528–534.

Corneo, G., Zardi, L., and Polli, E. (1972). Elution of human satellite DNAs on a methylated albumin kieselguhr chromatographic column: Isolation of satellite DNA 4. *Biochim. Biophys. Acta* **269**, 201–204.

Crain, W. R., Eden, F. C., Pearson, W. R., Davidson, E. H., and Britten, R. J. (1976a). Absence of short period interspersion of repetitive and non-repetitive sequences in the DNA of *Drosophila melanogaster*. *Chromosoma* **56**, 309–326.

Crain, W. R., Davidson, E. H., and Britten, R. J. (1976b). Contrasting patterns of DNA sequence arrangement in *Apis mellifera* (honeybee) and *Musca domestica* (housefly). *Chromosoma* **59**, 1–12.

Crick, F. H. C. (1971). General model for the chromosomes of higher organisms. *Nature (London)* **234**, 25–27.

Cutler, R. G., and Evans, J. E. (1967). Relative transcription activity of different segments of the genome throughout the cell division cycle of *Escherichia coli*. The mapping of ribosomal and transfer RNA and the determination of the direction of replication. *J. Mol. Biol.* **26**, 91–105.

Dahr R., Reddy, V. B., and Weissman, S. M. (1978). Nucleotide sequence of the DNA encoding the 5' terminal sequences of SV40 late mRNA. *J. Biol. Chem.* **253**, 612–620.

Davidson, E. H., Hough, B. R., Amenson, C. S., and Britten, R. J. (1973a). General interspersion of repetitive with non-repetitive sequence elements in the DNA of *Xenopus*. *J. Mol. Biol.* **77**, 1–23.

Davidson, E. H., Hough, B. R., Amenson, C. S., and Britten, R. J. (1973b). Arrangement and characterisation of repetitive sequence elements in animal DNAs. *Cold Spring Harbor Symp. Quant. Biol.* **38**, 295–301.

Davidson, E. H., Galau, G. A., Angerer, R. C., and Britten, R. J. (1975). Comparative aspects of DNA organisation in metazoa. *Chromosoma* **51**, 253–259.

Davies, J., and Nomura, M. (1972). The genetics of bacterial ribosomes. *Annu. Rev. Genet.* **6**, 203–234.

Dawid, I. B., Brown, D. D., and Reeder, R. H. (1970). Composition and structure of chromosomal and amplified ribosomal DNAs of *Xenopus laevis*. *J. Mol. Biol.* **51**, 341–360.

Doel, M. T., Houghton, M., Cook, E. A., and Carey, N. H. (1977). The presence of ovalbumin mRNA coding sequences in multiple restriction fragments of chicken DNA. *Nucleic Acids Res.* **4**, 3701–3712.

Doenier, R. C., Ohtsubo, E., Lee, H. J. and Davidson, N. (1974). Electron microscope heteroduplex studies of sequence relations among plasmids of *Escherichia coli*. VII. Mapping the ribosomal RNA genes of plasmid F14. *J. Mol. Biol.* **89**, 619–629.

Dott, P. J., Chuang, C. R., and Saunders, G. F. (1976). Inverted repeated sequences in the human genome. *Biochemistry* **15**, 4120–4125.

Doty, P., Marmur, J., Eigner, J., and Schildkraut, C. (1960). Strand separation and specific recombination in deoxyribonucleic acids: Physical chemical studies. *Proc. Natl. Acad. Sci. U.S.A.* **46**, 461–476.

Eden, F. C., Graham, D. E., Davidson, E. H., and Britten, R. J. (1977). Exploration of long and short repetitive sequence relationships in the sea urchin genome. *Nucleic Acids Res.* **4**, 1553–1567.

Efstratiadis, A., Kafatos, F. C., and Maniatis, T. (1977). The primary structure of rabbit β-globin mRNA as determined from cloned DNA. *Cell* **10**, 571–586.

Endow, S. A. (1977). Analysis of *Drosophila melanogaster* satellite IV with restriction endonuclease *Mbo* II. *J. Mol. Biol.* **114**, 441–449.

Filipski, J., Thiery, J.-P., and Bernardi, G. (1973). An analysis of the bovine genome by Cs_2SO_4-Ag^+ density gradient centrifugation. *J. Mol. Biol.* **80**, 177–197.

Finnegan, D. J., Rubin, G. M., Young, M. W., and Hogness, D. S. (1978). Repeated gene families in *Drosophila melanogaster. Cold Spring Harbor Symp. Quant. Biol.* **42**, 1053–1063.

Flamm, W. G., Bond, H. E., and Burr, H. E. (1966). Density–gradient centrifugation of DNA in a fixed angle rotor. *Biochim. Biophys. Acta* **129**, 310–319.

Flamm, W. G., McCallum, M., and Walker, P. M. B. (1967). The isolation of complementary strands from a mouse DNA fraction. *Proc. Natl. Acad. Sci. U.S.A.* **57**, 1729–1734.

Flamm, W. G., Walker, P. M. B., and McCallum, M. (1969). Some properties of the single strands isolated from the DNA of the nuclear satellite of the mouse (*Mus musculus*). *J. Mol. Biol.* **40**, 423–443.

Forsheit, A. B., Davidson, N., and Brown, D. D. (1974). An electron microscope heteroduplex study of the ribosomal DNAs of *Xenopus laevis* and *Xenopus mulleri. J. Mol. Biol.* **90**, 301–314.

Franke, W. W. (1974). Structure, biochemistry, and functions of the nuclear envelope. *Int. Rev. Cytol., Suppl.* **4**, 71–236.

Fry, K., and Salser, W. (1977). Nucleotide sequences of HS-α satellite DNA from kangaroo rat *Dipodomys ordii* and characterisation of similar sequences in other rodents. *Cell* **12**, 1069–1084.

Fry, K., Poon, R., Whitcombe, P., Idriss, J., Salser, W., Mazrimas, J., and Hatch, F. T. (1973). Nucleotide sequence of HS-β satellite DNA from kangaroo rat *Dipodomys ordii. Proc. Natl. Acad. Sci. U.S.A.* **70**, 2642–2646.

Gall, J. G., and Atherton, D. D. (1974). Satellite DNA sequences in *Drosophila virilis. J. Mol. Biol.* **85**, 633–664.

Gall, J. G., Cohen, E. H., and Polan, M. L. (1971). Repetitive DNA sequences in *Drosophila. Chromosoma* **33**, 319–344.

Gall, J. G., Cohen, E. H., and Atherton, D. D. (1973). The satellite DNAs of *Drosophila virilis. Cold Spring Harbor Symp. Quant. Biol.* **38**, 417–421.

Gallo, R. C., and Pestka, S. (1970). Transfer RNA species in normal and leukemic human lymphoblasts. *J. Mol. Biol.* **52**, 195–219.

Garon, C. F., Berry, K. W., and Rose, J. A. (1972). A unique form of terminal redundancy in adenovirus DNA molecules. *Proc. Natl. Acad. Sci. U.S.A.* **69**, 2391–2395.

Gelinas, R. E., and Roberts, R. J. (1977). One predominant 5'-undecanucleotide in adenovirus 2 late messenger RNA. *Cell* **11**, 533–544.

Georgiev, G. P. (1969). On the structural organisation of operon and the regulation of RNA synthesis in animal cells. *J. Theor. Biol.* **25**, 473–490.

Gilbert, W., and Maxam, A. (1973). The nucleotide sequence of the *Lac* operator. *Proc. Natl. Acad. Sci. U.S.A.* **70**, 3581–3584.

Gillis, M., De Ley, J., and De Cleene, M. (1970). The determination of molecular weight of bacterial genome DNA from renaturation rates. *Eur. J. Biochem.* **12**, 143–153.

Glover, D. M., and Hogness, D. S. (1977). A novel arrangement of the 18 S and 28 S sequences in a repeating unit of *Drosophila melanogaster* rDNA. *Cell* **10**, 167–176.

Glover, D. M., White, R. L., Finnegan, D. J., and Hogness, D. S. (1975). Characterisation of six cloned DNAs from *Drosophila melanogaster,* including one that contains the genes for rRNA. *Cell* **5**, 149–157.

Goldberg, R. B., Crain, W. R., Ruderman, J. V., Moore, G. P., Barnett, T. R., Higgins, R. C., Gelfand, R. A., Galau, G. A., Britten R. J., and Davidson, E. H. (1975). DNA sequence organisation in the genomes of five marine invertebrates. *Chromosoma* **51**, 225–251.

Graham, D. E., Neufeld, B. R., Davidson, E. H., and Britten, R. J. (1974). Interspersion of repetitive and non-repetitive DNA sequences in the sea urchin genome. *Cell* **1**, 127–137.

G ,ss, K., Probst, E., Schaffner, W., and Birnstiel, M. L. (1976a). Molecular analysis of the histone gene cluster of *Psammechinus miliaris:* I. Fractionation and identification of five individual histone mRNAs. *Cell* **8**, 455–469.

,ross, K., Schaffner, W., Telford, J., and Birnstiel, M. L. (1976b). Molecular analysis of the histone gene cluster of *Psammechinus miliaris:* III. Polarity and asymmetry of the histone-coding sequences. *Cell* **8**, 479–484.

Hahn, W. E., and Laird, C. D. (1971). Transcription of non-repeated DNA in mouse brain. *Science* **173**, 158–161.

Hamkalo, B. A., Miller, O. L., and Bakken, A. H. (1973). Ultrastructure of active eukaryotic genomes. *Cold Spring Harbor Symp. Quant. Biol.* **38**, 915–919.

Harel, J. N., Hanania, N., Tapiero, H., and Harel, L. (1968). RNA replication by nuclear satellite DNA in different mouse cells. *Biochem. Biophys. Res. Commun.* **33**, 696–701.

Hatch, F. T., and Mazrimas, J. A. (1974). Fractionation and characterisation of satellite DNAs of the kangaroo rat (*Dipodomys ordii*). *Nucleic Acids Res.* **1**, 559–575.

Hatch, F. T., Bodner, A. J., Mazrimas, J. A., and Moore, D. H., Jr. (1976). Satellite DNA and cytogenetic evolution. DNA quantity, satellite DNA and karyotypic variations in kangaroo rats (genus *Dipodomys*). *Chromosoma* **58**, 155–168.

Hennig, W. (1973). Molecular hybridisation of DNA and RNA *in situ. Int. Rev. Cytol.* **36**, 1–44.

Hörz, W., and Zachau, H. G. (1977). Characterisation of distinct segments in mouse satellite DNA by restriction nucleases. *Eur. J. Biochem.* **73**, 383–392.

Ikemura, T., and Nomura, M. (1977). Expression of spacer tRNA genes in ribosomal RNA transcription units carried by hybrid Col E1 plasmids of *Escherichia coli. Cell* **11**, 779–793.

Ilgren, C., Kirk, L. L., and Carbon, J. (1976). Isolation and characterisation of large transfer ribonucleic acid precursors from *Escherichia coli. J. Biol. Chem.* **251**, 922–929.

Jaskunas, S. R., Nomura, M., and Davies, J. (1974). Genetics of bacterial ribosomes. *In* "Ribosomes" (M. Nomura, A. Tissières, and P. Lengyel, eds.), pp. 333–368. Cold Spring Harbor Lab., Cold Spring Harbor, New York.

Jeffreys, A. J., and Flavell, R. A. (1977a). A physical map of the DNA regions flanking the rabbit β-globin gene. *Cell* **12**, 429–439.

Jeffreys, A. J., and Flavell, R. A. (1977b). The rabbit β-globin gene contains a large insert in the coding sequence. *Cell* **12**, 1097–1108.

Jelinek, W., Molloy, G., Fernandez-Munoz, R., Salditt, M., and Darnell, J. E. (1974). Secondary structure in heterogeneous nuclear RNA: Involvement of regions of repeated DNA sites. *J. Mol. Biol.* **82**, 361–370.

Jones, K. W. (1970). Chromosomal and nuclear location of mouse satellite DNA in individual cells. *Nature (London)* **225**, 912–915.

Jorgensen, P. (1976). A ribosomal RNA gene of *Escherichia coli* (rrnD) on λ daroE specialised transducing phages. *Mol. Gen. Genet.* **146**, 303–307.

Kedes, L. H., and Birnstiel, M. L. (1971). Reiteration and clustering of DNA sequences complementary to histone messenger RNA. *Nature (London), New Biol.* **230**, 165–169.

Kenerley, M. E., Morgan, E. A., Post, L., Lindahl, L., and Nomura, M. (1977). Characterisation of hybrid plasmids carrying individual ribosomal RNA transcription units of *Escherichia coli. J. Bacteriol.* **132**, 931–949.

Kersten, W., Kersten, H., and Szybalski, W. (1966). Physiochemical properties of complexes between DNA and antibiotics which affect RNA synthesis (actinomycin, daunomycin, cinerubin, nogalamycin, chromomycin, mithramycin and olivomycin). *Biochemistry* **5**, 236–244.

Kingsbury, D. T. (1969). Estimate of the genome size of various microorganisms. *J. Bacteriol.* **98**, 1400–1401.

Kit, S. (1961). Equilibrium sedimentation in density gradients of DNA preparations from animal tissues. *J. Mol. Biol.* **3**, 711–716.

Kleckner, N. (1977). Translocatable elements in procaryotes. *Cell* **11**, 11–23.

Klessing, D. F. (1977). Two adenovirus mRNAs have a common 5′ terminal leader sequence encoded at least 10 kb upstream from their main coding regions. *Cell* **12**, 9–21.

Laird, C. D., McConaughy, B. L., and McCarthy, B. J. (1969) On the rate of fixation of nucleotide substitutions in evolution. *Nature (London)* **224**, 149–154.

Lawn, R. M. (1977). Gene-sized DNA molecules of the *Oxytricha* macronucleus have the same terminal sequence. *Proc. Natl. Acad. Sci. U.S.A.* **74**, 4325–4328.

Lee, C. S., and Thomas, C. A., Jr. (1973). Formation of rings from *Drosophila* DNA fragments. *J. Mol. Biol.* **77**, 25–55.

Levine, A. J. (1976). SV40 and adenovirus early functions involved in DNA replication and transformation. *Biochim. Biophys. Acta* **458**, 213–241.

Lewin, B. (1975). Units of transcription and translation: Sequence components of heterogeneous nuclear RNA and messenger RNA. *Cell* **4**, 77–93.

McCarthy, B. J., and Duerksen, J. D. (1971). Fractionation of mammalian chromatin. *Cold Spring Harbor Symp. Quant. Biol.* **35**, 621–627.

McCarthy, B. J., and Farquhar, M. N. (1972). The rate of change of DNA in evolution. *Brookhaven Symp. Biol.* **23**, 1–43.

Manning, J. E., Schmid, C. W., and Davidson, N. (1975). Interspersion of repetitive and non-repetitive DNA sequences in the *Drosophila melanogaster* genome. *Cell* **4**, 141–155.

Manuelidis, L. (1977). A simplified method for preparation of mouse satellite DNA. *Anal. Biochem.* **78**, 561–568.

Marmur, J., and Doty, P. (1962). Determination of the base composition of deoxyribonucleic acid from its thermal denaturation temperature. *J. Mol. Biol.* **5**, 109–118.

Marmur, J., and Lane, D. (1960). Strand separation and specific recombination in deoxyribonucleic acids: Biological studies. *Proc. Natl. Acad. Sci. U.S.A.* **46**, 453–461.

May, E., Maizel, J. V., and Salzman, N. P. (1977). Mapping of transcription sites of Simian virus 40-specific late 16 S and 19 S mRNA by electron microscopy. *Proc. Natl. Acad. Sci. U.S.A.* **74**, 496–500.

Melli, M., Ginelli, E., Corneo, G., and diLernia, R. (1975). Clustering of the DNA sequences complementary to repetitive nuclear RNA of HeLa cells. *J. Mol. Biol.* **93**, 23–38.

Melli, M., Spinelli, G., Wyssling, H., and Arnold, E. (1977a). Presence of histone mRNA sequences in high molecular weight RNA of HeLa cells. *Cell* **11**, 651–661.

Melli, M., Spinelli, G., and Arnold, E. (1977b). Synthesis of histone messenger RNA of HeLa cells during the cell cycle. *Cell* **12**, 167–174.

Meyer, G. F., and Hennig, W. (1974). The nucleolus in primary spermatocytes of *Drosophila hydei. Chromosoma* **46**, 121–144.

Miyazawa, Y., and Thomas, C. A., Jr. (1965). Nucleotide composition of short segments of DNA molecules. *J. Mol. Biol.* **11**, 223–237.

Moar, M. H., Purdom, I. F., and Jones, K. W. (1975). Influence of temperature on the detectibility and chromosomal distribution of specific DNA sequences by *in situ* hybridisation. *Chromosoma* **53**, 345–359.

Molloy, G. R., Jelinek, W., Salditt, M., and Darnell, J. E. (1974). Arrangement of specific oligonucleotides within poly(A) terminated HnRNA molecules. *Cell* **1**, 43–53.

Morgan, E. A., Ikemura, T., and Nomura, M. (1977). Identification of spacer tRNA genes in individual ribosomal RNA transcription units of *Escherichia coli. Proc. Natl. Acad. Sci. U.S.A.* **74**, 2710–2714.

Mowbray, S. L., Gerbi, S. A., and Landy, A. (1975). Interdigitated repeated sequences in bovine satellite DNA. *Nature (London)* **253**, 367–370.

Nandi, U. S., Wang, J. C., and Davidson, N. (1965). Separation of deoxyribonucleic acids by Hg(II) binding and Cs_2SO_4 density–gradient centrifugation. *Biochemistry* **4**, 1687–1696.

Ohtsubo, E., Lee, H. J., Doenier, R. C., and Davidson, N. (1974). Electron microscope heteroduplex studies of sequence relations among plasmids of *Escherichia coli*. VI. Mapping of F14 sequences homologous to ϕ80dmetBJF and ϕ80dargECBH bacteriophages. *J. Mol. Biol.* **89**, 599–618.

Pace, N. R. (1973). Structure and synthesis of the ribosomal ribonucleic acid of prokaryotes. *Bacteriol. Rev.* **37**, 562–603.

Pardue, M. L., and Gall, J. G. (1970). Chromosomal localisation of mouse satellite DNA. *Science* **168**, 1356–1358.

Paul, J. (1972). General theory of chromosome structure and gene activation in eukaryotes. *Nature (London)* **238**, 444–446.

Peacock, W. J., Brutlag, D., Goldring, E., Appels, R., Hinton, C. W., and Lindsley, D. L. (1973). The organisation of highly repeated DNA sequences in *Drosophila melanogaster* chromosomes. *Cold Spring Harbor Symp. Quant. Biol.* **38**, 405–416.

Perlman, S., Phillips, C., and Bishop, J. O. (1976). A study of foldback DNA. *Cell* **8**, 33–42.

Philipson, L., Wall, R., Glickman, G., and Darnell, J. E. (1971). Addition of polyadenylate sequences to virus-specific RNA during adenovirus replication. *Proc. Natl. Acad. Sci. U.S.A.* **68**, 2806–2809.

Portmann, R., Schaffner, W., and Birnstiel, M. L. (1976). Partial denaturation mapping of cloned histone DNA from the sea urchin *Psammechinus miliaris. Nature (London)* **264**, 31–34.

Proudfoot, N. J. (1977). Complete 3′ non-coding region sequences of rabbit and human β-globin messenger RNAs. *Cell* **10**, 559–570.

Pyeritz, R. E., Lee, C. S., and Thomas, C. A., Jr. (1971). The cyclization of mouse satellite DNA. *Chromosoma* **33**, 284–296.

Rae, P. M. M. (1972). The distribution of repetitive DNA sequences in chromosomes. *Adv. Cell Mol. Biol.* **2**, 109–150.

Rae, P. M. M., and Franke, W. W. (1972). The interphase distribution of satellite DNA-containing heterochromatin in mouse nuclei. *Chromosoma* **39**, 443–456.

Reddy, V. B., Dahr, R., and Weissman, S. M. (1978). Nucleotide sequence of the genes for the simian virus 40 proteins VP2 and VP3. *J. Biol. Chem.* **253**, 621–630.

Reed, N. (1969). Fractionation of rat repeated sequences according to thermal stability. *Carnegie Inst. Wash., Yearb.* **68**, 386–388.

Reeder, R. H., Brown, D. D., Wellauer, P. K., and Dawid, I. B. (1976). Patterns of ribosomal DNA spacer lengths are inherited. *J. Mol. Biol.* **105**, 507–516.

Rees, H., and Jones, R. N. (1972). The origin of the wide species variation in nuclear DNA content. *Int. Rev. Cytol.* **32**, 53–92.

Ritossa, F. M., Atwood, K. C., Lindsley, D. L., and Spiegelman, S. (1966). On the chromosomal distribution of DNA complementary to ribosomal and soluble RNA. *Natl. Cancer Inst., Monogr.* **23**, 449–472.

Roberts, R. J., Arrand, J. R., and Keller, W. (1974). The length of the terminal repetition in adenovirus-2 DNA. *Proc. Natl. Acad. Sci. U.S.A.* **71**, 3829–3833.

Roberts, T. M., Lauer, G. D., and Klotz, L. C. (1976). Physical studies on DNA from "primitive" eucaryotes. *Crit. Rev. Biochem.* **3**, 349–449.

Rosbash, M., Ford, P. J., and Bishop, J. O. (1974). Analysis of the C-value paradox by molecular hybridisation. *Proc. Natl. Acad. Sci. U.S.A.* **71**, 3746–3750.

Rosbash, M., Campo, M. S., and Gummerson, K. S. (1975). Conservation of cytoplasmic RNA in mouse and rat. *Nature (London)* **258**, 682–686.

Russell, R. L., Abelson, J. N., Landy, A., Gefter, M. L., Brenner, S., and Smith, J. D. (1970). Duplicate genes for tyrosine transfer RNA in *Escherichia coli*. *J. Mol. Biol.* **47**, 1–13.

Ryskov, A. P., Saunders, G. F., Farashyan, V. R., and Georgiev, G. P. (1973). Double-helical regions in nuclear precursor of mRNA (pre-m-RNA). *Biochim. Biophys. Acta* **312**, 152–164.

Sakano, H., Yamada, S., Ikemura, T., Shimura, Y., and Ozeki, H. (1974). Temperature sensitive mutants of *Escherichia coli* for tRNA synthesis. *Nucleic Acids Res.* **1**, 355–371.

Sanger, F., Air, G. M., Barrell, B. G., Brown, N. L., Coulson, A. R., Fiddes, J. C., Hutchison, C. A., III, Slocombe, P. M., and Smith, M. (1977). Nucleotide sequence of bacteriophage φX174 DNA. *Nature (London)* **265**, 687–695.

Schachat, F. H., and Hogness, D. S. (1973). Repetitive sequences in isolated Thomas circles from *Drosophila melanogaster*. *Cold Spring Harbor Symp. Quant. Biol.* **38**, 371–381.

Schaffner, W., Gross, K., Telford, J., and Birnstiel, M. L. (1976). Molecular analysis of the histone gene cluster of *Psammechinus miliaris:* II. The arrangement of five histone-coding and spacer sequences. *Cell* **8**, 471–478.

Schildkraut, C. L., and Maio, J. J. (1969). Fractions of HeLa DNA differing in their content of Guanine + Cytosine. *J. Mol. Biol.* **46**, 305–312.

Schildkraut, C. L., Marmur, J., and Doty, P. (1961). The formation of hybrid DNA molecules and their use in studies of DNA homologies. *J. Mol. Biol.* **3**, 595–617.

Schildkraut, C. L., Marmur, J., and Doty, P. (1962). Determination of the base composition of deoxyribonucleic acid from its buoyant density in CsCl. *J. Mol. Biol.* **4**, 430–433.

Schlessinger, D. (1974). Ribosome formation in *Escherichia coli*. *In* "Ribosomes" (M. Nomura, A. Tissières, and P. Lengyel, eds.), pp. 393–416. Cold Spring Harbor Lab., Cold Spring Harbor, New York.

Schmid, C. W., and Deininger, P. L. (1975). Sequence organisation of the human genome. *Cell* **6**, 345–358.

Schmid, C. W., Manning, J. L., and Davidson, N. (1975). Inverted repeat sequences in the *Drosophila* genome. *Cell* **5**, 159–172.

Smith, M., Brown, N. L., Air, G. M., Barrell, B. G., Coulson, A. R., Hutchison, C. A., III, and Sanger, F. (1977). DNA sequence at the C termini of the overlapping genes A and B in bacteriophage φX174. *Nature (London)* **265**, 702–705.

Smith, M. J., Britten, R. J., and Davidson, E. H. (1975). Studies on nucleic acid reassociation kinetics: Reactivity of single-strand tails in DNA–DNA renaturation. *Proc. Natl. Acad. Sci. U.S.A.* **72**, 4805–4809.

Sommer, S., Salditt-Georgieff, M., Bachenheimer, S., Darnell, J. E., Furuichi, Y., Morgan, M., and Shatkin, A. J. (1976). The methylation of adenovirus-specific nuclear and cytoplasmic RNA. *Nucleic Acids Res.* **3**, 749–765.

Southern, E. M. (1970). Base sequence and evolution of guinea pig α-satellite DNA. *Nature (London)* **227**, 794–798.

Southern, E. M. (1971). Effects of sequence divergence on the reassociation properties of repetitive DNAs. *Nature (London), New Biol.* **232**, 82–83.

Southern, E. M. (1974). Eukaryotic DNA. *In* "Biochemistry of Nucleic Acids" (K. Burton, ed.), Vol. 6, pp. 101–139. Butterworth, London.

Southern, E. M. (1975). Long-range periodicities in mouse satellite DNA. *J. Mol. Biol.* **94**, 51–69.

Sutton, W. D., and McCallum, M. (1971). Mismatching and the reassociation rate of mouse satellite DNA. *Nature (London), New Biol.* **232**, 83–85.

Swift, H. H. (1950a). The desoxyribose nucleic acid content of animal nuclei. *Physiol. Zool.* **23**, 169–200.

Swift, H. H. (1950b). The constancy of desoxyribose nucleic acid in plant nuclei. *Proc. Natl. Acad. Sci. U.S.A.* **36**, 643–654.

Tartof, K. D. (1973). Regulation of ribosomal RNA gene multiplicity in *Drosophila melanogaster. Genetics* **73**, 57–71.

Tartof, K. D. (1975). Redundant genes. *Annu. Rev. Genet.* **9**, 355–385.

Tartof, K. D., and Dawid, I. B. (1976). Similarities and differences in the structure of X and Y chromosome rRNA genes of *Drosophila. Nature (London)* **263**, 27–30.

Thomas, C. A., Jr., Hamkalo, B. A., Misra, D. N., and Lee, C. S. (1970). Cyclization of eukaryotic deoxyribonucleic acid fragments. *J. Mol. Biol.* **51**, 621–632.

Thomas, C. A., Jr., Pyeritz, R. E., Wilson, D. A., Dancis, B. M., Lee, C. S., Bick, M. D., Huang, H. L., and Zimm, B. H. (1973). Cyclodromes and palindromes in chromosomes. *Cold Spring Harbor Symp. Quant. Biol.* **38**, 353–370.

Thomas, M., White, R. L., and Davis, R. W. (1976). Hybridisation of RNA to double-stranded DNA: Formation of R-loops. *Proc. Natl. Acad. Sci. U.S.A.* **73**, 2294–2298.

Tooze, J. (1973). "The Molecular Biology of Tumour Viruses." Cold Spring Harbor Lab., Cold Spring Harbor, New York.

Vinograd, J., and Hearst, J. E. (1962). Equilibrium sedimentation of macromolecules and viruses in a density gradient. *Fortschr. Chem. Org. Naturst.* **20**, 372–422.

Walbot, V., and Dure, L. S. (1976). Developmental biochemistry of cotton seed embryogenesis and germination. VII. Characterisation of the cotton seed genome. *J. Mol. Biol.* **101**, 503–536.

Walker, P. M. B. (1971). "Repetitive" DNA in higher organisms. *Prog. Biophys. Mol. Biol.* **23**, 145–190.

Wallace, H., and Birnstiel, M. L. (1966). Ribosomal cistrons and the nucleolus organizer. *Biochim. Biophys. Acta* **114**, 296–310.

Waring, M., and Britten, R. J. (1966). Nucleotide sequence repetition: A rapidly reassociating fraction of mouse DNA. *Science* **154**, 791–794.

Weinberg, E. S., Birnstiel, M. L., Purdom, I. F., and Williamson, R. (1972). Genes coding for polysomal 9 S RNA of sea urchins: Conservation and divergence. *Nature (London)* **240**, 225–228.

Weinberg, E. S., Overton, G. C., Schutt, R. H., and Reeder, R. H. (1975). Histone gene arrangement in the sea urchin. *Strongylocentrotus purpuratus. Proc. Natl. Acad. Sci. U.S.A.* **72,** 4815–4819.

Weinberg, R. A., and Penman, S. (1970) Processing of 45 S nucleolar RNA. *J. Mol. Biol.* **47,** 169–178.

Wellauer, P. K., Reeder, R. H., Carroll, D., Brown, D. D., Deutch, A., Higashinakagawa, T., and Dawid, I. B. (1974). Amplified ribosomal DNA from *Xenopus laevis* has heterogeneous spacer lengths. *Proc. Natl. Acad. Sci. U.S.A.* **71,** 2823–2827.

Wellauer, P. K., Dawid, I. B., Brown, D. D., and Reeder, R. H. (1976a). The molecular basis for length heterogeneity in ribosomal DNA from *Xenopus laevis. J. Mol. Biol.* **105,** 461–486.

Wellauer, P. K., Reeder, R. H., Dawid, I. B., and Brown, D. D. (1976b). The arrangement of length heterogeneity in repeating units of amplified and chromosomal ribosomal DNA from *Xenopus laevis. J. Mol. Biol.* **105,** 487–505.

Wells, R., Royer, H. D., and Hollenberg, C. P. (1976). Non-*Xenopus*-like DNA sequence organisation in the *Chironomus tentans* genome. *Mol. Gen. Genet.* **147,** 45–51.

Wells, R. D., and Blair, J. E. (1967). Studies of polynucleotides. LXXI. Sedimentation and buoyant density. Studies of some DNA-like polymers with repeating nucleotide sequences. *J. Mol. Biol.* **27,** 273–288.

Wensink, P. C., and Brown, D. D. (1971). Denaturation map of the ribosomal DNA of *Xenopus laevis. J. Mol. Biol.* **60,** 235–247.

Wesley, R. (1975). Inverted repetitious sequences in the macronuclear DNA of hypotrichous ciliates. *Proc. Natl. Acad. Sci. U.S.A.* **72,** 678–682.

Wetmur, J. G., and Davidson, N. (1968). Kinetics of renaturation of DNA. *J. Mol. Biol.* **31,** 349–370.

White, R. L., and Hogness, D. S. (1977). R loop mapping of the 18 S and 28 S sequences in the long and short repeating units of *Drosophila melanogaster* rDNA. *Cell* **10,** 177–192.

Wilson, D. A., and Thomas, C. A., Jr. (1974). Palindromes in chromosomes. *J. Mol. Biol.* **84,** 115–139.

Wimber, D. E., and Steffensen, D. M. (1970). Localisation of 5 S RNA genes on *Drosophila* chromosomes by RNA–DNA hybridisation. *Science* **170,** 639–641.

Wolfson, J., and Dressler, D. (1972). Adenovirus-2 DNA contains an inverted terminal repetition. *Proc. Natl. Acad. Sci. U.S.A.* **69,** 3054–3057.

Wu, J.-R., Pearson, W. R., Posakony, J. W., and Bonner, J. (1977). Sequence relationship between long and short repetitive DNA of the rat: A preliminary report. *Proc. Natl. Acad. Sci. U.S.A.* **74,** 4382–4386.

Young, M. W., and Hogness, D. S. (1978). A new approach for identifying and mapping structural genes in *Drosophila melanogaster. In* "Eukaryotic Genetic Systems" (G. Wilcox, J. Abelson, and C. F. Fox, eds.), ICN–UCLA Symposia on Molecular and Cellular Biology, Vol. 8, 315–331. Academic Press, New York.

Zimmer, C., and Luck, G. (1972). Stability and dissociation of the DNA complexes with distamycin A and netropsin in the presence of organic solvents, urea and high salt concentration. *Biochim. Biophys. Acta* **287,** 376–385.

2

Gene Reiteration and Gene Amplification

Adrian P. Bird

CELL BIOLOGY, VOL. 3

I. INTRODUCTION

The genome of an organism is more than a simple catalogue of coding sequences. It also contains some information about how gene products are to be deployed, both quantitatively and qualitatively, during the life cycle. Although little is known about the genetic control of qualitative differences between cells, we are beginning to understand some aspects of quantitative control. So far, much of this knowledge concerns the supply of RNA's or proteins that are required in large amounts over a short period of time. At the most basic level, this is achieved by providing the organism with multiple copies of the genes concerned. In other words, the need for more gene product is met by more copies of the gene.

A priori there are two ways in which the nuclear dosage of a gene may increase. One extreme is to carry a single gene copy in the germ line, but to generate multiple copies by amplification during each life cycle. The alternative is to incorporate multiple copies of the gene permanently into the germ cell genome, thus ensuring their distribution to all cells by mitosis. Examples of both these mechanisms will be covered in this chapter, as well as intermediate cases where amplification and multiple genomic copies are combined.

For a recent review on repeated genes, the reader is referred to Tartof (1975), and for reviews of gene amplification to Tobler (1975), Macgregor (1972), and Gall (1969).

II. ALTERNATIVES TO MULTIPLE GENES

Before discussing examples of multiple genes, it is worth considering the other means by which a cell or organism can respond to a demand for high rates of gene product synthesis. Perhaps the most important option is inherent in the mechanism of protein synthesis itself. By synthesizing a stable messenger RNA as an intermediate between DNA and protein, the template potential of a gene is greatly expanded. Messenger RNA may either accumulate while being actively translated, or it may be stored and brought into action all at once at some later time.

Given this inherent expansion step we may ask why multiple genes exist at all. The answer is that not all genes code for protein. Those coding instead for RNA synthesize no intermediate between the gene and its final product, and as a result their maximum rate of production is strictly limited by the number of gene copies. Looked at in this way, it is easy to rationalize the finding that almost all repeated genes code for an RNA product (28 S and 18 S ribosomal RNA, 5 S ribosomal RNA, transfer RNA). The only known exceptions, where repetition is also clearly related to high product demand, are the histone genes. The extenuating circumstances which apply in this case are discussed later (Section III,E).

Less clear-cut methods for raising the level of genetic activity are polyteny and polyploidy. Polyploid cells contain integral multiples of the haploid DNA content, making them triploid, tetraploid, hexaploid, etc., depending upon how many complete genomes are present. Polyploidy is sometimes found in somatic cells, such as the secretory cells of the silk gland in *Bombyx mori* (Suzuki *et al.,* 1972) and liver cells in a variety of animals. Occasionally it may be transmitted to subsequent generations as a new genome size. Polyteny also involves reduplication of the genome, although different parts are not always replicated to the same extent. Characteristically rounds of DNA replication occur without intervening chromatid separation and result in a giant, multichromatid chromosome. Polyteny is exclusively a somatic cell phenomenon found in organisms as diverse as the bean *Phaseolus coccineus* (Nagl, 1965), the ciliated protozoan *Stylonychia mytilus* (Ammermann, 1971), and the dipteran fly *Drosophila melanogaster* (Bridges, 1935).

At first sight both polyteny and polyploidy appear to be rather indiscriminate mechanisms for increasing the transcriptional capacity of a cell. Closer examination, however, suggests that this may be a simplistic interpretation. In fact, both kinds of cells are giants whose volume has increased in approximate proportion to their genome size. The number of genomes per unit volume may therefore be similar before and after the process, leaving transcriptional capacity essentially unchanged. It is possible that some other advantage attaches to having many genomes in one cell. Perhaps genetic activity is more easily coordinated when the genomes involved are bounded by the same nuclear membrane.

In multicellular organisms there is yet another method available for meeting an increased demand for a gene product: the number of cells devoted to its synthesis can be raised by cell division. Of course, this does not apply to proteins involved in basic cellular metabolism; these components must be available in quantities sufficient for the individual cell. Rather, it is products of importance to the organism as a whole, such as the production of hemoglobin, which can be simply stepped up by cell proliferation.

It is apparent from the above summary that the ways in which an organism can meet the need for high rates of genetic activity form a related hierarchy: more cells per organism, more genomes per cell, more genes per genome. In considering only the last of these, it is wise to keep in mind that this is only a part of the whole picture.

III. MULTIPLE GENES INTEGRATED IN THE GERM LINE GENOME

A. Which Genes Are Reiterated?

We have already noted that genes whose ultimate product is RNA cannot match the high rates of synthesis available to protein coding genes because they lack the intermediate mRNA step. Indeed, if a preparation of total cellular RNA is fractionated on an acrylamide gel, we find that the four major RNA peaks are all encoded by multiple genes (Fig. 1). Three of these major RNAs are structural components of the ribosome (28 S, 18 S and 5 S RNA), while the fourth is a heterogeneous population of transfer RNA's.

B. Nucleolar DNA—The 28 S and 18 S rRNA Genes

The nucleolus is the most prominent body in an interphase nucleus and so has always excited interest among cell biologists. Early work established that one or more chromosomal regions were invariably associated with the nucleolus, and these regions were christened "nucleolus organizer regions" (Heitz, 1931; McClintock, 1934). More recently, the powerful combination of genetics and nucleic acid hybridization (Gillespie and Spiegelman, 1965) related the genes for 28 S and 18 S rRNA with the nucleolus organizer (Ritossa and Spiegelman, 1965). In the RNA–DNA hybridization assay, DNA from the organism to be tested is denatured and immobilized on a nitrocellulose filter. The filter is then incubated with an excess of radioactive rRNA under conditions which promote annealing between rRNA and DNA sequences that code for rRNA (rDNA). Provided that the rRNA concentration is saturating, and in the presence of

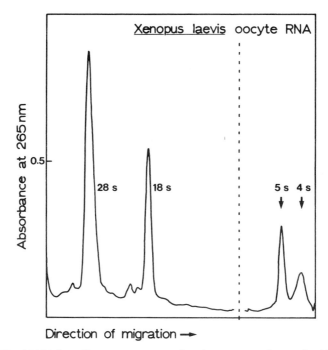

Fig. 1. Total RNA from *Xenopus* oocytes analyzed on a composite acrylamide gel. The major RNA species (28 S, 18 S, 5 S, and 4 S) are all encoded by reiterated genes. The dotted line marks the boundary between 2.3 and 15% acrylamide. (Figure kindly provided by Dr. P. J. Ford.)

appropriate controls, the amount of radioactivity bound by the filter gives the proportion of DNA complementary to (i.e., coding for) rRNA. Using this assay it was found that in *Drosophila melanogaster,* DNA from flies carrying 1, 2, 3, or 4 doses of the nucleolus organizer region hybridized rRNA in the ratio 1 : 2 : 3 : 4, respectively. A similar correlation between rDNA and the nucleolus was established for the African clawed frog, *Xenopus laevis.* DNA from frogs heterozygous for a deletion of the nucleolus organizer (Elsdale *et al.,* 1958) showed about half as much DNA complementary to rRNA as wild-type frogs, while DNA from embryos homozygous for the deletion showed no significant hybridization (Wallace and Birnstiel, 1966). The link between rDNA and the nucleolus was finally cemented by hybridization *in situ.* Here rRNA is annealed not to single-stranded DNA on a filter, but to DNA denatured in a fixed chromosome preparation (Gall, 1969; Gall and Pardue, 1969; John *et al.,* 1969). Subsequent autoradiography reveals that rRNA is bound only by the nucleolus organizer (Pardue, 1973).

These experiments, and others relating the nucleolus to rRNA synthesis (Brown and Gurdon, 1964), won general acceptance for the idea that rDNA is a key constituent of the chromosomal region responsible for nucleolus formation. More recent work has concentrated on the fine structure of purified rDNA. These experiments began with the realization that the buoyant density of *Xenopus laevis* rDNA differs from that of the rest of the DNA in cesium chloride solutions (Birnstiel *et al.*, 1966). Since that time elegant experimentation has established in ever increasing detail the arrangement of coding and noncoding sequences (Birnstiel *et al.*, 1968; Brown and Weber, 1968; Dawid *et al.*, 1970; Wensink and Brown, 1971; Wellauer *et al.*, 1974).

Figure 2 summarizes the results of this work. The rDNA of *Xenopus* is made up of alternating nontranscribed ''spacer'' sequences and sequences

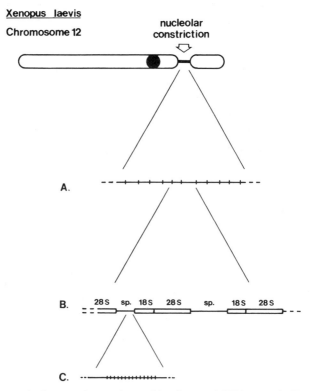

Fig. 2. The nucleolus organizer region and its ribosomal RNA genes in *Xenopus laevis*. (A) A few of the ~500 tandem repeated genes showing repeat units of different lengths interspersed with one another (see Section VII,C). (B) Simplified structure of the rDNA repeat unit showing two complete units which differ in their spacer (sp.) lengths. (C) Short tandem repeated sequences within the spacer (see Section VII,C).

which code for rRNA. The repeat unit is thus one coding sequence plus its adjacent spacer and is repeated several hundred times at each nucleolus organizer region (Fig. 2). These general features of rDNA organization appear remarkably constant in a wide variety of multicellular eukaryotes, although the average dimensions of the coding region and, more particularly of the spacer, do vary. Patterns of interspecies heterogeneity are considered in more detail in Section VII.

C. 5 S RNA Genes

5 S RNA is an integral component of the ribosome, and in certain "primitive" eukaryotes (e.g. yeast; Rubin and Sulston, 1973) the 5 S RNA genes are closely linked with the genes for the large ribosomal RNA's. Surprisingly, in higher eukaryotes this is not so, and 5 S RNA is encoded by gene clusters entirely separate from the nucleolus organizer. The reason for this separation is unknown. Best studied are the 5 S genes of Xenopus laevis. Hybridization to total DNA indicates 9000 to 24,000 genes per haploid chromosome set (Brown and Weber, 1968; Birnstiel et al., 1972). These are divided among at least 15 of the 18 chromosomes in clusters located exclusively at the chromosome tips or telomeres (Pardue et al., 1973). This pattern of chromosomal 5 S DNA sites is by no means universal. Another amphibian, the newt Notophthalmus viridescens, has four 5 S loci, none of which is telomeric (Pukkila, 1975), while Drosophila melanogaster has only one (Wimber and Steffensen, 1970). The average number of 5 S sequences per haploid complement also varies dramatically between species. Yeast (Retel and Planta, 1972) and Drosophila (Procunier and Tartof, 1975) both have about 200 5 S genes, or less than one-fiftieth of the Xenopus dose. At the other end of the scale, Notophthalmus possesses a massive 300,000 5 S genes in each chromosome complement; over three orders of magnitude more than Drosophila and yeast (Pukkila, 1975).

Like rDNA, the 5 S DNA of Xenopus laevis has been purified and found to consist of a string of alternating coding segments and nontranscribed spacers (Brown et al., 1971). It should be emphasized that only those genes which code for oocyte-specific variants of 5 S RNA have been isolated (Brownlee et al., 1972). Nevertheless, it seems likely that the structure established for this fraction will be of quite general occurrence, since it has also been identified in the distantly related fly Drosophila (Procunier and Tartof, 1976; Artavanis-Tsakonas et al., 1977).

D. Transfer RNA Genes

Transfer RNA genes are a more complex group than the various ribosomal genes, since forty or so codon classes are known, each encoded by a separate gene. In addition, it is not clear in all cases that tandem repeated sequences are responsible for their synthesis. One case in which reiterated tRNA genes have been identified is *Xenopus laevis* (Clarkson *et al.*, 1973), there being on average 200 clustered genes for each tRNA species. More recently, tandem repetition of the methionine initiator tRNA gene has been directly demonstrated by partial restriction endonuclease digestion (Clarkson and Kurer, 1976). It is not yet clear how universal this kind of repetition will prove to be. Hybridization of total tRNA to *Drosophila* (Ritossa *et al.*, 1966a) or yeast (Schweizer *et al.*, 1969) or human DNA (Hatlen and Attardi, 1971) gives average values of between 10 and 30 copies of each tRNA gene per haploid genome.

E. Histone Genes

We have seen that multiple genes are to be expected in those cases where RNA is the only gene product. Why, then, are the sea urchin histone genes present in 400- to 1000-fold reiteration? In part the answer appears to be that very large amounts of histone protein are required suddenly during early embryogenesis. There is another important point. Histone is not a catalytic protein, but is "soaked up" by newly synthesized DNA in a one to one mass ratio. When it is considered that the sea urchin cleavage nucleus synthesizes about 0.25 m of DNA every 30 min, the need for extreme amounts of histone is apparent. In support of this view, we find that a high proportion of blastula newly synthesized RNA is histone mRNA (Kedes and Gross, 1969) despite the presence of a considerable store of histone messenger in unfertilized eggs (Gross *et al.*, 1973). *Drosophila*, another organism whose cleavage stages are extremely rapid, again shows a high reiteration frequency of over 100 genes (Lifton *et al.*, 1978).

In contrast, organisms which do not divide so rapidly in early development, such as those possessing amniote eggs, do not show extensive histone gene reiteration. Thus, man and mouse possess only 10–20 histone genes per haploid set (Wilson *et al.*, 1974; Jacob, 1976). In the light of evidence for stage- and tissue-specific histone variants (Zweidler and Cohen, 1973; Ruderman and Gross, 1974; Cohen *et al.*, 1975), it is possible that these few genes are not functionally identical.

The case of *Xenopus laevis* provides a sobering example of the dangers of this kind of generalization. Here the rate of cell division during embryogenesis is once more very rapid, averaging 14 min per cell cycle from 50 to 5000 cells (Graham and Morgan, 1966). This, coupled with a large genome size compared to *Drosophila* and the sea urchin, argues for an enormous demand for histone at blastula. Yet the solution is different.

Each egg accumulates a large store of histone protein (Woodland and Adamson, 1976) in addition to large quantities of histone mRNA (Adamson and Woodland, 1974). Unlike the sea urchin, *de novo* synthesis of histone mRNA in cleaving embryos is not significant, and, correspondingly, the level of histone gene repetition is relatively low (Jacob *et al.*, 1976).

The organism in which histone gene fine structure has been most extensively studied is the sea urchin. Intensive work in a number of laboratories has established that histone DNA comprises tandem repeating units, each containing the genes for all five histones (reviewed in Kedes, 1976; Birnstiel and Chipchase, 1977). As sea urchins are not amenable to cytological or genetic studies, very little is known about chromosomal distribution of the histone repeat units. In this respect, of course, *Drosophila* is the organism of choice, and we now know that hDNA is localized at a region encompassing five polytene bands (Pardue *et al.*, 1977). Recent work has shown that here too each repeat unit codes for all five histones (Lifton *et al.*, 1978).

F. The Significance of Clustering

A striking feature of all the multiple genes considered so far is that they are not scattered about the genome independently, but instead form tightly knit clusters of like genes. One possible explanation for clustering is that tandem repeated genes most probably originated by serial duplications of what was initially a single copy sequence. Necessarily, multiple copies generated in this way will be clustered. There may also be positive selective advantages to clustering. For example, the maintainance of functional homogeneity among different members of the gene cluster in the face of mutation may require tandem repetition (see Section VII). Furthermore, tandem repetition may place all members of the gene cluster under common control such that they can be switched on or off coordinately (Finnegan *et al.*, 1978). This suggestion takes account of the observation that polytene chromosomes, lampbrush chromosomes, and perhaps also mitotic chromosomes (Benyajati and Worcel, 1976) are divided into units much larger than genes. It is possible that the structural changes which accompany transcription decondense whole units rendering all the genes therein transcribable. Clearly, if multiple genes were clustered in such a unit their transcription would be coordinated. Evidence from electron microscopy of active ribosomal genes is consistent with coordinate switching of all the repeat units in an array (McKnight and Miller, 1976).

By way of contrast, other kinds of multiple gene families have been discovered in *Drosophila melanogaster* (Finnegan *et al.*, 1978). The members of these families show sequence homogeneity, but unlike tandem genes they are scattered about the genome. Although their function is not yet known, it has been suggested that scattered multiple genes offer alternatives, whereby different cellular programs of genetic activity can include the same gene product without necessarily using the same gene copy (Finnegan *et al.*, 1978). Under this theory, dispersed multiple genes would confer flexibility of genetic programming rather than increased transcriptional potential.

IV. FLUCTUATIONS IN TANDEM GENE NUMBER

A. *Drosophila*—The *Bobbed* Locus

The existence of clustered multiple gene copies was first established by a combination of molecular hybridization and genetics. A natural extension of these experiments was to determine the exact number of genes in each cluster in the expectation that the number would be characteristic for a particular species. It soon became apparent, however, that the number of genes in a tandem array is subject to wide variability. Ritossa *et al.* (1966b) first established this in experiments with *bobbed* mutations of *Drosophila melanogaster,* and in so doing culminated a long series of purely genetic experiments (Stern, 1929; Spencer, 1944; Cooper, 1959). In phenotype, *bobbed* mutants display short thin bristles, a thin chitinous cuticle, low viability and fecundity, and slow development (Lindsley and Grell, 1968). Using RNA–DNA hybridization, Ritossa *et al.* (1966b) showed that chromosomes carrying the *bobbed* mutation bore fewer 28 S and 18 S rRNA genes than did wild-type chromosomes, and the number of genes decreased with an increasingly severe *bobbed* phenotype. In general the number of genes per nucleolus organizer varied continuously among both mutant and wild-type flies, with a cutoff point below which the *bobbed* phenotype was manifested. As well as demonstrating the lability of rRNA gene number per locus, these results beautifully account for the spectrum of *bobbed* phenotypes. Presumably, below a certain gene number, estimated as about 150 genes, some cells in the fly are unable to sustain the required rate of ribosome production, and the deformities characteristic of *bobbed* appear. The reduced rate of rRNA synthesis in various *bobbed* mutants broadly confirms this view (Mohan and Ritossa, 1970).

B. Amphibia

Evidence that rDNA amount varies in other species was foreshadowed by cytogenetic studies in amphibia. Although amphibia are short of conventional genetic markers, nucleolar variants were readily isolated and shown to be heritable (Humphrey, 1961). Furthermore, when lampbrush chromosomes were examined, mutants characterized by a small nucleolus also displayed a shorter nucleolar constriction than wild type (Callan, 1966). A direct link with ribosomal gene repetition was made in the toad *Bufo marinus* (Miller and Brown, 1969). Individuals were found to vary both in the lengths of their mitotic nucleolar constrictions and in the number of ribosomal genes as determined by RNA–DNA hybridization. Extending these findings, Miller and Knowland (1970) isolated "partial nucleolar mutants" from a population of wild frogs. Each mutant chromosome showed a reduced, but different, rDNA level, which was transmissible by genetic crosses in a Mendelian fashion. Similarly, Sinclair *et al.* (1974) studied ribosomal gene levels in four inbred stocks of the axolotl *Ambystoma mexicanum*. One stock had about one-third as much rDNA as the others and also displayed shorter nucleolar secondary constrictions.

Inconsistency in the chromosomal location of ribosomal genes has also been demonstrated. In a study of the newt *Notophthalmus viridescens*, Hutchison and Pardue (1975) found three chromosomal rDNA sites by *in situ* hybridization, but out of six males examined, none was homogygous for all three loci. The animals also showed a twofold range in the percent of total DNA complementary to rRNA. Similarly, when 5 S DNA sites were examined, some chromosomes apparently lacked 5 S sites that were clearly present on the homologous partner. Recent work on the Italian newt *Triturus vulgaris meridionalis* has emphasized the nonuniform chromosomal location of rDNA (Nardi *et al.*, 1977). When animals from different parts of Italy were tested by *in situ* hybridization with [³H]rRNA, all hybridized at the established nucleolus organizer region, but 13 out of 20 animals also show hybridization at a variety of other chromosomal sites. "Additional rDNA sites" of this kind were invariably heterozygous between homologues and did not appear to be restricted to particular chromosomes in the complement. In all the studies mentioned above, chromosomes from different tissues in the same animal showed identical rDNA patterns. Furthermore, when individuals with different rDNA patterns were crossed, the progeny displayed exactly the combinations to be expected if "additional rDNA sites" are passed on as Mendelian units (Batistoni *et al.*, 1978).

Evidence that widely different levels of rDNA redundancy can coexist in a wild interbreeding population came from *in situ* hybridization studies of plethodontid salamanders (Macgregor and Kezer, 1973; Hennen *et al.*, 1975). More often than not, homologous loci showed strongly different degrees of hybridization and, occasionally, absence of detectable hybridization at previously identified rDNA sites. In addition, salamanders chosen from two interbreeding populations were found to show marked heterogeneity in chromosomal hybridization accompanied by a 2.5- and 7.5-fold range in percentage of rDNA (Macgregor et al., 1977).

All these results are difficult to reconcile with another study of rDNA amount in amphibia (Buongiorno-Nardelli *et al.*, 1972). In these experiments, DNA from a variety of amphibian species was hybridized with rRNA, and the fraction of the genome coding for rRNA was calculated. Values were clustered at one, two, four, or eight times the lowest levels measured, suggesting that rDNA amount within a species is relatively constant and can only change during evolution by powers of two.

C. Plants

Intraspecies variation in ribosomal gene numbers has also been reported in a number of higher plants. In *Zea mays,* for example, different strains show a twofold range in percent of rDNA (Philips *et al.*, 1974; Ramirez and Sinclair, 1975). There is also variation between interbreeding individuals from the same strain in the onion *Allium cepa* (Maggini *et al.*, 1978). Members of two cultivars showed a 2- and 2.5-fold variation in percent of rDNA.

D. The Source of Variability

How does this variability in tandem gene number arise? Clearly it is not generated anew at each generation, since partial nucleolar deletions in both *Drosophila* and amphibia can be maintained by genetic crosses. On the other hand, unpredictable changes in ribosomal gene number have been observed at a low frequency. For example, populations of inbred flies can be shown in appropriate test crosses to harbor a significant fraction of *bobbed* chromosomes (Spencer, 1944; Ritossa *et al.*, 1966b). More directly, rDNA-deficient *bobbed* chromosomes maintained against wild-type partners can be shown to revert to wild type at a low rate (Bridges and Breheme, 1944).

These properties all fit with an unequal recombination mechanism as the source of variable redundancies. The tandem repeat structure lends itself to chromosome pairing and recombination in a variety of registers, only one of which can be considered equal. Crossover in any of the unequal registers would generate one chromosome with more than the starting number of tandem genes, and one with less (Fig. 3). This possibility has been tested in *Drosophila*. Schalet (1969) obtained two complementing X chromosomes bearing lethal deletions that overlapped either the left- or right-hand extremity of the nucleolus organizer. Male progeny of a female heterozyous for these deletions can only survive if a crossover event

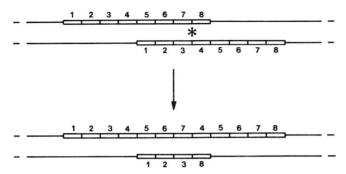

Fig. 3. Out of register crossover between identical arrays of tandem repeated genes to give unequal sized arrays. Each chromatid or homologous chromosome initially carries eight identical copies of the gene. Exchange occurs at the site marked by an asterisk.

occurs at the only region of homology between them, the nucleolus organizer. Such male recombinants were indeed observed, and often showed changes in intensity of the *bobbed* phenotype. It was inferred that the crossover resulted in a new dosage of ribosomal genes.

In theory unequal crossover need not be confined to homologous pairs of chromosomes at meiosis. Sister chromatids are generated at each DNA replication and are capable of recombination before chromosome separation at mitosis (Taylor, 1958). Also, functionally related tandem genes are often found at nonhomologous sites in the genome, and these may conceivably associate and recombine.

Despite the attractiveness of unequal crossing-over as an explanation for tandem gene variability, there are still profound uncertainties. Significant departures from this neat scheme have been observed. These are best documented in *Drosophila*, where sudden, predictable increases in rDNA amount occur in both somatic and germ line cells.

V. PREDICTABLE CHANGES IN rDNA REDUNDANCY IN SOMATIC CELLS

A. Ribosomal DNA Compensation in *Drosophila*

Drosophila genetics affords many opportunities for manipulating the nucleolus organizer-bearing sex chromosomes to give abnormal combinations. Assuming genome constancy in somatic cells we might expect that the number of ribosomal genes would be additive with increasing doses of the nucleolus organizer. Contrary to expectation, DNA from X/O flies has more than half as much rDNA as DNA from flies carrying two doses of the same X chromosome (Tartof, 1971). Since the extra rDNA in X/O individuals is not transmitted to the progeny, "rDNA compensation" appears to be exclusively a somatic phenomenon.

Interpretation of these experiments is hampered by the fact that the DNA tested came from whole flies. Adult *Drosophila* contain a significant fraction of polytene cells (Ashburner, 1970) in which the regulation of rDNA amount can be unusual (Hennig and Meer, 1971). With this in mind Spear and Gall (1973) again tested X/O flies for their rDNA content, but treated larval diploid cells (brains and imaginal discs) and polytene cells (salivary glands), separately. Whereas diploid rDNA content was closely proportional to the number of X chromosomes present, polytene cells showed the same fraction of rDNA regardless of their genetic constitution. This result immediately suggested an explanation for the rDNA compensation phenomenon. If polytene cells in the adult contribute the same amount of rDNA in either X/O or X/X flies, this would account for the nonadditivity of rDNA levels. The explanation is appealing because it avoids the need to postulate a compensation mechanism in diploid cells and because it explains why germ cells are exempt from the process. It also accounts for the finding that organisms incapable of polyteny, such as *Xenopus laevis* and *Zea mays,* show no evidence of compensation (Birnstiel *et al.,* 1966; Miller and Knowland, 1970; Philips *et al.,* 1974).

There are, however, limitations to the polytene theory of rDNA compensation (Tartof, 1973b). First, rDNA in polytene salivary glands is significantly underreplicated compared to euchromatic DNA and if this were also true of other polytene cells they would be numerically incapable of boosting the adult X/O rDNA content to the observed value. Second, there is the inconsistant behavior of the Y chromosome. Flies whose only rDNA is part of the Y chromosome do not appear to compensate (Tartof, 1971), yet salivary gland cells with the same genetic constitution do show independant replication of rDNA to the usual level (Spear, 1974).

B. Increased rDNA in Flax

Before leaving the subject of somatic variation in rDNA amount, it is necessary to mention the case of flax (Timmis and Ingle, 1973). Starting from a "plastic genotroph" it is possible, by appropriate treatment with fertilizers, to produce either large or small forms of the plant, both of which are stable. Large forms have a greater nuclear mass and DNA content than the smaller forms, and hybridization with rRNA shows that they also have about 50% more rDNA per nucleus. The increase is not sufficient by itself to account for the higher C value in large forms, suggesting that other DNA is also overrepresented.

VI. PREDICTABLE CHANGES IN rDNA REDUNDANCY THAT ARE TRANSMISSIBLE IN THE GERM LINE—rDNA MAGNIFICATION

Drosophila sex chromosomes that are deficient in rDNA generate the *bobbed* phenotype when homozygous or hemizygous. Normally such chromosomes arise and revert at a low rate consistent with a mechanism involving unequal recombination. However, in certain genetic combinations a *bobbed* chromosome can revert to wild type at a very much higher rate. The process is called "magnification" (Ritossa, 1968) and is confined to males. It begins when, for example, a *bobbed* X chromosome (X^{bb}) is placed opposite a Y chromosome carrying a severe *bobbed* mutation (Y^{bb-}). At first the resulting male fly is, as expected, strongly *bobbed* in phenotype. But if this genetic constitution is reestablished for a second generation, a high proportion of the male flies are either wild type, or phenotypically less *bobbed* than before. By the third generation, wild-type flies usually emerge. Each step in the loss of the *bobbed* condition is accompanied by an increasing percentage of rDNA in DNA from whole flies, until a wild-type complement is achieved. At this point magnification stops. Since the extra rDNA is associated with the X^{bb} chromosome, it appears that the presence of the Y^{bb-} chromosome somehow triggers a mechanism on the X^{bb} chromosome leading to a progressive increase in rDNA until a level capable of supporting normal growth is achieved. Not only the Y^{bb-} chromosome induces magnification. The phenomenon is also observed when a Y^{bb} chromosome is maintained with an X deleted for its ribosomal genes. In this case, though, it is the Y chromosome which magnifies (Boncinelli *et al.*, 1972).

That magnification occurs is not in doubt, but the mechanism is still far from settled. Ritossa (1972) has put forward a model in which unintegrated copies of the nucleolus organizer could pass between generations unstably in the germ line and crossover into the rDNA-deficient X^{bb} locus. In this way the model attempts to account for the uneven stability of newly magnified flies (see below). Tartof (1973b), on the other hand, is a proponent of unequal sister chromatid exchange as an explanation for magnification. In indirect support of the idea, he has found that when sister chromatid exchange is selected against, magnification is not found. The experiment makes use of ring-form X chromosomes carrying the *bobbed* mutation. Odd numbers of sister chromatid exchanges in a ring chromosome (1, 3, 5, etc.) would prevent segregation of the X chromatids at cell division and would therefore be lethal. Only germ cell X chromosomes which had exchanged an even number of times, or not at all, would survive. The lack of magnification in progeny of ring-X males may mean that the process depends on sister chromatid exchange. Another prediction of the unequal exchange model is that, since unequal exchange is a reciprocal event, decreases in rDNA would be expected to occur as well as increases. Indeed, such rDNA reduction was observed when a normal X chromosome was combined with a Y^{bb}, albeit at a very low frequency (Tartof, 1973b).

Both these models are severely taxed by the sheer range of unexplained observations. For example, while in some cases the rDNA accumulated during magnification is quite stable (Tartof, 1973b), in other cases it does not persist in genetic combinations which do not actively encourage magnification (Malva et al., 1972; Tartof, 1973a). Thus when an X^{bb} is combined with an X chromosome whose nucleolus is completely deleted for four successive generations, its rDNA content rises slowly from an estimated 160 to 263 genes. However, when this fourth generation chromosome is now combined with a normal Y, the number of genes drops to its original "premagnification" value (Tartof, 1973a). Instability has also been observed in newly magnified X^{bb} chromosomes induced by Y^{bb-}, though it gives way to stability after several generations (Henderson and Ritossa, 1970; Boncinelli et al., 1972).

The questions raised by differential stability of magnified rDNA are compounded by the finding that the bobbed phenotype can persist in flies apparently carrying more than enough genes to support a normal phenotype. One such case is the fourth generation X^{bb}/X^{NO-} combination discussed above (Tartof, 1973a). Another ineffectual increase in rDNA content has been observed in the first generation of a genotype prone to magnification (Ritossa et al., 1971). These male flies already show a marked increase in percent rDNA (so-called "premagnification"), and their sons will exhibit magnification, yet they themselves are still severely bobbed. Finally, the same phenomenon is seen during rDNA compensation when a bobbed locus is involved. Compensation can be demonstrated by RNA–DNA hybridization, yet it has no ameliorating effect on the bobbed phenotype (Ritossa, 1976).

It should be borne in mind that changes in rDNA redundancy in a particular chromosome are usually monitored experimentally by rRNA–DNA hybridization to whole fly DNA and are calculated by subtraction of the rDNA value on a known chromosome from the total measured redundancy. Both these procedures may exaggerate experimental uncertainties, and may account, for example, for the wide range of rDNA values measured on the Y^{bb-} chromosome: 0.093% (Ritossa, 1968), 0.036% (Tartof, 1973a), 0% (Spear, 1974). The last of these values was measured in larval diploid tissue.

Perhaps the most perplexing aspect of magnification is its relation to tandem gene number in normal cells. Is magnification an extreme case of some phenomenon which normally acts less spectacularly? Or is it an entirely abnormal process induced by particular combinations of defective chromosomes? At present we simply do not know.

VII. THE EVOLUTION OF TANDEM REPEATED GENES

A. Introduction

The examples discussed in Section III give general support to the view that a genome carries tandem multiple copies of a gene in response to selective pressure for more of that gene product. This seemingly simple device is apparently not without evolutionary complications. One problem is the maintenance of a stable level of redundancy as discussed in Section IV. Another complex, and perhaps related, question concerns the maintenance of functional homogeneity among a group of clustered, but separate, entities.

Evolution of rDNA has been examined experimentally by comparing two related species of frog, *Xenopus laevis* and *Xenopus borealis** (Brown *et al.*, 1972; Forsheit *et al.*, 1974). It has been shown that those parts of the rDNA repeat unit which code for product RNA are highly conserved both within and between species. However, the spacer DNA between coding sequences evolves in quite a different way. Within either species spacers form a closely related family, but between species there is insufficient similarity to permit the denatured sequences to reanneal with one another.

A similar situation has been found for 5 S DNA in these *Xenopus* species (Brown and Sugimoto, 1973a,b). The 5 S RNA genes are not confined to a single chromosomal region, but are clustered in separate groups at the tips of at least 15 chromosomes. Despite this, the spacer sequences of one species anneal with one another, and the melting temperature of the reassociated molecules suggests only about 4% mismatching. Between species the spacers are unrelated by the hybridization assay, and the average length of the *X. borealis* spacer is 2.5 times that of *X. laevis*.

---

* This species was previously identified as *Xenopus mulleri* in a number of publications. The error has since been corrected, and the species reidentified as *Xenopus borealis* (Brown *et al.*, 1977).

B. The Evolution of Repeated Genes Governed Only by Natural Selection

To begin with, we must consider the possibility that these results can be explained by natural selection alone. This view requires that any unselected parts of the cluster, such as nontranscribed spacers, nonessential parts of the gene, and whole repeat units in excess of the minimum requirements, will accumulate mutations over time and diverge. An important consequence is that the degree of sequence divergence between any two repeats in the same cluster and between two repeats from different clusters descended from a common ancestor will be the same.

The fact that *X. laevis* and *X. borealis* can produce viable hybrid progeny under artificial conditions clearly shows that they are descended from a common ancestor (Blackler, 1970). However, the fate of their rDNA since divergence presents a dilemma for the simple selectionist view. On one hand, their spacer sequences have diverged markedly, suggesting that the primary sequence in this part of the repeat is relatively free to change. While, on the other hand, spacer sequences within species are closely similar, implying sequence conservation. It seems that the genes in a cluster, or even in separate clusters, evolve not as independent entities but as a family, a tendency known as "horizontal evolution" (Brown *et al.,* 1972). This behavior is not confined to the ribosomal genes. Very highly repeated DNA sequences known collectively as satellite DNA's share many of the evolutionary properties of spacers. Satellite repeat units show great similarity within a species but often show considerable divergence between species (Hennig and Walker, 1970; Sutton and Mc-Callum, 1972). Neither satellites nor spacers are transcribed, and this together with their often rapid evolution suggests that primary sequence is not important. How then do they evolve as families? The simple selectionist view must suggest the untenable solution that hundreds of repeat units have by chance mutated in the same way, at the same positions, repeatedly.

C. Once per Generation Homogenization Mechanisms

In reviewing alternatives to the selection model, it is instructive to examine the historical development of our knowledge of repeated gene structure. At the time when the contradictions discussed above became known, all the available techniques indicated that rDNA repeat units within a species were identical. It was thus feasible to account for horizontal evolution by suggesting mechanisms specifically for homogenizing tandem repeat units at a stroke during each life cycle. Several such "correction mechanisms" were proposed in which one gene out of the array was either laterally amplified to replace the original cluster (see Brown *et al.*, 1972; Buongiorno-Nardelli *et al.*, 1972) or served as a template against which all the other genes in the array were corrected (Callan, 1967). All these models in their simplest forms were left stranded by the discovery that, on closer examination, genes within a cluster are not identical after all.

This new information resulted from the use of restriction endonucleases to investigate repetitive gene structure. Restriction endonucleases recognize a short sequence in DNA and, in the case of class II enzymes, make a double-stranded break at or near that site (Smith and Wilcox, 1970; Kelly and Smith, 1970; reviewed in Roberts, 1976). Thus a restriction site map of any homogeneous DNA fraction can be built up by treating the DNA with a restriction enzyme and analyzing the discrete fragments which are produced (Danna and Nathans, 1971). In *Xenopus* rDNA the recognition sequence for endonuclease EcoRI occurs twice per repeat unit (Morrow *et al.*, 1974). After digestion one of the resulting fragments comes from within the coding portion and is homogeneous in size. The other fragment contains the nontranscribed spacer and, surprisingly, shows marked length heterogeneity even when rDNA from one frog with a single nucleolus is tested (Wellauer *et al.*, 1974; Wellauer and Dawid, 1974). In the *X. laevis* population, spacers vary in length from 1000 to 5000 nucleotide pairs, and each nucleolus organizer carries a few spacer size classes from this distribution (Wellauer *et al.*, 1976b; Buongiono-Nardelli *et al.*, 1977). Figure 4 compares the pattern of repeat length in eight nucleolus organizers.

At first sight the fivefold range in rDNA spacer lengths does not accord well with the intraspecific homogeneity of spacer sequences by DNA annealing criteria (Dawid *et al.*, 1970). Recent results beautifully account for this disparity. It seems that spacers themselves contain short tandemly repeated sequences, and the primary cause of length heterogeneity is not the presence or absence of new sequences, but variation in the number of similar subrepeats (Wellauer *et al.*, 1976a; Botchan *et al.*, 1977). This is most clearly established for the oocyte-type 5 S genes of *Xenopus laevis*. Here repeat lengths are seen to vary in discrete quanta of about 15 nucleotides (Carroll and Brown, 1976a), and indeed a segment of the spacer is composed of a 15 nucleotide sequence tandemly repeated (Brownlee *et al.*, 1972). Direct sequencing of different spacer fragments purified by cloning in bacteria (Carroll and Brown, 1976a) shows that variable numbers of 15-mer repeats are primarily responsible for the length variation (Fedoroff and Brown, 1978).

Heterogeneity of repeat unit lengths has also been found in rDNA from mouse (Arnheim and Southern, 1977; Cory and Adams, 1977), man (Arn-

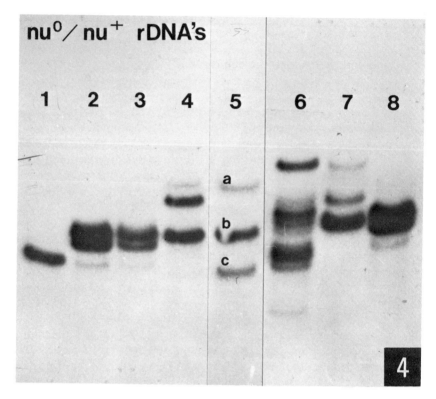

Fig. 4. Variability between rDNA repeat lengths in different nucleolus organizers. Each track analyzes one nucleolus because animals are heterozygous for an rDNA deletion (*nu*⁰). Blood cell DNA was digested with endonuclease *Hin*dIII, run out on an agarose gel, blotted, and hybridized with ³²P-rDNA. *Hin*dIII breaks once per repeat unit at a unique site. Apart from slots 2, 3, and 8, which are from animals carrying the same nucleolus organizer, there is pronounced variation between rDNA patterns. This variability is attributable to the nontranscribed spacer. Compare the internal homogeneity of slot 1 with the marked heterogeneity of slot 6. As an indication of size, bands 5a, b, and c are 13.8, 12.4, and 11.3 kb. (Figure kindly supplied by Dr. E. M. Southern and Dr. H. N. Kirkman.)

heim and Southern, 1977), and *Drosophila* (Glover and Hogness, 1977; White and Hogness, 1977; Wellauer and Dawid, 1977; Pellegrini *et al.*, 1977). The case of *Drosophila* is of particular interest, although of uncertain significance. Many rDNA repeat units on the X chromosome contain an insertion sequence within the 28 S gene. Length heterogeneity derives mainly from the presence or absence of the insertion, although there is also length variation in the spacer (Wellauer and Dawid, 1977). Not all tandem genes, however, are heterogeneous in length. For example, the histone genes of the sea urchin *Psammechinus* and the tRNA^{met} genes of *Xenopus* appear homogeneous in length despite the presence of noncoding DNA in the spacer (Shaffner *et al.*, 1976; Clarkson and Kurer, 1976).

When length heterogeneity is observed, there are usually only a few length classes in each individual. As a result, many repeat units must belong to each class and it is therefore possible to accommodate the observed heterogeneity by more elaborate formulations of the once per generation correction mechanisms. These formulations require that repeats of the same size result from correction of part, rather than all, of the gene array (Buongiorno-Nardelli *et al.*, 1977). Two predictions follow: first, that all repeats of like size will be clustered together and, second, that a combination of repeats will show change between generations owing to correction events.

Electron microscopy of heteroduplexes between a cloned spacer restriction fragment and single-stranded chromosomal rDNA long enough to contain several repeat units has shown convincingly that repeat units of the same length are not in fact clustered, but are interspersed with repeat units of other lengths (Wellauer *et al.*, 1976b). This finding has been confirmed by partial restriction endonuclease digestion of DNA from several animals with only one nucleolus per cell (E. M. Southern and H. N. Kirkman, unpublished observations). The results show that dimers of the rDNA repeat unit often contain spacers of different length. Reports that like repeats are clustered together (Buongiorno Nardelli *et al.*, 1977) are not confirmed by the restriction analysis, although there is evidence that different rDNA clusters show degrees of interspersion which sometimes deviate significantly from random (Wellauer *et al.*, 1976b).

Interspersion of variable repeat lengths also holds true for the 5 S genes in *Xenopus* which are found at multiple chromosomal loci. Restriction analysis of clones containing several contiguous repeat units (Carroll and Brown, 1976b) shows that even among 5 S repeat units from the same chromosomal locus, different spacer lengths are interspersed with one another.

Taken together these results argue strongly against once per generation correction mechanisms. This is reinforced by the finding that patterns of spacer heterogeneity are nearly always inherited intact in genetic crosses (Reeder *et al.*, 1976; see Section VII,G).

D. Unpredictable Homogenization Mechanisms

Having ruled out the once per generation correction models we are left with two kinds of explanation for the parallel evolution of gene clusters: sudden homogenization events which occur infrequently in the course of evolution ("saltatory events"; see Britten and Kohne, 1968), and frequent piecemeal events which do not homogenize at a stroke but which contribute to homogenization with time (sister chromatid exchange model; Smith, 1973). Neither explanation is easy to test, but already there are enough data for a preliminary evaluation.

E. Saltatory Replication

Saltatory replication is a blanket term for sudden homogenization mechanisms which occur irregularly over evolutionary time rather than in each life cycle. The molecular mechanism is not specified, but excision of a DNA segment followed by replication and reintegration is envisaged. The model would account for the horizontal evolution of $X.$ $laevis$ and $X.$ $borealis$ rDNA spacers in the following way. Consider the ancestral rDNA cluster. Since the last saltatory event the initially identical genes have accumulated mutations independently of one another and have given rise to a diverged, but related, family of repeats. Saltatory events occurring after the separation of $X.$ $laevis$ and $X.$ $borealis$ lines are unlikely to "amplify" the same repeat from within the diverged cluster. As a result, saltation will leave the two lines internally homogeneous but different from each other. The longer the two lines are separated, the more cycles of mutation and saltations will have occurred, and the bigger the difference between $X.$ $laevis$ and $X.$ $borealis$ rDNA families.

F. Piecemeal Homogenization—Unequal Sister Chromatid Exchange

The only well worked out example of this kind of model is unequal sister chromatid exchange (Smith, 1973; Edelman and Gally, 1970; Tartof, 1973b). Immediately following DNA replication in mitotic or meiotic cells, sister DNA molecules (i.e., sister chromatids) are in close proximity and probably remain so until chromosome separation takes place. Recombination between sisters can occur during this time, and has been visualized cytologically (Taylor, 1958; Prescott and Bender, 1963). As pointed out in Fig. 5, sister chromatid exchange should not be confused with the exchange between homologous chromosomes that occurs regularly at meiosis and less frequently in mitotic cells. Homologue exchange brings about new combinations of alleles on the chromosomes and can be detected genetically as an exchange of flanking markers. A sister chromatid exchange in which the sister molecules are perfectly aligned, however, leaves both products identical. Assuming that crossover only occurs at points which share sequence homology, it is apparent that two sister

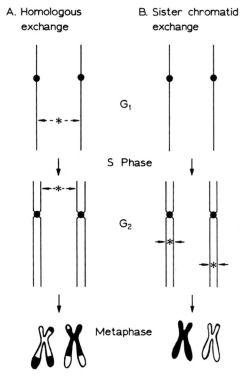

Fig. 5. Comparison between homologous exchange and sister chromatid exchange. (A) Exchange between homologues in the G_1 or G_2 phases of the cell cycle leads to characteristic changes in composition of the chromatids inherited by daughter cells. (B) Exchange between sister chromatids does not normally lead to changes in genetic composition because chromatids are the identical replication products of a single DNA molecule. Only if exchange is unequal can interchromatid differences arise.

clusters of tandemly repeated genes may crossover in many registers, only one of which leads to equal recombination. All unequal recombination will lead to one sister chromatid possessing more repeat units than the other (Fig. 3).

Smith (1973) has used a computer to follow a hypothetical array of 500 identical genes through a series of unequal crossovers. Each gene in the starting array was given a serial number and simple rules were applied for the degree of overlap between clusters and the minimum tolerated cluster size. After about 10,000 simulated crossover cycles, more than 90% of the genes in an array had the same serial number. In other words, unequal sister chromatid exchange tends to concentrate the descendants of one of the 500 starting repeats in the same array—it homogenizes the cluster.

At first sight this method of testing the mechanism seems to take an oversimplified view of the behavior of a tandem gene array in a real interbreeding population. This is because only one product of each crossover is followed. It seems that if we were to follow not only all the product chromosomes but all tandem gene arrays of that kind in the gene pool, then sister chromatid exchange would not lead to parallel evolution of their genes. Instead each array would tend to homogenize any one of the starting repeats independently of the others, leaving the repeat complexity in the population the same as at the outset. This criticism of the mechanism ignores "genetic drift." Random fluctuations in the frequency with which genes are passed from one generation to the next ensure that, even in the absence of selection, some genes are eliminated from the population and others become fixed. Ultimately all the genes of a particular type in the gene pool will be descended from one ancestral gene. For repetitive genes, this means that although different clusters will tend to homogenize independently, only one homogenizing cluster is destined to exist in a future population.

Not all aspects of the unequal crossover model have been fully worked out. In his original paper, Smith (1973) analyzed sister chromatid exchanges, rather than exchange between homologues, to simplify the computer analysis. It is, however, likely that unequal homologous exchanges, and even nonhomologous exchanges, could contribute to horizontal evolution. Unfortunately it is not possible to make convincing quantitative predictions based on the sister chromatid exchange model. Most of the parameters which affect the process are unknown (e.g., frequency of sister chromatid exchange in germ cells, frequency of homologous recombination and its effect on the rate of homogenization, ratio of equal to unequal crossovers, degree of sequence homology required for exchange, and rate of introduction of repeat length variants). Nevertheless, approximate calculations suggest that one unequal sister chromatid exchange per germ cell per generation would account for the observed resemblance between 100- to 1000-fold repeated tandem genes.

G. Unequal Exchange Versus Saltatory Replication

Both unequal exchange and saltatory replication explain the broad resemblance between ribosomal and 5 S genes within a species and the differences between species. By looking more closely at inter- and intraspecies heterogeneity it is possible to evaluate these two mechanisms more critically.

As we have seen, the primary source of intraspecific variation among repeat units is length variation caused by different numbers of spacer subrepeats. By contrast, qualitative sequence differences, while present, are not sufficient to interfere with reassociation of different denatured spacers. Unequal exchange accounts for the bias, since out of register crossovers between spacer subrepeats would automatically generate length variants. This paradox, whereby unequal sister chromatid exchange could both homogenize repeats and generate new length variants, is a natural consequence of the presence of short spacer subrepeats within the larger repeat unit.

Another feature of rDNA spacers in *Xenopus laevis* is the presence of higher-order periodicities in which the basic subrepeat of 15 nucleotides is itself part of a longer repeating structure (Botchan *et al.*, 1977). Satellite DNA's from various sources show identical properties, often displaying several ascending orders of repetition (Botchan, 1974; Southern, 1975a; Biro *et al.*, 1975). Again unequal exchange predicts just such higher-order periodicities in the course of repeated sequence evolution (Smith, 1973). In the case of mouse satellite, the observed frequency of higher repeats has been shown to fit with the expectations of unequal crossover (Southern, 1975a).

Saltatory replication does not accommodate these aspects of spacer evolution so smoothly. It is necessary to invoke occasional "lateral amplification" of spacer subrepeats both singly and in groups within one spacer, in addition to periodic excision, amplification, and reintegration of complete repeat units.

Although the weight of circumstantial evidence is so far on the side of homogenization by unequal recombination, there are some observations that do not fit neatly into this scheme. One example has been reported by Reeder *et al.* (1976) after studying the inheritance of rDNA spacer heterogeneity in *X. laevis*. These workers examined the progeny of four crosses involving animals heterozygous for a deletion of the ribosomal genes. *Eco*RI digestion of progeny rDNA permitted a comparison between parental and progeny spacers. Generally, spacer patterns were transmitted intact, except in 2 out of 51 progeny which showed significant departures from the paternal pattern. In both exceptional cases the change involved a measurable addition of one repeat size class to the parental cluster, but it was not clear whether unequal exchange or some "saltatory" process was responsible. It is significant that unequal crossover was considered the less likely explanation because it would be expected to introduce a variety of interspersed size classes rather than the one new spacer class that was observed. Another sudden change that is not readily explicable by unpredictable recombination events is magnification (see Section VI). The contending models invoke either abnormally high rates of unequal crossing-over between sister chromatids, or extrachromosomal rDNA replication followed by crossover into the rDNA deficient locus.

Note that the alternative explanations for these phenomena always involve recombination, but in one case it is between chromosomal clusters, and in the other case, between chromosomal and extrachromosomally replicated genes. The question is therefore whether or not extrachromosomal copies are involved. Extrachromosomal DNA existing as circular molecules has frequently been observed in eukaryotic somatic cells (Radloff *et al.*, 1967; Stanfield and Helinski, 1976). In *X. laevis* circular rDNA molecules have been found in blood and cultured kidney cells (Rochaix and Bird, 1975), and it has recently been claimed that testicular DNA of flies in the early stages of magnification contains circular DNA molecules (Graziani *et al.*, 1977). In all these cases, the presence of extrachromosomal genes is unexpected. Clearly, if repeated genes can, by accident or design, form extrachromosomal copies, and if these copies can reintegrate in the germ line, the course of repeated gene evolution will be affected.

To summarize, many features of tandem gene evolution are consistent with unequal exchange between sister chromatids or between chromosomes. Since much of our knowledge about fluctuations in tandem gene number is also consistent with unequal exchange (see Section IV), and since exchange itself is a well-known phenomenon, the mechanism must rank high as a possible cause of horizontal evolution. At the same time, sudden events have been observed that are difficult to explain by straightforward crossing-over. Some of these observations may be the result of small-scale saltatory events in which blocks of tandem genes are reintegrated after extrachromosomal replication. Further investigation of these phenomena will probably replace the inherently vague term "saltatory event" by specific molecular mechanisms.

VIII. AMPLIFICATION OF A UNIQUE GERM LINE GENE IN *TETRAHYMENA*

The evolutionary problems associated with tandem repeated genes are theoretically eliminated if a single gene is carried in the germ line and is amplified in the somatic cells of each generation. So far, this method for maintaining multiple genes has not been found in multicellular organisms. It does, however, occur in at least one unicellular eukaryote, the ciliated protozoan *Tetrahymena*. (Fig. 6)

In the vegetative phase *Tetrahymena* contains two nuclei, the micronucleus and the macronucleus. The micronucleus appears to be transcription-

Tetrahymena
pyriformis

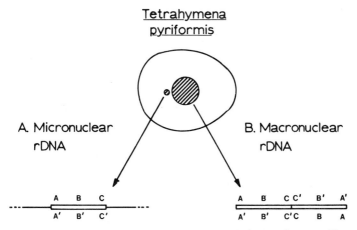

A. Micronuclear
rDNA

B. Macronuclear
rDNA

Fig. 6. Comparison of micro- and macronuclear rDNA's in *Tetrahymena*. The micronuclear genome contains a single integrated 25 S + 17 S coding sequence per haploid chromosome set. Macronuclear rDNA consists of multiple, extrachromosomal rDNA copies arranged in palindromic pairs. A', B', and C' are the sequences base paired with A, B, and C.

ally inactive, divides mitotically, undergoes meiosis during the sexual phase of the life cycle, and gives rise to the macronucleus. The macronucleus, on the other hand, is polyploid, is transcriptionally active, divides amitotically, and is generated anew following each sexual phase. Thus the micronucleus is a germ line nucleus, and the macronucleus is analogous to a somatic nucleus.

Considering first the micronucleus, saturation hybridization initially established a maximum of 20 ribosomal genes per haploid genome (Yao *et al.*, 1974). With the advent of restriction enzyme mapping and the blotting technique (Southern, 1975b), it was possible to investigate in more detail the arrangement of rDNA sequences and their linkage to other DNA in the chromosome (Yao and Gall, 1977). The results show that there is only one gene for the rRNA precursor in the micronucleus, and that it is integrated at a unique site in the chromosomal DNA (Fig. 6A).

Macronuclear rDNA differs from micronuclear rDNA in several important respects. First, it is amplified to a level of more than 200 gene copies per haploid DNA equivalent (Yao *et al.*, 1974), and all these copies are extrachromosomal (Gall, 1974; Engberg *et al.*, 1974; Yao and Gall, 1977). Second, structurally, the extrachromosomal rDNA molecules are paired inverted repeats of the chromosomal rDNA unit (Fig. 6B) (Karrer and Gall, 1976; Engberg *et al.*, 1976).

To explain this distinctive amplification procedure, Gall (1974) has suggested that *Tetrahymena* represents an evolutionary transition point between prokaryotes and multicellular eukaryotes. In common with pro-karyotes, there are few ribosomal genes in the germ line, but the greater cytoplasmic volume of a eukaryote demands a mechanism for boosting the ribosome supply.

Extrachromosomal rDNA has also been found in another ciliate, *Stylonichia* (Prescott *et al.,* 1973) and the slime mold, *Physarum* (Vogt and Braun, 1976). In the latter organism an inverted repeat structure for the rDNA has been established.

IX. AMPLIFICATION IN THE GERM CELLS OF MULTICELLULAR ORGANISMS

The genomes of multicellular eukaryotes contain tandem repeated ribosomal genes. Exact distribution of these multiple genes by mitosis presumably furnishes each somatic cell with sufficient rDNA for its needs. Nevertheless, germ cells, and in particular oocytes, often raise the level of rDNA per nucleus still further by amplification (Gall, 1969). In terms of cytology and molecular biology, the best studied case is *Xenopus laevis.* We shall now consider the known features of amplification in this or-ganism, following the life cycle of the germ cells (Fig. 7) rather than the chronological order in which discoveries were made.

A. Amplification in *Xenopus* Oocytes

1. The Earliest Events

The molecular events which accompany the initiation of amplification are still a subject of speculation. The reason for this is the inaccessibility of the very earliest detectable events, which occur about 3 weeks after fertilization when only 9–16 primordial germ cells are present (Kalt and Gall, 1974). By this stage the primordial germ cells have become estab-lished in the developing genital ridge, and *in situ* hybridization indicates that their rDNA content is similar to that in the surrounding somatic cells. Amplification coincides with a revival of mitotic activity in the germ cells as they enter the gonial phase.

Several kinds of models have been put forward to explain the molecular events at this stage. One extreme possibility is that amplified rDNA copies are never integrated into the chromosome, but pass episomally between generations (Wallace *et al.,* 1971; Brown and Blackler, 1972). There is now good reason to believe that this model is not correct. In a key experiment, Brown and Blackler (1972) observed that in the female progeny of crosses

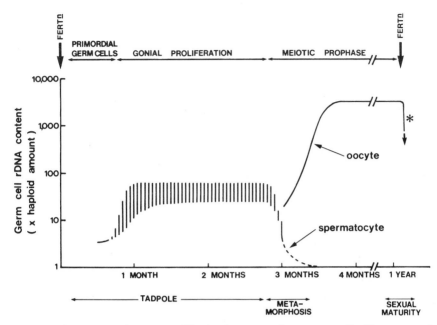

Fig. 7. Time course of rDNA amplification in *Xenopus laevis* germ cells. The amount of rDNA per cell is plotted against time. Note that the ordinate is a logarithmic scale. The sharp drop in rDNA at fertilization (asterisk) is thought to be due to dilution of the amplified rDNA copies by cell division rather than their destruction (see text).

between *Xenopus borealis* and *Xenopus laevis,* amplified rDNA is always of the *laevis* type regardless of which species was the maternal parent. The reason for the "dominance" of *laevis* is not known, but since an episome would be expected to show maternal inheritance, the inference is that amplified rDNA is generated from the chromosomes anew in each life cycle. In support of this conclusion, Wellauer *et al.* (1976b) have observed that amplified and chromosomal rDNA's in an individual female display a related range of repeat unit lengths.

How then might rDNA make the transition between chromosomal integration and the extrachromosomal state? One possibility is that a recombination event within the ribosomal gene array releases copies of the rDNA directly (Fig. 8A). This would explain the fact that amplified rDNA in oocytes is circular (Miller, 1964, 1966; Hourcade *et al.,* 1973a), but suffers from the disadvantage that the number of chromosomal genes would be reduced at each generation.

Other mechanisms propose that extrachromosomal rDNA copies arise by abnormal DNA replication. Figure 8B–D diagram the "reverse trans-

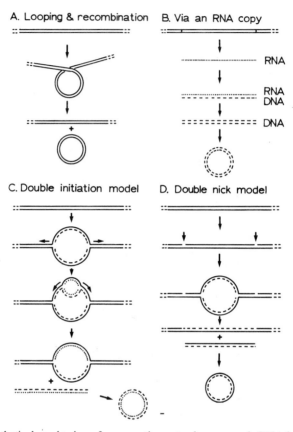

Fig. 8. Hypothetical mechanisms for generating extrachromosomal rDNA from the integrated chromosomal copies. (A) Looping and recombination. Sequence homology at the point where DNA strands cross leads to recombination and excision of a circular rDNA molecule. (B) Via an RNA copy. An RNA copy of the entire repeat unit is used as template for RNA-dependent DNA polymerase to give a double-stranded copy of the gene. (C) Double initiation model. DNA replication initiates bidirectionally to give a characteristic replication "bubble." After a short distance replication ceases, and a second bubble forms on one of the newly replicated strands. Replication proceeds to the extremities of the first bubble liberating a free linear rDNA copy. (D) Double nick model. Formation of a replication bubble between nicks on the same strand liberates a linear copy of the internick DNA.

criptase'' mechanism (Tocchini-Valentini and Crippa, 1971), ''the double initiation'' model (Sederoff and Tartof, cited in Tartof, 1975), and a ''double nick'' replication model. All require some provision for cyclization of the linear product, but all have the advantage that they leave the parental chromosome unchanged. At present there is no evidence for or against these models.

2. Amplified rDNA in Gonial Cells

After the initial amplification step, the primordial germ cells enter a proliferative phase and are termed gonia. The gonial nucleus is a large irregular body and is indistinguishable in males and females (Kalt, 1973). It may display up to 12 nucleoli and 5 to 40 times more rDNA than the surrounding somatic nuclei (Gall, 1969; Gall and Pardue, 1969). Electron microscopy provides direct evidence that this extra rDNA is extra-chromosomal, since ribosomal DNA from premeiotic gonads and from adult testes contains a high proportion of circular molecules (Bird, 1978). The presence of extrachromosomal rDNA in dividing gonia raises the question of how the level of amplification is maintained in an expanding cell population. There are two nonexclusive possibilities. Either the circular molecules replicate autonomously, or the population of free rDNA molecules is continually replenished from the chromosomal rDNA locus. It is not known whether one or both of these mechanisms operates, but the observation of a few circular replication intermediates among gonial rDNA suggests that autonomous replication can occur at this stage (Bird, 1978).

3. Meiotic Amplification

At meiosis the sexes differ for the first time. Early meiotic prophase is at first accompanied by a reduction in nuclear rDNA content (Kalt and Gall, 1974). In males the level remains low thereafter, but in females there is a dramatic second wave of amplification to about 2500 times the haploid rDNA content (Gall, 1969). It is this second wave of oocyte amplification which first excited interest in the process at the molecular level.

Cytological evidence for extrachromosomal DNA synthesis in early meiotic prophase had been available for many years (Painter and Taylor, 1942). Building on this early work, Gall (1968) showed that ovaries of newly metamorphosed frogs incorporate a high proportion of [^{3}H]thymidine into a DNA fraction which also hybridizes with rRNA. At the microscopic level, this synthesis correlates with the build up of a DNA-rich nuclear "cap" (Fig. 9) (Gall, 1968; Macgregor, 1968). The final link between the cap and rDNA was made possible by the development of *in situ* hybridization (Gall, 1969; Gall and Pardue, 1969; John *et al.*, 1969). Tritiated rRNA was hybridized to alkali-treated squash preparations of cap nuclei, and the hybridization assayed by autoradiography. The result unambiguously showed that the cap contains an amount of rDNA far in excess of either gonia or somatic cells.

Later studies have filled out the details of meiotic amplification. It begins about 1 week after premeiotic S phase, during the zygotene phase of

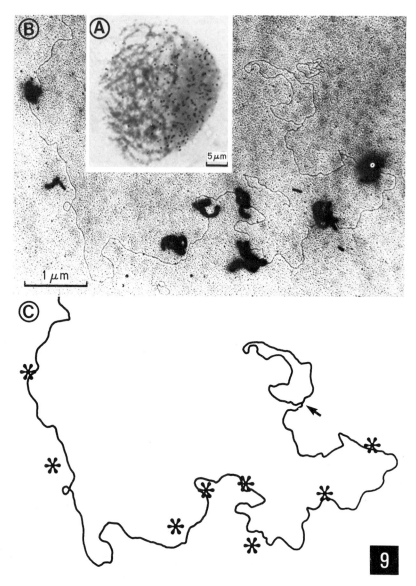

Fig. 9. The mechanism of meiotic amplification in *Xenopus* oocytes. (A) Light microscope autoradiograph of a pachytene oocyte nucleus labeled with tritiated thymidine. Silver grains show that DNA replication is going on in the "cap" of amplified rDNA. The unlabeled chromosomes are in the "bouquet" configuration with their tips grouped opposite the cap. (Figure taken from Bird and Birnstiel (1971) with permission from Springer-Verlag.) (B) Electron microscope autoradiograph of an rDNA molecule replicating by the rolling circle mechanism. Silver grains are seen over rDNA synthesized during a 3-hr incubation with tritiated thymidine and deoxycytidine. (C) Drawing of the same molecule with asterisks marking the silver grains. A replication fork (arrow) travels repeatedly round the rDNA circle generating a tail in which copies of the circle sequence are repeated in tandem. The circle contains one 28 S + 18 S rRNA gene plus nontranscribed spacer. (Figure adapted from Rochaix *et al.*, 1974, with permission of Academic Press.)

meiotic prophase. Synthesis of rDNA continues for 2–3 weeks in each oocyte, encompassing the whole of pachytene, and ceases as the cytological features of diplotene become evident (Bird and Birnstiel, 1971; Watson-Coggins and Gall, 1972). In this time the rDNA content of the nucleus rises to about 31 pg compared to a total chromosomal DNA content of 12 pg (four times the haploid value of 3 pg) (Macgregor, 1968). Thus almost three-quarters of the DNA in a *Xenopus* diplotene oocyte nucleus is extrachromosomal rDNA.

Coincident with this interest in early meiosis were important investigations of the rDNA from mature oocytes. Previous work by Miller (1964, 1966) had shown that the extrachromosomal multiple nucleoli of amphibian oocytes could be visualized as nucleoprotein rings in the light microscope. These rings were fragmented by treatment with DNase, but neither RNase nor Pronase destroyed their continuity. The inference was that multiple nucleoli contained rings of DNA at their cores. Taking advantage of the high buoyant density of rDNA in cesium chloride solution (Birnstiel *et al.,* 1966), Brown and Dawid (1968) analyzed the DNA from 10,000 *Xenopus* oocyte nuclei by ultracentrifugation. A prominent high density peak was indeed present, and it hybridized with rRNA. Furthermore, the amount of rDNA per nucleus in the mature oocyte agreed closely with the values obtained by Feulgen microspectrophotometry at the much earlier nuclear cap stage (Macgregor, 1968). Finally, Evans and Birnstiel (1968) sealed the connection between rDNA and multiple nucleoli by separating nucleoli from other oocyte constituents and showing that nucleolar DNA was rDNA.

4. Comparison between Amplified rDNA and Chromosomal rDNA

Amplified rDNA of *Xenopus laevis* has been rigorously compared with the somatic variety (Dawid *et al.,* 1970). The need for this comparison was occasioned by the higher buoyant density of amplified compared to somatic rDNA (Gall, 1968; Evans and Birnstiel, 1968). Nucleotide analysis, coupled with hybridization and thermal denaturation studies showed that the distinction is solely due to the presence of methyl groups on 13% of the cytidylic acid residues of somatic rDNA. Amplified rDNA, in contrast, is not detectably methylated.

Since the rDNA repeat units in one nucleolus organizer are usually heterogeneous in length, it is also possible to look for selectivity in the original choice of repeats for amplification. Analysis or rDNA from the whole ovary indicates that chromosomal repeats are not always equally represented among the amplified genes (Wellauer *et al.,* 1976b). This result has been extended by analyzing the amplified rDNA in single oocytes (Bird, 1978). Surprisingly, an oocyte amplifies only a small subset of the repeats present chromosomally, and different oocytes amplify different subsets (Fig. 10). Thus only a few repeat units are the source of amplified rDNA in the oocyte, and individual oocytes choose independently which repeats they will amplify.

In the female frog analyzed in this study, the spacers amplified most often were those most common in the chromosomes (Bird, 1978). However, in studies of the whole ovary, containing thousands of oocytes, this is often not found to be the case (Wellauer *et al.*, 1976b). Spacers which are a minor component of the chromosomal rDNA may sometimes predominate in total amplified rDNA from the same female. This difference cannot be completely explained by preferential selection of repeats from one of the homologous ribosomal gene clusters, since it is also seen in animals which possess only one nucleolus per diploid chromosome set (Reeder *et al.*, 1976). Rather it appears that certain repeats in a cluster are preferentially amplified—perhaps by virtue of their position in the tandem gene array. It should be borne in mind that we do not know whether extrachromosomal rDNA is generated only once, in the 3-week-old tad-

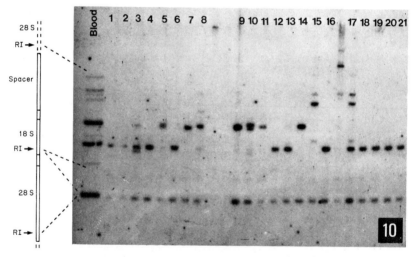

Fig. 10. *Eco*RI digestion of amplified rDNA in individual oocyte nuclei as compared with chromosomal (blood cell) rDNA. All oocytes and blood are from the same female *Xenopus laevis*. Each oocyte has amplified only a few of the repeats found in the chromosomes, and different oocytes have chosen different repeats. The figure is a 1-day autoradiograph of the blotted gel after hybridization to ^{32}P-rDNA. On the left is a schematic drawing of an rDNA repeat unit showing the fragments generated by *Eco*RI (RI) (Wellauer *et al.*, 1974) and their positions on the gel. Slots without DNA or showing incomplete digestion have been omitted from the numbering. (Figure taken from Bird, 1978, with permission of Cold Spring Harbor Laboratory, New York.)

pole, or whether the process occurs several times during gonial proliferation and at meiotic prophase.

5. The Mechanism of Meiotic Amplification

Considerably more is known about the molecular mechanism of meiotic amplification than about production of the first extrachromosomal rDNA copy. Ribosomal DNA purified from amplifying ovaries reveals a small proportion of tailed circles when examined in the electron microscope (Hourcade *et al.*, 1973a,b). The contour lengths of the circles suggest that they are monomers, dimers, trimers, etc., of the rDNA repeat unit, and this is confirmed by partial denaturation mapping. Furthermore, the tailed circle configuration suggests that the molecules are rolling circle intermediates in rDNA replication (Gilbert and Dressler, 1968). This conclusion has been shown to be correct by incubating the amplifying ovaries with [³H]thymidine and [³H]deoxycytidine prior to electron microscope autoradiography (Bird *et al.*, 1973; Rochaix *et al.*, 1974). The precursors preferentially enter tailed circles, showing that they are indeed replication intermediates, and the pattern of labeling in pulse and pulse–chase experiments conforms exactly with the expectation for a rolling circle mechanism (Fig. 9) The importance of rolling circle replication in meiotic rDNA synthesis has been borne out by studies of repeat length heterogeneity (Wallauer *et al.*, 1976b). Whereas chromosomal repeats of different lengths are intermingled, amplified rDNA comprises strings of identically sized units. The rolling circle mechanism would automatically generate such homogeneity (Gilbert and Dressler, 1968). Light microscope autoradiography of isolated high molecular weight rDNA is also consistent with the rolling circle mechanism (Bird, 1974). Replication points are not linked with one another, and chain growth is unindirectional rather than bidirectional as in somatic cells (Huberman and Riggs, 1968; Callan, 1972). The rate of replication is the same as for bulk chromosomal DNA, 10 μm or 30,000 nucleotides per hour (Bird, 1974).

It is known that mutants heterozygous for a deletion of the nucleolar DNA amplify to the same extent as wild-type frogs (Perkowska *et al.*, 1968). At first sight this suggests that oocytes are capable of measuring the amount of amplified rDNA and compensating for a shortage of chromosomal template. However, restriction enzyme analysis of amplified rDNA from single oocytes shows that even in normal animals very few chromosomal repeats contribute to amplification (Fig. 10) (Bird, 1978). It is, therefore, unlikely that the absence of a nucleolus affects the progress of amplification, and the concept of "compensation" need not be introduced. How amplification is terminated after the synthesis of 30 pg of rDNA remains a mystery.

6. Transcription of Amplified rDNA

Immediately following meiotic amplification, the oocyte enters dip-lotene, and the chromosomes adopt the lampbrush configuration. At the same time, the cap of amplified rDNA segregates into multiple nucleoli (Macgregor, 1968). Autoradiography after incubation with [³H]uridine shows that RNA synthesis is taking place in nucleoli at this time, and indeed nucleolar RNA synthesis is even detectable during meiotic amplifi-cation (A. P. Bird, unpublished observations). Nevertheless there is little mature newly synthesized 28 S and 18 S rRNA found in the cytoplasm (Mairy and Denis, 1971).

Ribosomal RNA synthesis begins in earnest at vitellogenesis, normally about 3 months after amplification. At this stage, multiple nucleoli become arranged near the nuclear membrane (Fig. 11) and intense 18 S and 28 S rRNA synthesis begins (Macgregor, 1967; Mairy and Denis, 1971). Direct visualization of rRNA synthesis in the multiple nucleoli shows coding regions saturated with RNA polymerases, each with an attached nascent rRNA precursor (Miller and Beatty, 1969; Scheer et al., 1977).

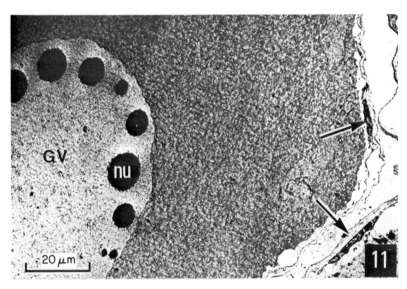

Fig. 11. Low-power electron micrograph of part of mature oocyte from *Xenopus laevis*. Amplified rDNA is located in multiple nucleoli (nu) and is being actively transcribed. The nucleoli are arranged close to the nuclear membrane which separates the oocyte nucleus or germinal vesicle (GV) from the egg cytoplasm. A layer of follicle cells surrounds each oocyte. Arrows mark two follicle cell nuclei for comparison with the germinal vesicle. (Figure kindly provided by George Duncan.)

7. The Ultimate Fate of Amplified rDNA

Oocyte maturation is accompanied by breakdown of the nuclear membrane and disintegration of the multiple nucleoli. It appears that nucleolar DNA becomes localized in Feulgen-positive granules, which then migrate to the animal pole of the egg (Brachet *et al.,* 1970; Hanocq *et al.,* 1974). As the first meiotic spindle forms, the granules vanish, but hybridization indicates that all amplified rDNA is still present (Brown and Dawid, 1968). Recently Thomas *et al.* (1977) have followed the level of rDNA in early embryos. No significant reduction is observed at least until the 16-cell stage. The results suggest that amplified genes are not destroyed at oocyte maturation, but neither are they replicated in the early embryo. Presumably they are progressively diluted as development proceeds.

8. The Synthesis of 5 S RNA during Oogenesis

Ribosomes contain 18 S, 28 S, and 5 S rRNA's in the ratio 1 : 1 : 1, but only the genes for 18 S and 28 S rRNA are amplified (Brown and Dawid, 1968). How then, does the oocyte meet its heavy 5 S RNA requirements? First, *Xenopus laevis* 5 S genes are repeated about 20,000 times per haploid genome compared to about 500 genes for rRNA. Thus the 4C oocyte contains some 80,000 5 S RNA genes. Second, there is reason to believe that most of these genes are reserved for use in the oocyte. Somatic cell 5 S RNA is homogeneous in sequence, whereas oocytes synthesize some somatic-type sequences plus a majority of oocyte-specific 5 S variants (Ford and Southern, 1973; Wegnez *et al.,* 1972). When a bulk preparation of 5 S genes is partially sequenced, only oocyte-type sequences are found (Brown *et al.,* 1972; Brownlee *et al.,* 1972), suggesting that the majority of 5 S genes code for oocyte-type RNA. Finally, the time span of active 5 S RNA synthesis is expanded. Previtellogenic oocytes synthesize comparatively little 18 S and 28 S rRNA, yet 5 S RNA synthesis is already occurring at a high rate at this stage (Ford, 1971).

We have already noted that the synthesis of structural RNA is slow compared to the potential rate of protein synthesis. Since amplification allows rRNA synthesis to transcend this limitation, 5 S RNA synthesis is the next most likely bottleneck in ribosome production. It is tempting to see the above-mentioned features of 5 S synthesis in *Xenopus* oocytes as a necessary adaptation to the occurrence of rDNA amplification.

B. Amplification in Insect Oocytes

Apart from the amphibia (and *Xenopus laevis,* in particular), amplification is also well documented in some species of insect. Particularly well-studied are oocytes of the cricket *Acheta domesticus,* which is a member of the order Orthoptera. The cytology of the process has been reported by a number of investigators (Lima-de-Faria *et al.,* 1968; Kunz, 1969; Cave and Allen, 1969), and it will be described here to illustrate the quite striking similarity to amphibian amplification.

Extra DNA is first detectable as several small Feulgen-positive bodies in oogonia. These bodies replicate and pass through oogonial mitoses, appearing ultimately as a single body at the leptotene stage of meiosis. During leptotene, the peripheral DNA body increases in size. Tritiated thymidine autoradiography shows that DNA synthesis is occurring within it and is not synchronous with premeiotic S phase. Synthesis is complete at pachytene, and the body assumes a less tightly packed, "pufflike" appearance. At this time, an RNA containing shell is formed around the body, and this is active in RNA synthesis. Synthesis continues into an arrested diplotene stage, and the DNA body eventually disperses.

The presence of nucleoli inside the body suggests that amplification of rDNA might be occurring, and this has been confirmed biochemically by Lima-de-Faria *et al.* (1969). DNA isolated from ovaries in which the DNA body is at its largest, is enriched for high density satellite compared with testis DNA. Hybridization shows that this satellite contains the ribosomal RNA genes. In addition, hybridization *in situ* shows that the DNA body is rich in rDNA (Cave, 1972; Ullman *et al.,* 1973). Finally, oocyte rDNA transcription has been visualized in the electron microscope (Trendelenburg *et al.,* 1973). Here, as in most other respects, *Acheta* amplification resembles that in *Xenopus.*

C. Low Levels of Amplification in Oocytes of Other Organisms

Ribosomal gene amplification of the kind discussed above has been shown in the oocytes of fish (Vincent *et al.,* 1969), a wide range of amphibia, and certain insects. In all these organisms it is manifested both cytologically, as a prominent "DNA body" or multiple nucleoli, and biochemically. Amplification on a far less dramatic scale has also been detected using only biochemical criteria in oocytes displaying a solitary nucleolus. In the phylum Mollusca, the surf clam appears to amplify rDNA by six- to eightfold (Brown and Dawid, 1968), and the related coot clam, by about twofold (Kidder, 1976). Also, an echiuroid worm, *Urechis caupo,* displays a sixfold amplification compared to sperm rDNA (Dawid and Brown, 1970). Such low increases are difficult to establish for two reasons. First, the oocytes concerned contain a significant fraction of mitochondrial DNA, whereas sperm do not. Correction of oocyte rRNA hybridization values is, therefore, required, and this may introduce errors. Second, the amount of rDNA per haploid genome can vary widely within a single species (see Section IV), making comparisons between the sperm of one individual and the oocytes of another uncertain. One of the few published cases in which amplification has been looked for but not found is the starfish, *Asterias forbesei* (Vincent *et al.*, 1969).

D. Amplification and Oocytes with Nutritive Cells

It is clear that amplification of nucleolar DNA in animal eggs is a widespread phenomenon. It argues for an unusually heavy requirement for the 28 S and 18 S rRNA components of ribosomes during the differentiation of these cells. Presumably the complicated amplification procedure is advantageous because it allows rRNA to be accumulated in a time commensurate with that required to accumulate the other constituents of the mature egg. Since multiple nucleoli contain rDNA equivalent to 3000–4000 chromosomal nucleolus organizers, compared to four nucleolus organizers in the meiotic chromosomes, we can conclude that in the absence of amplification, the mature oocyte would take many years, instead of several months, to differentiate.

The need for meiotic chromosomes to remain inviolate has meant that similar short cuts in the lampbrush stage are not feasible. In this respect the development of the "meroistic" ovary of insects, in which the transcriptional capacity of nurse cells is harnessed to the cause of egg development, represents a considerable advance. Nurse cells arise by a series of oogonial divisions and are closely connected with each growing oocyte. Their nuclei are often polyploid or polytene and engage in heavy RNA and ribosome synthesis on behalf of the egg (Ribbert and Bier, 1969). By contrast, oocyte nuclei remain relatively inactive in RNA synthesis, the chromosomes being condensed in a "karyosphere." As might be expected, amplification does not appear to be a feature of meroistic oocytes (Cave, 1974; Cave and Sixbey, 1976). The advantages which nutritive cells bestow on the organism can be appreciated by the difference in time required to complete oogenesis in the "panoistic" ovary of *Acheta* (100 days) and in the meroistic ovary of the blow fly *Calliphora* (6 days) (Ribbert and Bier, 1969).

Several variations on the meroistic process have been described. Young oocytes in Dytiscid water beetles (Bauer, 1933; Bier *et al.*, 1967) and in Tipulid flies (Lima-de-Faria and Moses, 1966; Bayreuther, 1956), while receiving RNA from nurse cells, nevertheless also possess a conspicuous DNA body which behaves in many ways like that in *Acheta*. The body gives rise to numerous nucleoli at diplotene and has been shown in Dytiscids to correspond to rDNA by cesium chloride ultracentrifugation and hybridization with rRNA (Gall *et al.,* 1969). More recently the amplified rDNA has been isolated and found to contain circular molecules such as those in *Xenopus* oocytes (Gall and Rochaix, 1974; Trendelenburg *et al.,* 1976).

Before leaving the subject of multicellular alternatives to amplification, it is necessary to describe an unusual variation found in the tailed frog, *Ascaphus truei* (Macgregor and Kezer, 1970). During the last three oogonial mitoses, all nuclei remain within the same cell and give rise to an eight-nucleate oocyte. Each nucleus amplifies less than 5 pg of DNA in contrast to the values per nucleus for *Xenopus, Bufo,* and *Necturus* of 27–30 pg (Macgregor, 1968; Gall, 1969; Brown and Dawid, 1968). However, the occurrence of amplification in all eight nuclei of an *Ascaphus* oocyte removes this disparity. Some time prior to ovulation, all but one of the nuclei disappear. This situation offers a primitive intermediate case between the self-sufficient oocyte exemplified by *Xenopus* and panoistic ovaries, and the nutritive cell—oocyte complex found in meroistic ovaries.

X. CONCLUDING REMARKS ON THE PECULIARITY OF RIBOSOMAL DNA

The reader will have noticed that the discussion of both repeated genes and gene amplification has been dominated by the 28 S and 18 S ribosomal RNA genes. This is partly due to the greater attention accorded to these genes by molecular biologists and partly to the fact that the behavior of rDNA is in some respects unusual. In conclusion it is worth summarizing some of the distinctive properties of rDNA.

The chromosomal location of ribosomal genes is betrayed by secondary constrictions in the metaphase chromosome set, although not all secondary constrictions are rDNA sites (Pardue, 1973). In addition, the nucleolar region in metaphase chromosomes reacts with stains such as ammoniacal silver, which leave the rest of the chromosomes unstained (Goodpasture and Bloom, 1975; Denton *et al.,* 1976). Another important generalization is that rDNA is often sandwiched between chromosome segments which are heterochromatic. For example, in *Drosophila,* rDNA is bounded on both sides by blocks of heterochromatin known to be rich in satellite DNA (Rae, 1970; Gall *et al.,* 1971; Botchan *et al.,* 1971). The generality of the relationship between nucleolar DNA and heterochromatin gives the impression that ribosomal genes are insulated from the euchromatic portions of the genome by blocks of transcriptionally inactive DNA.

The insulating quality of nucleolus-associated heterochromatin is nowhere more apparent than in *Drosophila* polytene cells. Here the euchromatic chromosome segments go through about nine doublings to give characteristic giant chromosomes (Rudkin, 1969). In contrast, the centromeric heterochromatin does not replicate significantly, and its constituent satellite DNA's are markedly underrepresented in polytene DNA (Rudkin, 1969; Gall *et al.,* 1971; Hennig, 1972). The ribosomal DNA, surrounded by nonreplicating chromosomal segments, replicates independently of both euchromatin and the satellites (Hennig and Meer, 1971; Spear and Gall, 1973).

Amplification is an extreme case of independent rDNA replication, and, despite many searches, there are no reliable instances of other genes being amplified in normal cells. In particular the 5 S genes, whose product is required in the same molar amounts as rDNA, appear to be incapable of amplification. This situation contrasts with the finding that some organisms amplify their oocyte ribosomal genes by only two- to eightfold (Brown and Dawid, 1968; Dawid and Brown, 1970; Kidder, 1976), suggesting that even relatively slight selective pressure for more rDNA can be met by amplification. Other distinctive phenomena are rDNA magnification (see Section VI), extrachromosomal rDNA in somatic cells of *Xenopus* and *Drosophila* (see Section VII), and intraspecies variation in the chromosomal location of rDNA (see Section IV). In these last two cases, however, it is not clear that the phenomena are exclusive to ribosomal genes.

The independent behavior of rDNA invites a reappraisal of some basic aspects of ribosomal gene organization. Most important is the question of linkage between rDNA and the remainder of the chromosome. Although the inheritance of a pattern of heterogeneous repeat lengths strongly suggests that rDNA is normally integrated into the chromosome (Reeder *et al.,* 1976), results of the kind discussed above raise the possibility that the relationship between nucleolar DNA and the chromosome is not a static one.

ACKNOWLEDGMENTS

I am very grateful to Chris Bostock, David Finnegan, and Ed Southern for critical readings of the manuscript.

REFERENCES

Adamson, E. D., and Woodland, H. R. (1974). Histone synthesis in early amphibian development: Histone and DNA synthesis are not coordinated. *J. Mol. Biol.* **88,** 263–278.

Ammermann, D. (1971). Morphology and development of the macronuclei of the ciliates *Stylonichia mytilus* and *Euplotes aediculatus*. *Chromosoma* **33**, 209–238.

Arnheim, N., and Southern, E. M. (1977). Heterogeneity of the ribosomal genes in mice and men. *Cell* **11**, 363–370.

Artavanis-Tsakonas, S., Shedl, P., Tschudi, C., Pirrotta, V., Steward, R., and Gehring, E. J. (1977). The 5 S genes of *Drosophila melanogaster*. *Cell* **12**, 1057–1067.

Ashburner, M. (1970). Function and structure of polytene chromosomes during insect development. *Adv. Insect Physiol.* **7**, 1–196.

Batistoni, R., Andronico, F., Nardi, I., and Barsacchi-Pilone, G. (1978). Chromosomal location of the ribosomal genes in *Triturus meridionalis* (Amphibia Urodela). III. Inheritance of the chromosomal sites for 18 S and 28 S ribosomal RNA. *Chromosoma* **65**, 231–240.

Bauer, H. (1933). Die wachsenden Oocytenberne einiger Insekten in ihrem Verhalten zur Nuklealfärburg. *Z. Zellforsch, Mikrosk. Anat.* **18**, 254–298.

Bayreuther, K. (1956). Die Oogenese der *Tipuliden*. *Chromosoma* **7**, 508–557.

Benyajati, C., and Worcel, A. (1976). Isolation, characterization, and structure of the folded interphase genome of *Drosophila melanogaster*. *Cell* **9**, 393–407.

Bier, K., Kunz, W., and Ribbert, D. (1967). Struktur und Funktion der Oocytenchromosomen und Nukleolen sowie der Extra-DNS während der Oogenese panoistischer und meroischer Insekten. *Chromosoma* **23**, 214–254.

Bird, A. P. (1974). Light microscope autoradiography of ribosomal DNA amplification in *Xenopus laevis*. *Chromosoma* **46**, 421–433.

Bird, A. P. (1978). A study of early events in ribosomal gene amplification. *Cold Spring Harbor Symp. Quant. Biol.* **42**, 1179–1183.

Bird, A. P., and Birnstiel, M. L. (1971). A timing study of DNA amplification in *Xenopus laevis* oocytes. *Chromosoma* **35**, 300–309.

Bird, A. P., Rochaix, J.-D., and Bakken, A. H. (1973). The mechanism of gene amplification in *Xenopus laevis* oocytes. In "Molecular Cytogenetics" (B. A. Hamkalo and J. Papaconstantinou, eds.), pp. 49–58. Plenum, New York.

Birnstiel, M. L., and Chipchase, M. (1977). Current work on the histone operon. *Trends Biochem. Sci.* **2**, 149–152.

Birnstiel, M. L., Wallace, H., Sirlin, J. L., and Fischberg, M. (1966). Localization of ribosomal DNA complements in the nucleolar organizer region of *Xenopus laevis*. *Natl. Cancer Inst., Monogr.* **23**, 431–447.

Birnstiel, M. L., Speirs, J., Purdom, I., Jones, K., and Loening, U. E. (1968). Properties and composition of the isolated ribosomal DNA satellite of *Xenopus laevis*. *Nature (London)* **219**, 454–463.

Birnstiel, M. L., Sells, B. H., and Purdom, I. F. (1972). Kinetic complexity of RNA molecules. *J. Mol. Biol.* **63**, 21–39.

Biro, P. A., Carr-Brown, A. C., Southern, E. M., and Walker, P. M. B. (1975). Partial sequence analysis of mouse satellite DNA. Evidence for short range periodicities. *J. Mol. Biol.* **94**, 71–86.

Blackler, A. W. (1970). The integrity of the reproductive cell line in the amphibia. *Curr. Top. Dev. Biol.* **5**, 71–87.

Boncinelli, E., Malva, C., Graziani, F., Polito, L., and Ritossa, F. M. (1972). rDNA magnification at the *bobbed* locus of the Y chromosome in *Drosophila melanogaster*. *Cell Differ.* **1**, 133–143.

Botchan, M. R. (1974). Bovine satellite I DNA consists of repetitive units 1,400 base pairs in length. *Nature (London)* **251**, 288–292.

Botchan, M. R., Kram, R., Schmid, C. W., and Hearst, J. E. (1971). Isolation and chromosomal localization of highly repeated DNA sequences in *Drosophila melanogaster*. *Proc. Natl. Acad. Sci. U.S.A.* **68**, 1125–1128.

Botchan, P., Reeder, R. H., and Dawid, I. B. (1977). Restriction analysis of nontranscribed spacers in *Xenopus laevis* ribosomal DNA. *Cell* **11**, 599–607.

Brachet, J., Hanocq, F., and Van Gansen, P. (1970). A cytochemical and ultrastructural analysis of *in vitro* maturation in Amphibian oocytes. *Dev. Biol.* **21**, 157–195.

Bridges, C. B. (1935). Salivary gland chromosome maps with a key to the banding of the chromosomes of *D. melanogaster*. *J. Hered.* **26**, 60–64.

Bridges, C. B., and Breheme, K. S. (1944). The mutants of *Drosophila melanogaster*. *Carnegie Inst. Wash., Publ.* No. 552.

Britten, R. J., and Kohne, D. E. (1968). Repeated sequences in DNA. *Science* **161**, 529–540.

Brown, D. D., and Blackler, A. W. (1972). Gene amplification proceeds by a chromosome copy mechanism. *J. Mol. Biol.* **63**, 75–83.

Brown, D. D., and Dawid, I. B. (1968). Specific gene amplification in oocytes. *Science* **160**, 272–280.

Brown, D. D., and Gurdon, J. B. (1964). Absence of ribosomal RNA synthesis in the anucleolate mutant of *X. laevis*. *Proc. Natl. Acad. Sci. U.S.A.* **51**, 139–146.

Brown, D. D., and Sugimoto, K. (1973a). 5 S DNAs of *Xenopus laevis* and *Xenopus mulleri:* Evolution of a gene family. *J. Mol. Biol.* **78**, 397–415.

Brown, D. D., and Sugimoto, K. (1973b). The structure and evolution of ribosomal and 5 S DNAs in *Xenopus laevis* and *Xenopus mulleri*. *Cold Spring Harbor Symp. Quant. Biol.* **38**, 501–505.

Brown, D. D., and Weber, C. S. (1968). Gene linkage by RNA–DNA hybridization: II. Arrangement of the redundant gene sequences for 28 S and 18 S rRNA. *J. Mol. Biol.* **34**, 681–698.

Brown, D. D., Wensink, P. C., and Jordan, E. (1971). Purification and some characteristics of 5 S DNA from *Xenopus laevis*. *Proc. Natl. Acad. Sci. U.S.A.* **68**, 3175–3179.

Brown, D. D., Wensink, P. C., and Jordan, E. (1972). A comparison of the ribosomal DNAs of *Xenopus laevis* and *Xenopus mulleri:* The evolution of tandem genes. *J. Mol. Biol.* **63**, 57–73.

Brown, D. D., Dawid, I. B., and Reeder, R. H. (1977). *Xenopus borealis* misidentified as *Xenopus laevis*. *Dev. Biol.* **59**, 266–267.

Brownlee, G. G., Cartwright, E., Mcshane, T., and Williamson, R. (1972). The nucleotide sequence of somatic 5 S RNA from *Xenopus laevis*. *FEBS Lett.* **25**, 8–12.

Buongiorno-Nardelli, M., Amaldi, F., and Lava-Sanchez, P. A. (1972). Amplification as a rectification mechanism for the redundant rRNA genes. *Nature (London), New Biol.* **238**, 134–137.

Buongiorno-Nardelli, M., Amaldi, F., Beccari, E., and Junakovic, N. (1977). Size of ribosomal DNA repeating units in *Xenopus laevis:* Limited individual heterogeneity and extensive population polymorphism. *J. Mol. Biol.* **110**, 105–118.

Callan, H. G. (1966). Chromosomes and nucleoli of the axolotl, *Ambystoma mexicanum*. *J. Cell Sci.* **1**, 85–108.

Callan, H. G. (1967). The organization of genetic units in chromosomes. *J. Cell Sci.* **2**, 1–7.

Callan, H. G. (1972). Replication of DNA in the chromosomes of eukaryotes. *Proc. R. Soc. London, Ser. B* **181**, 19–41.

Carroll, D., and Brown, D. D. (1976a). Repeating units of *Xenopus laevis* oocytetype 5 S DNA are heterogeneous in length. *Cell* **7**, 467–476.

Carroll, D., and Brown, D. D. (1976b). Adjacent repeating units in *Xenopus laevis* 5 S DNA can be heterogeneous in length. *Cell* **7**, 477–486.

Cave, M. D. (1972). Localization of ribosomal DNA within oocytes of the house cricket, *Acheta domesticus* (Orthoptera: Gryllidae). *J. Cell Biol.* **55**, 310–321.

Cave, M. D. (1974). Absence of rDNA amplification in a meroistic (telotrophic) ovary. *J. Cell Biol.* **63**, 53a.

Cave, M. D., and Allen, E. R. (1969). Syntheis of nucleic acids associated with a DNA-containing body in oocytes of *Acheta*. *Exp. Cell Res.* **58**, 201–212.

Cave, M. D., and Sixbey, J. (1976). Absence of ribosomal DNA amplification in a meroistic polytrophic ovary. *Exp. Cell Res.* **101**, 23–30.

Clarkson, S. G., and Kurer, V. (1976). Isolation and some properties of DNA coding for tRNA met from *Xenopus laevis*. *Cell* **8**, 183–195.

Clarkson, S. G., Birnstiel, M. L., and Serra, V. (1973). Reiterated transfer RNA genes of *Xenopus laevis*. *J. Mol. Biol.* **79**, 391–410.

Cohen, L., Newrock, K., and Zweidler, A. (1975). Stage specific switches in histone synthesis during embryogenesis of the sea urchin. *Science* **190**, 994–997.

Cooper, K. W. (1959). Cytogenetic analysis of major heterochromatic elements (especially Xh and Y) in *Drosophila melanogaster* and the theory of "heterochromatin." *Chromosoma* **10**, 535–588.

Cory, S., and Adams, J. M. (1977). A very large repeating unit of mouse DNA containing the 18 S, 28 S and 5.8 S rRNA genes. *Cell* **11**, 795–805.

Danna, K., and Nathans, D. (1971). Specific cleavage of simian virus 40 DNA by restriction endonuclease of *Haemophilus influenzae*. *Proc. Natl. Acad. Sci. U.S.A.* **68**, 2913–2917.

Dawid, I. B., and Brown, D. D. (1970). The mitochrondrial and ribosomal DNA components of oocytes of *Urechis caupo*. *Dev. Biol.* **22**, 1–14.

Dawid, I. B., Brown, D. D., and Reeder, R. H. (1970). Composition and structure of chromosomal and amplified ribosomal DNAs of *Xenopus laevis*. *J. Mol. Biol.* **51**, 341–360.

Denton, T. E., Howell, W. H., and Barrett, J. V. (1976). Human nucleolar organizer chromosomes: Satellite associations. *Chromosoma* **55**, 81–84.

Edelman, G. M., and Gally, J. A. (1970). Arrangement and evolution of eukaryotic genes. *In* "The Neurosciences: Second Study Program" (F. O. Schmitt, ed.), pp. 962–980. Rockefeller Univ. Press, New York.

Elsdale, T. R., Fischberg, M., and Smith, S. (1958). A mutation that reduces nucleolar number in *Xenopus laevis*. *Exp. Cell Res.* **14**, 642–643.

Engberg, J., Christiansen, G., and Leich, V. (1974). Autonomous rDNA molecules containing single copies of ribosomal RNA genes in the micronucleus of *Tetrahymena pyriformis*. *Biochem. Biophys. Res. Commun.* **59**, 1356–1365.

Engberg, J., Anderson, P., Leick, V., and Collins, J. (1976). Free ribosomal DNA molecules from *Tetrahymena pyriformis* GL are giant palindromes. *J. Mol. Biol.* **104**, 451–470.

Evans, D., and Birnstiel, M. (1968). Localization of amplified ribosomal DNA in the oocyte of *Xenopus laevis*. *Biochim. Biophys. Acta* **166**, 274–276.

Fedoroff, N. V., and Brown, D. D. (1978). The nucleotide sequence of oocyte 5 S DNA in *Xenopus laevis*. I. The AT-rich spacer. *Cell* **13**, 701–716.

Finnegan, D. J., Rubin, G. M., Young, M. W., and Hogness, D. S. (1978). Repeated gene families in *Drosophila melanogaster*. *Cold Spring Harbor Symp. Quant. Biol.* **42**, 1053–1063.

Ford, P. J. (1971). Non-coordinated accumulation and synthesis of 5 S ribonucleic acid by ovaries of *Xenopus laevis*. *Nature (London)* **233**, 561–564.

Ford, P. J., and Southern, E. M. (1973). Different sequence for 5 S RNA in kidney cells and ovaries of *Xenopus laevis*. *Nature (London), New Biol.* **241**, 7–10.

Forsheit, A. B., Davidson, N., and Brown, D. D. (1974). An electron microscope heteroduplex study of the ribosomal DNAs of *Xenopus laevis* and *Xenopus mulleri*. *J. Mol. Biol.*
90, 301–314.

Gall, J. G. (1968). Differential synthesis of the genes for ribosomal RNA during amphibian
oogenesis. *Proc. Natl. Acad. Sci. U.S.A.* **60,** 553–560.

Gall, J. G. (1969). The genes for ribosomal RNA during oogenesis. *Genetics* **61,** Suppl.,
121–131.

Gall, J. G. (1974). Free ribosomal RNA genes in the macronucleus of *Tetrahymena*. *Proc.
Natl. Acad. Sci. U.S.A.* **71,** 3078–3081.

Gall, J. G., and Pardue, M. L. (1969). Formation and detection of RNA–DNA hybrid
molecules in cytological preparations. *Proc. Natl. Acad. Sci. U.S.A.* **63,** 378–383.

Gall, J. G., and Rochaix, J. D. (1974). The amplified ribosomal DNA of Dytiscid beetles.
Proc. Natl. Acad. Sci. U.S.A. **71,** 1819–1823.

Gall, J. G., Macgregor, H. C., and Kidston, M. E. (1969). Gene amplification in the oocytes
of Dytiscid water beetles. *Chromosoma* **26,** 169–187.

Gall, J. G., Cohen, E. H., and Polan, M. L. (1971). Repetitive DNA sequences in
Drosophila. *Chromosoma* **33,** 319–344.

Gilbert, W., and Dressler, D. (1968). DNA replication: The rolling circle model. *Cold Spring
Harbor Symp. Quant. Biol.* **33,** 473–474.

Gillespie, D., and Spiegelman, S. (1965). A quantitiative assay for RNA–DNA hybrids with
DNA immobilized on a membrane. *J. Mol. Biol.* **12,** 829–842.

Glover, D. M., and Hogness, D. S. (1977). A novel arrangement of the 18 S and 28 S
sequences in a repeating unit of *Drosophila melanogaster* rDNA. *Cell* **10,** 167–176.

Goodpasture, C., and Bloom, S. E. (1975). Visualization of nucleolus organizer regions in
mammalian chromosomes using silver staining. *Chromosoma* **53,** 37–50.

Graham, C., and Morgan, R. W. (1966). Changes in the cell cycle during early amphibian
development. *Dev. Biol.* **14,** 439–460.

Graziani, F., Caizzi, R., and Gargano, S. (1977). Circular ribosomal DNA during ribosomal
magnification in *Drosophila melanogaster*. *J. Mol. Biol.* **112,** 49–64.

Gross, K. W., Ruderman, J., Jacobs-Lorena, M., Baglioni, C., and Gross, P. R. (1973).
Cell-free synthesis of histones directed by messenger RNA from sea urchin embryos.
Nature (London), New Biol. **241,** 272–274.

Hanocq, F., De Schutter, A., Hubert, E., and Brachet, J. (1974). Cytochemical and biochemical studies on progesterone-induced maturation in amphibian oocytes. DNA
synthesis. *Differentiation* **2,** 75–90.

Hatlen, L., and Attardi, G. (1971). Proportion of the HeLa genome complementary to
transfer RNA and 5 S RNA. *J. Mol. Biol.* **56,** 535–553.

Heitz, E. (1931). Nukleolen und Chromosomen in der Gattung *Vicia*. *Planta* **15,** 495–505.

Henderson, A., and Ritossa, F. (1970). On the inheritance of rDNA of magnified *bobbed* loci
in *Drosophila melanogaster*. *Genetics* **66,** 463–473.

Hennen, S., Mizuno, S., and Macgregor, H. C. (1975). *In situ* hybridization of ribosomal
DNA labelled with ^{125}iodine to metaphase and lampbrush chromosomes from newts.
Chromosoma **50,** 349–369.

Hennig, W. (1972). Highly repetitive DNA sequence in the genome of *Drosophila hydei*. II.
Occurrence in polytene tissues. *J. Mol. Biol.* **71,** 419–431.

Hennig, W., and Meer, B. (1971). Reduced polyteny of ribosomal RNA cistrons in giant
chromosomes of *Drosophila hydei*. *Nature (London), New Biol.* **233,** 70–72.

Hennig, W., and Walker, P. M. B. (1970). Variations in the DNA from two rodent families
(Cricetidae and Muridae). *Nature (London)* **225,** 915–919.

Hourcade, D., Dressler, D., and Wolfson, J. (1973a). The amplification of ribosomal RNA genes involves a rolling circle intermediate. *Proc. Natl. Acad. Sci. U.S.A.* **70**, 2926–2930.

Hourcade, D., Dressler, D., and Wolfson, J. (1973b). The nucleolus and the rolling circle. *Cold Spring Harbor Symp. Quant. Biol.* **38**, 537–550.

Huberman, J. A., and Riggs, A. D. (1968). On the mechanism of DNA replication in mammalian chromosomes. *J. Mol. Biol.* **32**, 327–341.

Humphrey, R. R. (1961). A chromosomal deletion in the Mexican axolotl (*Siredon mexicanum*) involving the nucleolus organizer and the gene for dark colour. *Am. Zool.* **1**, 361.

Hutchinson, N., and Pardue, M. L. (1975). The mitotic chromosomes of *Notophthalmus* (=*Triturus*) *viridescens:* Localization of C-banding regions and DNA sequences complementary to 18 S, 28 S, and 5 S ribosomal RNA. *Chromosoma* **53**, 51–69.

Jacob, E. (1976). Histone–gene reiteration in the genome of mouse. *Eur. J. Biochem.* **65**, 275–284.

Jacob, E., Malacinski, G., and Birnstiel, M. L. (1976). Reiteration frequency of the histone genes in the genome of the amphibian, *Xenopus laevis. Eur. J. Biochem.* **69**, 45–54.

John, H. A., Birnstiel, M. L., and Jones, K. W. (1969). RNA–DNA hybrids at the cytological level. *Nature (London)* **223**, 582–587.

Kalt, M. R. (1973). Ultra structural observations on the germ line of *Xenopus laevis. Z. Zellforsch. Mikrosk. Anat.* **138**, 41–60.

Kalt, M. R., and Gall, J. G. (1974). Observations on early germ cell development and premeiotic ribosomal DNA amplification in *Xenopus laevis. J. Cell Biol.* **62**, 460–472.

Karrer, K., and Gall, J. G. (1976). The macronuclear ribosomal DNA of *Tetrahymena pyriformis* is a palindrome. *J. Mol. Biol.* **104**, 421–453.

Kedes, L. H. (1976). Histone messengers and histone genes. *Cell* **8**, 321–331.

Kedes, L. H., and Gross, P. R. (1969). Synthesis and function of messenger RNA during early embryonic development. *J. Mol. Biol.* **42**, 559–575.

Kelly, T. J., and Smith, H. O. (1970). A restriction enzyme from *Haemophilus influenzae* II. Base sequence of the recognition site. *J. Mol. Biol.* **51**, 393–409.

Kidder, G. M. (1976). The ribosomal cistrons in clam gametes. *Dev. Biol.* **49**, 132–142.

Kunz, W. (1969). Die Entstehung multipler Oocytennukleolen aus akzessorischen DNS-Korpern bei *Gryllus domesticus. Chromosoma* **26**, 41–75.

Lifton, R., Karp, R., Goldberg, M., and Hogness, D. (1978). The histone genes of *Drosophila melanogaster. Cold Spring Harbor Symp. Quant. Biol.* **42**, 1047–1051.

Lima-de-Faria, A., and Moses, M. J. (1966). Ultrastructure and cytochemistry of metabolic DNA in *Tipula. J. Cell Biol.* **30**, 177–192.

Lima-de-Faria, A., Nilsson, B., Cave, D., Puga, A., and Jaworska, H. (1968). Tritium labelling and cytochemistry of extra DNA in *Acheta. Chromosoma* **25**, 1–20.

Lima-de-Faria, A., Birnstiel, M., and Jaworska, H. (1969). Amplification of ribosomal cistrons in the heterochromatin of *Acheta. Genetics* **61**(1), Suppl., 145–159.

Lindsley, D. L., and Grell, E. H. (1968). Genetic variations of *Drosophila melanogaster. Carnegie Inst. Wash., Publ.* No. 627.

McClintock, B. (1934). The relation of a particular chromosomal element to the development of the nucleoli in *Zea mays. Z. Zellforsch. Mikrosk. Anat.* **21**, 294–328.

Macgregor, H. C. (1967). Pattern of incorporation of [^{3}H]uridine into RNA of amphibian oocyte nucleoli. *J. Cell Sci.* **2**, 145–150.

Macgregor, H. C. (1968). Nucleolar DNA in oocytes of *Xenopus laevis. J. Cell Sci.* **3**, 437–444.

Macgregor, H. C. (1972). The nucleolus and its genes in amphibian oogenesis. *Biol. Rev. Cambridge Philos. Soc.* **47**, 177–210.

Macgregor, H. C., and Kezer, J. (1970). Gene amplification in oocytes with 8 germinal vesicles from the tailed frog, *Ascaphus truei* Steijneger. *Chromosoma* **29**, 189–206.

Macgregor, H. C., and Kezer, J. (1973). The nucleolar organizer of *Plethodon cinereus* (green). I. Location of the nucleolar organizer by *in situ* nucleic acid hybridization. *Chromosoma* **42**, 415–426.

Macgregor, H. C., Vlad, M., and Barnett, L. (1977). An investigation of some problems concerning nucleolus organizers in salamanders. *Chromosoma* **59**, 283–299.

McKnight, S. L., and Miller, O. L. (1976). Ultrastructural patterns of RNA synthesis during early embryogenesis of *Drosophila melanogaster*. *Cell* **8**, 305–313.

Maggini, F., Barsanti, P., and Marazia, T. (1978). Individual variation of the nucleolus organizer regions in *Allium cepa* and *Allium sativum*. *Chromosoma* **66**, 173–183.

Mairy, M., and Denis, H. (1971). Recherches biochimiques sur l'oogenèse, I. Synthese et accumulation du RNA pendant loogenèse du crapaud Sud-Africain *Xenopus laevis*. *Dev. Biol.* **24**, 143–150.

Malva, C., Graziani, F., Boncinelli, E., Polito, L., and Ritossa, F. (1972). Check of gene number during the process of rDNA magnification. *Nature (London), New Biol.* **239**, 135–137.

Miller, L., and Brown, D. D. (1969). Variation in the activity of nucleolar organizers and their ribosomal gene content. *Chromosoma* **28**, 430–444.

Miller, L., and Knowland, J. (1970). Reduction of ribosomal RNA synthesis and ribosomal RNA genes in a mutant of *Xenopus laevis* which organizes only a partial nucleolus. II. The number of ribosomal genes in animals of different nucleolar types. *J. Mol. Biol.* **53**, 329–338.

Miller, O. L. (1964). Extrachromosomal nucleolar DNA in amphibian oocytes. *J. Cell Biol.* **23**, 60a.

Miller, O. L. (1966). Structure and composition of peripheral nucleoli of salamander oocytes. *Natl. Cancer Inst., Monogr.* **23**, 53–66.

Miller, O. L., and Beatty, B. R. (1969). Visualization of nucleolar genes. *Science* **164**, 955–957.

Mohan, J., and Ritossa, F. M. (1970). Regulation of RNA synthesis and its bearing on the *bobbed* phenotype of *Drosophila melanogaster*. *Dev. Biol.* **22**, 495–512.

Morrow, J. F., Cohen, S. N., Chang, A. C. Y., Boyer, H. W., Goodman, H. M., and Helling, R. B. (1974). Replication and transcription of eukaryotic DNA in *Escherichia coli*. *Proc. Natl. Acad. Sci. U.S.A.* **71**, 1743–1747.

Nagl, W. (1965). Die Sat-riesenchomosomen der Kerne des Suspensors von *Phaseolus cooccineus* und ihre Verhalten während der Endomitose. *Chromosoma* **16**, 511–520.

Nardi, I., Barsacchi-Pilone, G., Batistoni, R., and Andronico, F. (1977). Chromosome location of the ribosomal genes in *Triturus vulgaris meridionalis* (Amphibia Urodela). II. Intraspecific variability in the number and position of chromosome loci for 18 S + 28 S ribosomal RNA. *Chromosoma* **64**, 68–84.

Painter, T. S., and Taylor, A. N. (1942). Nucleic acid storage in the toad's egg. *Proc. Natl. Acad. Sci. U.S.A.* **28**, 311–317.

Pardue, M. L. (1973). Localization of repeated DNA sequences in *Xenopus* chromosomes. *Cold Spring Harbor Symp. Quant. Biol.* **38**, 475–482.

Pardue, M. L., Brown, D. D., and Birnstiel, M. L. (1973). Localization of the genes for 5 S ribosomal RNA in *Xenopus laevis*. *Chromosoma* **42**, 191–197.

Pardue, M. L., Kedes, L. H., Weinberg, E. J., and Birnstiel, M. L. (1977). Localization of sequences coding for histone messenger RNA in the chromosomes of *D. melanogaster*. *Chromosoma* **63**, 135–151.

Pellegrini, M., Manning, J., and Davidson, N. (1977). Sequence arrangement of the rDNA of *Drosophila melanogaster*. *Cell* **10**, 213–224.

Perkowska, E., Macgregor, H. C., and Birnstiel, M. L. (1968). Gene amplification in the oocyte nucleus of mutant and wild-type *Xenopus laevis*. *Nature (London)* **217**, 649–650.

Philips, D. F., Weber, D. F., Klease, R. A., and Wang, S. S. (1974). The nucleolus organizer region of maize (*Zea mays* L.): Tests for ribosomal gene compensation or magnification. *Genetics* **77**, 285–297.

Prescott, D. M., and Bender, M. A. (1963). Autoradiographic study of the chromatid distribution of labelled DNA in two types of mamalian cells *in vitro*. *Exp. Cell Res.* **29**, 430–442.

Prescott, D. M., Murti, K. G., and Bostock, C. J. (1973). Genetic apparatus of *Stylonichia* sp. *Nature (London)* **242**, 576–600.

Procunier, J. D., and Tartof, K. D. (1975). Genetic analysis of the 5 S RNA genes in *Drosophila melanogaster*. *Genetics* **81**, 515–523.

Procunier, J. D., and Tartof, K. D. (1976). Restriction map of 5 S RNA genes of *Drosophila melanogaster*. *Nature (London)* **263**, 255–257.

Pukkila, P. J. (1975). Identification of the lampbrush chromosome loops which transcribe 5 S ribosomal RNA in *Notophthalmus (Triturus) viridescens*. *Chromosoma* **53**, 71–89.

Radloff, R., Bauer, W., and Vinograd, J. (1967). A dye–bouyant-density method for the detection and isolation of closed circular duplex DNA: The closed circular DNA in HeLa cells. *Proc. Natl. Acad. Sci. U.S.A.* **57**, 1514–1521.

Rae, P. (1970). Chromosoma distribution of rapidly annealing DNA in *Drosophila melanogaster*. *Proc. Natl. Acad. Sci. U.S.A.* **67**, 1018–1025.

Ramiriz, S. A., and Sinclair, J. H. (1975). Intraspecific variation of ribosomal gene redundancy in *Zea mays*. *Genetics* **80**, 495–504.

Reeder, R. H., Brown, D. D., Wellauer, P. K., and Dawid, I. B. (1976). Patterns of ribosomal spacer lengths are inherited. *J. Mol. Biol.* **105**, 507–516.

Retel, J., and Planta, R. J. (1972). Nuclear satellite DNAs of yeast. *Biochim. Biophys. Acta* **281**, 229–309.

Ribbert, D., and Bier, K. (1969). Multiple nucleoli and enhanced nucleolar activity in the nurse cells of the insect ovary. *Chromosoma* **27**, 178–197.

Ritossa, F. M. (1968). Unstable redundancy of genes for ribosomal RNA. *Proc. Natl. Acad. Sci. U.S.A.* **60**, 509–516.

Ritossa, F. M. (1972). Procedure for magnification of lethal deletions of genes for ribosomal RNA. *Nature (London)* **240**, 109–111.

Ritossa, F. (1976). The bobbed locus. *In* "The Genetics and Biology of *Drosophila*" (M. Ashburner and E. Novitski, eds.), Vol. 1B, pp. 801–846. Academic Press, New York.

Ritossa, F. M., and Spiegelman, S. (1965). Localization of DNA complementary to ribosomal RNA in the nucleolus organizer region of *D. melanogaster*. *Proc. Natl. Acad. Sci. U.S.A.* **53**, 737–745.

Ritossa, F. M., Atwood, K. D., Lindsley, D. L., and Spiegelman, S. (1966a). On the chromosomal distribution of DNA complementary to ribosomal and soluble RNA. *Natl. Cancer Inst., Monogr.* **23**, 449–472.

Ritossa, F. M., Atwood, K. C., and Spiegelman, S. (1966b). Molecular explanation of the *bobbed* mutants of *Drosophila* as partial deficiencies of "ribosomal" DNA. *Genetics* **54**, 819–834.

Ritossa, F., Malva, C., Boncinelli, E., Graziani, F., and Polito, L. (1971). On the first steps of rDNA magnification. *Proc. Natl. Acad. Sci. U.S.A.* **68**, 1580–1584.

Roberts, R. J. (1976). Restriction endonucleases. *Crit. Rev. Biochem.* **4**, 123–164.

Rochaix, J. D., and Bird, A. P. (1975). Circular ribosomal DNA and ribosomal DNA replication in somatic amphibian cells. *Chromosoma* **52**, 317–327.

Rochaix, J. D., Bird, A., and Bakken, A. (1974). Ribosomal RNA gene amplification by rolling circles. *J. Mol. Biol.* **87**, 473–487.

Ruderman, J. V., and Gross, P. R. (1974). Histones and histone synthesis in sea urchin development. *Dev. Biol.* **36**, 286–298.

Rudkin, G. T. (1969). Non-replicating DNA in *Drosophila*. *Genetics* **61**, Suppl., 227–238.

Schaffner, W., Gross, K. G., Telford, J., and Birnstiel, M. (1976). Molecular analysis of the histone gene cluster in *Psammechinus miliaris:* II. The arrangement of the five histone-coding and spacer sequences. *Cell* **8**, 471–478.

Schalet, A. (1969). Exchanges at the bobbed locus of *Drosophila melanogaster*. *Genetics* **63**, 133–153.

Scheer, V., Trendelenburg, M. F., Krohne, G., and Franke, W. (1977). Lengths and patterns of transcriptional units in amplified nucleoli of oocytes of *Xenopus laevis*. *Chromosoma* **60**, 147–167.

Schweizer, E., Mackechnie, C., and Halvorson, H. O. (1969). The redundancy of ribosomal and transfer RNA genes in *Saccharomyces cereviciae*. *J. Mol. Biol.* **40**, 261–277.

Sinclair, J. H., Carroll, C. R., and Humphrey, R. R. (1974). Variation in rDNA redundancy level and nucleolar organizer length of the Mexican axolotl. *J. Cell Sci.* **15**, 239–257.

Smith, G. P. (1973). Unequal crossover and the evolution of multigene families. *Cold Spring Harbor Symp. Quant. Biol.* **38**, 507–513.

Smith, G. P. (1978). What is the origin and evolution of repetitive DNAs? *Trends Biochem. Sci.* **3**, N34–N36.

Smith, H. O., and Wilcox, K. W. (1970). A restriction enzyme from *Haemophilus influenzae*. I. Purification and general properties. *J. Mol. Biol.* **51**, 379–391.

Southern, E. M. (1975a). Long range periodicities in mouse satellite DNA. *J. Mol. Biol.* **94**, 51–69.

Southern, E. M. (1975b). Detection of specific sequences among DNA fragments separated by gel electrophoresis. *J. Mol. Biol.* **98**, 503–517.

Spear, B. B. (1974). The genes for ribosomal RNA in diploid and polytene chromosomes of *Drosophila melanogaster*. *Chromosoma* **48**, 159–179.

Spear, B. B., and Gall, J. G. (1973). Independent control of ribosomal gene replication in polytene chromosomes of *Drosophila melanogaster*. *Proc. Natl. Acad. Sci. U.S.A.* **70**, 1359–1363.

Spencer, W. P. (1944). Isoalleles at the *bobbed* locus in *Drosophila hydei* populations. *Genetics* **29**, 520–536.

Stanfield, S., and Helinski, D. R. (1976). Small circular DNA in *Drosophila melanogaster*. *Cell* **6**, 333–345.

Stern, C. (1929). Uber die additive Wirckung multiplier Allele. *Biol. Zentralbl.* **49**, 261–290.

Sutton, W. D., and McCallum, M. (1972). Related satellites in the genus *Mus*. *J. Mol. Biol.* **71**, 633–656.

Suzuki, Y., Gage, L. P., and Brown, D. D. (1972). The genes for silk fibroin in *Bombyx mori*. *J. Mol. Biol.* **70**, 637–650.

Tartof, K. D. (1971). Increasing the multiplicity of ribosomal genes in *Drosophila melanogaster*. *Science* **171**, 294–297.

Tartof, K. D. (1973a). Regulation of ribosomal RNA gene multiplicity in *Drosophila melanogaster*. *Genetics* **73**, 57–71.

Tartof, K. D. (1973b). Unequal mitotic sister chromosomal exchange and disproportionate replication as mechanisms regulating ribosomal RNA gene redundancy. *Cold Spring Harbor Symp. Quant. Biol.* **38,** 491–500.

Tartof, K. D. (1975). Redundant genes. *Annu. Rev. Genet.* **9,** 355–385.

Taylor, J. H. (1958). Sister chromatid exchanges in tritium labelled chromosomes. *Genetics* **43,** 515–529.

Thomas, C., Hanocq, F., and Heilporn, V. (1977). Persistence of oocyte amplified rDNA during early development of *Xenopus laevis* eggs. *Dev. Biol.* **57,** 226–229.

Timmis, J. N., and Ingle, J. (1973). Environmentally induced changes in rRNA gene redundancy. *Nature (London), New Biol.* **244,** 235–236.

Tobler, H. (1975). Occurence and developmental significance of gene amplification. "Biochemistry of Animal Development," Vol. 3, pp. 91–143. Academic Press, New York.

Tocchini-Valentini, G. P., and Crippa, M. (1971). On the mechanism of gene amplification. *Lepetit Colloq. Biol. Med.* **2,** 237–243.

Trendelenburg, M. F., Scheer, U., and Franke, W. W. (1973). Structural organization of the transcription of ribosomal DNA in oocytes of the house cricket. *Nature (London), New Biol.* **245,** 167–170.

Trendelenburg, M. F., Scheer, V., Zentgraf, H., and Franke, W. W. (1976). Heterogeneity of spacer lengths in circles of amplified ribosomal DNA of two insect species, *Dytiscus marginalis* and *Acheta domesticus. J. Mol. Biol.* **108,** 453–470.

Ullman, J., Lima-de-Faria, A., Jaworska, H., and Bryngelsson, T. (1973). Amplification of ribosomal DNA in *Acheta.* V. Hybridization of RNA complementary to ribosomal DNA with pachytene chromosomes. *Hereditas* **73,** 13–24.

Vincent, W. S., Halvorsen, H. O., Chen, H.-R., and Shin, D. (1969). A comparison of ribosomal gene amplification in uni- and multi-nucleolate oocytes. *Exp. Cell Res.* **57,** 240–250.

Vogt, V. M., and Braun, R. (1976). Structure of ribosomal DNA in *Physarum polycephalum. J. Mol. Biol.* **106,** 567–587.

Wallace, H., and Birnstiel, M. L. (1966). Ribosomal cistrons and the nucleolar organizer. *Biochim. Biophys. Acta* **114,** 296–310.

Wallace, H., Morray, J., and Langridge, H. (1971). Gene amplification—An alternative theory to explain the synthesis of *Xenopus* nucleoli. *Nature (London)* **230,** 201–204.

Watson-Coggins, L., and Gall, J. G. (1972). The timing of meiosis and DNA synthesis during early oogenesis in the toad, *Xenopus laevis. J. Cell Biol.* **52,** 569–576.

Wegnez, M., Monier, R., and Denis, H. (1972). Sequence heterogeneity of 5 S RNA in *Xenopus laevis. FEBS Lett.* **25,** 13–20.

Wellauer, P. K., and Dawid, I. B. (1974). Secondary structure maps of ribosomal RNA and DNA. I. Processing of *Xenopus laevis* ribosomal RNA and structure of single-stranded ribosomal DNA. *J. Mol. Biol.* **89,** 379–395.

Wellauer, P. K., and Dawid, I. B. (1977). The structural organization of ribosomal DNA in *Drosophila melanogaster. Cell* **10,** 193–212.

Wellauer, P. K., Reeder, R. H., Carroll, D., Brown, D. D., Deutch, A., Higashinakagawa, T., and Dawid, I. B. (1974). Amplified ribosomal DNA from *Xenopus laevis* has heterogeneous spacer lengths. *Proc. Natl. Acad. Sci. U.S.A.* **71,** 2823–2827.

Wellauer, P. K., Dawid, I. B., Brown, D. D., and Reeder, R. H. (1976a). The molecular basis for length heterogeneity in ribosomal DNA from *Xenopus laevis. J. Mol. Biol.* **105,** 461–486.

Wellauer, P. K., Reeder, R. H., Dawid, I. B., and Brown, D. D. (1976b). The arrangement of length heterogeneity in repeating units of amplified and chromosomal ribosomal DNA from Xenopus laevis. *J. Mol. Biol.* **105,** 487–505.

Wensink, P. C., and Brown, D. D. (1971). Denaturation map of the ribosomal DNA of *X. laevis*. *J. Mol. Biol.* **60**, 235–248.

White, R. L., and Hogness, D. S. (1977). R-Loop mapping of the 18 S and 28 S sequences in the long and short repeating units of *Drosophila melanogaster* rDNA. *Cell* **10**, 177–192.

Wilson, M. C., Melli, M., and Birnstiel, M. L. (1974). Reiteration frequency of histone coding sequences in man. *Biochem. Biophys. Res. Commun.* **61**, 404–408.

Wimber, D. E., and Steffensen, D. M. (1970). Localization of 5 S RNA genes in *Drosophila* chromosomes by RNA–DNA hybridization. *Science* **170**, 639–641.

Woodland, H. R., and Adamson, E. D. (1976). The synthesis and storage of histones during the oogenesis of *Xenopus laevis*. *Dev. Biol.* **57**, 118–135.

Yao, M. C., and Gall, J. G. (1977). A single integrated gene for ribosomal RNA in a eukaryote, *Tetrahymena pyriformis*. *Cell* **12**, 121–132.

Yao, M. C., Kimmel, A., and Gorovsky, M. (1974). A small number of cistrons for ribosomal RNA in the germinal nucleus of a eukaryote, *Tetrahymena pyriformis*. *Proc. Nat. Acad. Sci. U.S.A.* **71**, 3082–3086.

Zweidler, A., and Cohen, L. H. (1973). Histone variants and new histone species isolated from mammalian tissues. *J. Cell Biol.* **59**, 378a.

3

Basic Enzymology of Transcription in Prokaryotes and Eukaryotes

Samson T. Jacob and Kathleen M. Rose

CELL BIOLOGY, VOL. 3

Copyright © 1980 by Academic Press, Inc.
All rights of reproduction in any form reserved.
ISBN 0-12-289503-7

I. INTRODUCTION

DNA-dependent RNA polymerase catalyzes the faithful transcription of DNA using ATP, GTP, CTP, and UTP as substrates. Since its discovery almost two decades ago by Weiss (1960), spectacular progress has been made in the elucidation of the general characteristics, subunit structure, and regulatory role of prokaryotic RNA polymerase. Difficulties in the quantitative solubilization of RNA polymerases from the nuclear matrix initially hampered studies on eukaryotic enzymes. The first successful extraction of RNA polymerases in high yield from rat liver nuclei (Jacob et al., 1968) paved the way to the characterization of these enzymes from a variety of eukaryotes.

Several fundamental differences in the template for RNA synthesis are evident between prokaryotes and nucleated organisms. For example, eukaryotes possess (a) approximately 10^3 times as much DNA per cell as prokaryotes; (b) a number of DNA molecules organized into discrete chromosomes; (c) a chromatin matrix consisting of DNA, histones, numerous nonhistone proteins, and RNA; and (d) large quantities of repetitive nucleotide sequences which lack structural genes. In spite of the disparity in the nature of the DNA template between pro and eukaryotes, the basic process of RNA synthesis has remained intact throughout evolution, and a number of similarities exist among all RNA polymerases. This chapter deals only with the general enzymatic proper-ties of prokaryotic and eukaryotic RNA polymerases. Other chapters in this volume will focus on the basic mechanism of DNA transcription and the role of chromatin components in transcription. Several excellent re-view articles on RNA polymerases have appeared in the past several years; the reader is asked to refer to these publications, contained in the list of General References at the end of this chapter, for more detailed information.

II. EUKARYOTIC RNA POLYMERASES

A. Multiple Forms and Their Intracellular Localization

In contrast to prokaryotes where a single DNA-dependent RNA polymerase is responsible for all RNA synthesis, eukaryotes contain sev-eral distinct RNA polymerases which are involved in the synthesis of discrete species of RNA. Three of these DNA-dependent enzymes are of nuclear origin, and one is localized within mitochondria. Plant cells con-tain an additional DNA-dependent RNA polymerase associated with chloroplasts. The nuclear enzymes can generally be resolved by DEAE-Sephadex chromatography. With increasing ionic strength, the enzymes elute in the following order: RNA polymerase I, RNA polymerase II, and, finally, RNA polymerase III (Jacob, 1973). Each individual class of nuclear enzyme has been further resolved into subspecies, either by additional ion exchange chromatography or by gel electrophoresis under nondenaturing conditions. Thus, class I RNA polymerase has been subfractionated into I_A and I_B, whereas class II and III enzymes have been resolved into three subspecies (II_A, II_B, II_0, and III_A, III_B, III_C, respectively).

Eukaryotic RNA polymerases are compartmentalized within the nucleus. Thus, polymerase I is localized in the nucleolus, whereas RNA polymerases II is exclusively found in the extranucleolar or nucleoplasmic fraction. Class III enzymes are found both in the nucleoplasmic and cytoplasmic fractions of the cell. One form of this enzyme (namely, III_B) is detected primarily in the cytosol, whereas III_A is confined to the nucleus (Seifart and Benecke, 1975). However, the majority of RNA polymerase III in the cytosol appears to have resulted from its release from the nucleus during tissue homogenization in the isotonic buffers (Austoker *et al.*, 1974; Lin *et al.*, 1976; Seifart and Benecke, 1975) which are commonly used for the preparation of cytosol fractions. Nevertheless, even when precautions are taken to minimize nuclear leakage, RNA polymerase III can still be detected in the cytosol fraction (Lin *et al.*, 1976). Thus, at present it cannot be ruled out that some RNA polymerase III may truly be a cytoplasmic (extramitochondrial) enzyme, although the function of such an enzyme population is not immediately apparent.

Each class of nuclear RNA polymerase occurs in two functionally distinct states (Yu, 1975). One population of enzyme is tightly bound to the chromatin and is designated as the "engaged" or "bound" form. The other population of RNA polymerase exists in the nucleoplasm in a freely diffusible or loosely bound state. The "bound" enzymes are retained in the nuclear matrix when the tissue is gently homogenized in hypotonic or isotonic buffers (Lin *et al.*, 1976), whereas the "free" enzymes are released to the cytosol fraction under such conditions. Recent studies tend to suggest that the "bound" form represents the transcriptionally active population of polymerase molecules (Matsui *et al.*, 1976; Leonard and Jacob, 1977). Most of the template-engaged polymerase II can be detected in structures designated as polynucleosomes, which have properties corresponding to euchromatin (Tata and Baker, 1978). Since polynucleosomes appear to contain the transcriptionally active DNA and most of the engaged RNA polymerase II, they may be an ideal system for studying transcription of structural genes *in vitro*. RNA polymerase II can be visualized in the transcriptionally active chromosome puffs of *Drosophila* polytene chromosomes (Jamrich *et al.*, 1977) using immunofluorescence techniques. It is of interest to note here that RNA polymerase from prokaryotes also exists in the cell in "free" and DNA-bound states. Analogous to chromatin preparations, a transcriptionally competent RNA polymerase–DNA complex can be isolated from *E. coli* by density gradient sedimentation (Pettijohn and Kamiya, 1967; Saitoh and Ishihama, 1977).

The DNA-dependent RNA polymerase associated with mitochondria is distinct from the nuclear enzymes (see Table I). Unlike the nuclear enzymes, mitochondrial RNA polymerase transcribes a single, generally circular, DNA molecule which is not associated with histones or other proteins. In fact, mitochondrial DNA has a greater resemblance to bacterial DNA than to nuclear DNA. The mitochondrial RNA polymerase is located within the inner mitochondrial membrane in close association with the mitochondrial DNA and/or membrane fraction. A distinct DNA-dependent RNA polymerase has also been identified in chloroplasts. Both mitochondrial and chloroplast RNA polymerases have been reviewed elsewhere (Jacob, 1973; Chambon, 1974).

B. Functional Role of Nuclear RNA Polymerases

Each RNA polymerase in eukaryotic cells is responsible for the synthesis of a particular type(s) of RNA. Ribosomal RNA (rRNA) is transcribed from nucleolar DNA as a large (45 S) RNA species containing the 28 S, 18 S, and 5.8 S rRNA components. Isolated nucleoli retain sufficient RNA polymerase I to synthesize rRNA *in vitro* (Jacob *et al.*, 1969; Ferencz and Seifart, 1975; Beebee and Butterworth, 1976; Mitsui *et al.*, 1977). Recent studies have also indicated a role for RNA polymerase I in the synthesis of some small nuclear or low molecular weight RNAs (Zieve *et al.*, 1977). It is not yet clear whether these small RNA species are synthesized by polymerase I in the nucleolus and then shuttled to the nucleoplasm or whether an extranucleolar population of polymerase I synthesizes these RNAs. Other low molecular weight RNAs in the cell are synthesized by RNA polymerase III; these include transfer RNA (tRNA), 5 S rRNA, and viral-specific low molecular weight RNA species (Roeder, 1976). RNA polymerase II catalyzes the synthesis of heterogeneous nuclear RNA (HnRNA) (Zybler and Penman, 1971), which has been shown, in several instances, to contain the precursor molecules for messenger RNAs (mRNAs) (Molloy and Puckett, 1976). Polymerase II has also been shown to be responsible for the synthesis of specific mRNA molecules (Suzuki and Giza, 1976; Bitter and Roeder, 1978).

Much of the evidence attributing synthesis of RNA types to a particular RNA polymerase has come from studies employing isolated nuclei and/or nucleoli which have retained the template-bound enzymes. The specificity of RNA polymerases in the synthesis of discrete RNA species has also been ascertained using exogenous RNA polymerase and isolated chromatin template containing little or no endogenous polymerase activity. For example, addition of RNA polymerase I, partially purified from a mammalian source, to isolated homologous chromatin or reconstituted chromatin can result in synthesis of rRNA-like products (Ballal *et al.*, 1977; Daubert *et al.*, 1977). However, it is not clear whether the added enzyme initiated new RNA chains or simply elongated the already initiated chains (Ballal *et al.*, 1977). Furthermore, transcription of homologous deproteinized DNA by the purified mammalian RNA polymerase I fails to yield rRNA-like products. In contrast, the ribosomal RNA gene containing the sequences for 28 S and 18 S rRNA, has been transcribed using "naked," deproteinized DNA and purified homologous RNA polymerase I from yeast. In particular, at low enzyme to DNA ratios, the purified yeast RNA polymerase I can transcribe native yeast DNA to produce a 42 S transcript (Van Keulen and Retel, 1977; Holland *et al.*, 1977). Thus, relatively accurate transcription of rRNA genes has been achieved from high molecular weight yeast DNA. However, it has not yet been rigorously proven that *only* rRNA chains have been initiated in this system.

As with rRNA, RNA polymerase II-specific products can be synthesized *in vitro* using chromatin as template. For example, the mRNAs for specific proteins such as globin (Steggles *et al.*, 1974; Wilson *et al.*, 1975) and ovalbumin (Towle *et al.*, 1977) are synthesized when exogenous RNA polymerase II is used for transcription of chromatin. However, in the case of globin gene transcription, a predominantly symmetrical mode of transcription (i.e., transcription of both strands of DNA duplex) occurs (Wilson *et al.*, 1975), implying that the factors required for asymmetrical transcription are either missing from the *in vitro* system or are not expressed under the conditions used for the assay. Contrary to globin mRNA synthesis, a significant proportion of sequences corresponding to ovalbumin gene are transcribed asymmetrically by polymerase II *in vitro*. Although globin and ovalbumin mRNAs can also be synthesized *in vitro* by *E. coli* RNA polymerase, recent evidence strongly suggests that the prokaryotic polymerase cannot always initiate at the correct promotor sites on mammalian chromatin. Thus, *in vitro* synthesis of appropriate mRNAs simply may be due to random initiation by this enzyme. *Esherichia coli* RNA polymerase can also catalyze RNA-dependent RNA synthesis using the endogenous mRNA as the template, which may lead to erroneous estimates of the newly formed transcripts unless proper precautions are taken

TABLE I

General Properties of DNA-Dependent RNA Polymerases[a]

Enzyme	Function	Molecular weight	Subunit structure	Sensitivity to inhibitors	
				Rifamycin	α-Amanitin
PROKARYOTES					
Bacteria —	Synthesis of all cellular RNA species	~500,000	Two large (155,000–165,000) and at least two small (40,000–90,000) subunits	Sensitive to low levels (<1 μg/ml)	Insensitive
Bacteriophages T4, SP01, SP82 —	Phage-specific RNA synthesis	~500,000	Host-modified subunits and/or additional polypeptides associated with host enzyme	Sensitive	?
Bacteriophages T7, gh-1 —	Phage-specific RNA synthesis	~100,000	Active enzyme consists of a single polypeptide	Sensitive only to polar rifamycin derivatives	?
EUKARYOTES (Cellular Location)					
Nucleolus I, (I$_A$, I$_B$)	Synthesis of rRNA precursors and	500,000–600,000	Two large subunits, 3–4 small subunits	Sensitive only to polar rifamycin	Insensitive, except yeast

118

		possibly some low molecular weight nonribosomal RNA's		(yeast contains 9 small subunits)	derivatives (>50 μg/ml)	which is sensitive to high levels (>100 μg/ml)
Nucleoplasm	II (II$_A$, II$_B$, II$_O$)	Synthesis of mRNA precursors	450,000–600,000	Two large and 3–7 small subunits	Sensitive to highly polar rifamycin derivatives	Completely sensitive to low levels (<1 μg/ml)
Nucleoplasm	III (III$_A$, III$_B$, III$_C$)	Synthesis of tRNA precursors, 5 S ribosomal rRNA's, viral-specific low molecular weight RNA's	~600,000	Two large and 8–11 small subunits	Sensitive to highly polar rifamycin derivatives	Completely sensitive to high levels (>100 μg/ml) except yeast, which is insensitive
Mitochondria	—	Transcription of mitochondrial DNA	~100,000	One polypeptide chain	Some preparations sensitive to low levels of parent rifamycin, while others are insensitive	Insensitive
Chloroplast	—	Transcription of chloroplast DNA	~500,000	Two large and several small subunits	Extracted enzyme insensitive	Insensitive

[a] For references, consult the text.

to alleviate this potential problem (Zasloff and Felsenfeld, 1977). It therefore seems evident that future investigations on the transcription of eukaryotic genes *in vitro* should preferably employ the eukaryotic RNA polymerases rather than *E. coli* RNA polymerase. This has been well illustrated by the observation that selective, accurate transcription of 5 S rRNA genes is achieved in isolated chromatin not by addition of *E. coli* polymerase, but only by exogenous RNA polymerase III (Parker and Roeder, 1977; Yamamoto *et al.*, 1977).

C. Differential Sensitivities of Nuclear RNA Polymerases to α-Amanitin

The nuclear RNA polymerases from most eukaryotes can be identified on the basis of their differential sensitivities to the mushroom toxin, α-amanitin. Class II RNA polymerase is sensitive to a very low dose (10^{-9}–$10^{-8} M$) of α-amanitin (Jacob *et al.*, 1970a,b; Kedinger *et al.*, 1970; Lindell *et al.*, 1970), whereas class III enzymes are inhibited at much higher levels (10^{-5}–$10^{-4} M$) (Weinmann and Roeder, 1974) of the toxin. Class I RNA polymerase is completely insensitive to the inhibitor. There are, however, few exceptions to the differential sensitivity of RNA polymerases I, II, and III to the amatoxin. For example, in yeast, class I enzymes are sensitive to high levels of α-amanitin, whereas class III RNA polymerases are insensitive to the toxin (Valenzuela *et al.*, 1976; Schultz and Hall, 1976). On the other hand, both RNA polymerases I and III from the insect *Bombyx mori* are resistant to α-amanitin at levels as high as 1 mg/ml (Roeder, 1976). Mitochondrial, chloroplast, and bacterial RNA polymerases are also resistant to this toxin (see Table I).

To date, polymerase II from all higher eukaryotes is inhibited by low concentrations of the mushroom poison. In view of the extremely small quantity of the toxin required for complete inhibition of RNA polymerase II, this inhibitor may be classified as an enzyme poison. Inhibition of RNA polymerase II and the resultant inhibition of mRNA and protein syntheses may, indeed, form the molecular basis of the mushroom poisoning in humans. Specific inhibition of mRNA synthesis by α-amanitin has found application in exploring the role of mRNA in cellular events (Jacob *et al.*, 1974; Biswas, 1978). Interestingly, α-amanitin can inhibit rRNA precursor synthesis *in vivo* (Jacob *et al.*, 1970c; Tata *et al.*, 1972; Sekeris and Schmid, 1972), presumably as a result of polymerase II inhibition. That is, the inhibition of rRNA synthesis by α-amanitin may be due to secondary responses which follow the inhibition of polymerase II-directed mRNA synthesis and the resultant inhibition of synthesis of "rapidly turning over" protein(s) controlling rDNA transcription (Jacob *et al.*, 1970c;

Lampert and Fiegelson, 1974; Lindell, 1976). Relative to inhibition of mRNA precursor synthesis, effects on rRNA precursor synthesis are, however, short-lived (Jacob *et al.*, 1970c).

Sensitivity of RNA polymerase II to α-amanitin can be affected by mutation (Chan *et al.*, 1972; Amati *et al.*, 1972; Somers *et al.*, 1975). Such mutants can be identified by the resistance of their RNA polymerase II to α-amanitin. RNA polymerase II from such cells binds the toxin with a lowered affinity (Lobban *et al.*, 1976; Ingles *et al.*, 1976). The α-amanitin-resistant mutation thus appears to involve a structural change in one of the subunits of RNA polymerase II. Relative to wild-type polymerase II, the amanitin-resistant enzyme possesses altered thermal denaturation characteristics (Lobban *et al.*, 1976) and template preferences (Bryant *et al.*, 1977). Another mutant (temperature sensitive), which appears defective in α-amanitin-sensitive RNA synthesis, has recently been isolated from a hamster cell line (Rossini and Baserga, 1978). At the nonpermissive temperature, the stimulation of α-amanitin sensitive RNA synthesis by serum is selectively inhibited and correlates well with RNA polymerase II activity. Furthermore, the mutant cells do not enter the S phase at the nonpermissive temperature. These data have prompted the investigators to conclude that RNA polymerase II activity is required for the entry of the cells into the S phase. These mutant cells provide a useful model for investigating the relationship between synthesis of ribosomal RNA and nonribosomal RNAs, since stimulation of ribosomal RNA synthesis is not affected at nonpermissive temperature.

α-Amanitin does not inhibit binding of RNA polymerase II to DNA, but rather primarily inhibits the elongation reaction (Jacob *et al.*, 1970b, Cochet-Meilhac and Chambon, 1974a). Studies using O-[^{3}H]methylde-methyl-γ-amanitin have shown that the equilibrium dissociation constants for complexes between the toxin and crude homogenates from various tissues or cells are identical to the equilibrium dissociation constant of pure polymerase II–amanitin complexes (Cochet-Meilhac and Chambon, 1974b). These data indicate that RNA polymerase II is the major cellular component to which this amatoxin binds. Although it had been previously shown that α-amanitin binds to RNA polymerase II in a 1 : 1 stoichiometric ratio with a binding constant of about $10^{-9} M$ for the complex formation, the exact determination of subunit(s) of the enzyme to which the toxin binds was unsuccessful owing to the dissociation of labeled amanitin from the subunits under the denaturing conditions used for gel electrophoresis. Recently, this problem has been circumvented by covalent coupling of [^{3}H]amanin to enzyme (amanin, a carboxylic acid, is a member of the family of amatoxins whose most prominent representative is α-amanitin) and affinity labeling by a water-soluble carbodiimide. Electrophoresis of the resultant conjugate under denaturing conditions has shown that the toxin binds to a large subunit (MW 140,000) of RNA polymerase II (Brodner and Wieland, 1976). The molecular mechanism of toxicity of α-amanitin must, therefore, reside in its selective binding to one of the large subunits of RNA polymerase II, which will lead to inhibition of mRNA synthesis.

III. PURIFICATION CHARACTERISTICS AND GENERAL PROPERTIES OF RNA POLYMERASES

A. Purification of Enzymes

A variety of methods have been used for the purification of DNA-dependent RNA polymerases from both prokaryotes and eukaryotes. In the case *E. coli,* they vary in ease, speed, yield, reproducibility, and final purity (Zillig *et al.,* 1970; Burgess, 1969; Berg *et al.,* 1971; Nüsslein and Heyden, 1972; Humphries *et al.,* 1973; Yarbrough and Hurwitz, 1974). Recently, two procedures have been developed for the rapid purification of *E. coli* RNA polymerase which recover nearly 50% of the enzyme in the final preparation. One of the procedures is an extension of the method introduced by Zillig *et al.* (1970). This involves initial removal of nucleic acid and the bulk of the protein from the cell lysate by fractionation with Polymin P. RNA polymerase can be completely precipitated with 0.35% Polymin P, whereas only 25–30% of the protein is precipitated at this step. Elution of the Polymin P precipitate with 0.5 M NaCl removes 50% of the extractable protein, while 1 M NaCl efficiently elutes the RNA polymerase. The enzyme is then purified by DNA–cellulose chromatography followed by high salt Bio-Gel A 5m chromatography (Burgess and Jendrisak, 1975).

Another procedure for rapid isolation of *E. coli* RNA polymerase takes advantage of the affinity of heparin for the enzyme. Two moles of this polyanion have been shown to bind to a mole of RNA polymerase (Zechel, 1971), one to the largest subunit, β', and the other to σ (Zillig *et al.,* 1971), the factor involved in selective transcription (see Section IV). The purification method (Sternbach *et al.,* 1975) consists of essentially three steps: (a) ammonium sulfate fractionation of cell lysates, (b) heparin–Sepharose chromatography, and (c) a high salt–sucrose gradient fractionation. In addition to a reduced number of steps in this scheme relative to conventional purifications, the final enzyme preparation is at least twice as active as other enzyme preparations (1960 versus 810 units/mg for the enzyme purified by Polymin P fractionation). The higher specific activity of RNA polymerase purified by heparin–Sepharose chromatography appears to be due to an unusually high content of σ factor in these enzyme preparations.

Unlike prokaryotic RNA polymerase, eukaryotic enzymes can be purified using isolated organelle (nuclei, nucleoli, mitochondria, and chloroplasts) as starting material. In itself, the preparation of the organelle prior to enzyme extraction represents a considerable purification. Owing to the diffusible nature of some of the nuclear enzymes, simultaneous quantitative recovery of all three RNA polymerases is not always feasible. The methods of solubilization and purification of mammalian RNA polymerases and difficulties encountered therein have been extensively reviewed recently (Jacob and Rose, 1978). Suffice it to say here that, in general, the three classes of nuclear enzymes can often be resolved by a single DEAE-Sephadex chromatographic column (Roeder and Rutter, 1969; Jacob, 1973). Extensive purification has been obtained by standard chromatographic and sizing techniques. The specific activities of highly purified eukaryotic RNA polymerases I, II, and III approach that of *E. coli* RNA polymerase when measured under similar conditions.

The column chromatographic fractionation of RNA polymerase on phosphocellulose deserves some comments. This ion exchange resin has been used extensively in the preparation of RNA polymerases from both prokaryotes and eukaryotes. The enzyme from *E. coli* loses the factor responsible for selective initiation, σ, on this column. Recently, it has been reported that if high concentrations of glycerol are present in the buffers for the phosphocellulose column, σ can be retained with the core RNA polymerase (Johnson *et al.*, 1971; Gonzalez *et al.*, 1977). Eukaryotic RNA polymerases do not appear to lose a σ-like factor upon phosphocellulose chromatography, although there are a few isolated reports relevant to loss of RNA "stimulating" factors during enzyme purification (Froehner and Bonner, 1973; Gissinger *et al.*, 1974; Goldberg *et al.*, 1977). The function and mechanism of action of these factors are at present unknown, as is their relationship to the cytoplasmic proteins which have been suggested to regulate the transcription of specific eukaryotic genes (Jacob *et al.*, 1970c; Lampert and Fiegelson, 1974; Lindell, 1976). The normal use of fairly high concentrations of glycerol (25–50%, v/v) in the purification of relatively labile eukaryotic RNA polymerases may explain retention of some subunits or factors throughout the purification. It is noteworthy that many of the highly purified preparations of nuclear RNA polymerases do, indeed, have a greater number of subunits than the bacterial enzymes (see Table I). In particular, the only enzyme (yeast RNA polymerase I) which has been shown to be capable of selective transcription on deproteinized DNA contains as many as 11 subunits (Valenzuela *et al.*, 1976). Variability in the number of subunits of eukaryotic RNA polymerases obtained from different laboratories may be partially due to different purification techniques, which have allowed retention of specific factors or subunits in some cases but not in others.

B. Stages of RNA Synthesis

The RNA synthesizing reaction catalyzed by DNA-dependent RNA polymerase is complex and occurs in several discrete steps. As elucidated in prokaryotic systems, these include (a) binding of enzyme to the DNA template or template site selection, (b) initiation of RNA synthesis with GTP or ATP, (c) RNA chain elongation using the four nucleoside triphosphates, and (d) chain termination and release (Chamberlin, 1974). Furthermore, each step itself appears to be multifaceted. The complexity of RNA synthesis has rendered its study extremely difficult, particularly in the case of eukaryotes, in which the DNA template is much more complex than in prokaryotes. The following section is a brief description of the reaction catalyzed by bacterial RNA polymerase.

The first step in the general transcription process is the interaction of RNA polymerase with DNA to form a binary complex. This complex is produced in the absence of nucleoside triphosphates and occurs primarily at promotor regions on the template. The formation of the binary complex leads to a limited opening of the DNA strands. The complex formed between enzyme and DNA is highly stable. The RNA polymerase in the binary complex protects a portion (about 40 base pairs) of the DNA from the action of pancreatic DNase. Isolation and characterization of this DNase-resistant fraction have facilitated identification of the RNA polymerase binding sites for several templates. Selection of biological binding sites (true promotors) is a key step in regulation of gene transcription; both repressors and positive control factors can operate at this step (Gilbert, 1976).

The second stage of the transcription process is initiation of RNA chains. The common mode of initiation involves the incorporation of an intact purine nucleoside triphosphate (pppX) to form the 5′-terminus of the RNA chain (Maitra and Hurwitz, 1965). In this reaction, two ribonucleoside triphosphates are coupled to give a dinucleotide tetraphosphate (Krakow *et al.*, 1976) as follows: $pppX + pppY \rightarrow pppXpY + pP_i$. The initiation of RNA chains *in vivo* occurs at specific promotor sites on the chromosome. However, initiation of RNA chains in regions other than the true promotors can occur. For example, when *E. coli* core polymerase (lacking the σ subunit, see Section IV) is employed for transcription, random initiation occurs. (Nonspecific initiation by eukaryotic polymerases often occurs when mammalian DNA is used as template. Invariably, these templates contain many single-strand breaks where RNA polymerases can bind. Consequently, studies using such DNA preparations can only provide an estimate of the activity of RNA polymerase rather than its capacity to bind to specific promotor regions and to initiate transcription of a specific gene selectively.)

During the elongation process, RNA polymerase, DNA, and nascent RNA are combined in a nondissociable ternary complex. This complex can be isolated by a variety of physical methods and used to study the elongation reaction specifically. The properties of the enzyme in the ternary complex are distinct from those of the binary complex. For example, the ternary complex is stable to elevated ionic strength. Enzyme in the ternary complex is resistant to inhibitors which interact only with free enzyme or which inhibit the initiation reaction. The affinity of σ subunit for the core RNA polymerase molecule is greatly reduced upon formation of the ternary complex. Apparently, σ is released from the core polymerase during elongation. The elongation reaction can be divided into at least five substeps (Krakow et al., 1976). These reactions are (a) binding of the nucleoside triphosphate, (b) formation of the phosphodiester bond, (c) release of the product pyrophosphate, (d) translocation of the incorporated nucleotide into the product terminus site, and (e) translocation of the enzyme to the next base on the template.

The last step in the transcription reaction is the selective termination of RNA chains and release from the DNA–polymerase complex. There is clear evidence that termination of RNA can occur at specific sites in vivo (Roberts, 1976). Although RNA polymerase itself can recognize some terminators in vitro, it fails to stop at others unless an accessory protein, the ρ factor (Roberts, 1969), is included. The ρ factor, which is an RNA-dependent nucleoside triphosphate phosphohydrolase, binds to the nascent RNA and utilizes ATP as an energy donor. The release of RNA from the template by ρ factor is accompanied by the cleavage of ATP to ADP and liberation of P_i (Lowery-Goldhammer and Richardson, 1974). In vivo, ρ is involved in termination of at least some of the transcripts which are terminated without ρ in vitro. This raises the possibility that ρ may be responsible for a majority of the transcription termination in the cell.

Following release of a discrete transcript and RNA polymerase from the template, the enzyme can then reassociate with σ factor and reinitiate at an available promotor. The transcription cycle can continue until the cell meets its demand for a specific RNA molecule.

C. General Reaction Characteristics

The basic RNA synthesizing reaction requires, in addition to enzyme, a polydeoxynucleotide template, divalent cation, and the four nucleoside triphosphates. The divalent cation requirement can be satisfied by Mg^{2+} or Mn^{2+}. In the case of RNA polymerase from prokaryotes, Mg^{2+} is generally used, because Mn^{2+} promotes DNA-independent homopolymer formation (Mehrota and Khorana, 1965). *Escherichia coli* RNA polymerase can also catalyze RNA-dependent RNA synthesis as demonstrated by the synthesis of "anti-mRNA" using endogenous globin mRNA associated with reticulocyte chromatin as the template (Zasloff and Felsenfeld, 1977). Eukaryotic RNA polymerases do not appear to utilize RNA as template. Generally, nuclear RNA polymerases are most active in the presence of Mn^{2+} following their solubilization from the chromatin matrix. This property usually persists throughout their purification. Of the three nuclear enzymes, RNA polymerase II shows the greatest preference for Mn^{2+}. In fact, the Mn^{2+}/Mg^{2+} activity ratio for this enzyme is up to tenfold higher than that for enzymes I and III. The optimal levels of Mg^{2+} and Mn^{2+} are quite varied, depending on the enzyme and template used, and range from 5 to 15 mM for Mg^{2+} and 1 to 5 mM for Mn^{2+}. The optimal concentrations of the divalent metal ions also depend on the total nucleoside triphosphate concentration in the incubation mixture.

The ionic strength of the reaction mixture has a dramatic effect on RNA synthesis. In the case of prokaryotic RNA polymerase, elevated ionic strength has been found to alter all the steps in the reaction (Chamberlin, 1974). The three nuclear RNA polymerases of eukaryotes exhibit characteristic salt optima when measured with a common template (usually calf thymus DNA). Typically, RNA polymerases I, II, and III are most active when the $(NH_4)_2SO_4$ concentrations in the assays are 30–60, 60–120, and 50–200 mM, respectively. These salt optima reflect only overall rates of RNA synthesis, rather than any specific aspect of the RNA synthesizing reaction. At higher salt concentrations, eukaryotic RNA polymerases cannot initiate on "naked" DNA. For example, rat liver or pig kidney RNA polymerases I and II are unable to initiate RNA synthesis at 150 and 400 mM $(NH_4)_2SO_4$, respectively. However, preinitiated enzymes can elongate at these same salt levels (Leonard and Jacob, 1977; Y. C. Lin and S. T. Jacob, unpublished observations).

The nature of the template used for transcription influences the overall reaction characteristics. For example, with denatured DNA as template, the total amount of RNA synthesized by *E. coli* RNA polymerase is decreased relative to double-stranded DNA, whereas the number of RNA chains is increased (Maitra and Hurwitz, 1967). Eukaryotic RNA polymerases I and II also initiate more RNA chains on denatured DNA than on native DNA. Single- or double-strand breaks in a duplex DNA result in increased transcription by *E. coli* "core" RNA polymerase (enzyme without σ factor) suggesting that core enzyme prefers initiation on single-strand regions or ends (Rosen and Rosen, 1969; Braun and Hechter, 1970). Although core RNA polymerase alone can transcribe intact (circular) double-stranded DNA, addition of σ factor greatly enhances transcription of such templates (5–75-fold). In eukaryotes, transcriptional activity measurements with purified enzymes are largely due to initiation of RNA chains at single-stranded regions or nicks on the DNA. Intact DNA's such as viral DNA are transcribed poorly by RNA polymerases I and II. However, RNA polymerase III is capable of transcribing fully intact double-stranded DNA's very efficiently (Hossenlopp *et al.*, 1975; Long *et al.*, 1976). It is possible that factors required for proper initiation of intact DNA are lost from RNA polymerase I and II during extensive purification.

On account of the complexity of the transcription process, measurement of the total amount of RNA synthesized *in vitro* in a given time has little meaning except for purification purposes. Some progress has been made in the analysis of the parameters involved in the individual reaction steps in prokaryotic systems. However, most studies dealing with the eukaryotic enzymes are still in the "black box" stage.

D. Subunit Structure of RNA Polymerases

1. Prokaryotes

Purification and characterization of the DNA-dependent RNA polymerase from a number of prokaryotes have been achieved in the past decade. The enzyme is large (MW 500,000) and consists of several polypeptide chains. RNA polymerase from *E. coli* has been studied most extensively and has the structure $\alpha_2\beta\beta'$ σ (holoenzyme) with the individual subunits having the following approximate molecular weights: α, 40,000; β, 155,000; β', 165,000; σ, 85,000–95,000 (Burgess, 1969). The σ subunit is easily separated from the core enzyme ($\alpha_2\beta\beta'$) by phosphocellulose chromatography (Burgess *et al.*, 1969) at low glycerol concentrations. At higher glycerol concentrations (50%), two fractions of RNA polymerase activity are observed, one of which contains σ (holoenzyme) and one which does not (core enzyme) (Rose, 1969; Johnson *et al.*, 1971; Gonzalez *et al.*, 1977). In the cell, σ is present in less than stoichiometric amounts relative to core polymerase (Ishihama *et al.*, 1976). Although the core enzyme is capable of faithful transcription of DNA, it initiates RNA chains indiscriminately. The function of σ is to facilitate the selection of true promoters for chain initiation (Burgess, 1971). Subsequent to chain initiation, σ is released from the core polymerase and can recombine with another core molecule for selective initiation (Burgess, 1969; Gerard *et al.*, 1972), possibly explaining why it is present in lesser molar quantities in the cell than the core enzyme.

In addition to the β, β', α, and σ subunits, varying levels of other polypeptides are associated with *E. coli* RNA polymerase. These include: omega (ω), MW 9000–12,000 (Burgess *et al.*, 1969; Berg *et al.*, 1971), and the X protein, MW 110,000 (Stetter and Zillig, 1974; Scheit and Stütz, 1975). As yet, no clear role in RNA synthesis has been elucidated for ω and X. In one instance, an altered σ subunit, σ', has been observed in RNA polymerase preparations (Fukuda *et al.*, 1974). Other proteins, such as the N protein (induced by coliphage λ infection), the M protein, ρ factor, and the cyclic AMP receptor protein may alter transcription by *E. coli* RNA polymerase. In general, these "regulatory" proteins do not copurify with bacterial RNA polymerase, although there is some evidence that N protein is closely associated with the β subunit of the enzyme (Ghysen and Pironio, 1972).

All prokaryotic RNA polymerases studied to date have similar structures and are comprised of several subunits. In an attempt to standardize the nomenclature for prokaryotic RNA polymerase, Zillig (Stetter and Zillig, 1974) proposed that the large polypeptides should be classified as β or β' on the basis of their functional role (see Section V), rather than size.

Bacteriophage infection generally results in either modification of the host RNA polymerase (see Section V,C) and/or production of a new phage-specific enzyme. Examples of the former case are infection in *E. coli* by phage T4 (Walter *et al.*, 1968; Travers, 1970; Stevens, 1972) and in *B. subtilis* by SP01 and SP82 phages (Spiegelman and Whitley, 1974; Fox and Pero, 1974; Duffy and Geiduschek, 1975). Examples of the latter are infection by coliphages T7 or T3 (Chamberlin *et al.*, 1970; Dunn *et al.*, 1970; Maitra, 1971) or *Pseudomonas* phage gh-1 (Towle *et al.*, 1975). The phage-specific RNA polymerases are generally comprised of a single polypeptide and exhibit a distinct preference for homologous phage DNA. One bacteriophage, N_4, contains a phage-specific RNA polymerase within the virion (Rothman-Denes and Schito, 1974). Thus, transcription of N_4 DNA does not require participation of the host RNA polymerase.

2. Eukaryotes

As in the case of prokaryotes, RNA polymerases from eukaryotes are also of large size and are composed of several subunits. Molecular weights of class I, II, and III RNA polymerases from several sources have been determined. They range from 400,000 to 650,000. In general, class III polymerase, with a molecular weight of more than 600,000, seems to be somewhat larger than enzymes I and II. Basically, all the eukaryotic RNA polymerases are comprised of two large subunits (130,000–220,000) and several (3–9) smaller polypeptides. Within an organism, each class of RNA polymerases has a unique subunit composition, the primary difference between the classes being in the molecular weight of the two large subunits (Sklar *et al.*, 1975). Indeed, two of the small subunits (molecular weight of approximately 25,000 and 16,000) appear to be common to all animal RNA polymerases (Kedinger *et al.*, 1974; Sklar *et al.*, 1975). The subunit composition of each class of polymerase has been phylogenetically conserved in that enzymes of the same class purified from several mammalian sources appear to have similar subunit compositions.

The subclasses of RNA polymerases also have distinct subunit compositions. Subclasses of enzyme generally differ from others in the same class by only one subunit. Interestingly, the subunits which are altered in the subclasses are different for each class of enzyme. Specifically, polymerase I_A from plasmacytoma 315 or rat liver is identical to I_B except that I_B contains a subunit of intermediate size (40,000–60,000) which is absent in I_A (Schwartz and Roeder, 1974; Matsui *et al.*, 1976); the largest subunit of II_A has a molecular weight of 214,000, while in II_B the largest subunit is 180,000 (Kedinger *et al.*, 1974); III_A differs from III_B by a change in one of the small subunits which has a molecular weight of 32,000 for the former enzyme and 33,000 for the latter (Sklar and Roeder, 1976). At present, little or no information is available as to the functional role of the individual subunits of eukaryotic RNA polymerases. Owing to the specificity of transcription of each enzyme class for a particular type of RNA, it seems plausible to speculate that such specificity might be determined by one of the subunits unique to each enzyme class, i.e., the larger subunits. Although each class of enzyme catalyzes the synthesis of different types of cellular RNA, it is not known whether the subclasses of enzymes are relegated to the synthesis of discrete species of RNA's.

The only RNA polymerase in eukaryotes which is apparently of small size is that present in mitochondria. This enzyme has been purified from a number of sources and appears to be composed of a single polypeptide with a molecular weight of 100,000 or less (Küntzel and Schäfer, 1971; Wu and David, 1972; Wintersberger, 1972). The properties of the mitochondrial enzyme are reminiscent of those of polymerases induced by T3, T7, and gh-1 bacteriophage infection, enzymes which are of small size and also highly selective in transcription of homologous template. Interestingly, RNA synthesis by some of the purified mitochondrial RNA polymerases (Küntzel and Schäfer, 1971; Scragg, 1971; Gallerani *et al.*,

1972; Reid and Parsons, 1971) is sensitive to rifamycin, the potent inhibitor of prokaryotic RNA polymerase, while the bacteriophage-specific enzymes are insensitive to this inhibitor. RNA polymerase from chloroplasts appears to be more complex than the mitochondrial enzyme. Analogous to the nuclear polymerases, it has a molecular weight of approximately 500,000. However, it is structurally distinct from RNA polymerase II from the same plant (Bottomley *et al.*, 1971).

In contrast to infection of bacteria by phages T3, T7, and gh-1, viral infection of mammalian cells does not appear to result in a small-sized viral-specific DNA-dependent RNA polymerase. Three types of response to viral infection have been described: (a) adequacy of host enzymes to transcribe viral genes [SV40 infection of HeLa cells (Hossenlopp *et al.*, 1975), polyoma transformation of BHK cells, (Cooper and Keir, 1975), adenovirus infection of KB cells (Austin *et al.*, 1973), and MuLV$_R$ infection of mouse spleen (Sethi and Gallo, 1975)]; (b) production of a viral-specific RNA polymerase of large size [vaccinia virus infection of HeLa cells (Nevins and Joklik, 1977)]. Although the vaccinia virus enzyme is of high molecular weight (425,000), it differs in its subunit structure from the host enzymes. (c) Increased activities and/or levels of host enzyme following infection [SV40 infection of CV-1 cells, (Righthand and Bagshaw, 1974) and Friend virus induction of mouse spleen (Babcock and Rich, 1973)]. The examples of elevated enzyme activity may be analogous to those observed after bacterial infection by T4, SP01, and SP82 phages, where host RNA polymerase is modified and/or associated with additional polypeptides.

IV. FUNCTION OF INDIVIDUAL SUBUNITS OF CORE
RNA POLYMERASE

As discussed in the previous sections, with the exception of mitochondrial RNA polymerase and certain phage-specific enzymes, DNA-dependent RNA polymerases from both prokaryotes and eukaryotes are multisubunit enzymes. Although the subunit compositions of RNA polymerases from many sources have been studied in some detail over the past decade, information concerning the functional role of the individual subunits in the transcription process is relatively scarce. Clearly, not all 4–11 subunits of purified RNA polymerases are necessary for RNA synthesis, since RNA polymerases containing only a single polypeptide chain do exist, for example in mitochondria (Küntzel and Schäfer, 1971) or as a result of infection by certain bacteriophages (Chamberlin *et al.*, 1970; Dunn *et al.*, 1971; Maitra, 1971; Towle *et al.*, 1975). However, the function of these small RNA polymerases is limited to transcription of specific templates. A divergence of function into several subunits may indicate that the more complex RNA polymerase molecules are subject to a wider variety of controls which can act via the individual polypeptide chains. Basically, three approaches have been used to elucidate the structure–function relationships of RNA polymerase subunits. These include (a) study of the binding sites of enzyme inhibitors and their effects on RNA synthesis (b) the properties of individual subunits produced after denaturation and the function of partially and/or completely reconstituted enzymes, and (c) the properties and subunit analyses of RNA polymerase mutants. All three approaches and, in some cases, a combination of techniques, have been used to study bacterial RNA polymerases, particularly the enzyme from *E. coli*. Data concerning the role of individual subunits in RNA polymerase from eukaryotes are very limited and, to date, consist primarily of the use of enzyme inhibitors. The following section will focus on basic properties of *E. coli* RNA polymerase, with brief references to other enzymes as necessary.

A. Studies Relevant to Individual Polypeptides

1. Subunit β'

Several lines of evidence indicate that the largest subunit of prokaryotic RNA polymerase, subunit β', is involved in binding to DNA. First, isolated β' has the capacity to bind template (Sethi and Zillig, 1970; Fukuda and Ishihama, 1974). However, since β' is also the most basic of the subunits, with lysine, histidine, and arginine contents totaling approximately 15 mole% (Fujiki and Zurek, 1975), its strong affinity for DNA is not an unexpected finding. This subunit also binds to the polyanion heparin (Zillig *et al.*, 1971). Although heparin apparently competes with DNA for template binding sites on RNA polymerase (Walter *et al.*, 1967), this polyanion does not bind specifically to β'. Chromatography of holoenzyme on a heparin–Sepharose matrix indicates that both σ factor and the $\beta' + \beta$ complex have strong affinities for the polyanion (Sternbach *et al.*, 1975). Heparin can also interact with eukaryotic RNA polymerases and is effective in preventing chain initiation (Cox, 1973; Ferencz and Seifart, 1975). Second, and more important in the designation of β' as the template-binding subunit is the observation that the complex $\beta\alpha_2$ does *not* bind DNA (Zillig *et al.*, 1971). Third, polydeoxy-4-thiothymidylic acid becomes attached to the β' subunit by photooxidation (Frishauf and Scheit, 1973). The linkage of this polydeoxynucleotide analogue can be overcome by increasing levels of template, again indicating that β' carries the template binding site. Fourth, DNA binding *in vitro* by RNA polymerase carrying a mutation in the β' subunit is altered in the mutant relative to wild-type enzyme (Khesin *et al.*, 1969; Panny *et al.*, 1974; Gross *et al.*, 1976). More recently, however, it has been reported that in *Salmonella typhimurium* a mutation in β' may affect RNA chain initiation and elongation rates in addition to DNA binding (Young and Wright, 1977). Further studies will be necessary to determine whether the alterations in initiation and elongation are the result of a direct involvement of β' in these processes or merely a result of an altered core enzyme configuration due to a mutation in β'.

2. Subunit β

Inhibitors of RNA synthesis which bind directly to the enzyme have served as valuable tools in studying the properties of the RNA polymerase molecule. The most widely studied inhibitor is the antibiotic rifamycin. As early as 1967 this antibiotic was shown to inhibit DNA-dependent RNA synthesis in a cell-free system containing *E. coli* RNA polymerase (Hartman *et al.*, 1967).

Although the parent compound rifamycin and closely related chemical derivatives, such as rifampicin and rifamycin SV, are very potent inhibitors of bacterial RNA polymerases, the nuclear enzymes from eukaryotes are insensitive to these antibiotics or are inhibited only at extremely high concentrations (Jacob *et al.*, 1968; Wehrli *et al.*, 1968). Interestingly, RNA synthesis in chloroplasts (Surzycki, 1969; Brown *et al.*, 1970) and rat liver mitochondria (Saccone *et al.*, 1971) is inhibited by rifampicin, suggesting that the RNA polymerases from these organelles may be structurally related to the bacterial enzyme. Although nuclear RNA polymerases from eukaryotes are resistant to inhibition by rifamycin itself, certain highly polar derivatives of the antibiotic, particularly those with lipophilic side chains, e.g., the 3-formyl-*O*-*n*-4-oximes of rifamycin SV, AF/013 and AF/08, can inhibit eukaryotic RNA polymerases. The concentrations of AF/013 required to inhibit eukaryotic RNA polymerases are generally severalfold higher than the concentration of rifamycin required to inhibit bacterial enzyme (for a review of inhibitors of eukaryotic RNA polymerases, see Lindell, 1977).

Rifamycin apparently binds to the β subunit of the bacterial RNA polymerase (Heil and Zillig, 1970), and resistance to the drug is a result of a mutation in this subunit. One mole of drug binds to 1 mole of enzyme. Bacterial RNA polymerase–rifamycin complexes are unable to initiate RNA synthesis (Sippel and Hartmann, 1968; Bautz and Bautz, 1970; Hinkle *et al.*, 1972). A similar phenomenon is observed for complexes of nuclear RNA polymerases and rifamycin derivatives (Meilhac and Chambon, 1973; Rose *et al.*, 1975). Both prokaryotic and eukaryotic RNA polymerases become resistant to the action of their respective inhibitors subsequent to the formation of an initiation complex. The similar modes of action of rifampicin on bacterial RNA polymerase and that of highly polar derivatives on nuclear enzymes suggests there may be a basic structural similarity between the prokaryotic and eukaryotic RNA polymerases. However, the fact that rifamycin itself cannot interact with eukaryotic enzymes indicates that the subunit to which this drug binds is not identical in prokaryotes and eukaryotes.

Two other antibiotics which bind to the β subunit of *E. coli* RNA polymerase are streptovaricin (Heil and Zillig, 1970) and streptolydigin (Iwakura *et al.*, 1973). Although binding sites for both rifamycin and streptolydigin are carried on the β subunit, they are clearly located on different parts of the β polypeptide (Ghysen and Pironio, 1972; Iwakura *et al.*, 1973). In contrast to the situation with *E. coli* enzyme, the streptolydigin- and rifamycin-binding sites of *Bacillus subtilis* RNA polymerase are carried on two separate polypeptides (Halling *et al.*, 1978). A mutation in the second largest polypeptide of the *B. subtilis* RNA polymerase is responsible for streptolydigin resistance, whereas a mutation in the largest subunit is responsible for rifamycin resistance. These studies suggest that, even among prokaryotes, divergence of function of individual polypeptides has occurred. Inhibition of RNA synthesis by streptolydigin occurs only when relatively high concentrations of this antibiotic (severalfold greater than rifamycin) are used. Streptolydigin does not inhibit initiation of RNA synthesis, but rather affects the elongation process (Siddhikol *et al.*, 1969; Schleif, 1969), apparently by reducing the rate of phosphodiester bond formation (Cassani *et al.*, 1971). These studies indicate that subunit β may play a role in the elongation reaction. It is of interest to note that high concentrations of the rifamycin derivative AF/013 can also inhibit the RNA chain elongation reaction catalyzed by rat liver polymerase II (Juhasz *et al.*, 1972; Rose *et al.*, 1975). Further evidence supporting a role for subunit β in RNA chain initiation and/or elongation is that β contains the binding site for nucleoside triphosphate substrates (Ishihama, 1972). Although both β and β' can bind the nucleoside analogue, 5-formyluridine triphosphate, this triphosphate binds only to β in the presence of template (Armstrong *et al.*, 1974), again indicating that β contains the substrate binding site.

3. Subunit α

A functional role for the smallest subunit of RNA polymerase, α, has not yet been fully elucidated. No inhibitors or antibiotics have been shown to interact directly with α. This subunit appears to be the most abundant polypeptide of RNA polymerase in the cell and can be found in free or dimeric form in cell extracts (Ito and Ishihama, 1975). Peptide mapping of α from a variety of bacteria suggests that this polypeptide is highly conserved (Lipkin *et al.*, 1976). Indirect evidence indicates that α may play a role in selection of promotor sites for transcription. This conclusion stems mainly from the finding that *E. coli* RNA polymerase molecules, modified as a result of bacteriophage T4 infection preferentially transcribe certain T4 genes. A major modification which occurs with T4 infection is ADP ribosylation of α.

Genetic studies in *E. coli* have indicated that the synthesis of β and β' subunits are closely coordinated. In fact, it appears that both genes are transcribed as a single polycistronic mRNA (Matzura *et al.*, 1971), sharing a common promotor. Synthesis of subunit α is not coordinated with β and β', and the gene for α, distant from the $\beta + \beta'$ transcription unit, is located in a group of ribosomal protein genes (Jaskunas *et al.*, 1975) with which it shares a common promotor (Jaskunas *et al.*, 1976). The location of the α gene amidst ribosomal genes suggests that its transcription may be closely coordinated with ribosomal protein synthesis. Recently, a mutation in the *rpoA* 109 locus, which codes for α, has been described (Fujiki *et al.*, 1976). This transversion, in which a histidine has been replaced by leucine, results in loss of capacity of the mutant strain to support growth of P2 phage. These studies suggest that a phage-specified protein must interact with α subunit to allow late phage expression. Sequence analysis of α (Ovchinnikov *et al.*, 1977) has permitted precise localization of the arginine residue which is ADP ribosylated upon T4 infection (Goff and Weber, 1971; Seifart *et al.*, 1971; Goff, 1974) at position 265 and of the *rpoA* 109 mutation at position 289 or 290. It thus appears that of the 328 amino acid residues of α, the region 265–290 is involved in expression of bacteriophage genes.

B. Reconstitution of RNA Polymerase

Although studies on individual subunits have led to some understanding of the basic properties of each polypeptide, such investigations do not comment on those properties which might require prior interaction with other subunits. Analysis of the sequence assembly of DNA-dependent RNA polymerase *in vitro* by mixing individual subunits obtained from dissociated enzyme has led to a partial understanding of the role of each subunit in the active enzyme. Reconstitution of active enzyme from subunits prepared by dissociation in $6\,M$ urea was first achieved in 1970 (Heil and Zillig, 1970; Lill and Hartmann, 1970). More recent work (Ishihama and Ito, 1972; Ishihama *et al.*, 1973; Yarbrough and Hurwitz, 1974; Palm *et al.*, 1975; Lill *et al.*, 1975; Saitoh and Ishihama, 1976), following the kinetics of subunit assembly, has established that RNA polymerase is reconstructed in the following manner

$$2\,\alpha \rightarrow \alpha_2 \xrightarrow{\beta} \alpha_2\beta \xrightarrow{\beta'} \alpha_2\beta\beta' \quad \text{(premature core)} \rightarrow E \quad \text{(active core)}$$

Reconstitution of an active RNA polymerase requires the presence of all subunits. The α subunit forms a complex with β prior to combination with β'. Using appropriate conditions, fully active enzyme can be reconstituted in the absence of σ subunit or DNA (Harding and Beychock, 1974; Saitoh and Ishihama, 1976). However, subunit σ and DNA help promote reconstitution of an active molecule. In the absence of σ, the primary complex of premature core enzyme can be activated by elevated temperature. Recent evidence (Taketo and Ishihama, 1977) indicates that the assembly of RNA polymerase *in vivo* occurs in a similar sequence to that observed *in vitro*.

Reconstitution studies have indicated that ω, a polypeptide of molecular weight 9000–12,000 found in varying amounts in preparations of purified RNA polymerase (Burgess *et al.*, 1969; Berg *et al.*, 1971), is not essential for enzyme activity *in vitro* (Heil and Zillig, 1970). Reconstitution studies have also demonstrated that the β' subunit carries the DNA binding site and determines ionic strength dependence and pH optima for the enzyme (Khesin *et al.*, 1969; Panny *et al.*, 1974; Young and Wright, 1977). Reconstitution of active core enzyme from the heterologous subunits obtained from *E. coli* and *Micrococcus luteus* RNA polymerases has been achieved (Lill *et al.*, 1975), indicating the close similarity of these enzymes. However, heterologous reconstitution of RNA polymerase subunits from *E. coli* and *Anacystis nidulans* does not yield active enzyme hybrids (Herzfeld and Kiper, 1976). These latter studies suggest a divergence of enzyme between bacteria and blue-green algae.

V. INTERRELATIONSHIP OF RNA POLYMERASES

A. RNA Polymerases as Metalloproteins

DNA-dependent RNA polymerases from all sources examined to date are zinc-containing proteins with metal contents ranging from 1 to 7 gm atoms/mole of enzyme (see Table II). Each mole of RNA polymerase from *E. coli* contains 2 moles of zinc (Scrutton *et al.*, 1971). At least one of the zinc ions is associated with the β' subunit. Replacement of zinc with cobalt yields an RNA polymerase which is fully active and yet initiates at different promotor regions on the template (Wu *et al.*, 1977).

These observations indicate that zinc is involved in DNA binding and further confirm a function for β' in template recognition. The second mole of zinc in *E. coli* RNA polymerase is probably associated with the β subunit or located between β and β' on the core enzyme (Wu *et al.*, 1977). Since zinc has been found to be associated with many nucleic acid-polymerizing enzymes, some of which do not bind to a template (Rose *et al.*, 1978), it is possible that the second mole of zinc in the *E. coli* enzyme plays a more general function in the RNA-synthesizing reaction. Zinc has been implicated in a number of RNA chain elongation reactions (Coleman, 1974; Lattke and Weser, 1977; Halling *et al.*, 1977; Rose *et al.*, 1978): it is, therefore, conceivable that zinc associated with subunit β might have such a role, as this subunit contains the nucleoside triphosphate-binding site.

The zinc content of RNA polymerases from two lower eukaryotes and one plant (see Table II) has also been investigated. We have recently observed that RNA polymerase I from a higher eukaryote (rat hepatoma) also contains zinc, with a metal content of approximately 2 gm atoms/mole of enzyme (K. M. Rose, I. L. Crawford, and S. T. Jacob, unpublished observations). As yet, no information is available as to which subunit(s) of eukaryotic RNA polymerase(s) contains the metal.

TABLE II

Zinc Content of DNA-Dependent RNA Polymerases

Enzyme	Zinc content (gm atom/mole)	Reference
Prokaryotes		
E. coli	2	Scrutton *et al.* (1971)
		Wu *et al.* (1977)
Coliphage T7	2–4	Coleman (1974)
B. subtilis	2	Halling *et al.* (1977)
Eukaryotes		
Euglena gracilis		
RNA polymerase I	2	Falchuk *et al.* (1977)
RNA polymerase II	2	Falchuk *et al.* (1976)
Yeast		
RNA polymerase I	2	Auld *et al.* (1976)
RNA polymerase II	1	Lattke and Weser (1976)
RNA polymerase III	2	Wandzilak and Benson (1978)
Wheat germ		
RNA polymerase II	7	Petranyl *et al.* (1977)

B. Immunological Characteristics

1. Prokaryotes

Preparation of antibodies to RNA polymerase and analysis of immunoprecipitates formed from such antibodies provide a powerful tool to evaluate structural–functional relationships among individual subunits of RNA polymerase and between enzymes from various sources. Antibodies prepared against *E. coli* holoenzyme ($\alpha_2\beta\beta'\sigma$) have been shown to react not only with holoenzyme and core enzyme but also against the individual subunits (Fukuda *et al.*, 1974). These studies indicate that at least some antigenic determinants of each subunit are exposed on the holoenzyme during antibody production. Fukuda *et al.* also found that antibodies against holoenzyme containing an altered σ (σ') are distinct from those against normal holoenzyme. Recently, Fukuda *et al.* (1977) have compared the antigenic relationship of RNA polymerase from a number of bacteria. RNA polymerases from *Salmonella typhimurium, Salmonella anatum, Serratia marcescens, Aerobacter aerogenes,* and *Proteus microbilis* are immunologically and structurally similar to enzyme from *E. coli*. In particular, the β subunit of the enzyme in all cases appears to be identical. In contrast, RNA polymerase from *B. subtilis* is immunologically and structurally distinct from *E. coli* enzyme, the major differences being in the large subunits. Other reports (Halling *et al.*, 1978) have also established a significant divergence in the β and β' subunits between these two bacteria.

2. Eukaryotes

Antibodies against nuclear RNA polymerases I and II have been prepared. Antiserum against RNA polymerase I_A purified from calf thymus (Kedinger *et al.*, 1974) can inhibit the RNA polymerase I activity from several mammalian sources, indicating antigenic similarity among these enzymes. *Escherichia coli* RNA polymerase activity is not affected by anti-polymerase I serum. Although the activity of purified RNA polymerase II is not inhibited by anti-polymerase I serum, partially purified fractions of polymerase II are sensitive to these antibodies. In a study of partially purified RNA polymerases I, II, and III, the order of inhibition by anti-polymerase I serum is I > III > II, indicating that enzymes I and III are more closely related antigenically than polymerases I and II (Hossenlopp *et al.*, 1975).

RNA polymerase II antibodies have been raised by injection of hens with purified calf thymus RNA polymerase II (Ingles, 1973). As with anti-polymerase I, anti-polymerase II inhibits RNA polymerase II activity from a number of mammalian sources, indicating antigenic homology between species. RNA polymerase II from eukaryotes more distant in the phylogenetic scale, such as *Xenopus laevis* and *Tetrahymena pyriformis,* is less sensitive to antibody neutralization. Anti-polymerase II serum can inhibit RNA polymerase I and III activities from the same source (Guialis *et al.,* 1977), albeit to a lesser extent than inhibition of enzyme II from other mammalian tissues. *Escherichia coli* RNA polymerase activity is not affected by anti-polymerase II serum, nor is RNA polymerase II affected by antiserum to *E. coli* enzyme (Ingles, 1973).

A sensitive radioimmunoassay has been developed to quantitate the amount of RNA polymerase II (Guialis *et al.,* 1977). This method of determining the number of enzyme molecules agrees well with the method based on the specific binding of labeled amanitin to polymerase II (Wieland and Fahrmeir, 1970; Cochet-Meilhac *et al.,* 1974b). Using the radioimmunoassay, it has been shown that the increased activity of amanitin-resistant polymerase II in amanitin-resistant hybrids of Chinese hamster ovary cells is due to accumulation of more toxin-resistant enzyme molecules (Guialis *et al.,* 1977).

Antibodies directed against hexaploid wheat RNA polymerase II react with the enzymes purified from dicotyledonous plants as well as monocotyledonous species, but they do not cross-react with yeast RNA polymerase II or *E. coli* RNA polymerase (Jendrisak and Guifoyle, 1978). Thus, not unexpectedly, plant class II enzymes are antigenically distinct from other RNA polymerases.

In summary, it appears that nuclear RNA polymerases of the same class are antigenically similar, although in species that are phylogenetically remote, this relationship diminishes or disappears. Furthermore, enzymes of the three classes within a species appear to be antigenically related, perhaps on account of the presence of some small molecular weight polypeptides which are highly conserved. Finally, bacterial RNA polymerase bears no antigenic similarity to nuclear RNA polymerases I or II.

C. Posttranslational Modifications

Modification of proteins by phosphorylation in the serine, threonine, and perhaps histidine residues has been reported in a number of systems. Both prokaryotic and eukaryotic RNA polymerases appear to undergo such a modification. Phosphorylation of σ factor in *E. coli* RNA polymerase has been correlated with its ability to recognize the promotor sites of T4 DNA (Martelo *et al.*, 1974). *Escherichia coli* core polymerase is phosphorylated as a result of bacteriophage T7 infection, with modification of both β' and β (Zillig *et al.*, 1975). In the case of T7 infection, phosphorylation of core polymerase may well be the primary factor in the shutdown of host gene transcription. Thus, as a result of bacteriophage infection, phosphorylation may play either a positive or negative role in control of gene expression.

Although there have been several reports on the phosphorylation of eukaryotic RNA polymerases *in vitro* using the endogenous protein kinases (Jacob and Rose, 1978), only one example of phosphorylation *in vivo* has been reported (Bell *et al.*, 1976). These investigators grew yeast in complete medium containing [^{32}P]phosphate and subsequently purified RNA polymerase I to homogeneity. Polyacrylamide gel electrophoresis under denaturing conditions revealed that five of the eleven subunits of the yeast polymerase I can be phosphorylated *in vivo*. The authors have speculated that phosphorylated RNA polymerase I may have an altered specific activity or turnover rate than the unphosphorylated enzyme. The phosphorylated enzyme may also interact with other specific nonhistones in a manner which might provide some control of transcription.

RNA polymerases can also be modified by ADP ribosylation. The α polypeptide(s) of *E. coli* RNA polymerase can be ADP ribosylated within 4 min after infection by bacteriophage T4 (Goff, 1974). The modification involves covalent attachment of one adenine nucleotide derived from NAD to one of the α polypeptides and results in an altered electrophoretic mobility of the subunit on polyacrylamide gels. Adenosine diphosphoribose is linked through its terminal ribose to the guanido nitrogen of arginine at the sequence threonine–valine–arginine, located at amino acid 265 in the α chain (Ovchinnikov *et al.*, 1977). T4 mutants which are unable to modify α still grow normally, indicating that ADP ribosylation is not crucial for phage production. However, the specificity of transcription by host enzyme appears to be altered upon ADP ribosylation of α (Mailhammer *et al.*, 1975).

Recently, RNA polymerase I from quail oviduct has been shown to be ADP ribosylated (Müller and Zahn, 1976), with a parallel reduction in the chromatin-associated RNA polymerase I activity. Progesterone administration *in vivo* decreases the specific activity of poly(ADP-ribose) polymerase, which is concomitant with increased RNA polymerase I activity in the quail oviduct nuclei (Müller and Zahn, 1976). It seems plausible that poly(ADP) ribosylation of RNA polymerase I may be one of the regulatory mechanisms by which the specificity of rDNA transcription is achieved.

VI. CONCLUSIONS

In spite of several disparities between prokaryotes and eukaryotes in the control of gene transcription, striking similarities exist in the basic enzymology of the transcriptive process. These include the following.

1. The RNA synthesizing reaction itself. Although the interaction between enzyme and template may be controlled by different mechanisms in nucleated and nonnucleated organisms, the basic catalytic properties of all RNA polymerases are remarkably similar.

2. The multisubunit structure of RNA polymerases. The majority of enzymes must transcribe information from complex templates and are comprised of several subunits. RNA polymerases which are restricted to transcription of relatively simple templates may be comprised of a single polypeptide chain. The existence of such enzymes in eukaryotes (mitochondria), as well as prokaryotes (after infection by certain bacteriophages), suggests that the capacity of a single polypeptide to catalyze DNA-dependent RNA synthesis has been preserved throughout evolution. The fact that most RNA polymerases from both prokaryotes and eukaryotes contain several subunits may well reflect the adaptations necessary to transcribe a variety of genetic information.

3. The metalloprotein nature of the enzymes. All RNA polymerases investigated to date appear to contain zinc. The presence of this metal in a number of other polymerases, e.g., RNA- and DNA-dependent DNA polymerases, tRNA nucleotidyltransferase, and poly(A) polymerase (Rose et al., 1978) suggests that zinc may be common to all nucleic acid polymerizing enzymes.

4. Sensitivity to selected drugs. The polyanion heparin appears to inhibit all DNA-dependent RNA polymerases by binding to the enzyme prior to RNA polymerase–DNA complex formation. Although rifamycin is a selective inhibitor of bacterial RNA polymerase, polar derivatives of this drug, e.g., AF/013, have the capacity to bind to all RNA polymerases. These studies tend to indicate some structural similarities between the DNA-dependent RNA polymerases from all sources.

5. Posttranslational modifications of RNA polymerases. Despite a limited amount of information on this subject, it is clear that RNA polymerase from both prokaryotes and eukaryotes can be modified posttranslationally and that these modifications can play a role in regulation of gene expression.

Calculations have been made as to the number of RNA polymerase molecules per cell. It has been estimated that an *E. coli* cell contains anywhere from 1300 (Burgess, 1971) to 4000 (Matzura *et al.*, 1973) molecules of enzyme. Based on amatoxin binding, each animal cell contains 4000 to 40,000 molecules of RNA polymerase II, depending on the metabolic state (Cochet-Meilhac *et al.*, 1974b). RNA polymerase II most probably represents 50–80% of the total nuclear RNA polymerases in animal cells; thus, an animal cell could contain up to 80,000 molecules of RNA polymerase. Although eukaryotes contain more enzyme molecules per cell, this is not the case if calculated per unit DNA. Since animal cells contain approximately 500 times more DNA than *E. coli,* the number of molecules of RNA polymerase in bacteria and animals might average 2000 and 160, respectively, for 3×10^6 base pairs of DNA. It must be borne in mind, however, that generally only 5–10% of the DNA in higher organisms is available for transcription at any particular time. Hence, the number of RNA polymerase molecules per unit of *transcriptionally active* DNA is similar in bacteria and animal cells.

Despite the manifestation of a number of common properties of RNA polymerases from most sources, it is clear that there are a several structural and immunological differences between these enzymes. Even within a single eukaryote, the multiple species of RNA polymerases are distinct. These structural variations between enzymes may represent the results of slow evolutionary diversion. Perhaps the most important consequence of such adaptations is that all RNA polymerases might not be subject to exactly the same controls.

Studies on the control of transcription in eukaryotes are still in the infant stage. No specific σ-like factors have been characterized in eukaryotes. Isolation and characterization of such regulatory factors in mammalian gene transcription will be a challenging problem for the future. Finally, the role of individual subunits of eukaryotic RNA polymerases, particularly of the larger subunits, in the transcriptional process must be elucidated.

ACKNOWLEDGMENTS

We are grateful to Mrs. Eileen Drust for her assistance in the preparation of this chapter. Studies carried out in our laboratory were supported by United States Public Health Service Grant CA-16438 and National Science Foundation Grants PCM 75-19768 and PCM 76-82224, American Cancer Society Institutional Grant IN-109, and a specialized Cancer Research Center Grant 1 P030 CA18450.

GENERAL REFERENCES

Biswas, B. B., Ganguly, A., and Das, A. (1975). Eukaryotic RNA polymerases and the factors that control them. *Prog. Nucl. Acid Res. Mol. Biol.* **15,** 145–184.

Burgess, R. R. (1971). RNA polymerase. *Annu. Rev. Biochem.* **40**, 711–740.
Chamberlin, M. J. (1974). Bacterial DNA-dependent RNA polymerase. *In* "The Enzymes" (P. D. Boyer, ed.), 3rd Ed., Vol. 10, pp. 333–374. Academic Press, New York.
Chamberlin, M. J. (1974). The selectivity of transcription. *Annu. Rev. Biochem.* **43**, 721–775.
Chambon, P. (1974). Eukaryotic RNA polymerases. *In* "The Enzymes" (P. D. Boyer, ed.), 3rd ed., Vol. 10, pp. 261–331. Academic Press, New York.
Chambon, P. (1975). Eukaryotic nuclear RNA polymerases. *Annu. Rev. Biochem.* **44**, 613–638.
Chambon, P., Gissinger, F., Kedinger, C., Mandel, J. L., and Meilhac, M. (1974). Animal nuclear DNA-dependent RNA polymerases. *In* "The Cell Nucleus" (H. Busch, ed.), pp. 270–307. Academic Press, New York.
Jacob, S. T. (1973). Mammalian RNA polymerases. *Prog. Nucl. Acid Res. Mol. Biol.* **13**, 93–126.
Jacob, S. T., and Rose, K. M. (1978). RNA polymerases and poly(A) polymerase from neoplastic tissues and cells. *Methods Cancer Res.* **14**, 191–241.
Lindell, T. J. (1977). Inhibitors of mammalian RNA polymerases. *Pharmacol. Ther. A* **2**, 195–225.
Losick, R., and Chamberlin, M., eds., (1976). "RNA Polymerase." Cold Spring Harbor Lab., Cold Spring Harbor, New York.
Sethi, V. S. (1971). Structure and function of DNA-dependent RNA polymerase. *Prog. Biophys. Mol. Biol.* **23**, 67–101.
Silvestri, L., ed. (1970). *Lepetit Colloq. Biol. Med.* **1**.
Transcription of Genetic Material (1970). *Cold Spring Harbor Symp. Quant. Biol.* **35**.

REFERENCES

Amati, P., Blasi, F., di Porzio, U., Riccio, A., and Traboni, C. (1975). Hamster α-amanitin-resistant RNA polymerase II able to transcribe polyoma virus genome in somatic cell hybrids. *Proc. Natl. Acad. Sci. U.S.A.* **72**, 753–757.
Armstrong, V., Sternbach, H., and Eckstein, F. (1974). Attempts to affinity label DNA-dependent RNA polymerase with formyl-UTP. *Fed. Eur. Biochem. Soc., Abstr. Commun., 9th Meet., Budapest* p. 176.
Auld, D. S., Atsuya, I., Campino, C., and Valenzuela, P. (1976). Yeast RNA polymerase I: A eukaryotic zinc metalloenzyme. *Biochem. Biophys. Res. Commun.* **69**, 548–554.
Austin, G. E., Bello, L. J., and Furth, J. J. (1973). DNA-dependent RNA polymerase of KB cells. Isolation of the enzymes and transcription of viral DNA. *Biochim. Biophys. Acta* **324**, 488–500.
Austoker, J. L., Beebee, T. J. C., Chesterton, C. J., and Butterworth, P. H. W. (1974). DNA-dependent RNA polymerase activity of Chinese hamster kidney cells sensitive to high concentrations of α-amanitin. *Cell* **3**, 227–236.
Babcock, F., and Rich, A. (1973). Deoxyribonucleic acid-dependent ribonucleic acid polymerases from murine spleen cells. Increased amounts of the nucleolar species in leukaemic tissue. *Biochem. J.* **133**, 797–804.
Ballal, N. R., Choi, Y. C., Mouche, R., and Busch, H. (1977). Fidelity of synthesis of preribosomal RNA in isolated nucleoli and nucleolar chromatin. *Proc. Nat. Acad. Sci. U.S.A.* **74**, 2446–2450.
Bautz, E. K. F., and Bautz, F. A. (1970). Initiation of RNA synthesis: The function of σ in the binding of RNA polymerase to promotor sites. *Nature (London)* **226**, 1219–1222.

Beebee, T. J. C., and Butterworth, P. H. W. (1976). The use of mercurated nucleoside triphosphate as a probe in transcription studies *in vitro*. *Eur. J. Biochem.* **66**, 543–550.

Bell, G. I., Valenzuela, P., and Rutter, W. J. (1976). Phosphorylation of yeast RNA polymerases. *Nature (London)* **261**, 429–431.

Berg, D., Barrett, K., and Chamberlin, M. (1971). Purification of two forms of *E. coli* RNA polymerase. *In* "Nucleic Acids," Part D (E. Grossman and K. Moldave, eds.), Methods in Enzymology, Vol. 21, pp. 506–519. Academic Press, New York.

Biswas, D. K. (1978). RNA–protein interactions in a cell-free system with isolated nuclei. *Biochemistry* **17**, 1131–1136.

Bitter, G. A., and Roeder, R. G. (1978). Transcription of viral genes by RNA polymerase II in nuclei isolated from adenovirus 2 transformed cells. *Biochemistry* **17**, 2198–2205.

Bottomley, W., Smith, H. J., and Bogorad, L. (1971). RNA polymerases of maize: Partial purification and properties of the chloroplast enzyme. *Proc. Natl. Acad. Sci. U.S.A.* **68**, 2412–2416.

Braun, T., and Hechter, O. (1970). Glucocorticoid regulation of ACTH sensitivity of adenyl cyclase in rat fat cell membranes. *Proc. Natl. Acad. Sci. U.S.A.* **66**, 995–1001.

Brodner, O. G., and Wieland, O. (1976). Identification of the amatoxin-binding subunit of RNA polymerase B by affinity labeling experiments. Subunit B_3—The true amatoxin receptor protein of multiple RNA polymerase B. *Biochemistry* **15**, 3480–3484.

Brown, R. D., Bastia, D., and Haselkorn, R. (1970). Effect of rifampicin on transcription in chloroplasts of *Euglena*. *Lepetit Colloq. Biol. Med.* **1**, 309–328.

Bryant, R. E., Adelberg, E. A., and Magee, P. T. (1977). Properties of an altered RNA polymerase II activity from an α-amanitin-resistant mouse cell line. *Biochemistry* **16**, 4237–4244.

Burgess, R. R. (1969). Separation and characterization of the subunits of ribonucleic acid polymerase. *J. Biol. Chem.* **244**, 6168–6176.

Burgess, R. R. (1971). RNA polymerase. *Annu. Rev. Biochem.* **40**, 711–740.

Burgess, R. R., and Jendrisak, J. J. (1975). A procedure for the rapid, large-scale purification of *Escherichia coli* DNA-dependent RNA polymerase involving Polymin P precipitation and DNA–cellulose chromatography. *Biochemistry* **14**, 4634–4638.

Burgess, R. R., Travers, A. A., Dunn, J. J., and Bautz, E. K. F. (1969). Factor stimulating transcription by RNA polymerase. *Nature (London)* **221**, 43–46.

Cassani, G., Burgess, R. R., Goodman, H. M., and Gold, L. (1971). Inhibition of RNA polymerase by streptolydigin. *Nature (London), New Biol.* **230**, 197–200.

Chamberlin, M. J. (1974). The selectivity of transcription. *Annu. Rev. Biochem.* **43**, 721–725.

Chamberlin, M. J., McGrath, J., and Waskell, L. (1970). New RNA polymerase from *Escherichia coli* infected with bacteriophage T₇. *Nature (London)* **228**, 227–231.

Chambon, P. (1974). Eucaryotic RNA polymerases. *In* "The Enzymes" (P. D. Boyer, ed.), 3rd Ed., Vol. 10, pp. 261–331. Academic Press, New York.

Chan, V. L., Whitmore, G. F., and Siminovitch, L. (1972). Mammalian cells with altered forms of RNA polymerase II. *Proc. Natl. Acad. Sci. U.S.A.* **69**, 3119–3123.

Cochet-Meilhac, M., and Chambon, P. (1974a). Animal DNA-dependent RNA polymerases. Mechanism of the inhibition of RNA polymerases B by amatoxins. *Biochim. Biophys. Acta* **353**, 160–184.

Cochet-Meilhac, M., and Chambon, P. (1974b). Animal DNA-dependent RNA polymerases. Determination of the cellular number of RNA polymerase B molecules. *Biochim. Biophys. Acta* **353**, 185–192.

Coleman, J. E. (1974). The role of Zn(II) in transcription by T₇ RNA polymerase. *Biochem. Biophys. Res. Commun.* **60**, 641–648.

Cooper, R. J., and Keir, H. M. (1975). Deoxyribonucleic acid-dependent ribonucleic acid polymerases from normal and polyoma-tranformed BHK-21/C13 cells. *Biochem. J.* **145**, 509–516.

Cox, R. (1973). Transcription of high-molecular-weight RNA from hen-oviduct chromatin by bacterial and endogenous form-B RNA polymerases. *Eur. J. Biochem.* **39**, 49–61.

Daubert, S., Peters, D., and Dahmus, M. E. (1977). Selective transcription of ribosomal sequences *in vitro* by RNA polymerase I. *Arch. Biochem. Biophys.* **178**, 381–386.

Duffy, J. J., and Geiduschek, E. R. (1975). RNA polymerase from phage SP01-infected and uninfected *Bacillus subtilis*. *J. Biol. Chem.* **250**, 4530–4541.

Dunn, J. J., Bautz, F. A., and Bautz, E. K. F. (1970). Different template specificities of phage T_3 and T_7 RNA polymerases. *Nature (London), New Biol.* **230**, 94–96.

Falchuk, K. H., Mazus, B., Ulpino, L., and Vallee, B. L. (1976). *Euglena gracilis* DNA-dependent RNA polymerase II: A zinc metalloenzyme. *Biochemistry* **15**, 4468–4475.

Falchuk, K. H., Ulpino, L., Mazua, B., and Vallee, B. L. (1977). *E. gracilis* RNA polymerase I: A Zinc metalloenzyme. *Biochem. Biophys. Res. Commun.* **74**, 1206–1212.

Ferencz, A., and Seifart, K. H. (1975). Comparative effect of heparin on RNA synthesis of isolated rat-liver nucleoli and purified RNA polymerase A. *Eur. J. Biochem.* **53**, 605–613.

Fox, T. D., and Pero, J. (1974). New phage-SP01-induced polypeptides associated with *Bacillus subtilis* RNA polymerase. *Proc. Natl. Acad. Sci. U.S.A.* **71**, 2761–2765.

Frishauf, A. M., and Scheit, K. H. (1973). Affinity labeling of *E. coli* RNA polymerase with substrate and template analogues. *Biochem. Biophys. Res. Commun.* **53**, 1227–1233.

Froehner, S. C., and Bonner, J. (1973). Ascites tumor ribonucleic acid polymerases. Isolation, purification and factor stimulation. *Biochemistry* **12**, 3064–3071.

Fujiki, H., and Zurek, G. (1975). The subunits of DNA-dependent RNA polymerase from *E. coli:* Amino acid analysis and primary structure of the N-terminal regions. *FEBS Lett.* **55**, 242–244.

Fujiki, H., Palm, P., Zillig, W., Calendar, R., and Sunshine, M. (1976). Identification of a mutation within the structural gene for the α subunit of DNA-dependent RNA polymerase of *E. coli*. *Mol. Gen. Genet.* **145**, 19–22.

Fukuda, R., and Ishihama, A. (1974). Subunits of RNA polymerase in function and structure; maturation *in vitro* of core enzyme from *Escherichia coli*. *J. Mol. Biol.* **87**, 523–540.

Fukuda, R., Iwakura, Y., and Ishihama, A. (1974). Heterogeneity of RNA polymerase in *Escherichia coli*. A new holoenzyme containing a new sigma factor. *J. Mol. Biol.* **83**, 353–367.

Fukuda, R., Ishihama, A., Saitoh, T., and Taketo, M. (1977). Comparative studies of RNA polymerase subunits from various bacteria. *Mol. Gen. Genet.* **154**, 135–144.

Gallerani, R., Saccone, P., Cantatore, P., and Gadaleta, M. N. (1972). DNA-dependent RNA polymerase from rat liver mitochondria. *FEBS Lett.* **22**, 37–40.

Gerard, G., Johnson, J., and Boezi, J. (1972). Release of the σ subunit of *Pseudomonas putida* deoxyribonucleic acid dependent ribonucleic acid polymerase. *Biochemistry* **11**, 989–997.

Ghysen, A., and Pironio, M. (1972). Relationship between the N function of bacteriophage λ and host RNA polymerase. *J. Mol. Biol.* **65**, 259–272.

Gilbert, W. (1976). Starting and stopping sequences for the RNA polymerase. *In* "RNA Polymerase" (R. Losick and M. Chamberlin, eds.), pp. 193–205. Cold Spring Harbor Lab., Cold Spring Harbor, New York.

Gissinger, F., Kedinger, C., and Chambon, P. (1974). Animal DNA-dependent RNA polymerases. General enzymatic properties of purified calf thymus RNA polymerases A_1 and B. *Biochimie* **56**, 319–333.

Goff, C. G. (1974). Chemical structure of a modification of the *Escherichia coli* ribonucleic acid polymerase alpha polypeptides induced by bacteriophage T₄ infection. *J. Biol. Chem.* **249**, 6181–6190.

Goff, C. G., and Weber, K. (1971). A T₄-induced RNA polymerase α subunit modification *Cold Spring Harbor Symp. Quant. Biol.* **35**, 101–108.

Goldberg, M., Perriard, J. C., and Rutter, W. J. (1977). A protein factor that stimulates the activity of DNA-dependent RNA polymerase I on double-stranded DNA. *Biochemistry* **16**, 1648–1654.

Gonzalez, N., Wiggs, J., and Chamberlin, M. J. (1977). A simple procedure for resolution of *Escherichia coli* RNA polymerase holoenzyme from core polymerase. *Arch. Biochem. Biophys.* **182**, 404–408.

Gross, G., Fields, D. A., and Bautz, E. K. F. (1976). Characterization of a ts beta mutant RNA polymerase of *Escherichia coli*. *Mol. Gen. Genet.* **147**, 337–341.

Guialis, A., Beatty, B. G., Ingles, C. J., and Crerar, M. M. (1977). Regulation of RNA polymerase II activity in alpha-amanitin resistant CHO hybrid cells. *Cell* **10**, 53–60.

Halling, S. M., Sanchez-Anzaldo, F. J., Fukuda, R. H., and Meares, C. F. (1977). Zinc is associated with the β subunit of DNA-dependent RNA polymerase of *Bacillus subtilis*. *Biochemistry* **16**, 2880–2884.

Halling, S. M., Burtis, K. C., and Doi, R. H. (1978). β' subunit of bacterial RNA polymerase is responsible for streptolydigin resistance in *Bacillus subtilis*. *Nature (London)* **272**, 837–839.

Harding, J. D., and Beychock, S. (1974). RNA polymerase assembly *in vitro*. Temperature dependence of reactivation of denatured core enzyme. *Proc. Natl. Acad. Sci. U.S.A.* **71**, 3395–3399.

Hartmann, G., Honikel, K. O., Knüsel, F., and Nüesch, J. (1967). The specific inhibition of the DNA-directed RNA synthesis by rifamycin. *Biochim. Biophys. Acta* **145**, 843–844.

Heil, A., and Zillig, W. (1970). Reconstitution of bacterial DNA-dependent RNA polymerase from isolated subunits as a tool for the elucidation of the role of the subunits in transcription. *FEBS Lett.* **11**, 165–168.

Herzfeld, F., and Kiper, M. (1976). The reconstitution of *Anacystis nidulans* DNA-dependent RNA polymerase from its isolated subunits. *Eur. J. Biochem.* **62**, 189–192.

Hinkle, D. C., Mangel, W. F., and Chamberlin, M. J. (1972). Studies of the binding of *Escherichia coli* RNA polymerase to DNA. The effect of rifampicin on binding and on RNA chain initiation. *J. Mol. Biol.* **70**, 209–220.

Holland, M. J., Hager, G. L., and Rutter, W. J. (1977). Transcription of yeast DNA by homologous RNA polymerases I and II: Selective transcription of ribosomal genes by RNA polymerase I. *Biochemistry* **16**, 16–24.

Hossenlopp, P., Wells, D., and Chambon, P. (1975). Animal DNA-dependent RNA polymerases. Partial purification and properties of three classes of RNA polymerases from uninfected and adenovirus-infected HeLa cells. *Eur. J. Biochem.* **58**, 237–251.

Humphries, P., McConnell, D. L., and Gordon, R. L. (1973). A procedure for the rapid purification of *Escherichia coli* deoxyribonucleic acid-dependent ribonucleic acid polymerase. *Biochem. J.* **133**, 201–203.

Ingles, C. J. (1973). Antigenic homology of eukaryotic RNA polymerases. *Biochem. Biophys. Res. Commun.* **55**, 364–371.

Ingles, C. J., Guialis, A., Lam, J., and Siminovitch, L. (1976). Alpha-amanitin resistance of RNA polymerase II in mutant Chinese hamster ovary cell lines. *J. Biol. Chem.* **251**, 2729–2734.

Ishihama, A. (1972). Subunits of ribonucleic acid polymerase in function and structure. Reversible dissociations of *Escherichia coli* ribonucleic acid polymerase. *Biochemistry* **11**, 1250–1258.

Ishihama, A., and Ito, K. (1972). Subunits of RNA polymerase in function and structure. Reconstitution of *E. coli* RNA polymerase from isolated subunits. *J. Mol. Biol.* **72,** 111–123.

Ishihama, A., Fukuda, R., and Ito, K. (1973). Subunits of RNA polymerase in function and structure. Enhancing role of sigma in subunit assembly of *E. coli* RNA polymerase. *J. Mol. Biol.* **79,** 127–136.

Ishihama, A., Taketo, M., Saitoh, T., and Fukuda, R. (1976). Control of formation of RNA polymerase in *Escherichia coli. In* "RNA Polymerase" (R. Losick and M. Chamberlin, eds.), pp. 485–502. Cold Spring Harbor Lab., Cold Spring Harbor, New York.

Ito, K., and Ishihama, A. (1975). Biosynthesis of RNA polymerase in *Escherichia coli.* Identification of intermediates in the assembly of RNA polymerase. *J. Mol. Biol.* **96,** 257–271.

Iwakura, Y., Ishihama, A., and Yura, T. (1973). RNA polymerase mutants of *Escherichia coli.* II. Streptolydigin resistance and its relation to rifampicin resistance. *Mol. Gen. Genet.* **121,** 181–196.

Jacob, S. T. (1973). Mammalian RNA polymerases. *Prog. Nucl. Acid Res. Mol. Biol.* **13,** 93–126.

Jacob, S. T., and Rose, K. M. (1978). RNA polymerases and poly(A) polymerase from neoplastic tissues and cells. *Methods Cancer Res.* **14,** 191–241.

Jacob, S. T., Sajdel, E. M., and Munro, H. N. (1968). Altered characteristics of mammalian RNA polymerase following solubilization from nuclei. *Biochem. Biophys. Res. Commun.* **32,** 831–838.

Jacob, S. T., Sajdel, E. M., and Munro, H. N. (1969). Regulation of nucleolar RNA metabolism by hydrocortisone. *Eur. J. Biochem.* **7,** 449–453.

Jacob, S. T., Sajdel, E. M., and Munro, H. N. (1970a). Different responses of soluble whole nuclear RNA polymerase and soluble nucleolar RNA polymerase to divalent cations and to inhibition by α-amanitin. *Biochem. Biophys. Res. Commun.* **38,** 765–770.

Jacob, S. T., Sajdel, E. M., and Munro, H. N. (1970b). Specific action of α-amanitin on mammalian RNA polymerase protein. *Nature (London)* **225,** 60–62.

Jacob, S. T., Sajdel, E. M., Muecke, W., and Munro, H. N. (1970c). Soluble RNA polymerases of rat liver nuclei: Properties, template specificity, and amanitin responses *in vitro* and *in vivo. Cold Spring Harbor Symp. Quant. Biol.* **35,** 681–691.

Jacob, S. T., Scharf, M. B., and Vesell, E. S. (1974). Role of RNA in induction of hepatic microsomal mixed function oxidases. *Proc. Natl. Acad. Sci. U.S.A.* **71,** 704–707.

Jamrich, M., Greenleaf, A. L., and Bautz, E. K. F. (1977). Localization of RNA polymerase in polytene chromosomes of *Drosophilia melanogaster. Proc. Natl. Acad. Sci. U.S.A.* **74,** 2079–2083.

Jaskunas, S., Burgess, R. R., and Nomura, M. (1975). Identification of a gene for the α subunit of RNA polymerase at the *Str-spc* region of the *Escherichia coli* chromosome. *Proc. Natl. Acad. Sci. U.S.A.* **72,** 5036–5040.

Jaskunas, S. R., Burgess, R. R., Lindahl, L., and Nomura, M. (1976). Two clusters of genes for RNA polymerase and ribosome components in *Escherichia coli. In* "RNA Polymerase" (R. Losick and M. Chamberlin, eds.), pp. 539–552. Cold Spring Harbor Lab., Cold Spring Harbor, New York.

Jendrisak, J., and Guifoyle, T. J. (1978). Eukaryotic RNA polymerases: Comparative subunit structures, immunological properties, and α-amanitin sensitivities of the class II enzymes from higher plants. *Biochemistry* **17,** 1322–1327.

Johnson, J. C., DeBacker, M., and Boezi, J. A. (1971). Deoxyribonucleic acid-dependent ribonucleic acid polymerase of *Pseudomonas putida. J. Biol. Chem.* **246,** 1222–1232.

Juhasz, P. P., Benecke, B. J., and Seifart, K. H. (1972). Inhibition of RNA polymerases from rat liver by the semi-synthetic rifampicin derivatives. *FEBS Lett.* **27,** 30–34.

Kedinger, C., Gniadowski, M., Mandel, J. L., Gissinger, F., and Chambon, P. (1970). α-Amanitin: A specific inhibitor of one of two DNA-dependent RNA polymerase activities from calf thymus. *Biochem. Biophys. Res. Commun.* **38**, 165–171.

Kedinger, C., Gissinger, F., and Chambon, P. (1974). Animal DNA-dependent RNA polymerases. Molecular structures and immunological properties of calf thymus enzyme A₁ and of calf thymus and rat liver enzymes B. *Eur. J. Biochem.* **44**, 421–436.

Khesin, R. B., Gorlenko, Z. M., Shemyakin, M. F., Stvolinsky, S. L., Mindlin, S. Z., and Ilyina, T. S. (1969). Studies on the function of the RNA polymerase components by means of mutations. *Mol. Gen. Genet.* **105**, 243–261.

Krakow, J. S., Rhodes, G., and Jovin, T. M. (1976). RNA polymerase: Catalytic mechanisms and inhibitors. *In* "RNA Polymerase" (R. Losick and M. Chamberlin, eds.) pp. 127–157. Cold Spring Harbor Lab., Cold Spring Harbor, New York.

Küntzel, H., and Schäfer, K. P. (1971). Mitochondrial RNA polymerase *Neurospora crassa*. *Nature (London), New Biol.* **231**, 265–269.

Lampert, A., and Fiegelson, P. (1974). A short lived polypeptide component of one of two discrete functional pools of hepatic nuclear α-amanitin resistant RNA polymerases. *Biochem. Biophys. Res. Commun.* **58**, 1030–1038.

Lattke, H., and Weser, U. (1976). Yeast RNA polymerase B: A zinc protein. *FEBS Lett.* **65**, 288–292.

Lattke, H., and Weser, U. (1977). Functional aspects of zinc in yeast RNA polymerase B. *FEBS Lett.* **83**, 297–300.

Leonard, T. B., and Jacob, S. T. (1977). Alterations in DNA-dependent RNA polymerases I and II from rat liver by thioacetamide. Preferential increase in the level of chromatin-associated nucleolar RNA polymerase IB. *Biochemistry* **16**, 4538–4544.

Lill, U. I., and Hartmann, G. R. (1970). Reactivation of denatured RNA polymerase from *E. coli*. *Biochem. Biophys. Res. Commun.* **39**, 930–935.

Lill, U. I., Behrendt, E. M., and Hartmann, G. R. (1975). Hybridization *in vitro* of subunits of the DNA-dependent RNA polymerase from *E. coli* and *M. luteus*. *Eur. J. Biochem.* **52**, 411–420.

Lin, Y.-C., Rose, K. M., and Jacob, S. T. (1976). Evidence for the nuclear origin of RNA polymerases identified in the cytosol: Release of enzymes from the nuclei isolated in isotonic sucrose. *Biochem. Biophys. Res. Commun.* **72**, 114–120.

Lindell, T. J. (1976). Evidence for an extranucleolar mechanism of actinomycin D action. *Nature (London)* **263**, 347–350.

Lindell, T. J. (1977). Inhibitors of mammalian RNA polymerases. *Pharmacol. Ther. A* **2**, 195–225.

Lindell, T. J., Weinberg, F., Morris, P. W., Roeder, R. G., and Rutter, W. J. (1970). Specific inhibition of nuclear RNA polymerase II by α-amanitin. *Science* **170**, 447–449.

Lipkin, V. M., Modyanov, N. N., Kocherginskaya, S. A., Chertov, O. Y., Nikiforov, V. G., and Lebedev, A. N. (1976). Comparison of peptide maps of α-subunits of RNA polymerases from microorganisms of enterobactereaceal family. *Bioorg. Khim.* **2**, 1174–1181.

Lobban, P. E., Siminovitch, L., and Ingles, C. J. (1976). The RNA polymerase II of an α-amanitin-resistant Chinese hamster ovary cell line. *Cell* **8**, 65–70.

Long, E., Dina, D., and Crippa, M. (1976). DNA-dependent RNA polymerase C from *Xenopus laevis* ovaries. Ability to transcribe intact double-stranded DNA. *Eur. J. Biochem.* **66**, 269–275.

Lowery-Goldhammer, C., and Richardson, J. P. (1974). An RNA-dependent nucleoside triphosphate phosphohydrolase (ATPase) associated with rho termination factor. *Proc. Natl. Acad. Sci. U.S.A.* **71**, 2003–2007.

Mailhammer, R., Yang, H. L., Reiness, G., and Zubay, G. (1975). Effects of bacteriophage

T_4, induced modification of *Escherichia coli* RNA polymerase on gene expression *in vitro*. *Proc. Natl. Acad. Sci. U.S.A.* **72**, 4928–4939.

Maitra, U. (1971). Induction of a new RNA polymerase in *Escherichia coli* infected with bacteriophage T3. *Biochem. Biophys. Res. Commun.* **43**, 443–450.

Maitra, U., and Hurwitz, J. (1965). The role of DNA in RNA synthesis. IX. Nucleoside triphosphate termini in RNA polymerase products. *Proc. Natl. Acad. Sci. U.S.A.* **54**, 815–822.

Maitra, U., and Hurwitz, J. (1967). The role of deoxyribonucleic acid in ribonucleic acid synthesis. *J. Biol. Chem.* **242**, 4897–4907.

Martelo, O. J., Woo, S. L. C., and Davie, E. W. (1974). Phosphorylation of *Escherichia coli* RNA polymerase by rabbit skeletal muscle protein kinase. *J. Mol. Biol.* **87**, 685–696.

Matsui, T., Onishi, T., and Muramatsu, M. (1976). Nucleolar DNA-dependent RNA polymerase from rat liver 2. Two forms and their physiological significance. *Eur. J. Biochem.* **71**, 361–368.

Matzura, H., Molin, S., and Maaløe, O. (1971). Sequential biosynthesis of the β and β' subunits of the DNA-dependent RNA polymerase from *Escherichia coli. J. Mol. Biol.* **59**, 17–25.

Matzura, H., Hansen, B. S., and Zeuthen, J. (1973). Biosynthesis of the β and β' subunits of RNA polymerase in *Escherichia coli. J. Mol. Biol.* **74**, 9–20.

Mehrota, B. D., and Khorana, H. G. (1965). Synthetic deoxyribopolynucleotides as templates for ribonucleic acid polymerase: The influence of temperature on template function. *J. Biol. Chem.* **240**, 1750–1753.

Meilhac, M., and Chambon, P. (1973). Animal DNA-dependent RNA polymerases. Initiation sites on calf-thymus DNA. *Eur. J. Biochem.* **35**, 454–463.

Mitsui, S., Fuke, M., and Busch, H. (1977). Fidelity of ribosomal ribonucleic acid synthesis by nucleoli and nucleolar chromatin. *Biochemistry* **16**, 39–44.

Molloy, G., and Puckett, L. (1976). The metabolism of heterogeneous nuclear RNA in animal cells. *Prog. Biophys. Mol. Biol.* **31**, 1–38.

Müller, W. E. G., and Zahn, R. K. (1976). Poly ADP-ribosylation of DNA-dependent RNA polymerase I from quail oviduct. Dependence on progesterone stimulation. *Mol. Cell. Biochem.* **12**, 147–159.

Nevins, J. R., and Joklik, W. K. (1977). Isolation and properties of the vaccinia virus DNA-dependent RNA polymerase. *J. Biol. Chem.* **252**, 6930–6938.

Nüsslein, C., and Heyden, B. (1972). Chromatography of RNA polymerase from *Escherichia coli* on single-stranded DNA–agarose columns. *Biochem. Biophys. Res. Commun.* **47**, 282–289.

Ovchinnikov, Y. A., Lipkin, V. M., Modyanov, N. N., Chertov, O. Y., and Smirnov, Y. V. (1977). Primary structure of alpha-subunit of DNA-dependent RNA polymerase from *Escherichia coli. FEBS Lett.* **76**, 108–111.

Palm, P., Heil, A., Boyd, D., Grampp, B., and Zillig, W. (1975). The reconstitution of *Escherichia coli* RNA polymerase from its isolated subunits. *Eur. J. Biochem.* **53**, 283–291.

Panny, R., Heil, A., Mazus, B., Palm, P., Zillig, W., Mindlin, S. Z., Ilyana, T. S., and Khesin, R. S. (1974). A temperature sensitive mutation of the beta'-subunit of DNA-dependent RNA polymerase from *E. coli* T16. *FEBS Lett.* **48**, 241–245.

Parker, C, S., and Roeder, R. G. (1977). Selective and accurate transcription of the *Xenopus laevis* 5 S RNA genes in isolated chromatin by purified RNA polymerase III. *Proc. Natl. Acad. Sci. U.S.A.* **74**, 44–48.

Petranyl, P., Jendrisak, J. J., and Burgess, R. R. (1977). RNA polymerase II from wheat germ containing tightly bound zinc. *Biochem. Biophys. Res. Commun.* **74**, 1031–1038.

Pettijohn, D., and Kamiya, T. (1967). Interaction of RNA polymerase with polyoma DNA. *J. Mol. Biol.* **29**, 275–295.

Reid, B. D., and Parsons, P. (1971). Partial purification of mitochondrial RNA polymerase from rat liver. *Proc. Natl. Acad. Sci. U.S.A.* **68**, 2830–2834.

Righthand, V. F., and Bagshaw, J. C. (1974). Increased DNA-dependent RNA polymerase activity from CV-1 cells productively infected by SV40. *Intervirology* **4**, 162–170.

Roberts, J. W. (1969). Termination factor for RNA synthesis. *Nature (London)* **224**, 1168–1174.

Roberts, J. W. (1976). Transcription termination and its control in *E. coli. In* "RNA Polymerase" (R. Losick and M. Chamberlin, eds.), pp. 247–271. Cold Spring Harbor Lab., Cold Spring Harbor, New York.

Roeder, R. G. (1976). Eukaryotic nuclear RNA polymerases. *In* "RNA Polymerase" (R. Losick and M. Chamberlin, eds.), pp. 285–329. Cold Spring Harbor Lab., Cold Spring Harbor, New York.

Roeder, R. G., and Rutter, W. J. (1969). Multiple forms of DNA-dependent RNA polymerase in eukaryotic organisms. *Nature (London)* **224**, 234–237.

Rose, K. M. (1969). Unprimed interdependent polymerization of ITP and CTP by RNA polymerase of *Pseudomonas putida*. M.S. Thesis, Michigan State Univ., East Lansing.

Rose, K. M., Ruch, P. A., and Jacob, S. T. (1975). Mechanism of inhibition of RNA polymerase II and poly(adenylic acid) polymerase by the O-n-octyloxime of 3-formylrifamycin SV. *Biochemistry* **14**, 3598–3603.

Rose, K. M., Allen, M. S., Crawford, I. L., and Jacob, S. T. (1978). Functional role of zinc in poly(A) synthesis catalyzed by nuclear poly(A) polymerase. *Eur. J. Biochem.* **88**, 29–36.

Rosen, O. M., and Rosen, S. M. (1969). Properties of an adenyl cyclase purified from frog erythrocytes. *Arch. Biochim. Biophys.* **131**, 449–456.

Rossini, M., and Baserga, R. (1978). RNA synthesis in a cell cycle-specific temperature sensitive mutant from a hamster cell line. *Biochemistry* **17**, 858–863.

Rothman-Denes, L. B., and Schito, G. C. (1974). Novel transcribing activities in N_4-infected *Escherichia coli. Virology* **60**, 65–72.

Saccone, C., Gallerani, R., Gadaleta, M. N., and Greco, M. (1971). The effect of α-amanitin on RNA synthesis in rat liver mitochondria. *FEBS Lett.* **18**, 339–341.

Saitoh, T., and Ishihama, A. (1976). Subunits of RNA polymerase in function and structure. VI. Sequence of the assembly *in vitro* of *Escherichia coli* RNA polymerase. *J. Mol. Biol.* **104**, 621–635.

Saitoh, T., and Ishihama, A. (1977). Biosynthesis of RNA polymerase from *Escherichia coli*. VI. Distribution of RNA polymerase subunits between nucleoid and cytoplasm. *J. Mol. Biol.* **115**, 403–416.

Scheit, K. H., and Stütz, A. (1975). The affinity of *E. coli* RNA polymerase to matrix bound rifamycin. *FEBS Lett.* **50**, 25–27.

Schleif, R. (1969). Isolation and characterization of streptolydigin resistant RNA polymerase. *Nature (London)* **223**, 1068–1069.

Schultz, L. D., and Hall, B. D. (1976). Transcription in yeast: Alpha-amanitin sensitivity and other properties which distinguish between RNA polymerases I and III. *Proc. Natl. Acad. Sci. U.S.A.* **73**, 1029–1033.

Schwartz, L. R., and Roeder, R. G. (1974). Purification and subunit structure of deoxyribonucleic acid-dependent ribonucleic acid polymerase I from the mouse myeloma MOPC 315. *J. Biol. Chem.* **249**, 5898–5906.

Scragg, A. H. (1971). Mitochondrial DNA-directed RNA polymerase from *Saccharomyces cerevisiae* mitochondria. *Biochem. Biophys. Res. Commun.* **45**, 701–706.

Scrutton, M. D., Wu, C.-W., and Goldthwait, D. A. (1971). The presence and possible role of zinc in RNA polymerase obtained from *Escherichia coli*. *Proc. Natl. Acad. Sci. U.S.A.* **68**, 2497–2501.

Seifert, K. H., and Benecke, B. J. (1975). DNA-dependent RNA polymerase C occurence and localization in various animal cells. *Eur. J. Biochem.* **53**, 293–300.

Seifert, W., Rabussay, D., and Zillig, W. (1971). On the chemical nature of alteration and modification of DNA dependent RNA polymerase of *E. coli* after T_4 infection. *FEBS Lett.* **16**, 175–179.

Sekeris, C. E., and Schmid, W. (1972). Action of α-amanitin *in vivo* and *in vitro*. *FEBS Lett.* **27**, 41–45.

Sethi, V. S., and Gallo, R. C. (1975). Deoxyribonucleic acid-dependent ribonucleic acid polymerases from spleen of uninfected and Rouscher murine leukemia virus-infected NIH Swiss mice. *Biochim. Biophys. Acta* **378**, 269–281.

Sethi, V. S., and Zillig, W. (1970). Dissociation of DNA-dependent RNA polymerase from *E. coli* in lithium chloride. *FEBS Lett.* **6**, 339–342.

Siddhikol, C., Erbstoeszer, J. W., and Weissblum, B. (1969). Mode of action of streptolydigin. *J. Bacteriol.* **99**, 151.

Sippel, A., and Hartmann, G. (1968). Mode of action of rifamycin on the RNA polymerase reaction. *Biochim. Biophys. Acta* **157**, 218–219.

Sklar, V. E. F., and Roeder, R. G. (1976). Purification and subunit structure of deoxyribonucleic acid-dependent ribonucleic acid polymerase III from the mouse plasmacytoma MOPC 315. *J. Biol. Chem.* **251**, 1064–1073.

Sklar, V. E. F., Schwartz, L. B., and Roeder, R. G. (1975). Distinct molecular structures of nuclear class I, II and III DNA-dependent RNA polymerases. *Proc. Natl. Acad. Sci. U.S.A.* **72**, 348–352.

Somers, D. G., Pearson, M. L., and Ingles, C. J. (1975). Isolation and characterization of an alpha-amanitin-resistant rat myoblast mutant cell line possessing alpha-amanitin-resistant RNA polymerase II. *J. Biol. Chem.* **250**, 4825–4831.

Spiegelman, G. B., and Whitley, H. R. (1974). Purification of ribonucleic acid polymerase from SP82-infected *Bacillus subtilis*. *J. Biol. Chem.* **249**, 1476–1482.

Steggles, A. W., Wilson, G. N., Kantor, J. A., Picciano, D. J., Falvey, A. K., and Anderson, W. F. (1974). Cell-free transcription of mammalian chromatin: Transcription of globin messenger RNA sequences from bone marrow chromatin with mammalian RNA polymerase. *Proc. Natl. Acad. Sci. U.S.A.* **71**, 1219–1223.

Sternbach, H., Engelhardt, R., and Lezius, A. G. (1975). Rapid isolation of highly active RNA polymerase from *Escherichia coli* and its subunits by matrix-bound heparin. *Eur. J. Biochem.* **60**, 51–55.

Stetter, K. O., and Zillig, W. (1974). Transcription in Lactobacillacae. DNA-dependent RNA polymerase from *Lactobacillus curvatus*. *Eur. J. Biochem.* **48**, 527–540.

Stevens, A. (1972). New polypeptides associated with DNA-dependent RNA polymerase of *Escherichia coli* after infection with bacteriophage T_4. *Proc. Natl. Acad. Sci. U.S.A.* **69**, 603–607.

Surzycki, S. J. (1969). Genetic functions of the chloroplast of *Chlamymonas reinhardi*: Effect of rifampin on chloroplast DNA-dependent RNA polymerase. *Proc. Natl. Acad. Sci. U.S.A.* **63**, 1327–1334.

Suzuki, Y., and Giza, P. E. (1976). Accentuated expression of silk fibroin genes *in vivo* and *in vitro*. *J. Mol. Biol.* **107**, 183–206.

Taketo, M., and Ishihama, A. (1977). Biosynthesis of RNA polymerase in *Escherichia coli*. V. Defects of the subunit assembly in a temperature-sensitive beta subunit mutant. *J. Mol. Biol.* **112**, 65–74.

Tata, J. R., and Baker, B. (1978). Enzymatic fractionation of nuclei. Polynucleosomes and RNA polymerase II as endogenous transcriptional complexes. *J. Mol. Biol.* **118**, 249–272.

Tata, J. R., Hamilton, M. J., and Shields, D. (1972). Effects of alpha-amanitin *in vivo* on RNA polymerase and nuclear RNA synthesis. *Nature (London), New Biol.* **238**, 161–164.

Towle, H. C., Jolly, J. F., and Boezi, J. A. (1975). Purification and characterization of bacteriophage gh-1 induced deoxyribonucleic acid-dependent ribonucleic acid polymerase from *Pseudomonas putida. J. Biol. Chem.* **250**, 1723–1733.

Towle, H. C., Tsai, M. J., Tsai, W. Y., and O'Malley, B. W. (1977). Effect of estrogen on gene expression in the chick oviduct. *J. Biol. Chem.* **252**, 2396–2404.

Travers, A. (1970). RNA polymerase and T4 development. *Cold Spring Harbor Symp. Quant. Biol.* **35**, 241–251.

Valenzuela, P., Hager, G. L., Weinberger, F., and Rutter, W. J. (1976). Molecular structure of yeast RNA polymerase III. Demonstration of the tripartite transcriptive system in lower eukaryotes. *Proc. Natl. Acad. Sci. U.S.A.* **73**, 1024–1028.

Van Keulen, H., and Retel, J. (1977). Transcription specificity of yeast RNA polymerase A. Highly specific transcription *in vitro* of the homologous ribosomal transcription units. *Eur. J. Biochem.* **79**, 579–588.

Walter, G., Zillig, W., Palm, P., and Fuchs, E. (1967). Initiation of DNA-dependent RNA synthesis and the effect of heparin on RNA polymerase. *Eur. J. Biochem.* **3**, 194–201.

Walter, G., Seifert, W., and Zillig, W. (1968). Modified DNA-dependent RNA polymerase from *E. coli* infected with bacteriophage T_4. *Biochem. Biophys. Res. Commun.* **30**, 240–247.

Wandzilak, T. M., and Benson, R. W. (1978). *Saccharomyces cerevisiae* DNA-dependent RNA polymerase III: A zinc metalloenzyme. *Biochemistry* **17**, 426–431.

Wehrli, W., Nüesch, J., Knüsel, F., and Staehlin, M. (1968). Action of rifamycins on RNA polymerase. *Biochim. Biophys. Acta* **157**, 215–217.

Weinmann, R., and Roeder, R. G. (1974). Role of DNA-dependent RNA polymerase III in the transcription of the tRNA and 5 S RNA genes. *Proc. Natl. Acad. Sci. U.S.A.* **71**, 1790–1794.

Weiss, S. B. (1960). Enzymatic incorporation of ribonucleoside triphosphates into the inter-polynucleotide linkages of ribonucleic acid. *Proc. Natl. Acad. Sci. U.S.A.* **46**, 1020–1030.

Wieland, T. H., and Fahrmeir, A. (1970). Constituents of amanita phalloides. Oxidation and reduction reactions at the α,δ-dihydroxyisoleucine side chain of O-methyl-α-amanitin. Methylalavamanitine, an nontoxic degradation product. *Justus Liebigs Ann. Chem.* **736**, 95–98.

Wilson, G. N., Steggles, A. W., and Neinhaus, A. W. (1975). Strand-selective transcription of globin genes in rabbit erythroid cells and chromatin. *Proc. Natl. Acad. Sci. U.S.A.* **72**, 4835–4839.

Wintersberger, E. (1972). Isolation of a distinct rifampicin-resistant RNA polymerase from yeast, *Neurospora,* and liver. *Biochem. Biophys. Res. Res. Commun.* **48**, 1287–1294.

Wu, C.-W., Wu, F. Y.-H., and Speckhard, D. C. (1977). Subunit location of the intrinsic divalent metal ions in RNA polymerase from *Escherichia coli. Biochemistry* **16**, 5449–5454.

Wu, G. J., and Dawid, I. B. (1972). Purification and properties of mitochondrial deoxyribonucleic acid dependent ribonucleic acid polymerase from ovaries of *Xenopus laevis. Biochemistry* **11**, 3589–3595.

Yamamoto, M., Jonas, D., and Seifert, K. (1977). Transcription of ribosomal 5 S RNA by

RNA polymerase C in isolated chromatin from HeLa cells. *Eur. J. Biochem.* **80**, 243–253.

Yarbrough, L. R., and Hurwitz, J. (1974). The isolation of subunits of deoxyribonucleic acid-dependent ribonucleic acid polymerase of *Escherichia coli. J. Biol. Chem.* **249**, 5400–5404.

Young, B. S., and Wright, A. (1977). Multiple effects of an RNA polymerase β' mutation on *in vitro* transcription. *Mol. Gen. Genet.* **155**, 191–197.

Yu, F.-L. (1975). An improved method for the quantitative isolation of rat liver nuclear RNA polymerases. *Biochim. Biophys. Acta* **395**, 325–336.

Zasloff, M., and Felsenfeld, G. (1977). Use of mercury-substituted ribonucleoside triphosphates can lead to artefacts in the analysis of *in vitro* chromatin transcripts. *Biochem. Biophys. Res. Commun.* **75**, 598–603.

Zechel, K. (1971). Transcription *in vitro:* Unter Suchrangen zur Binding und Initiation von RNA-Polymerase on DNA. Ph.D. Thesis, Univ. München, Munich.

Zieve, G., Benecke, B. J., and Penman, S. (1977). Synthesis of two classes of small RNA species *in vivo* and *in vitro. Biochemistry* **16**, 4520–4525.

Zillig, W., Zechel, K., and Halbwachs, H. J. (1970a). A new method of large scale preparation of highly purified DNA-dependent RNA polymerase from *E. coli. Hoppe-Seyler's Z. Physiol. Chem.* **351**, 221–224.

Zillig, W., Zechel, K., Rabussay, D., Schachner, M., Sethi, V. S., Palm, P., Heil, A., and Seifert, W. (1971b). On the role of different subunits of DNA-dependent RNA polymerase from *E. coli* in the transcriptive process. *Cold Spring Harbor Symp. Quant. Biol.* **35**, 47–58.

Zillig, W., Fujiki, H., Blum, W., Janekovic, D., Schweiger, M., Rahmsdorf, H. J., Ponta, H., and Hirsch-Kaufman, M. (1975). *In vivo* and *in vitro* phosphorylation of DNA-dependent RNA polymerase of *Escherichia coli* by bacteriophage-T4-induced protein kinase. *Proc. Natl. Acad. Sci. U.S.A.* **72**, 2506–2510.

Zybler, E., and Penman, S. (1971). Products of RNA polymerases in HeLa cell nuclei. *Proc. Natl. Acad. Sci. U.S.A.* **68**, 2861–2865.

4

Regulation of Transcription in Prokaryotes, Their Plasmids, and Viruses

Geoffrey Zubay

Copyright © 1980 by Academic Press, Inc.
All rights of reproduction in any form reserved.
ISBN 0-12-289503-7

I. THE SCOPE OF THIS CHAPTER

The beginning to the solution of the gene regulation problem was heralded by Jacob and Monod in their classic paper proposing the operon model of control (1961). Since this time it has become increasingly clear that progress in this area of biology requires a close interplay between the disciplines of genetics and biochemistry. Mutants isolated by genetic techniques indicate the number and location of genes involved in a particular regulatory event and facilitate the biochemical analysis of the corresponding gene products. Ideally, biochemical analyses are pursued from the whole cell down to a purified cell-free system using genetic correlates along the way. Usually two conditions when met constitute convincing evidence that a particular control mechanism is operative: (1) reconstruction of a regulatory system in a cell-free system with the alleged control factors present and (2) use of components from different genetic mutants showing the expected deviation from normal behavior seen *in vivo*.

At the present time a great deal is understood about the regulation of a limited number of gene systems such as the *lac* operon and the λ bacteriophage, and much information is being collected on a broad variety of gene types. It is essential that a limited number of systems be intensively investigated so that the full molecular details of the transcription process can be understood. It is also important that we expand the number of systems being studied, since new examples frequently illustrate basically new modes of regulation. A global picture of what is taking place in the whole cell can only be achieved by a broader knowledge of more gene systems.

In this chapter the subject of regulation of transcription in prokaryotes will be treated in toto. To pursue this goal effectively within a single chapter it is essential to be highly selective in the choice of gene systems to be discussed. A description of basically different modes of regulation as they affect individual genes and the way in which these modes are put together to give the complex patterns of expression existing in plasmids, bacteriophages, and whole cells will be emphasized. Consistent with these aims, some of the better understood systems will receive the most attention. Processing of RNA and protein and regulation of messenger RNA translation also play an important role in the ultimate expression of genes. However, these subjects will be discussed only in those instances where they definitely have a direct effect on the transcription process.

II. THE BASIC MECHANICS OF THE TRANSCRIPTION PROCESS

The enzyme responsible for RNA synthesis in most bacteria is a complex zinc metalloenzyme containing five protein subunits which in *E. coli* has a molecular weight of 476,000 daltons (for a review of the general properties of polymerase see Chamberlain, 1976). It contains two α subunits with molecular weights of 40,000, one β subunit with a molecular weight of 150,000, one β' subunit with a molecular weight of 160,000, and one σ subunit with a molecular weight of 86,000. The σ subunit is readily dissociable and appears to dissociate shortly after initiation of RNA synthesis. With σ the polymerase is referred to as the holoenzyme; without σ it is referred to as the apoenzyme or core enzyme. The subunits have been separated by urea or sodium dodecyl sulfate and resolved from each other by gel electrophoresis or chromatography. Reconstitution experiments suggest a stepwise assembly of the enzyme from its subunits as follows:

$$2\alpha \rightarrow \alpha_2 \xrightarrow{\beta} \alpha_2\beta \xrightarrow{\beta'} \alpha_2\beta\beta' \xrightarrow{\sigma} \alpha_2\beta\beta'\sigma$$

These steps indicate something of the structure of the enzyme. Thus, the α subunits are in close contact and probably related to each other by a dyad axis of symmetry if they obey the normal rules for dimer formation from identical subunits. The β subunit must be in contact with one or both α subunits. Reaction of holoenzyme with bifunctional cross-linking reagents has both confirmed and extended our topographical understanding of polymerase structure (Hillel and Wu, 1977). Such studies suggest the proximity of the following subunit pairs: $\beta\beta'$, $\beta\sigma$, $\beta'\sigma$, $\alpha\beta$, $\alpha\beta'$. Similar studies suggest that only the σ and β subunits come in direct contact with DNA at the initiation site for RNA synthesis (Hillel and Wu, 1978). Furthermore, σ binds to the nontranscribing DNA strand (C. W. Wu, personal communication) in the forward region where transcription is initiated. This leaves β to make the majority of contacts in the region upstream from the initiation point. A region of about 11 contiguous base pairs is believed to be unwound in the vicinity of the initiation site for transcription (Siebenlist, 1979). Other smaller polypeptides may be associated with the RNA polymerase, particularly in some phage-infected cells but no clear-cut function has been established for these. Genetic variants have been obtained for all of the subunits of the polymerase (Gross *et al.*, 1978). The genes for the β and β' subunits are adjacent to one another in the chromosome and appear to be cotranscribed, as might be expected from their equimolar appearance in the core enzyme. The genes for α and σ are located in different regions of the genome. With minor variations most prokaryotes which have been examined have a subunit structure similar to that of *E. coli* polymerase.

The normal polymerase structure can become significantly altered by certain types of phage infection or during sporulation in certain spore-forming bacteria, as discussed below. Whereas many phages use the normal or slightly modified host polymerase, some produce their own RNA polymerase, which is responsible for "late" viral transcription. For example, the T7-encoded RNA polymerase consists of a single polypeptide chain with a molecular weight of 70,000. The relative simplicity of this polymerase is consistent with its limited function; the enormous size and complexity of bacterial polymerase are probably a reflection of the large variety of genes it must recognize in a differential manner depending upon the physiologic state of the cell.

The process of transcription is commonly divided into three phases: initiation (which involves all those steps up to the formation of the first phosphodiester linkage) elongation, and termination (which involves a cessation of synthesis and release of the nascent RNA molecule from its complex with DNA and the polymerase). These three phases of polymerase action will be discussed in their simplest terms before turning to considerations of regulation.

A. Steps Involved in the Formation of an Initiation Complex

In vitro studies show that RNA polymerase can initiate RNA synthesis at a number of places other than true *in vivo* start sites. These places include single-strand regions on the DNA created by nicking of the template as well as a broad array of "weak" start sites where strong binding of polymerase to the template does not seem to be a prerequisite for initiation of RNA synthesis. Many early studies of polymerase binding and initiation were misguided because they were done with core polymerase. In the absence of σ factor, polymerase actually binds more strongly to DNA at many nonspecific sites. The binding was usually measured by a filter binding technique in which radioactively labeled DNA and polymerase were mixed and the retention of DNA on passing through a Millipore filter was used as a measure of whether or not it is bound to the polymerase. The discovery of σ factor by Burgess *et al.* (1969) was of paramount significance because correct promotor recognition only occurs in its presence. This was suggested from *in vitro* RNA synthesis studies using damaged calf thymus DNA and carefully prepared T7 bacteriophage DNA as templates. The former DNA was a fair template for core polymerase presumably because of the nicks it contained. The latter DNA was a poor template whose activity was improved ten to twentyfold if σ factor was present. Evidently, σ factor was indispensable to proper promotor recognition.

Subsequently, Bautz *et al.* (1969) showed that rifampicin inhibits initiation by binding to free RNA polymerase, thereby preventing formation of the initiation complex. They estimated the number of initiation sites on different viral templates by adding rifampicin after DNA–polymerase complex formation and measuring the number of γ-phosphate-labeled RNA's synthesized. One γ-labeled group should be formed for each rifampicin-resistant polymerase bound to an initiation site. The confidence produced by these early studies with σ factor led others to study interaction of holoenzyme with well-defined templates. Some of the more significant recent studies in this area have come from Chamberlain's and Schallers laboratories.

Most of Chamberlain's studies have been carried out with T7 DNA (Chamberlain, 1976). He distinguishes three types of binding of polymerase to DNA: nonspecific complexes, closed promotor complexes, and open promotor complexes. The first type of complex is formed with little site selectivity on the DNA and with little structural alteration of the DNA. The second type of complex is formed at or near the promotor with little structural alteration of the DNA. In contrast, the open promotor complex is formed at a specific site and is believed to involve the disruption of some hydrogen bonds in the base-paired structure of the DNA. Whereas all three structures are salt sensitive, formation of the open promotor complex is unique in being exquisitely temperature sensitive; it is not formed at low temperatures. This is believed to be due to the high activation energy necessary to bring about the necessary structural alterations in the DNA. This is true with a mesophile such as *E. coli* but obviously not true of cryophiles, which bear examination in this regard. Another characteristic of the open promotor complex is its ability to initiate RNA synthesis very rapidly when supplied with the four ribonucleoside triphosphates. By challenging such complexes with various concentrations of rifampicin and ribonucleoside triphosphates, it was possible to estimate that over 90% of the holoenzyme molecules in binary complexes with T7 DNA initiate an RNA chain with an apparent first-order rate constant for RNA chain initiation of 3.0 sec^{-1}. A controversial aspect of Chamberlain's interpretation concerns the distinction between a nonspecific closed complex and the specific closed promotor complex. The existence of the latter type of complex is based mainly on the observation that closed complexes formed at high salt can be converted to open promotor complexes without dissociation of the RNA polymerase. If the polymerase is able to traverse appreciable lengths of DNA in seeking a specific binding site, this conclusion would not be warranted. Further observations are needed to determine if specific metastable complexes at or near promotor sites can be formed without the structural alterations in the DNA believed to occur in the open promotor complex. Indeed the likelihood of this situation has been contested by Seeberg *et al.* (1977).

In Schaller's laboratory, RNA polymerase interaction with the double-stranded replicative form of bacteriophage fd has been studied (Seeburg *et al.*, 1977). The DNA was initially fragmented with restriction endonuclease II from *Haemophilus parainfluenzae* (*Hpa*II). The restriction fragments were reacted with *E. coli* RNA polymerase and the resulting complexes analyzed by filter binding followed by gel electrophoresis. Each of the five restriction fragments contain different promotors for the RNA polymerase, and through this type of analysis Seeberg *et al.* (1977) were able to measure rates of formation and dissociation for all five specific complexes. At 120 mM KCl the first-order rate constants for complex decay were determined to be 10^{-2} to 10^{-6} sec^{-1}. The second-order rate constants for complex formation were found to be about 10^6 to $10^7 M^{-1}$ sec^{-1}. From these values association constants for the individual promotors were calculated to range from 2×10^{-8} to $2 \times 10^{-11} M^{-1}$.

One of the most important conclusions of Seeburg *et al.* from these experiments was that the initiation efficiency of a promotor is determined by the rate of complex formation and not by the affinity for the enzyme. This follows quite naturally from the fact that a polymerase molecule strongly bound at the promotor locus is in a rapid start situation. Since the rapid start time is only a fraction of a second whereas the dissociation time is usually much longer, a rapid start complex once formed will almost always lead to initiation. If Seeburg *et al.* (1977) are correct, then the rate of formation of the rapid start complex is a more significant parameter than the association constant of the DNA–polymerase complex. Seeburg *et al.* (1977) also made cold stable complexes between polymerase and the fd fragments, which they felt should resemble Chamberlain's closed promotor complexes. However, they found a random distribution of polymerase on different DNA fragments and concluded that such cold stable complexes are not promotor specific and not an intermediate in promotor selection. It would appear that this important question raised by Schaller and Chamberlain over the existence of closed promotor complexes can only be resolved by further experiments. It should be emphasized that different conditions and different templates were used in the two laboratories.

The concept of a closed promotor complex is not unreasonable. Thus it seems quite likely that some recognition of the promotor takes place before any unwinding of the DNA. However, the experiments done thus far have not clearly demonstrated the existence of a closed promotor complex as a stable intermediate.

Seeburg *et al.* (1977) also found that the rate of formation and stability of open promotor complexes was enhanced in negatively supercoiled DNA. This is consistent with the findings of others that negatively supercoiled DNA makes a more effective template for RNA synthesis presumably because the unwinding of the DNA necessary to make an open promotor complex should be energetically favored by negative supercoiling.

All of the *in vitro* results on promotor complex formation described above were obtained at lower ionic strengths (0.12 *M* or less) than are believed to exist *in vivo* (~0.2 *M*). This is a limitation of the technique, since complexes are usually very unstable at higher ionic strengths which would be more characteristic of the intracellular environment. Perhaps this is an indication of the importance of supercoiling of the template or some other factors which are usually present *in vivo* that are lacking from the highly purified *in vitro* system. Schaller says that his rate measurements with fd DNA at 0.12 *M* fit very well to the ratios of RNA products observed in minicells (H. Schaller, personal communication). This suggests that the relative rates *in vitro* are reasonably accurate in the work on fd DNA even if the absolute values may differ from the *in vivo*.

The strong binding of RNA polymerase to the promotor site has been used to isolate and characterize those DNA regions in such a complex which are protected from the action of DNase. The polymerase-protected fragments usually constitute a region about 40 base pairs in length, including about 18 bases of the region coding for the 5′-end of the mRNA. The most common feature noticed in most promotor regions thus far isolated and examined is a region of 7 bases, with the most probable sequence TATPuATG, centered about ten bases before the transcription start. Most promotors examined have 5 out of 7 bases in common with this so-called Pribnow box. In the *lac* operon promotor two mutations in this region which cause a 10- to 25-fold enhancement in expression have been sequenced and found to contain single base alterations. The large effects of these mutations on expression suggests that sequence-specific contacts exist between the polymerase and the DNA in this region. A single mutation in the *lac* promotor a few bases to the left of the Pribnow box also has a large effect (fivefold increase) on expression, indicating that not only the Pribnow box is critical in determining the level of expression. This is also apparent from a sequence analysis of the *trp* promotor which has only three nucleotides in common with the Pribnow box. It seems likely that most fruitful correlations will be derived in the future from an analysis of the effect of base changes in the Pribnow box and surrounding area on the promotor strength.

It is expected that base alterations within the DNase protected region of the rapid start complex should in many cases be critical to the level of expression in transcription. Most revealing to the question of complex formation is the fact that regions outside the protected fragment are essential for the formation of the complex. Thus, if one releases the polymerase from the protected fragment, it will not rebind to form a rapid start complex. Studies on several promotors (λ, *lac, trp,* and fd) show that a region outside the protected fragment and 35 to 45 bases before the initiation point for messenger synthesis is essential to forming the rapid start complex. These findings have led to the suggestion that the initial recognition of the promotor region occurs outside the region of the rapid start complex. Such an entry box was originally thought to be the initial site recognized by the RNA polymerase, leading to a series of transitions and finally the stable rapid start complex.

Recently, Johnsrud (1978) and Siebenlist (1979) have used dimethyl sulfate as a probe for studying the interaction between polymerase and the promotor. This reagent methylates the N7 position on guanine or the N3 position on adenine in the double helix and in addition, the N1 position of adenine on the single strand. Protein bound to the double helix can either protect these positions or affect the reactivity to produce an enhanced methylation. The pattern of reactivity of the bases in the lactose promotor on binding polymerase suggests that the enzyme covers a region of at least 38 base pairs, stretching upstream from the origin of transcription. Inferred protein–DNA contacts occur mainly in the major groove of the DNA helix. The pattern of methylation of the T7 promotor–polymerase complex is consistent with the polymerase unwinding about 11 base pairs from the -9 to the $+2$ positions at the origin of transcription. Results of investigations on the polymerase–promotor complex with chemical probes are not entirely consistent with the earlier interpretations of the DNase protection experiments. These DNase protection experiments must be viewed with suspicion as there is no reason to believe that this reagent gives precise answers on the location of bound polymerase. On the one hand, a region where polymerase is not binding could give the impression of being protected from DNase action merely because of steric interference between the two proteins. This would lead to an overestimate for the region binding to polymerase. On the other hand, regions of DNA in direct contact with polymerase might in fact be susceptible to DNase action if they are not fully covered leading to an underestimate for bound DNA. Indeed protection experiments with dimethyl sulfate and other chemical reagents indicate that repressors, activators, and possibly polymerases make contacts mostly on one side of the double helix, leaving the phosphodiester linkages exposed on the other side. It seems likely that further results with the chemical probes will lead to a more precise picture on the how and where of polymerase binding.

The current model we favor for the rapid start complex for simple promotors not involving special activators is based on inspection of the sequences of known promotors and the chemical probe experiments discussed above. In this model there are two main areas of contact, one centered in the −35 region of the promotor with a favored sequence of TGTTGAC and one centered in the −10 region with the favored sequence TATpuATG. These two heptonucleotide regions will be referred to as polymerase binding site 1 (PBS1) or entry box and polymerase binding site 2 (PBS2) or Pribnow box. PBS2 also contains two or three bases in the unpaired region of the rapid start complex which extends from about −8 to +2 (see Siebenlist, 1979).

The attractiveness of this two-domain model for polymerase binding has led us to investigate the possibility that a similar situation may exist for eukaryotic RNA polymerase II. Inspection of all available sequences for promotors where the initiation site for mRNA transcription is known suggests that there may be two main domains of recognition. The first domain corresponding to PBS1 contains a decanucleotide centered at about −29 with the preferred sequence GPyATATAAGG; the second domain corresponding to PBS2 contains a pentanucleotide centered at −3 with the preferred sequence TGCTT. The data available on eukaryotic promotors are very limited, making these assignments very tentative and subject to a good deal of refinement.

Once the rapid start complex has been formed, initiation of RNA synthesis is rapid, taking only a fraction of a second. Initiation requires two ribonucleside triphosphates, which are presumed to bind partly to the polymerase and partly to one strand of the opened region of the template by Watson–Crick complementary base pairing. The point of initiation is fixed by the position of the polymerase on the template, but it is not always exact. For example, in the *lac* promotor mutant UV5 either a G or an A is the starting base; these are 8 and 9 bases from the center of the Pribnow box, respectively. Most RNA's initiate with a purine base A or G, and the product after dinucleotide cleavage preserves a triphosphate group on the 5'-end of the growing RNA according to the following reaction:

$$pppX + pppV \rightarrow pppXpV + ppi$$

B. Elongation

The function of elongation is to produce a continuous and faithful RNA replica of one of the template strands by the stepwise addition of nucleotide triphosphates to the 3'-OH tip of the growing RNA. One visualizes the double helix as being pried open in the direction of synthesis as it closes on those regions just copied. The rate of displacement of DNA–RNA hybrid from regions just transcribed may vary appreciably as discussed below. Studies with base analogues indicate that abnormal triphosphates, which make hydrogen-bonded base pairs with the same overall dimensions as Watson–Crick base pairs, are often acceptable substitutes for the normal triphosphates.

After the first phosphodiester linkage has been made, the characteristics of the DNA–polymerase complex change: it becomes completely resistant to rifampicin and also highly resistant to dissociation by other factors. Interestingly enough this extreme stability is not realized if single-strand DNA templates are used instead of helical templates. The σ factor is not required during elongation; spectroscopic experiments (Wu et al., 1975) show that it is normally released not coincident with but sometime after the first phosphodiester linkage is made. The released σ factor is reutilized by other core polymerases to initiate further synthesis. Elongation rates vary from about 20 to 60 nucleotides/sec for different genes under optimum growth conditions. Such rates can be achieved both in vivo or in vitro. Lowering the nucleotide concentrations significantly below 50 μM in vitro not only lowers the elongation rate but occasionally leads to premature arrest of RNA chain growth.

There are several unanswered questions about elongation for which one would like to have answers. First, one would like to know if elongation rates are sequence dependent. Although various suggestions have been made, there is a scarcity of actual observations. A priori one could imagine that sequences which bind strongly to polymerase or sequences which are difficult for polymerase to unwind would slow down the rate of transcription. Perhaps the best lead as to what types of sequences slow down rates of elongation will come from an analysis of sequences in and around bona fide termination sites, a topic which will be discussed in the next section. Clearly, terminator and attenuator sites (discussed below) signal stopping; in addition it seems likely that stopping or slowing down occur at other regions on the template. Second, one would like to know if elongation rates in some way are governed by translation rates so that the polymerase does not get too far ahead of the train of ribosomes on the track of growing messenger.

C. Termination and Antitermination*

The subject of termination was opened up by Roberts (1969), who succeeded in showing that an oligomeric protein named rho (monomer MW 50,000) catalyzed termination of "early right" (7 S) and "early left" (12 S) messenger synthesis in λ. In an *in vitro* system containing λ DNA and RNA polymerase two RNA's of about the correct size and sequence are made if rho is present. If rho is missing, synthesis continues to some indefinite point well past the normal termination points and the RNA and polymerases remain attached to the template. This experiment showed that rho, a protein found in uninfected cells, was necessary (at least *in vitro*) for normal termination of RNA synthesis and release of the nascent viral RNA from the template. After the Roberts experiment, scores of templates were examined under a variety of *in vitro* conditions to determine the effects of rho. In some cases rho is not required, in other cases it is required. Surprisingly, instances were also found where rho caused termination in the middle of a cistron. The question how many of these results were artifacts arising from a particular set of *in vitro* conditions could only be answered with the help of a genetic variant of rho. Recently the gene *rho* (formerly called *SuA*) has been identified in *E. coli* (Ratner, 1976). This finding will not only allow *in vivo* and *in vitro* comparisons to be made on the effects of normal and mutant rho on intercistronic regions but it suggests an explanation of the surprising intracistronic termination observed *in vitro* with rho (discussed below).

Until this reexamination of rho factor has been made, we must content ourselves with the preliminary evidence which is available. The comparison between the results with the closely related T3 and T7 bacteriophages illustrates a difficulty in assessing some of the presently available results. In both phages, transcription by host polymerase *in vivo* is limited to about 20% of the left-hand end of the genome. *In vitro* with T7 DNA and polymerase, transcription is terminated at approximately the same region, but with T3 DNA correct termination requires rho factor. In view of the genetic similarity of T3 and T7 one suspects, despite the *in vitro* results, that T7 DNA may in fact use rho as a catalyst for termination *in vivo*. *In vitro* some termination sites which do not appear to require the presence of rho for cessation of synthesis will not release the RNA or polymerase

---

* This subject has been reviewed by Adhya and Gottesman (1976).

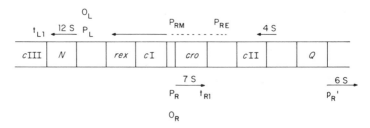

Fig. 1. Partial map of λ showing some genes, promotors, termination sites, and corresponding regions of transcriptions (arrows). See text for explanation.

from the termination point unless rho is added. This is the case for the *trp* attenuator discussed in the next section and clearly demonstrates one property of rho, namely, the catalysis of RNA release. In other *in vitro* situations termination and release of RNA can actually occur in the absence of rho, but still rho appears to increase the efficiency of the process. This is the case with λ6 S and λ45 RNA (see Fig. 1). Both of these RNA's are made *in vitro* with purified polymerase and released from the template in the absence of rho. The addition of rho stimulates synthesis of 6 S RNA by about 20% and 4 S by about 500%. This increased synthesis, particularly in the case of 4 S, is probably due to the rapid rho-catalyzed release, which unblocks the template making it available for further synthesis.

The facts just presented indicate that the dependence of termination on rho protein can vary from total to a catalytic enhancement. At present not a single termination site has been investigated which does not function more efficiently in the presence of rho. Rho-dependent termination can be conceptually divided into three events: (1) cessation of RNA synthesis, (2) release of the RNA from the template, and (3) release of the polymerase from the template. Little is known about step 3. It seems likely that step 1 is the first step in termination and does not involve rho at all but rather a sequence in the DNA which signals the polymerase to stop. The strength of the signal varies at different termination points. If the signal is strong, no further synthesis will occur, and in some cases (as with λ6 S and λ4 S discussed above) even release may occur. If the signal is weak (as with λ7 S and λ12 S discussed above), then the polymerase after pausing may proceed beyond the termination point with RNA synthesis. The pausing part of this model is supported by recent results of Rosenberg *et al.* (1978). They examined in kinetically controlled *in vitro* transcription reactions, the ability of RNA polymerase, initiated at P_R, the early right λ promotor, to move through the normally rho-dependent terminator λt_{R1} in the absence of rho. The results indicate that, even in the absence of rho, RNA polymerase undergoes a substantial pause within the t_{R1} site before it transcribes distal to this region.

It was mentioned earlier that rho can sometimes produce transcription termination in the intracistronic region of a message. In the case of polarity a nonsense mutation inserted into a gene leads to reduced expression of promotor distal genes of the same operon, resulting from a reduced synthesis of the corresponding RNA. In general, nonsense mutations produce a greater polar effect the closer they are to the beginning of a cistron. Since the immediate effect of the nonsense triplet is to inhibit the flow of ribosomes down the message, it was not clear why this should affect transcription, which precedes translation. The finding that *rho* mutants relieve polarity and that rho is encoded by a gene which suppresses polarity (*SuA* now called *rho*) has led to the proposal that rho causes polarity by a mechanism similar to that involved in normal termination (e.g., see Richardson, 1978). Polymerase reaches an intracistronic pause site. Since there are no ribosomes between the polymerase pause site and the last nonsense triplet, the intervening RNA is mostly exposed. If this exposed region of RNA is sufficiently large and contains the appropriate structure and sequence for rho attachment, then termination with release of RNA can result when rho comes in contact with the polymerase. In support of the idea that rho interacts directly with the RNA, Richardson has found that rho has a nucleotide triphosphatase activity which is dependent on the presence of RNA. Furthermore, the RNA must contain some C residues to bring about the reaction. From this it has been suggested that rho-catalyzed termination of RNA synthesis is an energy-dependent reaction involving ATP (or other XTP) hydrolysis. If this model for rho action is correct, then explanations have to be found for why RNA's such as ribosomal RNA are not prematurely cleaved since they do not carry ribosomes during transcription. Perhaps this is due to the binding of ribosomal proteins or the structure of the rRNA, making rho recognition difficult except at the normal termination site.

From the Richardson model it could be inferred that the role of RNA structure in rho action is complex. First, a sequence must exist which causes the polymerase to pause. Second, a sequence must exist on the upstream side of the pause site which is amenable to rho binding and rho-catalyzed RNA release at or near the immobilized polymerase. Third, for mRNA a sequence probably exists on the upstream side of the pause site that impedes ribosome flow. The third requirement can easily be met by the presence of one or more nonsense triplets judiciously placed. From a sequence analysis of some currently recognized termination sites, it has been suggested that a preferred termination structure contains a G- and C-rich region followed by a U-rich region at which termination occurs [for details see Adhya and Gottesman (1977)]. Adhya and Gottesman have suggested several possible explanations for these structural requirements. Frequently one also finds a sequence in the RNA close to the termination site amenable to formation of a hydrogen-bonded looped structure. Considering the varying dependence of different termination sites on rho and the scarcity of relevant sequence data, it is not possible to explain with any precision the relationship between the terminator proximal RNA structure and termination.

As λ infection progresses, termination at two early Rho-dependent locations ceases (see Fig. 1, $t_{L1} + t_{R1}$) so that transcription may proceed beyond these points. This change results from the synthesis of an antitermination factor, the N gene product, which arises from early leftward transcription. A number of experiments show that the action of N cannot be explained as a straightforward effect of N antagonizing rho at the termination site (Adhya et al., 1976). Thus, N acts as an antitermination factor only when RNA synthesis is initiated from P_R, the early right, or P_L, the early left promotor. Even termination signals not normally sensitive to N become sensitive when fused to these promoters. For instance when the gal operon is separated from prophage λ by about 15 bacterial genes (as it is in a λ lysogen), transcription from the λP_L promotor after prophage induction can override normal control mechanisms, including transcription termination signals that might be present in the bacterial genes located between the gal operon and the prophage λ. The N protein also relieves polarity in transcripts initiated at the P_R and P_L promotors. This type of phenomenon has led to the proposal that N action must begin at or near the P_R or P_L promotor. Recently, mutants have been isolated by Salstrom and Szybalski (1978) which shows that the λ leftward transcription starting from P_L contains a region downstream from the promotor which is essential for N-induced antitermination. The obvious but unproved model is that polymerase pauses in this critical region long enough to be modified by N protein, which may stay bound to the polymerase and prevent normal termination.

D. Attenuation

Another mode of transcription termination clearly influenced by rho and probably close in mechanism to that of polarity induced termination is called attenuation. Appreciable evidence has been obtained as to how attenuation works in the trp operon, but we are still some way from a complete understanding of the phenomenon. In the trp operon 161 bases in the RNA are transcribed (the so-called leader region) before reaching the translation region of the first large structural gene trp E. Under optimal growth conditions only about 1 out of 10 RNA's that begin at the initiation point for Trp-mRNA transcribes the complete message. The remaining chains are terminated between base 138 and 141.

In *rho* mutants producing defective rho factor termination in the leader region is attenuated; of the *rho* mutants tested so far the most potent gives a fourfold effect. Also, if growth conditions are such that the Trp-tRNA is undercharged with tryptophan, termination is greatly attentuated. Knowledge of the sequence of bases in the leader region has been crucial in suggesting possible mechanisms for the effects of both rho and Trp-tRNA on termination (Lee and Yanofsky, 1977). A possible initiation codon occurs at position 27 in the leader. That this region can serve as the initiation site for translation has been shown by fusing the *trp* leader to the *lac i* gene. A fused polypeptide containing seven amino acid residues coded for by the *trp* leader region was found to be at the amino-terminal end of the altered *lac* repressor protein (Schmeissner *et al.*, 1977). The importance of the leader peptide to the attenuation mechanism has received strong support from the finding that replacing the ATG initiator codon in the leader by ATA increases the efficiency of termination at the attenuator and lowers the increase in transcription beyond the attenuator under conditions of tryptophan starvation (Zurawski *et al.*, 1978). In phase with the ATG initiator at position 27–29 are tandemly arranged *trp* codons at positions 54–56 and 57–59 and a nonsense triplet at position 69–71 (see Fig. 2). If the ribosomes translating the leader region reach the nonsense triplet then termination would result from a rho-catalyzed reaction as it does in cases of polarity-induced termination discussed above. The termination point contains a string of U's preceded by a G-rich region, which is believed to be a favored pause site for polymerase (see previous section). The *rho* mutants terminate poorly at this point. In seeking an explanation for why a depletion of charged Trp-tRNA attenuates termination attention is immediately drawn to the tandemly located tryptophan codons in the 54–59 base region. Tryptophan is an infrequent codon in RNA, and two in tandem is most unusual. The effect of Trp-tRNA depletion would be to produce ribosome stalling before the first nonsense triplet. One must then find arguments why stalling of ribosomes at the tryptophan codons rather than at the nonsense triplet inhibits rho-induced termination in the leader. Lee and Yanofsky (1977) suggest that translation up to the nonsense triplet

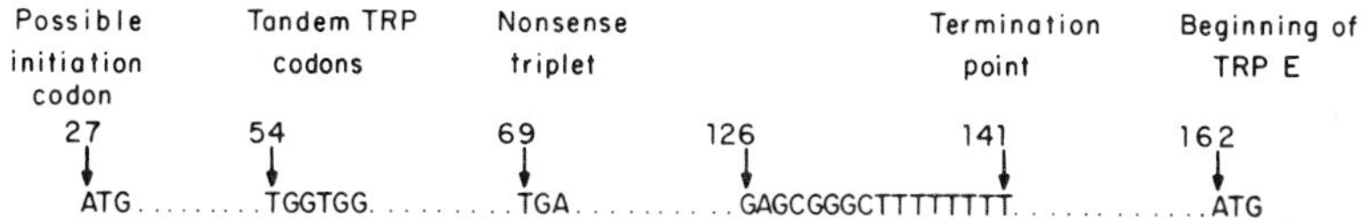

Possible initiation codon	Tandem TRP codons	Nonsense triplet		Termination point	Beginning of TRP E
27	54	69	126	141	162
↓	↓	↓	↓	↓	↓
ATG........	...TGGTGG........	...TGA........	...GAGCGGGCTTTTTTTTT...........		...ATG

Fig. 2. Sequence of bases in the *trp* leader region. See text for explanation.

might interfere with the formation of a complex hydrogen-bonded looped structure which could be formed between bases 74 to 78 and 115 to 119, which in turn might encourage a stem loop structure to form which is more favorable for polymerase pausing and rho-induced cleavage in the 138–141 region.

Additional support for the idea that the RNA secondary and tertiary structure is critical for attenuation comes from three further observations: (1) *Salmonella typhimurium*, which also shows attenuation in the *trp* operon, can form a very similar structure in this region; (2) mutants that alter the hydrogen bonding possibilities in this region lower attenuation (Stauffer *et al.*, 1978); and (3) substitution of ITP for GTP in the RNA, which should weaken the hydrogen bonding, eliminates attenuation *in vitro*. If should also be mentioned that high salt conditions (0.15 *M* Cl) inhibit rho action; this may be due to a stabilization of certain secondary structures. The argument that secondary or tertiary structure might interfere with rho-catalyzed cleavage was raised in a previous section to suggest how premature termination might be avoided in nascent ribosomal RNA. The model described for attenuation is most attractive in that it requires no special protein to explain the specificity. Rather, the specificity is explained by the sequence of bases in the RNA of the leader region. Consequently, the attenuation mechanism could readily be used to regulate other amino acid biosynthetic operons merely be replacing the tandem TRP condons by another set. In fact, there is excellent recent evidence that in both the phenylalanine and the histidine operons that a similar leader structure exists (Barnes, 1978; Di Nocera *et al.*, 1978; Zurowski *et al.*, 1978).

Although rho is most likely involved directly in termination of transcription in the leader region, *in vitro* experiments have not consistently confirmed this result. Thus, Lee and Yanofsky (1977) have found termination without rho, whereas Pannekoek *et al.* (1974) have found that high concentrations of rho are required. The assays used by these two groups may not be equally reliable. However, assuming that they are, we have sought an explanation of these results in the conditions used for RNA synthesis. The most notable difference is that Pannekoek *et al.* use 200 μM concentrations of all four triphosphates, whereas Lee and Yanofsky (1977) use 150 μM for three of the triphosphates and only 10-40 μM for the labeled triphosphate. Premature termination is a common problem (e.g., see Maizels, 1973) *in vitro*, which may result from using abnormal salt conditions or abnormally low concentrations of triphosphate. Thus, a pause site such as that found in the leader might become a termination site in the absence of rho if abnormally low concentrations of one or more triphosphates were used. Recently Platt has found that *in vitro*, using the conditions of Lee and Yanofsky (1977), in the absence of rho cessation of RNA synthesis occurs at the *trp* attenuator but no RNA release occurs unless rho is added (Platt, 1978), thus providing one more indication that Rho plays a role in attenuation.

E. Alternation

The term alternation is intended to designate the switching of RNA polymerase during elongation from transcribing one DNA strand to transcribing the complementary strand. Alternation occurs with surprising efficiency and regularity in an early region of the *lac* operon. Thus far it has only been observed *in vitro*. The reaction was recently discovered in transcription from restriction fragments of the *lac* operon using purified RNA polymerase (S. Mitra and G. Zubay, unpublished results). The product of the reaction is a single molecule of RNA which spontaneously adopts a double helix structure over part of its length connected by a short hairpin loop. The hairpin loop occurs at the region where polymerase switches from copying one DNA strand to copying the other strand. The double helix structure with the tight connecting hairpin loop is isolated intact from the total transcript by mild treatment with pancreatic RNase A, which selectively degrades single-stranded regions of the RNA. A family of RNase-resistant structures, very homogeneous in size, have been isolated, indicating that the sites of alternation and subsequent termination are very limited in number. Alternation is a basically different type of polymerase reaction which should be considered for its possible *in vivo* significance.

II. DIFFERENT WAYS OF REGULATING TRANSCRIPTION

In this section different types of control mechanisms will be considered in isolation from the standpoint of their biologic value and mode of operation. Examples discussed under each heading are selected because they are thought to be the best understood representation for a particular mode of regulation.

A. Gene Dosage

When growth conditions are such that a gene cannot supply the optimum amounts of a particular gene product even when functioning at its maximum rate, then there is a selective pressure for increasing the number of copies of that gene. For example, *E. coli* cells under conditions of rapid growth contain about 10^4 ribosomes. The maximum rate of reinitiation at the ribosomal gene promotor is about 1/sec. In rapid growth *E. coli* can duplicate once every 20 min, which would allow for the synthesis of only about 1200 ribosomes' worth of rRNA if there were only one gene for ribosomal RNA. In fact these are about seven copies for ribosomal RNA in the bacterial chromosome (Williams *et al.*, 1977). Evolutionary pressures have somehow led to gene duplication of rRNA genes to comply with the cells need for ribosomal RNA under conditions of rapid growth.

Possible mechanisms for this gene duplication will not be considered here. However, one might ask if under laboratory conditions it is possible to apply selective pressures so that the gene duplication process could be witnessed. Small extrachromosomal genetic elements supply numerous fitting examples. For instance, in *Streptococcus faecalis* resistance to the antibiotic tetracycline is due to the pAMα1 plasmid (Clewell and Yagi, 1977). Cultivation of bacteria harboring pAMα1 for a number of generations (40–50) in the presence of subinhibitory concentrations of tetracycline (Tc) results in the development of higher levels of resistance to the drug. This phenomenon is accompanied by an increase in size of the plasmid as shown by ultracentrifugation. Electron microscopy has been used to show that the increase in the size of the plasmid is due to the tandem duplication of a 2.6×10^6 dalton region containing the tetracycline-resistance gene. One or more tandem repeats have been observed. When the selective pressure for high levels of Tc resistance is removed, the plasmid gradually reverts to its original size with a single copy of the drug resistance gene. The instability of the multigene copy structure seen here is probably due to the tandem location of the duplicated genes.

In the case of rRNA genes discussed above, the copies have been dispersed. This gives the system greater stability in the absence of selective pressures. The gene dosage mechanism of regulating transcription is probably very widespread and given the patience could probably be demonstrated for almost any gene. In general, it is a slow response type of control requiring many generations and probably some selection to become effective even under the best of conditions.

B. Gene Transposition

The location of a gene or genetic element in the chromosome may be critical in determining its potential for expression. The interest in this type of control has been greatly stimulated by the discovery of insertion elements which mediate the necessary integration and excision process (see Bukhari *et al.*, 1977). In its simplest form the insertion sequence (IS) contains no known genes unrelated to insertion function and is usually shorter than 2 kilobases (kb). Insertion sequences are transposable segments of DNA that can insert into several sites in a genome by an unknown process which does not involve the use of the cells *rec* recombination system. More complex transposable elements known as transposons (Tn) behave formally like IS elements but contain additional genes unrelated to insertion function. The Tc resistance gene discussed in the previous section is a transposon. All transposons contain repetitious sequences at their ends. These are usually arranged in inverted order, but cases are known where the order is parallel. One of the first examples of how an insertion element could be used as a switch for turning a gene on and off was provided by Saedler *et al.* (1974). The 1400 kb insertion element IS2 readily integrates near the beginning of the *gal* operon. IS2 can be integrated in either of two orientations. When it is inserted in one way (orientation I), it prevents expression of the *gal* operon genes; when it is inserted in the opposite manner (orientation II) it allows for constitutive expression of the *gal* genes. By constitutive expression we mean expression which occurs unimpeded by the normal negative controls of the *gal* operon. To explain these results it has been hypothesized that IS2 contains a promotor which is oriented in the same direction as the *gal* operon in orientation II. Spontaneous conversion from orientation I to II occurs *in vivo*. The strong polar effect of IS2 in one orientation and its promotor capacity in the opposite orientation have been observed for a number of genes.

An example of clear biologic significance which most likely involves the use of an insertion element is provided by the phenomenon of phase variation in *Salmonella*. *Salmonella* strains switch from making one type of flagellar antigen to making another with a small probability which is two to three orders of magnitude higher than the frequency of mutation. Flagellar antigens are specified by two genes designated *H1* and *H2*. The expression of these genes is regulated so that only one gene activity, or phase, is expressed in an individual cell at any given time. By heteroduplex mapping of the appropriate closed fragments, Zieg *et al.* (1977) have shown that there is a DNA segment adjacent to the *H2* gene which can undergo inversion. A correlation was demonstrated between the phase state of the *H2* gene and the sequence of the adjacent segment. The explanation for the action of this segment could be similar to that used to explain the effect of IS2 inversion on *gal* operon expression.

Given the appropriate selective pressures, the effects of insertion elements could make themselves felt very quickly indeed. Within a population of microorganisms with no selective pressure one would expect roughly a 50–50 distribution in the orientation of a particular IS element at a particular gene site. With a selective pressure favoring one orientation the cells containing that form should quickly "take over" the culture. If the opposite selective pressure was subsequently exerted the cell containing the IS elements oriented in the opposite direction (arising spontaneously at a rate of about one in a thousand) should "take over" the culture with a slight delay because of their low number in the population. From the preceding considerations it seems likely that insertion sequences and transposons could be quite useful in a variety of situations where biphasic control of expression would be desirable.

C. Polymerase Modification

The existing structure of the cellular RNA polymerase could be directly modified so that the recognition pattern for different promotors is changed. Such structural modifications can be brought about by subunit replacement or by subunit alteration involving covalent or noncovalent changes. Suggestive evidence that polymerase modifications play a significant physiologic role under unusual growth conditions or after virus infection has been found.

Evidence that the λN gene product affects the λ transcription pattern by preventing termination is discussed in Sections II,C, and V,D. The physiologic significance of this reaction seems incontestable (see Section V,D). Proof that the antitermination activity of N results from binding to RNA polymerase is lacking (Epp and Pearson, 1976, personal communication). Whereas it has been shown that the N protein or a part of the N protein binds to polymerase, it is not clear that this is how the protein affects the transcription pattern *in vivo* or *in vitro*.

When T4 bacteriophage infects *E. coli,* there is an abrupt change in metabolism which favors synthesis of progeny phage. At the level of transcription there is an inhibition of the initiation of host mRNA synthesis, while completion of already initiated host mRNA molecules goes on. At the same time synthesis of a discrete set of T4 mRNA's, known as "immediate early" is begun. As infection progresses different classes of T4 RNA's are synthesized; these RNA's are known by a complex terminology which is intended to signify their order of appearance such as "delayed early," "quasi-lates," "anti-lates," and "true lates." All phage transcription is sensitive to rifamycin, arguing strongly that the host polymerase or a derivative thereof with the host β subunit is involved. Analysis of host polymerase after T4 infection indicates chemical modification of the existing polymerase subunits and association with some new small phage-encoded polypeptides. The virus-induced chemical modifications of polymerase have been carefully studied in Zillig's laboratory (Skorko *et al.,* 1977). An enzyme injected during infection brings about rapid "alteration," which involves covalent binding of ADP-ribose to one of the two α subunits present in the core enzyme. A fraction of the σ, β, and β' subunits also becomes labeled with ^{32}P. Alteration is reversible and appears only transiently *in vivo*.

The second change of RNA polymerase ("modification") involves most probably ADP-ribosylation of all α subunits. Unlike alteration, modification is irreversible and requires T4-specific protein synthesis; it is one-half complete about 2 min after infection. The ability of normal and T4-modified polymerase to transcribe different classes of genes has been analyzed in a DNA-directed cell-free system for protein synthesis. Here gene expression is measured indirectly by determining the amount of gene-related protein that is synthesized (Mailhammer *et al.*, 1975). This system is composed of DNA, a so-called S-30 cell-free extract containing all the macromolecular components necessary for RNA and protein synthesis, and all the necessary salts and substrates. For the purposes of this particular study the S-30 extract was made from a rifampicin-sensitive strain. Normal or T4-modified polymerase was isolated from a rifampicin-resistant strain and introduced into the cell-free system in the presence of rifampicin. Under these conditions all transcription results from the added purified polymerase. In this way the difference in proteins synthesized in the presence of normal or T4-modified polymerase can be studied in isolation. Such studies show that T4-modified polymerase is greatly reduced in its ability to express host genes, such as the *lac* operon, the *trp* operon, and the Su_{III}^{+} Tyr-tRNA gene. Polymerase modification was not sufficient to bring about synthesis of any nonearly T4 gene products, but in conjunction with crude protein extracts from late T4-infected cells, the modification enhanced the expression of certain nonearly gene products. *In vivo* it has been shown that T4 gene products from genes 33, 45, and 55 are required for late transcription (e.g., see Rabussay and Geiduschek, 1977a). The active fraction in late T4-infected extracts necessary to bring about limited nonearly gene expression may involve these three proteins, which are known to bind to the RNA polymerase, and it may involve other factors necessary to bring about the limited nonearly protein synthesis seen *in vitro*.

A more detailed analysis of the effects of these proteins both *in vitro* and *in vivo* is necessary for a fuller understanding of the situation. In the *in vitro* system host polymerase shows a strong preference for T4 DNA compared to host DNA. This may explain the early type of inhibition of host RNA synthesis which is proportional to the multiplicity of infection but insensitive to viral protein synthesis.

A number of other lytic bacteriophages (e.g., T5, SP82, and SP01) appear to produce proteins *in vivo* which bind to the host polymerase, thereby modifying the transcriptional specificity (e.g., see Rabussay and Geiduschek, 1977a; Losick and Pero, 1977).

Sporulation is limited to certain microorganisms that undergo profound alterations during a relatively inactive metabolic period. In the soil bacterium *B. subtilis* this leads to loss of most cytoplasm and the formation of a heavy protective wall as well as antibiotic synthesis. Rifampicin-resistant mutants of *B. subtilis* have been isolated which are unable to sporulate, indicating the importance of certain features of polymerase structure to the sporulation process. Furthermore, the onset of sporulation is associated with a marked decrease in the activity of the α subunit of RNA polymerase and with the appearance of at least two sporulation-specific polypeptides that bind to core enzyme. Evidence for a sporulation-specific inhibitor of α activity has come from an examination of RNA polymerase in crude extracts of sporulating cells (e.g., see Losick and Pero, 1977). Two polypeptides with molecular weights of 85,000 and 27,000 not present in vegetative cells have been found in association with the polymerase of sporulating bacteria. Detailed studies of the promotor specificity of the vegetative and sporulation polymerases are underway in a number of laboratories.

The polymerase modifications discussed thus far involve irreversible or not readily reversible changes in the structure of the enzyme. This is correlated with physiologic situations which are irreversible in the case of viral infection and not readily reversible in the case of sporulation.

There are indications that polymerase can be modified in a highly reversible manner for purposes of modulating the rate of stable RNA synthesis according to the rate of protein synthesis. Guanosine 5'-diphosphate 3'-diphosphate (ppGpp), a compound discovered to be naturally occurring by Cashel and Gallant (1969), appears to be a pleiotropic effector of transcription present in bacteria. This compound is produced during a stalling reaction of protein synthesis on the ribosome, and the amount that is made is roughly proportional to the amount of uncharged tRNA on the ribosome. The intracellular concentration of ppGpp is about 50 μM when amino acids are abundant and as high as 500 μM when one or more amino acids are lacking. *In vivo* the so-called stringent response brought about by amino acid starvation results in severe reduction in the synthesis of the stable RNA's, rRNA, and tRNA. The ppGpp concentration rise precedes the reduction of RNA synthesis upon amino acid starvation. This is true in wild-type bacteria or in mutants known as *relA* that accumulte ppGpp at a much slower rate.

In mutants with much slower breakdown rates of ppGpp, readdition of the deprived amino acids does not allow an immediate resumption of normal rates of rRNA synthesis because the ppGpp concentration takes longer to fall to its basal level of about 50 μM. The effect of ppGpp on the expression of a number of different genes has been studied in the DNA-directed system for RNA and protein synthesis similar to that used in the T4 studies discussed above (e.g., see Reiness *et al.*, 1975). It was found that ppGpp inhibits rRNA synthesis and expression of the *E* gene of the *arg ECBH* operon, whereas it stimulates expression of many other operons including the *ara, lac* and *trp* operons. Transfer RNA synthesis as measured by the synthesis of Su_{III}^+ Tyr-tRNA synthesis is only slightly inhibited by ppGpp. From *in vivo* studies we would have expected both rRNA and tRNA synthesis to be strongly inhibited by ppGpp. The weak inhibition of tRNA synthesis seen *in vitro* may be explained by the use of two promotors for this gene *in vitro*, one of which is stimulated and the other of which is inhibited by ppGpp (H.-L. Yang and G. Zubay, unpublished results).

Among the other genes studied for ppGpp effects *in vitro*, it is of particular interest to note that the synthesis of some ribosomal proteins is severely inhibited by 300 μM ppGpp (Lindahl *et al.*, 1976). This correlates with *in vivo* evidence which shows a parallel between the stringent response for both rRNA and ribosomal proteins; these results also lend credence to the use of the *in vitro* approach for assessing the effects of ppGpp on gene expression. The concept of the function of ppGpp that emerges from the *in vitro* studies is that of a general transcriptional effector which exerts its influence in a gene-specific manner. Much recent work has been done in more purified cell-free systems in an attempt to pinpoint the site of action of ppGpp. Such systems usually contain a suitable template for transcription of rRNA, RNA polymerase, and sometimes partially purified fractions from whole cell extracts. No consistent or convincing picture has emerged from these studies, since the amount of transcription of rRNA is greatly reduced in the absence of unknown factors from whole cell extracts and both the amount of transcription and the effect of ppGpp vary greatly with the ionic strength used. It does not seem likely that this situation will be clarified until mutants in the target site for ppGpp action have been isolated. Then a comparison of *in vitro* and *in vivo* results with components obtained from normal and mutant cells should lead to a firm conclusion as to the site of action of ppGpp.

D. Polymerase Replacement

When the lytic bacteriophage T7 infects *E. coli,* transcription takes place from the so-called left-hand end of the genome, and 20% of the DNA is transcribed as a single RNA molecule. Among the four genes located in this region the only one indispensable for virus growth encodes an RNA polymerase. This polymerase is far simpler than the host polymerase, being composed of a single protein subunit with a molecular weight of 107,000. The T7 polymerase is resistant to inhibition by rifampicin and streptolydigin, two antibiotics that inhibit bacterial RNA polymerase. The T7 polymerase transcribes the r strand of T7 DNA, initiates RNA chains only with pppG, and elongates these chains very rapidly *in vitro;* an average rate of 230 nucleotides/sec has been measured at 37°C, which is about five times faster than the rate of transcription of the left-hand end of T7 RNA by *E. coli* polymerase. The late transcription pattern, which arises mainly from the right-hand 80% of the genome, results exclusively from the action of the T7 RNA polymerase. Unlike early transcription, the late transcription pattern appears complex, with as many as seven distinct promotors and two different termination sites being recognized. Not surprisingly a complex array of RNA's is produced in late transcription. It is presumed that these late promotors are structurally related since they are all recognized exclusively by the T7 polymerase. As might be expected the early promotors recognized by *E. coli* polymerase cannot be recognized by T7 polymerase. The situation with T7 represents an extreme case where, to bring about specific promotor recognition, a new polymerase is created. This is also true of the closely related T3 phage.

As in the case of T4, T7 phage infection results in the inhibition of host RNA synthesis. In the latter case the host polymerase is completely turned off by proteins encoded by the early T7 genes (Hesselbach and Nakada, 1977). Since host polymerase is not needed for viral replication after early transcription, it is not surprising that T7 takes this more obvious route to inhibiting host transcription.

Transcription from the double-stranded DNA bacteriophage N4 is resistant to rifampicin even when the drug is added prior to infection. Apparently N4 does not use host polymerase even during the early stages of transcription. Two classes of RNA are synthesized from N4 DNA after infection, one which does not require translation of the phage genome after infection and one which does. The first class of RNA is believed to be synthesized by a virion-encapsulated RNA polymerase which is injected into the cell with the DNA (Falco *et al.,* 1977). The second class of RNA's is presumed to be transcribed by another virus-encoded polymerase made after infection.

E. Structural Alteration in the Template

Modifications in the primary, secondary, and tertiary structure of already formed DNA can exert an influence on the polymerase recognition and transcription properties of a gene. It is highly likely that some of these changes in structure are important *in vivo*.

Primary structure changes to be considered as significant to some events in transcription include transient single-strand cleavage and base modification. Single-strand cleavage is known to encourage transcription *in vitro* at otherwise inactive initiation sites. Mature T5 bacteriophage contains five single-strand breaks in one of the DNA strands. The best known cases where base modifications affect the transcription pattern involve the hydroxymethylation of cytosine, which is an aid to late transcription in T4 phage and probably in other T-even phages. However, it is not clear how base modification affects transcription in such cases.

Gross changes in secondary structure involving the transition from a duplex to a single-stranded structure may play a role in regulating transcription. In certain viruses such as T4 some late genes are only transcribed at the time when the DNA is being replicated. Since replication involves the transient appearance of single-stranded DNA, a possible explanation for this is that some promotors are only recognized when the template is in the single-stranded form. However, recently some doubt has been cast on the importance of concomitant DNA synthesis to late viral gene expression by the observation *in vitro* that whereas prior DNA replication seems to be essential for late gene expression, it is not essential for the two processes to be occurring simultaneously (Rabussay and Geiduschek, 1977b). More convincing evidence that single-stranded DNA can be important in some forms of transcription comes from work on the phage N4. The N4 phage has already been discussed in Section III,D, because of the remarkable occurrence of two viral-encoded RNA polymerases. One of these viral polymerases is found associated with the mature virion. This polymerase is responsible for early viral RNA synthesis (class I RNA) after infection. *In vitro* this polymerase will only transcribe from single-stranded DNA. This requirement for a single-stranded template *in vitro* suggests that at least part of the viral DNA exists in a single-stranded form when it is active in early transcription *in vivo,* although direct proof of this is lacking.

It has been known for several years that the introduction of negative supercoils into the DNA structure can substantially stimulate the *in vitro* template activity for transcription. Comparison of the RNA's made from relaxed and supercoiled DNA indicates that negative supercoiling increases the activity of already functioning promotors and leads to active initiation from previously inactive regions. Supercoiling also appears to be important in DNA replication and recombination reactions. The most likely explanation for the effect of negative supercoiling on transcription is that the formation of rapid start complexes requires a partial unwinding of the double helix, which should introduce positive supercoils into a closed covalent circular double helix molecule (cccDNA). If the cccDNA is in an activated supercoiled state, the energy for unwinding the double helix could in part be supplied by the partial relaxation of this supercoiled structure. Direct evidence that the main effect of negative supercoiling on stimulating RNA synthesis is on the formation of the rapid start complex and not on elongation has come from the measurement of the various parameters in purified *in vitro* systems (e.g., see Richardson, 1975; Seeburg *et al.*, 1977).

DNA which has been isolated from whole cells usually contains a high degree of negative supercoiling when examined *in vitro*. Even the *E. coli* chromosome when carefully isolated without removing the RNA and protein is in a highly supercoiled state (Worcel and Burgi, 1972). This is most intriguing and surprising, since even a single nick in one strand of the chromosome should allow the DNA to adopt the more thermodynamically stable structure with no supercoils. Evidently the complex of DNA with RNA and protein introduces sufficient local constraints on axial rotation to allow the majority of the chromosome to adopt a highly supercoiled state. This picture of the *E. coli* chromosome is reinforced by Kavenoff and Bowen (1976), whose electron microscopic observations of carefully prepared chromosomes indicate the presence of about 100 supercoiled loops of DNA radiating from a central core of RNA. The mystery as to how the high energy supercoiled structure is formed and maintained *in vivo* has been partly answered by Gellert *et al.* (1976), who discovered the enzyme gyrase, which converts relaxed closed circular DNA into the high energy negatively supercoiled structure in an ATP-dependent reaction.

The significance of gyrase and supercoiling for various transcription processes can now be assessed with the help of the antibiotics novobiocin and coumermycin, which specifically inhibit the gyrase. *Escherichia coli* cells are normally sensitive to novobiocin. Insensitive mutants have modified gyrase protein. Novobiocin studies show that not only host survival but also the replication of many phages is dependent upon the gyrase. Whether the crucial reactions involve DNA or RNA synthesis or both is not known at this time. The large effect of supercoiling on transcription patterns necessitates that much further work on the regulation of transcription be done so that the effects of supercoiling can be ascertained. The importance of such studies cannot be overemphasized, since supercoiling may have both qualitative and quantitative effects depending upon the gene system.

Recently the effect of supercoiling on the expression of a number of genes has been examined in a coupled cell-free system (Yang *et al.*, 1979;

H. L. Yang and co-workers, unpublished observations) for transcription and translation. This has been done by using the same DNA either in the supercoiled or in the relaxed circular configuration. The relaxed circular DNA will normally supercoil in this system as a result of DNA gyrase action. However, the action of gyrase can be blocked by gyrase inhibitors such as novobiocin. The results show a wide range of gene sensitivities to supercoiling. Negative supercoiling never inhibits gene expression but the level of stimulation which it causes varies from negligible to tenfold or more depending upon the gene. Small super-coiled DNAs active in transcription lose most of their supercoiling as a result of DNA unwinding and DNA–RNA hybrid formation. This raises a most sophisticated consideration about gene activity for genes which are highly dependent on supercoiling. It seems likely that such a gene located near another very active gene will show a lower level of expression because of the decrease in regional supercoiling caused by DNA–RNA hybrid formation. Two genes interacting in this way could be quite far apart and still affect each others' activity. All that is required is that the two genes be situated in the same supercoiled "unit" of DNA. Small plasmids probably constitute a single unit in which the restricted axial rotation essential to maintaining the supercoiled state results from the closed, covalent circular structure. Chromosomal DNA clearly must contain many such units in which the restricted axial rotation essential to maintaining the supercoiled state results from the closed, covalent circular structure.

F. Positive or Negative Control Factors Other Than Polymerase

The initiation locus for transcription contains a rapid start binding site for RNA polymerase and frequently a binding site(s) for one or more regulator proteins. Regulator proteins that function by binding to the DNA either inhibit or facilitate polymerase interaction with its binding site.

1. Negative Control Systems

In *negative control* systems the regulator protein *inhibits* polymerase interaction. Some negative control regulator proteins bind strongly by themselves and are called *repressors*. The *lac* operon is regulated in part by a repressor protein. This operon is not expressed as long as the *i* gene repressor (a highly specific repressor encoded by the *i* gene) binds to a segment of the initiation locus. In order for the *lac* operon to be expressed, the repressor must be removed from the initiation locus. Spontaneous dissociation of the repressor is very small because of the low dissociation constant for this complex (around 10^{-11} mole/liter under physiological conditions. Appreciable dissociation requires direct interaction of repressor with a small molecule modulator known as the *antirepressor* (or inducer). Binding of the antirepressor to the repressor is believed to produce an allosteric transition that drastically lowers the affinity of the repressor for the initiation site on the DNA. For the *lac* operon a derivative of lactose (allolactose) or various synthetic derivatives such as isopropyl-β-D-thiogalactoside (IPTG) serve as antirepressors. A negative control regulator protein that must combine with one or more molecules in order to function is referred to as an *aporepressor*. An aporepressor is converted to active repressor by combination with a *corepressor*. The *trp* operon consists of a cluster of five structural genes involved in the biosynthesis of tryptophan. This operon is repressed at one end by the binding of an aporepressor–corepressor complex. The aporepressor is a protein encoded by a distantly located *trp R* gene and the corepressor is tryptophan.

The selective advantages of the two control systems just described seems clear. The *lac* operon, which encodes enzymes for metabolizing lactose, is silent in the absence of lactose. If insufficient lactose is present, the enzymes would serve no purpose. The *trp* operon is inactive at high levels of tryptophan. As long as there is an adequate supply of tryptophan, there is no need for the enzymes responsible for the biosynthesis of tryptophan.

The genetic and biochemical evidence in support of the repressor model for control of the *lac* operon is overwhelming. This system will be discussed in some detail since more is known about it than any other negatively controlled system. The genetic map of the *lac* region was shown to contain three contiguous structural genes, Z, Y, and a, with an adjacent control locus known as the promotor–operator region. Another control locus or gene called i is located close to the operator (O) locus. The three structural genes code for different proteins involved in lactose metabolism. In particular, the Z gene codes for the enzyme β-galactosidase (β-gal), which hydrolyzes lactose to its component monosaccharides.

Expression of the Z, Y, and a genes is very low unless lactose or a synthetic analogue of lactose such as IPTG is present in the growth medium. Under these conditions one sees a thousandfold increase of *lac* operon proteins in the cell. In the absence of lactose or IPTG, large quantities of the *lac* proteins can also occur by mutation of O to O^c or i^+ to i^-. This suggests that O and i are involved in regulation. In partial diploids (merodiploids), it was found that O^c is cis dominant to O^+. Thus, the O locus affects only those genes to which it is in direct apposition. In contrast, i^+ is dominant to i^- in the cis or trans position. The results of these two dominance tests led Jacob and Monod to the operon model of control, which stipulates that the i gene makes a diffusible product (repressor) that normally binds to the O locus (operator), preventing expression of the operon. The small molecules lactose or IPTG disrupt this repressor–operator complex by binding to the repressor, thereby permitting expression of the operon.

Further understanding of operon control required that cell-free biochemical techniques be used to dissect from the cell each control factor, study it in isolation, and demonstrate the effects which occur when the control factors are recombined in the purified state. In both crude and purified cell-free systems it has been shown that repressor blocks transcription. Also, in specially contrived situations where the initiation site is moved far to the left of the repressor binding site, it has been shown that repression is far less efficient in inhibiting transcription. Evidently repressor is less efficient at blocking elongation than initiation. Other experiments have been done to show that antirepressor IPTG binds to repressor. Indeed, it was through use of radioactively labeled IPTG as a detector that repressor was first purified by Gilbert and Müller-Hill (1967).

Interaction studies have shown that repressor binds to DNA in general but several orders of magnitude more strongly to DNA containing the O site. DNA containing operator constitutive (O^c) alterations binds repressor poorly, and under all conditions antirepressor dissociates the complex of DNA and repressor. As the ionic strength is increased, the strength of the repressor–operator complex is greatly reduced, showing that electrostatic interaction between the negatively charged phosphates of the DNA and positively charged groups on the protein contributes substantially to the binding. Binding to single-stranded DNA is poor and does not show significant site specificity. This plus the fact that binding is very rapid and complete at low temperature argues against major structural changes in the DNA as a result of repressor binding. It seems likely that repressor binds to the double helix without significant disruption of the double helix base-paired structure. This does not rule out minor conformational changes, such as slight bends or change in pitch of the double helix to obtain an optimal fit between the interacting sites on the DNA and the repressor (e.g., see Goeddel et al., 1978).

Numerous quantitative relationships were established in the cell-free system which would be difficult or impossible to establish in whole cells. Thus, it was shown by varying repressor concentration that one repressor molecule is sufficient for inhibiting gene expression. It was also shown that in the presence of excess repressor that gene expression is directly proportional to the square of the IPTG concentration. The latter fact suggests that derepression requires two antirepressor molecules. It has been proposed that the tetrameric repressor molecule contains two operator binding sites and four inducer binding sites (Zubay and Lederman, 1969;

Zubay and Chambers, 1971) and that each operator binding site is sensitive to two of the inducer binding sites so that the tetrameric molecule can be visualized as two functionally separate parts. Most arrangements for tetramers would lead to more than one operator binding site from symmetry considerations alone. Thus, tetramers can be arranged with tetrahedral symmetry or other symmetries which give three orthogonally related dyad axes. The sugar–phosphate backbond of the DNA has a dyad axis, and the alignment of one of the protein dyad axes with that of the DNA might produce the best interaction. Furthermore, if the sequence of bases in the operator region on one DNA chain were the same as that in the other DNA chain, a perfect dyad axis would exist for the operator region of the DNA. It is probable that such an arrangement would give the best interaction with a repressor molecule of similar symmetry.

I would not be surprised if most repressor–operator complexes obeyed these symmtry rules. Indeed all sequenced promotors that interact with or are believed to interact with repressors have substantial dyad symmetries in the presumptive repressor binding region o. The concept would appear to have enormous predictive value for the analysis of promotors whose behavior is less well known. Thus if one finds substantial segments of dyadic symmetry in the region of the promotor there is a high probability that this is indicative of a repressor binding site. If this dyadic rule is universal, then all repressors in the active form should be multimers containing an even number of identical subunits, for only with such structures are dyad axes in the protein possible. In the DNA not all bases in the region of interaction need show the symmetry for two reasons: (1) only a fraction of the bases interact directly with repressor and (2) the repressor site overlaps regions in the DNA serving other functions, such as the polymerase binding site or the initiation point for transcription.

Sequence studies have yielded most important information about the binding sites for polymerase and repressor. The sequence of bases in the promotor–operator region of the *lac* operon and the surrounding areas are indicated in Fig. 3. In this figure the bases are numbered from position $+1$, which indicates the initiation point for transcription (-1 is the first base away from the transcriptional start). Protection experiments in which operator containing DNA complexed with repressor is digested with DNase indicate that repressor binds between base pairs -5 to $+21$. In the discussion above (section IIA) arguments were raised that the polymerase binding site is situated approximately between bases -38 and $+2$. Clearly the repressor and the polymerase binding sites have considerable overlap so that the binding of one protein could preclude the binding of the other. In Fig. 3 the repressor binding sites for three other operons *trp, λP_R* and λP_L are also indicated. The position and size of the repressor binding sites

Fig. 3. DNA sequences in the region of the repressor and polymerase binding sites for the *trp, lac, *λP_R and λP_L promotor–operators. The polymerase rapid start complex forms between about -40 and $+10$ where $+1$ is the initiation point for transcription. Regions containing Pribnow boxes (PBS2) and entry boxes (PBS1) are indicated by brackets. Approximate repressor binding sites are bracketed. Bases showing dyadic symmetry about approximate centers of repressor binding sites are overlined and underlined. Repressor binding sites are defined by regions which are protected by repressor from DNase action; they also include sites of mutants to operator constitutivity ($O^+ \rightarrow O^c$). See text for further explanation.

vary somewhat, but in all cases there is considerable overlap with the polymerase binding sites. Whereas there are other places that repressor could bind to prevent initiation, this may be the most common situation. It does create one obvious complication, which is that the nucleotide sequences which interact specifically with polymerase and repressor must be intermingled to some extent.

In the case of the *lac* and *trp* repressor, there is an additional complication. A portion of the transcribed region is also included in the repressor binding site. In the region of the *lac* operator we find a striking display of dyadic symmetry. Twenty-eight out of 34 base pairs in the -7 to $+28$ region form a perfect dyad. Two interpretations have been offered for the existence of this symmetry: the one made above that the DNA binding site on repressor should have a dyad axis of symmetry because it is composed of an even number of identical subunits. Such a structure should interact optimally with DNA that has a similar dyad axis of symmetry. An alternative hypothesis suggested by Gierer (1966) is that the twofold symmetry in the sequence represents the availability of a second arrangement of the DNA, in which the two strands are pulled out and rewound around themselves to form a cruciform structure, the complementary base pairs interacting within each strand separately. Such a structure could be stabilized by binding to the repressor protein and would permit the protein to interact with DNA in a complex three-dimensional way. Although the Gierer model would create a structure with many potential advantages for interaction, it is considered an unlikely possibility. The extremely rapid reaction of repressor with operator-containing DNA even at 0°C argues against such radical alterations in the DNA structure.

Gilbert et al. (1977) have reviewed two types of evidence relating to the importance of various positions in the operator site to repressor binding: (1) base changes brought about by mutation that alter the position of the interaction of the repressor and (2) a determination of which bases in the sequence the repressor protects against chemical attack. These studies do not directly support the arguments about symmetrical interaction but suggest that the bases toward the center of the sequence (from +3 to +19 in Fig. 3) are most influential in repressor binding. Sometime ago Lin and Riggs (1974) showed that the *lac* repressor binds tenfold more tightly to bromouracil-substituted *lac* operator than to unsubstituted *lac* operator. Most recently Ogata and Gilbert (1977) have obtained more direct evidence that most of the bromouracil-substituted thymines in the central 22 base region of the operator can be cross-linked to repressor. Taken together these observations extend the range over which close contacts between repressor and operator can be demonstrated and suggests the importance of DNA major groove contacts since this is where the bromo substituent is located.

Evidence that dyadic symmetry is an important factor in the repressor binding site comes from parallel studies on the *trp*, λP_R, and λP_L repressor binding sites. In all cases substantial dyadic symmetry is found in the operator binding sites (see Fig. 3 and Ptashne, 1975; Bennet et al., 1978). A full evaluation of the importance of the symmetric and asymmetric regions in the DNA to repressor binding will require much further work and possibly new approaches, such as X-ray crystallography.

One of the most useful conclusions to come out of the Gilbert approach to studying repressor–promotor interaction is that all regions of the DNA which interact with repres: r are located on one side of the double helix (Goeddel et al., 1978). This gives the picture of a side-by-side interaction rather than of a protein wrapping itself around the double helix. This type of situation also may be a general characteristic of activator protein–DNA interactions as discussed below for CAP.

The simple negative control systems considered here appear to function in a very simple and efficient manner by binding at a site overlapping the polymerase binding site. Frequently, negative control regulator proteins are present at a concentration of only ten to twenty copies per cell. This is all that is needed because they bind strongly and selectively. *A priori* the greatest limitations of the negative control system are twofold, the necessity for binding close to the polymerase binding site and the impossibility of using two negative control systems in tandem when it is desirable to have alternative or combined signals for turning on a particular gene or operon. In part this may account for the evolution of more complex positive control systems in the prokaryotic cell.

2. Positive Control Systems

In the simplest positive control systems the regulator protein facilitates polymerase interaction with the DNA. Inherent in the concept of the positive control system is the notion that the DNA sequence is contrived so that the RNA polymerase molecule by itself has difficulty forming a rapid start initiation complex. A positive control protein that binds strongly without a cofactor is called an *activator*. A small molecule modulator that prevents the binding of the activator is called an *antiactivator*. A positive control protein which cannot function by itself is called an *apoactivator* and the necessary attendant molecule(s) known as *coactivator(s)*.

The ara C protein and the CAP protein are parts of two positive control systems which have received a great deal of attention [for a recent review of work on CAP see Zubay (1979)]. Both of these proteins are apoactivators, and the coactivators are L-arabinose and cAMP, respectively. In this section we will concentrate on the mechanism underlying the action of CAP, the catabolite gene activator protein; the ara C protein will be discussed at a later point.

The discovery of CAP involved close collaboration between geneticists and biochemists (Zubay *et al.,* 1970; Zubay, 1969). First, it was found that cAMP reverses the effect glucose has on inhibiting the synthesis of β-galactosidase (β-gal) and a multitude of other enzyme systems associated with catabolism. Second, it was found that cAMP increases the synthesis of β-gal in a DNA-directed system for protein synthesis by 20- to 30-fold. From this it was inferred that there must be a positive control protein which reacted with cAMP to stimulate *lac* expression as well as expressions of a number of other genes associated with catabolism that were inhibited when glucose was present in the growth media. Selection techniques were devised to detect a mutant that could not make catabolic enzymes even when cAMP was present. Such a mutant was used as the source of cell-free extract and found to be defective in the *in vitro* synthesis of β-gal even when cAMP was present. This defect was corrected by adding extract from wild-type cells from which the active factor CAP was isolated.

The CAP protein (encoded by the *crp* gene) is a dimer composed of identical monomers with a molecular weight of 22,000. CAP binds cAMP and binds to DNA when cAMP is present. CAP has no visible affinity for RNA polymerase when the two proteins are mixed together in solution. Mutants in the *lac* promotor which eliminated the stimulation of β-gal synthesis by the cAMP–CAP complex showed that a site necessary for CAP interaction was separate from the polymerase initiation site and further removed from the structural genes. Binding studies (Majors, 1975; Mitra *et al.*, 1975) have shown that whereas CAP has a general affinity for DNA which is greatly augmented in the presence of cAMP, it binds preferentially to fragments of DNA containing the *lac* promotor–operator region. A general model for CAP action was proposed (Zubay and Chambers, 1971) that CAP functions by complexing to a segment of the promotor adjacent to RNA polymerase, thereby facilitating the formation of the RNA polymerase initiation complex.

The synthesis of β-galactosidase from the *lac* operon *in vitro* was found to be proportional to the concentration of CAP and to the square of the concentration of cAMP, suggesting the involvement of two cAMP molecules in the formation of an active CAP complex. Since CAP is a dimer and therefore probably has dyadic symmetry, the presence of two cAMP binding sites on the molecule seems highly likely. By arguments identical to those used in the prediction of dyadic symmetry in the repressor binding site, it was predicted that the CAP binding site on DNA would show dyadic symmetry (Zubay and Chambers, 1971). The approximate location of the CAP binding site was defined by a series of genetic deletions to be in the -50 to -80 base pair region. Inspection of the sequence in this region reveals a 14 base region between -55 and -68 (see Fig. 4) which shows dyadic symmetry for 12 of the 14 base pairs. This is believed to be the site of CAP interaction. Gilbert (1976) suggested this location from data which showed that single base pair changes that block the function of CAP at the *lac* promoter involve GC to AT transitions at positions -66 and -59, respectively (indicated in Fig. 4); these sites are symmetrically disposed about the dyad axis of symmetry. The sequence in the *gal* promoter shows some similarities in the -60 region (see Fig. 4). Twelve out of 16 bases in the -60 region are located on a dyad axis of symmetry. Only the outer three base pairs are both on the dyad symmetry axis and similar in composition to those found in the presumptive CAP site in *lac* (see Fig. 4).

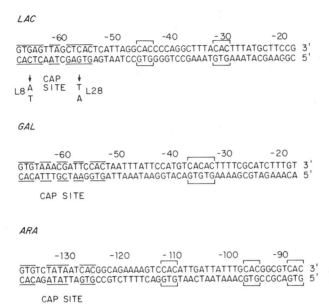

Fig. 4. Selected regions of the *lac*, *gal*, and *ara* promotor–operator regions including presumptive CAP binding sites. Sequence data from Dickson *et al.* (1975) for *lac*, Musso *et al.* (1977) for *gal*, and B. Smith and R. Schleif (unpublished results) for *ara*. Regions showing dyadic symmetry about the center of the presumptive binding sites are overlined and underlined. Numbering of bases is consistent with that used in Fig. 3. Regions containing reiteration of the right most triplet in the binding site are indicated by brackets. See text for further explanation.

Based on this it seems likely that the critical sequence in a CAP site is

$$
\begin{array}{l}
\text{GTG——————————CAC} \\
\text{CAC——————————GTG}
\end{array}
$$

In *lac* there are 8 base pairs in between this critical sequence, whereas in *gal* there are 10. Recently it has been contested that this site in the *gal* operon is the CAP binding site (Di Lauro *et al.*, 1979). Further work is necessary to resolve this point. Recently sequence data for the promotor–operator region of the *ara* operon (Smith and Schleif, 1978; Greenfield *et al.*, 1978) has been obtained. In the region of the *ara* promotor from −138 to −125 there is a 14 base pair segment with 10 of the 14 base pairs located on a dyad axis of symmetry, and the outer 3 bases show the same structure as the presumptive CAP binding site in *lac*. Thus this may be the CAP binding site in the *ara* operon. Greenfield *et al.* (1978) have made the same proposal. In *lac* and *gal* the function of CAP binding is probably to facilitate formation of the rapid start complex. In the case of *ara* the function of CAP binding is probably to facilitate entry of the ara C regulatory protein to its positive control site as explained below. In order to understand why *lac* and *gal* require CAP for activation, it seems useful to compare the sequences in these promotors with those in promotors which are activated by polymerase alone. It is most striking that in the -35 region required for polymerase complex formation one finds the heptanucleotide TGTTGAC in *trp*, λP_R and λP_L, whereas in *lac* only 4 bases agree with this sequence and in *gal* the agreement is even poorer (see Figs. 3 and 4). A *lac* mutant called UV5 shows a high level of CAP-independent expression; it contains two changes at positions -8 and -7 in the Pribnow box (PBS2, which produce the ideal sequence for this polymerase binding site. It seems likely that CAP facilitates formation of the polymerase–DNA complex in cases where the PBS1 sequence is appreciably different from the optimum sequence; how CAP does this is a matter of speculation. As in the case of repressor, CAP binding is very rapid even at the lowest temperatures, arguing against radical structural alteration of the DNA as a result of CAP binding (S. Mitra and G. Zubay, unpublished results). However, the detection of a specific complex of CAP with double helix DNA does not preclude some conformational alteration of this complex prior to polymerase binding. This is difficult to assess partly because polymerase does not form a rapid start complex even after CAP binding unless the temperature is raised.

One further similarity in the three CAP-sensitive promoters is the reiteration of the right half of the presumptive CAP binding site $_{GTG}^{CAC}$ (see Fig. 4). Whether this is relevant or not to CAP action is unclear, since no known mutants eliminating these reiterations have been characterized. If the reiterated triplets in question turn out to have no affect on operon activity, then simple explanations for CAP action may be adequate. These can be conceptually divided into two categories: In the first, CAP binding brings about limited structural alteration of the DNA facilitating polymerase binding. In the second, CAP binding stabilizes polymerase binding to an adjacent site by virtue of an affinity between the two proteins (suggested by Gilbert, 1976). I favor the Gilbert suggestion, particularly in view of the results of Siebenlist (1979) mentioned above which show that the only unpaired bases in the rapid start complex are in the -8 to $+2$ region of the promotor.

Most recently John Majors (personal communication) has significantly advanced our understanding of how CAP binds at the *lac* promoter. The alkylating agents, dimethyl sulfate (DMS) and ethylnitrosourea (EtNU) were used as probes for determining close contacts between CAP and DNA. DMS reacts with the adenine N3 and the guanine N7 positions whereas EtNU reacts with the phosphates. Experiments with the DMS probe showed that CAP binding protected the two outer guanines on each side of the binding site from methylation; if premethylated these groups strongly inhibited CAP binding. The pattern of protection observed demonstrates symmetrical interaction in two adjacent large groves of the DNA. Experiments with EtNU both confirm the importance of the symmetry and show that, like *lac* repressor, CAP interacts with only one side of the DNA helix over the region covered by the symmetry axis.

G. Factors Affecting Termination

Termination could be used to regulate transcription under conditions where termination is provisional. The most likely genes to be regulated by a provisional termination signal are those which occur after the signal. Regulation of termination was first demonstrated in the case of phage λ (see Section II,C). Early right and early left transcription in λ are terminated by rho (see Fig. 1). This termination event is antagonized by the N gene product of λ. In the presence of adequate N protein, transcription continues beyond the provisional termination signals. This has been demonstrated both *in vivo* and *in vitro*. The antitermination activity of N relates not only to the termination site but to the promotor. Thus, rho-sensitive termination signals not normally sensitive to the antitermination effects of N become sensitive if the genes are artificially rearranged so that initiation is from the P_R or P_L promotors. Genetic deletion studies indicate that N action depends on a region downstream from the P_L promotor and just to the right of the gene *cro,* which is the first gene transcribed from the early right promotor P_R. Models for what happens in these regions to make N an effective antiterminator at some distance have been discussed in Section II,C.

The importance of N protein in antitermination is supported by the observation that λN^- mutants, which do not grow on wild-type cells, can grow on certain *rho* mutants. The λQ gene is required for late transcription to the right in λ. It has been suggested by Roberts (1975) that the Q gene product also functions by acting as an antiterminator of rho. It is believed the Q protein prevents termination of the $\lambda 6$ S RNA (see Fig. 1), thereby stimulating late transcription. Roberts says this proposal requires further investigation.

H. Factors Affecting Attenuation

Attenuation is defined as termination of RNA synthesis before the usual termination site. Under conditions where attenuation is operative, the affirmative decision to make an mRNA can be made at some point downstream from the initiation site as well as at the promotor. In this event, a nascent transcript is prematurely terminated. *A priori* it would appear that the advantage of such a situation is that it allows for greater amplification in the regulatory response and the use of different regulatory devices responsive to different needs of the cell. Since most of the evidence for an attenuation device comes from studies of the *trp* operon, we will focus on this system here. In the *trp* operon an attenuation site which leads to premature termination is located about 141 nucleotides from the initiation point. A description of this system as well as a mechanism involving rho and Trp-tRNA in the termination process has been given in Section II,D. When attenuation is optimal, only about one out of ten RNA's initiated at the promotor elongate beyond 141 nucleotides. Of those chains which do elongate beyond this point, most are believed to terminate at the end of the *trp* operon. What are the potential advantages of the attenuator in this case? First, it allows for greater overall amplification in the RNA synthesis. For example, if the amplification factor at the attenuator is 10 as indicated above and if the amplification factor at the promotor is also 10 then the overall amplification factor in going from a fully repressed to a fully derepressed level of Trp-mRNA synthesis should be 100. Second, the use of two controls in series allows one to use different signals to stimulate a positive response. At the promotor the tryptophan concentration is the signal, whereas at the attenuator the charged Trp-tRNA concentration appears to be the signal. Assuming that the main if not the only function of tryptophan in *E. coli* is to make up part of a protein, these would appear to be very similar signals. However, one can conceive of circumstances where despite a reasonable concentration of tryptophan the charged Trp-tRNA supply is abnormally low. This would serve as a signal for the cell to synthesize more tryptophan by encouraging the synthesis of Trp-mRNA synthesis beyond the attenuator site.

The model described for attenuation in Section II,D is elegant in its simplicity. The specificity of the attenuation depends upon the location of two tandemly located codons for tryptophan in a special leader sequence of bases in the RNA and otherwise uses factors (Trp-tRNA, rho factor, and RNA polymerase) that are already present for other purposes. The general mechanism could be readily extended to other amino acid operons merely by replacing the trp codons by two codons for another amino acid. In this connection one is reminded of the enormous amount of skillful investigation conducted by Phillip Hartman's and Bruce Ames' laboratories in search of a classic repressor gene for the histidine operon. Not only has no such repressor ever been found but everything points to charged His-tRNA as being the key factor in control (e.g., Brenner and Ames, 1971). This is exactly what one would predict if there were no classic regulation at the promotor but only an attenuator site for control downstream from the initiation site. Recent evidence on the existence of an attenuator in the histidine operon has been cited in Section II,D. The so-called operator constitutive mutants of the histidine operon which lead to "derepressed" synthesis of the histidine biosynthetic enzymes could be reinterpreted as mutants in the attenuator region. Whereas a detailed model for regulation by attenuation has been made for an amino acid operon, it is not difficult to imagine that attenuation could be involved in quite unrelated situations using somewhat different control factors.

I. Factors Affecting Polarity

Certain structural gene mutations not only affect the gene in which they occur but also limit the expression of other promotor distal genes. Genes between the mutation and the promotor are not affected. Such mutations are called polar. Nonsense mutants, insertions, or deletions often lead to polarity. In Section II,C, it was pointed out that *rho* mutants greatly alleviated polarity and therefore that the mechanism underlying polarity might be similar in some respects to the mechanism of rho-catalyzed termination and attenuation. If this is the case, the reduced expression of operator distal genes after the polarity mutation might be expected to result primarily from a reduced synthesis of RNA after the polar mutation. A gradient of polarity exists along the gene such that a nonsense mutation close to the beginning of a gene has a much greater polarity than the same mutation near the beginning of the next gene. The gradient is not a completely smooth affair, and it has been suggested that this is due to intracistronic translational restarts (e.g., see Zipser, 1970). This is consistent with the involvement of rho. If the ribosome is able to reattach a short distance beyond the nonsense mutation, it does not leave much unprotected single strand RNA for rho to attack. The type of polarity brought about by mutations from normal to nonsense triplets is a rare and random affair. By contrast the polarity brought about by insertion or translocation of insertion sequences (discussed in Section III,B) occurs frequently and is nonrandom, making it much more amenable to use in the systematic regulation of transcription. Again, that polarity is involved is indicated by the requirement for rho. Examples of such cases were given in Section III,B. It seems likely that many more examples of this type of regulation will be discovered in the future.

There is another type of polarity phenomenon sometimes called "natural polarity," which may be important in regulating gene expression in multicistronic operons. It is believed that in some operons, such as *trp* and *gal,* all genes within the operon are equally expressed, suggesting that they are transcribed and translated by the same amount. By contrast there is some indication that the *his* and the *lac* operons show a "natural polarity," i.e., there is less protein made per gene length for promotor distal than for promotor proximal genes. It has been hypothesized (Ames and Hartman, 1963) that in such cases the order of genes in the operon might be determined by the relative quantities of each enzyme needed to maintain a smooth flow of intermediates through the pathway. Natural polarity probably results from an event which takes place in the intercistronic region. Since detailed studies have not been made of the RNA synthesized for such operons, it is not possible to say whether the lowered expression of promotor distal genes is due to a falling off of ribosomes at the intercistronic regions or a rho-induced termination of transcription or both. Current evidence favors the former event, since the relative amount of *lac* operon encoded proteins is unaffected by alterations in rho (J. Beckwith, personal communication). If rho was involved in this sort of polarity one would have expected a relative increase in the promotor distal gene products in cells containing defective rho.

An artificially contrived situation involving the juxtapositioning of the *gal* operon at different distances from the λP_L promotor may serve as a useful model for understanding some aspects of polarity (Adhya *et al.,* 1976). The closer the *gal* operon is brought to the λP_L promotor the more effect N protein is at stimulating *gal* expression. Some aspects of this situation have been discussed in Section II,C. Any genes which are read from either λP_L or λP_R show some escape from rho-induced termination in the presence of the λ N protein. However, this effect of N is not 100% with the consequence that each time a termination signal is reached there is a possibility of some read through and some termination. This leads to a decreasing gradient of RNA synthesis as the distance from P_L and the number of intervening termination signals increases.

IV. HOW DIFFERENT COMBINATIONS OF FACTORS ARE PUT TOGETHER TO CONTROL SPECIFIC GENES

Despite the limited state of knowledge about regulation of transcription that is currently available, it is clear that the expression of most genes or operons results from more than one controlling factor. These factors are put together in combinations that suit the cell or viral needs for the particular gene products. Thus far, almost every system which has been investigated has brought to light new possibilities and new principles. Different gene systems have evolved in their own way, each taking advantage of the large variety of mechanisms that are intrinsically available to regulate transcription. In this section a select group of systems will be discussed which illustrate the use of different combinations of devices for transcriptional control.

A. The *lac* Operon

The *lac* operon is composed of a control region and three structural genes, Z, y and a, which code for the three enzymes β-galactosidase, β-galactoside permease, and thiogalactoside transacetylase, respectively. The function of this system is to concentrate lactose and β-galactosides from the growth medium (y gene function) and to hydrolyze them to monosaccharides (Z gene function). No definite role has been assigned to thiogalactoside transacetylase.

Monosaccharides comprise an important source of carbon–carbon bonds for the synthesis of other compounds and energy production. In the *lac* operon there is evidence for three transcriptional control elements which comes partly from *in vivo* and partly from *in vitro* observations. The small molecule modulators for these controls are lactose, cAMP, and ppGpp. Lactose (or rather a derivative of lactose known as allolactose) is the antirepressor for the highly specific *i* gene repressor, which affects the activity of the *lac* operon by binding to the operator region of the operon. This repressor binding site partially overlaps the polymerase binding site as discussed above (see Fig. 3 and 5). Significant activation of the *lac* operon requires removal of the repressor from its operator binding site. Somewhat further removed from the structural genes is a binding site for CAP (see Figs. 4 and 5). The CAP protein in combination with cAMP is a general activator for many catabolite-sensitive genes, probably hundreds in *E. coli*. In cAMP⁻ (cya⁻) or CAP⁻ (crp⁻) strains *lac* expression only reaches about 5% of its maximum. In the presence of CAP and high levels of cAMP, the expression of the operon increases about twenty-fold *in vivo* and *in vitro*. As far as is known, CAP concentration is fairly constant and insensitive to growth conditions, maintaining a level of about 1000 molecules per cell. By contrast the cAMP concentration varies from about 10^{-4} to 10^{-7} M depending upon the growth conditions. The cAMP concentration is controlled by factors which are most sensitive to the level of glucose and nitrogen source in the growth media; the higher the glucose level the lower the cAMP level and the lower the level of expression of the *lac* operon and other catabolite sensitive genes which are controlled by cAMP. If both lactose and glucose are present in the growth media, the level of cAMP is low. The instruction to the *lac* operon seems to be that if a simpler carbon source is available, use it first and do not waste energy making enzymes for lactose utilization. Finally, ppGpp stimulates the expression of the *lac* operon (this has only been shown *in vitro*). It cannot be said exactly why this is so or what physiologic value it has. The

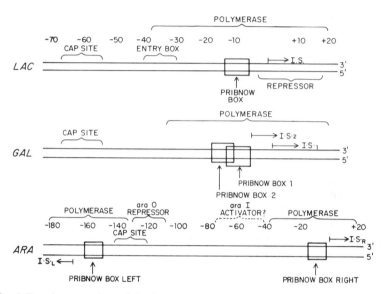

Fig. 5. Regulatory regions of the *lac, gal,* and *ara* operons. Base numbering system is according to convention used in Fig. 3. Scale used for *ara* operon is twice that used for *lac* and *gal.* Location of initiation sites indicated as I.S. Location of Pribnow boxes (PBS2) is obvious from sequence data and initiation site data. Location of CAP binding sites has been discussed in Section III, F. Location of entry box (PBS1) is indicated for *lac* but not for *gal* or *ara,* since the sequence data for *gal* and *ara* does not indicate a typical entry box. See text for further explanation.

concentration of ppGpp is inversely proportional to the gross rate of cellular RNA synthesis. The ppGpp modulator has a stimulatory, inhibitory, or negligible effect, depending upon the gene or operon. For the *lac* operon, the effect is a two- to threefold stimulation of expression either in the crude or purified *in vitro* systems.

Recent *in vitro* studies indicate the tertiary structure of the DNA strongly affects *lac* operon expression (Yang *et al.,* 1979). These studies involve the use of linear λ lac DNA in a coupled transcription–translation system. When novobiocin is present (2 μg/ml), only 20% of the normal level of β-gal is obtained. Since novobiocin inhibits gyrase, the enzyme that causes supercoiling, the inference from these results is that supercoiled DNA is at least five times more effective in *lac* operon expression than DNA lacking supercoiling.

B. The *ara* Operon

The *ara* operon contains three structural genes, *ara B, ara A,* and *ara D,* whose gene products can bring about the stepwise conversion of

L-arabinose into D-xylulose 5-phosphate. The regulatory locus is bounded by the *ara B* gene on one side and by the *ara C* gene on the other side (see Fig. 5). *In vivo* the expression of the operon requires L-arabinose, and it is inhibited by glucose, indicating a catabolite repression effect and suggesting the involvement of the cAMP–CAP complex. Genetic studies on whole cells have shown that the ara C protein is required for expression. Further work has revealed the presence of two sites in the control region, one where ara C protein acts as an activator and one where the ara C protein acts like a repressor (Englesberg *et al.*, 1969). The order of sites is known to be repressor site (*ara O*), activator site (*ara I*), and transcription start site (*I.S.*). Wilcox *et al.* (1974) demonstrated that the *ara C* gene and the nearby *ara BAD* operon are transcribed in opposite directions. The promotors for these genes are near to each other, separated only by a 167 base region between the structural genes for *ara C* and *ara B* (see Fig. 5).

In the coupled cell-free system for DNA-directed protein synthesis it has been shown that optimal expression of the *BAD* operon requires the ara C protein, ppGpp, cAMP–CAP, and L-arabinose; the synthesis of functional ara C protein has also been demonstrated (Zubay, 1973). One-third of maximal expression of the operon is achieved in the absence of cAMP, indicating that at high levels of ara C protein and ppGpp the need for the cAMP–CAP system can be partly overcome (Yang and Zubay, 1973). Based on this finding a model was proposed which placed the *ara I* site close to the polymerase binding site suggesting the order from left to right, CAP site, *ara I* site, and polymerase binding site (see Fig. 5). If this is correct, the function of CAP in the *ara* operon would most likely be to facilitate the binding of the ara C protein at the *ara I* site, which in turn would facilitate the binding of polymerase.

The regulation of *ara C* expression has also been studied, and it has been found that the expression of this gene *in vivo* is inhibited by ara C protein and stimulated to a small extent (two- to threefold) by cAMP–CAP (Casadaban, 1976). This is quite remarkable in view of the fact that the presumptive CAP binding site overlaps the polymerase binding site. Perhaps the two proteins bind on opposite sides of the DNA. The closeness of the control loci for *ara C* and *ara BAD* and the similarity in the control machinery suggest that there may be overlapping use of the regulatory region and some interference effects. In support of this, Hirsch and Schleif (1977) have shown in a purified transcription system that optimum activity of the *ara C* gene requires cAMP–CAP and an inactive *BAD* operon. If *ara C* protein is present with cAMP–CAP, the main operon (*BAD*) is turned on and the activity of the *ara C* gene is greatly reduced. According to Smith and Schleif (1978) the control loci for the two promotors must be located in a 167 base region. A model for the control region based on the aforegoing considerations is presented in Fig. 5. The location of the Pribnow boxes for leftward and rightward transcription is obvious from the sequence data. The location of the repressor binding site for ara C protein (*ara O*) has been determined by Smith and Schleif (1978). The location of the CAP site is predicted by arguments presented in Section III,F,2. The location of the activation site for the ara C protein (*ara I*) is unknown except that it is believed to be somewhere between the CAP site and the polymerase binding site. The repressor binding site overlaps the CAP binding site, and therefore repressor may inhibit gene expression by preventing CAP binding. The CAP site is strategically located so that it could facilitate binding of the ara C protein. It is abnormally close to the alleged leftward initiation site ($I.S._L$), which may account for the smaller stimulation of this promotor by cAMP–CAP. It is not immediately obvious why *BAD* activity should inhibit *C* expression as claimed by Hirsch and Schleif (1977).

In the *lac* and *trp* operons there is no evidence for modulation of repressor synthesis. Under all conditions of growth where measurements have been made about 10–20 copies of repressor are produced. This is sufficient to give optimum repression so there is no need to make any more. The situation is quite different for operons regulated by specific activators. In such instances optimum expression requires the binding of the activator at the promotor, and the rate of formation of such a complex should depend upon the concentration of the activator protein. Thus, many more copies of an activator protein would probably be necessary to produce optimum expression. Economy dictates that a feedback control of regulatory protein synthesis would prevent excessive synthesis of the regulatory protein when it is not needed.

The use of two controls, cAMP–CAP and L-arabinose–ara C, to regulate the *ara* operon illustrates the same principle used in *lac* operon control. There is no need to make these enzymes unless L-arabinose is present and then only if glucose, the simpler-to-use carbon source, is lacking. The observation that partial escape from cAMP–CAP control results when ppGpp is present has only been made *in vitro*, but it might suggest that in the presence of L-arabinose and ppGpp, it is sometimes beneficial to have expression of the operon even when the glucose level is adequate to satisfy the energy and carbon needs. In this regard it might be pointed out that the end product of the metabolic pathway defined by the *ara* operon is D-xylulose 5′-phosphate, which is an intermediate in the multipurpose phosphogluconate pathway.

Lis and Schleif (1973) discovered that at intermediate intracellular concentrations of cAMP the *lac* operon is seven to eight times more active than the *ara* operon. This means that a higher level of cAMP is necessary to form the CAP complex at the *ara* promotor than at the *lac* promotor. Such a difference might serve a useful function. A C_6 compound resulting from lactose hydrolysis is a more efficient source of energy than a C_5 compound such as L-arabinose. If both lactose and arabinose were available for carbon and energy sources, the lactose would probably be metabolized first.

There are many similarities between the controls used to regulate D-serine deaminase (Dsdase) synthesis and those used to regulate the *ara BAD* operon. For optimal expression *in vivo* and *in vitro* cAMP–CAP and the protein encoded by the *dsdC* gene are required. It is believed that the *dsdC* gene product is a specific apoactivator which functions in conjunction with the coactivator D-serine (Bloom *et al.,* 1975; Heincz and McFall, 1978a,b; Heincz, Keller and McFall, 1979). The *dsdC* gene product gives suboptimal expression of the *dsdase* gene in the absence of cAMP. Possibly the function of this is to prevent toxic levels of D-serine from accumulating or to provide a nitrogen source when D-serine is not needed as a carbon source.

C. The *gal* Operon

The *gal* operon is an example of an operon providing alternative modes of activation. In the case of the *lac* and the *ara* operons usually two conditions have to be met to activate transcription: the sugar lactose or arabinose must be present in the growth medium and glucose must be absent. To ensure this mechanism of activation the operons employ two control switches operating in series as described above. Both switches must be in the ''on'' position for optimal transcription. The metabolic principle underlying this type of control is that the cell uses the simplest sugar available to generate energy and as a carbon source.

In some cases the enzymes of an operon might be useful for quite different metabolic functions. When this happens, we might expect to find alternative mechanisms of activating the operon. Tyler *et al.* (1974) have suggested that the *hut* operon, which encodes enzymes that catabolize histidine, is such a case. They have hypothesized that this operon is responsive to the cells' need for carbon or nitrogen compounds with cAMP–CAP serving as the activation system when carbon is in demand and with glutamine synthetase serving as the activation system when nitrogen is in demand.

The *gal* operon, which encodes three enzymes that metabolize galactose, may exemplify another and possibly better understood case involving alternative modes for activation. The three structural genes of the *gal* operon *gal E, gal T,* and *gal K* code for the enzymes UDP-galactose-4-epimerase, galactose-1-P-uridyltransferase, and galactose kinase, respectively. These enzymes convert galactose into UDP-gal, a form which is useful either for catabolism when galactose is needed as a carbon or energy source, or for anabolism when galactose is needed as a substrate in cell envelope synthesis. Adjacent to the three structural genes of the *gal* operon is a complex promotor region containing binding sites for the specific *gal* repressor, CAP, and RNA polymerase. The sequence of bases for the control region and the first part of the operon are known and illustrated in part in Fig. 4 (Musso *et al.*, 1977). In the -53 to -68 nucleotide region 13 out of 16 base pairs show a dyad axis of symmetry. The outer 3 bases have the same sequence as the hypothesized binding site for CAP in the *lac* operon. They are located at about the same distance from the initiation site (S_1), which makes it seem likely that this is the CAP binding site. Mutants in the repressor binding site showing poor repressor binding (O^c's) have thus far been found in the -60 to -66 base region (Adhya and Miller, 1979), which suggests that the *gal* repressor binding site overlaps the CAP binding site rather than the polymerase binding site.

Musso *et al.* (1977) have demonstrated *in vitro* (but not *in vivo*) the existence of two initiation sites S_1 and S_2 for transcription (see Fig. 5). These two start sites are only five nucleotides apart. Centered about 9 bases before each start site are sequences representative of two overlapping Pribnow boxes, possibly indicating the existence of two rapid start binding sites for RNA polymerase. Transcription from each of the initiation sites responds to different regulatory mechanisms. In the presence of cAMP and CAP, transcription initiates only at S_1. In the absence of cAMP or CAP, transcription starts from S_2. The *gal R* repressor inhibits transcription from both initiation sites. This is somewhat surprising in view of the fact that the repressor binding site in *gal* is located at some distance from the polymerase binding site as already indicated. Since it strongly overlaps the CAP binding site, it is easy to see how it interferes with initiation at S_1, which requires cAMP–CAP. The situation at S_2 is less clear. The S_2 site is 5 nucleotides closer to the repressor binding site so that repressor may interfere with the polymerase "entry box" for S_2 initiation. Another provocative observation of Musso *et al.* (1977) is that most but not all initiation from S_2 stops after transcription of the first 7 nucleotides. This is reminiscent of an attenuator mechanism of control and leads one to wonder if an undiscovered antitermination factor exists for control from S_2. Clearly, further information is necessary before we can say that the S_1 and S_2 initiation sites are definitely correlated with the catabolic and anabolic needs, respectively, of the cell for galactose.

D. The *trp* Operon

The *trp* operon uses two regulators, one before and one after initiation. The *trp* operon contains a cluster of five structural genes involved in the synthesis of four enzymes which bring about the conversion of anthranilic acid to tryptophan in five discrete steps. For details and references, refer to the review by Platt (1978). The main use of the end product of this operon is in protein synthesis, and consequently the function of the enzymes using anthranilate in this pathway should be to satisfy the cells tryptophan needs for protein synthesis. In spite of this, the control apparatus for the *trp* operon appears to contain two switches, one at the promotor (see Section II,C) and one at the attenuator, which is located (see Section II,D and III,H) about 140 bases downstream from the promotor. These two regulatory mechanisms for the *trp* operon have been discussed above, and we shall not elaborate on them here. The control switch at the promoter involves the *trp R* aporepressor and corepressor tryptophan. The control switch at the attenuator involves rho factor and charged tryptophanyl-tRNA. Rho is also involved in termination at the end of the operon, but as far as we know this does not serve a regulatory function. If the concentration of the amino acid in the cell is low, the promotor is active. However, if the concentration of charged Trp-tRNA is high, about 8 out of 10 of the nascent RNA chains will not proceed beyond the attenuator site. Optimal transcription of the entire operon requires a low level of both tryptophan and the charged tRNA.

There are at least two conceivable advantages for having these two control switches in series: (1) For a system with two metabolic switches the amplification factor between full on and full off should be the product of what one could achieve with only one switch. (2) Under some conditions for example, very rapid protein synthesis, it is possible that a higher level of amino acid is required to keep the tRNA at an adequately charged level (suggested by C. Yanofsky). The structure of the *trp* promotor is surprising in that the region around -10 has only 3 out of 7 bases in agreement with the favored sequence for the Pribnow box (see Fig. 3). Most highly active promotors have at least 5 out of 7. In spite of this the *trp* promotor is quite active. Clearly it is not possible to understand the efficiency of the *trp* promotor in terms of the Pribnow box alone. As noted before, the -35 region, believed to be important in making a rapid start complex with RNA polymerase, has the same heptanucleotide sequence in *trp*, λP_R and λP_L. Genetic studies reviewed by Platt (1978) suggest that the -38 to -31 region is important for promotor function and that the -20 to -5 region is most important in making a rapid start complex.

E. The lambda C_I Gene

The lambda C_I gene is a complex system involving both positive and negative controls. When λ bacteriophage DNA infects a sensitive host cell, it either undergoes replication or lysogenizes by integrating into the host chromosome as a prophage. In the lysogenic state certain conditions lead to activation of the prophage and subsequent vegetative replication. The concentration of the $\lambda\,C_I$ gene is a decisive factor involved in phage replication, lyogeny, and induction of the prophage. In view of the complex roles played by the $\lambda\,C_I$ product it is not surprising that regulation of $\lambda\,C_I$ expression should be complex. Many aspects of the control process governing C_I expression are still poorly understood. However, it is worthy of discussion as an example of a gene whose control is subjected to a variety of signals. A most excellent recent summary of this subject is given in a paper by Weisberg *et al.* (1977); this review should be seen for details and references.

The λ mutants which fail to lysogenize or which lysogenize poorly have been classified into four groups, *cI, cII, cIII,* and *cY*. The *cI* mutants never lysogenize and cannot be maintained in the lysogenic state. By contrast *cII, cIII,* or *cY* mutants lysogenize occasionally and as such they are indistinguishable from wild-type lysogens. Temperature-sensitive mutants of *cI* can lysogenize at permissive temperatures but invariably lyse at nonpermissive temperatures, showing that *cI* is continuously required to maintain the lysogenic state. Abundant genetic and biochemical evidence has shown that the cI protein acts as a repressor to block early λ functions by binding at the O_L and O_R operators (see Fig. 1). The first evidence that cI protein induces its own synthesis came from studies of a lysogen which bears a temperature-sensitive gene for cI and other defects which prevent phage induction at high temperatures. At permissive temperatures this phage shows immunity to superinfection by exogenous phage, demonstrating the presence of cI protein. If this lysogen is maintained at nonpermissive temperatures for some time, it shows no immunity at nonpermissive temperatures and the return to immunity at low temperatures is very slow. Further genetic studies indicate that repressor promotes its own synthesis by binding at O_R. This and other experiments suggest the existence of a leftward promotor near O_R, which is referred to as the promotor for repressor maintenance, P_{RM}. However, there is some controversy whether the stimulation of repressor synthesis at P_{RM} results from a transcriptional or posttranscriptional effect. The O_R region is complex, containing at least three sites for binding cI protein, and there are indications from *in vitro* studies that at low repressor concentration *cI* expression is stimulated whereas at high repressor concentrations, where most of the binding sites are occupied, it is repressed.

The decision between the lysogenic and lytic pathways made after λ infection is the result of a number of phage and host gene products that affect *cI* expression. Lambda mutants in *cII, cIII,* and *cY* make very little

cI product after infection and consequently they rarely lysogenize. Currently, it is believed that there are two promotors which are involved in cI repressor synthesis, P_{RM} discussed above and P_{RE} (see Fig. 1), so named because it is believed to be involved in repressor synthesis soon after infection. The activity of P_{RM} is controlled by cI and cro. The activity of P_{RE} is directly modulated by the products of the cII and $cIII$ genes and indirectly by the product of the cro gene, which affects the synthesis of cII and $cIII$ gene products. The cII and $cIII$ gene products are neither required nor synthesized in established lysogens. The action of cII and $cIII$ products is seen after infection of a sensitive cell as a transient stimulation of cI gene expression. Infection with $\lambda cIII^+ cII^+$ results in a rate of repressor synthesis five to ten times higher per cI gene copy than in established lysogens. In addition to cII and $cIII$ gene products, the intactness of a DNA site called cY is required for rapid repressor synthesis soon after infection. The cis dominance of this mutation suggests cY is a promotor which has been called P_{RE} in Fig. 1. The establishment of immunity also requires the presence of a functional N gene. Lambda N^- mutants synthesize only 7% as much repressor protein as wild-type phage. Since N function is known to be required for the synthesis of most phage proteins, e.g., cII and $cIII$ as well as for normal phage DNA replication, the requirement for N product could be indirect.

The cII and $cIII$ mutants synthesize at least two late proteins, endolysin and tail antigen, sooner than wild-type virus. The delay in rightward late gene expression, promoted by cII and $cIII$ products, appears to be mediated through the postulated P_{RE} promotor, since it is not seen in cY mutants. It has been suggested that the leftward transcription from P_{RE} directly depresses rightward transcription from P_R, which in turn limits the expression of Q whose product is required for late gene expression.

The gene cro is located in the early rightward transcript of λ (see Fig. 1) and therefore does not require N for synthesis. Cro mutants do not form plaques on sensitive hosts as the lytic functions appear to be repressed. This property appears to be associated with a continuous elevated rate of repressor synthesis after infection. This contrasts with wild-type phage, which switch from the P_{RE}-dependent, high-level rate of repressor synthesis to the low-level P_{RM}-dependent rate at about 10 min after infection. It appears, therefore, that the accumulation of cro product is responsible for this shut-off of P_{RE}. Further genetic studies indicate that cro acts by blocking the synthesis of cII and $cIII$. This is followed by the rapid inactivation of cII and $cIII$ products, which leads to a shutting down of transcription from P_{RE}. Direct biochemical evidence that cro acts as a repressor by binding to O_L and O_R operators has recently been obtained by Takeda $et\ al.$ (1977). Since cro acts as a repressor by binding to O_L and O_R, it mimics repressor action to some extent. At least one significant difference in the action of cI and cro exists, however. As mentioned above the binding of cI repressor to O_R at low levels stimulates cI expression from P_{RM}. Cro binding reduces cI expression from P_{RM} even in the presence of active repressor.

All of the evidence that P_{RE} is actually a site for the initiation of RNA synthesis is indirect. It has been suggested that P_{RE} is an attenuator site downstream from another promotor stimulating leftward transcription (Szybalski, 1977). Whether P_{RE} is a true promotor, attenuator, or for that matter an alternator will probably be resolved by direct biochemical evidence as to where leftward transcription from P_{RE} originates.

F. Divergent Transcription

Two cases of operons with transcription emanating in diverging directions from the interior of the operon are well documented, the *bio* operon (Ketner and Campbell, 1975) and the *arg ECBH* cluster (Jacoby, 1972; Elseviers *et al.*, 1972). In the negatively controlled *bio* operon there are five structural genes (*A, B, F, C, D*) essential for the synthesis of biotin with a promotor–operator region located between the *A* and *B* genes. Leftward transcription yields the *A* gene product and rightward transcription yields the *B, F, C,* and *D* gene products in the indicated order. The operon is negatively controlled by a specific aporepressor protein and the corepressor biotin. Recent evidence indicates that the true corepressor of the biotin operon may be the acylanhydride of biotin-AMP. (M. Eisenberg and O. Prakash, personal communication). Constitutive operator mutants for *A* gene expression are also constitutive for *B, F, C,* and *D* expression, which suggests that there may be only one repressor binding site for the entire gene cluster. The closeness of the control regions for leftward and rightward transcription is also indicated by the finding that point mutants that increase leftward transcription four- to sixfold diminish rightward transcription by two-thirds. One is reminded of the *in vitro* experiments on the *ara* operon (see Section IV,B), where it was stated that rightward transcription of the *ara BAD* cluster inhibits leftward transcription of *ara C*. In this case also the regulatory genes for the two transcripts are very close together. In the biotin operon it is believed that there are two promotors P_L and P_R which transcribe divergently from two partially overlapping face-to face promotors (Otsuka and Abelson, 1978). The putative operator contains an imperfect dyad axis of symmetry that partially overlaps the promotors.

A situation similar to the one seen with the *bio* operon exists in the *arg ECBH* cluster. This grouping contains four of the nine known genes for arginine biosynthesis (Kelker *et al.*, 1976). The promotor–operator region for the cluster is located between the *E* and *C* genes with leftward transcription yielding the *C*, *B*, and *H* gene products. The conclusions which can be drawn from measurements of *arg* gene products in strains bearing various regulatory mutations is very similar to the situation with the *bio* operon.

Divergent transcription as seen in *bio* and *arg* provides another variation in operon structure. Because two promotors are involved with different intrinsic association constants for polymerase, one anticipates that expression of the two transcripts could be quite different and noncoordinate at different levels of derepression. This would appear to be one possible advantage of this type of operon arrangement, but there may be others. Clearly a great deal is to be learned by future studies of the *arg* and *bio* systems.

V. OVERALL PATTERNS OF TRANSCRIPTIONAL CONTROL OBSERVED IN DIFFERENT GENE SYSTEMS

There are basically three different types of chromosomes that replicate in prokaryotes: the cellular chromosome, the viral chromosome, and the plasmid. The plasmid and the host chromosome can coexist as independent replicating units seemingly indefinitely. By contrast, most virus chromosomes do not coexist indefinitely with the host. Lysogenic virus can enter either a lytic or a prophage state. In the prophage state they are integrated into the host chromosome and duplicate as the host duplicates. In the lytic state they duplicate rapidly and kill the host cell, which they usually lyse, releasing infectious particles. So far as we know lysogenic viruses in the prophage state and plasmids utilize the unmodified host RNA polymerase. Any virus which enters the irreversible lytic cycle either modifies or replaces the host polymerase for its own transcription. In this section certain generalizations will be discussed relating to the patterns of transcriptional control seen in cells, viruses, and plasmids.

A. Organization of Controls in the Cellular Chromosome

The single *E. coli* chromosome contains enough DNA for about 3000 genes. Every bacterial cell is genetically totipotent and responds rapidly to changes in environment. This requires all the gene regulation responses to be readily reversible in either the on or off direction. A system delicately poised for rapid response probably contains sufficient quantities of many regulator proteins, which may not be present at optimum concentrations when needed. At any given time only a small fraction (about 5–10%) of the total genes are turned on, and it appears that this is determined primarily not by the concentration of regulator proteins but by the concentrations of small molecule effectors. These small molecule effectors include coactivators, corepressors, antirepressors, and antiactivators, which interact directly with the regulator proteins, as well as those small molecule effectors that interact directly with the RNA polymerase.

Many genes are subject to multiple controls, and in such cases organization into a hierarchy is usually evident. This can be illustrated by reviewing the situation with respect to the *lac* operon. In the *lac* operon there is evidence for three control elements. The small molecule modulators for these controls are lactose, cAMP, and ppGpp. Lactose (allolactose) is the antirepressor for the highly specific *i* gene repressor, which specifically affects the activity of the *lac* operon. The repressor binding site partially overlaps the polymerase binding site as discussed above. Somewhat further removed from the structural genes is a binding site for CAP. CAP in combination with cAMP is a general activator for many catabolite-sensitive genes, probably hundreds, in *E. coli*. Although CAP can bind in the presence of repressor, the latter must be removed before the *lac* operon can be activated. Finally, ppGpp stimulates transcription of the *lac* operon by binding to the polymerase. The concentration of ppGpp is inversely proportional to the gross rate of cellular RNA synthesis. The ppGpp can have a stimulatory, inhibitory or negligible effect on transcription depending upon the gene or operon. For *lac,* the effect is stimulatory at least *in vitro.*

During its normal life cycle an organism will grow in an appropriate culture medium until it has produced the maximum number of cells it is capable of forming under the particular conditions. The limitation may be set by the supply of oxygen, the pH, the carbon or energy source, or by other nutritional factors. As the culture stops growing it enters the so-called stationary phase. This may be followed eventually by cell death or sporulation in select groups of microorganisms. The transition to sporulation involves major morpholgical and metabolic changes frequently including the production of antibiotics. It seems likely that these changes are accompanied by a drastic alteration in the pattern of transcription. In this regard it is particularly noteworthy that in *B. subtilis* the onset of sporulation is associated with both a marked decrease in the activity of the σ subunit of RNA polymerase and with the appearance of at least two sporulation-specific polypeptides that bind to core enzyme (Losick and Pero, 1977). The precise regulatory significance of these changes in the RNA polymerase and the significance of other factors important to sporulation remains a fertile field for investigation.

B. Organization of Controls in DNA Bacteriophages Containing 30 to 60 Genes

For most regulated bacterial genes it is the concentration of small molecule effectors that are the determining factors in whether or not a gene will be expressed. In marked contrast, small molecule effectors appear to play a minor role in viral development. Following infection, controls in the form of phage-encoded regulatory proteins are introduced in a sequential and irreversible manner. When T bacteriophages infect *E. coli,* vegetative phage replication ensues and ultimately the cells lyse (for an excellent recent review of this subject see Rabussay and Geiduschek, 1977a). In the case of T4 infection, modification of the host polymerase is a key factor in the regulation of transcription, whereas with T7, replacement of the host polymerase by a phage-encoded enzyme is a key factor in the regulation.

T4 infection begins with the transcription of about 25 early genes by *E. coli* polymerase, mainly from the l strand. Products of early transcription shut down host transcription and translation. Some small polypeptide products of early synthesis become associated with host polymerase. Other products lead to modification of the subunits of host polymerase, including changes in size, adenylation, and phosphorylation (Skorko *et al.,* 1977). Further transcription results principally from the r strand by the modified host polymerase. Presumably, the various modifications of the *E. coli* polymerase change its affinity for the different initiation sites so that the appropriate genes are transcribed when they are needed.

T7 infection begins with transcription of about 20% of the DNA from the end of one strand by *E. coli* polymerase. This transcription leads to the synthesis of a phage-encoded polymerase that transcribes the remaining 80% of the same DNA strand. The *E. coli* polymerase cannot transcribe the late genes of the T7 phage. Furthermore, other factors encoded by early T7 genes inhibit the transcription of host chromosome by *E. coli* polymerase.

The λ bacteriophage encodes at least four regulator proteins, cI, N, cro, and Q and possibly cII and cIII (for a review of this subject see Weisberg *et al.,* 1977). This bacteriophage can exist in an active or an inactive form in *E. coli.* In the inactive form the λ chromosome is integrated into the host chromosome. This inactive state is maintained by the presence of the *cI* repressor, which binds to the so-called early right (O_R) and early left (O_L) initiation loci and also modulates its own synthesis. The repressor is synthesized in small amounts when λ is in the prophage state from the P_{RM} promotor, and together with the tandemly located *rex* gene represent the only λ genes that are known to be active at this time. Various disturbances which interfere with host DNA metabolism trigger a series of events that result in inactivation of the cI protein. This is the first irreversible step leading to λ replication and cell death. Once the cI repressor has been released from the two operators O_R and O_L, transcription begins at the corresponding promotors P_R and P_L. The early right and early left transcription units are quite short as a result of rho-sensitive termination points. Early left transcription leads to synthesis of the regulator protein N, and early right transcription leads to the synthesis of the regulator protein cro (see Fig. 1). By a mechanism which has been discussed above, N leads to antitermination so that transcription continues in the early left and early right cistrons to points considerably beyond the original terminator points. The extended transcription from the right transcriptional unit results in the synthesis of Q protein. This protein may be an antiterminator that turns on late rightward transcription of λ. In the meantime, the buildup of the cro gene product diminishes transcription from P_L and P_R probably by binding to sites within O_L and O_R. The cro gene product also inhibits any transcription of the cI gene from the P_{RM} promotor (promotor for repressor maintenance) by binding to the nearby O_R operator. Following cro-induced repression of P_R most of the late rightward transcription begins at a promotor following the Q gene and known as P'_R. Undoubtedly, further aspects of λ development will be discovered in the near future.

C. Organization of Controls in Plasmids

Plasmids are extrachromosomal genetic elements capable of semiautonomous replication in a cell (for a general review of this subject see Falkow, 1975). All bacterial plasmids in the mature form have been characterized as covalently closed circular molecules of duplex DNA. Plasmids are commonly divided into two classes. Those that promote conjugal transfer of DNA are known as conjugative plasmids, of which the F and R plasmids are examples. Those incapable of promoting conjugation are known as nonconjugative plasmids, of which the colicinogenic factor Col E1 is an example. Even plasmids of the latter class can be transferred with the help of other plasmids or by artificial means. For example, Col E1 plasmid is efficiently transferred by an F-containing strain. The size of plasmid DNA's varies from about 2 to 100×10^6 daltons. Conjugative plasmids are usually larger.

The genes of plasmids may be broadly classified into three types, those required for autonomous replication, those required for cell-to-cell transfer, and, finally, those genes required for neither of the above but which give the plasmid some of its obvious phenotypic properties such as drug resistance or toxin production. Each of these gene types is designed to carry out a specific function. The fact that the different sets of plasmid genes require special conditions for expression suggests that special regulatory mechanisms exist to control the activities of those genes either singly or coordinately.

It seems likely that replication of plasmids is subject to more than one type of control. Plasmids can be considered in two categories according to the number of copies of plasmid DNA formed in the host cell. Those existing in one or a few copies are called stringent (standing for stringent control of DNA synthesis); those existing in multiple copies are called relaxed. Usually, conjugative plasmids show stringent control, whereas nonconjugative plasmids show relaxed control. Exceptions to this generalization exist such as plasmid R6K, which is conjugative, but present in multiple copies. Some relaxed plasmids, such as Col E1, show a spectacular increase in number when chloramphenicol is added to a growing culture. This drug inhibits host chromosome replication but not Col E1 plasmid replication. A number of studies have been done to determine the mechanism of plasmid replication as well as the dependence of replication on genes needed for host chromosome replication. In no cases have precise mechanisms for control of replication been determined.

Another phenomenon indicative of the regulation of replication is the so-called plasmid incompatibility. Originally F incompatability referred to the fact that in male strains of *E. coli* the F factor exists either in an integrated state in Hfr strains or in an autonomous state F$^+$, but it cannot be established in both states simultaneously. It has been shown that this is a result of the failure of the F factor to replicate in the autonomous state in the Hfr. So-called *inc* mutants have been isolated where this property of incompatibility is lost. Incompatibility is a common property of plasmids. Indeed it has been used as a basis for classification; those plasmids which are incompatible with each other seem to be closely related and are classified in the same incompatibility group.

The genes required for conjugal transfer appear to be coordinately regulated. Since conjugal genes serve one function, that of transferring DNA from one cell to another, it would be wasteful to have some active and some dormant. Genetic analysis shows that the F factor carries an operon with about twelve genes coding for DNA transfer properties, the so-called *tra* operon (Helmuth and Achtman, 1975). Genetic and biochemical studies indicate that these genes are controlled in a positive way by the *tra J* gene, which is situated at one end of the operon (Willetts, 1977).

Except for F factors, fertility properties of most conjugatable plasmids are generally repressed in the cell, but repressed mutants that transfer with a high frequency have been isolated. Studies on regulatory mutants isolated for F and R plasmids have led to a model for control of the transfer genes. In this model the R factor product FIN (product of the *fin* gene) and the F factor product P$_F$ (product of the *tra P* gene) interact to form the F transfer inhibitor. This acts to prevent the synthesis or function of the product (J_F) of the F gene *tra J*, the absence of which in turn prevents the synthesis or function of several if not most of the transfer (*tra*) gene products. Other control systems modulating transfer have been described (e.g., see Gasson and Willetts, 1975).

Cyclic AMP and CAP protein appear to play an important role in regulating the expression of a number of plasmid genes involved in drug resistance and bacteriocin production. Such a mechanism has been demonstrated most completely for chloramphenicol acetyltransferase, the enzyme carried by a number of plasmids, which is responsible for inactivating chloramphenicol. Thus, de Crombrugghe *et al.* (1973) have shown that a DNA-directed cell-free system for synthesis of this enzyme is stimulated fifteenfold by cAMP and CAP.

Whereas many of the plasmid-encoded enzymes conferring drug resistance are believed to be synthesized continuously, synthesis of the protein or proteins conferring tetracycline resistance by the transposon Tn 10 is strongly induced by the drug tetracycline or closely related analogues (Yang *et al.*, 1976). It appears that a plasmid-encoded repressor is involved which is antagonized by tetracycline (Yang *et al.*, 1976). A similar, but evolutionarily unrelated, type of inducible tetracycline resistance probably exists in the plasmid PSC 101 (Tait and Boyer, 1978).

Clearly much more work must be done on plasmids before many useful generalizations can be made about the types of transcriptional control mechanisms used. At this stage a hazy picture emerges of a system symbiotic with the host which uses the host RNA polymerase in an unmodified form in conjunction with various positive and negative control systems. The most unusual and perhaps unique type of regulation seen is that involved in incompatiability. Unfortunately very little is known about the mechanism of regulation underlying the incompatibility phenomenon.

VI. CONCLUDING REMARKS

The analysis of gene expression has been enormously accelerated by two recent technological breakthroughs: chemical methods for sequence determination in DNA and biochemical techniques which permit the capture of genes or fragments of genes of interest in small hybrid plasmids. In the future, most genes of interest will be available in biochemical amounts and their sequences known before attacking the question how they are regulated. This will permit enormous short cuts to be made in their analysis. As indicated in this chapter, it is already possible in many cases to identify promotors, termination sites, CAP binding sites, and repressor binding sites. This type of information, combined with what we know about ribosome recognition sites and initiation and termination sites for translation, will permit informed guesses to be made about possible gene products. More subtle questions about what controls the level of gene expression can now be asked with greater facility.

In concluding this chapter it seems appropriate to make some remarks about the future of research in this area. Exploration of controls used in different systems has revealed a large variety of control mechanisms, and every system which has been explored in sufficient detail so far has increased our basic knowledge as to how these controls function. Obviously there are far too many systems which could be explored than there are explorers, and we must set a practical limit somewhere. The practical limit should probably be that new systems only be entered into either if an unusual type of control process appears to be involved, as in the regulation of incompatibility between closely related plasmids, or if it is of great interest in some other regard, as in the case of the control of expression of hybrid plasmids, which is of great practical importance but whose study will not necessarily lead to the establishment of many new basic principles.

The most difficult thing to predict is the discovery of new types of control mechanisms. These will probably appear as the result of serendipity in the course of systematic investigations. While there is a lot of filling in to do before we can discuss the regulation of even a single bacteria, virus, or plasmid in molecular terms, the basic advances in our understanding of transcriptional controls will involve mostly an investigation of the structural details of RNA polymerase and a limited number of regulating proteins and how they interact with their allosteric effectors and DNA. It seems likely that these advances will come about largely as the result of two approaches which have contributed little to our understanding of regulation thusfar: (1) crystallographic determination of the protein structures and (2) analysis of the effects of directed changes of the nucleotide sequence in the control region.

ACKNOWLEDGMENTS

This work was supported by grants from the National Institutes of Health (5RO1-AI-13277, 2RO1-GM-16648). I am grateful to the following colleagues for advice given during the course of preparing this chapter: S. Adhya, J. R. Beckwith, M. Cashel, M. Chamberlain, M. Gottesman, W. Szybalski, T. Platt, J. P. Richardson, J. Roberts, H. Schaller, and C. Yanofsky. Many others showed their generosity by contributing reprints, preprints and unpublished observations of their works. Members of my laboratory including Huey-Lang Yang, Sudha Mitra, Simon Chen, and David Foster helped me to form a seminar study group in which a number of the works reported here were considered. The assistance in manuscript preparation given by Ms. Victoria Chien and Linda Sproviero is appreciated.

REFERENCES

Adhya, S., and Gottesman, M. (1977). Control of transcription termination. *Annu. Rev. Biochem.* **47**, 967–996.

Adhya, S., and Miller, W. (1979). Modulation of the two promoters of the galactose operon of *Escherichia coli. Nature* (*London*) **279**, 492–494.

Adhya, S., Gottesman, M., de Crombrugghe, and Court, D. (1976). Transcription termination regulates gene expression. *In* "RNA Polymerase" (R. Losick and M. Chamberlain, eds.), pp. 719–773. Cold Spring Harbor Lab., Cold Spring Harbor, New York.

Ames, B. N., and Hartman, P. E. (1963). The histidine operon. *Cold Spring Harbor Symp. Quant. Biol.* **28**, 349–355.

Barnes, W. (1978). DNA sequence from the histidine operon control region: Seven histidine codons in a row. *Proc. Natl. Acad. Sci. U.S.A.* **75**, 4281–4285.

Bautz, E. F., Bautz, E. F., and Dunn, J. (1969). *E. coli* factor: A positive control element in phage T4 development. *Nature* (*London*) **223**, 1022–1024.

Bennett, G. N., Schweingruber, M. E., Brown, K. D., Squires, C., and Yanofsky, C. Y. (1973). Nucleotide sequence of the promoter–operator region of the tryptophan operon of *Escherichia coli. Proc. Natl. Acad. Sci. U.S.A.* **73**, 2351–2355.

Bloom, F. R., McGall, E., Young, M. C., and Carothers, A. M. (1975). Positive control in the D-serine deaminase system of *Escherichia coli* K-12. *J. Bacteriol.* **121**, 1092–1101.

Brenner, M., and Ames, B. N. (1971). The histidine operon and its regulation. *In* "Metabolic Regulation V" (H. Vogel, ed.), pp. 349–387. Academic Press, New York.

Bukhari, A. I., Shapiro, J. A., and Adhya, S. L. (1977). "DNA Insertion Elements, Plasmids and Episomes." Cold Spring Harbor Lab., Cold Spring Harbor, New York.

Burgess, R. R., Travers, A. A., Dunn, J. J., and Bautz, E. K. F. (1969). Factor stimulating translation by RNA polymerase. *Nature* (*London*) **221**, 43–45.

Casadaban, M. J. (1976). Regulation of the regulatory gene for the arabinose pathway, ara C. *J. Mol. Biol.* **104**, 557–566.

Cashel, M., and Gallant, J. (1969). Two compounds implicated in the function of the RC gene of *Escherichia coli. Nature* (*London*) **221**, 838–841.

Chamberlain, M. J. (1976). Interaction of RNA polymerase with the DNA template. *In* "RNA Polymerase" (R. Losick and M. Chamberlain, eds.), pp. 159–181. Cold Spring Harbor Lab., Cold Spring Harbor, New York.

Clewell, D. B., and Yagi, Y. (1977). Amplification of the tetracycline resistance determinant on plasmid pAMα1 in *Streptococcus faecalis. In* "DNA Insertion Elements, Plasmids and Episomes" (A. I. Bukhari, J. A. Shapiro and S. L. Adhya, eds.) pp. 235–246. Cold Spring Harbor Lab., Cold Spring Harbor, New York.

de Crombrugghe, B., Pastan, I., Shaw, W. V., and Rosner, J. L. (1973). Stimulation by cyclic AMP and ppGpp of chloramphenicol. *Nature* (*London*), *New Biol.* **241**, 237–239.

Dickson, R. C., Abelson, J., Barnes, W. M., and Reznikoff, W. S. (1975). Genetic regulation: The lac control region. *Science* **187**, 27–35.

DiLauro, R., Taniguchi, T., Musso, R., and de Crombrugghe, B. (1979). Unusual location and function of the operator in the *E. coli* galactose operon. *Nature* (*London*) **279**, 494–500.

DiNocera, P. P., Blasi, F., DiLauro, R., Frunzio, R., and Bruni, C. R. (1978). Nucleotide sequence of the attenuator region of the histidine operon of *Escherichia coli* K-12. *Proc. Natl. Acad. Sci. U.S.A.* **75**, 4276–4280.

Elseviers, D., Cunin, R., Glansdorff, N., Baumberg, S., Ashcroft, E. (1972). Control regions within the arg ECBH gene cluster of *Escherichia coli* K12. *Mol. Gen. Genet.* **117**, 349–366.

Englesberg, E., Squires, C., and Meronk, F. (1969). The L-arabinose operon in *Escherichia coli* B/r: A genetic demonstration of two functional states of the product of a regulator gene. *Proc. Natl. Acad. Sci. U.S.A.* **62**, 1100–1107.

Epp, C., and Pearson, M. L. (1976). Association of bacteriophage lambda N gene protein with *E. coli* RNA polymerase. *In* "RNA Polymerase" (R. Losick and M. Chamberlain, eds.) pp. 667–692. Cold Spring Harbor Lab., Cold Spring Harbor, New York.

Falco, S. C., Laan, K. V., and Rothman-Denes, L. B. (1977). Virion-associated RNA polymerase required for bacteriophage N4 development. *Proc. Natl. Acad. Sci. U.S.A.* **74**, 520–523.

Falkow, S. (1975). "Infectious Multiple Drug Resistance." Pion Limited, London.

Gasson, A., and Willetts, N. (1975). Five control systems preventing transfer of *Escherichia coli* K-12 sex factor F. *J. Bacteriol.* **122**, 518–525.

Gellert, M., Mizuuchi, K., O'Dea, H., and Nash, H. A. (1976). DNA gyrase: An enzyme that introduces superhelical turns into DNA. *Proc. Natl. Acad. Sci. U.S.A.* **73**, 3872–3876.

Gierer, A. (1966). Model for DNA and protein interactions and the function of the operator. *Nature (London)* **212**, 1480–1481.

Gilbert, W. (1976). Starting and stopping sequences. *In* "RNA Polymerase" (R. Losick and M. Chamberlain, eds.), pp. 198. Cold Spring Harbor Lab., Cold Spring Harbor, New York.

Gilbert, W., and Muller-Hill, B. (1967). The *lac* operator is DNA. *Proc. Natl. Acad. Sci. U.S.A.* **58**, 2415–2419.

Gilbert, W., Majors, J., and Maxam, A. (1977). How proteins recognize DNA sequences. *Life Sci. Res. Rep.* **4**, 167–178.

Goeddal, D. V., Yansura, D. G., and Caruthers, M. H. (1978). How *lac* repressor recognizes *lac* operator. *Proc. Natl. Acad. Sci. U.S.A.* **75**, 3578–3582.

Greenfield, L., Boone, T., and Wilcox, G. (1978). DNA sequence of the *ara* BAD promoter in *Escherichia coli* B/r. *Proc. Natl. Acad. Sci. U.S.A.* **75**, 4724–4728.

Gross, C., Hoffman, J., Ward, C. Hager, D., Burdick, G., Berger, H., and Burgess, R. (1978). Mutation affecting thermostability of of sigma subunit of *Escherichia coli* RNA polymerase lies near the dna G locus at about 66 min on the *E. coli* genetic map. *Proc. Natl. Acad. Sci. U.S.A.* **75**, 427–431.

Helmuth, R., and Achtman, M. (1975). Operon structure of DNA transfer cistrons on the F sex factor. *Nature (London)* **257**, 652–656.

Heincz, M. C., and McFall, E. (1978a). Role of the dsdC activator in regulation of D-serine deaminase synthesis. *J. Bacteriol.* **137**, 96–103.

Heincz, M. C., and McFall. E. (1978b). Role of small molecules in regulation of D-serine deaminase synthesis. *J. Bacteriol.* **137**, 104–110.

Heincz, M. C., Kelker, N. E., and McFall, E. (1978). Positive control of D-serine deaminase synthesis *in vitro*. *Proc. Natl. Acad. Sci.* U.S.A. **75**, 1695–1699.

Hesselbach, B. H., and Nakada, D. (1977). "Host shutoff" function of bacteriophage T7: Involvement of T7 gene *Z* and gene *0.7* in the inactivation of *Escherichia coli* RNA polymerase. *J. Virol.* **24**, 736–745.

Hillel, Z., and Wu, C. W. (1977). Subunit topography of RNA polymerase from *Escherichia coli*. A cross-linking study with bifunctional reagents. *Biochemistry* **16**, 3334–3342.

Hillel, Z., and Wu, C.-W. (1978). Photochemical crosslinking studies on the interaction of *Escherichia coli* RNA polymerase with T7 DNA. *Biochemistry* **17**, 2954–2961.

Hirsch, J., and Schleif, R. (1977). The *ara C* promotor: Transcription, mapping and interaction with the *ara BAD* promotor. *Cell* **11**, 545–550.

Jacob, F., and Monod, J. (1961). Genetic regulatory mechanisms in the synthesis of proteins. *J. Mol. Biol.* **3**, 318–356.

Jacoby, G. A. (1972). Control of the argECBH cluster in *Escherichia coli*. *Mol. Gen. Genet.* **117**, 337–348.

Johnsrud, L. (1978). Contacts between *Escherichia coli* RNA polymerase and in *lac* operon promoter. *Proc. Natl. Acad. Sci. U.S.A.* **75**, 5314–5318.

Kavenoff, R. and Bowen, B. (1976). Electron microscopy of membrane-free folded chromosomes from *Escherichia coli*. *Chromosoma* **59**, 89–101.

Kelker, N. E., Mass, W. K., Yang, H.-L., and Zubay, G. (1976). *In vitro* synthesis and repression of argininosuccinase in *Escherichia coli* K12; partial purification of the arginine repressor. *Mol. Gen. Genet.* **144**, 17–20.

Ketner, C., and Campbell, A. (1975). Operator and promotor mutations affecting divergent transcription in the bio gene cluster of *Escherichia coli*. *J. Mol. Biol.* **96**, 13–27.

Lee, F., and Yanofsky, C. (1977). Transcription termination at the *trp* operon attenuators of *Escherichia coli* and *Salmonella typhimurium:* RNA secondary structure and regulation of termination. *Proc. Natl. Acad. Sci. U.S.A.* **74**, 4365–4369.

Lin, S. Y., and Riggs, A. D. (1974). Photochemical attachment of *lac* repressor to bromodeoxyuridine-substituted *lac* operator by ultraviolet radiation. *Proc. Natl. Acad. Sci. U.S.A.* **71**, 947–951.

Lindahl, L., Post, L., and Nomura, M. (1976). DNA-dependent *in vitro* synthesis of ribosomal proteins, protein elongation factors, and RNA polymerase subunit α: Inhibition by ppGpp. *Cell* **9**, 439–448.

Lis, J. T., and Schleif, R. (1973). Different cyclic AMP requirements for induction of the arabinose and lactose operons of *Escherichia coli*. *J. Mol. Biol.* **79**, 149–162.

Losick, R., and Pero, J. (1977). Regulatory subunits of RNA polymerase. *In* "RNA Polymerase" (R. Losick and M. Chamberlain, eds.) pp. 227–246. Cold Spring Harbor Lab., Cold Spring Harbor, New York.

Mailhammer, R., Yang, H.-L., Reiness, C., and Zubay, G. (1975). Effects of bacteriophage T4-induced modification of *Escherichia coli* RNA polymerase on gene expression *in vitro*. *Proc. Natl. Acad. Sci. U.S.A.* **72**, 4928–4932.

Maizels, N. (1973). The nucleotide sequence of the lactose messenger ribonucleic acid transcribed from the UV5 promotor mutant of *Escherichia coli*. *Proc. Natl. Acad. Sci. U.S.A.* **70**, 3585–3589.

Majors, J. (1975). Specific binding of CAP factor to *lac* promoter DNA. *Nature (London)* **256**, 672–673.

Mitra, S., Zubay, G., and Landy, A. (1975). Evidence for the preferential binding of the catabolite gene activator protein (CAP) to DNA containing the *lac* promoter. *Biochem. Biophys. Res. Commun.* **67**, 857–863.

Musso, R. E., DiLauro, R., Adhya, S., and de Crombrugghe, B. (1977). Dual control for transcription of the galactose operon by cyclic AMP and its receptor protein at two interspersed promoters. *Cell* **12**, 847–854.

Ogata, R., and Gilbert, W. (1977). Contacts between the *lac* repressor and thymines in the *lac* repressor. *Proc. Natl. Acad. Sci. U.S.A.* **74**, 4973–4976.

Otsuka, A., and Abelson, J. (1978). The regulatory region of the biotin operon in *Escherichia coli*. *Nature (London)* **276**, 689–694.

Pannekoek, H., Perbal, B., and Pouwels, P. H. (1974). The specificity of transcription *in vitro* of the tryptophan operon of *Escherichia coli*. II. The effect of Rho factor. *Mol. Gen. Genet.* **132**, 291–306.

Platt, T. (1978). Regulation of gene expression in the tryptophan operon of *Escherichia coli*.

In "The Operon" (J. H. Miller and W. S. Reznikoff, eds.) pp. 263–302. Cold Spring Harbor Lab., Cold Spring Harbor, New York.

Ptashne, M. (1975). Repressors, operators, and promoters in bacteriophage lambda. *Harvey Lect.* **69**, 143–171.

Rabussay, D., and Geiduschek, E. P. (1977a). Regulation of gene action in the development of lytic bacteriophages. *Comprehensive Virol.* **8**, 1–196.

Rabussay, D., and Geiduschek, E. P. (1977b). Phage T4-modified RNA polymerase transcribes T4 late genes *in vitro*. *Proc. Natl. Acad. Sci. U.S.A.* **74**, 5305–5309.

Ratner, D. (1976). Evidence that mutations in the *suA* polarity suppressing gene directly affect termination factor rho. *Nature (London)* **259**, 151–153.

Reiness, C., Yang, H.-L., Zubay, G., and Cashel, M. (1975). Effects of guanosine tetraphosphate on cell-free synthesis of *Escherichia coli* ribosomal RNA and other gene products. *Proc. Natl. Acad. Sci. U.S.A.* **72**, 2881–2885.

Richardson, J. P. (1975). Initiation of transcription by *Escherichia coli* RNA polymerase from supercoiled and non-supercoiled bacteriophage PM2 DNA. *J. Mol. Biol.* **91**, 477–487.

Richardson, J. P. (1978). RNA synthesis termination factor rho. *Proc. FEBS Meet. 11th.* **43**, 153–162.

Roberts, J. (1975). Transcription termination and late control in phage lambda. *Proc. Natl. Acad. Sci. USA* **72**, 3300–3304.

Roberts, J. (1969). Termination factor for RNA synthesis. *Nature (London)* **224**, 1168–1171.

Rosenburg, M., Court, D., Shimatake, H., and Wulff, D. L. 1978). A *ρ*-dependent termination site in the gene coding for tyrosine t RNA su₃ of *Escherichia coli*. *Nature (London)* **272**, 414–422.

Saedler, H., Reif, H. J., Hu, S., and Davidson, N. (1974). IS2, a genetic element for turn-off and turn-on of gene activity in *E. coli*. *Mol. Gen. Genet.* **132**, 265–289.

Salstrom, J. S., and Szybalski, W. (1978). Coliphage λnut L⁻: A unique class of mutants defective in the site of *N* utilization for antitermination of leftward transcription. *J. Mol. Biol.* **124**, 195–221.

Schmeissner, V., Ganem, D., and Miller, J. H. (1977). Genetic studies of the lac repressor. II. Fine structure deletion map of the lac *I* gene, and its correlation with the physical map. *J. Mol. Biol.* **109**, 303–326.

Seeburg, P. H., Nüsslein, C., and Schaller, H. (1977). Interaction of RNA polymerase with promoters from bacteriophage fd. *Eur. J. Biochem.* **74**, 107–113.

Siebenlist, U. (1979). RNA polymerase unwinds an 11-base pair segment of a phage T7 promoter. *Nature (London)* **279**, 651–652.

Skorko, R., Zillig, W., Rohrer, H., Fujiki, H., and Mailhammer, R. (1977). Purification and properties of the NAD⁺: protein ADP-ribosyl-ADP-ribosyltransferase responsible for the T4-phage-induced modification of the α subunit of DNA-dependent RNA polymerase of *Escherichia coli*. *Eur. J. Biochem.* **79**, 55–66.

Smith, B. R., and Schleif, R. (1978). Nucleotide sequence of the L-arabinose regulatory region of *Escherichia coli* K12. *J. Biol. Chem.* **253**, 6931–6935.

Stauffer, G. V., Zurawski, G., and Yanofsky, C. (1978). Single basepair alterations in the *Escherichia coli trp* operon leader region that relieve transcription termination at the *trp* attenuator. *Proc. Natl. Acad. Sci. U.S.A.* **75**, 4833–4837.

Szybalski, W. (1977). Initiation and regulation of transcription and DNA replication in coliphage lambda. *In* "Regulatory Biology" (J. C. Copeland and G. A. Mayluff, eds.), pp. 2–45. Ohio State Univ. Press, Columbus.

Tait, R. C., and Boyer, H. W. (1978). On the nature of tetracycline resistance controlled by the plasmid pSC101. *Cell* **13**, 73–81.

Takeda, Y., Folkmanis, A., and Echols, H. (1977). Cro regulatory protein specified by bacteriophage λ. Structure, DNA-binding, and repression of RNA synthesis. *J. Biol. Chem.* **252**, 6177–6183.

Tyler, B., Deleo, H. B., and Magasanik, B. (1974). Activation of transcription of hut DNA by glutamine synthetase. *Proc. Natl. Acad. Sci. USA* **71**, 225–229.

Weisberg, R. A., Gottesman, S., and Gottesman, M. E. (1977). Bacteriophage λ: The lysogenic pathway. *Comprehensive Virol.* **8**, 197–258.

Wilcox, G., Boulter, J., and Lee, N. (1974). Direction of transcription of the regulatory gene *ara C* in *Escherichia coli* B/r. *Proc. Natl. Acad. Sci. U.S.A.* **71**, 3635–3639.

Willetts, N. (1977). The transcriptional control of fertility in F-like plasmids. *J. Mol. Biol.* **112**, 141–148.

Williams, B. G., Blattner, F. R., Jaskunas, S. R., and Nomura, M. (1977). Insertion of DNA carrying ribosomal protein genes of *Escherichia coli* into Charon vector phages. *J. Biol. Chem.* **252**, 7344–7354.

Worcel, A., and Burgi, E. (1972). On the structure of the folded chromosome of *Escherichia coli. J. Mol. Biol.* **71**, 127–147.

Wu, C.-W., Yarbrough, L. R., Hillel, Z., and Wu, F. Y. (1975). Sigma cycle during *in vitro* transcription: Demonstration by nanosecond fluorescence depolarization spectroscopy. *Proc. Natl. Acad. Sci. U.S.A.* **72**, 3019–3023.

Yanofsky, C. Personal communication.

Yang, H.-L., and Zubay, G. (1973). Synthesis of the arabinose operon regulator protein in a cell-free system. *Mol. Gen. Genet.* **122**, 131–136.

Yang, H.-L., Levy, S., and Zubay, G. (1976). Synthesis of an R plasmid protein associated with tetracycline resistance is negatively regulated. *Proc. Natl. Acad. Sci. U.S.A.* **73**, 1509–1512.

Yang, H.-L., Heller, K., Gellert, M., and Zubay, G. (1979). Differential sensitivity of gene expression to inhibitors of DNA gyrase. *Proc. Natl. Acad. Sci. U.S.A.* **76**, 3304–3308.

Zieg, J., Silverman, M., Hilmen, M., and Simon, M. (1977). Recombinational switch for gene expression. *Science* **196**, 170–172.

Zipser, D. (1970). Polarity and translational punctuation. In "The Lactose Operon" (J. R. Beckwith and D. Zipser, eds.), pp. 221–232. Cold Spring Harbor Laboratory, Cold Spring Harbor, New York.

Zubay, G. (1969). In "The Role of Adenyl Cyclase and Cyclic 3′5′-AMP in Biological Systems" (T. W. Rall, M. Rodbell, and P. Conliffe, eds.), pp. 231–235. Natl. Inst. Health, Bethesda, Maryland.

Zubay, G. (1979). The isolation and properties of CAP, the catabolite gene activator. In "Methods in Enzymology," Vol. *65* (in press).

Zubay, G., and Lederman, M. (1969). DNA-directed peptide synthesis, VI. Regulating the expression of the *lac* operon in a cell-free system. *Proc. Natl. Acad. Sci. U.S.A.* **62**, 550–554.

Zubay, G., and Chambers, D. A. (1971). Regulating the *lac* openon. In "Metabolic Regulation V" (H. Vogel, ed.) p. 340. Academic Press, New York.

Zubay, G., Schwartz, D., and Beckwith, J. (1970). Mechanism of activation of catabolite-sensitive genes: A positive control system. *Proc. Natl. Acad. Sci. U.S.A.* **66**, 104–110.

Zurawski, G., Brown, K., Killingly, D., and Yanofsky, C. (1978a). Nucleotide sequence of the leader region of the phenylalanine operon of *Escherichia coli. Proc. Natl. Acad. Sci. U.S.A.* **75**, 4271–4275.

Zurawski, G., Elseviers, D., Stauffer, G. V., and Yanofsky, C. (1978b). Translational control of transcription termination at the attenuator of the *Escherichia coli* operon. *Proc. Natl. Acad. Sci. U.S.A.* **75**, 5988–5992.

5

Structural Manifestation of Nonribosomal Gene Activity

J.-E. Edström

I. INTRODUCTION

The borderline between structural and chemical analysis of the genetic material was perhaps distinct a few years ago, today this is not so. In this chapter, information obtained by light and electron microscopy on essentially unfragmented genetic material will be used as the framework for analysis of its structure–function relations.

215

Copyright © 1980 by Academic Press, Inc.
All rights of reproduction in any form reserved.
ISBN 0-12-289503-7

In structuring this chapter, I have considered it important to distinguish between transcription in somatic cells and meiotic transcription as displayed by the lateral loops of lampbrush chromosomes. The justification of this distinction will be introduced later. The major part of this chapter, dealing with somatic transcription, is divided into sections dealing with nonpolytene and polytene genetic material. This is mainly a practical subdivision based on different types of analytical methods that can be used with the two types of material. In polytene chromosomes, it is often possible to relate defined structural modifications to chemical and functional parameters at recognized genetic loci, which is rarely the case for nonpolytene material. This is my excuse for devoting most space to polytene material with which I am also more familiar.

II. CHROMATIN IN NONPOLYTENE SOMATIC CELLS

A. Gross Structural Correlates of Transcriptive Activity

In nonpolytene interphase cells, chromatin can exist in either a compact form (condensed chromatin or heterochromatin) or in a dispersed form (euchromatin) with a much lower fibril density and DNA content per unit volume. In nuclei containing both types of chromatin, the heterochromatin is preferentially localized around the nucleolus and on the inside of the nuclear envelope. It has long been known from a large number of studies that the ratio of euchromatin to heterochromatin in a cell increases with increasing protein synthetic activity. This is dramatically shown when relatively heterochromatic lymphocytes are stimulated to blastic transformation (Frenster, 1974). The fact that the relative amounts of the two types of chromatin are interdependent indicates that they are different modifications of the same cellular component. Early studies showed the euchromatin to incorporate RNA precursors and the heterochromatin to be relatively inactive in transcription (Hsu, 1962; Frenster et al., 1963; Littau et al., 1964; Granboulan and Granboulan, 1965). Nevertheless, all euchromatin may not be immediately engaged in transcription. Karasaki (1965) and Milner and Hayhoe (1968) concluded that the highest incorporation occurs in the border area between euchromatin and heterochromatin. Fakan and Bernhard (1971), Nash et al. (1975), and Fakan et al. (1976) using very short pulses of labeled RNA precursors showed that the initial incorporation occurs preferentially in this border area and that the label redistributes to remaining parts of the euchromatin after a chase in nonradioactive medium. Concomitantly with this redistribution, the size distribution of the labeled RNA changes from a spectrum characteristic for heterogeneous nuclear RNA (hnRNA) toward lower molecular weight values (Fakan *et al.*, 1976). The inner euchromatic areas may thus be the site of processing of RNA rather than of its synthesis.

The histone–DNA ratio is similar in the two types of chromatin, but there is an excess of nonhistone proteins in the euchromatin (Frenster, 1965), a feature that has a parallel during transcription in polytene chromosomes (see Section III,C).

Even if condensed interphase chromatin is considered to be inactive in RNA synthesis, erythrocyte nuclei that appear to be completely heterochromatic have been reported to synthesize both low and high molecular weight RNA (Zentgraf *et al.*, 1975). This has a parallel in the transcription of 5 S and 4 S RNA in mitotic chromosomes (Zylber and Penman, 1971), which may be thought of as the most extreme form of heterochromatin.

In the context of the present discussion, one of the most interesting examples of the correlation between structure and activity is the heterochromatinization of one of the female X chromosomes in mammals (Lyon, 1961). There is ample evidence that heterochromatinization of the X chromosome is paralleled by genetic inactivity for most of the alleles it carries (see Lyon, 1972, for review).

In coccids, including the mealybug, the paternal set of chromosomes becomes heterochromatic in males at an early embryonic stage, which has a parallel in nonexpression of paternally derived genetic markers (Brown and Nur, 1964; Brown, 1969). The relation between the state of the chromatin and genetic activity is further illustrated by the reactivation in some tissues of the paternal set both structurally and genetically (Nur, 1967).

B. Activity Correlates at the Fine Structural Level

1. Studies on Sectioned Material

The structural complexity of the genetic material and the difficulties of distinguishing between DNA and RNA have been severe obstacles for electron microscopic (EM) studies on sectioned material. Nucleic acids have been classified by nuclease digestions, but results are often inconclusive because of interference by fixatives on enzyme action and the low effect of such treatments on overall electron density (for discussion, see Bouteille *et al.*, 1974). The DNA bleaching technique of Bernhard (1969), which leaves ribonucleoprotein (RNP) with contrast, has been important for distinguishing between DNP and RNP (Fig. 1).

Structural components related to genetic activity, detected by this technique are the so-called perichromatin fibrils, as a rule 30–50 Å thick

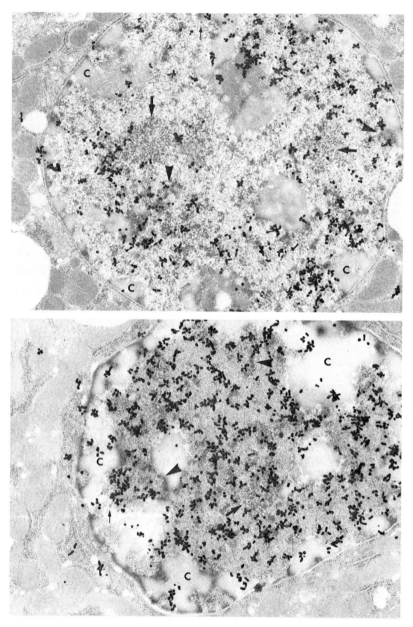

Fig. 1. Nuclei of rat hepatocytes labeled in culture with tritiated uridine, prepared according to the EDTA technique, and subjected to high resolution autoradiography. Upper picture after 5 min of labeling, lower after 5 min of labeling followed by 4 hr of nonradioactive chase. Arrowheads, perichromatin fibrils; large arrows, clusters of interchromatin granules; small arrows, perichromatin granules; c, condensed chromatin regions. ×23,000. (From Fakan *et al.*, 1976.)

and sometimes partially wrapped up into granules (Bernhard, 1969). They can be made to disappear by ribonuclease digestion following protease treatment (Monneron and Bernhard, 1969). The fibrils are found mainly in the border area between heterochromatin and euchromatin, but are present also in other regions of the euchromatin. Extranucleolar incorporation of RNA precursors was found to occur initially over these structures (Fakan and Bernhard, 1971; Nash *et al.*, 1975; Fakan *et al.*, 1976) (Fig. 1). Their occurrence is positively correlated with changes in the state of transcriptive activity (Petrov and Bernhard, 1971; Nash *et al.*, 1975). Bernhard and co-workers have suggested that the perichromatin fibrils in the border region between euchromatin and heterochromatin represent the RNP which is in the process of being formed as RNA is transcribed.

The relation to transcription is less clear for another nonbleached structure, the perichromatin granule (Fig. 1), which is 350–450 Å in diameter, surrounded by a halo, and has a structure suggesting that it is formed of a coiled fiber. It resembles in some respects Balbiani ring granules that are characteristic of certain chromosome puffs (Vazques-Nin and Bernhard, 1971). Perichromatin granules are present at the periphery of condensed chromatin, sometimes well within the euchromatic areas and close to the nuclear pores, where they can appear as if engaged in a process of unravelling, with a filamentous protrusion toward a pore (Monneron and Bernhard, 1969). In contrast to Balbiani ring granules they are not, as a rule, observed outside the nuclear envelope.

A third group of structures is represented by the so-called interchromatin granules (Fig. 1) that are 200–300 Å in diameter and often are seen interconnected by filaments, possibly forming a network (Monneron and Bernhard, 1969). They do not seem to be the site of transcription, and only a very weak labeling with radioactive RNA precursors can be observed several days after precursor administration (Fakan and Bernhard, 1973). The function of these structures is unknown, but the network they seem to participate in may represent the same structure as described by Miller *et al.* (1978). Miller *et al.* demonstrated an intranuclear network to which newly synthesized hnRNA makes connections and suggested that it serves as a framework for the processing of this RNA.

2. Studies on Spread Material

The chromatin spreading technique introduced by Miller and Beatty (1969) transforms the complex three-dimensional architecture to a much simpler two-dimensional representation and often allows a distinction between DNP and RNP simply from topographical relations. Most of this kind of work so far has dealt with preribosomal or lampbrush loop transcription. Studies on somatic nonribosomal transcription processes are scarce, and with one exception do not give examples of defined gene activities.

The basic model of a transcription complex is the classic, almost self-explanatory "Christmas tree" configuration first described for ribosomal RNA (rRNA) genes in amphibian oocytes (Miller and Beatty, 1969). The highly illustrative character of these rRNA transcription figures is due to the multiplicity of transcribing polymerases, which results in clear nascent RNA size gradients, and to the fact that they are genetically and biochemically defined.

The abundance of gene products in a cell varies over several orders of magnitude, with the majority of genes represented by only a few transcribed sequences per cell (Bishop et al., 1974). Intense transcription comparable to that of rRNA units, therefore, occurs only exceptionally. Furthermore, unprocessed primary transcripts have, as a rule, not been identified for nonribosomal genetic units. Although RNP chains attached to DNA filaments can be observed in mammalian cells (Hamkalo and Miller, 1973; Puvion-Dutilleul et al., 1977), the absence of visible gradients (due to the large distances between the growing RNA chains on the template) do not as a rule permit nonribosomal transcription units to be characterized.

Embryonic material from insects with a more intense transcriptive activity offers better possibilities than adult tissues (Foe et al., 1976; Laird et al., 1976; McKnight and Miller, 1976). Intensely transcribing units (fibril gradients) were described in syncytial cleavage nuclei from Drosophila embryos (McKnight and Miller, 1976). During this early stage, with rapid cleavage of nuclei, interphase lasts only 3.4 min. Possibly as a result of a limited transcription time, the fibril gradients are small, covering less than 2 μm of chromatin, which itself is densely packed with polymerases. The general appearance of the gradients agrees with the "Christmas tree" arrangement characteristic of the amphibian oocyte nucleoli. The configuration of the RNP filaments varies between the gradients. Some contain knoblike figures throughout their length; in others a distinct secondary structure is lacking. The density of polymerases is about 30/μm of chromatin, compared to about 50/μm for rRNA transcription units. During a later developmental stage, when the cell cycle lengthens (cellular blastoderm), a second type of transcription units appears, covering 3–6 μm and usually with longer spacings between the polymerases (Fig. 2).

Laird et al. (1976) also studied the cellular blastoderm stage in Drosophila and Oncopeltus. On the basis of number of nucleosomes per length unit, they determined the DNA "packing ratio," i.e., the length of B form DNA to chromatin contour length, to be about 1.6 for active nonribosomal units (as compared to 1.1–1.2 for active ribosomal units).

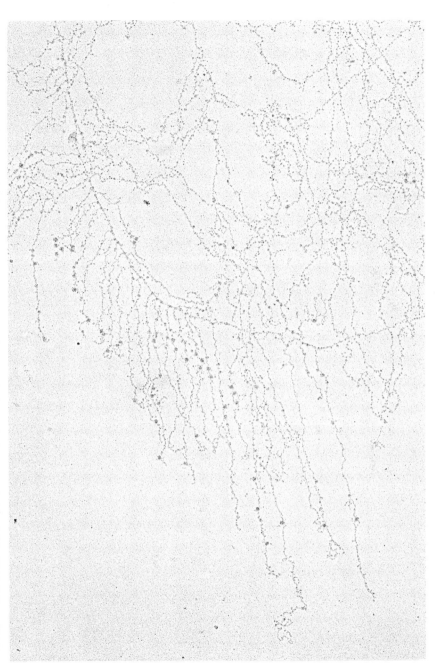

Fig. 2. Nonnucleolar RNP fibril gradient from the cellular blastoderm stage of *Drosophila melanogaster* embryos. (From McKnight and Miller, 1976.)

Furthermore, they derived the size of a series of individual transcription units from the distribution and size of fibrils in RNP fibril gradients. They could then relate RNP fibril length to length of the transcribed DNA and found a highly variable foreshortening of RNA, in the range of 4–23 times reduction in the length of the RNP fibril as compared to the length of the transcribed DNA. Furthermore, they found the size range for the transcription units to be similar to the range for DNA contents in individual *Drosophila* chromomeres, suggesting that most of the chromomere is transcribed. This conclusion may, however, be questioned. When transcribing polymerases are scarce, the likelihood that no growing RNA chain is present in an active transcription unit at a given moment increases as the size of the unit decreases. In other words an active unit may display an RNA chain only intermittently, particularly if it is small. Such units could be underrepresented in a statistical analysis. Furthermore, as shown by the results of McKnight and Miller (1976), the fibril gradient in a given situation may not be representative for the overall range of transcription unit sizes. Firm conclusions can be drawn when an identified primary transcription unit is localized to an identified chromomere. Such a result has been achieved for the BR 2 transcript of *Chironomus tentans*. The results indicate that the transcript occupies only a minor part of the chromomere (Daneholt *et al.*, 1977b). Not all fibril gradients follow the Christmas tree pattern, and more complex transcription-processing patterns can be found (Laird *et al.*, 1976).

In studies on embryonic chromatin from *Oncopeltus*, Foe *et al.* (1976) pointed out a possible fundamental difference between rRNA and non-ribosomal chromatin. The former is free of nucleosomes during transcription and during a phase immediately preceding transcription, whereas the nonribosomal chromatin contained nucleosomes. They suggested the designation rho (ρ) chromatin for the former and nu (ν) chromatin for the latter. Depending on the activity of the ν chromatin, the number of nucleosomes per length unit will vary. In inactive chromatin the DNA packing ratio is 2.2; with increasing activity this ratio can decrease to a value of about 1.6 in *Oncopeltus*. Probably the value is even lower in intensely transcribing units such as the syncytial transcription units of *Drosophila* embryos. The interspersed arrangement of nucleosomes and fibrils carrying polymerases are illustrated by micrographs of lampbrush loops during low transcriptory activity (Scheer, 1978), which may here serve to illustrate the arrangement in principle also during somatic nonribosomal transcription (Fig. 3).

The transcription unit of the silk fibroin gene has been visualized and is found to be densely packed with polymerases (McKnight *et al.*, 1976). If the same DNA packing ratio as determined for rRNA genes it assumed,

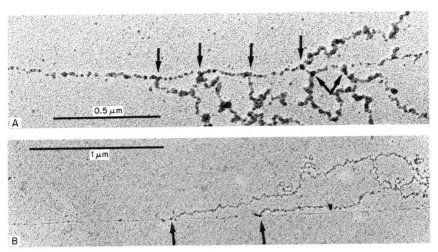

Fig. 3. Transcriptional complexes from lampbrush chromosome loops of *Triturus* oocytes in stages of reduced transcriptional activity, untreated (A) and treated (B) with 0.5% Sarkosyl to remove nucleosomes. Arrows indicate basal polymerase-containing particles. (From Scheer, 1978.)

i.e., 1.12, a good agreement is obtained with biochemical determinations of the gene size (Lizardi and Brown, 1975).

III. POLYTENE CHROMOSOMES

A. Structural Aspects

Polytene chromosomes result from the repeated replication of interphase chromosomes without intervening mitosis with the resulting daughter chromosomes remaining laterally aligned. This results in microscopically visible, cross-banded extended bodies (Fig. 4). On the level of the individual chromosome, the band is equivalent to the chromomere. Polytene chromosomes are best known from the larval salivary glands of the Diptera (e.g., *Drosophila, Chironomus, Sciara*), but they also occur in other dipteran tissues. They can also be found in other groups of organisms, i.e., Collembola (Cassagnau, 1971), ciliates (Ammermann, 1971), and some angiosperms (Nagl, 1969).

Accepting the present concept that the interphase chromosome is one continuous DNA double helix running the length of the chromosome (Kavenoff and Zimm, 1973), the overall DNA packing ratio (extended B form DNA length divided by the chromosome length) in dipteran polytene

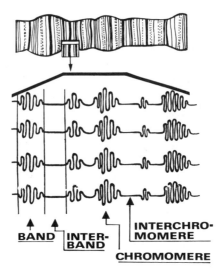

Fig. 4. Diagram of the relation between a polytene chromosome and its constituent chromosomes (chromatids).

chromosomes has been estimated to about 50 (Beermann, 1972). In *Drosophila* the average length of bands is similar to that of interbands or about 0.12 μm (Beermann, 1972). Swift (1962) has determined the DNA concentration in interbands of *Sciara coprophila* using a sensitive Feulgen modification and his data suggest a DNA packing ratio of 1–2. The ratio is, however, unlikely to be much less than 2 because of the packing effect on DNA of nucleosomes. On the assumption that the upper level applies in Swift's determinations, there should be on the average about 700 base pairs (bp) of DNA per interband. This should be compared to an average DNA content per chromomere for the *Drosophila* X chromosome of about 30,000 bp (Rudkin, 1965). The DNA packing ratio can be calculated to be about 100 in the bands, or about 50 times as much as in interbands. A ratio of 1 : 60 for the DNA concentration in interbands to bands in *Drosophila* is mentioned by Swift (1965).

The parts of polytene chromosomes in which bands and interbands alternate are designated euchromatic. In many species there are additional components of the chromosome set. Thus, in *Drosophila* the polytene chromosomes are joined together at the centromeric regions by a compact mass of darkly staining material without band structure. This is designated α-heterochromatin after Heitz (1934) and is transcriptionally inactive (Lakhotia and Jacob, 1974); it is underreplicated during polytenization (Heitz, 1934; Berendes and Keyl, 1967; Mulder *et al.*, 1968; Rudkin, 1969) and contains simple sequence DNA (Gall *et al.*, 1971). Between the α-heterochromatin and the euchromatic chromosome arms there are short intervening nonbanded parts of the chromosome arms called β-heterochromatin (Heitz, 1934), which are more fibrogranular in texture and less dense than the α-heterochromatin. The α-heterochromatin is equivalent to the centromeric heterochromatin of mitotic chromosomes, but a structure corresponding to β-heterochromatin is not seen in mitotic chromosomes (see Sokoloff and Zacharias, 1977, for discussion). β-Heterochromatin incorporates RNA precursors and contains RNP granules (Lakhotia and Jacob, 1974). Its DNA hybridizes *in situ* to mRNA (Spradling *et al.*, 1975); it undergoes replication during polytenization (Plaut, 1963); and it has been shown to contain moderately repetitive sequences (Wensink *et al.*, 1974; Renkawitz, 1978). Few, if any, structural genes have been localized by genetic analysis to the β-heterochromatin (see Beermann, 1962, for review). A diagram of the relation between the different forms of chromatin in polytene and mitotic chromosomes is given in Fig. 5. It should be stressed, finally, that heterochromatin can be located also in noncentromeric regions, interspersed with euchromatin (intercalary heterochromatin), at the telomeres, and that in several cases whole polytene chromosomes may be heterochromatic (see Beermann, 1962, for review).

B. Band–Interband Organization in Relation to Gene Distribution

One of the important foundations for gene expression studies involving polytene chromosomes was established when Beermann (1952), Mechelke (1953), and Breuer and Pavan (1955) showed that the differences between polytene chromosomes of cells in different tissues or different stages of development are due to functional modifications rather than to changes in the basic pattern of bands. This suggests that the band–interband pattern is related to the stable underlying informational content in DNA. This view acquired even more significance when it became clear that bands

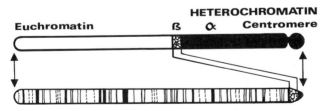

Fig. 5. Diagram of the relative distribution of euchromatin and heterochromatin in a *Drosophila* polytene chromosome and the corresponding mitotic chromosome. (Redrawn from Gall *et al.*, 1971.)

may, at least in certain instances, represent single units of function. Support for this view came from deletion studies which localized allelic mutants in the same band–interband region (reviewed in Beermann, 1972). Relations approaching 1 : 1 between complementation units, defined as operational units for genes, and bands (or interbands) were found in studies where small sections of the *Drosophila* genome were mapped for an increasing number of mutants (Judd *et al.*, 1972; Shannon *et al.*, 1972; Hochman, 1973). Complementation units are nonoverlapping on the genetic map and each is the site of mutants that do not complement each other in the trans position. The identification of bands with such units nevertheless has ambiguities. Thus, there may be some uncertainty in clearly defining the number of bands, of screening for all mutations, and obtaining enough mutations to saturate the investigated region. There are also functional uncertainties. Mutations with polar effects may increase the apparent size of a unit and lower the apparent number of genes. Intragenic complementation, on the other hand, may increase the number of units. For a recent review of band–gene relations, see Ashburner (1976).

The problems involved also are highlighted by the variation in band numbers between species, apparently unrelated to genetic complexity. Band numbers have been estimated to about 5000 for *Drosophila melanogaster* (Beermann, 1972), 3500 for *D. hydei* (Berendes *et al.*, 1973) and 1900 for *Chironomus tentans* (Beermann, 1952)—all of which can be expected to have similar numbers of genes. A parallel exists for the number of meiotic chromomeres, for which similar large differences may exist in closely related species (Section IV,B).

A possible exception to the band constancy rule has recently been preliminary reported (Ribbert, 1977). In the dipteran *Calliphora* polyteny can be induced in the nurse cells by inbreeding and selection. The nurse cells, which provide the oocyte with RNA, exhibit a band pattern which shows overall differences from the pattern in the polytene chromosomes of the pupal bristle-forming cells.

There has been much discussion as to whether structural genes are located in interbands or bands, a discussion stimulated by the apparent discrepancy between the high DNA content in the bands and the apparent single gene content there, and also by demonstrations that large parts of the bands can be deleted without any apparent functional effects (Sorsa *et al.*, 1973). On the other hand, interband DNA contents of 1000 bp or less may not be able to accommodate genetic units, neither in the concept of transcription units (Laird *et al.*, 1976) nor as complementation units. Ritossa (1976) related the sizes of different complementation units which have known polypeptide products to their recombinational map lengths and estimated the sizes of several complementation units to be, as a rule, many thousand base pairs. This is far too large to be contained in interbands, but is of a size more typical for chromomeric DNA contents. The discrepancy between the size of an average messenger or protein and units of this size could be accounted for by processing, involving excision and ligation, of a long primary transcript.

Whereas most of the DNA of eukaryotes cannot easily be accounted for in terms of genes for somatic functions, it is nevertheless present in the somatic cells in most eukaryotes and may partially be transcribed into hnRNA. In the hypotrichous ciliates a mechanism has developed in which only a few percent of the DNA sequences present in the micronucleus (representing the germ line) are retained in the macronucleus for the vegetative functions (Prescott and Murti, 1973; Ammermann *et al.*, 1974). The macronucleus anlagen contains polytene chromosomes with bands and interbands such as in the Diptera. During the phase of DNA elimination each band, altogether about 10,000, becomes surrounded by a membrane (Ammermann, 1971) and DNA is eliminated more or less uniformly from all the bands (Prescott and Murti, 1973). Thus, most of the chromomeric DNA sequences do not seem to be necessary for the somatic functions, and strains devoid of micronuclei grow well for many generations. Probably, however, the eliminated DNA is expressed in the conjugation process and may have an equivalent in DNA expressed during the meiotic transcription in metazoa. A possible parallel is the situation in gall midges in which most of the germ line DNA is eliminated from somatic cells and in which the eliminated DNA appears to be predominantly active in transcription during meiosis (Section IV,A).

C. Transcription

1. Structure and Histochemistry

a. **Light Microscopy.** Transcription in polytene chromosomes usually is associated with local swellings of the bands, so-called puffs. There are also situations where transcription occurs in the absence of visible puff formation. Here the discussion will be limited to morphological and histochemical aspects of puffing, whereas puff induction will be treated in this volume, Chapter 7 by Peter Rae.

Puffing implies an unraveling of the compacted DNP filaments in the chromomere. As a consequence, the region is either broadened and lengthened in most cases or the extended DNA projects laterally in loops which together form a large ringlike structure called a Balbiani ring (BR) (Beermann, 1952) (Fig. 6). Balbiani rings show greater transcriptional

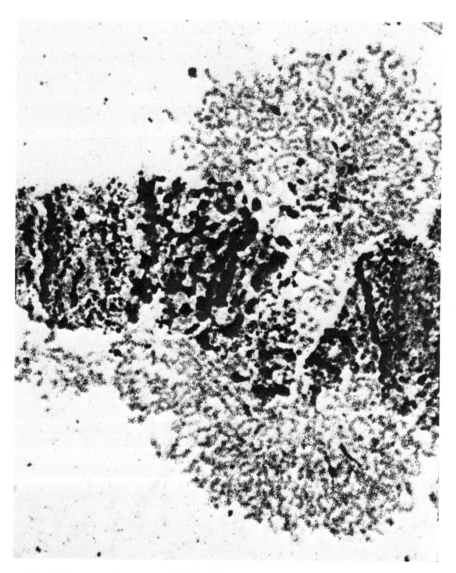

Fig. 6. Electron micrograph of Balbiani ring 2 from *C. tentans*. ×6500. Courtesy of J. Hyde.

activity than non-BR puffs (Pelling, 1964). Balbiani rings do not exist in all Diptera; they have been found in low numbers, and mainly but not only, in the salivary glands of the nematoceran Diptera which include chironomids and simuliids. They are also present in Collembola salivary glands (Cassagnau, 1971). A similar structure has been reported to exist as one class of puffs in *Sciara* (Gabrusewycz-Garcia and Garcia, 1974), although these puffs do not equal the nematoceran BR in size and transcriptory activity.

Although an ordinary puff may obscure the banded structure in a large segment of a polytene chromosome, there is evidence from structural studies that as a rule it originates in a single band (Clever, 1961; Pelling, 1964; Sorsa, 1969). An origin of a DNA puff in a single band has also been described (Breuer and Pavan, 1955). Also a BR can originate in a single band, as is the case for BR 4 (Beermann, 1961) and BR 6 (Beermann, 1973) in *Chironomus pallidivittatus* salivary gland chromosomes (Fig. 7) and a BR in the Malpighian tubules from *C. tentans* (Beermann, 1952). In other cases, a more complex origin is indicated. BR 1 in both *C. tentans* and *C. pallidivittatus* originates in two bands separated by three intervening bands, and the two components have even become separated in the latter species by translocations to give two separate BR's (Beermann, 1962). A complex origin also has been reported for a BR in *Smittia* (Bauer, 1957, cited in Beermann, 1962). In *Acricotopus* the BR's arise regularly from two to three closely adjacent thick bands (Mechelke, 1959). BR 2 in *Chironomus* was originally described as originating in a single, rather thin band in chromosome 4 (Beermann, 1952). The origin later was reassigned to a neighboring thick band (Daneholt *et al.*, 1977b; Beermann, unpublished observations). These chromomeres contain about five times more DNA than an average sized *Chironomus* chromomere (Table I). All in all it appears, therefore, that Balbiani rings often have their origin in blocks of DNA that are larger than the usual chromomeres.

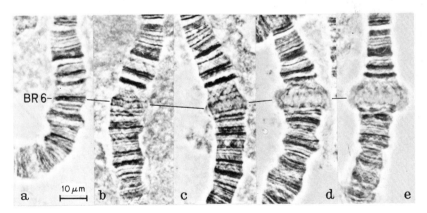

Fig. 7. Segments of chromosome 3 from *C. pallidivittatus*, showing the Balbiani ring 6 band at increasing degrees of puffing, from (a) nonpuffed to (e) maximal puffing. (From Beermann, 1973.)

TABLE I

DNA and RNA Content per Chromomere from Polytene and Lampbrush Chromosomes

	DNA (10^3 bp)	DNA packing ratio[a]	RNA (kb)	RNA/DNA (w/w)	
Chironomus tentans					
BR 1			561[b]		
BR 2	470[c]		1638[b]	1.74	(22.1[d])
BR 3			318[b]		
Average non-BR puff	100[e]		23[b]	0.11	
Drosophila melanogaster X chromosome	30[f]	100[g]			
Triturus oocyte lampbrush chromosomes	6000[h]	1000[i]	120,000[h]	10[j]	(200[k])
Notophtalmus oocyte lampbrush chromosomes	4000[l]				

[a] See Section I,B.

[b] Edström *et al.* (1978), values from cells with chromosomes having 8000–16,000 chromatids; calculated on the assumption of an average chromatid number of 12,000.

[c] Derksen, Wieslander, and Daneholt, to be published.

[d] Value in transcribing part, maximal value, based on assumption of only one transcription unit, 37,000 bp long (Case and Daneholt, 1978).

[e] Daneholt and Edström (1967).

[f] Rudkin (1965).

[g] Beermann (1972).

[h] Calculated from a diploid DNA content of 67 pg (Edström and Gall, 1963) and a chromomere number of 5000 (Callan, 1963).

[i] Sommerville *et al.* (1978b).

[k] Calculated on the assumption that 5% of the total DNA is transcribing and present in the

[j] Calculated on the assumption that 5% of the total DNA is transcribing and present in the loops (Sommerville and Malcolm, 1976).

[l] Calculated from a diploid DNA content of 89 pg (Edström and Gall, 1963) and a chromomere number of 10,000 (Gall, 1963).

With the exception of a group of puffs among the sciarid flies, there is no evidence for changes in the amount of DNA during puffing. The concentration (but not the amount) of DNA per band decreases as the result of unraveling, and Beermann and Bahr (1954) have calculated that the minimal length of extended chromatin in BR 2 of *C. tentans* is 5 μm, representing a length increase of five times. Values between 2 and 12 μm of extended puff chromatin have been observed in *D. hydei* by Berendes (1969).

Puffs vary in size, but the RNA concentration has been judged to be similar in puffs of different size from *C. tentans* (Pelling, 1964). This would suggest that bands puff in proportion to the amounts of RNP present. The incorporation of RNA precursors within a chromosome set has, furthermore, been found to occur roughly in proportion to the size of the individual puff (Pelling, 1964; Zhimulev and Belyaeva, 1975). Bands not visibly puffed may show evidence for transcription in the form of RNP granules (Gabrusewycz-Garcia and Garcia, 1974), and visible puff formation, therefore, is probably not a very sensitive indicator of total transcription. About 50% more bands are seen by autoradiography to incorporate RNA precursors than the number of bands observed to puff in *C. tentans* (Pelling, 1964). Also in *Drosophila* nonpuffed bands (and interbands) have been found to be labeled after RNA precursor administration (Zhimulev and Belyaeva, 1975; Bonner and Pardue, 1976). In *Drosophila* chromosomes stretched with a micromanipulator, the number of bands labeled by RNA precursors exceeded by several hundred percent the number of visible puffs in the corresponding chromosome segments (Ananiev and Barsky, 1978), and altogether close to 40% of the total number of bands were labeled. Sequences coding for histone mRNA in *D. melanogaster* are located in a banded region without puffs, although there occurs transcription in the region (Pardue *et al.,* 1977).

In addition to DNA and histones, puffs contain nonhistone protein with specific staining properties (Clever, 1961; Holt, 1970). Such protein has also been detected in visibly nonpuffed bands and during puff induction before the visible appearance of the puff (Berendes, 1968).

In the case of bands containing the 5 S genes, there is neither structural nor autoradiographic evidence for transcription (Sorsa, 1973; Alonso and Berendes, 1975). This may be a consequence of small transcript size and may be expected to apply in other similar instances.

b. Electron Microscopy. Puffs are observed to contain 100–500 Å RNP granules, the average size being characteristic for a given puff (Beermann and Bahr, 1954; Swift, 1965; Sorsa, 1969; Gabrusewycz-Garcia and Garcia, 1974; Skaer, 1977). Granules belonging to different size classes within a given puff also have been observed (Gabrusewycz-Garcia and Garcia, 1974). The RNP is not always regularly granular but may in some cases be more or less fibrillar (Swift, 1965; Sorsa, 1969). Granules in the larger size range, i.e., 300–500 Å, are characteristic of some but not all BR's (Beermann and Bahr, 1954; Stevens and Swift, 1966). BR 3 in *C. tentans* contains granules 130–200 Å in diameter (Sass, 1978). Since, furthermore, non-BR puffs may contain granules of the large size characteristic for most BR's, a BR cannot be defined on the basis of the size of the RNP granules. Puff granules have been observed to be attached by a stalk to what is probably DNP (Beermann and Bahr, 1954; Stevens and Swift, 1966; Gabrusewycz-Garcia and Garcia, 1974; Daneholt, 1975; Sass, 1978).

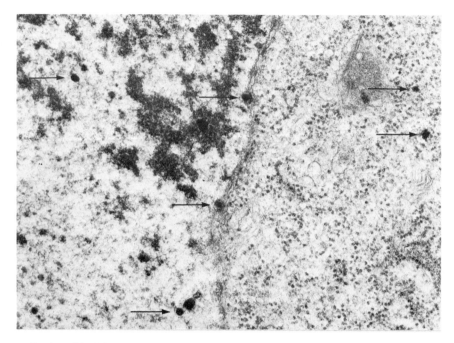

Fig. 8. Balbiani ring granules (arrows) from salivary gland cells of *C. tentans*, as seen in the nucleus (left), traversing the nuclear envelope, and in the cytoplasm (right). Compare the size of the granules with that of the ribosomes in the cytoplasm. × 49,000. Courtesy of J. Thyberg.

Distinct BR granules are seen in the nuclear sap, in nuclear pores, and in the cytoplasm outside of the nuclear envelope (Stevens and Swift, 1966) (Fig. 8).

A particularly large RNP aggregate, 1000–3000 Å in diameter, has been observed in *D. hydei* in a chromosome region which can be activated by vitamin B_6 (Derksen *et al.*, 1973). It also can be observed in the nuclear sap. It consists of a central core surrounded by 300 Å RNP particles and may be a storage form for the latter, the RNA of which may be a 40 S RNA that can be isolated from this puff (Bisseling *et al.*, 1976). Particles of this character seem to be of exceptional occurrence, but a similar complex has been observed in one region of *D. virilis* salivary gland chromosomes (Swift, 1965).

RNP particles also are found in puffs in which there is a differential synthesis of DNA. In *Rhynchosciara* and *Sciara* a few puffs that form shortly before metamorphosis show disproportionate synthesis of DNA (Breuer and Pavan, 1955; Rudkin and Corlette, 1957; Crouse and Keyl,

1968). These puffs also synthesize RNA, with a peak of synthetic activity after or simultaneously with the period of maximal DNA synthesis (Gabrusewycz-Garcia, 1964; Pavan and Da Cunha, 1969). During early stages, the DNA puffs are characterized by a coarse and dense chromatin network which is later loosened up and acquires RNP granules. These granules are on the average smaller and more homogeneous than those of the RNA puffs of the same organism (Gabrusewycz-Garcia and Garcia, 1974).

In certain situations puffs may show a morphological connection with nucleolar formation, the functional significance of which is not yet understood. In *Sciara* the nucleolar organizer region, which is localized on the X chromosome, extends fine strands of DNA throughout the nuclear sap (Gerbi, 1971). Bodies with ultrastructural characteristics of nucleolar material are formed in close association with DNA puffs and heterochromatic bands (Gabrusewycz-Garcia and Kleinfeld, 1966; Gabrusewycz-Garcia, 1972).

2. Interband Transcription

It is quite clear that numerous polytene chromosome bands in dipteran chromosomes must be involved in transcription. The amount of RNA complementary to the DNA in *Drosophila* (Levy *et al.*, 1976), the size of the transcripts in *Drosophila* (Lengyel and Penman, 1975; Laird *et al.*, 1976) and in *Chironomus* (Daneholt, 1972; Egyházi, 1974), and the pattern of labeling of RNA in polytene chromosomes all require an involvement of at least part of the chromomeric DNA in transcription. Since the interband DNA represents a considerably smaller fraction of the genome, its engagement is more difficult to assess, although several claims have been made that such an involvement exists.

The evidence for interband transcription has to be evaluated against what is known regarding the relative distribution of DNA between bands and interbands. The concentration of DNA in interbands is at most 5% of that in the bands, and probably even less (Section III,A). The concentration of histones is likely to parallel that of the DNA (Gorovsky and Woodward, 1967). Consequently at least 95% of the total volume in the interbands cannot be accounted by the DNA–histone chromatin element. It is possible that this component is nuclear sap permeating the chromosomes. The polytene chromosomes in *C. tentans* occupy about half of the nuclear volume. This suggests that of the order of one-fourth of the nuclear sap might be localized in the interbands.

The possible sap content of interband regions seriously complicates the interpretation of almost all the evidence in favor of interband transcription. This is not only because the nuclear sap contains RNP granules but also because it becomes labeled soon after RNA precursor administration due to the short times required for the synthesis of low molecular weight 4 S and 5 S RNA's. Furthermore, in contrast to general belief, 4 S RNA is not immediately transferred to the cytoplasm. For *Chironomus* newly synthesized 4 S RNA has been shown to be retained in the nucleus for 30–60 min before being exported to the cytoplasm (Edström and Tanguay, 1974), probably as a consequence of the fact that it undergoes intranuclear maturation (Egyházi *et al.*, 1969; Lönn, 1977a). The 4 S RNA accounts for an appreciable part of the total RNA synthesis in these cells. Thus, even very short pulses of RNA precursors, are likely to result in some autoradiographic labeling of the nuclear sap and hence of the interbands. Until experiments of this kind are supplemented with controls involving inhibition of RNA polymerase II, they should be evaluated with great caution. Ananiev and Barsky (1978), using micromanipulator-stretched chromosomes did not find unequivocal evidence for interband labeling that could be attributed to newly synthesized RNA in *Drosophila*. Interbands surrounded by two unlabeled bands were never labeled.

A problem in evaluating the role of interband DNA in transcription is the very low concentration as compared to band DNA (Section III,A). It may thus not be easy to exclude, by autoradiographic procedures, that this DNA is transcribed more or less in proportion to its relative concentration. All that can be said at present is that there is no satisfactory autoradiographic evidence that a significant transcription occurs in interbands.

Evidence for interband transcription based on the presence of RNP granules (Skaer, 1977) is complicated by similar difficulties i.e., the possibility exists that granules are present because of an nonspecific permeation of nuclear sap. Interband granules, many of which appear to be similar to 350–450 Å perichromatin granules, must represent the transcripts of appreciable parts of DNA. Thus, the 300 Å granules of the vitamin B_6 puff in *D. hydei* in all probability contain a 40 S RNA (Bisseling *et al.*, 1976), i.e., about 10,000 nucleotides, and the BR granules, similar in size to perichromatin granules, must contain on the order of 40,000 nucleotides (Case and Daneholt, 1978). The average DNA content of an interband in *Drosophila* is, however, likely to be less than 1000 bp. It is consequently not likely that the interband DNA can contribute significantly to the visible RNP material which appears to be present in interbands. If, on the other hand, one assumes that interband DNA is part of a transcription unit large enough to produce these granules, in other words is part of a unit that extends into an adjoining band, one could then accept the view that interband RNP granules have originated *in situ*. In such a case, the interband RNP should, however, always be accompanied by considerably larger numbers of band- or puff-localized RNP particles. This, however, is not observed (Skaer, 1977), and it appears, therefore, highly improbable that the presence of RNP in interbands is a morphological correlate of interband transcription, but is much more likely to be due to permeating nuclear sap containing RNP granules.

One indirect piece of evidence for interband transcription derives from the distribution of RNA polymerase II studied by indirect immunofluorescence in squashed polytene chromosomes. The enzyme was found to be present not only in puffs but also, more weakly, in interbands (Plagens *et al.*, 1976; Jamrich *et al.*, 1977). However, in view of the possible sap content of the interbands and also the nonspecific binding properties of RNA polymerase II, these results cannot be considered as strong support for interband transcription.

There is, however, at least one instance where interbands may be genetically active. The 5 S RNA genes in *Drosophila* occupy two closely adjoining bands, and there is good evidence that the 5 S genes occur in one continuous cluster (Artavanis-Tsakonas *et al.*, 1977), suggesting that the interband region between the two bands is occupied by potentially active 5 S sequences.

3. Chemical Aspects

a. **RNA.** RNA in active chromosome regions has been characterized with regard to content per puff (Edström *et al.*, 1978), base composition (Edström and Beermann, 1962), size distribution (Daneholt, 1972; Bisseling *et al.*, 1976; Serfling, 1976), sequence arrangement (Lambert, 1972; Wobus and Serfling, 1977), and rate of transcription (Egyházi, 1974, 1975). Most of the work has been carried out on large puffs such as the Balbiani rings (Table I). BR 2 of *C. tentans,* which was measured in salivary glands at the end of larval development, contains on the average 11 pg of locally derived RNA (Edström et al., 1978), but values up to 30 pg were recorded for individual rings. If one takes in account the size of the finished transcript [37 kilobases (kb) (Case and Daneholt, 1978)] and the fact that the electrophoretic size distribution of this RNA (Daneholt, 1972; Egyházi, 1975) suggests that most of it exists as a series of growing chains, it is possible to estimate the spacing between the growing chains, if all the RNA derives from one transcription unit per chromomere. For the most RNA-rich rings a maximal density of growing chains would be about $20/\mu m$ if one assumes a DNA packing ratio of 1.2, which is similar to the polymerase density of syncytial transcription units (McKnight and Miller, 1976). For the majority of the rings, the density would be lower, however, and this would be even more so in case the RNA derives from several transcription units.

BR 1 and BR 3 contain on the average about 4 and 2 pg RNA synthe-sized *in situ*, respectively, whereas the corresponding value for an average non-BR puff is 0.15 pg RNA. There is thus between 75 and 15 times more RNA per BR than per non-BR puff.

The size of the transcripts derived from BR 1 are similar to that of BR 2 RNA (Egyházi, 1976). The size of BR 3 RNA is not known. For non-BR puff RNA in *C. tentans,* the electrophoretic size distribution and the rate by which chromosomal RNA label is removed from the puffs after admin-istration of an inhibitor of initiation (Egyházi, 1974) suggest that the size distribution is of the same order as that for BR 1 and BR 2 RNA, and also similar to the size distribution for the hnRNA of mammalian cells. Assum-ing idealized simple transcription units of the order of 30–45 kb, it can be calculated that the average polymerase density is quite low in non-BR puffs, only about 1 per 6 μm of chromatin. This means that many rela-tively small transcription units, even if active, may not display any nas-cent RNA chains at any given moment (cf. II,B).

Transcription products are unlikely to be stored in the BR after transcription, since the RNA disappears from the puffs as soon as the largest nascent RNA chains are eliminated after inhibition of transcription initiation (Egyházi, 1975). There are, however, documented cases of RNP aggregates, such as occur for the vitamin B_6 puff in *D. hydei* (Derksen *et al.,* 1973), where finished products accumulate in the puff (Bisseling *et al.,* 1976).

With regard to the subsequent fate of the transcribed RNA, the ultra-structural appearance of BR granules in the sap and outside of the nuclear envelope has a chemical correlate in the form of BR 2 RNA in the cyto-plasm of a size not measurably different from that of the transcript in the nuclear sap (Daneholt *et al.,* 1977b). The total nuclear sap compartment (including sap permeating the chromosomes) contains about ten times as much BR-derived RNA as is attached to the BR. The nucleus has about 5% of the total cellular content of BR-derived RNA (Edström *et al.,* 1978). BR granules occur in the cytoplasm in the neighborhood of the nuclear envelope (Swift, 1965; Stevens and Swift, 1966) but are not the only struc-tural correlate of intact BR RNA, which is present in polysomes (Daneholt *et al.,* 1977a) and distributed throughout the whole cytoplasm (Lönn, 1978) in close association to the endoplasmic reticulum (Lönn, 1977b).

The vitamin B_6 puff product can also be traced in the nuclear sap as a 40 S molecule (Bisseling *et al.,* 1976), which is a parallel to the ultrastructural appearance of the large composite RNP granules typical for the puff. These aggregates seem to be dissolved into simpler components before export to the cytoplasm (Derksen *et al.,* 1973), but the cytoplasmic counterpart is not known with certainty although a cytoplasmic polyadenylated 15 S RNA shows *in situ* hybridization to the vitamin B_6 puff (Lubsen *et al.*, 1978).

Export to the cytoplasm of puff products not measurably different in size from the transcripts may be a peculiarity of certain BR products. In addition to BR 2 RNA, it applies to BR 1 RNA (Edström and Rydlander, 1976) and has been reported for BR products from *C. thummi* (Serfling, 1976). That condition may, however, not be characteristic for transcripts from non-BR puffs. A 35 S RNA component exists in nuclei and cytoplasm of *C. tentans* salivary gland cells and appears to be derived from higher molecular weight precursors originating in non-BR puffs (Egyházi, 1978). Furthermore, differences between nuclear and cytoplasmic RNA in size and sequence complexity also suggest that size processing is the rule for most nuclear products.

b. Proteins. As has been pointed out previously, the spacing between nascent RNP chains is very large in most active chromosome bands. It is, therefore, not to be expected that the nucleosome density or histone concentration would be measurably affected during puff formation. This expectation is supported by histochemical measurements with the alkaline fast green method, which show that the histone content of bands appears to be unchanged during puffing (Gorovsky and Woodward, 1967). Histones also have been investigated by immunofluorescence and appear to be present in proportion to DNA concentration (Desai *et al.*, 1972; Alfageme *et al.*, 1976; Plagens *et al.*, 1976).

There is an increase during puffing in the amount of total protein (Holt, 1971) and also of material which has staining properties of nonhistone protein (Clever, 1961; Holt, 1970). Since it does not appear possible to label nonhistone puff proteins with radioactive amino acids injected shortly before puff induction (Pettit and Rasch, 1966; Holt, 1970) or by administering labeled amino acids several hours before puff formation (Holt, 1970), these proteins probably preexist in the cell.

RNA polymerase II can be detected by immunofluorescence in puffs (and interbands) (Fig. 9) but is absent in compact bands (Plagens *et al.*, 1976; Jamrich *et al.*, 1977; Greenleaf *et al.*, 1978). The fluorescence is not always proportional to puff size (Greenleaf *et al.*, 1978) nor is this to be expected, since polymerase density may be only one factor influencing the size of the puff, transcript size being another. Little is known about other proteins associated with the appearance of puffs. One protein fraction, designated ρ and having a size between 80,000–110,000 daltons, has been localized by immunofluorescence in *Drosophila* third instar salivary gland polytene chromosomes in puffs and in a number of nonpuffed bands, most of which puff at some stage during third instar or prepupal development

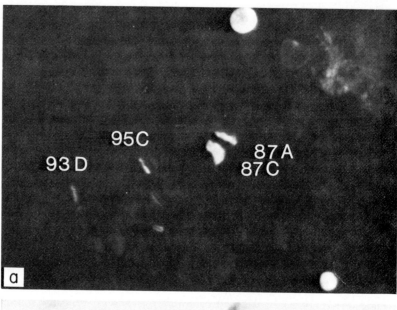

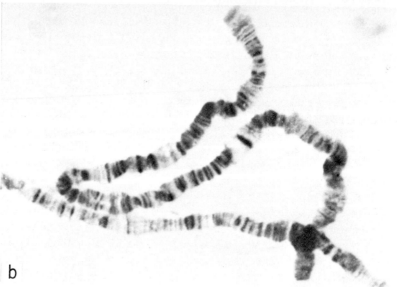

Fig. 9. Indirect immunofluorescent visualization of RNA polymerase II in heat-shocked *Drosophila* larva (a) and phase contrast view (b) of the same specimen after orcein staining. The fluorescent heat-shock puffs are localized on chromosome arm 3 R. (From Jamrich, 1978.) (Courtesy of M. Jamrich.)

(Silver and Elgin, 1977). It may be that this protein has a role in maintaining the chromatin in a configuration that makes it available for transcription. Since visible puffing is, however, not a very sensitive criterion for an active chromosome band, it is also possible that the ρ protein is connected with ongoing transcriptive activity in those unpuffed bands in which it is localized. Proteins known to be associated with growing RNA chains have not yet been characterized in chromosome puffs. Nevertheless, in other systems, a small set of proteins has been characterized associating with hnRNA, some of which may constitute the homologue of puff RNA (Samarina *et al.*, 1968; Beyer *et al.*, 1977; Martin *et al.*, 1977).

4. Genetic Correlates of Puffs

Chromosome puffs in *Diptera* are entities that make it possible to correlate a change in structure from condensed band to a puff with the onset of a defined genetic function, rather than solely with transcriptive activity. Work with this aim has focused on two types of gene products, secretory proteins and proteins formed in response to heat shock.

In salivary glands BR's are tissue specific and therefore have been assumed to be involved in coding for secretory proteins. Correlative evidence for an involvement of a Balbiani ring in the determination of secretory protein was obtained in *Acricotopus* by Baudisch and Panitz (1968) and in *C. tentans* by Pankow *et al.* (1976). More direct evidence for a genetic role for a defined BR has come from work by Beermann (1961) and Grossbach (1969). In these cases, the inheritance of a type of secretory granule or of a defined secretory protein was followed in hybrids between sibling *Chironomus* species in which salivary gland cells have slightly different heritable sets of secretory proteins. The presence of small genome segments containing defined BR's could be correlated with the presence of defined secretory protein fractions. Still, a BR may be more complex genetically than a single coding unit (cistron) for a secretory protein. The complex origin of certain BR's in chironomids has already been discussed. Additional complexity is seen in a certain strain of *C. thummi* which develops a new BR without any measurable change in the pattern of secretory proteins (Wobus *et al.*, 1971). Furthermore, different chironomids with different numbers of BR's in the salivary gland chromosomes have spectra of secretory protein components which contain similar numbers of proteins (Wobus *et al.*, 1970).

The ease of genetic manipulation in *Drosophila* has made it possible to delineate with a high degree of precision the role of a specific puff. Korge (1977) has located a genetic unit that has many characteristics of a structural gene for a defined secretory protein in a particular chromosome segment in the *Drosophila* X chromosome. This segment probably contains only one band and displays a larval puff, the presence of which is correlated with the appearance of the secretory protein. Nevertheless, if rather complex assumptions are made in this case as well as for similar correlations (Korge, 1975; Akam et al., 1978), these puffs do not necessarily have to contain the structural genes for the polypeptide chains. A compelling formal demonstration of such a role would probably necessitate an identification of the relevant puff transcripts.

The DNA puffs of the sciarids are of special interest as a potential example of amplification of structural genes. Such a mechanism may have wide biological implications, since it has recently been demonstrated to occur for folate reductase genes in sarcoma cells (Schimke et al., 1977). In the case of the DNA puffs, there is evidence that the formation of these structures is correlated with particular genetic activities. Thus, the appearance of a poly(A)-containing 16 S RNA is correlated with the appearance temporally and regionally in *Rhynchosciara* salivary glands of a specific DNA puff (Okretic et al., 1977). A similar correlation of a defined protein with this puff and for another protein with another DNA puff also is observed (Winter et al., 1977a). Both proteins appear in the gland secretion (Winter et al., 1977b).

Exposure of *Drosophila* larvae to high temperatures or to treatments which interfere with oxidative metabolism leads to the appearance of a small number of new puffs in the salivary gland chromosomes, the so-called heat-shock puffs (Fig. 8), and a concomitant regression of normal puffs (Ritossa, 1962, 1964; Ashburner, 1970; Berendes, 1972). Heat shock leads to the production of at least six new proteins and to a cessation of most of the labeling of the proteins previously being produced (Tissières et al., 1974; McKenzie et al., 1975; Lewis et al., 1975; Koninkx, 1976). These proteins also appear in *Drosophila* tissue culture cells and other tissues (McKenzie et al., 1975; Lewis et al., 1975). The similarity in numbers of proteins and puffs at first suggested a simple relation between them. McKenzie et al. (1975) found that the dominant mRNA fraction following heat shock was of a size sufficient to code for the dominant protein fraction and that this RNA hybridized in situ to two puffs, 87 A and 87 B (later redesignated 87 C). The latter puff previously had been found to be the one most vigorously incorporating tritiated uridine (Tissières et al., 1974). Spradling et al. (1975) found that the total mRNA from heat-shocked tissue culture cells hybridized in situ to six bands known to puff during heat shock, whereas an additional three bands known to form heat-shock puffs (Ashburner, 1970) did not label. They also detected significant hybridization over a tenth locus not previously being known to be involved in the heat-shock response. A confirmation that at least these ten bands are involved in the heat-shock response was obtained in in situ hybridization studies of heat-shock RNA from imaginal discs (Bonner and Pardue, 1976) when all ten bands hybridized heat-shock RNA. In a later paper Spradling *et al.* (1977) reported on studies in which separated mRNA fractions were hybridized individually to chromosome squashes. Each of three different mRNA fractions hybridized predominantly a single heat shock locus, i.e. the bands 63 C, 95 D, and 67 B. Mirault *et al.* (1977) in *in vitro* translation studies could assign a protein to two of these three species, i.e., those originating in 63 C and 95 D. In these two cases, consequently, the gross features have been outlined in the gene expression chains. The relation of the two bands 87 A and 87 C to mRNA is more complex. Both Spradling *et al.* (1977) and Henikoff and Meselson (1977) found that the dominating mRNA fraction showed sequence complementarity toward both 87 A and 87 C and that the amount of hybridization is two to three times greater in 87 C as compared to 87 A. The 87 C band furthermore showed complementarity toward a less abundant mRNA fraction (Spradling *et al.,* 1977) and toward heterodisperse nuclear RNA (Henikoff and Meselson, 1977). Mirault *et al.* (1977) using *in vitro* translation identified the product of the dominating mRNA fraction hybridizing 87 A and 87 C as one or two 70,000 dalton proteins, dominating in the heat-shock response. The results from *in situ* hybridization indicating multiple representation of the gene(s) for the 70,000 dalton protein in 87 A and 87 C may explain the finding by Ish-Horowicz *et al.* (1977) that no induced heat-shock protein is missing from embryos homozygous for deletions spanning the 87 C locus, whereas the 70,000 dalton component is lacking when deletions cover the whole 87 A–87 C region. Two similar but not identical cloned segments of *Drosophila* DNA, complementary to the mRNA fraction that directs the synthesis of the 70,000 dalton protein, were found by Schedl *et al.* (1978) to hybridize to 87 A and 87 C, but other aspects of their arrangement are unknown. Tandemly repeated sequences hybridizing to other mRNA fractions induced by heat shock are also present in both 87 C and 87 A in arrangements that are now being detailed by cloning techniques (Lis *et al.,* 1978).

5. Genetic Role of BR 1 and BR 2 in Chironomus tentans

The large size of BR 1 and BR 2 in *C. tentans* have made them suitable objects for microchemical analysis and the exploration of their genetic role. Their high molecular weight 75 S RNA is transferred to the cytoplasm without measurable size reduction (Daneholt *et al.,* 1977b) and is derived from two or three bands, but the BR 2 puff is usually dominating. *In vivo* 50–100% of the RNA which is transcribed in the BR's is also exported to the cytoplasm (Edström *et al.,* 1977), where it is present at a concentration of 2–4% (Case and Daneholt, 1978; Edström *et al.,* 1978), predominantly localized in polysomes (Daneholt *et al.*, 1977a; Wieslander and Daneholt, 1977). Grossbach (1973) has suggested that a 0.5×10^6 dalton secretory protein constitutes the translation product because of its size and since it accounts for about 50% of the total secretory protein synthesis. Before this protein can be accepted as representative in size for the primary translation product, it has to be shown, however, that the large size is not a result of posttranslatory modifications, as stressed by Grossbach (1973).

IV. CHROMOSOMES DURING MEIOSIS

A. Structural Aspects

It is not possible in this chapter to cover the subject of transcription during meiosis in much detail. My intention is to review only certain aspects of interest for comparisons with somatic transcription. For a current review of meiotic transcription in amphibian oocytes, where it is best known, the reader is referred to Sommerville (1977).

Meiotic transcription has been most studied in the amphibian primary oocyte in the diplotene stage, when the meiotic chromosomes appear as bivalents with chiasmata. The basic chromosome structure revealed by light and electron microscopic observations and enzyme digestions has been described by Callan (1963) and Gall (1963). The chromosome, which is duplicated at this stage with two chromatids, is a linear structure in which a single DNA duplex presumably runs in each chromatid from one end of the chromosome to the other (Fig. 10). The DNP filament forms a large number of thickenings, or chromomeres, which can be visualized as resulting from a local folding of the DNP fiber very much like in polytene chromomeres, although the DNA packing ratio is much higher in amphibian oocyte chromomeres (Table I). The chromomeres of the two chromatids appear morphologically as one unit, like the interchromomeric fiber, although the latter can be demonstrated to consist of two DNA fibers (Gall, 1963). From each of the two fused chromomeres, loops project in a symmetrical fashion, different chromomeres having loops of different appearance. In amphibia the loops can be dozens of micrometers in circumference and can be visualized as an unraveled part of the chromomere. The loop is believed to derive from one end of the chromomere, toward an interchromomeric segment (Old *et al.*, 1977). The sizes of the loops are dependent on the phase of meiosis, and they are generated from loop-free chromomeres in early diplotene, probably by an unraveling which occurs only at one of the insertions (MacGregor and Andrews, 1977), since the other end may be defined by insertion close to an inter-

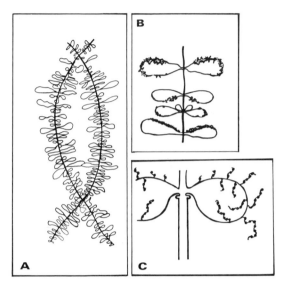

Fig. 10. Schematic representation of a lampbrush chromosome bivalent with two cross-overs (A), a series of lampbrush loops (B), and a single loop with nascent RNP (C).

chromomeric region. Each loop is covered by RNP material. Often, but not always, this material shows a continuous increase in diameter from one insertion to the other, creating a thin insertion end and a thick insertion end. The chromomeres and interchromomeric parts are free of RNP (Mott and Callan, 1975). Incorporation of RNA precursors occurs on the loops (Gall and Callan, 1962; Snow and Callan, 1969) in a manner similar to incorporation into puffs. The loops with their RNP or loop matrix give the chromosomes an appearance responsible for the designation lampbrush chromosomes.

Lampbrush chromosomes fitting this general observation are also known in oocytes of several other genera (see Davidson, 1968, for review). In many cases, loops cannot be distinctly observed in the light microscope, perhaps because the size of the loops may be related to the DNA content (Callan, 1963). In these cases the chromosomes can be observed with a lateral indistinctness or "fuzziness" accompanied by RNA labeling over fuzzy regions. With some evident exceptions, lampbrush chromomeres are likely to be of general occurrence in oocytes. One such exception are meroistic insect oocytes, e.g., *Drosophila,* in which nurse cells provide the oocyte with RNA. It is possible that a lampbrush transcriptional pattern is taken over by the nurse cells in these cases. In the insects *Musca* and *Calliphora,* the nurse cells may become polytenic. In these cells, a generalized diffuse pattern of RNA precursor incorporation has been observed rather than the discontinuous type characteristic of polytene chromosomes in other tissues (Ribbert and Bier, 1969). In gall midges the characteristic meiotic transcription in the oocytes (to the extent that it can be studied in these organisms with very small genomes) takes place in extra chromosomes (E chromosomes) that are eliminated from the somatic cells, whereas the chromosomes to be retained in somatic cells are compact and show no evidence of RNA synthesis (Kunz, 1970; Kunz et al., 1970).

Lampbrush chromsomes may also occur generally in spermatocytes, although they are more difficult to study than in oocytes because loop development is less pronounced. They have been observed by light microscopy in Orthoptera (Henderson, 1964) and by electron microscopy in the mouse (Kierszenbaum and Tres, 1974) and Chironomus (Keyl, 1975). The Y chromosome of Drosophila forms loops similar to lampbrush loops, but it is uncertain whether it is homologous to other meiotic lampbrush chromosomes. A lateral fuzziness of spermatocyte bivalents is of common occurrence, suggesting that the presence of lateral loops is a general phenomenon during meiosis. Lampbrush loop activity is already present in pachytene in spermatocytes. Because this is before the phase of chromosome repulsion, the two loops from each chromosome project in the same direction (Henderson, 1971) rather than in opposite directions as in oocytes. According to Henderson (1971), the second meiotic prophase also shows lampbrush loops in Orthoptera. In spermatocytes, usually only the paired autosomes display the lampbrush condition. In Orthoptera the X univalent is transcriptionally inactive and condensed (Henderson, 1964), and this also applies to the condensed XY bivalent of mammalian spermatocytes, the sex vesicle (Monesi, 1965; Kierszenbaum and Tres, 1974). This is probably not due to an inherent inability of the X to participate in meiotic transcription since, by all evidence, the corresponding XX bivalent is transcriptionally active in oocytes (Ohno et al., 1961; Kunz, 1967).

B. Transcription

1. Structure

The RNP covering the loops in amphibian oocyte lampbrush chromosomes may vary in gross appearance, i.e., in texture, thickness, and distribution. Homologous loops are usually similar, but heterozygosity in loop morphology may occur. There are also instances of large heteromorphic chromosome segments probably related to sex determination. In many instances, the RNP forms a single gradient in thickness along the loop, but other arrangements, such as tandemly arranged gradients or a more uniform thickness of RNP (Callan, 1963), may occur. RNP consisting of interconnected 200–300 Å particles have been observed in electron micrographs of sectioned material (Malcolm and Sommerville, 1974; Mott and Callan, 1975). In formamide these structures can be made to appear as a pearl necklace with the RNP granules interconnected by a thin RNA fiber (Malcolm and Sommerville, 1977). This arrangement is general for the lampbrush loops, and individuality in appearance occurs as a result of different patterns of aggregation of the individual RNP particles (Mott and Callan, 1975).

In spread preparations of oocyte lampbrush chromosomes, RNP fibril gradients can be seen in which the DNP axis is densely covered by transcribing polymerases (Hamkalo and Miller, 1973; Angelier and Lacroix, 1975; Scheer et al., 1976; Scheer, 1978). The matrix units vary in size but may occur in lengths corresponding to tens of micrometers (Fig. 11). Fibril gradients may occur as solitary units, but tandem arrangements can also be seen, with units of different size and even different polarities (Angelier and Lacroix, 1975; Scheer et al., 1976). The transcribing polymerases as a rule fill all the space on the chromatin axis. During regression of the loops, the density of RNP fibrils decreases, and nucleosomes appear (Fig. 3) (Scheer, 1978).

Spread spermatocytes from *Chironomus* show chromosomes with a lampbrushlike organization (Keyl, 1975). About 98% of the DNA is calculated to lie in the loops, which are covered by an amorphous material the nature of which is unclear. In *Drosophila* XO spermatocytes, the density of RNP filaments is below that of amphibian lampbrush loops, but nevertheless considerably higher than in most transcriptional complexes from somatic cells (Glätzer, 1975). Long, branched transcripts can be observed in arrangements of a size suggesting that they represent transcripts of very large pieces of DNA, possibly entire chromomeres. Similar transcription complexes have been observed in mouse spermatocytes (Kierszenbaum and Tres, 1974).

2. Chemical Aspects

a. RNA. The RNP of amphibian lampbrush chromosomes has a much lower RNA–protein ratio than RNP in chromatin of somatic nuclei. It can be calculated from the data of Izawa et al. (1963) that the RNA–protein ratio is less than 0.02 in the former, and Sommerville (1973) gives a figure of less than 0.03 for the RNA component of the loop RNP. The overall RNA–DNA ratio is about 10 (Izawa et al., 1963; Edström and Gall, 1963), but since only 5% of the DNA is exposed in the loops (Sommerville et al., 1978b), the overall RNA–DNA ratio in active parts should be about 200, which is even higher than for the BR 2 transcription unit (see Table I). The

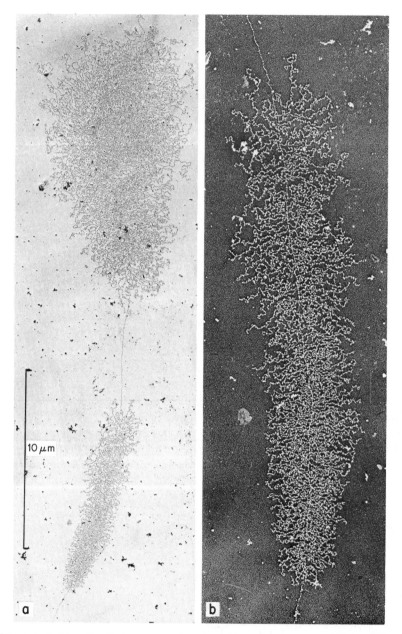

Fig. 11. Part of lampbrush chromosome loop from *Pleurodeles* oocytes with two different sized transcriptional complexes (a) and a single transcription complex (b). (From Angelier and Lacroix, 1975.)

RNA of the loops has a base composition similar to that of DNA in *Triturus* oocytes (Edström and Gall, 1963). After shedding, i.e., after reaching the nuclear sap, the RNA composition undergoes drastic changes, resulting in high uridylic acid contents to an extent that varies for different cells (Edström and Gall, 1963; Sommerville, 1973).

The RNA in amphibian lampbrush chromosomes is complementary to about 5% of the genome (Sommerville and Malcolm, 1976), and only 3% of the RNA shows sequence homology with cDNA prepared from messengerlike RNA from the same cells. From the length of the fibril gradients, which may correspond to lengths of up to about 100 μm of nascent transcripts, RNA of giant size would be expected. There are no methods to determine the size of such large RNA which may, however, not necessarily exist *in vivo*, since the RNA molecules may be nicked. Large-size RNA molecules, sedimenting in the range of 50 S to 100 S (Malcolm and Sommerville, 1977) and measuring 3×10^6 to 10×10^6 daltons in denaturing formamide gels (Denoulet *et al.*, 1977), have been obtained.

The cDNA produced from RNA templates with properties of mRNA hybridizes mainly with unique sequences and corresponds to about 10,000 different mRNA sequences (Rosbash *et al.*, 1974; Sommerville and Malcolm, 1976), which is about the same as the number of lampbrush loops. This degree of informational diversity is similar in two amphibian species with widely different amounts of DNA in their genomes (Rosbash *et al.*, 1974).

RNA precursors as a rule are incorporated following a pulse along the whole extent of the matrix-covered loop, but in exceptional cases, such as the giant granular loops of *Triturus*, pulse labeling is seen only at the thin insertion end. In the latter case label spreads over the whole loop during a period of 10 days following the pulse (Gall and Callan, 1962). It is not known how this movement occurs, since it seems to be the rule that loops are stationary (Malcolm and Sommerville, 1974; Old *et al.*, 1977).

b. Proteins. More than 95% of loop RNP is protein in amphibian oocytes, consisting of a large number of molecular species in the range of 10,000 to 150,000 daltons (Sommerville and Hill, 1973; Scott and Sommerville, 1974; Maundrell, 1975; Malcolm and Sommerville, 1977). This is in marked contrast to the composition of RNP in hnRNA in somatic nuclei where only a few protein species are found (Section III,C). Most of the proteins are present in virtually all loops in proportion to the amounts of covering RNP, as judged by immunofluorescence (Sommerville *et al.*, 1978a). One particular protein is present in only a small number of loops of unknown function present in several chromosomes. Oocytes contain a 40 S to 42 S RNP particle consisting of 5 S and 4 S RNA and two distinct proteins. One of the latter is present in putative 5 S RNA loops, and the other is absent from this loop pair but present in a set of loops which could, therefore, represent 4 S RNA genes (Sommerville *et al.*, 1978a).

3. Functional Aspects

The chromomeres in amphibian lampbrush chromosomes occur in numbers of several thousand per chromosome set, and values in the range of 4000–10,000 have been given (Callan, 1963; Gall, 1963; Vlad and Macgregor, 1975). The numbers are thus about twice as high as for dipteran chromomeres of polytene chromosomes (Section III,B). Since this difference is, nevertheless, moderate and could reflect differences in functional complexity and numbers of genes, the possibility should be entertained that the two types of chromomeres may be homologous. Support for this view is given by Keyl (1975), who finds the number of loops in *Chironomus* spermatocyte chromosomes not significantly different from the number of polytene chromosome chromomeres in the same organism. As in the case of Diptera, even closely related amphibian species may show significant differences in chromomere numbers (Vlad and Macgregor, 1975), suggesting that an identification of chromomeres with single genes, even if feasible in some cases, may not always be possible. Since chromomeres may occasionally be the site of more than one pair of loops (Sommerville, 1977) and in view of the complex arrangement of RNP fibril gradients that sometimes occurs in the loops (Section I,B), this is hardly surprising.

It can be calculated that the average chromomere in *Triturus* contains 6000 kb and in *Notophtalmus* 4000 kb of DNA. This is based on genome sizes of 33.5 and 44.5 pg of DNA (Edström and Gall, 1963) and chromomere numbers of 5000 and 10,000, respectively (Callan, 1963; Gall, 1963). For dipteran polytene chromosomes values of 30 kb per chromomere (Rudkin, 1965) and 100 kb per chromomere (Daneholt and Edström, 1967) have been obtained. The amphibian chromomeres thus contain about 100 times more DNA than they do in Diptera (Table I). Even if only 5% is transcriptionally active in loops, or 250 kb, the transcription templates are enormous compared to the lengths of mRNA. Thus, even if amphibian loops were to be sites of single genes, as a rule only about 1% of the transcription can be accounted for in mRNA production, unless the genes are reiterated. The value obtained by nucleic acid hybridization (Sommerville and Malcolm, 1976) is only slightly higher (3%).

Chromomeres, as reflected in loop morphology phenotypes, behave as heritable units. Heterozygosity in loop morphology can occur, and loop phenotypes are transmitted according to Mendelian principles and distributed in populations according to Hardy–Weinberg rules (Callan and Lloyd, 1960). The character of the RNP matrix is, furthermore, charac-

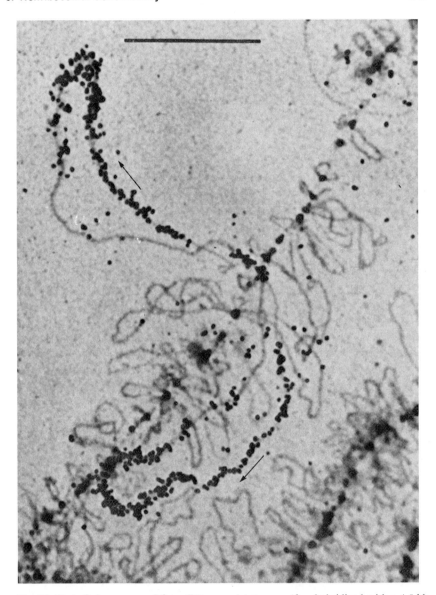

Fig. 12. Part of chromosome I from *Triturus cristatus carnifex*, hybridized with a 4.5 kb pair fragment of the histone gene cluster of *Strongylocentratus purpuratus*, containing the coding regions for histones H2b, H4, and H1. The conservation of histone sequences during evolution permits the use of echinoderm DNA for hybridizing newt histone RNA sequences. Scale, 20 μm. Arrows indicate direction of transcription. (From Old *et al.*, 1977.)

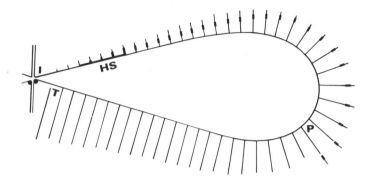

Fig. 13. Diagram of the transcription of sequences complementary to sea urchin histone genes in a *Triturus* lampbrush loop. The loop DNA is drawn thicker in the region containing the histone sequences (HS). The RNA transcripts project from the loop, the length of the RNA arbitrarily drawn contracted to one tenth of the transcribed DNA; I indicates the initiation site and T, termination. Thickened portions of the RNA chains represent histone sequences, and P indicates where processing of transcripts occurs, liberating histone messenger RNA from the loop matrix. To the left is shown a portion of the nontranscribed main chromosome axis, including a chromomere. (Redrawn from Old *et al.,* 1977.)

teristic for the whole loop in spite of its enormous length (Callan and Lloyd, 1960).

When radioactive DNA complementary to mRNA is used to hybridize to transcripts *in situ,* transcripts of sequences with known somatic function are shown to be present on lampbrush loops (Fig. 12). This applies to 5 S RNA sequences (Pukkila, 1975), histone messenger (Old *et al.,* 1977), and globin mRNA sequences (Old and Callan, unpublished data, cited in Sommerville, 1977). Observations on histone gene transcription indicates that the processing events may differ in different histone loops, that there may be extremely long transcription templates, and that processing may commence during transcription (Old *et al.,* 1977) (Fig. 13). The significance of this, as well as the long uninterrupted transcription of 5 S gene batteries (Pukkila, 1975), is not yet understood; nor is it understood why globin genes are transcribed in nonerythroid cells. It is, therefore, important to point out that the demonstration of transcription of sequences known to have mRNA capabilities is not equivalent to the demonstration of the corresponding genetic function.

C. *Drosophila* Y Chromosome Loops

In all *Drosophila* species investigated so far the Y chromosome develops about half a dozen long, RNP covered loops (Fig. 14). They differ in appearance and degree of development in different *Drosophila* species,

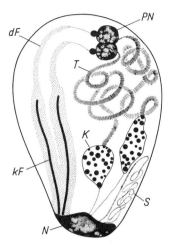

Fig. 14. Diagram of the Y chromosome arrangement in primary spermatocytes of *Drosophila hydei*. The two chromatids run in parallel having the beginning and the end in the mass of nucleolar material. The different loops are designed kF, compact threads; dF, diffuse threads; PN, pseudonucleolus; T, tubular ribbons; K, clubs; S, nooses. N stands for nucleolus. (From Hennig, 1967.)

and are particularly well developed in *D. hydei,* where each loop pair has a characteristic morphology (Meyer, 1963). Barely visible pairs of lampbrush loops develop in *Drosophila* spermatocyte autosomes (Hess, 1973), but the Y chromosome loops are extremely large, having lengths of the order of 100 μm (Hennig *et al.,* 1973). Like amphibian lampbrush loops, they are covered with RNP and incorporate RNA precursors (Hennig, 1967, 1968). Very large nascent transcripts which are similar to amphibian lampbrush loops in the density of the RNP fibrils can be observed in spread spermatocyte nuclei (Hennig *et al.,* 1973). In contrast to amphibian loops, the Y chromosome loops do not as a rule originate with both insertions from a chromomere, but rather the loops are in most cases parts of a more or less continuous extended structure without a basal chromomere. Another difference is that the chromatids (thence the loops) are unpaired, except at the centromere (Hennig *et al.,* 1973). Translocation experiments (Hess, 1967) and loop deletions that give rise to sterility with structural derangements in the developing sperm characteristic for each type of deletion (Meyer, 1968) have been used to map Y chromosome loop functions. Temperature-sensitive mutations also exist, an indication that the role of these loops is exerted via proteins (Leoncini, 1977). At the present time it may perhaps be advisable not to equate these structures with the lampbrush loops characteristic of paired chromosomes during meiosis.

V. COMPARATIVE ASPECTS

A. Polytene versus Nonpolytene Chromatin

The extent of structural modification of the genetic material resulting from transcriptional activity is a product not only of the number of active loci but also of the transcriptory characteristics of each locus, such as the size of transcripts and the density of growing RNP fibrils. In polytene chromosomes, the effect of these factors is to some extent displayed separately. Thus, the number of active loci is seen as a spectrum of band activities (puffing and RNA incorporation), whereas the density of growing fibrils and the size of the transcripts determine the total RNP content in the active locus. If the product of the latter two factors is sufficiently large, the result is a visible puff. These structural changes occur in chromosome regions that are banded and are called euchromatic. The term euchromatin in this context has little to do with the current activity state, but is related to the basic arrangement of the genetic material.

In the chromatin of somatic cells without polytene chromosomes, on the other hand, the relation between the factors enumerated above and the structural manifestations resulting from transcription is probably a function of the concerted action of these factors such that they cannot be distinguished individually. It is well known that transcriptory activity on the whole is accompanied by loosening of chromatin, which also is called euchromatin. In this case, the term has a different meaning than in polytene chromosome material, since it is not related to the basic chromosome organization but to an activity state. The amount of euchromatin is, however, not only a function of the number of active genes but also a function of the intensity of gene expression. Thus, certain differentiated nuclei, e.g., those in large neurons, may be entirely euchromatic, whereas others, e.g., in adjoining glial cells, may be mainly heterochromatic. Such large differences are likely to be related to differences in total amount of transcriptive activity rather than number of active genes. The euchromatin in nonpolytene cell chromatin can probably not be equated with puffs in polytene chromosomes, because (as discussed in Section I,B) the nascent RNA in nonpolytene nuclei appears to be localized mainly in the border region between heterochromatin and euchromatin. Most of the euchromatin, although dispersed, may therefore not be directly involved in transcription.

At a finer structural level there is probably no fundamental difference between polytene and nonpolytene structural activity correlates. Large RNP granules are seen in association with the chromatin as well as in the nuclear sap in both types. In nonpolytene cells most of the transcription is associated with perichromatin fibrils, which can be partially granular and which probably have their counterpart in polytene chromosome puff RNP granules, which often are partially fibrillar. The perichromatin fibrils are probably the homologues of the puff granules, and whether the difference in appearance is an effect of the DNA bleaching technique or reflects an *in vivo* difference may be a question of minor importance.

B. Somatic versus Meiotic Transcription

The appearance of transcriptive processes during meiotic prophase is to some extent influenced by the fact that the chromatin has to be organized into chromosomes of dimensions that permit the orderly meiotic structural chromosome rearrangements. There are good reasons, however, to assume that differences between somatic and meiotic transcription may be more deep-seated than merely a result of this structural difference.

One problem in evaluating differences between the two types of transcription is that practically all knowledge regarding meiotic transcription derives from amphibian oocytes, particularly those with high genome values of DNA. This may in itself give the transcription certain atypical characteristics. However, it is doubtful whether the differences between somatic and meiotic transcription can have this explanation alone.

In spite of indirect evidence for a role of lampbrush loops in a traditional type of gene expression, many characteristics of lampbrush loop transcription are not easily understood. On a gross level, an important characteristic of meiotic transcription is that as a rule it occurs in meiotically paired chromosomes, not in univalents. It is difficult to see why this should be so if normal somatic gene expression is the main feature of meiotic transcription. The inactive X chromosome in the male probably has the same density of somatically active genes as other chromosomes. It is consequently possible that the bulk of meiotic transcription is somehow related to chromosome pairing rather than somatic gene expression.

A case difficult to explain in a somatic sense is the meiotic transcription in gall midge meiocytes, which seems to take place in the E chromosomes. The E chromosomes are eliminated from somatic cells and cannot contain somatically active genetic material, whereas the somatic chromosomes appear to be transcriptionally inactive in meiocytes.

An important characteristic of meiotic transcription is that it occurs in thousands of loci. Such widespread transcription could possibly be understood in oocytes as an accumulation of messages in preparation for early embryogenesis. If such a generalized transcription occurs also in spermatocytes, about which less is known but which seems plausible from their lampbrush loop state, its occurrence would be more difficult to understand, since the sperm contains little or no RNA. Several properties of the transcripts, such as size, transcription intensity, RNP composition, time spent on chromosomes, and posttranscriptional processing appear to be markedly different than in somatic transcription.

Parallels can be drawn between the chromomere organization visible during meiosis and in the dipteran polytene chromomeres. As far as can be decided at present, they may well be homologous structures, although this certainly cannot be considered established. The structural modifications accompanying transcriptory activity may, however, be different. It is characteristic of lampbrush loops that the unwound DNA retains its two insertions close together in the chromomere. This way of looping out is not entirely representative for puffing. The chromomere length is changed by puffing, which would not be the case if real lampbrush loops were formed. Whether the difference is trivial and simply due to a necessity for meiotic chromosomes not to change length due to transcriptory activity, or whether it has more deep-seated causes, remains to be seen. An attempt has been made to explain several characteristics of meiotic transcription against the need for a mechanism operating during meiosis, lowering the cost of gene substitution in eukaryotes (Edström, 1976).

REFERENCES

Akam, M. E., Roberts, D. B., Richards, G. P., and Ashburner, M. (1978). *Drosophila:* The genetics of two major larval proteins. *Cell* **13**, 215–225.

Alfageme, C. R., Rudkin, G. T., and Cohen, L. H. (1976). Locations of chromosomal proteins in polytene chromosomes. *Proc. Natl. Acad. Sci. U.S.A.* **73**, 2038–2042.

Alonso, C., and Berendes, H. D. (1975). The location of 5 S (ribosomal) RNA genes in *Drosophila hydei. Chromosoma* **51**, 347–356.

Ammermann, D. (1971). Morphology and development of the macronuclei of the ciliates *Stylonychia mytilus* and *Euplotes aediculatus. Chromosoma* **33**, 209–238.

Ammermann, D., Steinbrück, G., von Berger, L., and Hennig, W. (1974). The development of the macronucleus in the ciliated protozoan *Stylonychia mytilus. Chromosoma* **45**, 401–429.

Ananiev, E. V., and Barsky, V. E. (1978). Localization of RNA synthesis sites in the 1 B–3 C region of the *Drosophila melanogaster* X chromosome. *Chromosoma* **65**, 359–371.

Angelier, N., and Lacroix, J. C. (1975). Complexes de transcription d'origines nucléolaire et chromosomique d'ovocytes de *Pleurodeles waltii* et *P. poireti. Chromosoma* **51**, 323–335.

Artavanis-Tsakonas, S., Schedl, P., Tschudi, C., Pirrotta, V., Steward, R., and Gehring, W. J. (1977). The 5 S genes of *Drosophila melanogaster. Cell* **12**, 1057–1067.

Ashburner, M. (1970). Patterns of puffing activity in the salivary gland chromosomes of *Drosophila. Chromosoma* **31**, 356–376.

Ashburner, M. (1976). Aspects of polytene chromosome structure and function. *In* "Organization and Expression of Chromosomes" (V. G. Allfrey, E. K. F. Bautz, B. J. McCarthy, R. T. Schimke, and A. Tissières, eds.), pp. 81–95. Dahlem Konf., Berlin.

Baudisch, W., and Panitz, R. (1968). Kontrolle eines biochemischen Merkmals in den Speicheldrüsen von *Acricotopus lucidus* durch einen Balbiani Ring. *Exp. Cell Res.* **49**, 470–476.

Beermann, W. (1952). Chromomerenkonstanz und spezifische Modifikation der Chromosomenstruktur in der Entwicklung und Organdifferenzierung von *Chironomus tentans*. *Chromosoma* **5**, 139–198.

Beermann, W. (1961). Ein Balbiani-Ring als Locus einer Speicheldrüsen-Mutation. *Chromosoma* **12**, 1–25.

Beermann, W. (1962). Riesenchromosomen. *In* "Protoplasmologia" (M. Alfert, H. Bauer, and C. V. Harding, eds.), Vol. VID, pp. 1–165. Springer-Verlag, Berlin and New York.

Beermann, W. (1972). Chromomeres and genes. *In* "Results and Problems in Cell Differentiation" (W. Beermann, J. Reinert, and H. Ursprung, eds.), Vol. IV, pp. 1–33. Springer-Verlag, Berlin and New York.

Beermann, W. (1973). Directed changes in the pattern of Balbiani ring puffing in *Chironomus:* Effects of sugar treatment. *Chromosoma* **41**, 297–326.

Beermann, W., and Bahr, G. F. (1954). The submicroscopic structure of the Balbiani ring. *Exp. Cell Res.* **6**, 195–201.

Berendes, H. D. (1968). Factors involved in the expression of gene activity in polytene chromosomes. *Chromosoma* **24**, 418–437.

Berendes, H. D. (1969). Induction and control of puffing. *Ann. Embryol. Morphol., Suppl.* **1**, 153–164.

Berendes, H. D. (1972). The control of puffing in *Drosophila hydei*. *In* "Results and Problems in Cell Differentiation" (W. Beermann, J. Reinert, and H. Ursprung, eds.), Vol. IV, pp. 181–207. Springer-Verlag, Berlin and New York.

Berendes, H. D., and Keyl, H.-G. (1967). Distribution of DNA in heterochromatin and euchromatin of polytene nuclei of *Drosophila hydei*. *Genetics* **57**, 1–13.

Berendes, H. D., Alonso, S., Helmsing, P. J., Leenders, H. J., and Derksen, J. (1973). Structure and function in the genome of *Drosophila hydei*. *Cold Spring Harbor Symp. Quant. Biol.* **38**, 645–654.

Bernhard, W. (1969). A new staining procedure for electron microscopical cytology. *J. Ultrastruct. Res.* **27**, 250–265.

Beyer, A. L., Christensen, M. E., Walker, B. W., and LeStourgeon, W. M. (1977). Identification and characterization of the packaging proteins of core 40 S hnRNP particles. *Cell* **11**, 127–138.

Bishop, J. O., Morton, J. G., Rosbash, M., and Richardson, M. (1974). Three abundance classes in HeLa cell messenger RNA. *Nature (London)* **250**, 199–204.

Bisseling, T., Berendes, H. D., and Lubsen, N. H. (1976). RNA synthesis in puff 2-48BC after experimental induction in *Drosophila hydei*. *Cell* **8**, 299–304.

Bonner, J. J., and Pardue, M. L. (1976). The effect of heat shock on RNA synthesis in *Drosophila* tissues. *Cell* **8**, 43–50.

Bouteille, M., Laval, M., and Dupuy-Coin, A. M. (1974). Localization of nuclear functions as revealed by ultrastructural autoradiography and cytochemistry. *In* "The Cell Nucleus" (H. Busch, ed.), Vol. 1, pp. 3–71. Academic Press, New York.

Breuer, M. E., and Pavan, C. (1955). Behaviour of polytene chromosomes of *Rhynchosciara angelae* at different stages of larval development. *Chromosoma* **7**, 371–386.

Brown, S. W. (1969). Developmental control of heterochromatization in coccids. *Genetics* **61**, Suppl., 191–198.

Brown, S. W., and Nur, U. (1964). Heterochromatic chromosomes in the coccids. *Science* **145**, 130–136.

Callan, H. G. (1963). The nature of lampbrush chromosomes. *Int. Rev. Cytol.* **15**, 1–34.

Callan, H. G., and Lloyd, L. (1960). Lampbrush chromosomes of crested newts *Triturus cristatus* (Laurenti). *Philos. Trans. R. Soc. London, Ser. B* **243**, 135–219.

Case, S. T., and Daneholt, B. (1978). The size of the transcription unit in Balbiani ring 2 of *Chironomus tentans* as derived from analysis of the primary transcript and 75 S RNA. *J. Mol. Biol.* **124**, 223–241.

Cassagnau, P. (1971). Les chromosomes salivaires polytenes chez *Bilobella grassei* (Denis) (Collemboles: Neanuridae). *Chromosoma* **35**, 57–83.

Clever, U. (1961). Genaktivitäten in den Riesenchromosomen von *Chironomus tentans* und ihre Beziehungen zur Entwicklung. I. Genaktivierungen durch Ecdyson. *Chromosoma* **12**, 607–675.

Crouse, H. V., and Keyl, H.-G. (1968). Extra replications in the "DNA-puffs" of *Sciara coprophila*. *Chromosoma* **25**, 357–364.

Daneholt, B. (1972). Giant RNA transcript in a Balbiani ring. *Nature (London), New Biol.* **240**, 229–232.

Daneholt, B. (1975). Transcription in polytene chromosomes. *Cell* **4**, 1–9.

Daneholt, B., and Edström, J.-E. (1967). The content of deoxyribonucleic acid in individual polytene chromosomes of *Chironomus tentans*. *Cytogenetics* **6**, 350–356.

Daneholt, B., Andersson, K., and Fagerlind, M. (1977a). Large-sized polysomes in *Chironomus tentans* salivary glands and their relation to Balbiani ring 75 S RNA. *J. Cell Biol.* **73**, 149–160.

Daneholt, B., Case, S. T., Derksen, J., Lamb, M. M., Nelson, L., and Wieslander, L. (1977b). The size and chromosomal location of the 75 S RNA transcription unit in Balbiani ring 2. *Cold Spring Harbor Symp. Quant. Biol.* **42**, 867–876.

Davidson, E. H. (1968). "Gene Activity in Early Development." Academic Press, New York.

Denoulet, P., Muller, J.-P., and Lacroix, J.-C. (1977). RNA metabolism in amphibian oocytes. I. Chromosomal pre-messenger RNA synthesis. *Biol. Cell.* **28**, 101–108.

Derksen, J., Berendes, H. D., and Willart, E. (1973). Production and release of a locus-specific ribonucleoprotein product in polytene nuclei of *Drosophila hydei*. *J. Cell Biol.* **59**, 661–668.

Desai, L. S., Pothier, L., Foley, G. E., and Adams, R. A. (1972). Immunofluorescent labelling of chromosomes with antisera to histones and histone fractions. *Exp. Cell Res.* **70**, 468–471.

Edström, J.-E. (1976). Meiotic versus somatic transcription with special reference to Diptera. *In* "Organization and Expression of Chromosomes" (V. G. Allfrey, E. K. F. Bautz, B. J. McCarthy, R. T. Schimke, and A. Tissières, eds.), pp. 301–316. Dahlem Konf., Berlin.

Edström, J.-E., and Beermann, W. (1962). The base composition of nucleic acids in chromosomes, puffs, nucleoli, and cytoplasm of *Chironomus* salivary gland cells. *J. Biophys. Biochem. Cytol.* **14**, 371–380.

Edström, J.-E., and Gall, J. G. (1963). The base composition of ribonucleic acid in lampbrush chromosomes, nucleoli, nuclear sap, and cytoplasm of *Triturus* oocytes. *J. Cell Biol.* **19**, 279–284.

Edström, J.-E., and Rydlander, L. (1976). Identification of cytoplasmic RNA from individual Balbiani rings. *Biol. Zentralbl.* **95**, 521–530.

Edström, J.-E., and Tanguay, R. (1974). Cytoplasmic ribonucleic acids with messenger characteristics in salivary glands of *Chironomus tentans*. *J. Mol. Biol.* **84**, 569–583.

Edström, J.-E., Ericson, E., Lindgren, S., Lönn, U., and Rydlander, L. (1977). Fate of Balbiani ring RNA *in vivo*. *Cold Spring Harbor Symp. Quant. Biol.* **42**, 877–884.

Edström, J.-E., Lindgren, S., Lönn, U., and Rydlander, L. (1978). Balbiani ring RNA content and half-life in nucleus and cytoplasm of *Chironomus tentans* salivary gland cells. *Chromosoma* **66**, 33–44.

Egyházi, E. (1974). A tentative initiation inhibitor of chromosomal heterogeneous RNA synthesis. *J. Mol. Biol.* **84**, 173–183.

Egyházi, E. (1975). Inhibition of Balbiani ring RNA synthesis at the initiation level. *Proc. Natl. Acad. Sci. U.S.A.* **72**, 947–950.

Egyházi, E. (1976). Quantitation of turnover and export to the cytoplasm of hnRNA transcribed in the Balbiani rings. *Cell* **7**, 507–515.

Egyházi, E. (1978). Kinetic evidence that a discrete messenger-like RNA is formed by post-transcriptional size reduction of heterogeneous nuclear RNA. *Chromosoma* **65**, 137–152.

Egyházi, E., Daneholt, B., Edström, J.-E., Lambert, B., and Ringborg, U. (1969). Low molecular weight RNA in cell components of *Chironomus tentans* salivary glands. *J. Mol. Biol.* **44**, 517–532.

Fakan, S., and Bernhard, W. (1971). Localization of rapidly and slowly labelled nuclear RNA as visualized by high resolution autoradiography. *Exp. Cell Res.* **67**, 129–141.

Fakan, S., and Bernhard, W. (1973). Nuclear labelling after prolonged ^{3}H-uridine incorporation as visualized by high resolution autoradiography. *Exp. Cell Res.* **79**, 431–444.

Fakan, S., Puvion, E., and Spohr, G. (1976). Localization and characterization of newly synthesized nuclear RNA in isolated rat hepatocytes. *Exp. Cell Res.* **99**, 155–164.

Foe, V. E., Wilkinson, L. E., and Laird, C. D. (1976). Comparative organization of active transcription units in *Oncopeltus fasciatus*. *Cell* **9**, 131–146.

Frenster, J. H. (1965). Nuclear polyanions as de-repressors of synthesis of ribonucleic acid. *Nature (London)* **206**, 680–683.

Frenster, J. H. (1974). Ultrastructure and function of heterochromatin and euchromatin. *In* "The Cell Nucleus" (H. Busch, ed.), Vol. 1, pp. 565–580. Academic Press, New York.

Frenster, J. H., Allfrey, V. G., and Mirsky, A. E. (1963). Repressed and active chromatin isolated from interphase lymphocytes. *Proc. Natl. Acad. Sci. U.S.A.* **50**, 1026–1032.

Gabrusewycz-Garcia, N. (1964). Cytological and autoradiographic studies in *Sciara coprophila* salivary gland chromosomes. *Chromosoma* **15**, 312–344.

Gabrusewycz-Garcia, N. (1972). Further studies of the nucleolar material in salivary gland nuclei of *Sciara coprophila*. *Chromosoma* **38**, 237–254.

Gabrusewycz-Garcia, N., and Garcia, A. M. (1974). Studies on the fine structure of puffs in *Sciara coprophila*. *Chromosoma* **47**, 385–401.

Gabrusewycz-Garcia, N., and Kleinfeld, R. G. (1966). A study of the nucleolar material in *Sciara coprophila*. *J. Cell Biol.* **29**, 347–359.

Gall, J. G. (1963). Chromosomes and cytodifferentiation. *In* "Cytodifferentiation and Macromolecular Synthesis" (M. Locke, ed.), pp. 119–143. Academic Press, New York.

Gall, J. G., and Callan, H. G. (1962). H^3-uridine incorporation in lampbrush chromosomes. *Proc. Natl. Acad. Sci. U.S.A.* **48**, 562–570.

Gall, J. G., Cohen, E. H., and Polan, M. L. (1971). Repetitive sequences in *Drosophila*. *Chromosoma* **33**, 319–344.

Gerbi, S. A. (1971). Localization and characterization of the ribosomal RNA cistrons in *Sciara coprophila*. *J. Mol. Biol.* **58**, 499–511.

Glätzer, K. H. (1975). Visualization of gene transcription in spermatocytes of *Drosophila hydei*. *Chromosoma* **53**, 371–379.

Gorovsky, M. A., and Woodward, J. (1967). Histone content of chromosomal loci active and inactive in RNA synthesis. *J. Cell Biol.* **33**, 723–728.

Granboulan, N., and Granboulan, P. (1965). Cytochimie ultrastructurale du nucléole. II.

Etude des sites de synthèse du RNA dans le nucléole et le noyau. *Exp. Cell Res.* **38**, 604–619.

Greenleaf, A. L., Plagens, U., Jamrich, M., and Bautz, E. K. F. (1978). RNA polymerase B (or II) in heat induced puffs of *Drosophila* polytene chromosomes. *Chromosoma* **65**, 127–136.

Grossbach, U. (1969). Chromosomen-Aktivität und biochemische Zelldifferenzierung in den Speicheldrüsen von *Camptochironomus*. *Chromosoma* **28**, 136–187.

Grossbach, U. (1973). Chromosome puffs and gene expression in polytene cells. *Cold Spring Harbor Symp. Quant. Biol.* **38**, 619–627.

Hamkalo, B. A., and Miller, O. L., Jr. (1973). Electron microscopy of genetic activity. *Annu. Rev. Biochem.* **42**, 379–396.

Heitz, E. (1934). Über α- und β-Heterochromatin sowie Konstanz und Bau der Chromomeren bei *Drosophila*. *Biol. Zentralbl.* **54**, 588–609.

Henderson, S. A. (1964). RNA synthesis during male meiosis and spermiogenesis. *Chromosoma* **15**, 345–366.

Henderson, S. A. (1971). Grades of chromatid organization in mitotic and meiotic chromosomes. I. The morphological features. *Chromosoma* **35**, 28–40.

Henikoff, S., and Meselson, M. (1977). Transcription at two heat shock loci in *Drosophila*. *Cell* **12**, 441–451.

Hennig, W. (1967). Untersuchungen zur Struktur und Funktion des Lampenbürsten-Y-Chromosoms in der Spermatogenese von *Drosophila*. *Chromosoma* **22**, 294–357.

Hennig, W. (1968). Ribonucleic acid synthesis of the Y chromosome of *Drosophila hydei*. *J. Mol. Biol.* **38**, 227–239.

Hennig, W., Meyer, G. F., Hennig, I., and Leoncini, O. (1973). Structure and function of the Y chromosome of *Drosophila hydei*. *Cold Spring Harbor Symp. Quant. Biol.* **38**, 673–683.

Hess, O. (1967). Complementation of genetic activity in translocated fragments of the Y chromosome in *Drosophila hydei*. *Genetics* **56**, 283–295.

Hess, O. (1973). Local structural variations of the Y chromosome of *Drosophila hydei* and their correlation to genetic activity. *Cold Spring Harbor Symp. Quant. Biol.* **38**, 663–671.

Hochman, B. (1973). Analysis of a whole chromosome in *Drosophila*. *Cold Spring Harbor Symp. Quant. Biol.* **38**, 381–389.

Holt, T. K. H. (1970). Local protein accumulation during gene activation. I. Quantitative measurements on dye binding capacity at subsequent stages of puff formation in *Drosophila hydei*. *Chromosoma* **32**, 64–78.

Holt, T. K. H. (1971). Local protein accumulation during gene activation. II. Interferometric measurements of the amount of solid material in temperature induced puffs of *Drosophila hydei*. *Chromosoma* **32**, 428–435.

Hsu, T. C. (1962). Differential rate in RNA synthesis between euchromatin and heterochromatin. *Exp. Cell Res.* **27**, 332–334.

Ish-Horowicz, D., Holden, J. J., and Gehring, W. J. (1977). Deletions of two heat-activated loci in *Drosophila melanogaster* and their effects on heat-induced protein synthesis. *Cell* **12**, 643–652.

Izawa, M., Allfrey, V. G., and Mirsky, A. E. (1963). Composition of the nucleus and chromosomes in the lampbrush stage of the newt oocyte. *Proc. Natl. Acad. Sci. U.S.A.* **50**, 811–817.

Jamrich, M. (1978). Lokalisierung von Histon- und Nichthistonproteinen auf den chromosomen und dem Chromatin von *Drosophila melanogaster*. Ph.D. Thesis, Univ. Heidelberg, Heidelberg.

Jamrich, M., Greenleaf, A. L., and Bautz, E. K. F. (1977). Localization of RNA polymerase in polytene chromosomes of *Drosophila melanogaster*. *Proc. Natl. Acad. Sci. U.S.A.* **74**, 2079–2083.

Judd, B. H., Shen, M. W., and Kaufman, T. C. (1972). The anatomy and function of a segment of the X chromosome of *Drosophila melanogaster*. *Genetics* **71**, 139–156.

Karasaki, S. (1965). Electron microscopic examination of the sites of nuclear RNA synthesis during amphibian embryogenesis. *J. Cell Biol.* **26**, 937–958.

Kavenoff, R., and Zimm, B. H. (1973). Chromosome-sized DNA molecules from *Drosophila*. *Chromosoma* **41**, 1–27.

Keyl, H.-G. (1975). Lampbrush chromosomes in spermatocytes of *Chironomus*. *Chromosoma* **51**, 75–91.

Kierszenbaum, A. L., and Tres, L. L. (1974). Transcription sites in spread meiotic prophase chromosomes from mouse spermatocytes. *J. Cell Biol.* **63**, 923–935.

Koninkx, J. F. J. C. (1976). Protein synthesis in salivary glands of *Drosophila hydei* after experimental gene induction. *Biochem. J.* **158**, 623–628.

Korge, G. (1975). Chromosome puff activity and protein synthesis in larval salivary glands of *Drosophila melanogaster*. *Proc. Natl. Acad. Sci. U.S.A.* **72**, 4550–4554.

Korge, G. (1977). Direct correlation between a chromosome puff and the synthesis of a larval salivary protein in *Drosophila melanogaster*. *Chromosoma* **62**, 155–174.

Kunz, W. (1967). Funktionsstrukturen im Oocytenkern von *Locusta migratoria*. *Chromosoma* **20**, 332–370.

Kunz, W. (1970). Genetische Aktivität der Keimbahnchromosomen während des Eiwachstums von Gallmücken (Cecidomyiidae). *Verh. Dtsch. Zool. Ges.* **64**, 42–46.

Kunz, W., Trepte, H.-H., and Bier, K. (1970). On the function of the germ line chromosomes in the oogenesis of *Wachtliella persicariae* (Cecidomyiidae). *Chromosoma* **30**, 378–406.

Laird, C. D., Wilkinson, L. E., Foe, V. E., and Chooi, W. Y. (1976). Analysis of chromatin-associated fiber arrays. *Chromosoma* **58**, 169–192.

Lakhotia, S. C., and Jacob, J. (1974). EM autoradiographic studies on polytene nuclei of *Drosophila melanogaster*. II. Organization and transcriptive activity of the chromocentre. *Exp. Cell Res.* **86**, 253–263.

Lambert, B. (1972). Repeated DNA sequences in a Balbiani ring. *J. Mol. Biol.* **72**, 65–75.

Lengyel, J., and Penman, S. (1975). hnRNA size and processing as related to different DNA content in two dipterans: *Drosophila* and *Aedes*. *Cell* **5**, 281–290.

Leoncini, O. (1977). Temperatursensitive Mutanten im Y-Chromosom von *Drosophila hydei*. *Chromosoma* **63**, 329–357.

Levy, B. W., Johnson, C. B., and McCarthy, B. J. (1976). Diversity of sequences in total and polyadenylated nuclear RNA from *Drosophila* cells. *Nucleic Acids Res.* **3**, 1777–1789.

Lewis, M., Helmsing, P. J., and Ashburner, M. (1975). Parallel changes in puffing activity and patterns of protein synthesis in salivary glands of *Drosophila*. *Proc. Natl. Acad. Sci. U.S.A.* **72**, 3604–3608.

Lis, J. T., Prestidge, L., and Hogness, D. S. (1978). A novel arrangement of tandemly repeated genes at a major heat shock site in *D. melanogaster*. *Cell* **14**, 901–919.

Littau, V. C., Allfrey, V. G., Frenster, J. H., and Mirsky, A. E. (1964). Active and inactive regions of nuclear chromatin as revealed by electron microscope autoradiography. *Proc. Natl. Acad. Sci. U.S.A.* **52**, 93–100.

Lizardi, P. M., and Brown, D. D. (1975). The length of the fibroin gene in the *Bombyx mori* genome. *Cell* **4**, 207–215.

Lönn, U. (1977a). Exclusive nuclear location of precursor 4 S RNA *in vivo*. *J. Mol. Biol.* **112**, 661–666.

Lönn, U. (1977b). Direct association of Balbiani ring 75 S RNA with membranes of the endoplasmic reticulum. *Nature (London)* **270**, 630–631.

Lönn, U. (1978). Delayed flow through cytoplasm of newly synthesized Balbiani ring 75 S RNA. *Cell* **13**, 727–733.

Lubsen, N. H., Sondermeijer, P. J. A., Pages, M., and Alonso, C. (1978). *In situ* hybridization of nuclear and cytoplasmic RNA to locus *2-48BC* in *Drosophila hydei*. *Chromosoma* **65**, 199–212.

Lyon, M. F. (1961). Gene action in the X-chromosome of the mouse (*Mus musculus* L.). *Nature (London)* **190**, 372–373.

Lyon, M. F. (1972). X-chromosome inactivation and developmental patterns in mammals. *Biol. Rev. Cambridge Philos. Soc.* **47**, 1–35.

Macgregor, H. C., and Andrews, C. (1977). The arrangement and transcription of "middle repetitive" DNA sequences on lampbrush chromosomes of *Triturus*. *Chromosoma* **63**, 109–126.

McKenzie, S. L., Henikoff, S., and Meselson, M. (1975). Localization of RNA from heat-induced polysomes at puff sites in *Drosophila melanogaster*. *Proc. Natl. Acad. Sci. U.S.A.* **72**, 1117–1121.

McKnight, S. L., and Miller, O. L., Jr. (1976). Ultrastructural patterns of RNA synthesis during early embryogenesis of *Drosophila melanogaster*. *Cell* **8**, 305–319.

McKnight, S. L., Sullivan, N. L., and Miller, O. L. (1976). Visualization of the silk fibroin transcription unit and nascent silk fibroin molecules on polyribosomes of *Bombyx mori*. *Prog. Nucleic Acid Res. Mol. Biol.* **19**, 313–319.

Malcolm, D. B., and Sommerville, J. (1974). The structure of chromosomederived ribonucleoprotein in oocytes of *Triturus cristatus carnifex*. *Chromosoma* **48**, 137–158.

Malcolm, D. B., and Sommerville, J. (1977). The structure of nuclear ribonucleoprotein of amphibian oocytes. *J. Cell Sci.* **24**, 143–165.

Martin, T., Billings, P., Pullman, J., Stevens, B., and Kinniburgh, A. (1977). Substructure of nuclear ribonucleoprotein complexes. *Cold Spring Harbor Symp. Quant. Biol.* **42**, 899–909.

Maundrell, K. G. (1975). Proteins of the newt oocyte nucleus: Analysis of the non-histone proteins from lampbrush chromosomes, nucleoli and nuclear sap. *J. Cell Sci.* **17**, 579–588.

Mechelke, F. (1953). Reversible Strukturmodifikationen der Speicheldrüsenchromosomen von *Acricotopus lucidus*. *Chromosoma* **5**, 511–543.

Mechelke, F. (1959). Beziehungen zwischen der Menge der DNS und dem Ausmass der potentiellen Oberflächenentfaltung von Riesenchromosomenloci. *Naturwissenschaften* **46**, 609.

Meyer, G. F. (1963). Die Funktionsstrukturen des Y-Chromosoms in den Spermatocyten-Kernen von *Drosophila hydei*, *D. neohydei*, *D. repleta* und einigen anderen *Drosophila*-Arten. *Chromosoma* **14**, 207–255.

Meyer, G. F. (1968). Spermiogenese in normalen und Y-Defizienten Männchen von *Drosophila melanogaster* und *D. hydei*. *Z. Zellforsch. Mikrosk. Anat.* **84**, 141–175.

Miller, O. L., and Beatty, B. R. (1969). Visualization of nucleolar genes. *Science* **164**, 955–957.

Miller, T. E., Huang, C.-Y., and Pogo, A. O. (1978). Rat liver nuclear skeleton and ribonucleoprotein complexes containing hnRNA. *J. Cell Biol.* **76**, 675–691.

Milner, G. R., and Hayhoe, F. G. (1968). Ultrastructural localization of nucleic acid synthesis in human blood cells. *Nature (London)* **218**, 785–787.

Mirault, M.-E., Goldschmidt-Clermont, M., Moran, L., Arrigo, A. P., and Tissières, A. (1977). The effect of heat shock on gene expression in *Drosophila melanogaster*. *Cold Spring Harbor Symp. Quant. Biol.* **42**, 819–827.

Monesi, V. (1965). Differential rate of ribonucleic acid synthesis in the autosomes and sex chromosomes during male meiosis in the mouse. *Chromosoma* **17**, 11–21.

Monneron, A., and Bernhard, W. (1969). Fine structural organization of the interphase nucleus in some mammalian cells. *J. Ultrastruct. Res.* **27**, 266–288.

Mott, M. R., and Callan, H. G. (1975). An electron microscope study of the lampbrush chromosomes of the newt *Triturus cristatus*. *J. Cell Sci.* **17**, 241–261.

Mulder, M. P., van Duijn, P., and Gloor, H. J. (1968). The replicative organization of DNA in polytene chromosomes of *Drosophila hydei*. *Genetica* **39**, 385–428.

Nagl, W. (1969). Puffing of polytene chromosomes in a plant (*Phaseolus vulgaris*). *Naturwissenschaften* **56**, 221–222.

Nash, R. E., Puvion, E., and Bernhard, W. (1975). Perichromatin fibrils as components of rapidly labelled extranucleolar RNA. *J. Ultrastruct. Res.* **53**, 395–405.

Nur, U. (1967). Reversal of heterochromatization and the activity of the paternal chromosome set in the male mealy bug. *Genetics* **56**, 375–389.

Ohno, S., Kaplan, W. D., and Kinosita, R. (1961). X-chromosome behavior in germ and somatic cells of *Rattus norvegicus*. *Exp. Cell Res.* **22**, 535–544.

Okretic, M. C., Penoni, J. S., and Lara, F. J. S. (1977). Messenger-like RNA synthesis and DNA chromosomal puffs in the salivary glands of *Rhynchosciara americana*. *Arch. Biochem. Biophys.* **178**, 158–165.

Old, R. W., Callan, H. G., and Gross, K. W. (1977). Localization of histone gene transcripts by *in situ* hybridization. *J. Cell Sci.* **27**, 57–79.

Pankow, W., Lezzi, M., and Holderegger-Mähling, I. (1976). Correlated changes of Balbiani ring expansion and secretory protein synthesis in larval salivary glands of *Chironomus tentans*. *Chromosoma* **58**, 137–153.

Pardue, M. L., Kedes, L. H., Weinberg, E. S., and Birnstiel, M. L. (1977). Localization of sequences coding for histone messenger RNA in the chromosomes of *Drosophila melanogaster*. *Chromosoma* **63**, 135–151.

Pavan, C., and da Cunha, A. B. (1969). Chromosomal activities in *Rhynchosciara* and other Sciaridae. *Annu. Rev. Genet.* **3**, 425–450.

Pelling, C. (1964). Ribonucleinsäure-synthese der Riesenchromosomen. Autoradiographische Untersuchungen an *Chironomus tentans*. *Chromosoma* **15**, 71–122.

Petrov, P., and Bernhard, W. (1971). Experimentally induced changes of extranucleolar ribonucleoprotein components of the interphase nucleus. *J. Ultrastruct. Res.* **35**, 386–402.

Pettit, B. J., and Rasch, R. W. (1966). Tritiated histidine incorporation into *Drosophila* salivary gland chromosomes. *J. Cell. Comp. Physiol.* **68**, 325–334.

Plagens, U., Greenleaf, A. L., and Bautz, E. K. F. (1976). Distribution of RNA polymerase on *Drosophila* polytene chromosomes as studied by indirect immunofluorescence. *Chromosoma* **59**, 157–165.

Plaut, W. (1963). On the replicative organization of DNA in the polytene chromosome of *Drosophila melanogaster*. *J. Mol. Biol.* **7**, 632–635.

Prescott, D. M., and Murti, K. G. (1973). Chromosome structure in ciliated protozoans. *Cold Spring Harbor Symp. Quant. Biol.* **38**, 609–618.

Pukkila, P. J. (1975). Identification of the lampbrush chromosome loops which transcribe 5 S ribosomal RNA in *Notophtalmus* (*Triturus*) *viridiscens*. *Chromosoma* **53**, 71–89.

Puvion-Dutilleul, F., Bernadac, A., Puvion, E., and Bernhard, W. (1977). Visualization of two different types of nuclear transcription complexes in rat liver cells. *J. Ultrastruct. Res.* **58**, 108–117.

Renkawitz, R. (1978). Characterization of two moderately repetitive DNA components localized within the β-heterochromatin of *Drosophila hydei*. *Chromosoma* **66**, 225–236.

Ribbert, D. (1977). Nonhomologizable chromomere patterns of experimentally induced polytene chromosomes in the germline and that of true somatic cells in the fly *Calliphora erythrocephala*. *Abstr. Helsinki Chromosome Conf.* p. 152.

Ribbert, D., and Bier, K. (1969). Multiple nucleoli and enhanced nucleolar activity in the mouse cells of the insect ovary. *Chromosoma* **27**, 178–197.

Ritossa, F. (1962). A new puffing pattern induced by temperature shock and DNP in *Drosophila*. *Experientia* **18**, 571–573.

Ritossa, F. (1964). Experimental activation of specific loci in polytene chromosomes of *Drosophila*. *Exp. Cell Res.* **35**, 601–607.

Ritossa, F. M. (1976). Eukaryotic gene unit. In "Organization and Expression of Chromosomes" (V. G. Allfrey, E. K. F. Bautz, B. J. M. McCarthy, R. T. Schimke, and A. Tissières, eds.), pp. 153–166. Dahlem Konf., Berlin.

Rosbash, M., Ford, P. J., and Bishop, J. O. (1974). Analysis of the C-value paradox by molecular hybridization. *Proc. Natl. Acad. Sci. U.S.A.* **71**, 3746–3750.

Rudkin, G. T. (1965). The relative mutabilities of DNA in regions of the X chromosome of *Drosophila melanogaster*. *Genetics* **52**, 665–681.

Rudkin, G. T. (1969). Non-replicating DNA in *Drosophila*. *Genetics* **61**, Suppl., 227–238.

Rudkin, G. T., and Corlette, S. L. (1957). Disproportionate synthesis of DNA in a polytene chromosome region. *Proc. Natl. Acad. Sci. U.S.A.* **43**, 964–968.

Samarina, O. P., Lukanidin, E. M., Molnar, J., and Georgiev, G. P. (1968). Structural organization of nuclear complexes containing DNA-like RNA. *J. Mol. Biol.* **33**, 251–263.

Sass, H. (1978). Untersuchungen zur Struktur und Funktion der Balbianiringe von *Chironomus tentans*. Dissertation, Eberhard-Karls-Univ., Tübingen.

Schedl, P., Artavanis-Tsakonas, S., Steward, R., Gehring, W. J., Mirault, M.-E., Goldschmidt-Clermont, M., Moran, L., and Tissières, A. (1978). Two hybrid plasmids with *D. melanogaster* DNA sequences complementary to mRNA coding for the major heat shock protein. *Cell* **14**, 921–929.

Scheer, U. (1978). Changes of nucleosome frequency in nucleolar and non-nucleolar chromatin as a function of transcription: An electron microscopic study. *Cell* **13**, 535–549.

Scheer, U., Franke, W. W., Trendelenburg, M. F., and Spring, H. (1976). Classification of loops of lampbrush chromosomes according to the arrangement of transcriptional complexes. *J. Cell Sci.* **22**, 503–520.

Schimke, R. T., Alt, F. W., Kellems, R. E., Kaufman, R., and Bertino, J. R. (1977). Amplification of dihydrofolate reductase genes in methotrexate-resistant cultured mouse cells. *Cold Spring Harbor Symp. Quant. Biol.* **42**, 649–657.

Scott, S. E. M., and Sommerville, J. (1974). Location of nuclear proteins on the chromosomes of newt oocytes. *Nature (London)* **250**, 680–682.

Serfling, E. (1976). The transcripts of Balbiani rings from *Chironomus thummi*. *Chromosoma* **57**, 271–283.

Shannon, M. P., Kaufman, T. C., Shen, M. W., and Judd, B. H. (1972). Lethality patterns and morphology of selected lethal and semilethal mutations in the zeste-white region of *Drosophila melanogaster*. *Genetics* **72**, 615–638.

Silver, L. M., and Elgin, S. C. R. (1977). Distribution patterns of three subfractions of *Drosophila* nonhistone chromosomal proteins: Possible correlations with gene activity. *Cell* **11**, 971–983.

Skaer, R. J. (1977). Interband transcription in *Drosophila*. *J. Cell Sci.* **26**, 251–266.

Snow, M. H., and Callan, H. G. (1969). Evidence for a polarized movement of the lateral loops of newt lampbrush chromosomes during oogenesis. *J. Cell Sci.* **5**, 1–25.

Sokoloff, S., and Zacharias, H. (1977). Functional significance of changes in the shape of the polytene X chromosome in *Phryne*. *Chromosoma* **63**, 359–384.

Sommerville, J. (1973). Ribonucleoprotein particles derived from the lampbrush chromosomes of newt oocytes. *J. Mol. Biol.* **78**, 487–503.

Sommerville, J. (1977). Gene activity in the lampbrush chromosomes of amphibian oocytes.

In "Biochemistry of Cell Differentiation II" (J. Paul, ed.), Vol. XV, pp. 79–156. Univ. Park Press, Baltimore, Maryland.

Sommerville, J., and Hill, R. J. (1973). Proteins associated with the heterogeneous nuclear RNA in newt oocytes. *Nature (London), New Biol.* **245**, 104–106.

Sommerville, J., and Malcolm, D. B. (1976). Transcription of genetic information in amphibian oocytes. *Chromosoma* **55**, 183–208.

Sommerville, J., Crichton, C., and Malcolm, D. (1978a). Immunofluorescent localization of transcriptional activity on lampbrush chromosomes. *Chromosoma* **66**, 99–114.

Sommerville, J., Malcolm, D. B., and Callan, H. G. (1978b). The organization of transcription on lampbrush chromosomes. *Philos. Trans. R. Soc. London, Ser. B* **283**, 359–366.

Sorsa, M. (1969). Ultrastructure of puffs in the proximal part of chromosome 3 R in *Drosophila melanogaster. Ann. Acad. Sci. Fenn., Ser. A4,* **150**, 1–21.

Sorsa, V. (1973). Ultrastructure of the 5 S RNA locus in salivary gland chromosomes of *Drosophila melanogaster. Hereditas* **74**, 297–301.

Sorsa, V., Green, M. M., and Beermann, W. (1973). Cytogenetic fine structure and chromosomal localization of the white gene in *Drosophila melanogaster. Nature (London)* **245**, 34–37.

Spradling, A., Penman, S., and Pardue, M. K. (1975). Analysis of *Drosophila* mRNA by *in situ* hybridization: Sequences transcribed in normal and heat-shocked cultured cells. *Cell* **4**, 395–404.

Spradling, A., Pardue, M. L., and Penman, S. (1977). Messenger RNA in heat-shocked *Drosophila* cells. *J. Mol. Biol.* **109**, 559–587.

Stevens, B. J., and Swift, H. (1966). RNA transport from nucleus to cytoplasm in *Chironomus* salivary glands. *J. Cell Biol.* **31**, 55–78.

Swift, H. (1962). Nucleic acids and cell morphology in dipteran salivary glands. *In* "The Molecular Control of Cellular Activity" (J. M. Allen, ed.), pp. 73–125. McGraw-Hill, New York.

Swift, H. (1965). Molecular morphology of the chromosome. *In Vitro* **1**, 26–49.

Tissières, A., Mitchell, H. K., and Tracy, U. M. (1974). Protein synthesis in salivary glands of *Drosophila melanogaster:* Relation to chromosome puffs. *J. Mol. Biol.* **84**, 389–398.

Vazques-Nin, G., and Bernhard, W. (1971). Comparative ultrastructural study of perichromatin- and Balbiani ring granules. *J. Ultrastruct. Res.* **36**, 842–860.

Vlad, M., and Macgregor, H. C. (1975). Chromomere number and its genetic significance in lampbrush chromosomes. *Chromosoma* **50**, 327–347.

Wensink, P. C., Finnegan, D. J., Donelson, J. E., and Hogness, D. S. (1974). A system for mapping DNA sequences in the chromosomes of *Drosophila melanogaster. Cell* **3**, 315–325.

Wieslander, L., and Daneholt, B. (1977). Demonstration of Balbiani ring RNA sequences in polysomes. *J. Cell Biol.* **73**, 260–264.

Winter, C. E., de Bianchi, A. G., Terra, W. R., and Lara, F. J. S. (1977a). Relationships between newly synthesized proteins and DNA puff patterns in the salivary glands of *Rhynchosciara americana. Chromosoma* **61**, 193–206.

Winter, C. E., de Bianchi, A. G., Terra, W. R., and Lara, F. J. S. (1977b). The giant DNA puffs of *Rhynchosciara americana* code for polypeptides of the salivary gland secretion. *J. Insect Physiol.* **23**, 1455–1459.

Wobus, U., and Serfling, E. (1977). The repetition frequency of DNA in Balbiani ring 2 of *Chironomus thummi. Chromosoma* **64**, 279–286.

Wobus, U., Panitz, R., and Serfling, E. (1970). Tissue specific gene activities and protein in the *Chironomus* salivary gland. *Mol. Gen. Genet.* **107**, 215–223.

Wobus, U., Serfling, E., and Panitz, R. (1971). Salivary gland proteins of a *Chironomus thummi* strain with an additional Balbiani ring. *Exp. Cell Res.* **65**, 240–246.

Zentgraf, H., Scheer, U., and Franke, W. W. (1975). Characterization and localization of the RNA synthesized in mature avian erythrocytes. *Exp. Cell Res.* **96**, 81–95.

Zhimulev, I. F., and Belyaeva, E. S. (1975). ^{3}H-uridine labelling patterns in the *Drosophila melanogaster* salivary gland chromosomes X, 2R and 3L. *Chromosoma* **49**, 219–231.

Zylber, E. A., and Penman, S. (1971). Synthesis of 5 S and 4 S RNA in metaphase-arrested HeLa cells. *Science* **172**, 947–949.

6

The Expression of Animal Virus Genes

Raymond L. Erikson

I. INTRODUCTION

In the context of this chapter the expression of animal virus genes will be discussed in terms of the molecular mechanisms involved in the generation of viral messenger RNA and in the production of mature functional viral proteins. Many viruses produce several monocistronic mRNA's which program the synthesis of proteins that require very little processing to yield mature viral proteins. At the other extreme, viruses have evolved that produce only a single species of mRNA in the cytoplasm of infected cells with only a single site for the initiation of protein synthesis. Consequently, a large polyprotein precursor is synthesized which is rapidly cleaved in the infected cell to generate all the virus-encoded proteins.

CELL BIOLOGY, VOL. 3

Intermediate modes of expression also are found whereby each of a few subgenomic mRNA's codes for a subset of viral proteins. This rich diversity of expression will be illustrated by the discussion of a number of well-studied viral systems. Because of limited space, the replication of viral genomes will not be considered in this chapter. References will be kept to a minimum, but the reader will find additional references in the citations that are given.

II. THE RNA-CONTAINING VIRUSES

Animal viruses whose genetic information is encoded in single-stranded RNA are usually classed as positive- or negative-strand viruses. By convention (Baltimore, 1971), viral mRNA found on polysomes is said to have positive strand polarity, since its sequences program protein biosynthesis, that is, it is the "sense" strand. When a virus produces mRNA of the same polarity as that of the RNA found in the virus particle, it is designated a positive-strand virus. Many virus particles contain RNA with polarity opposite to that of their mRNA, and these are designated negative-strand viruses. These virus particles usually contain a polymerase which transcribes the RNA of the virion into mRNA soon after infection. This is necessary because uninfected cells have not been shown to contain RNA-dependent RNA polymerases (Baltimore, 1971).

A. The Positive-Strand Viruses

1. Picornaviruses

Picornavirus is a widely used term for a group of small, ether-resistant, polyhedral, RNA-containing viruses. These viruses as a group were among the first to be studied in detail at the molecular level. The many important observations made with these viruses subsequently influenced the thinking of cell biologists about other experimental systems. For example, the demonstration that a polycistronic viral RNA genome is translated into a polyprotein which is subsequently cleaved into mature functional polypeptides was of considerable importance (Summers and Maizel, 1968; Holland and Kiehn, 1968; Jacobson and Baltimore, 1968).

The capsids of purified picornaviruses contain nearly equal numbers of four polypeptides with approximate molecular weights of 32,000–35,000, 28,000–30,000, 23,000–27,000, and about 7000 (Rueckert, 1971). Recently a low molecular weight polypeptide covalently linked to the 5'-end of the poliovirus genome has been reported (Lee *et al.*, 1977). Virions contain one molecule of single-stranded RNA with a molecular weight of about 2.4×10^6 to 2.6×10^6, approximately 7500 nucleotides, which could encode over 200,000 daltons of protein, or about 5 to 8 average-sized proteins. The molecule is linear and some sequence information has been obtained for poliovirus RNA (Nomoto *et al.,* 1977).

Protein-free genomic RNA is infectious and contains polyadenylic acid at the 3'-end, two characteristics that suggest that it may act as mRNA. Direct evidence now available shows that the 5'-terminal sequence of poliovirion RNA is identical to that of poliovirus mRNA (Nomoto *et al.,* 1977), and, in addition, virion RNA has been translated in cell-free extracts into virus-specific polypeptides similar to those found in infected cells (Villa-Komaroff *et al.,* 1975; Pelham, 1978). Therefore, genomic RNA is equivalent to mRNA and has been designated the plus strand according to convention.

Evidence on the molecular mechanism of expression of picornaviruses comes from two equally important sources: (1) studies on infected tissue culture cells, and (2) studies on the translation of viral RNA in cell-free extracts.

Presumably the first step in the infectious process after penetration and uncoating is the association of parental virion RNA with cellular ribosomes, although no direct demonstration of this event in the infected cell has yet been made. As far as is known, the only viral messenger RNA found in infected cells throughout the infectious cycle is the same size as virion RNA, therefore all virus-specific proteins must be translated from one species of messenger. Initial analyses of virus-specific proteins in infected cells presented an apparent paradox because the sum of their molecular weights exceeded the coding capacity of the viral RNA (Maizel and Summers, 1968). This situation was soon clarified when subsequent studies led to the idea that the viral mRNA was translated into a giant polyprotein which was cleaved to mature functional proteins. This notion was supported by the identification, in cells treated with amino acid analogues to reduce protease activity, of a polypeptide larger than 200,000 daltons, a size which would approximate the total coding capacity of the viral genome (Jacobson *et al.,* 1970). However, in general, the primary products are smaller, and most likely cleavage occurs during translation so that the uncleaved product is seldom observed (Butterworth and Rueckert, 1972).

The drug pactamycin prevents the initiation of protein synthesis while allowing the completion of peptides already in the process of chain elongation. Consequently, during pactamycin treatment nucleotide sequences more distant from an initiation site will yield more polypeptide product than those close to an initiation site. Genetic maps constructed for picornaviruses using pactamycin in this way show that the genes for all polypeptide products can be ordered in a single linear array (Rekosh, 1972; Butterworth and Rueckert, 1972).

When viral RNA is used as message for cell-free translation it stimulates incorporation of amino acids such as methionine as well as ^{35}S from N-formyl[^{35}S]methionyl-tRNA$_f^{Met}$ (Villa-Komaroff *et al.*, 1975) which donates methionine only at the N-terminus of polypeptides. A single tryptic peptide containing N-formyl[^{35}S]methionine was found among all the virus-specific polypeptides translated *in vitro,* suggesting that initiation of protein synthesis may occur at only one site. The *in vitro* polypeptides are similar to the *in vivo* polypeptides in amino acid sequence, indicating that the single initiation peptide observed *in vitro* is likely to correspond to authentic initiation *in vivo.* Taken together, these results demonstrate that picornavirus expression occurs by translation of genome-like RNA on cellular ribosomes from a single initiation site near the 5'-end of the RNA molecule. Translation of the entire length of the message occurs without internal initiation at each gene, and the product is cleaved into mature functional polypeptides.

The origin of the proteases that carry out the cleavage of viral proteins is unclear, but of some interest. Similar cleavage patterns are observed with the same virus in different host cells (Kiehn and Holland, 1970). Perhaps the specificity of cleavage occurs because of the evolution of specific viral amino acid sequences that are susceptible to proteases resident in a number of host cells. On the other hand, a recent experiment carried out by Pelham (1978) suggests that translation of viral RNA in cell-free extracts results in the synthesis of a protease which cleaves newly synthesized viral polypeptides. Such a result raises the possibility that, after a certain nucleotide sequence has been translated, the amino acid sequences produced have a protease activity which acts intramolecularly. Whatever the case may be, research on this aspect of viral expression may prove to be very interesting in the near future.

Among other virus-specific proteins, a viral RNA replicase must be synthesized. However, despite a great deal of mostly unpublished effort, a purified enzyme has not been obtained to date. Synthesis of progeny proceeds through the use of a negative-strand template. As with most viruses, the details of assembly of progeny RNA into mature virus are only vaguely understood.

2. Togaviruses

Another positive-strand RNA-containing virus group has recently been renamed togaviruses. Most of the well-studied members of this group were originally referred to as arboviruses (*ar*thropod-*bo*rne). Specific examples most frequently described in the literature are Sindbis virus and Semliki Forest virus (SLV), and the information presented here will concern these two viruses, although much of the information may prove to describe togaviruses generally.

The virions are different from picornaviruses in that they contain a ribonucleoprotein core surrounded by an envelope. The envelope contains a lipid bilayer with two virus-encoded glycoproteins, termed E_1 and E_2, with molecular weights of approximately 52,000. The core contains one viral protein, molecular weight 30,000, and one single-stranded RNA molecule which has a molecular weight of about 4.2×10^6 and a sedimentation coefficient of 42 S. The protein-free RNA is infectious and polyadenylylated and, in this respect, is similar to picornavirus RNA. The size of the RNA indicates that it has nearly twice the information content of picornavirus RNA.

In contrast to picornaviruses, two forms of virus-specific RNA are found in infected cells: one is genome length, 42 S, and the other is 26 S with a molecular weight of 1.8×10^6.

It has been shown that the 26 S mRNA contains sequences that are present at the 3'-end of the 42 S viral RNA (Kennedy, 1976) and that they are both synthesized from a 42 S negative RNA strand found in infected cells (Simmons and Strauss, 1972). Recent studies (Brzeski and Kennedy, 1978) utilizing uv inactivation kinetics of RNA synthesis have shown that 26 S mRNA is generated by internal transcriptive initiation at a point about two-thirds of the way from the 3'-end of the 42 S negative-strand template. On an average, it appears that each negative-strand RNA template has one transcriptase engaged in producing a 42 S RNA molecule and three engaged in producing 26 S RNA (Fig. 1). These results eliminate the possibility that the 26 S RNA is produced by nuclease cleavage of a full-length 42 S RNA with the remainder of the RNA being discarded.

The 26 S species has been shown to direct the synthesis of virion structural proteins (Clegg and Kennedy, 1975). This RNA is a polycistronic message, containing a single initiation site as measured by the incorporation of [^{35}S]methionine from N-formyl[^{35}S]methionyl-tRNA$_f^{Met}$ (Cancedda et al., 1975), and it directs the synthesis of a soluble core protein and two membrane-bound glycoproteins (Wirth et al., 1977). The 26 S mRNA is translated on membrane-bound polyribosomes; the core protein is synthesized first and is released by proteolytic cleavage during synthesis on the cytoplasmic side of endoplasmic reticulum membranes, while newly synthesized envelope glycoproteins are found in the lipid bilayer in a protease-resistant form. This process is diagrammed in Fig. 2. It has been suggested that the nascent glycoprotein chains are responsible for the tight attachment of the 26 S mRNA–polyribosome complex to the mem-

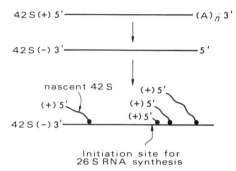

Fig. 1. Diagram of the expression of the Sindbis virus RNA genome. This virus, which is one example of togaviruses, contains an infectious positive RNA genome. A virus-encoded polymerase(s) is synthesized in infected cells and results in the generation of an RNA complementary to virion RNA. From this negative strand a 42 S RNA of genome size is made, which presumably serves either for progeny virus formation or as mRNA for nonstructural proteins. Structural proteins, the core and the two envelope glycoproteins E_1 and E_2, are translated from a 26 S membrane-bound mRNA.

branes perhaps via hydrophobic N-termini. The N-termini are inserted into the lipid bilayer during synthesis resulting in the removal of the viral glycoproteins from the cytoplasmic side of the lipid bilayer. Studies such as these have obvious implications for the synthesis of nonviral glycoproteins as well. The sequence at the 5'-end of the 42 S mRNA serves as message for other viral proteins, such as the virus-specific RNA-dependent RNA polymerase (Clewley and Kennedy, 1976), of which perhaps fewer molecules are required.

The scheme for genome expression evolved by togaviruses, such as Sindbis virus and SLV, entails the production of increased amounts of message for the subset of virus-specific proteins needed in large amounts. Although there are certain similarities in the structure of the genomes of picornaviruses and togaviruses, picornaviruses, in contrast, apparently produce equimolar amounts of all virus-encoded polypeptides.

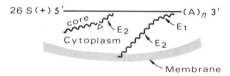

Fig. 2. Diagram of the translation of 26 S Sindbis virus mRNA. The mRNA is polycistronic for structural proteins and the gene order is 5'-core-E_2-E_1-$(A)_n$3'. The core protein is cleaved immediately after synthesis and remains in the cytoplasm, ribosomes continue and synthesize envelope proteins which are immediately inserted into the endoplasmic reticular membrane resulting in a membrane-bound polyribosome complex.

3. Retroviruses

RNA-containing tumor viruses, such as the avian, murine, and feline sarcoma viruses, which have been placed in a group known as retrovirus, have been under intensive study for several years. They have been the object of such interest not only because they cause malignant disease but also because it is now well accepted that viral genes exist as RNA in virions and also as DNA integrated in host cell chromosomes.

The study of the expression of viral genes in the integrated state yields information on the expression of cellular genes as well. Virus-specific mRNAs must be transported from the nucleus to the cytoplasm and in this respect have pathways of processing that are analogous to those of normal cellular genes as well as those of DNA viruses which replicate in the nucleus. The machinery is thus unlike that of most RNA-containing viruses that replicate in the cytoplasm. In the context of this chapter, space limitations permit only a short outline of the essential virological aspects of gene expression and these, in turn, are limited to consideration of the RNA tumor viruses that require no helper viruses for a replicative cycle. The best characterized of these viruses are the avian sarcoma viruses (ASV), and the details mentioned here are derived from studies with ASV; however, it is likely that many of the events described for these viruses will be found to be true for most RNA tumor viruses.

Avian sarcoma virions include an envelope which contains a virus-encoded glycoprotein(s). Within the envelope is a ribonucleoprotein core containing four virus-encoded proteins that probably play a structural role, as well as the DNA polymerase which transcribes the viral RNA into the DNA which is subsequently integrated into the host cell chromosome. The genome of ASV is diploid, consisting of identical single-stranded RNA molecules, 10,000 nucleotides in length, held together by hydrogen bonds near their 5'-termini. The diploid nature of the genome, which is probably the case for all retroviruses, is unique among viruses and may have evolved to permit reverse transcription, but no evidence is available on this point.

Several genes have been identified and mapped on the ASV genome: *gag* (group-specific antigens), encoding the four structural proteins of the virion core; *pol,* encoding the viral RNA-directed DNA polymerase; *env,* encoding the glycoprotein(s) of the viral envelope; and *src,* which is responsible for neoplastic transformation of the host cell (Vogt, 1977). The order of these genes on viral RNA is 5'-*gag-pol-env-src*-poly(A) 3' (Duesberg *et al.,* 1976). There is also an RNA sequence, denoted *com,* near the poly(A) which encodes no known product. Although the genome is polycistronic, the expression of viral genes is noncoordinate. For example, several thousand molecules of structural proteins appear in the viral core, whereas only a few DNA polymerase molecules are found there (Davis and Rueckert, 1972). This circumstance prevails in the entire infected cell as well.

Radiolabeled DNAs complementary to each of the viral genes have been prepared and used for molecular hybridization with RNA from infected cells to detect virus-specific poly(A)-containing mRNAs which are polyribosome-associated and have the same polarity as virion RNA. Very little negative-strand RNA is detectable in infected cells. The results of these experiments reveal three distinct virus-specific mRNA's: 38, 28, and 21 S in size. The gene composition of 38 S mRNA is 5'-gag-pol-env-src-com-poly(A), that of 28 S mRNA is 5'-env-src-com-poly(A), and that of 21 S mRNA is 5'-src-com-poly(A) (Weiss et al., 1977; Hayward, 1977). The mechanism for generation of the subgenomic mRNAs is at present unknown, although it may be surmised that they are generated by processing of a full-length genome transcript because it has been shown, for example, that each subgenomic mRNA contains at its 5'-end a sequence of RNA at least 100 nucleotides in length which is found only at a single site at the 5'-end of genome-length RNA. Presumably this 100-nucleotide segment of RNA is "spliced" onto the sequences comprising the message during the processing of full-length 38 S RNA into the mature functional messages (Krzyzek et al., 1978). (Splicing is discussed in more detail below.)

The primary translation product of the gag gene from 38 S RNA, both in vitro and in vivo, is a polypeptide designated Pr76, which is apparently cleaved by proteases present in infected cells to the four proteins which compose the viral core (von der Helm and Duesberg, 1975; Vogt et al., 1975). No viral mRNA with the pol gene at its 5'-end has so far been identified, but a translation product of 38 S viral RNA has been identified both in vivo and in vitro which has a molecular weight of 180,000 and which contains the antigenic determinants and tryptic peptides characteristic of viral DNA polymerase (Purchio et al., 1977). It seems likely, therefore, that pol is expressed by the uninterrupted translation of 38 S message into a read-through translation product containing both the gag and pol gene products and that functional polymerase is generated by protease cleavage of this 180,000-dalton polypeptide (Oppermann et al., 1977).

On the other hand, the env gene product is translated from subgenomic RNA. In cells infected with an ASV mutant with a deletion in the src gene, 21 S RNA contains env rather than src near the 5'-end, and this RNA from infected cells, when microinjected into cells, results in detectable glycoprotein synthesis (Stacey, et al., 1977).

The translation of src is similar to that of env in that its subgenomic message contains the src gene at its 5'-end. The product of the src gene is a phosphoprotein of molecular weight 60,000 (Brugge *et al.,* 1978; Purchio *et al.,* 1978). Furthermore, subgenomic 21 S viral RNA is translated *in vitro* into a polypeptide similar in structure and function to that synthesized in transformed cells (Erikson *et al.,* 1978).

Therefore, avian sarcoma viruses have evolved a unique mechanism of expression which permits gene products to be produced at different levels. Similarities do exist between the structure of ASV messages and those of other positive-strand RNA viruses; however, with the exception of *pol,* the genes expressed are located at the 5'-end of a mRNA. The expression of *pol* probably occurs through the uninterrupted translation of the first two genes of 38 S RNA. The *gag* gene products, the virion core proteins, and functional DNA polymerase are generated by proteolytic cleavage of a precursor polypeptide; in contrast, the mature *env* and *src* gene products appear to arise through specific initiation and termination of protein synthesis. An outline of the steps in the life cycle of ASV is shown in Fig. 3.

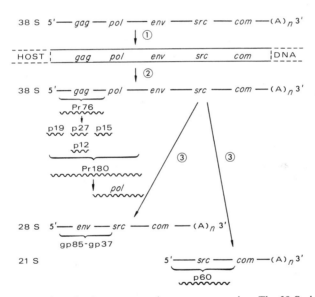

Fig. 3. Representation of avian sarcoma virus gene expression. The 38 S viral genome is reverse-transcribed into a closed circular duplex DNA molecule and integrated into host DNA (1). The gene order of integrated DNA is colinear with viral RNA. Following integration and cell division viral DNA is transcribed into 38 S genome-length RNA, which is polyadenylylated and capped. The structure of "caps" is described below (2). Translation of 38 S RNA is described in the text. Concomitant with, or subsequent to, 38 S RNA synthesis virus-specific RNA is processed into subgenomic mRNAs (3). The 5'-terminal gene is expressed in each case to generate the viral envelope glycoproteins gp85 and gp37, as well as the transforming gene product p60 from subgenomic mRNA. The *gag* and *pol* genes are expressed as described in the text. No attempt is made to illustrate the short spliced sequence on each subgenomic mRNA.

B. The Negative-Strand Viruses

1. Paramyxoviruses

This large group of viruses includes parainfluenza, Newcastle disease (NDV), Sendai, and mumps viruses. Other viruses, such as measles, form a distinct subgroup of paramyxoviruses based on antigenicity and certain structural features of the virion, but less is known about the expression of their viral genomes. More complete information is available on the replication of NDV and Sendai virus, and the details presented here are drawn from work on these viruses. A number of important properties of these viruses, such as their ability to cause persistent infection of cultured cells and their possible involvement in chronic diseases, are beyond the scope of this chapter.

Paramyxovirus virions contain a ribonucleoprotein nucleocapsid enclosed in an envelope. There are two glycoproteins in the envelope, one, termed HN, has both hemagglutinating and neuraminidase activities, and the second, termed F, is associated with virion to cell fusion and the hemolysis of red blood cells. These two proteins are of great biological interest and importance from the standpoint of virus replication. For example, under certain conditions, Sendai virions are formed which contain a biologically inactive precursor to F, termed F_0. These virions lack fusion properties (Scheid and Choppin, 1974), but the cleavage of F_0 by proteases yields a new amino terminus which is very hydrophobic and which may confer new biological properties to the virions permitting interaction of the viral membrane with the host cell membrane during infection and fusion (Gething et al., 1978). There is also a nonglycosylated protein (M) located immediately inside the viral envelope. The nucleocapsid protein is designated NP and there are two other internal proteins (P and L) which may be involved in RNA polymerase activity. The estimated total molecular weight of these structural proteins is nearly 500,000 and, in addition, two nonstructural virus-specific proteins with molecular weights of 36,000 and 22,000 have been found in Sendai-infected cells (Lamb et al., 1976).

The RNA isolated from virions has a sedimentation coefficient of 50–57 S (Duesberg and Robinson, 1965). All the available information from several different investigators shows that paramyxovirus genomes consist of one single-stranded RNA molecule with a molecular weight of approximately 5.5 to 6.0 × 10^6. It very quickly became apparent from early molecular studies that the RNA found on the polyribosomes of infected cells was complementary to virion RNA. This was probably the first well-documented case of this phenomenon (Bratt and Robinson, 1967). This result, coupled with the lack of infectivity of protein-free viral RNA, indicated that the RNA packaged in the virion was not the "sense" strand in contrast to all viruses discussed up to this point. Furthermore, virus-specific mRNA was not of genome length, but most of it was, rather, a heterogeneous group about 18 S in size. In molecular hybridization experiments, this RNA proved to be complementary to about 60% of the viral genome, representing, therefore, a combined molecular weight of 3.4×10^6. These species have been translated in cell-free extracts to yield five of the known virion polypeptides (Weiss and Bratt, 1976; Clinkscales *et al.*, 1977). These results prove the messenger function of the complementary RNA. There is also a 35 S species of mRNA which has not as yet been reported to have been translated *in vitro*. However, the fact that it is associated with polyribosomes and is poly(A)-containing makes it a candidate message, very likely for the L protein which has a molecular weight of 180,000, bringing the total weight of the polypeptide products to about 500,000.

Although it was not the first reported case of a virus which contained an RNA transcriptase, NDV virions were shown to have a low level of enzymatic activity and followed the pattern expected of negative-strand RNA viruses, namely, to carry the enzyme required to transcribe virion RNA into mRNA. The viral polypeptide responsible for the enzymatic activity has not been unequivocally identified, but transcriptive complexes usually contain the NP and L polypeptides plus at least one other virion protein (Buetti and Choppin, 1977). Perhaps the interaction of several components is necessary for accurate transcription. Whether the subgenomic messages are generated by cleavage of a precursor or by independent initiations is unknown at this time.

The expression of this negative-strand RNA virus involves transcription, by a virion-associated polymerase, of virion RNA into subgenomic mRNAs that are probably monocistronic because they are about the size required for the proteins produced and because in pulse-chase experiments no evidence for a high molecular polypeptide precursor of virus-specific proteins has been uncovered (Lamb and Choppin, 1977). Consequently, the paramyxoviruses have evolved a mechanism for genomic expression which results in the production of subgenomic monocistronic mRNAs which are translated into viral proteins. Figure 4 diagrams the production of paramyxovirus mRNAs.

2. Rhabdoviruses

Rhabdoviruses, particularly the various types of vesicular stomatitis virus (VSV), have been the subject of intensive investigation at the molecular level for the past several years. These viruses tend to be bullet

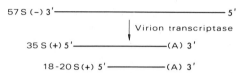

Fig. 4. Diagram of paramyxovirus gene expression. The viral genome has negative-strand polarity and the virus contains a polymerase. A 35 S virus-specific mRNA is found in infected cells as well as five species of 18–20 S monocistronic mRNA's.

shaped, somewhat unusual for an enveloped animal virus. Inside the viral envelope is a ribonucleoprotein core which contains a single molecule of single-stranded noninfectious RNA. The infectious virion is approximately 180 nm in length and about 65 nm in diameter at the blunt end. Typically, in virus stocks passed at high multiplicity, short truncated particles appear which contain shortened viral RNA. These particles have been termed defective-interfering particles because they will not replicate upon single infection of cells, but require that the cell be simultaneously infected with a normal-length particle as well. However, they do have biological activity because they interfere with the replication of infectious particles and reduce the yields of normal virus (Huang, 1973).

Infectious virions yield a noninfectious RNA molecule of 4.0 to 4.5×10^6 daltons depending on the serotype under study. The five well-characterized Indiana serotype VSV proteins have a total molecular weight of about 380,000, accounting for nearly all of the coding capacity of the genome (Wagner *et al.*, 1972). However, the New Jersey serotype which has a 4.5×10^6 dalton genome may encode another gene product (Pringle *et al.*, 1971; Franze-Fernandez and Banerjee, 1978). The virion polypeptides are designated L, large (190,000 daltons); G, glycoprotein (65,000 daltons); N, nuclear capsid (50,000 daltons); M, membrane, (29,000 daltons). Another polypeptide, NS for nonstructural (42,000 daltons) is a phosphoprotein of which very little is incorporated into virions, although it is present in large quantities in infected cells. Initially it was suggested that this protein may have been a contaminant in purified virus preparations; however, it now appears that, along with L and N, NS plays a significant role in RNA transcription (Imblum and Wagner, 1975) and is, no doubt, an essential virion component.

The noninfectious nature of protein-free viral RNA and the finding that mRNA in infected cells is complementary to virion RNA (Huang *et al.*, 1970), as is paramyxovirus mRNA, led to the discovery of an RNA-dependent RNA polymerase in vesicular stomatitis virions by Baltimore *et al.* (1970), a finding which, in turn, led to a framework for describing the expression of viral genes as outlined by Baltimore (1971). The enzymatic activity can be demonstrated after the virion is partially disrupted with a nonionic detergent. In the presence of four nucleoside triphosphates, several size classes of RNA complementary to virion RNA are synthesized and are polyadenylylated by another enzymatic activity present in the virion (Banerjee and Rhodes, 1973). RNAs synthesized *in vivo* and *in vitro* are similar in size and coding capacity. Studies on the *in vitro* translation of VSV mRNA show that 17 and 14.5 S mRNA's specifically code for the G and the N polypeptides, respectively. Another species, 12 S in size, which consists of two mRNAs encodes the NS and M polypeptides (Knipe *et al.*, 1975). A 31 S mRNA also observed may be the message for the L protein (Morrison *et al.*, 1974). There is probably little or no post-translation cleavage of VSV proteins, and the size of the mRNA's and their respective translation products suggests that VSV generates subgenomic messages which are monocistronic.

Studies with VSV mRNA's, carried out in parallel with reovirus as discussed below, revealed the surprising new fact that these viral mRNAs contained a 5'-terminal structure having the form $G(5')ppp(5')Ap$ . . . (Abraham & Banerjee, 1976; Abraham *et al.*, 1975a,b) which was also methylated in the presence of S-adenosylmethionine. The terminal guanosine is added posttranscriptionally with GTP as the substrate. Therefore, vesicular stomatitis virions contain enzymatic activities for RNA transcription and for processing, such as "capping" and polyadenylylation, although these activities have not been specifically associated with any particular viral polypeptide. Subsequently, these types of 5'-terminal structures have been found on most eukaryotic mRNA's (Perry *et al.*, 1975). Studies on the origin of the phosphate in the triphosphate group of the "cap" showed that both the α- and β-phosphates of GTP were incorporated into the blocked structure and that only one phosphate could have been derived from a transcriptional event (Shatkin, 1976).

Ball and White (1976) studied the expression of VSV genes in a system which linked transcription and translation. Detergent-activated VSV was added to a cell-free extract of mouse L cells in which the mRNA's synthesized were immediately translated. The synthesis of each virus-specific polypeptide was measured as a function of uv light dose needed to block the production of each polypeptide in order to obtain an apparent target size for each viral gene. It was found that the target size for the N protein gene corresponded to the size of its mRNA, but that the target size for other genes appeared to increase in a manner proportional to the sum of their molecular weights. Taken together, the results suggested that transcription began at a single point on the viral genome and proceeded in the order $3'$-N-NS-M-G-L-$5'$. These data, and those subsequently obtained by Abraham and Banerjee (1976), are consistent with the interpretation that the genome is transcribed from a single initiation site and that

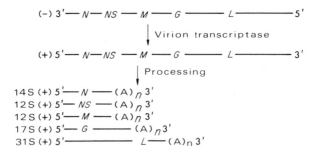

Fig. 5. Diagram of rhabdovirus gene expression. This negative-strand virus contains enzymatic activities able to carry out the transcription of the genome, beginning from a single initiation site, and the subsequent processing by cleavage, capping, and polyadenylylation of the transcript to monocistronic mRNAs for each virion protein.

the RNA is processed after synthesis, resulting in the generation of monocistronic mRNA's with a 5'-monophosphate. Simultaneous with processing, or soon thereafter, the 5'-end is capped with GDP and methylated. Finally, there is concomitant polyadenylylation of the 3'-ends to produce a mature functional message. Whether or not all these enzymatic functions are virally encoded is not known at this time, although if they are, some of the viral proteins would necessarily carry out two functions because most of the coding capacity of the RNA has been accounted for. It is of considerable interest that, to produce its mRNA, a relatively simple RNA virus has evolved a rather elaborate mechanism which bears a resemblance to that used for cellular mRNA synthesis. Figure 5 presents a summary of the data described above.

3. Myxoviruses

Influenza viruses are the only viruses presently assigned to this group. These viruses, of which there are three unrelated serological types, designated A, B and C, have been isolated from a variety of animals, and because their role in human disease is of obvious importance they have been the focus of a great deal of experimental investigation. In the past few years, a combination of genetic and biochemical analyses has resulted in excellent descriptions of individual viral genes and their products (Palese, 1977).

From the early analyses of influenza virus, it appeared that several different sizes of rather heterogeneous RNA were extracted from purified virions, each of which appeared too small to code for all of the structural polypeptides (Duesberg and Robinson, 1967; Duesberg, 1968). However, it had been suggested previously that the high rates of recombination observed between viral mutants may be explained by the reassortment of a segmented genome from each of two parents to yield recombinant progeny virus (Hirst, 1962). Indeed, improved fractionation techniques using polyacrylamide gels have subsequently shown that there are eight reproducibly observed RNA species present in influenza virions, and that their small size is not due to endonuclease cleavage of a high molecular weight genome (McGeoch et al., 1976; Palese, 1977).

The use of polyacrylamide gel electrophoresis has also permitted the identification of eight proteins in purified virions (Ritchey et al., 1977). There are three polypeptides, P1, P2, and P3, in the molecular weight range of 90,000. The glycoprotein responsible for the hemagglutinating activity (HA) of the virus has a molecular weight of about 80,000. This glycoprotein is often cleaved proteolytically in cells to yield two disulfide-linked polypeptides, HA_1 and HA_2 (Lazarowitz et al., 1971), which under certain conditions results in an increase of virus infectivity (Lazarowitz and Choppin, 1975). The other envelope glycoprotein, which has a molecular weight of 58,000, has neuraminidase activity (NA). A smaller polypeptide of 27,000 daltons is found on the internal surface of the viral envelope, and a single protein (NP) of molecular weight 65,000 is located in the ribonucleoprotein core. Finally, a virally coded nonstructural virus-specific protein (NS) with a molecular weight of 23,000 is found in the nucleus of infected cells. All molecular weights presented here are approximate because of strain differences of a few thousand daltons. Variations are also found in the size of individual RNA segments in different strains. Using the RNA segments as markers for different strains, and strain-specific antigenic differences, a genetic map of the influenza virus genome has been obtained that permits assignment of each gene product to a specific RNA segment (Palese, 1977).

Although there were early reports that influenza may be a positive-strand virus, there is now general agreement that virion RNA has negative polarity. RNA from infected cell polysomes protects virion RNA in molecular hybridization experiments and programs the synthesis of virus-specific polypeptides in cell-free extracts (Etkind and Krug, 1975). Polyribosome-associated virus-specific RNA is poly(A)-containing, whereas virion RNA is not, and purified virions contain the RNA polymerase required to generate mRNA after infection (Chow and Simpson, 1971). All of these results show that influenza is a negative-strand virus.

Consistent with the correlations in size of the genomic RNA segments and the viral proteins, it appears likely that each genomic segment is transcribed into a message of nearly identical size. There is evidence that transcriptional controls exist because the mRNAs for NP and NS proteins are produced in greater amounts in infected cells, as are these proteins

(Etkind *et al.*, 1977). All virus-specific proteins are synthesized from monocistronic mRNA, and there is no detectable posttranslational cleavage in order to generate mature structural proteins (Ritchey *et al.*, 1977).

As pointed out by Palese (1977), the detailed molecular description of the expression of the influenza virus genome now permits the design of experiments to elucidate which gene products are most important in determining the virulence of various viral strains. In addition, since human pandemics have been postulated to be caused by viruses generated by recombination between strains of human and animal origin, molecular studies on new isolates may quickly identify their origin and lead to more rational vaccine design.

It appears from the foregoing that negative-strand viruses utilize most of their genetic content to encode proteins which enter progeny virus. Several of these viruses, such as VSV, contain a number of enzymatic activities that can be understood only if a single viral polypeptide has more than one function or, if during maturation, the virion sequesters host enzymes for its own devices. On the other hand, positive-strand viruses, such as picornaviruses and togaviruses, use only a fraction of their information for structural proteins, while the enzymes utilized for replication of viral RNA and/or synthesis of mRNA remain associated with the infected cells.

C. Double-Stranded RNA-Containing Viruses

Reoviruses

The best-studied example of a virus with a double-stranded RNA genome is reovirus (*r*espiratory *e*nteric *o*rphan) isolated from humans. Similar viruses have been isolated from plants, numerous animals, insects, and other invertebrates (Joklik, 1974). The result showing that the virus contained double-stranded RNA (Gomatos and Tamm, 1963) along with the fact that it is easy to grow resulted in a number of highly productive investigations on this virus.

This virus is not coated by a membrane, but rather possesses two capsid shells composed of protein subunits. It is relatively stable and under proper conditions can be readily purified in good yields with very low particle to plaque forming unit ratio, an unusual high efficiency for an animal virus. Reovirus polypeptides are saddled with rather awkward designations, based on relative mobility on polyacrylamide gels, a parameter which is not consistent in different buffer systems (Shatkin and Both, 1976), consequently the literature is difficult to follow on this matter. A Tris-glycine buffered polyacrylamide system resolves four proteins present in the core, λ_1, λ_2, μ, and σ, with apparent molecular weights of 153,000, 148,000, 79,000, and 54,000, respectively, while μ_2, σ_2, and σ_4, with molecular weights of 73,000, 52,000, and 43,000, respectively, are found in the outer shell.

As was the case with influenza virus, RNA from virions consistently was obtained in several different sizes, the largest of which was too small to encode all the viral structural proteins. Eventually it became clear that reovirus contains a segmented genome composed of ten double-stranded RNA fragments (Shatkin *et al.*, 1968) which are present in virions prior to extraction (Millward and Graham, 1970). Furthermore, a diphosphate at the 5'-end of each segment suggests that their synthesis is independently initiated (Miura *et al.*, 1974).

The ten genomic RNA segments fall into three size classes: three large segments with molecular weights 2.3 to 2.7 $\times$ 10^6; three medium, 1.3 to 1.6 $\times$ 10^6; and four small, 0.6 to 0.9 $\times$ 10^6. When analyzed on polyacrylamide gels, different types of reovirus show RNA segments with unique profiles. Also, different types of reovirus show unique antigenic determinants and patterns of pathogenicity which can be used as genetic markers in the analysis of recombinants derived from mixed infections with two reovirus types. Analysis of the origin of the RNA segments in these recombinants permits the assignation of each gene product to a specific RNA segment (Weiner *et al.*, 1977; Sharpe *et al.*, 1978). These data also show that recombinants in reovirus crosses arise through the physical reassortment of the genome segments.

Several enzymatic activities are associated with reovirus cores, one of which is RNA polymerase (Borsa and Graham, 1968; Shatkin and Sipe, 1968). Both in infected cells (Ward *et al.*, 1972) and *in vitro* the viral genome is transcribed into ten mRNA species, the sizes of which correspond to those of the genome, indicating that a mRNA is produced for each segment. These mRNAs lack the conventional poly(A)-containing 3' sequences, but are associated with polysomes in the infected cell. Moreover, ten polypeptide gene products have been detected in infected cells and by cell-free translation of RNA transcribed *in vitro* (Both *et al.*, 1975). The sizes of these ten polypeptides closely correspond to those expected if each mRNA were monocistronic. Probably some posttranslation cleavage occurs in infected cells, but the total molecular weight change is small (Shatkin and Both, 1976).

In addition to RNA polymerase activity, reovirus cores contain three other enzymatic activities of importance to mRNA synthesis. These include a capping activity that results in the addition of GTP to the 5'-end of mRNA in an inverted 5' linkage G(5')ppp(5')GpC . . . both *in vivo* and *in vitro*, and the presence of two methylating activities results in a final structure m^7 G(5')ppp(5')G_p^mC . . . . Enzymatic activity is lost when attempts are made to separate it from cores, consequently no functions have been assigned to particular viral polypeptides (Shatkin and Both, 1976). The function of the novel cap structure is still not clear, but it appears to improve ribosome binding and translation in cell-free extracts (Shatkin and Both, 1976) as shown by direct comparisons of messages with and without caps. This novel and surprising development concerning the structure of reovirus mRNA was soon shown to be generally true for other viral as well as cellular mRNA's (Perry *et al.*, 1975; Perry, 1976; Perry & Kelley, 1976; Rottman et al., 1974). This is another striking important example of an observation first elucidated using viral systems, which subsequently had considerable impact on other aspects of the study of gene expression.

In summary, the reovirus genome is composed of segments, each of which is transcribed into a monocistronic mRNA which is not polyadenylylated. These mRNA's are modified at the 5'-termini and translated into proteins which undergo relatively little posttranslational cleavage.

Table I contains a collection of the data on all the RNA-containing viruses mentioned above. It must be emphasized that the molecular weights should be viewed as only approximate. The reader will find in any particular reference values that depend on the type of analysis performed and on the molecular weight markers employed.

III. THE DNA-CONTAINING VIRUSES

A. Double-Stranded DNA-Containing Viruses

1. Papova- and Adenoviruses

Although the structure of the papova- and adenovirus genomes is very different, it seems reasonable to discuss these two groups together because of certain similarities in gene expression. Both viruses replicate their genome in the nucleus and thus face the problem, as the host cell does, of moving viral mRNA's to the cytoplasm for translation. Beginning in the spring of 1977 both viruses were the focus of research efforts which resulted in totally new and unexpected insights into eukaryotic gene expression, namely, that DNA sequences are often not colinear with their mature mRNAs. This result was first documented (Berget *et al.*, 1977; Chow *et al.*, 1977) for mRNA's coding for the late adenovirus polypeptide. After the slightest of pauses while the waves subsided, numerous investigators were able to extend these observations to other viruses as well as to normal cellular genes, thus providing another illustration of the seminal role virus research plays in producing new insights concerning cell biology in general.

The genome of adenovirus is a linear double-stranded DNA molecule of 20 to 25 × 10^6 daltons (Green *et al.*, 1967), about 35,000 base pairs, which is large enough to encode 30–40 polypeptides. The virus life cycle is typically described as consisting of an early phase, before viral DNA synthesis begins, and a late phase, after the onset of DNA replication. Adenovirus DNA is transcribed in the nucleus of infected cells by an α-amanitin-sensitive polymerase of host origin. Protein synthesis is required in the infected cell for the switch from early to late transcription, and for the onset of viral DNA synthesis, perhaps because viral-encoded polypeptides are required for these events.

The analysis of expression of various genes located at different positions on the genome is based to a large extent on physical maps of the genome generated by cleavage with restriction endonucleases. The DNA fragments are ordered along the genome and are used in nucleic acid hybridization experiments with stable virus-specific RNA transcripts found in the cytoplasm of infected cells. Detailed maps of the regions of viral DNA complementary to early and late mRNA are available (Flint, 1977). The details of these maps are beyond the scope of this chapter; nevertheless, it is of interest to note that the general patterns are not simple. For example, early mRNA is transcribed from four different regions of the genome, two on each strand (Fig. 6). The actual number of nucleotides expressed as early mRNA depends on the strain and varies from 20 to 40%. At late times after infection, early mRNA continues to be synthesized, while there is increased synthesis of late mRNA, most of which is transcribed from only one strand.

Specific mRNA's can be selected by hybridization to restriction endonuclease fragments of viral DNA and used for cell-free translation (Lewis *et al.*, 1976). Viral polypeptides made in this way can be matched with those synthesized *in vivo* and, consequently, genes for specific polypeptides can be mapped within the genome. A large fraction of the adenovirus genome has been described in this way (Flint, 1977).

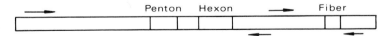

Fig. 6. Early adenovirus mRNA's. The arrows represent the location and direction of synthesis of early mRNA from DNA sequences in the adenovirus genome. The penton, hexon, and fiber are three adenovirus genes for structural proteins, which are transcribed late and are used here to orient the genome.

TABLE I

Expression of the Genes of RNA-Containing Viruses

Genome	mRNA (s)		Translation products and molecular weights
	Size	Comments	
Picornavirus Size: 35 S Molecular weight: 2.4 to 2.6 × 10⁶ Nucleotides: 7500 Polarity: Positive	35 S	Polycistronic	Polyprotein (> 200,000)
Togavirus Size: 42 S Molecular weight: 4.2 × 10⁶ Nucleotides: 12,500 Polarity: Positive	42 S 26 S	Polycistronic Polycistronic, produced in greater quantities than 42 S	Nonstructural polypeptides Structural proteins for viral core and envelope
Retrovirus Size: 38 Sᵃ Molecular weight: 3 × 10⁶ ᵃ Nucleotides: 10,000ᵃ Polarity: Positive	38 S 28 S 21 S 35 S	Polycistronic, only the first two genes translated Polycistronic, probably only *env* gene translated Only *src* gene translated Subgenomic	Internal core proteins (76,000); joint core–polymerase product (180,000) Viral glycoprotein(s) Viral transforming protein (60,000)
Paramyxovirus Size: 50–57 S Molecular weight: 5.5 to 6.0 × 10⁶ Nucleotides: 17,500 Polarity: Negative	18–22 S	Five subgenomic species	L (> 160,000) P (~ 75,000, Sendai) HN (67,000) Fo (65,000) NP (56,000) M (41,000)
Rhabdovirus Size: 42 S Molecular weight: 4.0 to 4.5 × 10⁶	31 S 17 S 14.5 S	All are monocistronic, poly-adenylylated and generated by cleavage of a precursor	L (190,000) G (69,000) N (50,000)

Nucleotides: 13,000
Polarity: Negative

12.0 S   NS (42,000)
12.0 S   M (29,000)

Myxovirus (Influenza)

Size: 8 segments

Molecular weight: Total $\sim 5.7 \times 10^6$

All are monocistronic transcripts of each genome segment

Same size as genome segments

Nucleotides[b]:

1. 3500	P1	
2. 3000	P2	(80,000–94,000)
3. 2950	P3	
4. 2450	HA	($\sim$80,000)
5. 2000	NP	($\sim$65,000)
6. 1720	NA	(58,000)
7. 1080	M	($\sim$55,000)
8. 870	NS	($\sim$23,000)

Reovirus

Size: 10 segments

Molecular weight: Total $\sim 15 \times 10^6$

Polarity: Negative

All are monocistronic transcripts of each genome segment

Same size as genome segments

Nucleotides[b]:

1		λ_1 (153,000)
2	About 4500 base pairs	λ_2 (148,000)
3		λ_3 (143,000)
4		μ_1 (79,000)
5	About 2300 base pairs	μ_2 (77,000)
6		μ_3 (72,000)
7		σ_1 (54,000)
8	About 1200 base pairs	σ_2 (52,000)
9		σ_3 (49,000)
10		σ_4 (43,000)

Polarity: Double-stranded

[a] Size of subunit.
[b] Actual sizes vary with strain.

Since adenovirus is a tumor virus, it is of some interest to explore this special aspect of gene expression. Viral DNA sequences from the left-hand 12–14% of the genome are the only ones reproducibly found in transformed cells and, therefore, are candidates for containing genes which encode transforming proteins. About one-half of these sequences are commonly expressed as mRNA in transformed cells (Flint *et al.*, 1976). Furthermore, only viral DNA fragments isolated from the left-hand end are able to transform cells, thus providing more direct evidence concerning their role in transformation (Graham *et al.*, 1974). The information encoded in these adenovirus genes is also expressed during normal lytic growth of the virus; consequently, the transformation of cells may be only incidental to a function evolved for virus replication.

When abundant late viral mRNA found on polyribosomes was hybridized to viral DNA, the structures observed in the electron microscope showed short sequences of about 160 nucleotides at the 5'-end of mRNA which did not hybridize to the adjacent DNA which constituted the structural gene (Berget *et al.*, 1977). Instead, the 5'-terminal sequences of the mRNAs were transcribed from a region of the genome remote from the structural gene and were then joined to the principal mRNA sequence to generate a mature functional message (Chow *et al.*, 1977), as shown in Fig. 7. At this time there is no published evidence that the short splice is translated into protein, leaving the role of these RNA sequences unclear.

At late times after infection, host protein synthesis is depressed and a total of 22 virus-specific polypeptides is observed in infected cells. However, all of them have not been clearly shown to be virus-encoded. Thirteen of these proteins have molecular weights similar to those found in purified virus. There is evidence of proteolytic processing of some virus-specific polypeptides; however, several proteins are not altered after synthesis, and there is no evidence for any very large precursors for several virion proteins (Lewis *et al.*, 1974).

Papovaviruses, as exemplified by simian virus 40 (SV40) and polyomavirus, have also been intensively studied in part because they are tumor viruses, but also, because of their small size, they seemed to offer the possibility that the expression of all their genes could be described completely. It seems as though virologists are very close to realizing such

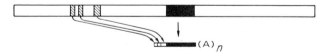

Fig. 7. Example of splicing of late adenovirus hexon mRNA. The mature mRNA contains, in addition to sequences which encode the polypeptide (shaded region), sequences from three regions of the genome which are not contiguous.

a goal. The genome is a closed circular DNA with a molecular weight of about 3.0×10^6 as judged by physical measurements (Weil and Vinograd, 1963), and the more than 5000 base pairs in the molecule have been nearly completely sequenced (Fiers et al., 1978; Reddy et al., 1978).

Prior to viral DNA synthesis about one-half of the genome is transcribed into early mRNA from one strand, while late mRNA is copied from the other strand after DNA synthesis has been initiated (Lindstrom and Dulbecco, 1972; Sambrook et al., 1972). Late mRNA encodes the three major virion capsid proteins VP1, VP2, and VP3, which have molecular weights of 45,000, 39,000 and 27,000, respectively (Salzman and Khoury, 1974). This poly(A)-containing mRNA is observed to sediment at 16 S and 19 S, but it appears that the 16 S species is derived from the 19 S species by processing in the cytoplasm of infected cells (Aloni et al., 1975).

With the aid of the DNA sequence, and some knowledge of the amino acid composition of VP1, VP2, and VP3, predictions have been made about the location of the genes for each of these structural proteins (Reddy et al., 1978). It appears that almost all of the DNA sequences which encode VP3 lie within those which are also used to code for VP2. Since VP3 may not be derived by protease cleavage of VP2 (Cole et al., 1977) there is probably an independent initiation event for the translation of VP3 at an AUG codon which is read as an internal methionine for VP2. Consequently, two proteins are ultimately synthesized from one strand of DNA by a mechanism other than the processing of a high molecular weight precursor.

In contrast, the initiation codon for VP1 lies within the DNA sequence which codes for the carboxy-termini of VP2 and VP3, but in a different reading frame (Contreras et al., 1977). Such complexities may have been anticipated from studies on the genome of ϕX174 (Sanger et al., 1977), but the occurence of 5'-terminal sequences in the mRNA for these proteins which are not transcribed from DNA sequences immediately adjacent to those which encode the protein sequences was not foreseen from bacteriophage studies. As is the case for adenovirus, these spliced sequences are short, about 100–200 nucleotides in length, and are apparently not translated, leaving their function unclear.

In understanding the expression of viral genes, splicing of RNA transcribed from two noncontiguous DNA sequences became more complex and interesting as the study of early papovavirus mRNAs and polypeptides unfolded. One early virus-specific protein is T antigen which has an apparent molecular weight of about 90,000–100,000 (Tjian et al., 1978), and which would require nearly all of the information content of the early gene region of papovaviruses for its production. However, three apparent paradoxes developed when it was found that (i) deletions within the early region of the genome still permitted expression of an unaltered T antigen and (ii) a second small t antigen, molecular weight 17,000, observed in infected cells (Sleigh *et al.*, 1978; Schaffhausen *et al.*, 1978), was found to be translated from an mRNA nearly equal in size to the mRNA for large T antigen (Paucha *et al.*, 1978). Both proteins share antigenic determinants and N-terminal amino acid sequences and, therefore, appear to be encoded by overlapping DNA sequences, but since some deletion mutants show normal large T while they fail to express small t, it appears that certain sequences code for large T only (Sleigh *et al.*, 1978). (iii) Sequence analysis of viral DNA showed that there are translation termination codons in all three reading frames within the early gene coding region.

The possibility of spliced messages provided a way out of this dilemma and, indeed, evidence that this occurs has been presented (Aloni *et al.*, 1977; Berk and Sharp, 1978). The mRNA for small t is 2500 nucleotides in length and consists of 630 nucleotides at the 5'-end spliced to 1900 nucleotides, while the mRNA for large T is 2200 nucleotides in length and consists of 330 nucleotides at the 5'-end spliced to 1900 nucleotides. The 5'-ends of both mRNAs are the same and are translated in the same reading frame up to the point of the first splice, but beyond the point of the splice, translation of small t continues in a reading frame that may contain a codon which results in termination, whereas translation of large T antigen continues in a reading frame which generates the larger protein, as shown in Fig. 8. This result suggests an explanation for the fact that certain deletions in the T antigen gene do not affect the size of the polypeptide product. It also contrasts to late structural spliced mRNA's in that the spliced sequences are translated. The molecular mechanism whereby two

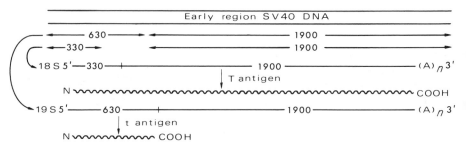

Fig. 8. Expression of early SV40 genes. The same DNA sequence is transcribed into two different spliced mRNA's, an 18 S RNA which is translated into large T antigen, as well as a 19 S RNA which is translated into small t antigen. The two proteins have the same amino acid sequence at their amino terminus.

mRNA's are spliced from the transcripts of a single DNA sequence is unknown at this time, but the evolution of such a mechanism indicates still another process that permits a virus maximum utilization of a small genome. One can also appreciate from the foregoing outline that nearly every detail concerning the expression of the SV40 and polyomavirus genomes may shortly be available.

2. Herpesviruses

Herpesviruses have been isolated from a large number of species, and among the viruses classified in this group there are no doubt a variety of life styles. In man there are several distinct herpesviruses, herpes simplex virus (HSV) 1 and 2, as well as varicella-zoster, Epstein-Barr, and cytomegalovirus. It is beyond the scope of this chapter to examine differences in the molecular expression of these virus systems nor is it possible to discuss the very interesting capacity of these viruses to establish persistent infections. Nearly all of the information presented here has been obtained in studies on HSV 1 or 2.

Herpesviruses are large enveloped viruses about 100 nm in diameter each of which contains a linear, double-stranded DNA genome. As do papova- and adenoviruses, they replicate in the nucleus.

The DNA genome of all the herpesviruses is considerably larger than that of papova- or adenoviruses; estimates by various investigators range from 82 to 100 $\times$ 10^6 daltons in size. Considerable progress has been made in describing the molecular anatomy of the HSV genome (Jones, *et al.,* 1977; Skare and Summers, 1977). Kinetic reannealing experiments yield complexity measurements consistent with physical size determinations (Frenkel and Roizman, 1971). Consequently, these viruses may encode at least 100 proteins of average size.

Infection of cells with herpesviruses results in the controlled appearance of virus-specific polypeptides with concomitant decline in cellular protein synthesis (Powell and Courtney, 1975). There is a set of polypeptides, termed α, which is synthesized early and which do not require prior virus-specific protein synthesis. A second set, termed β, is synthesized only after the synthesis of α polypeptides and, finally, one or more β polypeptides are required for the onset of γ polypeptide synthesis (Pereira *et al.,* 1977). Over 50 virus-specific proteins have been identified in infected cells, many of which are over 100,000 daltons in size. These polypeptides alone have a total molecular weight of 4 $\times$ 10^6 and account for about 75% of the coding capacity of the viral genome. However, it is not clear that all of these are virus-encoded because specific antisera are not available nor has cell-free translation of purified viral mRNAs been accomplished.

Thymidine kinase and DNA polymerase are among the enzymes that are probably encoded by HSV (Kit *et al.*, 1974). The major virion polypeptides can be identified by polyacrylamide gel electrophoresis, and there may be as many as 30 with a surprising number showing molecular weights greater than 100,000 (Roizman and Furlong, 1974). Relatively few studies by pulse-chase analysis have been undertaken to determine if large molecular weight precursors are synthesized, but translation of mRNAs from infected cells in cell-free extracts, where little proteolytic processing occurs, shows that a number of cell-free products comigrate with virus-specific polypeptides from infected cells (Preston, 1977), suggesting that at least some viral products undergo little processing.

Transcriptional studies on the HSV genome are not available in the detail of those on papova- or adenoviruses, but the general outlines exist. Prior to DNA synthesis about 20% of the viral genome is transcribed, the remainder of the genome being transcribed after the onset of DNA synthesis (Swanstrom and Wagner, 1974; Ben-Zeev & Becker, 1977). By 2 hr after infection, viral poly(A)-containing mRNA accounts for 60 to 75% of the newly synthesized mRNA on polyribosomes, and by 6 hr postinfection most newly synthesized mRNA is viral, demonstrating how quickly virus-specific events replace those related to cellular functions (Stringer *et al.*, 1977). Size analysis of the RNA transcripts from the nucleus indicates that they are larger than cytoplasmic viral RNA, suggesting that some processing similar to that of host cell RNA may occur; however, at this time no information is available on the possible occurrence of herpesvirus-specific spliced mRNAs.

3. Poxviruses

Poxviruses were among the first DNA-containing viruses to be investigated at the molecular level. They have the unusual property, for a DNA virus, of replicating in the cytoplasm. Furthermore, infection rapidly reduces host cell macromolecular synthesis, which permits some studies to be carried out more easily. However, the very large genome, with a molecular weight over 1.2×10^8, and the potential to produce about 150 proteins makes detailed molecular analysis of gene expression extremely complex. Nevertheless, a number of interesting phenomena have been described.

Poxviruses are enveloped and appear as oval or brick-shaped particles about 250 nm in size. Soon after infection the outer envelope is removed to yield core particles (Dales, 1963). Similar core particles can also be produced *in vitro* by treatment of purified virions with a nonionic detergent. These cores were shown by Kates and McAuslan (1967) and Munyon *et al.* (1967) to contain RNA polymerase. This enzyme, which transcribes viral DNA, was the first virion-associated enzyme described which clearly had an essential role in the life cycle of the virus, although the general significance of the discovery may not have been appreciated at the time. Detergent-activated virions also are able to polyadenylylate viral RNA (Kates and Beeson, 1970a,b; Moss *et al.*, 1973).

Subsequently, poxvirus mRNA was shown to be capped and methylated by enzymatic activity resident in virion cores (Wei and Moss, 1974; Boone & Moss, 1977). Only a portion of the viral DNA is transcribed by virion polymerase (Kates and Beeson, 1970a,b), but the actual fraction expressed varies depending on methods used for virus purification (Boone and Moss, 1978). These latter studies also showed two types of transcription controls. First, early viral mRNA, made prior to DNA synthesis, may be a subset of the total virus-specific mRNA made after the onset of DNA synthesis. Late mRNA protects additional DNA sequences in nucleic acid hybridizations, but probably contains early mRNA because no additional sequences are protected by the addition of early mRNA to late mRNA during reannealing. Another type of control is exhibited in abundance levels for specific mRNA sequences. There was a single abundance class for early mRNA, but two classes differing by 11-fold for late mRNA were found.

Early mRNA from infected cells probably corresponds to that RNA synthesized by detergent-activated cores as shown by coupled transcription–translation systems (Cooper and Moss, 1978). Vaccinia virus cores were placed in a cell-free lysate able to translate mRNA, and the polypeptides synthesized programmed by the virus-specific mRNA were then analyzed by polyacrylamide gel electrophoresis. At least 23 *in vitro* products comigrated with early virus-specific proteins present in cells 2 hr postinfection. The expression of at least some of these proteins is probably necessary for the initiation of viral DNA synthesis and late transcription, since the inhibition of early protein synthesis prevents the onset of viral DNA synthesis.

The size of the primary RNA transcripts can be deduced from ultraviolet light inactivation experiments in the coupled transcription–translation system. These experiments are analogous to those Ball and White (1976) carried out with VSV. Pelham (1977) irradiated cores prior to their addition to the cell-free extract and studied kinetics of inactivation of the synthesis of vaccinia proteins. He found that there was a linear relationship between the molecular weights of the proteins made *in vitro* and the rate of uv inactivation of their synthesis. From these data he suggested that tandem polycistronic transcription of mRNA for these early proteins was unlikely, because if such transcription did occur there would be proteins whose synthesis would be more sensitive to ultraviolet light than expected from their size. It is likely that among these early proteins are enzymes such as thymidine kinase (Kit *et al.*, 1977) which is a virus-encoded enzyme as are DNA polymerase and polynucleotide ligase (Moss, 1974).

The question of protease cleavage of primary translation products has also been studied for poxvirus-specific proteins, and at least three major late proteins are formed by cleavage of high molecular weight precursors (Katz and Moss, 1970a,b; Moss and Rosenblum, 1973).

The expression of the poxvirus genome is complex and is only beginning to be understood, but it is known that early mRNA representing about 14% of the genome is synthesized and processed by enzymes present in the virion. After early gene expression, DNA synthesis is initiated and late transcription begins. Late mRNA translation results in synthesis of virion-associated enzymes and viral structural proteins.

B. Single-Stranded DNA-Containing Viruses

Parvoviruses

Parvoviruses contain the smallest genome of any known animal virus. Perhaps, therefore, it is not surprising that a major group of these viruses, the adenovirus-associated viruses, are replication defective and have an absolute requirement for simultaneous infection with an adenovirus. However, others, including a large number isolated from rodents, are not defective despite their small genomes.

Parvoviruses are small, about 20 nm in size, and are very stable (Rose, 1974). They contain 20–25% DNA. The DNA from the virions is a linear single-stranded structure which ranges in different viruses between 1.4×10^6 to 1.7×10^6 in molecular weight. Therefore, the virus may encode only about 150,000 daltons of protein if only one reading frame in the genome is expressed.

Even nondefective parvoviruses replicate best in cells that are rapidly dividing. Indeed, there is evidence that for their replication these viruses require one or more cellular functions produced at specific stages in the cell cycle (Tennant and Hand, 1970). Infection of synchronized cells suggests that parvoviruses require a factor made about 5 hr after S phase begins (Rhode, 1973).

Three proteins are usually found in parvovirions with molecular weights which vary with the virus studied, but which fall into three size classes; A (72,000 to 92,000), B (62,000 to 79,000), and C (55,000 to 66,000). The total molecular weight of A, B, and C exceeds the coding capacity of the genome. Tattersall *et al.* (1977) demonstrated convincingly, however, that these polypeptides are all related and that B is probably derived from A by proteolytic cleavage and that C is similarly derived from B. Since these polypeptides are the same in virus grown in two different host cells, it is likely that they are virus-encoded (Tattersall *et al.*, 1976).

Little is known about the number or sequence complexity of virus-specific RNA's. The predominant viral RNA is about 18 S in size, of polarity opposite to that of the genome, corresponding to a molecular weight of 6.5 to 7.5 × 10⁵. This RNA, if it consisted of only a single class of RNA, would account for no more than one-half of the genome (Salzman and Redler, 1974). Additional studies to determine how much of the genome such RNA protects after reannealing have not been reported. Furthermore, since the DNA replicates via the synthesis of a duplex, molecular studies also must be done to see if virus-specific RNA can be found which has the same polarity as virion DNA.

IV. REMARKS

The events described in this chapter omit subtle interactions of viruses and their hosts. The production of virus-specific mRNAs has been described as well as the translation products of these RNAs. This seemed necessary because, with few exceptions, the two processes are closely linked. When a viral mRNA is present in infected cells, it is translated, and there is little evidence for translational control among the viruses discussed. However, additional important considerations, such as the posttranslational modification of viral proteins by glycosylation or phosphorylation, have not been discussed, although a great deal is known in this area. Fascinating aspects of the maturation of enveloped viruses have been described elsewhere but were not discussed here, although they are closely related to polypeptide synthesis. Furthermore, discussion of the functions of certain tumor virus-encoded polypeptides which are involved in cell transformation is beyond the scope of this chapter.

A large number of virus groups were surveyed in order to make the reader aware of the variety of life styles that have evolved. We should also expect that additional examples of this variety and complexity remain to be found. Additional experiments on well-studied systems and investigations on newly discovered viruses will, in the future, expand that presented here.

REFERENCES

Abraham, G., Rhodes, D. P., and Banerjee A. K. (1975a). Novel initiation of RNA synthesis *in vitro* by vesicular stomatitis virus. *Nature (London)* **255**, 37–40.

Abraham, G., Rhodes, D. P., and Banerjee, A. K. (1975b). The 5' terminal structure of the methylated mRNA synthesized *in vitro* by vesicular stomatitis virus. *Cell* **5**, 51–58.

Abraham, G., and Banerjee, A. K. (1976). The nature of the RNA products synthesized *in vitro* by subviral components of vesicular stomatitis virus. *Virology* **71**, 230–241.

Aloni, Y., Shani, M., and Reuveni, Y. (1975). RNAs of simian virus 40 in productively infected monkey cells: Kinetics of formation and decay in enucleate cells. *Proc. Natl. Acad. Sci. U.S.A.* **72**, 2587–2591.

Aloni, Y., Dhar, R., Laub, O., Horowitz, M., and Khoury, G. (1977). Novel mechanism for RNA maturation: The leader sequences of simian virus 40 mRNA are not transcribed adjacent to the coding sequences. *Proc. Natl. Acad. Sci. U.S.A.* **74**, 3686–3690.

Ball, L. A., and White, C. N. (1976). Order of transcription of genes of vesicular stomatitis virus. *Proc. Natl. Acad. Sci. U.S.A.* **73**, 442–446.

Baltimore, D. (1971). Expression of animal virus genomes. *Bacteriol. Rev.* **35**, 235–241.

Baltimore, D., Huang, A. S., and Stampfer, M. (1970). Ribonucleic acid synthesis of vesicular stomatitis virus, II. An RNA polymerase in the virion. *Proc. Natl. Acad. Sci. U.S.A.* **66**, 572–576.

Banerjee, A. K., and Rhodes, D. P. (1973). *In vitro* synthesis of RNA that contains polyadenylate by virion-associated RNA polymerase of vesicular stomatitis virus. *Proc. Natl. Acad. Sci. U.S.A.* **70**, 3566–3570.

Ben-Zeev, A., and Becker, Y. (1977). Requirement of host cell RNA polymerase II in the replication of herpes simplex virus in α-amanitin-sensitive and -resistant cell lines. *Virology* **76**, 246–253.

Berget, S. M., Moore, C., and Sharp, P. A. (1977). Spliced segments at the 5' terminus of adenovirus 2 late mRNA. *Proc. Natl. Acad. Sci. U.S.A.* **74**, 3171–3175.

Berk, A. J., and Sharp, P. A. (1978). Spliced early mRNAs of simian virus 40. *Proc. Natl. Acad. Sci. U.S.A.* **75**, 1274–1278.

Boone, R. F., and Moss, B. (1977). Methylated 5'-terminal sequences of vaccinia virus mRNA species made *in vivo* at early and late times after infection. *Virology* **79**, 67–80.

Boone, R. F., and Moss, B. (1978). Sequence complexity and relative abundance of vaccinia virus mRNAs synthesized *in vivo* and *in vitro*. *J. Virol.* **26**, 554–569.

Borsa, J., and Graham, A. F. (1968). Reovirus: RNA polymerase activity in purified virions. *Biochem. Biophys. Res. Commun.* **33**, 895–901.

Both, G. W., Lavi, S., and Shatkin, A. J. (1975). Synthesis of all the gene products of the reovirus genome *in vivo* and *in vitro*. *Cell* **4**, 173–180.

Bratt, M. A., and Robinson, W. S. (1967). Ribonucleic acid synthesis in cells infected with Newcastle disease virus. *J. Mol. Biol.* **23**, 1–21.

Brugge, J. S., Erikson, E., Collett, M. S., and Erikson, R. L. (1978). Peptide analysis of the transformation-specific antigen from avian sarcoma virus-transformed cells. *J. Virol.* **26**, 773–782.

Brzeski, H., and Kennedy, S. I. T. (1978). Synthesis of alphavirus-specified RNA. *J. Virol.* **25**, 630–640.

Buetti, E., and Choppin, P. W. (1977). The transcriptase complex of the paramyxovirus SV5. *Virology* **82**, 493–508.

Butterworth, B. E., and Rueckert, R. R. (1972). Gene order of encephalomyocarditis virus as determined by studies with pactamycin. *J. Virol.* **9**, 823–828.

Cancedda, R., Villa-Komaroff, L., Lodish, H. F., and Schlesinger, M. (1975). Initiation sites for translation of Sindbis virus 42 S and 26 S messenger RNAs. *Cell* **6**, 215–222.

Chow, N. L., and Simpson, R. W. (1971). RNA-dependent RNA polymerase activity associated with virions and subviral particles of myxoviruses. *Proc. Natl. Acad. Sci. U.S.A.* **68**, 752–756.

Chow, L. T., Gelinas, R. E., Broker, T. R., and Roberts, R. J. (1977). An amazing sequence arrangement at the 5' ends of adenovirus 2 messenger RNA. *Cell* **12**, 1–8.

Clegg, J. C. S., and Kennedy, S. I. T. (1975). Translation of Semliki Forest virus intracellular 26 S RNA: characterization of the products synthesized *in vitro*. *Eur. J. Biochem.* **53**, 175–183.

Clewley, J. P., and Kennedy, S. I. T. (1976). Purification and polypeptide composition of Semliki Forest virus RNA polymerase. *J. Gen. Virol.* **32**, 395–411.

Clinkscales, C. W., Bratt, M. A., and Morrison, T. G. (1977). Synthesis of Newcastle disease virus polypeptides in a wheat germ cell-free system. *J. Virol.* **22**, 97–101.

Cole, C. N., Landers, T., Goff, S. P., Manteuil-Brutlag, S., and Berg, P. (1977). Physical and genetic characterization of deletion mutants of simian virus 40 constructed *in vitro*. *J. Virol.* **24**, 277–294.

Contreras, R., Rogiers, R., Van de Voorde, A., and Fiers, W. (1977). Overlapping of the VP_2-VP_3 gene and the VP_1 gene in the SV40 genome. *Cell* **12**, 529–538.

Cooper, J. A., and Moss, B. (1978). Transcription of vaccinia virus mRNA coupled to translation *in vitro*. *Virology* **88**, 149–165.

Dales, S. (1963). The uptake and development of vaccinia virus in strain L cells followed with labeled viral deoxyribonucleic acid. *J. Cell Biol.* **18**, 51–72.

Davis, N. L., and Rueckert, R. R. (1972). Properties of a ribonuceloprotein particle isolated from Nonidet P-40-treated Rous sarcoma virus. *J. Virol.* **10**, 1010–1020.

Duesberg, P. H. (1968). The RNAs of influenza virus. *Proc. Natl. Acad. Sci. U.S.A.* **59**, 930–937.

Duesberg, P. H., and Robinson, W. S. (1965). Isolation of the nucleic acid of Newcastle disease virus (NDV). *Proc. Natl. Acad. Sci. U.S.A.* **54**, 794–800.

Duesberg, P. H., and Robinson, W. S. (1967). On the structure and replication of influenza virus. *J. Mol. Biol.* **25**, 383–405.

Duesberg, P. H., Wang, L. H., Mellon, P., Mason, W. S., and Vogt, P. K. (1976). Towards a complete genetic map of Rous sarcoma virus. *Animal Virology ICN-UCLA Symp. Mol. Cell. Biol.* **4**, 107–125.

Erikson, E., Collett, M. S., and Erikson, R. L. (1978). *In vitro* synthesis of a functional avian sarcoma virus transforming-gene product. *Nature (London)* **274**, 919–921.

Etkind, P. R., and Krug, R. M. (1975). Purification of influenza viral complementary RNA: Its genetic content and activity in wheat germ cell-free extracts. *J. Virol.* **16**, 1464–1475.

Etkind, P. R., Buchhagen, D. L., Herz, C., Broni, B. B., and Krug, R. M. (1977). The segments of influenza viral mRNA. *J. Virol.* **22**, 346–352.

Fiers, W., Contreras, R., Haegeman, G., Rogiers, R., Van de Voorde, A., Van Heuverswyn, H., Van Herreweghe, J., Volckaert, G., and Ysebaert, M. (1978). Complete nucleotide sequence of SV40 DNA. *Nature (London)* **273**, 113–120.

Flint, J. (1977). The topography and transcription of the adenovirus genome. *Cell* **10**, 153–166.

Flint, S. J., Sambrook, J., Williams, J. F., and Sharp, P. A. (1976). Viral nucleic acid sequences in transformed cells: IV. A study of the sequences of adenovirus 5 DNA and RNA in four lines of adenovirus 5-transformed rodent cells using specific fragments of the viral genome. *Virology* **72**, 456–470.

Franze-Fernandez, M. T., and Banerjee, A. K. (1978). *In vitro* RNA transcription by the New Jersey serotype of vesicular stomatitis virus. I. Characterization of the mRNA species. *J. Virol.* **26**, 179–187.

Frenkel, N., and Roizman, B. (1971). Herpes simplex virus: Genome size and redundancy studied by renaturation kinetics. *J. Virol.* **8**, 591–593.

Gething, M. J., White, J. M., and Waterfield, M. D. (1978). Purification of the fusion protein of Sendai virus: Analysis of the NH_2-terminal sequence generated during precursor activation. *Proc. Natl. Acad. Sci. U.S.A.* **75**, 2737–2740.

Gomatos, P. J., and Tamm, I. (1963). The secondary structure of reovirus RNA. *Proc. Natl. Acad. Sci. U.S.A.* **49**, 707–714.

Graham, F. L., Abrahams, P. J., Mulder, C., Heijneker, H. L., Warnaar, S. O., DeVries, F. A. J., Fiers, W., and Van der Eb, A. J. (1974). Studies on *in vitro* transformation by DNA and DNA fragments of human adenoviruses and simian virus 40. *Cold Spring Harbor Symp. Quant. Biol.* **39**, 637–650.

Green, M., Piña, M., Kimes, R., Wensink, P. C., MacHattie, L. A., and Thomas, C. A., Jr. (1967). Adenovirus DNA, I. Molecular weight and conformation. *Proc. Natl. Acad. Sci. U.S.A.* **57**, 1302–1309.

Hayward, W. S. (1977). Size and genetic content of viral RNAs in avian oncovirus-infected cells. *J. Virol.* **24**, 47–63.

Hirst, G. K. (1962). Genetic recombination with Newcastle disease virus, polioviruses and influenza. *Cold Spring Harbor Symp. Quant. Biol.* **27**, 303–309.

Holland, J. J., and Kiehn, E. D. (1968). Specific cleavage of viral proteins as steps in the synthesis and maturation of enteroviruses. *Proc. Natl. Acad. Sci. U.S.A.* **60**, 1015–1022.

Huang, A. S. (1973). Defective interfering viruses. *Annu. Rev. Microbiol.* **27**, 101–117.

Huang, A. S., Baltimore, D., and Stampfer, M. (1970). Ribonucleic acid synthesis of vesicular stomatitis virus. III. Multiple complementary messenger RNA molecules. *Virology* **42**, 946–957.

Imblum, R. L., and Wagner, R. R. (1975). Inhibition of viral transcriptase by immunoglobulin directed against the nucleocapsid NS protein of vesicular stomatitis virus. *J. Virol.* **15**, 1357–1366.

Jacobson, M. F., and Baltimore, D. (1968). Polypeptide cleavages in the formation of poliovirus proteins. *Proc. Natl. Acad. Sci. U.S.A.* **61**, 77–84.

Jacobson, M. F., Asso, J., and Baltimore, D. (1970). Further evidence on the formation of poliovirus proteins. *J. Mol. Biol.* **49**, 657–669.

Joklik, W. K. (1974). Reproduction of reoviridae. *In* "Comprehensive Virology", vol. 2 (H. Fraenkel-Conrat and R. R. Wagner, eds.), pp. 231–334. Plenum, New York.

Jones, P. C., Hayward, G. S., and Roizman, B. (1977). Anatomy of herpes simplex virus DNA. VII. αRNA is homologous to noncontiguous sites in both the L and S components of viral DNA. *J. Virol.* **21**, 268–276.

Kates, J., and Beeson, J. (1970a). Ribonucleic acid synthesis in vaccinia virus. I. The mechanism of synthesis and release of RNA in vaccinia cores. *J. Mol. Biol.* **50**, 1–18.

Kates, J., and Beeson, J. (1970b). Ribonucleic acid synthesis in vaccinia virus. II. Synthesis of polyriboadenylic acid. *J. Mol. Biol.* **50**, 19–33.

Kates, J. R., and McAuslan, B. R. (1967). Poxvirus DNA-dependent RNA polymerase. *Proc. Natl. Acad. Sci. U.S.A.* **58**, 134–141.

Katz, E., and Moss, B. (1970a). Formation of vaccinia virus structural polypeptide from a higher-molecular-weight precursor: Inhibition by rifampicin. *Proc. Natl. Acad. Sci. U.S.A.* **66**, 677–684.

Katz, E., and Moss, B. (1970b). Vaccinia virus structural polypeptide derived from a high-molecular-weight precursor: Formation and integration into virus particles. *J. Virol.* **6**, 717–726.

Kennedy, S. I. T. (1976). Sequence relationships between the genome and the intracellular RNA species of standard and defective-interfering Semliki Forest virus. *J. Mol. Biol.* **108**, 491–511.

Kiehn, E. D., and Holland, J. J. (1970). Synthesis and cleavage of enterovirus polypeptides in mammalian cells. *J. Virol.* **5**, 358–367.

Kit, S., Leung, W. C., Jorgensen, G. N., Trkula, D., and Dubbs, D. R. (1974). Thymidine kinase isozymes of normal and virus-infected cells. *Cold Spring Harbor Symp. Quant. Biol.* **39**, 703–715.

Kit, S., Jorgensen, G. N., Liav, A., and Zaslavsky, V. (1977). Purification of vaccinia virus-induced thymidine kinase activity from [^{35}S]methionine-labeled cells. *Virology* **77**, 661–676.

Knipe, D., Rose, J. K., and Lodish, H. F. (1975). Translation of individual species of vesicular stomatitis viral mRNA. *J. Virol.* **15**, 1004–1011.

Krzyzek, R. A., Collett, M. S., Lau, A. F., Perdue, M. L., Leis, J. P., and Faras, A. J. (1978). Evidence for splicing of avian sarcoma virus 5'-terminal genomic sequences onto viral-specific RNA in infected cells. *Proc. Natl. Acad. Sci. U.S.A.* **75**, 1284–1288.

Lamb, R. A., and Choppin, P. W. (1977). The synthesis of Sendai virus polypeptides in infected cells. II. Intracellular distribution of polypeptides. *Virology* **81**, 371–381.

Lamb, R. A., Mahy, B. W. J., and Choppin, P. W. (1976). The synthesis of Sendai virus polypeptides in infected cells. *Virology* **69**, 116–131.

Lazarowitz, S. G., and Choppin, P. W. (1975). Enhancement of the infectivity of influenza A and B viruses by proteolytic cleavage of the hemagglutinin polypeptide. *Virology* **68**, 440–454.

Lazarowtiz, S. G., Compans, R. W., and Choppin, P. W. (1971). Influenza virus structural and nonstructural proteins in infected cells and their plasma membranes. *Virology* **46**, 830–843.

Lee, Y. F., Nomoto, A., Detjen, B. M., and Wimmer, E. (1977). A protein covalently linked to poliovirus genome RNA. *Proc. Natl. Acad. Sci. U.S.A.* **74**, 59–63.

Lewis, J. B., Anderson, C. W., Atkins, J. F., and Gesteland, R. F. (1974). The origin and destiny of adenovirus proteins. *Cold Spring Harbor Symp. Quant. Biol.* **39**, 581–590.

Lewis, J. B., Atkins, J. F., Baum, P. R., Solem, R., Gesteland, R. F., and Anderson, C. W. (1976). Location and identification of the genes for adenovirus type 2 early polypeptides. *Cell* **7**, 141–151.

Lindstrom, D. M., and Dulbecco, R. (1972). Strand orientation of simian virus 40 transcription in productively infected cells. *Proc. Natl. Acad. Sci. U.S.A.* **69**, 1517–1520.

McGeoch, D., Fellner, P., and Newton, C. (1976). Influenza virus genome consists of eight distinct RNA species. *Proc. Natl. Acad. Sci. U.S.A.* **73**, 3045–3049.

Maizel, J. V., Jr., and Summers, D. F. (1968). Evidence for differences in size and composition of the poliovirus-specific polypeptides in infected HeLa cells. *Virology* **36**, 48–57.

Millward, S., and Graham, A. F. (1970). Structural studies on reovirus: Discontinuities in the genome. *Proc. Natl. Acad. Sci. U.S.A.* **65**, 422–429.

Miura, K. I., Watanabe, K., Sugiura, M., and Shatkin, A. J. (1974). The 5'-terminal nucleotide sequences of the double-stranded RNA of human reovirus. *Proc. Natl. Acad. Sci. U.S.A.* **71**, 3979–3983.

Morrison, T., Stampfer, M., Baltimore, D., and Lodish, H. F. (1974). Translation of vesicular stomatitis messenger RNA by extracts from mammalian and plant cells. *J. Virol.* **13**, 62–72.

Moss, B. (1974). Reproduction of poxviruses. In "Comprehensive Virology", vol. 3 (H. Fraenkel-Conrat and R. R. Wagner, eds.), pp. 405–474. Plenum, New York.

Moss, B., and Rosenblum, E. N. (1973). Protein cleavage and poxvirus morphogenesis: Tryptic peptide analysis of core precursors accumulated by blocking assembly with rifampicin. J. Mol. Biol. 81, 267–269.

Moss, B., Rosenblum, E. N., and Paoletti, E. (1973). Polyadenylate polymerase from vaccinia virions. Nature (London), New Biol. 245, 59–63.

Munyon, W., Paoletti, E., and Grace, J. T., Jr. (1967). RNA polymerase activity in purified infectious vaccinia virus. Proc. Natl. Acad. Sci. U.S.A. 58, 2280–2287.

Nomoto, A., Kitamura, N., Golini, F., and Wimmer, E. (1977). The 5'-terminal structures of poliovirion RNA and poliovirus mRNA differ only in the genome-linked protein VPg. Proc. Natl. Acad. Sci. U.S.A. 74, 5345–5349.

Oppermann, H., Bishop, J. M., Varmus, H. E., and Levintow, L. (1977). A joint product of the genes gag and pol of avian sarcoma virus: A possible precursor of reverse transcriptase. Cell 12, 993–1005.

Palese, P. (1977). The genes of influenza virus. Cell 10, 1–10.

Paucha, E., Mellor, A., Harvey, R., Smith, A. E., Hewick, R. M., and Waterfield, M. D. (1978). Large and small tumor antigens for simian virus 40 have identical amino termini mapping at 0.65 map units. Proc. Natl. Acad. Sci. U.S.A. 75, 2165–2169.

Pelham, H. R. B. (1977). Use of coupled transcription and translation to study mRNA production by vaccinia cores. Nature (London). 269, 532–534.

Pelham, H. R. B. (1978). Translation of encephalomyocarditis virus RNA in vitro yields an active proteolytic processing enzyme. Eur. J. Biochem. 85, 457–462.

Pereira, L., Wolff, M. H., Fenwick, M., and Roizman, B. (1977). Regulation of herpesvirus macromolecular synthesis. V. Properties of α polypeptides made in HSV-1 and HSV-2 infected cells. Virology 77, 733–749.

Perry, R. P. (1976). Processing of RNA. Annu. Rev. Biochem. 45, 605–629.

Perry, R. P., and Kelley, D. E. (1976). Kinetics of formation of 5' terminal caps in mRNA. Cell 8, 433–442.

Perry, R. P., Kelley, D. E., Friderici, K., and Rottman, F. (1975). The methylated constituents of L cell messenger RNA: Evidence for an unusual cluster at the 5' terminus. Cell 4, 387–394.

Powell, K. L., and Courtney, R. J. (1975). Polypeptides synthesized in herpes simplex virus type 2-infected HEp-2 cells. Virology 66, 217–228.

Preston, C. M. (1977). The cell-free synthesis of herpesvirus-induced polypeptides. Virology 78, 349–353.

Pringle, C. R., Duncan, I. B., and Stevenson, M. (1971). Isolation and characterization of temperature-sensitive mutants of vesicular stomatitis virus, New Jersey serotype. J. Virol. 8, 836–841.

Purchio, A. F., Erikson, E., and Erikson, R. L. (1977). Translation of 35 S and of subgenomic regions of avian sarcoma virus RNA. Proc. Natl. Acad. Sci. U.S.A. 74, 4661–4665.

Purchio, A. F., Erikson, E., Brugge, J. S., and Erikson, R. L. (1978). Identification of a polypeptide encoded by the avian sarcoma virus src gene. Proc. Natl. Acad. Sci. U.S.A. 75, 1567–1571.

Reddy, V. B., Thimmappaya, B., Dhar, R., Subramanian, K. N., Zain, B. S., Pan, J., Ghosh, P. K., Celma, M. L., and Weissman, S. M. (1978). The genome of simian virus 40. Science 200, 494–502.

Rekosh, D. (1972). Gene order of the poliovirus capsid proteins. J. Virol. 9, 479–487.

Rhode, S. L., III. (1973). Replication process of the parvovirus H-1. I. Kinetics in a parasynchronous cell system. *J. Virol.* **11**, 856–861.

Ritchey, M. B., Palese, P., and Schulman, J. L. (1977). Differences in protein patterns of influenza A viruses. *Virology* **76**, 122–128.

Roizman, B., and Furlong, D. (1974). The replication of herpesviruses. *In* "Comprehensive Virology", vol. 3 (H. Fraenkel-Conrat and R. R. Wagner, eds.), pp. 229–403. Plenum, New York.

Rose, J. A. (1974). Parvovirus reproduction. *In* "Comprehensive Virology", vol. 3 (H. Fraenkel-Conrat and R. R. Wagner, eds.), pp. 1–61. Plenum, New York.

Rottman, F., Shatkin, A. J., and Perry, R. P. (1974). Sequences containing methylated nucleotides at the 5′ termini of messenger RNAs: Possible implications for processing. *Cell* **3**, 197–199.

Rueckert, R. (1971). Picornaviral architecture. *In* "Comparative Virology" (K. Maramorosch and E. Kurstak, eds.), pp. 255–306. Academic Press, New York.

Salzman, N. P., and Khoury, G. (1974). Reproduction of papovaviruses. *In* "Comprehensive Virology", vol. 3 (H. Fraenkel-Conrat and R. R. Wagner, eds.), pp. 63–141. Plenum, New York.

Salzman, L. A., and Redler, B. (1974). Synthesis of viral-specific RNA in cells infected with the parvovirus, Kilham rat virus. *J. Virol.* **14**, 434–440.

Sambrook, J., Sharp, P. A., and Keller, W. (1972). Transcription of simian virus 40 I. Separation of the strands of SV40 DNA and hybridization of the separated strands to RNA extracted from lytically infected and transformed cells. *J. Mol. Biol.* **70**, 57–71.

Sanger, F., Air, G. M., Barrell, B. G., Brown, N. L., Coulson, A. R., Fiddes, J. C., Hutchison, C. A., III, Slocombe, P. M., and Smith, M. (1977). Nucleotide sequence of bacteriophage φX174 DNA. *Nature (London)* **265**, 687–695.

Schaffhausen, B. S., Silver, J. E., and Benjamin, T. L. (1978). Tumor antigen(s) in cells productively infected by wild-type polyoma virus and mutant NG-18. *Proc. Natl. Acad. Sci. U.S.A.* **75**, 79–83.

Scheid, A., and Choppin, P. W. (1974). Identification of biological activities of paramyxovirus glycoproteins. Activation of cell fusion, hemolysis, and infectivity by proteolytic cleavage of an inactive precursor protein of Sendai virus. *Virology* **57**, 475–490.

Sharpe, A. H., Ramig, R. F., Mustoe, T. A., and Fields, B. N. (1978). A genetic map of reovirus. I. Correlation of genome RNAs between serotypes 1, 2, and 3. *Virology* **84**, 63–74.

Shatkin, A. J. (1976). Capping of eucaryotic mRNAs. *Cell* **9**, 645–653.

Shatkin, A. J., and Both, G. W. (1976). Reovirus mRNA: Transcription and translation. *Cell* **7**, 305–313.

Shatkin, A. J., and Sipe, J. D. (1968). RNA polymerase activity in purified reoviruses. *Proc. Natl. Acad. Sci. U.S.A.* **61**, 1462–1469.

Shatkin, A. J., Sipe, J. D., and Loh, P. (1968). Separation of ten reovirus genome segments by polyacrylamide gel electrophoresis. *J. Virol.* **2**, 986–991.

Simmons, D. T., and Strauss, J. H. (1972). Replication of Sindbis virus. II. Multiple forms of double-stranded RNA isolated from infected cells. *J. Mol. Biol.* **71**, 615–631.

Skare, J., and Summers, W. C. (1977). Structure and function of herpesvirus genomes. II. *Eco*RI, *Xba*1 and *Hin*dIII endonuclease cleavage sites on herpes simplex virus type 1 DNA. *Virology* **76**, 581–595.

Sleigh, M. J., Topp, W. C., Hanich, R., and Sambrook, J. F. (1978). Mutants of SV40 with an altered small t protein are reduced in their ability to transform cells. *Cell* **14**, 79–88.

Stacey, D. W., Allfrey, V. G., and Hanafusa, H. (1977). Microinjection analysis of envelope-glycoprotein messenger activities of avian leukosis viral RNAs. *Proc. Natl. Sci. U.S.A.* **74**, 1614–1618.

Stringer, J. R., Holland, L. E., Swanstrom, R. I., Pivo, K., and Wagner, E. K. (1977). Quantitation of herpes simplex virus type 1 RNA in infected HeLa cells. *J. Virol.* **21**, 889–901.

Summers, D. F., and Maizel, J. V., Jr. (1968). Evidence for large precursor proteins in poliovirus synthesis. *Proc. Natl. Acad. Sci. U.S.A.* **59**, 966–971.

Swanstrom, R. I., and Wagner, E. K. (1974). Regulation of synthesis of herpes simplex type 1 virus mRNA during productive infection. *Virology* **60**, 522–533.

Tattersall, P., Cawte, P. J., Shatkin, A. J., and Ward, D. C. (1976). Three structural polypeptides coded for by minute virus of mice, a parvovirus. *J. Virol.* **20**, 273–289.

Tattersall, P., Shatkin, A. J., and Ward, D. C. (1977). Sequence homology between the structural polypeptides of minute virus of mice. *J. Mol. Biol.* **111**, 375–394.

Tennant, R. W., and Hand, R. E., Jr. (1970). Requirement of cellular synthesis for Kilham rat virus replication. *Virology* **42**, 1054–1063.

Tjian, R., Fey, G., and Graessmann, A. (1978). Biological activity of purified simian virus 40 T antigen proteins. *Proc. Natl. Acad. Sci. U.S.A.* **75**, 1279–1283.

Villa-Komaroff, L., Guttman, N., Baltimore, D., and Lodish, H. F. (1975). Complete translation of poliovirus RNA in a eukaryotic cell-free system. *Proc. Natl. Acad. Sci. U.S.A.* **72**, 4157–4161.

Vogt, P. K. (1977). Genetics of RNA tumor viruses. *In* "Comprehensive Virology", vol. 9 (H. Fraenkel-Conrat and R. R. Wagner, eds.), pp. 341–455. Plenum, New York.

Vogt, V. M., Eisenman, R., and Diggelmann, H. (1975). Generation of avian myeloblastosis virus structural proteins by proteolytic cleavage of a precursor polypeptide. *J. Mol. Biol.* **96**, 471–493.

Von der Helm, K., and Duesberg, P. H. (1975). Translation of Rous sarcoma virus RNA in a cell-free system from ascites Krebs II cells. *Proc. Natl. Acad. Sci. U.S.A.* **72**, 614–618.

Wagner, R. R., Prevec, L., Brown, F., Summers, D. F., Sokol, F., and MacLeod, R. (1972). Classification of rhabdovirus proteins: A proposal. *J. Virol.* **10**, 1228–1230.

Ward, R., Banerjee, A. K., LaFiandra, A., and Shatkin, A. J. (1972). Reovirus-specific ribonucleic acid from polysomes of infected L cells. *J. Virol.* **9**, 61–69.

Wei, C. M., and Moss, B. (1974). Methylation of newly synthesized viral messenger RNA by an enzyme in vaccinia virus. *Proc. Natl. Acad. Sci. U.S.A.* **71**, 3014–3018.

Weil, R., and Vinograd, J. (1963). The cyclic helix and cyclic coil forms of polyoma viral DNA. *Proc. Natl. Acad. Sci. U.S.A.* **50**, 730–738.

Weiner, H. L., Drayna, D., Averill, D. R., Jr., and Fields, B. N. (1977). Molecular basis of reovirus virulence: Role of the *S1* gene. *Proc. Natl. Acad. Sci. U.S.A.* **74**, 5744–5748.

Weiss, S. R., and Bratt, M. A. (1976). Comparative electrophoresis of the 18–22 S RNAs of Newcastle disease virus. *J. Virol.* **18**, 316–323.

Weiss, S. R., Varmus, H. E., and Bishop, J. M. (1977). The size and genetic composition of virus-specific RNAs in the cytoplasm of cells producing avian sarcoma-leukosis viruses. *Cell* **12**, 983–992.

Wirth, D. F., Katz, F., Small, B., and Lodish, H. F. (1977). How a single Sindbis virus mRNA directs the synthesis of one soluble protein and two integral membrane glycoproteins. *Cell* **10**, 253–263.

7

Aspects of Cytoplasmic and Environmental Influences on Gene Expression

Peter M. M. Rae

301

I. INTRODUCTION

The purpose of this chapter is to describe in a fairly general way some diverse instances of manifest cytoplasmic or extracellular influence on nuclear gene expression. The following discussions of natural and experimental situations that demonstrate the cytoplasm-to-nucleus component of nucleocytoplasmic interaction are not comprehensive. Rather, a selection has been made of examples I believe are particularly instructive or are otherwise important for their potential for further studies on the modulation of nuclear activity by cytoplasmic factors.

Other chapters in this treatise, especially those in this volume, discuss gene expression and its regulation in more or less specific molecular terms. To an extent this is also the case for the section of this chapter dealing with steroid hormone influences on target cells. The bulk of this chapter, however, is designed to complement such aspects of cell biology by summarizing results of cytological, developmental, and genetic, as well as biochemical, studies that have provided important demonstrations of the effects of the cellular milieu on patterns of gene expression.

Prominent among these have been experimental studies on the effects on the composition and activities of a nucleus, from one cell type or stage, of its being confronted with a cytoplasm having another orientation, such as in transplantations of amoeba nuclei from cells in one stage of the cell cycle to those in another, transplantations of somatic nuclei into oocytes or unfertilized eggs, and fusions between heterospecific and/or heterotypic vertebrate cells. Similarly outstanding have been studies on the determination of the course of differentiation in early embryos by cytoplasmic factors preformed in oocytes and specifically positioned during egg maturation or shortly after fertilization. Finally, there is a great deal of specific information on nearly all phases of steroid hormone induction of gene expression in receptive cells. The influence of steroid hormones on genetic activity involves interaction between hormone and a cytoplasmic receptor and the association of a resultant complex with chromatin, so that this system provides particularly illustrative examples of extracellular influences on gene expression acting through the cytoplasm. The chapter closes with a brief discussion of the recently discovered phenomenon of the development of drug resistance in mammalian cell lines through an increase, in surviving cells, in the number of genes coding for the enzyme subject to inhibition by the drug. Certainly this is not a routine response of nuclei to environmental factors, as it appears that rare events (gene duplications) are selected for by constant exposure of cells to gradually increasing concentrations of the drug in the medium, but it is a remarkable example of evolution in action and is amenable to detailed study.

In the interest of brevity, and so that the text of this chapter does not continually stray from the charge given by its title, many obvious examples and other elegant studies relevant to the theme have been either omitted entirely or only referred to in passing. To compensate for this, I have attempted to cite reviews and original articles that have particularly comprehensive discussions.

II. CYTOPLASMIC INFLUENCES ON GENE EXPRESSION IN TRANSPLANTED NUCLEI

A. Nuclear Transplantations in Amoeba

Much of what is known of gross molecular aspects of nucleocytoplasmic interactions has come from nuclear transplantation and other microsurgical studies on amoeba carried out over the past twenty years or so by Goldstein and Prescott and their associates. *Amoeba proteus* has provided indications of a quantitative influence of cytoplasm on the timing of mitosis in amputated cells (Prescott, 1956) and a possible qualitative influence on the timing and course of DNA synthesis in transplanted nuclei (Prescott and Goldstein, 1967). Further, studies of the fates of labeled macromolecules (protein and RNA) introduced into amoebae via nuclear transplantation (by which binucleate cells are generated) have revealed the existence of proteins that concentrate in nuclei but also shuttle between nucleus and cytoplasm (Legname and Goldstein, 1972) and of low molecular weight RNA species with similar properties (Goldstein and Ko, 1974). The roles of these migrating macromolecules are not known, but their demonstration in amoeba has provided a background for, and corroboration of, evidence of similar nucleocytoplasmic interactions from other nuclear transplantation and cell fusion studies to be discussed in following sections. A review of work on amoeba may be found in Goldstein (1974).

B. Transplantation of Nuclei into Unfertilized Eggs

Nuclear transplantations involving the insertion of a somatic cell nucleus into an enucleated unfertilized egg were done classically to demonstrate that nuclei do not irreversibly lose genetic information, or totipotency, during the course of somatic differentiation. As early as 1952, Briggs and King successfully grew differentiating embryos from *Rana pipiens* eggs whose chromosomes were replaced with single diploid nuclei from embryos, and in 1962, Gurdon was able to grow a few adult *Xenopus laevis* from enucleated eggs that were given intestinal epithelium nuclei

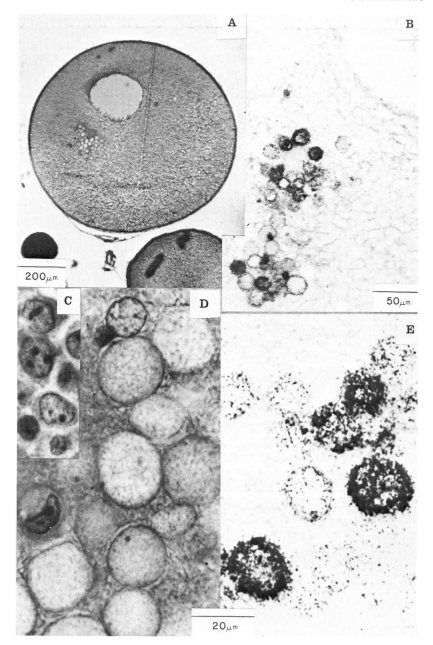

Fig. 1. Changes in the appearance of HeLa cell nuclei upon injection into the cytoplasm of *Xenopus laevis* oocytes parallel morphological changes undergone by other types of somatic nuclei upon injection into unfertilized amphibian eggs. (A) Section through an oocyte four

from swimming tadpoles. These results have been profoundly significant to the fields of cell and developmental biology, and among their implications is that a nucleus from a differentiating or differentiated cell is effectively reprogrammed to behave as a zygotic nucleus by the cytoplasm of the egg. It is some of the changes the nucleus undergoes that is the subject of this section.

1. Responses of Transplanted Nuclei to Egg Cytoplasm

The basic plan of transplantation experiments takes advantage of the position of the presumptive maternal chromosomes in the mature unfertilized egg. The chromosomes are in meiotic metaphase II, and they are at the surface of the egg near the animal pole. They may thus be removed from the egg with a flick of a glass needle, or they may be destroyed by ultraviolet microbeam irradiation. The donor nucleus, prepared by gentle shear lysis of cells and necessarily surrounded by a relatively minute amount of donor cell cytoplasm, is then microinjected into the activated recipient egg. Upon injection, donor nuclei undergo several obvious and subtle changes in morphology, content, and metabolism (Fig. 1).

Most clearly, nuclei swell tremendously after injection. Intestinal epithelium nuclei increase in volume about thirtyfold, from 160 μm^3 to about 4500 μm^3 (Gurdon and Brown, 1965). "Differentiated" (postblastula) nuclei lose nucleoli, and these changes are associated with gross changes in patterns of RNA synthesis. Brown and Littna (1964) showed that in normal *Xenopus* embryo development, the gross patterns of RNA synthesis are the following: unfertilized eggs make small amounts of heterogeneous RNA, cleavage embryos make heterogeneous (nonribosomal) RNA and transfer RNA, and only at gastrula is ribosomal RNA synthesis detectable. The latter is correlated with the appearance of the nucleolus, but not until the neurula stage is ribosomal RNA synthesis predominant. Gurdon and Brown (1965) then showed that ribosomal RNA synthesis in nuclei from neurula embryos is shut down after transplantation into mature eggs. The nuclei assumed the egg pattern of RNA synthesis, then as development of transplantation embryos proceeded, the daughter nuclei undertook the normal developmental phases of RNA synthesis.

---

days after injection, showing the giant oocyte nucleus (germinal vesicle) and several swollen HeLa nuclei. (C) HeLa nuclei in an oocyte 10 min after injection, and (D) 4 days after injection. By 4 days, the injected nuclei have swollen, the chromatin has dispersed, and nucleoli have disappeared. (B) and (E) are autoradiograms of oocytes sectioned 4 days after injection with HeLa nuclei prelabeled with [^{3}H]thymidine, showing the stability of DNA in the swollen nuclei. Reprinted with permission from Gurdon, J. B. (1976). Injected nuclei in frog oocytes: Fate, enlargement, and chromatin dispersal. *J. Embryol. Exp. Morphol.* **36,** 523–540.

Another relatively evident change that donor nuclei undergo is the uptake and loss of proteins after transplantation. Arms (1968) and Ecker and Smith (1971) showed autoradiographically that oocyte cytoplasms prelabeled with leucine rapidly transfer substantial amounts of label into injected cell nuclei during the period of nuclear swelling. The stability of much of this uptake was shown by Ecker and Smith, who transferred blastula nuclei which developed from such transplants into other, unlabeled, enucleated eggs, and followed the distribution of label to daughter nuclei during subsequent cleavages. The converse experiment, showing loss of proteins from nuclei of late blastulae and gastrulae when they are injected into enucleate eggs, was done by DiBerardino and Hoffner (1975). They determined cytochemically, and autoradiographically with [^{3}H]lysine or [^{3}H]leucine, that acidic proteins were preferentially lost from donor nuclei, while basic proteins (presumably mostly histones) were largely retained. The influx and egress of proteins is consistent with an exchange of nuclear proteins which results in a reprogramming of the donor nucleus.

The third general change in behavior of donor nuclei in eggs concerns DNA synthesis. Arms (1968) showed that during the influx of proteins, transplanted nuclei commence DNA synthesis. Gurdon (1967) demonstrated that the induction of DNA synthesis can occur not only in transplanted embryonic nuclei but also in such nuclei as those from adult brain, 99% of which nuclei do not normally take up [^{3}H]thymidine into DNA. Interestingly, the induction of DNA synthesis in transplanted nuclei does not occur in oocytes before they have been activated to mature, and it is not dependent upon protein synthesis. This point will be raised again later.

2. An Ooplasmic Determinant of Embryonic Nuclear Activity

Nuclear transplantation has been used in the study of a specific instance of an oocyte cytoplasmic factor influencing subsequent nuclear activities in embryogenesis. In the axolotl, *Ambystoma mexicanum*, there is a simple recessive maternal effect gene o (ova deficient; Humphrey, 1966). Females homozygous for o produce eggs which always arrest at gastrula, irrespective of the genotype of the male with which an o/o female is mated. The mutation leads to a cytoplasmic deficiency of a protein or proteins (termed o^+ substance), as demonstrated by the correction of subsequent developmental arrest by injection of nucleoplasm from an o^+/o^+ or an o^+/o germinal vesicle or of cytoplasm from a wild type or heterozygous mature egg into a mutant egg; moreover, the factor has the heat lability and trypsin sensitivity properties of a protein (Briggs and Justus, 1968). The effect of the lack of o^+ substance appears first in blastulae, where RNA synthesis is not detectable in offspring of o/o mothers. The arrest at gastrulation indicates that blastulae nuclei are nonfunctional, since even enucleate eggs will cleave and form a blastopore.

That the o$^+$ substance is involved in a stable, heritable, activation of nuclei during blastulation was shown in nuclear transplantation experiments (Brothers, 1976). Nuclei from normal mid- to late blastulae function in the normal development of larvae when transplanted into enucleated o/o eggs, but nuclei from early blastulae do not. Further, serial transplantation of late blastula nuclei into o/o eggs, and subsequent transfer of early blastula nuclei from these into other o/o eggs demonstrated that the activation was stable over several cell generations. The converse experiment, that of injecting o$^+$ eggs with nuclei from cleavage stages and early, middle, and late blastulae of embryos resulting from a cross involving an o/o female, indicated that nuclei from mutant eggs are receptive to the o$^+$ substance through early blastula. After this stage, unknown and irreversibly damage is done (nuclei from abnormal mid- and late blastulae could not support development). Brothers (1976) has concluded that the o$^+$ substance is a regulatory protein, needed to activate the capacity of nuclei to express genes required for normal development from gastrulation.

C. Transplantation of Nuclei into Immature Oocytes

Interspecific nuclear transplants into *oocytes* have also been performed and have produced interesting results. It should first be pointed out, as an indication of the strong influence of cytoplasmic factors on nuclear activities, that rapidly dividing blastula nuclei, for example, active in DNA synthesis and relatively inactive in RNA synthesis will *reverse* these features upon injection into oocytes, so that their metabolic activity becomes consonant with that of the resident germinal vesicle (the giant oocyte nucleus) (Gurdon, 1968).

In terms of influences of oocyte cytoplasm on the expression of specific genes in transplanted somatic nuclei, evidence from interspecific transplants where both species- and tissue-specific protein synthesis could be monitored has indicated that there are factors in ooplasm that reprogram introduced nuclei to undertake the activities of germinal vesicles. Etkin (1976) injected *Ambystoma texanum* liver nuclei into either nucleated or enucleated *A. mexicanum* (axolotl) oocytes, then assayed proteins synthesized over time after injection for the presence of an enzyme specific for *A. texanum* liver [an electrophoretic variant of alcohol dehydrogenase (ADH)] and one normally present in both oocytes and liver of *A. texanum* [a species-specific variant of lactate dehydrogenase (LDH)]. He found that within 1 week of culture, recipient axolotl oocytes contained *texanum*-specific LDH, but that even after 3 weeks no ADH was detectable despite the fact that the ADH : LDH activity ratio in liver is about 0.75 : 1. The experiments suggest that transplanted nuclei continue to carry out functions common to oocytes and liver cells, but that a tissue-specific function of liver nuclei is not maintained in their new environment.

A converse experiment, demonstrating the *activation* by oocyte cytoplasm of genes not normally expressed in somatic nuclei, was performed by DeRobertis and Gurdon (1977). They injected nuclei from cultured *Xenopus* kidney cells into *Pleurodeles* oocytes, then compared two-dimensional gel electropherograms of ^{14}C-labeled proteins synthesized in injected oocytes with proteins normally synthesized in *Xenopus* cultured cells, *Xenopus* oocytes, and *Pleurodeles* oocytes. DeRobertis and Gurdon estimated that among the proteins synthesized in *Xenopus* oocytes, at least sixteen do not appear in kidney cells, and at least eight kidney proteins are unique to that tissue; in addition, at least fifteen *Xenopus* oocyte-specific proteins can be distinguished from *Pleurodeles* oocyte proteins. With these data as references, DeRobertis and Gurdon determined that among the proteins newly synthesized in injected *Pleurodeles* oocytes, at least six were *Xenopus* specific. Three of these were proteins common to *Xenopus* oocytes and kidney cells, and most important, the other three were *Xenopus* oocyte specific. No proteins whose synthesis was directed by the injected *Xenopus* kidney nuclei were kidney specific, so that *Pleurodeles* ooplasm effectively redirected the synthetic patterns of the *Xenopus* kidney cell nuclei to conform to those of oocyte nuclei (at least in terms of the production of translatable messenger RNA). All of these nuclear transplantation experiments point to the important conclusion that oocyte cytoplasm modulates nuclear activity and contains factors necessary for the regulation of nuclear activities through early embryogeny. This point will be developed in the following section.

III. CYTOPLASMIC DETERMINANTS IN EMBRYOGENY

It has been evident since the latter part of the nineteenth century that normal development in diverse metazoans depends on cytoplasmic factors present in the unfertilized egg and on the integrity of their topological relationships. Studies bearing on this have been discussed at length by Davidson (1976), and in this section attention is given to a few particularly instructive examples of localized cytoplasmic determinants of morphogenesis.

A. Pole Plasms in Diptera

1. Germ Cell Determinants in Drosophila

By the eighth cleavage division in early *Drosophila* embryogenesis nuclei reach the posterior pole plasm. Immediately, about ten pole cells form apart from the remaining embryonic syncytium (which by the twelfth or thirteenth division produces the cellular blastoderm). During this time before gastrulation the pole cells undergo a pair of divisions, then during gastrulation they are included within the posterior midgut invagination (Fig. 2). Subsequently, they migrate through the midgut layer of cells and reach the embryonic gonads, where they establish the germ line (Sonnenblick, 1950). If the pole plasm is leaked out of a puncture or is destroyed by cauterization or ultraviolet irradiation before cleavage nuclei reach the posterior end of an embryo, normal flies will develop but they will be sterile [see Eddy (1975) for a review of early experimental studies on pole plasm].

Establishment of the fate of pole cells in embryogenesis proceeded in large part by following basophilic granules that are in the egg polar plasm and are incorporated into the pole cells (Counce, 1963). These polar granules were first examined in the electron microscope by Mahowald (1962; 1968; see Fig. 3). In the early embryo, the granules are discrete

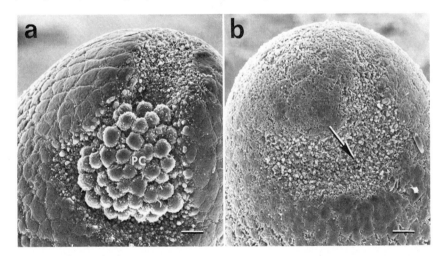

Fig. 2. Scanning electron micrographs of the posterior tip of *Drosophila melanogaster* embryos at the beginning of gastrulation. (a) Embryo showing approximately 45 pole cells (PC) in the posterior midgut invagination. (b) Embryo which had been treated with 7200 ergs cm⁻² of uv irradiation during the cleavage stage, showing no pole cells in the invagination (arrow). × 725. Reprinted with permission from Mahowald *et al.* (1976).

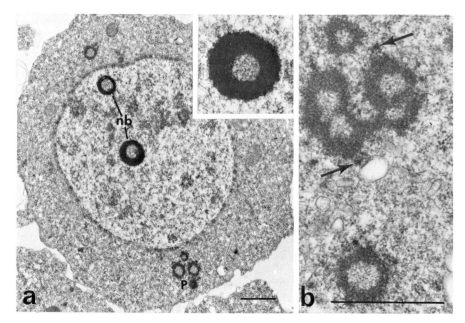

Fig. 3. (a) Survey of a pole cell in *Drosophila melanogaster* to show the structure of polar granules (P) during gastrulation. Occassionally the polar granules reaggregate into large granules but they are always dispersed in the cytoplasm. The nucleus contains two nuclear bodies (nb). × 11,000. Insert: nuclear body. × 37,000. (b) Higher magnification of polar granules in *D. melanogaster* during gastrulation. The substructure of the granules remains an interwoven mat of fibrils. Small electron-dense particles (arrows), approximately 50 nm in diameter, are found adjacent to the granules during gastrulation. Possibly they are early stages in the fragmentation of the granules, which results in their becoming associated with the nuclear envelope. × 37,000. Reprinted with permission from Mahowald *et al.* (1976).

bodies, about 0.25 to 1 μm in diameter, comprised of meshworks of inter-woven fibrils about 150 Å in thickness. In the mature unfertilized egg, the granules are associated with mitochondria, but shortly after fertilization they fragment into smaller spherical or rod-shaped structures. Ribosomes become associated with the periphery of each granule, and the granules spread around and into the surrounding ooplasm. Just prior to pole cell formation the granules become clumped together in groups, then at the time of pole cell formation there is a fragmentation of the polar granules into small, dense clumps. During the blastoderm stage, the polar granules aggregate into a number of clusters of spherical granules 0.5 to 0.75 μm in diameter, which may have a hollow core (Fig. 3b). At the blastoderm stage, also, about four spherical, "hollow" bodies appear in the nuclei of pole cells (Fig. 3a). They persist through gastrulation, but are not evident when the pole cells migrate from the gut to the embryonic gonad. Similarly, the cytoplasmic polar granules retain the blastoderm structure until the pole cells migrate, at which time the granules become associated with the outer aspect of the nuclear envelope as fibrous bodies or ''nuage''-like material (Mahowald, 1971a).

Light microscopic cytochemical studies on polar granules, with azure B at pH 4, and electron microscopic cytochemistry with indium trichloride (Mahowald, 1971b) have shown that polar granules in mature eggs are rich in RNA (as well as protein). Also, ultraviolet irradiation spectral studies showed a peak of sterility induction at 254 nm (Poulson and Waterhouse, 1960). Mahowald found that following fertilization, the ability of the polar granules to bind indium or to stain with basophilic dyes decreased until the blastoderm stage, when they ceased to react. This suggested that at the latter stage RNA was no longer present, and Mahowald proposed that the granules contain messenger RNA that is extracted from them and translated during the organization of pole cells.

More direct studies on the function of pole plasm and polar granules have involved transplantation experiments. Illmensee and Mahowald (1974) induced pole cells in the *anterior* of embryos by the injection of pole plasm at that site. The experimentally induced pole cells had nuclear and cytoplasmic granules at the blastoderm stage, and Illmensee and Mahowald showed that these could act as functional germ cells by transplanting them into the posterior of other blastoderm stage embryos of a different genotype, growing these to adults, and demonstrating that some progeny of recipients had a genotype that must have resulted from germ cell activity of donated ''anterior pole cells.'' These experiments were a clear and direct demonstration of the existence of ooplasmic germ cell determinants.

Following the establishment of this procedure, and considering the fact that several *Drosophila* species differ with respect to the morphology of polar granules (Mahowald, 1968), Mahowald *et al.* (1976) undertook a set of experiments that would allow the determination of some properties of polar granules and of whether or not they have genetic continuity. The experiments involved the injection of pole plasm from cleavage stage *Drosophila immigrans* embryos into the anterior tip of genetically marked *D. melanogaster* cleavage embryos (mwh e⁴). The *immigrans* granules are large, and semicrystalline, and are easily distinguished from *D. melanogaster* granules. In the induced anterior pole cells, the morphology of both cytoplasmic and nuclear polar granules was of the *immigrans* type. The structure of the nuclear granules was particularly interesting because it suggested that they are the synthetic product of injected *immigrans* pole plasm (perhaps translated from polar granule messenger RNA).

The anterior pole cells induced by the injection of *immigrans* pole plasm were able to function as germ cells, as judged from the results of injection of the induced mwh e⁴ pole cells into the posterior of blastoderm stage y sn³ mal hosts that had been irradiated posteriorly at cleavage. One-third of the developed flies that were fertile expressed the donor phenotype in crosses with mwh e⁴ partners. Thus, there is no particular species specificity to the induction of functional pole cells by pole plasm. Three of the five flies that showed the donor phenotype were females, and this permitted a study of polar granule morphology in F_1 progeny. From earlier morphological studies (Mahowald, 1971a), it had appeared possible that there was a cytoplasmic continuity of germ plasm granules, the germane observation being that polar granules seemed to be transformed into perinuclear fibrous bodies ("nuage") in the germ line, so that material related to granules might never have disappeared from the cytoplasm. However, polar granules in progeny of the females mentioned above were always of the *melanogaster* type, discounting direct continuity of the granules themselves.

What may be said of the development of germ plasm in *Drosophila?* Mahowald (1962) had shown that definitive polar granules appear during the midvitellogenesis stage of oocyte maturation (stage 10). Illmensee *et al.* (1976) then injected pole plasm from wild type *D. melanogaster* oocytes at different maturation stages into the anterior tip of mwh e⁴ embryos in early cleavage, and subsequently examined the recipient embryos by electron microscopy for the presence of pole cells (having both polar granules and nuclear bodies). Their findings are summarized in the tabulation below.

Source of polar plasm	Embryos examined	Percentage with pole cells
Unfertilized eggs	9	22
Oocyte stage		
14	8	38
13	5	80
12	7	0
10 and 11	5	0

A determinative capacity is reached between stages 12 and 13, despite the fact that granules are present in stage 10 oocytes. Observations which may bear on the acquisition of the capacity by oocytes are the following: At stages 11 to 12, RNA synthesis in the nurse cell chamber stops, the chamber breaks down, and there is a great increase in oocyte RNA content. Protein synthesis in the egg continues until stage 13, and the only change between stages 12 and 13 with regard to the granular material seems to be one of amount. It thus appears that functional polar granules are assembled in more than one step during oocyte development, with the granules visible as early as stage 10 serving as foci for the organization of components transported or synthesized at later stages.

2. Anterior Polar Determinants

Cytoplasmic determination of anterior structures in dipteran embryos is a phenomenon basically similar to that of posterior pole cell determination in that the integrity of anterior egg cytoplasm is essential for the subsequent development of anterior embryonic structures; while there are no polar granules or other distinguishing features in this cytoplasm, there is evidence of the involvement of localized ribonucleoprotein in anterior determination. For example, in the midge *Smittia,* experimental injury of the anterior pole of cleavage embryos (before nuclei have reached the region) frequently results in the development of a "double abdomen," where head, thorax, and anterior abdominal segments do not appear but are replaced by a mirror image of the apparently normal posterior abdomen. Kalthoff *et al.* (1977) combined centrifugal stratification of embryo components with ultraviolet microbeam irradiation to determine the nature and location of cytoplasmic anterior determinants. The outcome of their experiments was that double abdomen embryos were produced at greatest frequency after ultraviolet irradiation of an anterior cytoplasmic layer from which organelles larger than ribosomes had been excluded. Other studies on sensitivities of anterior determinants to ultraviolet light (Kalthoff, 1971a) and to ribonuclease (Kandler-Singer and Kalthoff, 1976) have pointed to a ribonucleoprotein composition for the determinative factor(s). These might be positioned by nurse cells during oogenesis, as these cells are adjacent to the anterior, but not the posterior, pole of the developing oocyte (Kalthoff, 1971b).

In *Drosophila,* too, it is the anterior pole of the developing egg that is in proximity with nurse cells (Gill, 1964), and there is evidence that the maternal genotype alone determines the establishment of anterior structures in developing embryos. Bull (1966) correlated the appearance of bicaudal (double abdomen) embryos with a particular chromosome 2 carried in a laboratory stock, finding that the bicaudal effect was expressed when this chromosome was homozygous or in flies heterozygous for this chromosome and a second chromosome deficient for the *vestigial* locus. She further determined that the bicaudal phenotype was strictly a property of the maternal rather than the embryonic genotype, in that double abdomen embryos and embryos with other anterior defects appeared only when the mother had the above chromosome constitution, regardless of the paternal genotype.

As was clear from studies on posterior pole plasm, experimental and genetic studies of the anterior pole of *Drosophila* and other dipteran embryos thus indicate a prelocalization of cytoplasmic cortical factors that operate during embryogeny in the determination of patterns of nuclear and cell differentiation. Indeed, there is likely an array of preformed cytoplasmic determinants of morphogenesis (Sander, 1976), and Schubiger *et al.* (1977) have demonstrated with ligation experiments that for the development of larval structures between those most anterior and posterior in *Drosophila*, interaction among some such factors is necessary as early as the cleavage syncytium stage, before the development of the cellular blastoderm.

B. The Polar Lobe in *Ilyanassa*

Similar, and in some respects more detailed, information on cytoplasmic determinants of nuclear activities in embryogenesis has come from classic and recent studies on the role of cytoplasmic lobes that appear transiently during the first few cleavage divisions of embryos of certain annelids and molluscs, such as the mud snail *Ilyanassa* (Fig. 4). At the first

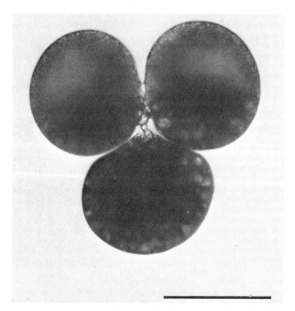

Fig. 4. The egg of the mud snail *Ilyanassa* at the "trefoil" stage of first cleavage. The upper spheres are the AB and CD blastomeres, and the lower one is the polar lobe. The bar represents 100 μm. Reprinted with permission from Clement and Tyler (1967). Copyright 1967 by the American Association for the Advancement of Science.

cleavage, producing the AB and CD blastomeres, the lobe appears as a large protuberance at the vegetal pole (trefoil stage) then is resorbed into the CD blastomere. The lobe reappears at the second cleavage and becomes part of the D blastomere, which is thus much larger than the other three, and the D quadrant of early embryos is always distinctive. Classic experiments by Clement (1952) and others evaluating the effects on development of removal of the polar lobe at the trefoil stage revealed that the lobe (which contains no nucleus) contains essential determinative information. Delobed embryos continue to develop, and at a rate comparable with that of normal embryos, but they are grossly deficient in that they lack eyes, velum, shell, foot, heart, and intestine, and they fail to develop larval axes [see also Guerrier *et al.* (1978) for *Dentalium*]. Studies on the effects of removal of D macromeres at successive cleavage stages (Clement, 1962), where more of the mentioned structures appeared in embryos as the 1d, 2d, 3d, and 4d daughters of the D, 1D, 2D, and 3D macromeres, respectively, were allowed to become incorporated into the embryo, have suggested that in normal embryos various pieces of information in the polar lobe are specifically partitioned among the several cells in the D quadrant lineage.

The nature and mechanism(s) of action of morphogenetic determinants in the *Ilyanassa* polar lobe are not known, but there are good indications that a unique population of preformed (maternal) messenger RNA is present in the trefoil stage polar lobe, and that components of the lobe have at least a quantitative effect on RNA synthesis in embryonic nuclei. Clement and Tyler (1967) compared the rates of protein synthesis in whole eggs, lobeless eggs, and isolated lobes of *I. obsoleta* and found that trefoil lobe cytoplasm incorporates about one-eighth as much labeled amino acid into protein as does an intact egg. It does so for at least a day after detachment, so that preformed messenger RNA is translated as early as the first cleavage stage and as a population is relatively stable. As regards qualitative aspects of protein synthesis in normal and lobeless embryos, Newrock and Raff (1975) determined that the electrophoretic patterns of proteins synthesized in normal and delobed embryos during the first 48 hr of development (past gastrulation) are significantly different. In experiments involving inhibition of new RNA synthesis throughout early development with actinomycin D (which shuts down embryonic transcription but has no noticeable effect on the first 2 days of development), Newrock and Raff found that while there is substantial synthesis and translation of RNA in untreated early embryos, *much* of the difference between patterns of proteins synthesized in normal and delobed embryos is due to polar lobe-specific translation of preformed mRNA.

Some effects of polar lobe determinants, be they products of the aforementioned site-specific translation or other factors, on gene expression (transcription) later in development have been measured by Davidson *et al.* (1965). They determined that normal and lobeless embryos synthesize comparable amounts of RNA during the first few cleavage stages, but that between about 20 and 72 hr after trefoil, the rate of RNA synthesis ([³H]uridine uptake into RNA during a 2-hr pulse at various intervals) in lobeless embryos gradually declines to a value nearly one-half that in normal embryos, despite the fact that at every stage there is the same number of cells in each type. These results point to the conclusion that because normal and lobeless embryos contain the same number of nuclei, the observed difference in RNA synthesis between the two must be due to at least quantitative effects of polar lobe material on transcription at later stages in development.

C. Cortical Determinants in an Ascidian

A classic and truly remarkable example of the "mosaic egg," in which the egg cytoplasm is segregated into regions each of which is destined to become uniquely incorporated into one of the several cell lineages, is found in the ascidians (Conklin, 1905). In the fertilized egg of *Cynthia partita,* five zones are distinguishable in that each has a pigment content different from any other, and in his studies of cell lineage during early embryogenesis Conklin determined that the five regions segregate into the following five larval cell types: tail muscle, coelomic mesoderm, notochord and neural plate, endoderm, and ectoderm, so that the egg pigmentation pattern is effectively a fate map (Conklin's illustrations of early embryogeny are reproduced in Davidson, 1976).

Whittaker (1973) took advantage of this fact (namely, that in early embryos cells destined to give rise to various differentiated structures can be identified by their pigment content) in asking if a relationship could be drawn between cytoplasmic segregation in eggs and the subsequent induction of the synthesis of tissue-specific enzymes. It had been determined previously by Whittaker and others that relatively early in development acetylcholinesterase activity appears and is localized in tail muscle cells, and tyrosinase activity appears uniquely in the brain pigment cells. Whittaker (1973) was able to demonstrate that in cleavage-arrested *Ciona intestinalis* embryos (in which nuclear divisions proceed in the absence of cell division), blastomeres representing lineages of the two tissues specifically produced the respective enzymes after obligatory RNA synthesis, and at times corresponding to the temporal program of normal development.

Whittaker's elegant experiments were based on the following: (1) in normal embryos, acetylcholinesterase activity is first detectable in the presumptive muscle cells of the early neurula (at 8 hr of development) and tyrosinase activity appears at the tailbud stage (9 hr), and (2) treatment of embryos at various cleavage stages with inhibitors of cytokinesis such as cytochalasin B arrested embryos at a given cell number but permitted nuclear divisions. With regard to the appearance and distribution of acetylcholinesterase in cytochalasin-treated embryos arrested at various cell numbers, Whittaker found that the positions and the maximum number of cells in a given embryo that reacted positively to a histochemical test for activity of this enzyme corresponded exactly with the muscle cell lineage. As noted, acetylcholinesterase activity appears at 8 hr of development and tyrosinase at 9 hr. If protein synthesis is inhibited with puromycin from 1 hr before the activities are normally demonstrable, they will not appear; if RNA synthesis is inhibited with actinomycin for 3 hr before the normal appearance of the respective enzyme activities, such activities are not detectable (however, administration of actinomycin at 2 hr or less before the normal appearance does permit the scheduled appearance of activities). The simplest and very appealing interpretation of Whittaker's results is that cytoplasmic factors, positioned in the fertilized egg, segregate during cleavage and ultimately signal tissue-specific RNA synthesis and subsequent protein synthesis using this RNA as template.

IV. BEHAVIOR OF NUCLEI IN SOMATIC CELL HETEROKARYONS

Because cell hybrids and their use in somatic cell genetics are discussed in Volume I of this series (Ringertz and Ege, 1978), this section is essentially limited to a description of the effects of incorporation into heterokaryons on the inactive nuclei of avian erythrocytes, and of interactions in homokaryons between similar nuclei in different stages of the cell cycle. The fusion of chick erythrocytes with HeLa cells or mouse fibroblasts will be concentrated upon because erythrocytes are highly and terminally differentiated, and changes their nuclei undergo after fusion are striking. For a thorough discussion of the subject, the reader is referred to the monograph by Ringertz and Savage (1976).

A. Reactivation of Avian Erythrocyte Nuclei in Heterokaryons

Cell fusions are mediated by agents such as irradiated Sendai virus or polyethylene glycol, which cause formation of cell bridges and subsequent total fusion between cells in contact with one another; during the process, erythrocytes are hemolyzed so that practically only their plasma mem-

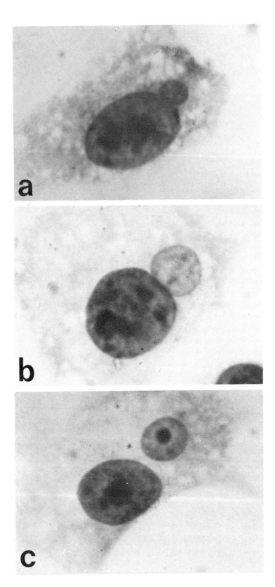

Fig. 5. Reactivation of erythrocyte nuclei in heterokaryons. The heterokaryons were made by fusing 12-day-old embryo erythrocytes with mouse A_9 cells. (a) Twelve hours after fusion the erythrocyte nucleus is still small, condensed, and darkly staining. (b) By 48 hr after fusion the erythrocyte nucleus has expanded. (c) By 4 days after fusion the erythrocyte nucleus has developed a large, round nucleolus. Reprinted with permission from Sidebottom (1974).

branes and nuclei are involved in the fusion. Once in heterokaryons, chicken erythrocyte nuclei [which otherwise synthesize no DNA and very little RNA (e.g., Zentgraf *et al.*, 1975)] are activated to synthesize RNA at a high rate, ultimately directing the synthesis of chick-specific proteins, and to synthesize DNA (reviewed by Sidebottom, 1974).

The first obvious consequence of cell fusion is that erythrocyte nuclei, which are normally highly pycnotic, undergo enlargement (Harris, 1965). Over the period of 2 or 3 days, nuclear volume increases up to about twentyfold, and dry mass increases up to about eightfold. During this time the chromatin disperses, and near the end of this period a nucleolus appears (Fig. 5). Dispersal of the chromatin is accompanied by conformational changes which result in its having a greater affinity for the DNA intercalating dye acridine orange and a lower thermal stability than the chromatin in highly condensed nuclei (Bolund *et al.*, 1969). The DNA in such erythrocyte chromatin is thus considered to be more accessible to modulators of genetic activity than is DNA in normal erythrocyte chromatin, and the chromatin is described as being "activated." Such activation is manifested by an influx of host proteins into the erythrocyte nucleus and by RNA and DNA synthesis. Such events are similar to those that occur when differentiated nuclei are transplanted into eggs or oocytes (Section II,B,1).

Activated erythrocyte nuclei undertake RNA synthesis soon after swelling begins, and the rate of RNA synthesis is correlated with the increase in size of the nucleus (Harris, 1967). During the first couple of days postfusion, the newly synthesized RNA is heterodisperse, and it does not enter the cytoplasm (Harris *et al.*, 1969). At day 3 after fusion, when the nucleolus appears, ribosomal RNA synthesis is detectable, and this RNA, along with heterogeneous RNA, appears in the cytoplasm. This point will be discussed later in the section. DNA synthesis in erythrocyte nuclei begins around 24 hr after fusion, so that by 48 hr the DNA content of a nucleus is doubled (i.e., the nuclei are in G_2, having started in G_0).

The protein influx is measured as a substantial increase in nuclear dry mass. The proteins are nucleus specific, and they may be newly synthesized, or a class of shuttling proteins (Goldstein, 1974) and thus of host nucleus origin, or both. Ringertz *et al.* (1971) elegantly demonstrated the nuclear specificity in experiments involving indirect immunofluorescence, where HeLa cell–chick erythrocyte heterokaryons were challenged in cytological preparations with sera which contained antibodies against nucleolar protein, nucleoplasmic protein, and cytoplasmic protein. HeLa (human) proteins were assayed for in hen erythrocyte nuclei using the following antisera: anti-human nucleolus serum was from an individual with lupus erythematosus; anti-human nucleoplasm serum was from a person with an erythematosus facial lesion; and anti-human cytoplasm serum was from a patient with severe rheumatoid arthritis. When these sera were presented to cytological preparations of heterokaryons that were fixed at different days postfusion and immunofluorescence was measured, Ringertz *et al.* found that human nucleolar antigen was detectable in chick nuclei within 1 hr or so, while human nucleoplasm antigen appeared within about 12 hr. Human cytoplasm antigens did not enter the chick nuclei in detectable quantities, so that the movement of human nuclear proteins into erythrocyte nuclei was specific. Sera were also prepared which contained antibodies against chicken-specific nucleolar proteins. In cytological immunofluorescence tests on heterokaryons, these proteins appeared in both chick and HeLa nuclei about 12 hr postfusion.

The appearance of chick-specific nucleolar proteins follows the appearance of a nucleolus in the erythrocyte nucleus of a heterokaryon. At about the same time, other evidence of gene expression in the activated erythrocyte nucleus emerges (Harris and Cook, 1969). At the time of cell fusion, chick-specific cell surface antigens are naturally a component of the heterokaryon cell membrane. For hours after fusion the amount of these antigens decreases fairly rapidly until nucleoli make their appearance in the chick nuclei. At this time, chick surface antigens reappear as part of the heterokaryon cell membrane. Similarly, in heterokaryons involving mouse A_9 cells, which do not have the enzyme inosinic acid pyrophosphorylase, the enzyme becomes detectable after the appearance of the chick nucleolus, as can be measured by the capacity of heterokaryons to incorporate tritiated hypoxanthine into RNA.

The importance of the presence of a nucleolus in activated erythrocyte nuclei to the onset of detectable chick-specific protein synthesis is not understood. There is certainly a correlation between the appearance of a nucleolus and the appearance of nonribosomal (and presumably informational) RNA in the heterokaryon cytoplasm (see above), and Deák *et al.* (1972) demonstrated in ultraviolet microbeam irradiation studies that the integrity of a nucleolus was essential for the *de novo* appearance of chick proteins in heterokaryons. However, there are no indications from other systems that synthesis, transport, or translation of messenger RNA requires the coordinate activity of nucleoli. This paradox, which remains, is discussed by Deák *et al.* (1972).

B. Modulation of Differentiated Functions in Heterotypic Hybrid Cells

The interspecific activation of chick erythrocyte nuclei by host proteins in heterokaryons may be limited to general ("housekeeping"-type) cell functions, since it has not been shown to extend to the conversion of the chick nucleus to a state where it directs the synthesis of a differentiated cell product. Neither chick hemoglobin nor a histospecific protein characteristic of the host but coded for by a chick gene are produced by gene activity in chick nuclei in heterokaryons, so that the level of activation reaches neither redetermination of erythropoietic function nor reprogramming to differentiated functions consonant with those of host cells. Ringertz *et al.* (1972) fused chick erythrocytes with rat myoblasts and myotubes, and screened heterokaryons for the presence of chick myosin. None was detected by immunological criteria, although in control experiments where chick *myoblasts* were fused with rat myotubes, both rat and chick myosins were synthesized (myoblasts do not synthesize myosin and other proteins characteristic of muscle, while myotubes do).

On the other hand, studies on interspecific hybrids between cells of cultured lines that manifest distinct differentiated functions have shown that in *synkaryons* (mononucleate hybrid cells having chromosomes of both fusion partners) the genome of one "parent" can be induced to undertake some, but not all, differentiated activities of the other (reviewed in Davidson, 1974). For example, Peterson and Weiss (1972) fused rat hepatoma cells [which produce serum albumin, tyrosine aminotransferase (TAT) and liver (type B) fructose-1,6-diphosphate aldolase] with mouse fibroblasts (which have a very low level of TAT and synthesize neither serum albumin nor type B aldolase) and determined immunologically that in some resultant clones both rat and mouse albumin genes were expressed (the albumins of the two species are serologically distinguishable). However, the hybrids failed to synthesize detectable amounts of TAT and type B aldolase. These observations make the point that factors responsible for the *induction* of a differentiated function such as serum albumin synthesis are active across species lines; they also indicate that the *maintenance* of other differentiated functions in hybrids such as these (synthesis of hepatoma TAT and type B aldolase) is under a control separate from that leading to induction. Other studies beyond the scope of this chapter (e.g., Brown and Weiss, 1975) suggest that dosage of the respective parental genes in hybrid clones is a determinant of the maintenance of expression.

C. Fusions of Cells in Different Stages of the Cell Cycle

With regard to DNA synthesis in nuclei of fused cells, it was shown by Johnson and Harris (1969a) that in HeLa homokaryons formed from asynchronous cells there is an imposition of synchrony of DNA synthesis (as judged by autoradiography of fused cells pulsed with [^{3}H]thymidine at

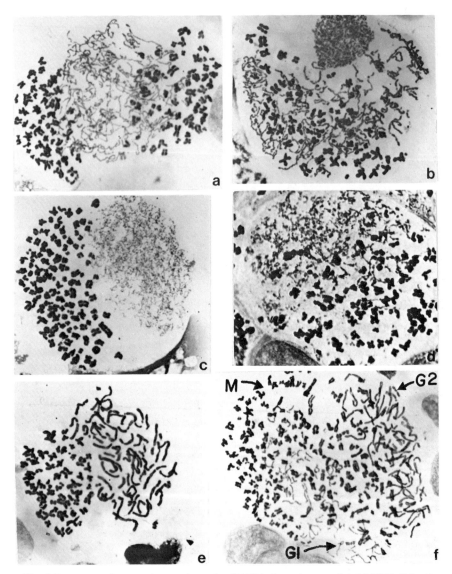

Fig. 6. Premature chromosome condensation (PCC) in interphase nuclei. (a) PCC of the G_1 nucleus in a heterophasic M/G_1 cell at 30 min after fusion. Note the long and slender G_1 chromosomes with single chromatids; the more fully condensed chromosomes are from the mitotic cell blocked with colcemid. (b) Heterophasic cell formed by the fusion of two G_1 cells with one mitotic cell. PCC can be observed in the G_1 nuclei, one of which still shows its nuclear outline. (c) Premature condensation of S chromatin 45 min after fusion between mitotic and S phase cells. Chromatin of the S nucleus presents a fragmented appearance. (d) Premature chromosome condensation in a nucleus, blocked at the G_1–S boundary by double

1-day intervals) and of mitosis. This is similar to the gross activation of erythrocyte nuclei in HeLa–chick cell fusions, in which both HeLa and chick nuclei synthesize DNA essentially in synchrony (Johnson and Harris, 1969b).

In examining more closely the imposition of synchrony on DNA synthesis in HeLa cell homokaryons, Rao and Johnson (1970) fused cells from different populations of *previously* synchronized cells. They found that when cells in S phase were fused with those in G_1, DNA synthesis was prematurely induced in the latter nuclei. The "G_1" nuclei continued to synthesize DNA as the "S" nuclei proceeded into G_2, but mitosis was synchronous (and somewhat later than in control S/S fusions). Presumably, a diffusible component of S phase cells promotes DNA synthesis in cells otherwise in G_1. On the other hand, when S phase cells were fused with cells in G_2, the latter did not reinitiate DNA synthesis; rather, when these nuclei entered mitotic prophase, chromatin in the S phase nuclei condensed and synthesis decayed. Rao and Johnson described this as "premature chromosome condensation," and found further that when cells in G_1 and G_2 as well as in S were fused with cells actually in mitosis, nuclei of all three were induced to form chromosomes prematurely (Fig. 6). While the "G_2 chromosomes" were essentially normal, chromatin in G_1 nuclei condensed into single chromatids, and S phase chromatin condensed less completely into a reticulum of large and small fragments interspersed with barely condensed chromatin. There is thus evidently a diffusible inducer of chromatin condensation in mitotic cells that is capable of interacting productively (and perhaps directly, as condensation occurs within 30 min of fusion) with chromatin at any stage of the cell cycle.

V. STEROID HORMONE INFLUENCES ON TARGET CELL NUCLEI

There are two general sorts of hormone–cell interactions in higher metazoa. One involves interactions of peptide hormones, such as insulin

thymidine treatment, after fusion with a mitotic cell. Note that the pattern of condensation at upper left is intermediate between that of G_1 and S nuclei. (e) PCC of the G_2 nucleus in an M/G_2 heterophasic cell. The more condensed chromosomes are from the mitotic cell treated with colcemid for 19 hr. (f) A heterophasic cell formed by the fusion between G_1, G_2, and mitotic cells. Note the differences between G_1, G_2, and metaphase (M) chromosomes. Reprinted with permission from Johnson, R. T. and Rao, P. N. (1970). Mammalian cell fusion: Induction of premature chromosome condensation in interphase nuclei. *Nature* (*London*) **226**, 717–722.

and glucagon, with specific receptor proteins (usually glycoproteins) that are integral components of plasma membranes of target cells (Kahn, 1976). These complexes activate or stimulate adenylate cyclase in the case of glucagon, for example, or depress adenylate cyclase activity and stimulate guanylate cyclase in the case of insulin. Since such hormones act at the level of the plasma membrane, they do not directly influence the nuclear metabolism of receptor-containing cells [see Tager and Steiner (1974) and Cuatrecasas (1974) for reviews on peptide hormones and their receptors].

The other form of hormone–cell interaction is germane to this chapter, and involves the actual entry of steroid hormones, such as estradiol, progesterone, or testosterone, into target cells, where they bind to and modify specific receptor proteins, forming complexes that then enter target cell nuclei and modulate patterns of transcription (Gorski and Gannon, 1976).

A. Sex Hormones in Vertebrates

A paradigm of steroid hormone–target cell interaction is estrogen and its effects on the mammalian uterus (Jensen and DeSombre, 1973). Estrogen is secreted by ovarian follicles; the chief target tissue is the uterus, which rapidly takes up and concentrates the hormone. Jensen and Jacobson (1960) found that a single injection of $[^3H]\beta$-estradiol into an immature female rat actually resulted in the rapid uptake of the steroid by a variety of tissues (muscle, kidney, liver, vagina, and uterus) but that the initial rate of uptake differed in increasing order of the organs as listed, and that the retention by each organ as a function of time after a single injection decayed with very different kinetics depending on whether or not the organ was a physiologically responsive target. For example, uptake of the hormone analogue by kidney and liver peaked within 20 min after injection and retention decayed exponentially thereafter, while uptake by vagina and uterus peaked more slowly and more broadly, and retention of the hormone dropped gradually over a period of hours after the single injection. Thus, the feature that distinguishes target from nontarget tissues is not so much uptake of a steroid hormone but its retention. Jensen *et al.* (1967) also found that specific retention of hormone took place in organ culture and tissue homogenates, and that it could be competitively inhibited *in vitro* by certain estradiol analogues which physiologically inhibit the uterotrophic action of estrogen (principally endometrium proliferation).

1. Cytoplasmic and Nuclear Hormone–Receptor Complexes

The target-specific retention of estradiol is not sensitive to inhibitors of RNA or protein synthesis, but is due to the presence of a tissue-specific, preformed protein, as Toft and Gorski (1966) first showed. They treated immature rat uteri with [³H]estradiol, homogenized the organ, prepared a 105,000 g supernatant, then centrifuged this into a sucrose gradient. They found a sharp peak of radioactivity at 8 S, which was later found to be reversibly convertible in KCl or NaCl concentrations of 0.15 M or more to material that sedimented at 4 S. This was the first good evidence for the presence in target cells of a hormone receptor.

Early cell fractionation experiments such as these had also shown that up to 70 or 80% of the [³H]estradiol was located in a low speed pellet of homogenates. This was thought to be an aggregation artifact until autoradiographic studies by Stumph and Roth (1966) on sections of intact tissues showed accumulation of [³H]estradiol in nuclei. It was then found that nuclei could be extracted with 0.3–0.4 M KCl to release estradiol-binding material that sedimented in sucrose gradients at 5 S. In *in vivo* and *in vitro* studies of the uptake and binding of estradiol by uterine tissues, Gorski *et al.* (1968) and Jensen *et al.* (1968) showed that the nuclear estradiol–receptor complex is derived from the cytoplasmic (cytosol) complex and is not a consequence of estradiol transfer from one receptor to another. The evidence on which this conclusion was based is the following: (1) 8 S protein was constitutive in immature uterine cytosol, and reacted directly with estradiol in a temperature-independent way both *in vivo* and *in vitro*. The 5 S binding protein, on the other hand, was not detectable except where estradiol was incubated with uterine homogenates or with nuclei *plus* cytosol at temperatures between 25° and 37°C. (2) Prior heating of a cytosol fraction to 45°C destroyed 8 S estradiol binding and prevented the ability of the cytosol to promote the appearance of 5 S complex. (3) If intact immature uteri were exposed at 2°C to [³H]estradiol, the steroid accumulated as 8 S complex and was extranuclear by autoradiography. If the uteri were warmed to 37°C, redistribution of the steroid took place to give a predominantly nuclear location of the tritium, and this could be extracted as a 5 S complex. Finally (4), the appearance of 5 S complex was at the expense of 8 S cytosol complex, and reestablishment of the latter was dependent on protein synthesis.

As noted, the 8 S cytosol receptor referred to in the preceding paragraph exists as such under conditions of low ionic strength and is convertible to a 4 S moiety when the salt concentration is elevated to around 0.15 M or greater. Because this higher ionic strength is more physiological, it is considered that 4 S reflects the native size of the receptor, and that the 8 S structure is an aggregate (e.g., Yamamoto and Alberts, 1972).

DNA has been implicated in the conversion of the 4 S cytosol receptor complex to the 5 S nuclear receptor complex because DNase treatment releases the 5 S complex from nuclei (Harris, 1971). Yamamoto and Alberts (1972) investigated the role of DNA in the *in vitro* conversion of 4 S complex to 5 S complex by DNA–cellulose chromatography. They incubated rat uterus extracts with [³H]estradiol, passed the material into a column, then compared the sedimentation value of complex in unfractionated extract with the value of complex bound to DNA–cellulose in $0.15 M$ NaCl and eluted with $0.4 M$ NaCl. By exposure to DNA, estradiol complex was converted from 4 S to 5 S, and the 5 S form was found to bind about 15 times more strongly to DNA than unconverted 4 S material (which could be eluted from the column with $0.25 M$ NaCl).

In terms of molecular changes which underlie the conversion of 4 S complex to 5 S complex, Yamamoto and Alberts (1972) found from gel filtration and sucrose gradient centrifugation studies that there is a change in both size and shape. The 4 S complex has a molecular weight of 60,000 and is a prolate ellipsoid with a length to width ratio of 3.8 : 1, while the 5 S complex has a molecular weight of 104,000 and a length to width ratio of 5.4 : 1. Evidently, conversion involves the addition of a second subunit to the 4 S complex.

2. Interaction of Hormone–Receptor Complexes with Target Cell Chromatin

Having determined (Yamamoto and Alberts, 1972) that the conversion of 4 S receptor requires the presence of hormone and involves an interaction with DNA, Yamamoto and Alberts (1974) investigated the nature and specificity of estradiol-dependent binding of receptor to DNA by use of a technique termed "sedimentation partition chromatography," which allows a measurement of equilibrium dissociation constants. They found that irrespective of the kind of DNA used [rat uterus DNA, *E. coli* DNA, or poly(dAT)], the molar dissociation constant (K_{RD}) for receptor binding was on the order of $5 \times 10^{-4} M$ DNA sites (base pairs). Since the number of sites in poly(dAT) is the same as the number of base pairs, this meant that receptor binding to DNA had no nucleotide sequence specificity (although receptor discriminated between DNA and double-stranded RNA), so that there could be an essentially unlimited number of low affinity receptor binding sites in chromatin. Yamamoto and Alberts also determined that there was no difference between purified DNA and uterine chromatin with respect to receptor binding. These results were consistent with other observations *in vitro* and *in vivo* that indicated a nonspecificity of receptor binding in nuclei. For example, Chamness *et al.* (1973) showed with several organisms that nuclei of both target and nontarget tissues effectively bind hormone–receptor complex. Williams and Gorski (1972) investigated the equilibrium distribution of estradiol in cytosol and nuclear fractions of uteri exposed to different estradiol concentrations. They found that at estradiol concentrations between 1 and 10 nM (bracketing the physiological range) there was always about five times as much estradiol in the nuclear compartment as in the cytosol, so that the proportion of hormone-bound 4 S and 5 S receptors was constant irrespective of the amount of hormone. Also, the proportion did not change during washes of the uteri during which half the [³H]estradiol was removed from the tissue. The constancy of the proportion of nucleus-bound to cytoplasmic receptor complex as hormone levels change demonstrates a rapid equilibrium between the two, and a nonsaturability (nonspecificity) of nuclear binding sites.

As is noted below (Section V,A,3), steroid hormones clearly induce specific nuclear responses in target tissues. In order to resolve the paradox presented by this and the experimental results just summarized, Yamamoto and Alberts (1975) proposed that high affinity (specific) 5 S receptor binding sites exist in the genome of an organism, but that they are undetectable under the assay conditions that have been applied. In formulating their hypothesis, Yamamoto and Alberts drew a parallel between this system and the interaction of *E. coli* lactose repressor protein with *E. coli* DNA [where the repressor protein binds to the appropriate DNA operator sequence with a K_{diss} of 2 to $6 \times 10^{-12} M$, but also binds to many other DNA's with a K_{diss} of around $10^{-3} M$ (Lin and Riggs, 1975)]. The data of Williams and Gorski (1972) permit the conclusion that in the rat genome there are in excess of 30,000 low affinity 5 S receptor binding sites (and probably many more); however Yamamoto and Alberts have calculated that from one to 1000 specific sites in a genome for the binding of receptor (with K_{RD}'s of 10^{-12} to $10^{-9} M$) could be operative but masked by low affinity binding.

While Yamamoto and Alberts and their associates have concluded that steroids mediate tissue-specific responses through direct interaction of their receptors with DNA, O'Malley and his co-workers have promoted the concept that target tissue-specific nonhistone chromosomal proteins are intermediaries ("acceptors") in the specific association of receptors with appropriate genes. From studies on progesterone receptor in chicken oviduct (a specific target tissue) and its interactions with nuclei from various organs, O'Malley *et al.* (1971) reported that chick oviduct nuclei were capable of binding larger amounts of hormone–receptor complex than nuclei from heart, intestine, lung, or spleen. This apparent tissue preference prompted the "nuclear acceptor hypothesis," by which target cell chromatin contains "acceptor sites with a specific affinity for the receptor molecules" (O'Malley and Means, 1974). As DNA itself did not demonstrate such specificity, O'Malley *et al.* (1972) dissected chick oviduct chromatin biochemically by removing various protein fractions and assaying the residue for ability to bind [³H]progesterone upon incubation with oviduct cytosol pretreated with the radioactive hormone. They found that the presence of a particular acidic (nonhistone) protein fraction was necessary for a binding level that approached that of unfractionated chromatin, and that the active component was absent from nonhistones of nontarget chromatin. This fraction contained proteins that were soluble in 2 M NaCl, 5 M urea, pH 8.5, and comprised 35% of all acidic proteins. The conclusion drawn from such experiments was that a particular acidic protein component of oviduct chromatin contained specific acceptor molecules that promoted tissue-specific transcriptional responses to hormone treatment. Some of the results of such studies on progesterone–receptor interaction with nuclei and chromatin are at variance with results obtained with other systems (e.g., Chamness et al., 1973; discussed by Yamamoto and Alberts, 1976), so that progesterone–chick oviduct interactions may not be generalizable to other hormone–target cell systems.

3. Metabolic Responses in Target Nuclei

An early response of target cells to hormone treatment is an *apparent* large burst of RNA synthesis (measured as incorporation of radioactive precursor into RNA) that occurs within minutes and rapidly subsides (e.g., Means and Hamilton, 1966). However, this is not a net change in rate of synthesis but is due to an increased uptake by the cells of RNA precursor upon hormone administration; an actual increase in RNA synthesis in response to hormone does not take place until some hours later (Billing et al., 1969). It has been demonstrated, however, that there is a transient, threefold increase in endogenous RNA polymerase II (nucleoplasmic enzyme) activity in isolated uterine nuclei that peaks at 30 min after an injection of estradiol into immature rats and subsides to the basal level within 2 hr. This burst is followed closely by a twofold increase in polymerase I (nucleolar enzyme) activity that is stable for at least 12 hr after a single hormone injection, and a fourfold increase in polymerase II activity that peaks at 6 hr and slowly decays (Glasser et al., 1972). The initial and transient increase in polymerase II activity is not dependent on protein synthesis (is insensitive to cycloheximide) and is considered to be a primary response to the hormone. The stimulation of polymerase I activity and the later increase in polymerase II activity will not occur in the presence of cycloheximide; they are thus secondary responses that may depend on protein synthesis directed by RNA transcribed during the early burst of activity, but not to the extent that the increases are the result of new enzyme synthesis (Borthwick and Smellie, 1975).

The increases in RNA polymerase activity appear to be the result of increased template activity of the chromatin in target cell nuclei. For example, it has been reported that there is a large burst, then a rapid decline, of histone H4 acetylation in rat uteri that precedes the initial increase in polymerase II activity and is insensitive to pretreatment with cycloheximide or actinomycin D (Libby, 1972). The independence of this response from RNA and protein synthesis categorizes it as a primary one. O'Malley's group (e.g., Schwartz *et al.*, 1975) has determined that *E. coli* RNA polymerase is more active on chromatin prepared from chick oviducts pretreated with estradiol than on chromatin from untreated chick oviducts, and Tsai *et al.* (1975) reported that changes in the *in vitro* template activity of oviduct chromatin as a function of time after the estrogen analogue diethylstilbestrol was administered to (or withdrawn from) chicks paralleled the changes in amount of nucleus-bound estrogen receptor. Specifically, they presaturated various chromatin preparations with *E. coli* RNA polymerase in the absence of added nucleoside triphosphates (generating "preinitiation complexes"), then added triphosphates, rifampicin, and heparin, and allowed transcription to occur for several minutes at 37°C. Since rifampicin inhibits the initiation of transcription and heparin and rifampicin inhibit reinitiation, incorporation of precursors into RNA was a measure of the number of stable initiation complexes formed during the preincubation of chromatin with *E. coli* polymerase. Tsai *et al.* also measured the amount of estrogen receptor in aliquots of nuclei from which chromatin was prepared by assaying for [³H]estradiol binding ability, and found, as noted, that there was a correlation between the number of transcription initiations in chromatin and the number of receptors bound to nuclei. Because *E. coli* RNA polymerase does not initiate faithfully in chromatin and uses both sense and nonsense polynucleotide chains as template (Wilson *et al.*, 1975), and because the number of detected initiation sites was in excess of 25,000 per picogram of DNA (>60,000/chick diploid nucleus), the conclusion that is permitted from these studies is that hormone–receptor interaction with chromatin results in a fairly general alteration of chromatin structure that favors transcription.

Included among regions of chromatin that are subject to modification by hormone–receptor complexes are obviously those that contain genes coding for target tissue-specific proteins; whether hormone–receptor complexes interact with appropriate sites in the genome through high affinity binding to DNA or through acceptor proteins, the induction or stimulation of specific messenger RNA and protein synthesis has been demonstrated in systems in which a few species of polypeptides are produced in abundance in target cells. In the context of sex hormones, the induction by estradiol of vitellogenin (the precursor of egg yolk proteins) synthesis in amphibian and avian liver has become an important and amenable system for study (Tata, 1976), but the chick oviduct has been the principal subject of research over the past decade or so.

Tubular gland cells that comprise the great majority of cells in the magnum portion of the avian oviduct respond to estrogen administration *in vivo* by undergoing proliferation and by synthesizing large amounts of the major egg white protein ovalbumin and of coanlbumin (Palmiter, 1972). Exposure of estrogen-pretreated oviduct to progesterone results in the stimulation of avidin synthesis (Chan *et al.*, 1973). The experimental system is immature chick oviduct (before gonads have developed to the point of steroid production), where response to hormone administered orally or by injection mimics the course of events in older hens. By about 18 hr after the initiation of estrogen treatment, ovalbumin synthesis is detectable, and the rate of its synthesis increases to a plateau at about 10 days. If chicks are withdrawn from hormone treatment, ovalbumin synthesis declines to an undetectable level, but if hormone is readministered ovalbumin synthesis is underway after a lag of only 3 hr and is maximal by 4 days (Palmiter, 1972). Ovalbumin synthesis accounts for more than half of all protein synthesis in maximally induced oviducts, and as discussed by Palmiter (1975), the production of 3×10^{19} molecules of ovalbumin per day in an oviduct (640,000 molecules/min/cell) is the consequence of the presence of about 100,000 ovalbumin messenger RNA ($mRNA_{ov}$) molecules per cell [up from less than ten mRNA molecules detectable in nuclei of uninduced oviduct cells by hybridization of radioactive DNA complementary to purified $mRNA_{ov}$ (McKnight *et al.*, 1975)]. In turn, this concentration of $mRNA_{ov}$ is calculated to be the result of a steady state synthetic rate of 34 molecules/min/cell, corresponding to the transcriptior of a single ovalbumin gene per haploid genome at about 35% of the theoretically maximum rate (Palmiter, 1975).

The kinetics of the appearance and accumulation of ovalbumin and conalbumin messenger RNA's after hormone induction have been studied *in vivo* (e.g., Palmiter *et al.*, 1976) and in oviduct explant cultures (McKnight, 1978). In oviducts of chicks secondarily stimulated with estrogen (after withdrawal from primary stimulation, see above) $mRNA_{ov}$ is detectable after a lag of about 3 hr by complementary DNA hybridization or by the capacity of RNA to direct the synthesis of ovalbumin in reticulocyte lysates. Conalbumin mRNA ($mRNA_{con}$) and conalbumin itself appear with practically no lag, and this likely has to do with the facts that the basal level of conalbumin synthesis in withdrawn chicks is relatively high and that this synthesis is preferentially stimulated by low doses of estrogen. The 3-hr lag in the appearance of $mRNA_{ov}$ after hormone treatment is not yet understood. That some protein synthesis is necessary for the stimulation of $mRNA_{ov}$ synthesis is strongly suggested by McKnight's demonstration *in vitro* that if cycloheximide or puromycin was added together with hormone to explant cultures neither $mRNA_{ov}$ nor $mRNA_{con}$

was produced, but if the drug was removed at a point within the normal lag phase, the induction of the mRNA's occurred on schedule. On the other hand, while Palmiter's *in vivo* studies on the effects of protein synthesis inhibitors on the appearance of $mRNA_{ov}$ indicated partial suppression, the results could also be interpreted as being due to toxic side effects, and he concluded that intervening protein synthesis is not essential for estrogen stimulation of $mRNA_{ov}$ production. Instead, Palmiter *et al.* (1976) have proposed that hormone–receptor complex binding to high affinity (productive) sites in chromatin (Yamamoto and Alberts, 1974) is at some distance from the gene specifically stimulated, and that the lag phase (3 hr in the case of ovalbumin) represents the length of time necessary for the receptor complex (or another molecule or an alteration in chromatin structure) to linearly traverse the distance between the binding site and the gene. The acceptance of this "receptor translocation" model, or of the model in which protein synthesis of an unknown nature is a primary consequence of hormone induction upon which the activation of such specific genes as that for ovalbumin depends, awaits studies that reconcile apparently disparate data. Studies on ecdysone induction of gene expression summarized in Section V,C bear on this issue, and favor the latter model.

4. Androgen Effects on Target Tissues

Studies on the effects of testosterone on gene activity in such target tissues as testis and prostate have closely paralleled those described above for estrogens in their target cells. For example, Fang *et al.* (1969) and Mainwaring (1969) detected cytoplasmic dihydroxytestosterone (DHT) receptor in prostate tissue. The receptor had properties similar to those described above for female hormone receptors. Fang and Liao (1971) and Mainwaring and Peterken (1971) determined that the hormone–receptor complex was translocated to cell nuclei, and that "nuclear acceptor proteins" were a component of the nonhistone protein fraction of chromatin. In a fashion similar to that of O'Malley *et al.* (1972), Klyzsejko-Stefanowicz *et al.* (1976) assayed various deproteinized and reconstituted chromatins for the ability to bind DHT—receptor complex, and found that a nonhistone fraction with affinity for DNA was most responsible for binding of the complex to chromatin; further, the binding was enhanced when proteins in this fraction were phosphorylated. The influence of androgens on RNA synthesis in target tissue was examined by Parker and Mainwaring (1977), who compared the complexities of polyadenylated RNA in prostates of normal and castrated rats by RNA–DNA hybridization kinetics analysis. They found that a class of abundant nonribosomal RNA in normal prostate was greatly reduced in prostates of castrated animals and concluded that testosterone regulates the concentration (and perhaps the synthesis) of particular poly(A) + RNA's in target cells.

B. Glucocorticoid Induction of Specific Gene Expression in Mammalian Cells

Glucocorticoids are a class of adrenocorticosteroids that includes cortisone. Dexamethasone is a synthetic glucocorticoid that has been used to advantage in diverse studies of hormone-mediated regulation of genetic activity. Treatment of cultured rat hepatoma cells with 10^{-6} M dexamethasone elicits the novel appearance of fewer than ten protein species against a background of more than a thousand species that are synthesized both before and after hormone is included in the medium (Ivarie and O'Farrell, 1978). Among these few proteins that are specifically newly synthesized in response to the hormone analogue is tyrosine aminotransferase (TAT), the steroid-mediated induction of which had been studied extensively in Tomkins' laboratory (Tomkins et al., 1966). The induction of TAT synthesis upon dexamethasone administration is rapid, having a lag phase of less than 30 min; decay of the rate of synthesis upon removal of the hormone is also rapid, as TAT messenger RNA ($mRNA_{TAT}$) has a short half-life of about 1–1.5 hr (Steinberg et al., 1975). The induction of TAT synthesis is recognized as a primary response to exposure of cells to the hormone not only because of its suddenness, but principally because induction occurs in the absence of protein synthesis (Peterkofsky and Tomkins, 1967).

The participation of hormone receptors in the dexamethasone-mediated stimulation of TAT synthesis is in a manner that is completely analogous to the estrogen receptor situation described above (Section V,A). Studies of the properties of receptors in various mouse lymphoma cell lines that are resistant to toxic levels of dexamethasone have genetically dissociated the functions of cytoplasmic binding of hormone to receptor, translocation of this complex to the nucleus, and binding of the complex to DNA (e.g., Gehring and Tomkins, 1974). Thus, the allosteric interaction of steroid hormones with constitutive cytoplasmic receptors to promote their productive interaction with the genome may be a general phenomenon.

Dexamethasone will also induce mouse mammary tumor virus RNA synthesis in cultures of mouse mammary carcinomas. Mouse mammary tumor virus (MMTV) is an RNA virus that replicates using a DNA copy that is integrated into the host cell genome, and while cultures of mouse mammary carcinomas spontaneously release MMTV, treatment with dexamethasone results in an up to twentyfold increase in the cellular concentration of MMTV RNA. Such an increase in amount of MMTV RNA is due to a hormone-mediated enhancement of viral RNA synthesis (Ringold *et al.*, 1977). Once again, this involvement of hormone in the stimulation of specific RNA synthesis is via a receptor that has high affinity for glucocorticoids and an activation-dependent affinity for DNA. Stimulation of MMTV RNA synthesis is a primary response to hormone treatment, as it requires neither DNA nor protein synthesis (Ringold *et al.*, 1975).

C. Gene Activation by Ecdysone in *Drosophila*

The giant polytene chromosomes of dipteran salivary gland cells have been paramount in studies of many aspects of chromosome structure and function; a most important property of these chromosomes has been that loci of genetic activity can be visualized in cytological preparations as distinctive "puffs" that stand out against a background of bands of condensed chromatin (chromomeres) and interbands. The puffs are considered to arise from one or a few bands, and are sites of intense RNA synthesis [reviewed by Beerman, 1972, and in chapter 5 of this volume (Edstrom, 1979)]. Ecdysone, the "molting hormone" is secreted primarily by the larval ring gland into the hemolymph, and a high titer late in the last larval instar induces pupariation. Effects at other stages of *Drosophila* development have been described by Garen *et al.* (1977).

In the midge *Chironomus tentans* (Clever, 1966), and in *Drosophila melanogaster* (Ashburner, 1972), events surrounding pupariation include specific changes in the puffing patterns of salivary gland chromosomes. These changes are dependent upon exposure of the glands to ecdysone, and can be mimicked earlier in the last larval instar by experimental treatment with the hormone. When the ecdysone titer rises sharply, puffs present through much of the prepupal instar regress, and a number of new puffs appear in a sequence by which they may be categorized as *early puffs* (within minutes) or *late puffs* (after a lag of a few hours). The appearance of the late puffs is accompanied by regression of the early puffs, and Clever and Romball (1966) showed with *Chironomus* that the appearance of both early and late puffs in response to ecdysone treatment is sensitive to actinomycin D, while cycloheximide prevents the appearance of the late, but not the early, puffs. The strong implication of this observation is that the appearance of late puffs depends upon protein synthesis directed by RNA generated in the early puffs.

Ashburner (1972) determined that explanted *Drosophila* salivary glands respond to ecdysone treatment in a way that parallels *in vivo* reactions to the hormone. Subsequently, Ashburner *et al.* (1973) detailed the sequence of puffing events in salivary chromosomes *in vitro* upon exposure to hor-

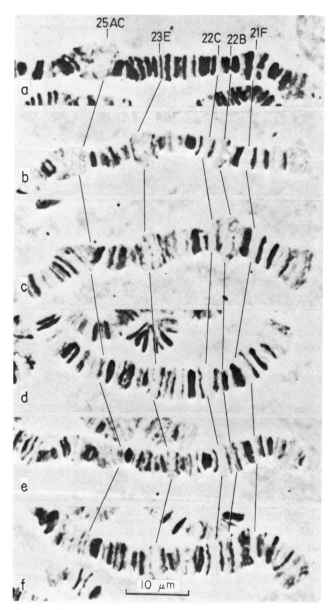

Fig. 7. Ecdysone-induced changes in the pattern of puffs in the distal region of chromosome 2L in cultured third instar larval salivary glands of *Drosophila melanogaster*. 25AC is an example of an intermolt puff which regresses upon exposure of the gland to ecdysone; 23E is induced very rapidly by the hormone and regresses after a period of activity; 22C is a puff which *in vivo* is largest at the time of puparium formation and *in vitro* is largest between 8 and

mone, and the effects of inhibition of protein synthesis on this sequence [see also Ashburner (1974), Ashburner and Richards (1976), and Richards (1976a,b)]. Of the half-dozen early puffs and the 100+ late puffs that are induced by ecdysone treatment, emphasis was given to analysis of the behavior of three prominent early puffs (23E on chromosome 2 and 74E and 75B on chromosome 3) that appear within 10 min of ecdysone administration, and five late puffs (62E, 78D and 63E on the left arm of chromosome 3 are maximally puffed at 5, 6, and 8 hr, respectively; 82F on the right arm of 3 is maximal at 10 hr; and 22C on chromosome 2 is maximal at 8 hr). As an example, puff activities in chromosome 2 are illustrated in Fig. 7.

The size attained by early puffs (a rough measure of synthetic activity) is dependent upon the dose of ecdysone over the broad range between 10^{-9} and $10^{-6} M;$ late puffs will not appear at doses below $5 \times 10^{-8} M$, and puffing is maximal above $2 \times 10^{-7} M$, so that their response is essentially all or none. The activity of early puffs requires constant exposure to hormone, and if ecdysone is washed from the glands they will regress prematurely; washing also results in the premature induction of some late puffs (see below). Induction of five of the six early puffs is unaffected by treatment of the glands with inhibitors of protein synthesis, but no late puffs will appear. Further, the normal regression of early puffs at about 4 hr is blocked by such inhibitors, so that protein synthesis is required for both this event and the subsequent appearance of late puffs.

From data such as these, Ashburner et al. (1973) have formulated the hypothesis that early and late puffs alike are regulated by both ecdysone (as an ecdysone–receptor complex) and the products of early puffs, and that these are antagonists that compete for binding with particular chromosomal loci (Fig. 8). According to the hypothesis, early puff regions bind and respond to ecdysone–receptor, with the magnitude of the response reflecting the cellular concentration of ecdysone (see Section V,A,2); receptor also binds to late puff regions, but nonproductively. The protein products of early puffs that are induced by ecdysone–receptor induce the late puffs, but only after their concentration has reached such a level that they can effectively displace ecdysone–receptor that is already bound at late puff regions. The products of early genes also compete with ecdysone–receptor at the early puff sites, and upon displacement of the latter the puff is no longer subject to hormone stimulation and it regresses.

10 hr of culture. (a) Squash of an unincubated control salivary gland. (b) Squash of a gland incubated for 1 hr in $5 \times 10^{-6} M$ ecdysterone. (c) Four hours in $1 \times 10^{-6} M$ hormone. (d) Eight hours in $2.5 \times 10^{-7} M$. (e) Ten hours in $5 \times 10^{-6} M$. (f) Twelve hours in $5 \times 10^{-4} M$. Reprinted with permission from Ashburner (1972).

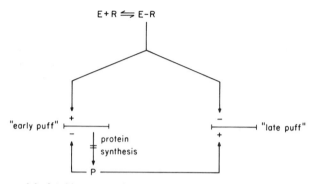

Fig. 8. The model of Ashburner *et al.* (1973) of the control of puffing at early and late puff sites in the salivary chromosomes of *Drosophila*. Ecdysone (E) binds to a receptor molecule to form the E–R complex. This complex acts positively at early puff sites and negatively at late puff sites. The induced product (P) of an early puff site acts positively at late sites and negatively at the early site (see the text). The synthesis of P is sensitive to inhibitors of protein synthesis. Redrawn from Ashburner *et al.* (1973), with permission.

VI. "GENE AMPLIFICATION" IN CULTURED MAMMALIAN CELLS

Mouse and hamster cell lines resistant to the drug methotrexate have been established, and in the murine case at least, resistance is the consequence of a heritable increase in the number of genes coding for the enzyme affected by the drug (Schimke *et al.*, 1978).

Methotrexate is an analogue of folic acid, and it is an inhibitor of dihydrofolate reductase. Cells treated with methotrexate are thus unable to generate tetrahydrofolate, which is a cofactor in the biosynthesis of purine nucleosides, thymidylate, and glycine and methionine, and therefore are incapable of growth. However, cell lines resistant to methotrexate have been selected by growing cultures in gradually increasing concentrations of the drug. In such lines of murine and hamster cells there is an up to 300-fold increase in dihydrofolate reductase activity, and this is due to an increased rate in enzyme synthesis over that in sensitive cell lines (Alt *et al.*, 1976; Hanggi and Littlefield, 1976). The increase in enzyme activity is directly correlated with increased amounts of dihydrofolate reductase messenger RNA (Kellems *et al.*, 1976; Chang and Littlefield, 1976).

There are two forms of resistance to high levels of methotrexate in mouse cell lines. Some lines are resistant only while there is a maintenance of the drug in the medium, and revert to sensitivity over a number of generations in medium free of methotrexate (*revertant*). Others are *stable* in that cells will continue to produce high levels of dihydrofolate reductase in the absence of methotrexate. In order to determine gene number in the various stable, revertant, and sensitive lines, Alt *et al.* (1978) prepared radioactive DNA (cDNA) highly enriched in sequences complementary to dihydrofolate reductase messenger RNA and used this as a probe for the gene in DNA reassociation kinetics measurements. With mouse liver DNA as the reaction driver, radioactive cDNA formed duplexes at a rate indicating that the gene for dihydrofolate reductase is a unique (single copy) sequence in the normal mouse genome. Hybridization of the probe with highly purified DNA from stable resistant and revertant lines, however, displayed increased rates of annealing which paralleled relative amounts of dihydrofolate reductase enzyme and its messenger RNA in the various lines (i.e., there were from about 10 to

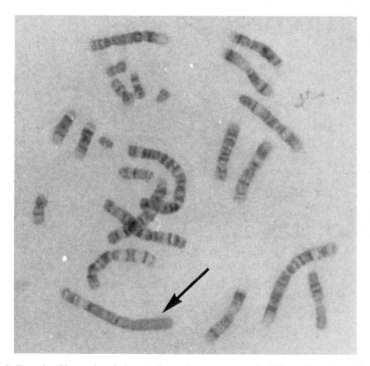

Fig. 9. Trypsin–Giemsa banded metaphase chromosomes of a Chinese hamster cell of line DC-3F/A3, showing a homogeneously staining region (HSR) in one of the chromosomes (arrow). This long chromosome is a marker for the cell line, which developed a 108,400-fold resistance to amethopterin (an antifolate drug) and a 170-fold increase in dihydrofolate reductase activity after exposure to the drug. Reprinted with permission from Biedler and Spengler (1976). Copyright 1976 by the American Association for the Advancement of Science.

about 200 copies of the dihydrofolate reductase gene per genome in the several revertant and stable resistant lines). The obvious conclusion from such studies is that cells which survive treatment with high doses of methotrexate do so because levels of dihydrofolate reductase sufficient to overcome the inhibition are the consequence of an increase in gene number.

Information on the physical state of these additional copies of the gene is not published at this writing, so that nothing definite may be said of the mechanism of their production—be it tandem duplication or extrachromosomal amplification as occurs with genes for ribosomal RNA in some organisms [see Chapter 2 of this volume (Bird, 1979)]. A cytogenetic observation provides some circumstantial support for the consideration that tandem gene multiplication may be involved. Both mouse (Biedler *et al.*, 1965) and hamster (Biedler and Spengler, 1976) cell lines resistant to methotrexate have distinctive karyotypes in that a specific chromosome in each has a long region with unusual staining properties (nonbanding with the Giemsa technique) that is not observed in cells having low dihydrofolate reductase levels (Fig. 9).

Schimke *et al.* (1978) have noted that the selection of methotrexate resistant cell lines must be a gradual process, where drug concentration is increased incrementally over many cell generations. They estimate that one in 10^6 to 10^7 cells survives each stepwise increase in methotrexate concentration, so that the multiplication process is the product of several rare events. It is likely, as Alt *et al.* (1978) suggest, that the few hundred dihydrofolate reductase genes in highly resistant cells are the accumulation, over generations, of spontaneous multiplications that were selected for their survival value. There is as yet no information on why some resistant lines are stable and others are not.

VII. CONCLUSIONS

The attempt has been made in this chapter to point out instances of cytoplasmic and environmental influences on gene expression that illustrate the diversity of both biological phenomena bearing on this subject and the methods with which they are studied.

Classic experiments in developmental biology have provided the knowledge that factors positioned in the unfertilized egg are major contributors to the determination of distinctive nuclear activities in embryonic cells incorporating different segments of ooplasm. The application of cytochemistry and biochemistry to the study of developing systems has led to the concept that these factors are preformed ribonucleoproteins, of which the operative component appears to be RNA, possibly messenger coding for proteins with regulatory functions. One of several important questions as yet unanswered asks the nature of the circumstances, or the clock, that determine when in early development such factors exert their influence. Other cytoplasmic factors having immediate general and specific regulatory effects on nuclear activities in oocytes and mature eggs of amphibia have been demonstrated in nuclear transplantation experiments, and the evidence suggests these are proteins formed during oocyte maturation. The modes of action of these cytoplasmic controllers of gene expression or chromosome replication are not understood (indeed, none have been identified at any level save the o^+ substance in axolotl), and they are expected to be complex, since genes are turned both on and off in such experimental situations.

Nuclear transplants involving unicellular eukaryotes and cell fusions involving higher metazoans have also indicated that components of cytoplasm (some perhaps transient) have the capacity to modulate nuclear activity. In heterokaryons involving avian erythrocyte nuclei, activation of these nuclei appears to accompany the influx of nuclear proteins from the host cell, but, again, the identification of proteins actually involved in the activation of chromatin and determination of the nature of the interaction(s) has yet to be made. Interestingly, the activation of erythrocyte nuclei across distant species lines in heterokaryons seems to involve expression of genes coding for housekeeping proteins, but not for histospecific proteins of either partner in the heterokaryon. A paucity of detailed information is also the case for the experimental demonstrations that cells in different stages of the cell cycle have soluble components that upon cell fusion impose properties of these stages on nuclei of cells in other stages. However, an interesting feature of this phenomenon is that the imposition is heirarchical, with cells in later stages having dominance over cells in earlier stages.

Studies of steroid hormone-mediated induction of specific genetic activity in target cells have provided perhaps the greatest amount of information on cytoplasmic (and environmental) regulation of gene expression. There are certainly still large gaps in our knowledge, but the process whereby systemic hormone is sequestered in target cells by specific receptors, and the receptors are modified by this association to a state in which they bind to and modify chromatin, is comparatively well understood. Less clear, indeed the subject of controversy, is the nature of the interaction between receptor complex and chromatin that leads to the transcription of specific genes at a high rate, but it is apparent that interactions between receptor complexes and a genome may promote target cell-specific transcription directly (independent of concomitant protein synthesis) or through induced (protein) intermediaries that act at sites in the genome removed from the loci of primary receptor–chromatin interaction. An environmental influence on gene expression that has not been covered in this chapter is the chemical induction of erythropoietic differentiation, particularly synthesis of hemoglobins, in cultured mouse cells transformed by Friend leukemia virus (e.g., Singer *et al.*, 1974). A wide variety of compounds such as dimethyl sulfoxide and other polar molecules, short chain fatty acids, and purines initiate erythroid differentiation, but they can have different results. For example, the relative rates of synthesis of two β-hemoglobin chains, and the relative concentrations of their mRNA's, differ between cells induced by polar compounds and by fatty acids (Nudel *et al.*, 1977). While butyrate treatment induces hemoglobin synthesis in Friend erythroleukemic cells and also affects levels of histone acetylation in these and other types of cells (e.g., Candido *et al.*, 1978), the relationship between these phenomena and the nature of the induction are not known. This is a potentially very important system for the clarification of the spectrum of processes that direct gene expression.

REFERENCES

Alt, F. W., Kellems, R. E., and Schimke, R. T. (1976). Synthesis and degradation of folate reductase in sensitive and methotrexate-resistant lines of S-180 cells. *J. Biol. Chem.* **251**, 3063–3074.

Alt, F. W., Kellems, R. E., Bertino, J. R., and Schimke, R. T. (1978). Selective multiplication of dihydrofolate reductase genes in methotrexate-resistant variants of cultured murine cells. *J. Biol. Chem.* **253**, 1357–1370.

Arms, K. (1968). Cytonucleoproteins in cleaving eggs of *Xenopus laevis*. *J. Embryol. Exp. Morphol.* **20**, 367–374.

Ashburner, M. (1972). Patterns of puffing activity in the salivary gland chromosomes of *Drosophila*. VI. Induction by ecdysone in salivary glands of *Drosophila melanogaster* cultured *in vitro*. *Chromosoma* **38**, 255–281.

Ashburner, M. (1974). Sequential gene activation by ecdysone in polytene chromosomes of *Drosophila melanogaster*. II. The effects of inhibition of protein synthesis. *Dev. Biol.* **39**, 141–157.

Ashburner, M., and Richards, G. (1976). Sequential gene activation by ecdysone in polytene chromosomes of *Drosophila melanogaster*. III. Consequences of ecdysone withdrawal. *Dev. Biol.* **54**, 241–255.

Ashburner, M., Chihara, C., Meltzer, P., and Richards, G. (1973). Temporal control of puffing activity in polytene chromosomes. *Cold Spring Harbor Symp. Quant. Biol.* **38**, 655–662.

Beermann, W. (1972). Chromomeres and genes. *In* "Developmental Studies on Giant Chromosomes: Results and Problems in Cell Differentiations" (W. Beermann, ed.), Vol. 4, pp. 1–33. Springer-Verlag, Berlin and New York.

Biedler, J. L., and Spengler, B. A. (1976). Metaphase chromosome anomaly: Association with drug resistance and cell-specific products. *Science* **191**, 185–187.

Biedler, J. L., Albrecht, A. M., and Hutchinson, D. J. (1965). Cytogenetics of mouse leukemia L1210. I. Association of a specific chromosome with dihydrofolate reductase activity in amethopterin-treated sublines. *Cancer Res.* **25,** 246–257.

Billing, R. J., Barbiroli, B., and Smellie, R. M. S. (1969). The mode of action of oestradiol. I. The transport of RNA precursors into the uterus. *Biochim. Biophys. Acta* **190,** 52–59.

Bird, A. P. (1979). Gene amplification. *In* "Cell Biology: A Comprehensive Treatise" (L. Goldstein and D. M. Prescott, eds.), Vol. 3, Academic Press, New York.

Bolund, L., Ringertz, N. R., and Harris, H. (1969). Changes in the cytochemical properties of erythrocyte nuclei reactivated by cell fusion. *J. Cell Sci.* **4,** 71–87.

Borthwick, N. M., and Smellie, R. M. S. (1975). The effects of oestradiol-17β on the ribonucleic acid polymerases of immature rabbit uterus. *Biochem. J.* **147,** 91–101.

Briggs, R., and Justus, J. T. (1968). Partial characterization of the component from normal eggs which corrects the maternal effect of gene *o* in the Mexican axolotl (*Ambystoma mexicanum*). *J. Exp. Zool.* **147,** 105–116.

Briggs, R., and King, T. J. (1952). Transplantation of living nuclei from blastula cells into enucleated frogs' eggs. *Proc. Natl. Acad. Sci. U.S.A.* **38,** 455–463.

Brothers, A. J. (1976). Stable nuclear activation dependent on a protein synthesized during oogenesis. *Nature (London)* **260,** 112–115.

Brown, D. D., and Littna, E. (1964). RNA synthesis during the development of *Xenopus laevis,* the South African clawed toad. *J. Mol. Biol.* **8,** 669–687.

Brown, J. E., and Weiss, M. C. (1975). Activation of production of mouse liver enzymes in rat hepatoma–mouse lymphoid cell hybrids. *Cell* **6,** 481–494.

Bull, A. L. (1966). *Bicaudal,* a genetic factor which affects the polarity of the embryo in *Drosophila melanogaster. J. Exp. Zool.* **161,** 221–242.

Candido, E. P. M., Reeves, R., and Davia, J. R. (1978). Sodium butyrate inhibits histone deacetylation in cultured cells. *Cell* **14,** 105–113.

Chamness, G. C., Jennings, A. W., and McGuire, W. L. (1973). Oestrogen receptor binding is not restricted to target nuclei. *Nature (London)* **241,** 458–460.

Chan, L., Means, A. R., and O'Malley, B. W. (1973). Rates of induction of specific translatable messenger RNAs for ovalbumin and avidin by steroid hormones. *Proc. Natl. Acad. Sci. U.S.A.* **70,** 1870–1874.

Chang, S. E., and Littlefield, J. W. (1976). Elevated dihydrofolate reductase mRNA levels in methotrexate-resistant BHK cells. *Cell* **7,** 391–396.

Clement, A. C. (1952). Experimental studies on germinal localization in *Ilyanassa.* I. The role of the polar lobe in determination of the cleavage pattern and its influence in later development. *J. Exp. Zool.* **121,** 593–625.

Clement, A. C. (1962). Development of *Ilyanassa* following removal of the D macromere at successive cleavage stages. *J. Exp. Zool.* **149,** 193–215.

Clement, A. C., and Tyler, A. (1967). Protein-synthesizing activity of the anucleate polar lobe of the mud snail *Ilyanassa obsoleta. Science* **158,** 1457–1458.

Clever, U. (1966). Gene activity patterns and cellular differentiation. *Am. Zool.* **6,** 33–41.

Clever, U., and Romball, C. G. (1966). RNA and protein synthesis in the cellular response to a hormone, ecdysone. *Proc. Natl. Acad. Sci. U.S.A.* **56,** 1470–1476.

Conklin, E. G. (1905). The organization and cell lineage of the ascidian egg. *J. Acad. Nat. Sci. Philadelphia* **13,** 1–118.

Counce, S. J. (1963). Developmental morphology of polar granules in *Drosophila* including observations on pole cell behavior and distribution during embryogenesis. *J. Morphol.* **112,** 129–145.

Cuatrecasas, P. (1974). Membrane receptors. *Annu. Rev. Biochem.* **43,** 169–214.

Davidson, E. H. (1976). "Gene Activity in Early Development." Academic Press, New York.

Davidson, E. H., Haslett, G. W., Finney, R. J., Allfrey, V. G. and Mirsky, A. E. (1965). Evidence for prelocalization of cytoplasmic factors affecting gene activation in early embryogenesis. *Proc. Natl. Acad. Sci. U.S.A.* **54**, 696–704.

Davidson, R. L. (1974). Gene expression in somatic cell hybrids. *Annu. Rev. Genet.* **8**, 195–218.

Deák, I., Sidebottom, E., and Harris, H. (1972). Further experiments on the role of the nucleolus in the expression of structural genes. *J. Cell Sci.* **11**, 379–391.

DeRobertis, E. M., and Gurdon, J. B. (1977). Gene activation in somatic nuclei after injection into amphibian oocytes. *Proc. Natl. Acad. Sci. U.S.A.* **74**, 2470–2474.

DiBerardino, M. A., and Hoffner, N. J. (1975). Nucleo-cytoplasmic exchange of nonhistone proteins in amphibian embryos. *Exp. Cell Res.* **94**, 235–252.

Ecker, R. E., and Smith, L. D. (1971). The nature and fate of *Rana pipiens* proteins synthesized during maturation and early cleavage. *Dev. Biol.* **24**, 559–576.

Eddy, E. M. (1975). Germ plasm and the differentiation of the germ cell line. *Int. Rev. Cytol.* **43**, 229–280.

Edstrom, J.-E. (1979). Structural manifestations of gene activity. *In* "Cell Biology: A Comprehensive Treatise" (L. Goldstein and D. M. Prescott, eds.), Vol. 3, Academic Press, New York.

Etkin, L. D. (1976). Regulation of lactate dehydrogenase (LDH) and alcohol dehydrogenase (ADH) synthesis in liver nuclei, following their transfer into oocytes. *Dev. Biol.* **52**, 201–209.

Fang, S., and Liao, S. (1971). Androgen receptors. *J. Biol. Chem.* **246**, 16–24.

Fang, S., Anderson, K. M., and Liao, S. (1969). Receptor proteins for androgens. *J. Biol. Chem.* **244**, 6584–6595

Garen, A., Kauvar, L., and Lepesant, J.-A. (1977). Roles of ecdysone in *Drosophila* development. *Proc. Natl. Acad. Sci. U.S.A.* **74**, 5099–5102.

Gehring, U., and Tomkins, G. M. (1974). A new mechanism for steroid unresponsiveness: Loss of nuclear binding activity of a steriod hormone receptor. *Cell* **3**, 301–306.

Gill, K. S. (1964). Epigenetics of the promorphology of the egg in *Drosophila melanogaster*. *J. Exp. Zool.* **155**, 91–104.

Glasser, S. R., Chytil, F., and Spelsberg, T. C. (1972). Early effects of oestradiol-17β on the chromatin and activity of the deoxyribonucleic acid-dependent ribonucleic acid polymerases (I and II) of the rat uterus. *Biochem. J.* **130**, 947–957.

Goldstein, L. (1974). Movement of molecules between nucleus and cytoplasm. *In* "The Cell Nucleus" (H. Busch, ed.), Vol. I, pp. 387–438. Academic Press, New York.

Goldstein, L., and Ko, C. (1974). Electrophoretic characterization of shuttling and non-shuttling small nuclear RNAs. *Cell* **2**, 259–269.

Gorski, J., and Gannon, F. (1976). Current models of steroid hormone action: A critique. *Annu. Rev. Physiol.* **38**, 425–450.

Gorski, J., Toft, D., Shyamala, G., Smith D., and Notides, A. (1968). Hormone receptors: Studies on the interaction of estrogen with the uterus. *Recent Progr. Hormone Res.* **24**, 45–72.

Guerrier, P., van den Biggelaar, J. A. M., van Dongen, C. A. M., and Verdonk, N. H. (1978). Significance of the polar lobe for the determination of dorsoventral polarity in *Dentalium vulgare* (da Costa). *Dev. Biol.* **63**, 233–242.

Gurdon, J. B. (1962). Adult frogs derived from the nuclei of single somatic cells. *Dev. Biol.* **4**, 256–273.

Gurdon, J. B. (1967). On the origin and persistence of a cytoplasmic state inducing nuclear DNA synthesis in frogs' eggs. *Proc. Natl. Acad. Sci. U.S.A.* **58**, 545–552.

Gurdon, J. B. (1968). Changes in somatic cell nuclei inserted into growing and maturing amphibian oocytes. *J. Embryol. Exp. Morphol.* **20**, 401–414.

Gurdon, J. B., and Brown, D. D. (1965). Cytoplasmic regulation of RNA synthesis and nucleolus formation in developing embryos of *Xenopus laevis*. *J. Mol. Biol.* **12**, 27–35.

Hanggi, U. J., and Littlefield, J. W. (1976). Altered regulation of the rate of synthesis of dihydrofolate reductase in methotrexate-resistant hamster cells. *J. Biol. Chem.* **251**, 3075–3080.

Harris, G. S. (1971). Nature of oestrogen specific binding sites in the nuclei of mouse uteri. *Nature (London), New Biol.* **231**, 246–248.

Harris, H. (1965). Behavior of differentiated nuclei in heterokaryons of animal cells from different species. *Nature (London)* **206**, 583–588.

Harris, H. (1967). The reactivation of the red cell nucleus. *J. Cell Sci.* **2**, 23–32.

Harris, H., and Cook, P. R. (1969). Synthesis of an enzyme determined by an erythrocyte nucleus in a hybrid cell. *J. Cell Sci.* **5**, 121–134.

Harris, H., Sidebottom, E., Grace, D. M., and Bramwell, M. E. (1969). The expression of genetic information: A study with hybrid cells. *J. Cell Sci.* **4**, 499–526.

Humphrey, R. R. (1966). A recessive factor (*o*, for ova deficient) determining a complex of abnormalities in the Mexican axolotl (*Ambystoma mexicanum*). *Dev. Biol.* **13**, 57–76.

Illmensee, K., and Mahowald, A. P. (1974). Transplantation of posterior polar plasm into *Drosophila*. Induction of germ cells at the anterior pole of the egg. *Proc. Natl. Acad. Sci. U.S.A.* **71**, 1016–1020.

Illmensee, K., Mahowald, A. P., and Loomis, M. R. (1976). The ontogeny of germ plasm during oogenesis in *Drosophila*. *Dev. Biol.* **49**, 40–65.

Ivarie, R. D., and O'Farrell, P. H. (1978). The glucocorticoid domain: Steroid-mediated changes in the rate of synthesis of rat hepatoma proteins. *Cell* **13**, 41–55.

Jensen, E. V., and Jacobson, H. I. (1960). *In* "Biological Activities of Steroids in Relation to Cancer" (G. Pincus and E. P. Vollmer, eds.), pp. 161–178. Academic Press, New York.

Jensen, E. V., and DeSombre, E. R. (1973). Estrogen–receptor interaction. *Science* **182**, 126–134.

Jensen, E. V., DeSombre, E. R., Hurst, D. J., Kawashima, T., and Jungblut, P. W. (1967). Estrogen–receptor interactions in target tissues. *Arch. Anat. Microsc.* **56**, Suppl., 547–569.

Jensen, E. V., Suzuki, T., Kawashima, T., Stumpf, W. E., Jungblut, P. W., and DeSombre, E. R. (1968). A two-step mechanism for the interaction of estradiol with rat uterus. *Proc. Natl. Acad. Sci. U.S.A.* **59**, 632–638.

Johnson, R. T., and Harris, H. (1969a). DNA synthesis and mitosis in fused cells. I. HeLa homokaryons. *J. Cell Sci.* **5**, 603–624.

Johnson, R. T., and Harris, H. (1969b). DNA synthesis and mitosis in fused cells. II. HeLa–chick erythrocyte heterokaryons. *J. Cell Sci.* **5**, 625–643.

Kahn, C. R. (1976). Membrane receptors for hormones and neurotransmitters. *J. Cell Biol.* **70**, 261–286.

Kalthoff, K. (1971a). Photoreversion of UV induction of the malformation "double abdomen" in the egg of *Smittia* spec. (Diptera, Chironomidae). *Dev. Biol.* **25**, 119–132.

Kalthoff, K. (1971b). Position of targets and period of competence for UV-induction of the malformation "double abdomen" in the egg of *Smittia* spec. (Diptera, Chironmidae). *Wilhelm Roux Arch. Entwicklungsmech. Org.* **168**, 63–84.

Kalthoff, K., Hanel, P. and Zissler, D. (1977). A morphogenetic determinant in the anterior pole of an insect egg (*Smittia* spec., Chironomidae, Diptera). *Dev. Biol.* **55**, 285–305.

Kandler-Singer, I., and Kalthoff, K. (1976). RNase sensitivity of an anterior morphogentic determinant in an insect egg (*Smittia* spec., Chironomidae, Diptera). *Proc. Natl. Acad. Sci. U.S.A.* **73**, 3739–3743.

Kellems, R. E., Alt, F. W., and Schimke, R. T. (1976). Regulation of folate reductase synthesis in sensitive and methotrexate-resistant sarcoma 180 cells. *J. Biol. Chem.* **251**, 6987–6993.

Klyzesejko-Stefanowicz, L., Chiu, J. F., Tsai, Y.-H., and Hnilica, L. S. (1976). Acceptor proteins in rat androgenic tissue chromatin. *Proc. Natl. Acad. Sci. U.S.A.* **73**, 1954–1958.

Legname, C., and Goldstein, L. (1972). Proteins in nucleocytoplasmic interactions. VI. Is there an artifact responsible for the observed shuttling of proteins between cytoplasm and nucleus in *Amoeba proteus*? *Exp. Cell Res.* **75**, 111–121.

Libby, P. R. (1972). Histone acetylation and hormone action. Early effects of oestradiol-17β on histone acetylation in rat uterus. *Biochem. J.* **130**, 663–669.

Linn, S.-Y., and Riggs, A. D. (1975). The general affinity of *lac* repressor for *E. coli* DNA: Implications for gene regulation in prokaryotes and eukaryotes. *Cell* **4**, 107–111.

McKnight, G. S. (1978). The induction of ovalbumin and conalbumin mRNA by estrogen and progesterone in chick oviduct explant cultures. *Cell* **14**, 403–413.

McKnight, G. S., Pennequin, P., and Schimke, R. T. (1975). Induction of ovalbumin mRNA sequences by estrogen and progesterone in chick oviduct as measured by hybridization to complementary DNA. *J. Biol. Chem.* **250**, 8105–8110.

Mahowald, A. P. (1962). Fine structure of pole cells and polar granules in *Drosophila melanogaster*. *J. Exp. Zool.* **151**, 201–215.

Mahowald, A. P. (1968). Polar granules of *Drosophila*. II. Ultrastructural changes during early embryogenesis. *J. Exp. Zool.* **167**, 237–262.

Mahowald, A. P. (1971a). Polar granules of *Drosophila*. III. The continuity of polar granules during the life cycle of *Drosophila*. *J. Exp. Zool.* **176**, 329–344.

Mahowald, A. P. (1971b). Polar granules of *Drosophila*. IV. Cytochemical studies showing loss of RNA from polar granules during early stages of embryogenesis. *J. Exp. Zool.* **176**, 345–352.

Mahowald, A. P., Illmensee, K., and Turner, F. R. (1976). Interspecific transplantation of polar plasm between *Drosophila* embryos. *J. Cell Biol.* **70**, 358–373.

Mainwaring, W. I. P. (1969). A soluble androgen receptor in the cytoplasm of rat prostate. *J. Endocrinol.* **45**, 531–541.

Mainwaring, W. I. P., and Peterken, B. M. (1971). A reconstituted cell-free system for the specific transfer of steroid receptor complexes into nuclear chromatin isolated from rat ventral prostate gland. *Biochem. J.* **125**, 285–295.

Means, A. R., and Hamilton, T. H. (1966). Early estrogen action: Concomitant stimulations within two minutes of nuclear RNA synthesis and uptake of RNA precursors by the uterus. *Proc. Natl. Acad. Sci. U.S.A.* **56**, 1594–1598.

Newrock, K. M., and Raff, R. A. (1975). Polar lobe specific regulation of translation in embryos of *Ilyanassa obsoleta*. *Dev. Biol.* **42**, 242–261.

Nudel, U., Salmon, J. E., Terada, M., Bank, A., Rifkind, R. A., and Marks, P. A. (1977). Differential effects of chemical inducers on expression of β-globin genes in murine erythroleukemia cells. *Proc. Natl. Acad. Sci. U.S.A.* **74**, 1100–1104.

O'Malley, B. W., and Means, A. R. (1974). Female steroid hormones and target cell nuclei. *Science* **183**, 610–620.

O'Malley, B. W., Toft, D. O., and Sherman, M. R. (1971). Progesterone-binding components of chick oviduct. II. Nuclear components. *J. Biol. Chem.* **246**, 1117–1122.

O'Malley, B. W., Spelsberg, T. C., Schrader, W. T., Chytil, F., and Steggles, A. W. (1972).

Mechanisms of interaction of a hormone–receptor complex with the genome of a eukaryotic target cell. *Nature (London)* **235**, 141–144.

Palmiter, R. D. (1972). Regulation of protein synthesis in chick oviduct. I. Independent regulation of ovalbumin, conalbumin, ovomucoid and lysozyme induction. *J. Biol. Chem.* **247**, 6450–6459.

Palmiter, R. D. (1975). Quantitation of parameters that determine the rate of ovalbumin synthesis. *Cell* **4**, 189–197.

Palmiter, R. D., Moore, P. B., Mulvihill, E. R., and Emtage, S. (1976). A significant lag in the induction of ovalbumin messenger RNA by steroid hormones: A receptor translocation hypothesis. *Cell* **8**, 557–572.

Parker, M. G., and Mainwaring, W. I. P. (1977). Effects of androgens on the complexity of poly(A) RNA from rat prostate. *Cell* **12**, 401–407.

Peterkofsky, B., and Tomkins, G. M. (1967). Evidence for the steroid induced accumulation of tyrosine aminotransferase messenger RNA in the absence of protein synthesis. *Proc. Natl. Acad. Sci. U.S.A.* **60**, 222–228.

Peterson, J. A., and Weiss, M. C. (1972). Expression of differentiated functions in hepatoma cell hybrids: Induction of mouse albumin production in rat hepatoma–mouse fibroblast hybrids. *Proc. Natl. Acad. Sci. U.S.A.* **69**, 571–575.

Poulson, D. F., and Waterhouse, D. F. (1960). Experimental studies on pole cells and midgut differentiation in Diptera. *Aust. J. Biol. Sci.* **13**, 541–567.

Prescott, D. M. (1956). Relation between cell growth and cell division. II. The effect of cell size on cell growth rate and generation time in *Amoeba proteus*. *Exp. Cell Res.* **11**, 86–98.

Prescott, D. M., and Goldstein, L. (1967). Nuclear–cytoplasmic interaction in DNA synthesis. *Science* **155**, 469–470.

Rao, P. N., and Johnson, R. T. (1970). Mammalian cell fusion. Studies on the regulation of DNA synthesis and mitosis. *Nature (London)* **225**, 159–164.

Richards, G. (1976a). Sequential gene activation by ecdysone in polytene chromosomes of *Drosophila melanogaster*. IV. The mid-prepupal period. *Dev. Biol.* **54**, 256–263.

Richards, G. (1976b). Sequential gene activation by ecdysone in polytene chromosomes of *Drosophila melanogaster*. V. The late prepupal puffs. *Dev. Biol.* **54**, 264–275.

Ringertz, N. R., and Ege, T. (1978). Use of mutant, hybrid, and reconstructed cells in somatic cell genetics. *In* "Cell Biology: A Comprehensive Treatise" (L. Goldstein and D. M. Prescott, eds.), Vol. 1, pp. 191–234. Academic Press, New York.

Ringertz, N. R., and Savage, R. E. (1976). "Cell Hybrids." Academic Press, New York.

Ringertz, N. R., Carlsson, S. A., Ege, T., and Bolund, L. (1971). Detection of human and chick nuclear antigens in nuclei of chick erythrocytes during reactivation in heterokaryons with HeLa cells. *Proc. Natl. Acad. Sci. U.S.A.* **68**, 3228–3232.

Ringertz, N. R., Carlsson, S. A., and Savage, R. E. (1972). Nucleocytoplasmic interactions and the control of nuclear activity. *Adv. Biosci.* **8**, 219–234.

Ringold, G. M., Yamamoto, K. R., Tomkins, G. M., Bishop, J. M., and Varmus, H. E. (1975). Dexamethasone-mediated induction of mouse mammary tumor virus RNA: A system for studying glucocorticoid action. *Cell* **6**, 299–305.

Ringold, G. M., Yamamoto, K. R., Bishop, J. M., and Varmus, H. E. (1977). Glucocorticoid-stimulated accumulation of mouse mammary tumor virus RNA: Increased rate of synthesis of viral RNA. *Proc. Natl. Acad. Sci. U.S.A.* **74**, 2879–2883.

Sander, K. (1976). Specification of the basic body pattern in insect embryogenesis. *Adv. Insect Physiol.* **12**, 125–238.

Schimke, R. T., Alt, F. W., Kellems, R. E., Kaufman, R., and Bertino, J. R. (1978). Amplification of folate reductase genes in methotrexate-resistant cultured mouse cells. *Cold Spring Harbor Symp. Quant. Biol.* **42**, 649–657.

Schubiger, G., Moseley, R. C., and Wood, W. J. (1977). Interaction of different egg parts in determination of various body regions in *Drosophila melanogaster*. *Proc. Natl. Acad. Sci. U.S.A.* **74**, 2050–2053.

Schwartz, R. J., Tsai, M.-J., Tsai, S. Y., and O'Malley, B. W. (1975). Effect of estrogen on gene expression in the chick oviduct. V. Changes in the number of RNA polymerase binding and initiation sites in chromatin. *J. Biol. Chem.* **250**, 5175–5182.

Sidebottom, E. (1974). Heterokaryons and their uses in studies of nuclear function. *In* "The Cell Nucleus" (H. Busch, ed.), Vol. I, pp. 439–469. Academic Press, New York.

Singer, D., Cooper, M., Maniatis, G. M., Marks, P. A., and Rifkind, R. A. (1974). Erythropoietic differentiation in colonies of cells transformed by Friend virus. *Proc. Natl. Acad. Sci. U.S.A.* **71**, 2668–2670.

Sonnenblick, D. P. (1950). The early embryo of *Drosophila melanogaster*. *In* "Biology of Drosophila" (M. Demerec, ed.), pp. 62–167. Wiley, New York.

Steinberg, R. A., Levinson, B. B., and Tomkins, G. M. (1975). Kinetics of steroid induction and deinduction of tyrosine aminotransferase synthesis in cultured hepatoma cells. *Proc. Natl. Acad. Sci. U.S.A.* **72**, 2007–2011.

Stumpf, W. E., and Roth, L. J. (1966). High resolution autoradiography with dry mounted, freeze dried frozen sections. Comparative study of six methods using two diffusible compounds ^{3}H-estradiol and ^{3}H-mesobilirubinogen. *J. Histochem. Cytochem.* **14**, 274–287.

Tager, H. S., and Steiner, D. F. (1974). Peptide hormones. *Annu. Rev. Biochem.* **43**, 509–538.

Tata, J. R. (1976). The expression of the vitellogenin gene. *Cell* **9**, 1–14.

Toft, D., and Gorski, J. (1966). A receptor molecule for estrogens: Isolation from the rat uterus and preliminary characterization. *Proc. Natl. Acad. Sci. U.S.A.* **55**, 1574–1581.

Tomkins, G. M., Thompson, E. B., Hayashi, S., Gelehrter, Y., Granner, D., and Peterkofsky, B. (1966). Tyrosine transaminase induction in mammalian cells in tissue culture. *Cold Spring Harbor Symp. Quant. Biol.* **31**, 349–360.

Tsai, S. Y., Tsai, M. J., Schwartz, R., Kalimi, M., Clark, J. H., and O'Malley, B. W. (1975). Effects of estrogen on gene expression in chick oviduct: Nuclear receptor levels and initiation of transcription. *Proc. Natl. Acad. Sci. U.S.A.* **72**, 4228–4232.

Whittaker, J. R. (1973). Segregation during ascidian embryogenesis of egg cytoplasmic information for tissue-specific enzyme development. *Proc. Natl. Acad. Sci. U.S.A.* **70**, 2096–2100.

Williams, D., and Gorski, J. (1972). Kinetic and equilibrium analysis of estradiol in uterus: A model of binding-site distribution in uterine cells. *Proc. Natl. Acad. Sci. U.S.A.* **69**, 3464–3468.

Wilson, G. N., Steggles, A. W., and Nienhuis, A. W. (1975). Strand-selective transcription of globin genes in rabbit erythroid cells and chromatin. *Proc. Natl. Acad. Sci. U.S.A.* **72**, 4835–4839.

Yamamoto, K. R., and Alberts, B. M. (1972). *In vitro* conversion of estradiol-receptor protein to its nuclear form: Dependence on hormone and DNA. *Proc. Natl. Acad. Sci. U.S.A.* **69**, 2105–2109.

Yamamoto, K. R., and Alberts, B. M. (1974). On the specificity of the binding of the estradiol receptor protein to deoxyribonucleic acid. *J. Biol. Chem.* **249**, 7076–7086.

Yamamoto, K. R., and Alberts, B. M. (1975). The interaction of estradiol–receptor protein with the genome: An argument for the existence of undetected specific sites. *Cell* **4**, 301–310.

Yamamoto, K. R. and Alberts, B. M. (1976). Steroid receptors: Elements for modulation of eukaryotic transcription. *Annu. Rev. Biochem.* **45**, 721–746.

Zentgraf, H., Scheer, U., and Franke, W. W. (1975). Characterization and localization of the RNA synthesized in mature avian erythrocytes. *Exp. Cell Res.* **96**, 81–95.

8

Molecular Aspects of the Regulation of Eukaryotic Transcription: Nucleosomal Proteins and Their Postsynthetic Modifications in the Control of DNA Conformation and Template Function

Vincent G. Allfrey

347

Copyright © 1980 by Academic Press, Inc.
All rights of reproduction in any form reserved.
ISBN 0-12-289503-7

I. INTRODUCTION

Current views of transcriptional control in eukaryotic cells generally recognize dual aspects of protein involvement—participation in the regulation of chromatin structure, largely attributed to histones, and selection of particular DNA sequences for template function in the RNA polymerase reaction, attributed to nonhistone chromosomal proteins. In fact, such distinctions are no longer clear, as histones and nonhistones both play a role in chromatin structure and both influence its transcriptional activity in interdependent ways. This chapter focuses on recent developments in this rapidly expanding area. It will emphasize that transcriptional control is both positive and negative, and that there are close correlations between structure and function in transcribing and nontranscribing regions of the chromatin. In addition, it will introduce wherever possible, a third major variable: modulation of DNA-binding affinity and chromosomal protein interactions by postsynthetic modifications of polypeptide chains. It will be shown that enzymatic reactions, such as acetylation and phosphorylation, which alter the charge and structure of histones and a variety of other nuclear proteins, represent a dynamic interface between chromatin and its environment and permit nearly instantaneous adaptation of the genome to changing hormonal and developmental stimuli. In addition, this interface offers a highly sensitive target to carcinogens which can act both directly and indirectly to modify the structure of key chromosomal proteins, as well as the DNA of the affected cell.

Progress in the field of nuclear proteins, chromatin structure, and gene regulation has accelerated rapidly as new experimental procedures have been brought to bear on biological systems selected for their remarkable advantages in visualization, activation, or fractionation of the genome. This brief overview draws upon the recent literature to illustrate some major developments and to consider their implications for models of transcriptional control. The primary focus will be on the nucleosome, the fundamental unit of chromatin structure, and its modification by enzymes responsive to physiological control systems.

A. Transcriptional Control in Eukaryotes

1. Differential Gene Expression

It is now well established that the DNA sequence complexity of the eukaryotic genome is never fully expressed in the messenger RNA population of a single cell type. The mouse genome, for example, has a DNA complexity sufficient to encode approximately 10^6 messenger RNA's (mRNA's) of average size, yet hybridization studies using mRNA's and copy DNA's (cDNA's) indicate that mouse liver and kidney have a sequence complexity corresponding to about 1 to 2×10^4 different mRNA's (Hastie and Bishop, 1976). In mouse brain, the total poly(A)+mRNA is complementary to only 3.8% of the nonrepeated DNA sequences (Bantle and Hahn, 1976). The corresponding figure for rat liver poly(A)+mRNA is 2.8% of the total single-copy DNA, enough to encode about 30,000 distinct RNA sequences with an average length of about 1800 nucleotides (Savage *et al.,* 1978).

It is clear that such figures, even when doubled to account for asymmetric transcription, represent minimal estimates of DNA template function because the sequence complexity of nuclear RNA is far greater than that of the cytoplasmic messenger fraction. In mouse brain, total nuclear RNA is about 5.6 times more complex than the poly(A)+mRNA, and the sequence diversity of poly(A)-containing heterogeneous nuclear RNA (hnRNA) is three to four times greater than that of the poly(A) mRNA in the cytoplasm (Bantle and Hahn, 1976). Similarly, the polyadenylated nuclear RNA of *Drosophila* cells has a complexity five to ten times greater than that of cytoplasmic mRNA (Levy-W. *et al.,* 1976). Based on nuclear RNA complexity, recent estimates of the percent transcription of the unique DNA sequences of the rat genome in different tissues range from 9.2% in the thymus to 31.2% in the brain (Chikaraishi *et al.,* 1978).

A large proportion of nuclear RNA sequences homologous to single-copy DNA are held in common in different tissues (Chikaraishi *et al.,* 1978), and a similar overlap is seen in cytoplasmic mRNA's. For example, comparisons of the liver and oviduct of the hen suggest that 85% of the different mRNA sequences are present in both tissues (Axel *et al.,* 1976). The total poly(A)+RNA sequence overlap between functionally distinct regions of the brain may exceed 80% (Kaplan *et al.,* 1978). The overlap deserves particular attention in considering mechanisms of transcriptional control because it suggests that many common regulatory elements are likely to be present in different tissues of the organism.

The programming of differentiation during embryonic development involves sequential changes in mRNA diversity and proportions. The polysomal mRNA at the gastrula stage of sea urchin embryogenesis has been estimated to contain some 14,000 transcripts, representing about 2.7% of the total DNA sequence complexity of the organism (Galau *et al.,* 1974). Comparisons of the gastrula mRNA's with those of other embryonic stages and adult tissues suggest that each stage of differentiation involves the activation of several thousand structural genes, and that the programming of early embronic development may require the activation of 20,000–30,000 diverse DNA sequences (Galau et al., 1976).

The characteristics of individual cell types reflect the accumulation and utilization of abundant RNA species which code for the proteins necessary to express those characteristics, and definitive changes can be seen in the mRNA population during the differentiation of individual organs. The sequence complexity and frequency distributions of pancreatic RNA's are profoundly altered during embryonic development. Between 14 and 20 days of gestation in the rat, the RNA's characteristic of the adult pancreatic acinar cell are increased several hundredfold in the embryonic pancreas, in parallel with the increased rate of synthesis of pancreas-specific proteins (Harding et al., 1977). The amplification of mRNA's coding for secretory enzymes continues until amylase mRNA, for example, comprises about 30% of the total poly(A)+mRNA of the adult pancreatic acinar cell. The tissue specificity of the amplification is seen in the failure of amylase cDNA to hybridize with the RNA's of brain or kidney (Harding and Rutter, 1978). It is highly unlikely that cytoplasmic mRNA's coding for other typical pancreatic secretory proteins, such as lipase A and carboxypeptidase A, are produced in other tissues in appreciable amounts, since the corresponding enzyme activities are not detectable (Bradshaw and Rutter, 1972). Thus, specialization of the pancreatic acinar cell is largely attributable to selective activation of a limited gene set coding for the secretory enzymes, while those genes remain essentially silent in other tissues, such as liver, kidney, or brain. Many such hybridization studies, using mRNA's and cDNA's, confirm the view that divergent pathways of differentiation require both positive and negative transcriptional control mechanisms, and further evidence for this viewpoint will now be presented.

2. Negative and Positive Control in Eukaryotic Transcription

In bacteria and viruses, negative transcriptional control is well established as a major regulatory mechanism, the most notable example being suppression of RNA synthesis from the *lac* operon of *E. coli* by the *lac* repressor protein. Binding of the *lac* repressor through its basic aminoterminal region to the operator DNA has been analyzed in detail (Gilbert and Müller-Hill, 1966, 1967; Lin and Riggs, 1975; Barkley *et al.*, 1975; Kao-Huang *et al.*, 1977; Ogata and Gilbert, 1978). The system provides a model of great significance for transcriptional control in higher organisms. It establishes the fact that a DNA-binding protein may have very high specificity and that its effects may be reversible (Barkley *et al.*, 1975). Moreover, the *lac* operon is subject to positive control by the cyclic

AMP-binding protein (Eron and Block, 1971), and the repressor prevents the binding of the latter and RNA polymerase to *lac* promoter sites.

There are also many clear examples of negative transcriptional control in eukaryotic cells. A suppression of RNA synthesis is seen in the loss of maternal (mature oocyte) RNA sequences during development of the sea urchin embryo from the 16-cell to the gastrula stage (Hough-Evans *et al.*, 1977); in the cessation of synthesis of mRNA's coding for early embryonic histone types (Grunstein, 1978); in the 60-fold reduction in the number of cytoplasmic poly(A)+mRNA species during erythroid cell maturation (Hsu *et al.*, 1978); in the species-specific suppression of histone H1 and H2B production in mouse–human hybrid cells (Ajiro *et al.*, 1978); and in the selective inhibition of the mouse nucleolus organizer in hybrids of human fibrosarcoma cells and mouse peritoneal macrophages (Miller *et al.*, 1976).

Positive transcriptional control in eukaryotes—defined here as the inductive process leading to the appearance of "new" RNA species or to a selective amplification of particular transcripts—is a common event in embryonic development. Messenger RNA's coding for histone variants, such as the late β, γ, and δ subtypes of histone H2A are not detectable in the early sea urchin embryo. The polysomal mRNA of the early blastula directs the synthesis of the predominant histone variant H2Aα, but mRNA from later embryonic stages (mesenchyme blastula and gastrula) directs the synthesis of H2A$_{\beta\gamma\delta}$, while H2A$_\alpha$ synthesis is diminished (Newrock *et al.*, 1978). [The size of histone gene transcripts also varies in different embryonic stages (Kunkel *et al.*, 1978).] The histone genes of the sea urchin exist in multiple copies, each organized into a repeating structure and containing coding sequences for all five major histone classes (Kedes, 1976; Sures *et al.*, 1978; Overton and Weinberg, 1978). The changing frequencies of histone variant mRNA's during embryonic development suggest that transcription can be coordinated at multiple loci, and that control over these reiterated genes is both positive and negative.

The genes coding for protamine, which exist in single copies (Sakai *et al.*, 1978), are also subject to positive control. During spermiogenesis in the trout, protamine messengers are synthesized in the nuclei of the primary spermatocytes and progressively accumulate in the cytoplasm of the maturing cells. Protamine mRNA's are not present in the trout liver in significant amounts, and, more significantly, they are not detectable in testes cells at developmental stages preceding the primary spermatocyte (Iatrou and Dixon, 1978).

Similar evidence for positive transcriptional control during cell differentiation is provided by studies of hemoglobin messenger RNA synthesis during the maturation of erythroid cells. The globin genes exist in single copies; they are not amplified in mouse erythroid cells, and there is no elimination of globin genes in nonerythroid tissues (Packman *et al.*, 1972; Harrison *et al.*, 1974). Fetal erythroid precursor cells contain little if any globin mRNA, but when cultured in the presence of erythropoietin they accumulate 1800 molecules of globin mRNA per cell (Ramirez *et al.*, 1975). Similarly, as rabbit erythroblasts undergo maturation, globin mRNA levels increase from 10,000 to approximately 23,000 copies per cell (Clissold *et al.*, 1977). When murine leukemia cells infected with Friend leukemia virus are cultured in the presence of dimethyl sulfoxide (DMSO) they differentiate to erythroblasts. Little or no hybridizable globin mRNA is found in uninduced cells, but cells cultured in the presence of DMSO accumulate globin mRNA's and synthesize hemoglobin (Ross *et al.*, 1972). The high specificity of globin gene activation is indicated by hybridization experiments showing no major differences in the expression of several thousand other polyadenylated mRNA's in DMSO-treated cells (Minty *et al.*, 1978).

While studies of gene activation have generally focused on the accumulation of mRNA's which code for proteins characteristic of a differentiated cell type, such as globin mRNA's, it should be noted that such genes may not be completely suppressed in other cell types. Estimates of globin mRNA per cell nucleus range from about 20 molecules in fibroblasts and lymphoma cells to about 80 molecules in hepatocytes and brain cells. These figures are to be compared with the 140,000 copies of globin mRNA in the cytoplasm of the red cell. It is significant also, that the majority of the globin mRNA sequences in nonerythroid tissues are restricted to the cell nucleus (Humphries *et al.*, 1976). Whether these RNA sequences remain permanently locked in the high molecular weight nuclear precursor of globin mRNA (Kwan *et al.*, 1977), whether they eventually serve as templates for globin synthesis, or whether they are selectively degraded remains to be determined. In any case, the results indicate that suppression of globin gene function is not absolute, and they suggest that similar low levels of expression may be true for many other genes. This compounds the sequence complexity of cellular RNA populations, in addition to the overlap of gene expression representing the synthesis of mRNA's needed for the "housekeeping" functions common to all cells.

Positive transcriptional control over the synthesis of specific mRNA sequences is dramatically illustrated in the response of various cell types to hormonal stimulation. Estrogens are necessary for the growth and differentiation of the tubular gland cells of the avian oviduct, in which they regulate the synthesis of egg white proteins, such as ovalbumin, conalbumin, or ovomucoid. The concentration of translatable or hybridizable ovalbumin and ovomucoid mRNA sequences varies during primary stimulation with estrogens, after hormone withdrawal, and again following secondary stimulation (Means *et al.*, 1972; Palmiter, 1973; Palmiter and Carey, 1974; Harris *et al.*, 1975; McKnight *et al.*, 1975; O'Malley *et al.*, 1978; Palmiter *et al.*, 1978; Tsai *et al.*, 1978). During primary estrogen treatment the concentration of ovalbumin mRNA (mRNA$_{ov}$) increases from essentially zero to about 50,000 molecules per cell (Harris *et al.*, 1975). During estrogen withdrawal, the mRNA$_{ov}$ content drops precipitously, leaving only about 60 mRNA$_{ov}$ molecules per tubular gland cell, and there is no detectable ovalbumin synthesis in the withdrawn state (McKnight *et al.*, 1975). Readministration of a single dose of estrogen to withdrawn chicks leads to a 1000-fold increase in mRNA$_{ov}$ concentration (Harris *et al.*, 1975).

The activation of chromatin by steroid hormones has been analyzed in terms of hormone receptor function, transport of the hormone–receptor complex to the nucleus, and increased rates of RNA chain initiation (O'Malley *et al.*, 1978). The effects of estrogens and other hormones upon nucleosome structure in their target cells will be considered in Section II.

Estrogen effects on RNA synthesis are particularly evident for those RNA sequences present in great abundance, such as mRNA$_{ov}$ (Monahan *et al.*, 1976) and the ovomucoid messenger (Tsai *et al.*, 1978), but estrogen withdrawal is reported to have little effect on the total mRNA complexity (Cox, 1977).

The same gene in different target tissues may respond differently to steroid hormones. An interesting example is seen in the induction of conalbumin and transferrin synthesis in the chick oviduct and liver, respectively. Despite the difference in nomenclature, both polypeptides appear to be identical products of a single gene locus acting in the two target tissues. Measurements of conalbumin messenger RNA by cDNA hybridization showed a tenfold increase in the oviduct of estrogen-treated animals, but only a twofold increase in transferrin messenger RNA in the liver. Progesterone also stimulated the production of conalbumin mRNA in the oviduct, but it had no effect on transferrin mRNA synthesis in the liver (Lee *et al.*, 1978). The finding that the transferrin gene, which occurs in only one copy per haploid genome, responds differently to the same hormone in different tissues raises many interesting questions about the organization of the gene in different cell types, possible variations in chromatin structure, and competing regulatory systems in the two target cells.

There are many other examples of hormonal stimulation of mRNA synthesis in hepatic cells. Hydrocortisone increases mRNA levels for tyrosine aminotransferase (Nickol *et al.*, 1978) and also induces the synthesis of the mRNA for tryptophan oxygenase (Schutz *et al.*, 1975). An important point is that the effects of steroids on the liver provide examples of both positive and negative transcriptional control. There is an extraordinary increase in vitellogenin mRNA content in the livers of cockerels treated with estradiol-17β (Tata, 1976); Deeley *et al.*, 1977), but estrogen administration to male rats suppresses the synthesis of the mRNA for α 2u globulin (Kurtz *et al.*, 1976). Conversely, androgens increase the levels of α 2u globulin mRNA in castrated male rats. In minimal deviation hepatomas, the α 2u globulin gene is silenced (Feigelson and Kurtz, 1978).

Analyses of steroid hormone effects on transcription have generally stressed the appearance of messenger RNA species, but the induction of the "new" proteins specified by those mRNA's is often coupled to an increase in the protein synthetic capacity of the responding cell. Ribosomal RNA synthesis is stimulated when prostatic cells respond to testosterone (Liao *et al.*, 1966) and when cells of the toad bladder are stimulated by aldosterone (Wilce *et al.*, 1976). Even a simple eukaryote, *Achlya ambisexualis,* responds to a steroid sex hormone (antheridol) with an increased rate of ribosomal RNA synthesis (Timberlake, 1976) and a commensurate synthesis of ribosomal proteins (Michalski, 1978). As in other cases of chromatin response to androgens [e.g., testosterone effects on the rat prostate (Nyberg *et al.*, 1976)], the state of the chromatin in male strains of *Achlya* is modified, and the number of RNA polymerase initiation sites is increased after exposure to antheridol (Sutherland and Horgen, 1977). In all of these systems, an immediate effect of hormonal stimulation is a modification of nucleosome structure, as will be discussed in Section II.

In the developmental program of many insects, such as *Drosophila melanogaster,* various organs respond differently to the steroid hormone ecdysone, and the activities of different chromosomal regions become evident in the "puffing" patterns of polytene chromosomes. The specificity of the hormone effects in different tissues has been made evident by *in situ* hybridization techniques. When newly synthesized salivary gland RNA was added to the chromosomes of other organs, it was found that the salivary RNA did not hybridize detectably to the site of the major ecdysone-induced puff in wing imaginal disc chromosomes, confirming the dissimilar effects of hormone action in different tissues. Moreover, hybridization of ^{3}H-RNA from the salivary gland to the sites of ecdysone-induced puffs in salivary chromosomes could be detected only if the RNA was labeled in the presence of ecdysone (Bonner and Pardue, 1977). This is a clear indication that the hormone selectively activates transcription at regions of the chromosome where the DNA assumes a more extended configuration. This change in chromatin structure in the puff appears to correlate with a selective loss of histone H1 (Jamrich *et al.*, 1977).

Similar precision in gene activation is seen in the response of *Drosophila* chromosomes to elevated temperature. Heating the cells to 37°C causes characteristic puffs to appear, while preexisting puffs regress. *In situ* hybridization studies show that one of the major heat-shock loci, *87B*, hybridizes selectively to an induced polysomal mRNA which codes for the 70,000 dalton heat-shock protein (McKenzie *et al.*, 1975; Mirault *et al.*, 1978). Regression of the normal developmental puffs after heat treatment is accompanied by the disappearance of the corresponding mRNA's (Spradling *et al.*, 1977; Mirault *et al.*, 1978).

Emphasis has been placed on these hormonal and temperature-dependent responses for several reasons; first, to emphasize the point that transcriptional control in eukaryotes is both positive and negative; second, to stress the dual aspects of genetic control in metazoans—selective activation of different genes in different cell types and the overlapping synthesis of mRNA's needed for housekeeping functions in all cell types; third, to point out that transcriptional control extends to reiterated sequences, such as ribosomal and histone genes, as well as to unique DNA sequences; and finally, to provide the background for the subsequent discussions of the control of chromatin structure and function by DNA-binding proteins. Many of the effects of hormones and other developmental stimuli can be related to changes in chromosomal protein structure and metabolism which ultimately modify the conformation and nucleosomal organization of the DNA strand.

Before analyzing these changes in molecular terms, some further points should be made about the interactions between nucleus and cytoplasm in establishing the transcriptional program of differentiating cells. Factors that control the onset, duration, and end of specific gene functions can be shown to originate in the cytoplasm. When salivary gland nuclei of *Drosophila* are transplanted into the egg cytoplasm at different stages, the puffing pattern of the polytene chromosomes varies with the developmental state of the egg (Kroeger, 1960). Similarly, genes that are suppressed in *Xenopus* kidney cells are reactivated when kidney nuclei are microinjected into the oocyte (De Robertis and Gurdon, 1977). Conversely, the genes that are normally expressed in the kidney but not expressed in oocytes become inactivated in the oocyte cytoplasm. The reactivation of transcription in transplanted nuclei does not require a new round of DNA synthesis and, thus, cannot be attributed to selective gene amplification (De Robertis and Gurdon, 1977). [However, in some developmental programs, such as differentiation of chick neural retina and cartilage, there is evidence for tissue-specific amplification of particular DNA sequences (Strom *et al.*, 1978).]

The results of nuclear transplantation experiments in *Drosophila* and in

Xenopus are fully consistent with classic embryological tenets that the morphogenetic patterns of early development are controlled by the cytoplasm of the fertilized egg. Studies of *Ambystoma* mutants confirm the need for cytoplasmic proteins in nuclear activation. Cytoplasmic deficiency of a protein synthesized during oogenesis in *Ambystoma* leads to developmental arrest at gastrulation. The injection of cytoplasm from a normal mature egg into a mutant egg corrects the deficiency (Briggs and Cossens, 1966). Gastrulation requires the presence of a nucleus (Gurdon, 1974). In the *Ambystoma* mutant the nucleus is functional, but proteins present in the normal egg cytoplasm are essential for the activation of the nuclear genes required for gastrulation and organogenesis (Brothers, 1976). How this is achieved remains to be determined, but the entry of ooplasmic proteins into a microinjected nucleus is known to be highly selective (De Robertis *et al.*, 1978), and many of the proteins targeted to the nucleus have DNA-binding properties, as will be discussed for proteins of the high-mobility group (HMG) class in Section III. The presence of DNA-binding proteins which regulate transcription in oocyte nuclei is strongly indicated by the fact that free DNA's, such as 5 S DNA and ribosomal DNA are transcribed with high fidelity when injected into *Xenopus* oocyte nuclei, initiating RNA synthesis at the right sequences in the correct DNA strand, and synthesizing RNA's of the correct length (Brown and Gurdon, 1977; Trendelenburg and Gurdon, 1978). [However, when the ribosomal DNA of *Dytiscus* is injected into *Xenopus* oocyte nuclei, some rDNA molecules initiate abnormally and the transcripts may fail to terminate properly (Trendelenburg *et al.*, 1978). This may indicate some species specificity in the factors involved in RNA polymerase attachment and chain termination.]

It is significant that genes in the suppressed state can be reactivated by proteins from another cell type and even from a different species. This is particularly evident in cell-fusion experiments involving avian erythrocytes, which have little transcriptional activity (Zentgraf *et al.*, 1975), and mammalian cells. When the nucleus of a chick erythrocyte is introduced into the cytoplasm of a HeLa cell it resumes the synthesis of RNA (Harris, 1965). Reactivation of the erythrocyte nucleus leads to the synthesis of specific avian enzymes, such as inosinic acid pyrophosphorylase (Harris and Cook, 1969), as well as to the synthesis of chick ribosomal RNA (Bramwell, 1978). These events are accompanied by changes in the morphology, nonhistone protein content, and DNA accessibility in the reactivated nucleus (Bolund *et al.*, 1969; Appels *et al.*, 1975). Human proteins are known to enter the avian nucleus prior to the onset of RNA synthesis (Appels *et al.*, 1975), and one may assume that these proteins (and other factors entering the nucleus) are effective in modifying and reactivating the chromatin of the distant species.

B. Structural Correlates of Gene Activity

The introduction of techniques, such as high-resolution electron microscopy, autoradiography, and *in situ* hybridizations employing radioactive mRNA's, cDNA's, and cloned DNA sequences of known function, has provided new insights into the relationship between the structure of the chromatin fiber and its template activity in transcription. The lampbrush chromosomes of amphibian oocytes provide a particularly revealing model of the structural correlates of gene control. Ribonucleic acid synthesis in lampbrush chromosomes takes place on extended DNA-containing "loops" which project laterally from many sites along the chromosome axis (Gall and Callan, 1962; Callan, 1963; Izawa *et al.,* 1963; Miller, 1965; Miller *et al.,* 1972; Malcolm and Sommerville, 1974; Mott and Callan, 1975; Scheer *et al.,* 1976). The complete chromosome complement contains at least 10^4 loop pairs bearing nascent RNA chains. This represents transcription of only about 4% of the total DNA, the great majority of which occurs in the dense axial chromomeres as tightly packed and transcriptionally inert superhelical coils (Mott and Callan, 1975). Estimates of the proportion of DNA present in the loops (5%) (Callan, 1963) are in good agreement with results showing that somewhat over 4% of the nonrepetitive DNA sequences hybridize with oocyte nuclear RNA (Sommerville and Malcolm, 1976). In some instances, the functions of particular loops have been characterized; e.g., loops which hybridize selectively to 5 S RNA (or its complementary DNA) have been mapped in the lampbrush chromosomes of *Notopthalmus*. The reiterated 5 S RNA coding sequences are scattered on 4 chromosomes in 34 loops which range in size from 15 to 200 μm. It is significant that no hybridization to the adjoining dense centromeric chromatin was observed in any of these regions (Pukkila, 1975). The scattering of the 5 S genes on different loops in different chromosomes suggests that transcriptional control over these reiterated DNA sequences involves a coordinate action of diffusible control factors. These factors may be short-lived, as synthesis of the oocyte-type 5 S RNA is suppressed in the somatic cells of amphibians (Ford and Southern, 1973).

The histone genes have also been localized to particular loops of newt lampbrush chromosomes by *in situ* hybridization techniques (Old *et al.,* 1977). A careful examination of the loops after hybridization to cloned histone DNA sequences reveals that the histone coding region of each particular loop occurs at a fixed position around the loop axis. This is important, because it rules out earlier models of transcription in which the chromomeric DNA was spooled out and reeled in during the course of RNA synthesis, a process which would place the histone genes at different positions in the loop. Transcription of the histone genes occurs unidirectionally, a single DNA strand acting as the template, but the histone mRNA transcription unit is only a fraction of the total length of the loop, and the surrounding DNA sequences are not transcribed, despite the fact that they are extended. It follows that the extension of the DNA strand is not a sufficient cause for transcription, and that initiation and termination require other forms of control.

As a result of the intense activity of the histone loops, the *Xenopus* oocyte accumulates thousands of copies of mRNA's coding for early embryonic histones (Levenson and Marcu, 1976), but their translation is delayed until after fertilization takes place. The storage of "masked" messenger RNA's for long periods prior to their utilization is an important aspect of translational control in early embryogenesis (Gross *et al.*, 1973). However, activation of cytoplasmic protein synthesis after fertilization can occur in the absence of a nucleus (Gurdon, 1974) and, thus, it does not appear to be under immediate genetic influence. This control mechanism will not be considered in the present analysis of transcriptional control in the nucleus.

As maturation of the oocyte proceeds, the chromosome loops retract and the incorporation of RNA precursors is drastically curtailed (Davidson *et al.*, 1964). Suppression of RNA synthesis by actinomycin D also leads to retraction of the loops (Izawa *et al.*, 1963). Similar correlations between compaction of the DNA and its inactivity in RNA synthesis have been noted in somatic cell nuclei of mammals (Littau *et al.*, 1964; Granboulan and Granboulan, 1966), in plant (*Trillium*) microspores (Kemp, 1966), and in the polytene chromosomes of *Drosophila* (Lakhotia and Jacob, 1974).

The transcription unit in lampbrush chromosomes appears as an array of lateral ribonucleoprotein fibrils of increasing length attached to the axial DNA strand by RNA polymerase molecules (Mott and Callan, 1975; Scheer *et al.*, 1976). Loops may contain one or more functional units and differ in the placement of transcription units along the loop. In some loops the transcribing regions are closely apposed, while in others they are separated by fibril-free and presumably nontranscribed DNA sequences (Fig. 1). The latter have a beaded, nucleosomal conformation, which is not evident in the transcription unit (Scheer *et al.*, 1976). The spacing and polarity of the transcription units in some loops indicates that continuous travel of the RNA polymerase complexes into spacer regions or subsequent genes does not occur, and that detachment of the RNA polymerases must take place at the termination site.

In some complex loops, adjacent transcriptional units differ in the polarity of transcription. This is evident in the opposite progression of RNP fibril lengths along the DNA axis (Fig. 2). Alternative transcription from

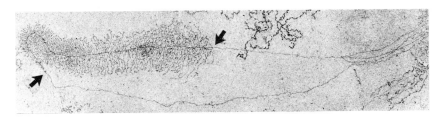

Fig. 1. Electron micrograph of a lampbrush chromosome "loop" of an amphibian oocyte (*Triturus cristatus*). The loop (total length 53 μm) contains one functional transcription unit with ribonucleoprotein fibrils of increasing length attached to the axial DNA strand. The arrows indicate initiation and termination sites of the transcription matrix. Fibril-free (apparently nontranscribed) regions 25 and 13 μm in length, respectively, occur at each end of the transcription unit. These nontranscribed regions are studded with nucleosomes. × 7500. (Photograph courtesy of Drs. U. Scheer and H. Zentgraf.)

one or the other DNA strand is implied (Scheer *et al.*, 1976). A reversed polarity of transcription is seen on palindromic DNA sequences, such as those which encode the ribosomal RNA's of *Physarum* (Molgaard *et al.*, 1976; Vogt and Braun, 1976; Grainger and Ogle, 1978; Campbell *et al.*, 1979) or *Tetrahymena* (Karrer and Gall, 1976). A more detailed analysis of the *Physarum* rDNA molecule and the structural differences between its transcribing and spacer regions will be presented in Section II.

Although this chapter is not primarily concerned with the processing of RNA transcripts or with the final disposition of the message, observations on lampbrush chromosomes provide important clues to this important aspect of transcription. The primary transcripts on the chromosome loops are generally much longer than the average-size messenger RNA in the cytoplasm (Miller and Hamkalo, 1972; Sommerville and Malcolm, 1976). In fact, some messengerlike sequences are apparently present in large

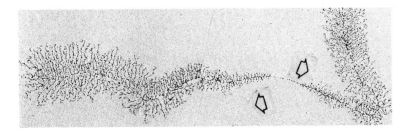

Fig. 2. Electron micrograph of a lampbrush chromosome "loop" of *Acetabularia cliftonii*. Two transcription units showing opposite polarity of transcription, as judged by the gradients in RNP fibril lengths, are arranged head-to-head about a short "spacer." Arrows indicate the direction of transcription. × 17,500. (Photograph courtesy of Drs. U. Scheer and H. Zentgraf.)

primary transcripts up to 6×10^4 nucleotides long (Sommerville and Malcolm, 1976). This is in accord with evidence that the sequence complexity of embryonic nuclear RNA's greatly exceeds that of cytoplasmic RNA's (Hough *et al.*, 1975) and with many recent observations on the presence of intervening sequences in structural genes. The chick ovalbumin gene, for example, is split into seven messenger-coding sequences separated by six intervening sequences. All of these sequences are contained in a chromosomal DNA region of 6000 nucleotide pairs, which is more than three times longer than the ovalbumin message (Mandel *et al.*, 1978). It is significant that the boundaries between the various coding and intervening sequences share direct or similar repeats which define common excision-ligation points (Breathnach *et al.*, 1978; Caterall *et al.*, 1978). Processing of long primary transcripts to mRNA's could involve elimination of the intervening sequences by mechanisms similar to those employed in prokaryotes for the processing of preribosomal RNA's. In *E. coli*, the DNA sequences flanking the 16 S rRNA gene are potentially capable of hybridizing to each other to form a stem consisting of 26 base pairs (Young and Steitz, 1978). The complementary sequences in the primary transcript are presumed to form a double-stranded stem containing the ends of the single-stranded 1700 nucleotide 16 S rRNA chain. It is known that both proximal and distal RNase III cuts of of the 16 S rRNA precursor occur in the base-paired sequences. Endonuclease action at those points would free the 16 S ribosomal RNA from the long primary transcript. Analogous mechanisms for the elimination of intervening sequences in eukaryotes may be postulated to involve cleavage of "snap-back," self-annealing sequences in hnRNA. Heterogeneous nuclear RNA contains many sequences that are transcribed from "inverted repeat" DNA, while cytoplasmic mRNA contains many fewer such sequences (Jelinek, 1977; Jelinek *et al.*, 1978), as would be expected if they were removed during hnRNA processing by endonuclease cleavage.

The processing of nuclear transcripts may also require the participation of small nuclear RNA's, if the analogy with prokaryotes is extended. A small 350-nucleotide RNA is needed in the processing of bacterial suppressor tRNA's by RNase P (Stark *et al.*, 1978). A low molecular weight nuclear RNA which hydrogen bonds to nuclear poly(A)+RNA and is not present in poly(A)-terminated mRNA's in the cytoplasm has been described (Jelinek and Leinwand, 1978). Whether other small nuclear RNA's play a role in the processing of primary transcripts is not known, but the fact that they remain associated with chromatin even when transcription is suppressed argues against their association with the primary transcript (Goldstein *et al.*, 1978).

The processing of hnRNA by the excision of intervening sequences joined at their ends by base pairing would be expected to generate small loops or open circles containing the introns. This may account for the presence of many small, unattached ring structures along the lateral RNP fibrils of *Triturus* lampbrush chromosomes (Scheer *et al.*, 1976).

Electron microscopy of the ribosomal genes in amphibians [*Xenopus* and *Notopthalmus* (*Triturus*)] (Miller and Beatty, 1969; Franke *et al.*, 1976; Scheer *et al.*, 1976; Woodcock *et al.*, 1976; Franke *et al.*, 1978; Scheer, 1978), in insects (*Drosophila* and *Oncopeltus*) (Meyer and Hennig, 1974; Foe *et al.*, 1976; McKnight and Miller, 1976; Foe, 1978; McKnight *et al.*, 1978), and in unicellular organisms [*Amoeba proteus* (Murti and Prescott, 1978) and *Physarum polycephalum* (Grainger and Ogle, 1978)] has provided some of the most graphic and significant information on the organization of transcriptionally active chromatin. In all systems, the activity of the reiterated ribosomal genes is readily visualized in positively stained chromatin spreads (Miller and Beatty, 1969; Miller and Bakken, 1972). The transcription units (often called "matrix units") in amphibian nucleoli are generally seen as tandem arrays of RNP fibrils of increasing length attached to an axial DNA strand by closely apposed RNA polymerase molecules. Each transcription unit is discrete, separated from the next unit by a DNA "spacer" which is usually fibril-free. In *Xenopus*, the ribosomal genes occur in tandem arrays of uniform polarity, but in *Physarum* the ribosomal genes occur at opposite ends of a long central spacer, and transcription proceeds in opposite directions toward each end of the palindromic rDNA molecule (Grainger and Ogle, 1978).

Changes in the ultrastructure of ribosomal chromatin take place during periods of altered ribosomal RNA synthesis and are particularly evident at progressive stages of amphibian oogenesis (Scheer *et al.*, 1976; Franke *et al.*, 1978; Scheer, 1978) and embryogenesis in insects (Foe *et al.*, 1976; McKnight and Miller, 1976; Foe, 1978; McKnight *et al.*, 1978). When the ribosomal genes are fully active, the strands of nucleolar chromatin appear smooth and are devoid of nucleosomes in the transcribed units (Scheer, 1978). This nonnucleosomal character also extends into the spacer regions which contain "prelude" complexes (Franke *et al.*, 1976). When the ribosomal genes are inactivated at later stages of oogenesis, nucleolar chromatin progressively assumes a beaded appearance, with nucleosomes first becoming manifest in the extended fibril-free and apparently nontranscribed regions. In full-grown oocytes, when ribosomal RNA synthesis is shut down, most of the nucleolar chromatin is beaded and indistinguishable from inactive, nonnucleolar chromatin (Scheer, 1978). The absence of beaded nucleosomal arrays in actively transcribing chromatin containing ribosomal genes has been noted repeatedly in amphibians and insects (Foe *et al.*, 1976; Woodcock *et al.*,

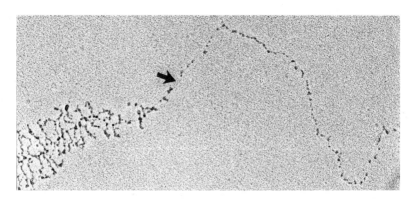

Fig. 3. Electron micrograph of a ribosomal transcription unit and contiguous "spacer" in the nucleolar chromatin of *Triturus alpestris*. The DNA in the nontranscribed "spacer" is organized into a beaded nucleosomal configuration. DNA in the transcription unit and adjacent to its origin occurs as an extended, nucleosome-free fiber. The arrow indicates an unbeaded DNA sequence immediately adjacent to the transcription unit, which is studded with RNA polymerase molecules and attached transcripts. × 60,000. (Photograph courtesy of Drs. U. Scheer and H. Zentgraf.)

1976; Foe, 1978; Scheer, 1978; Franke *et al.*, 1978). Failure to observe a beaded conformation is in accord with measurements of the length of the transcription unit. In the ribosomal genes of *Oncopeltus,* the DNA has a calculated compaction ratio of 1.2 μm of B structure DNA per micrometer of chromatin (Foe *et al.*, 1976). It follows that the DNA cannot be coiled around nucleosomes in the usual two turns per nucleosome. The DNA sequences between the ribosomal transcription units of *Oncopeltus* generally have a beaded morphology and show a correspondingly higher compaction ratio of about 2.3 (Foe *et al.*, 1976; Foe, 1978). Similar differences in chromatin organization in active and inactive regions are also seen in the extrachromosomal rDNA of *Physarum polycephalum*. The transcription units at each end of the linear rDNA molecule contain DNA in the extended configuration (Grainger and Ogle, 1978; Johnson *et al.*, 1978b) while DNA in the nontranscribed central spacer occurs in a beaded (Grainger and Ogle, 1978) and relatively nuclease-resistant conformation (Johnson *et al.*, 1978a,b).

Comparable observations have been made on nonnucleolar chromatin. In retracting loops of lampbrush chromosomes, reduced transcriptional activity is associated with the reappearance of beaded subunits along the DNA strand. The results are consistent with the view that nucleosomes, as beaded structures, are absent from heavily transcribed chromatin regions, and that they are reformed after gene inactivation (Scheer, 1978).

Whether nucleosomes are visible in a transcription unit appears to depend upon the rate of transcription in that region of the chromatin. In genes that are transcribed very slowly (as judged by the sparse distribution of lateral RNP fibrils) the nascent RNP chains may be attached to beaded chromatin strands. Since the nucleosome is an impediment to transcription, as judged by biochemical studies (see Section II), one may conclude that mechanisms exist for facilitating the passage of the RNA polymerase without removing the nucleosome from the DNA template. This may involve a reversible change in nucleosomal conformation to permit a relaxation of constraints on the DNA template during passage of the polymerase and rapid reformation of the beaded structure after traverse by the enzyme. (Mechanisms which would permit rapid changes in the structure of the nucleosome will be considered in Section II.)

Electron microscopy of ribosomal transcription complexes at different stages in oogenesis and embryonic development reveals that individual transcription units, even adjacent ones, may differ with regard to their content of nascent rRNA chains and in the frequency and distribution of RNA polymerase molecules on the DNA template (McKnight and Miller, 1976; Foe *et al.,* 1976; Foe, 1978; McKnight *et al.,* 1978). This is an important point because it means that the rate of production of ribosomal RNA's is not simply controlled by the number of gene copies or by the total concentration of the polymerase. Biochemical studies lend further support to this conclusion. In *Physarum polycephalum,* for example, measurements of rDNA content show no differences between the transcriptionally active microplasmodial form and the transcriptionally inert microsclerotial form (Ryser and Braun, 1974), and in *Dictyostelium discoideum* changes in ribosomal RNA synthesis during development are not simply correlated with levels of RNA polymerase I activity (Pong and Loomis, 1973). It follows that other control mechanisms regulate the binding, progression, and release of the polymerase in the rDNA transcription unit.

The distribution of RNA polymerases on chromatin is not random. Immunofluorescence studies of salivary gland chromosomes in *Drosophila* show that RNA polymerase II (B) is present almost exclusively in the puffs and interband regions. Heat-shock treatment alters the puffing pattern and results in a redistribution of RNA polymerase molecules (Jamrich *et al.,* 1977). The localization of the polymerases in the interband regions is in accord with autoradiographic studies of [3H]uridine incorporation which showed that the primary uptake of the RNA precursor takes place in the diffuse chromatin adjacent to the bands (Zhimulev and Belyaeva, 1975).

In summary, transcriptionally active regions of chromatin differ in conformation and composition from nontranscribed regions. Highly active transcription units, such as those seen in the loops of lampbrush chromosomes and in the nucleoli of early oocytes, or in the puffs of insect polytene chromosomes, contain DNA in a highly extended configuration. Many RNA polymerase molecules bearing nascent RNA chains (and associated proteins) occupy the extended DNA template. The density of the polymerases in ribosomal DNA may exceed 50 polymerase molecules per micrometer (McKnight and Miller, 1976), and the number decreases when ribosomal RNA synthesis is suppressed (Foe *et al.*, 1976). The DNA strand in the transcription unit is not beaded in the typical nucleosomal morphology but is relatively smooth. However, its diameter (about 70 Å) exceeds that of a free DNA strand (20 Å), presumably due to the presence of associated proteins. Nontranscribed chromatin, after spreading, displays a beaded morphology with about 34 nucleosomes/μm, with the beads joined by a 32 Å thick filament, and free of attached RNA polymerase molecules (Foe *et al.*, 1976). In slowly transcribing regions of chromatin, a beaded conformation may exist within a transcription unit, suggesting that nucleosomes may reform quickly after traverse by the RNA polymerase.

The physical state and the activity of chromatin change in response to a wide variety of hormonal and developmental stimuli. Such changes require alterations in the conformation of the DNA template and in its accessibility to regulatory factors involved in chain initiation and termination and to RNA-polymerizing enzymes. Evidence that these alterations are achieved by modulation of the binding between DNA and chromosomal proteins will now be presented.

II. CONTROL OF DNA CONFORMATION

The magnitude of the problem in organizing the eukaryotic genome can be illustrated by data on the human lymphocyte nucleus, which contains about 7×10^{-12} g of DNA which, if fully extended, would be equivalent to over 2 m (215 cm) of DNA double helix! Most of the DNA is organized into 200–250 Å fibers with an average mass of 5.95×10^{-16} g/μm (Goulomb and Bahr, 1974). The total length of chromatin fiber in the lymphocyte nucleus, as determined by quantitative electron microscopy, is 7.59 cm, and the DNA packing ratio (length of DNA: length of chromatin fiber is about 28.3 : 1. From the packing ratio and the mass of 1 μm of DNA (3.26×10^{-18} g), one can calculate that each micrometer of chromatin fiber contains about 1×10^{-16} g of DNA. It follows that the DNA is associated with far more than an equal amount of protein and RNA. Much of this may be peripheral to the DNA strand, like the RNP fibrils in the transcription units, while other components, such as the histones and HMG proteins (see Section II,A) are bound to DNA as integral parts of the chromatin strand. Still other components, including the RNA polymerases, have transient associations with the DNA strand.

The problem of compacting 2 m of human DNA into a lymphocyte nucleus only 6 μm in diameter is solved by a series of DNA-coiling mechanisms, beginning at the level of the nucleosome.

A. The Proteins of Nucleosomes and Their Role in Chromatin Organization

The primary control over DNA conformation is exerted by its interactions with basic proteins: histones in somatic cells and protamines in many types of sperm cells. (Because the substitution of protamines for histones in spermatogenesis correlates with extreme chromatin compaction and the loss of RNA synthetic activity, the protamines will not be considered further in this analysis of chromatin organization and function in somatic cells.)

The basic unit of chromatin structure is the nucleosome, a flattened nucleohistone bead about 110 Å in diameter and 57 Å in height (Finch *et al.*, 1977). Electron micrographs of spread chromatin fibers show linear, regularly-spaced arrays of nucleosomes distributed along a 30 Å filament at a frequency of about 34 nucleosomes/μm (Woodcock, 1973; Olins and Olins, 1974; Langmore and Wooley, 1975; Oudet *et al.*, 1975; Woodcock *et al.*, 1976; Foe *et al.*, 1976). The electron microscopic visualization of chromatin organization as a periodic repeating structure resembling beads-on-a-string is confirmed by many biochemical and biophysical studies. It has led to the prevailing view that the DNA strand is periodically wrapped around clusters of histones, each cluster comprising an octamer made up of two molecules each of the four major histone classes: H2A, H2B, H3, and H4 (Kornberg, 1974, 1977; Noll, 1974; Van Holde *et al.*, 1974; Joffe *et al.*, 1977).

Coiling of the DNA strand around the histone cores (Van Holde *et al.*, 1974; Baldwin *et al.*, 1975; Pardon *et al.*, 1976; Hjelm *et al.*, 1977) represents the first level of compaction of the genetic material in interphase chromatin. In all species examined, the nucleosome core particles contain a uniform length of DNA (140–145 nucleotide pairs) (Axel, 1975; Sollner-Webb and Felsenfeld, 1975; Thomas and Kornberg, 1975; Compton *et al.*, 1976; Noll, 1976; Shaw *et al.*, 1976; Thomas and Kornberg, 1976; Whitlock and Simpson, 1976a). Coiling of the DNA around the octameric histone complex results in a fivefold shortening of its contour length (Oudet *et al.*, 1975) and also alters its topological state by the introduction of supercoils

(Germond *et al.*, 1975; Camerini-Otero *et al.*, 1978). Approximately 1.75 turns of DNA of contour length about 425 Å envelop the octameric histone complex: (H2A)$_2$ (H2B)$_2$ (H3)$_2$ (H4)$_2$ (Finch and Klug, 1978). On the average nucleosomes each contain one molecule of histone H1 (Goodwin *et al.*, 1977a) associated with DNA sequences adjoining the nucleosome core (Baldwin *et al.*, 1975; Shaw *et al.*, 1976; Varshavsky *et al.*, 1976; Whitlock and Simpson, 1976b; Noll and Kornberg, 1977; Finch and Klug, 1978). Binding of H1 to a 20 nucleotide pair segment of DNA in the linker region (Noll and Kornberg, 1977) would complete two turns of the double helix around the nucleosome, one turn for each 80 nucleotide pairs.

H1 is the most complex histone, considerably longer than the other histones and differing from them in its clustered distributions of basic amino acids. It has a high potential for self-aggregation and, most significantly, has a capacity for cross-linking chromatin strands (Littau *et al.*, 1965; Bradbury *et al.*, 1973) and bridging distant DNA sequences in viral chromatin (Griffith and Christiansen, 1978). There are many indications that H1 plays a key role in the control of higher orders of chromatin structure (Littau *et al.*, 1965; Bradbury *et al.*, 1973; Renz *et al.*, 1977; Thoma and Koller, 1977; Worcel and Benyajati, 1977; Griffith and Christiansen, 1978). The selective removal of histone H1 from liver chromatin leads to a major structural transition in which the 200 Å fibers disperse to form a beaded strand of nucleosomes separated from each other by distances of 150–200 Å (Thoma and Koller, 1977). The assembly of nucleosomes into 200 Å chromatin fibers probably involves a cooperative interaction of histone H1 with the nucleosome chain. This generates assemblies (200 Å "superbeads") (Renz *et al.*, 1977, 1978) containing eight to ten nucleosomes in discrete groupings (Stratling *et al.*, 1978). This additional compaction, added to the fivefold shortening of DNA contour length in coiling about the nucleosome, would provide a linear packing ratio for DNA in the native chromosome fiber of about 25 to 1, in good agreement with the packing ratio of 28 to 1 measured in human lymphocyte chromatin (Golomb and Bahr, 1974). It has also been proposed that histone H1 can stabilize each nucleosome by binding to the linker DNA on either side of the core particle (Gaubatz *et al.*, 1978). In either case, removal of H1 would be expected to release constraints on the DNA strand, and one might expect that the removal of H1 would correlate with gene activity. It has been noted that H1 is absent from the puffs of *Drosophila* polytene chromosomes (Jamrich *et al.*, 1977).

Although the nucleosome "cores" from diverse species contain a uniform length of DNA, the average lengths of DNA linkers between the beads may vary from species to species (Compton *et al.*, 1976; Johnson *et al.*, 1976; Morris, 1976a; Noll, 1976; Thomas and Furber, 1976; Lohr *et al.*, 1977), from tissue to tissue (Morris, 1976b), and in different cell types within a tissue (Thomas and Thompson, 1977; Todd and Garrard, 1977). The significance of the variability in spacer lengths remains unclear, but the spacing of nucleosomes in avian erythrocytes is known to be altered as the erythrocyte-specific histone, H5, accumulates (Weintraub, 1978). Since H5 is a variant of histone H1, this suggests that differences in the H1 complement in different cell types may affect the spacing of the chromatin subunits. Histone H1 has long been known to comprise a mixture of closely related proteins (Kincade and Cole, 1966; Kincade, 1969; Panyim and Chalkley, 1969; Marks *et al.*, 1975; Hohmann, 1978) that are present in various proportions in different tissues, and subject to multiple forms of phosphorylation and ADP-ribosylation. (The postsynthetic modifications will be considered in Sections II,F and H). Therefore, the potential exists for extensive modulation of chromatin organization by the use of different structural variants and combinations of histone H1.

While the varying proportions of histone H1 subtypes may be related to the variability in spacer lengths between nucleosomes, there are also indications of specific associations with particular DNA sequences, such as the association of different H1 variants with various satellite DNA's in *Drosophila virilis* (Blumenfeld *et al.*, 1978).

The histones of the nucleosome core (H2A, H2B, H3, and H4) occur in fixed relative proportions (Bustin *et al.*, 1977; Kornberg, 1977; Joffe *et al.*, 1977). Two molecules of each occur in the octameric complex. The parallel distributions of the core histones, as visualized by the binding of antibodies to individual histone classes in nucleosomes (Bustin *et al.*, 1976) and in polytene chromosomes (Kurth *et al.*, 1978), suggests that this simple, uniform stoichiometry exists for nucleosomal core proteins throughout the genome. However, the situation is complicated by the existence of histone variants and a wide range of postsynthetic modifications. Two subtypes of histone H3 are present in bovine tissues (Marzluff *et al.*, 1972), and multiple forms of histones H2A, H2B, and H3 have been identified in the sea urchin embryo (Newrock *et al.*, 1978b), in *Drosophila* embryos (Alfagame *et al.*, 1974), and in mammalian cells (Franklin and Zweidler, 1977). Different H4 mRNA's are produced after hatching of *Lytechinus* (Grunstein, 1978) and, in *Stronglylocentrotus*, different mRNA's for variants of histones H2A and H2B dominate in the early and late blastula (Newrock *et al.*, 1978a). The significance of the changes in proportions of histone variants is not known, but it is clear that all nucleosomes cannot be identical in this regard. A further complication arises in the postsynthetic modifications of the core histones by acetylation, phosphorylation, methylation, ADP-ribosylation, and isopeptide branching. These modifications will be considered subsequently; they are mentioned here to emphasize the potential for heterogeneity in nucleosome structure, even if a uniform set of histone variants occurs in the octamer.

The formation and stability of the nucleosome depend upon specific interactions between the component histone classes. The specificity of these interactions and their effects on DNA conformation have been investigated by the use of physical probes, such as fluorescence anisotropy and lightscattering (D'Anna and Isenberg, 1974; Li, 1977), nuclear magnetic resonance (Lewis *et al.*, 1975; Moss *et al.*, 1976a,b), X-ray diffraction (Kornberg and Thomas, 1974, Boseley *et al.*, 1976; Bradbury *et al.*, 1978), circular dichroism (Whitlock and Simpson, 1976a; Fasman, 1977), and electron microscopy (Germond *et al.*, 1975; Woodcock and Frado, 1978; Simpson *et al.*, 1978). The main result to be stressed here is that the histone complex involves interactions between structured segments of the apolar C-terminal regions of the histones to form a globular core, leaving the basic NH_2-terminal regions free to interact with the DNA. (Further evidence for DNA binding to the amino-terminal regions of the histones will be considered subsequently.)

The arrangement of histones within the nucleosome has been investigated using protein cross-linking reagents capable of spanning the distances between neighboring lysine residues (Olins and Wright, 1973; Ilyin *et al.*, 1974; Chalkley and Hunter, 1975; Hyde and Walker, 1975; Thomas and Kornberg, 1975; Van Lente *et al.*, 1975; Jackson, 1978) or bridging histones by the activation of their tyrosine residues (Martinson and McCarthy, 1975, 1976; Martinson *et al.*, 1976). The so-called zero-length cross-linking reagents, tetranitromethane and ultraviolet irradiation, reveal the very close associations between the histones of the core particle. Two major dimers are produced: H2A–H2B and H2B–H4 (Martinson and McCarthy, 1976; Martinson *et al.*, 1976). Cross-linking with carbodiimide reveals an H3–H4 dimer (Bonner and Pollard, 1975). Separate domains are involved in the association of a given histone (H2B, for example) with different neighboring histones (such as H2A and H4). Significantly, analysis of the sites of cross-linking within the histone chains indicate that the interactions primarily involve the apolar regions of the molecule.

Physical probes of histone mixtures in solution have also indicated preferred interactions between the same pairs of histones: H2A–H2B, H2B–H4, and H3–H4 (D'Anna and Isenberg, 1974; Lewis *et al.*, 1975). H2A–H2B dimers and H3–H4 tetramers form readily in solution (Kornberg and Thomas, 1974; Moss *et al.*, 1976a,b), and at high salt concentrations all four nucleosomal histones assemble to form heterotypic tetramers (H2A)(H2B)(H3)(H4) (Weintraub *et al.*, 1975). Heterotypic tetramers also appear during dissociation of nucleosome cores at high ionic strengths, and they form an equilibrium with the octamer (Chung *et al.*,

1978). The free tetramers become unstable as the salt concentration is lowered, but they become stabilized by association with DNA.

The presence of heterotypic tetramers is in accord with evidence that the nucleosome possesses an axis of dyad symmetry (Finch and Klug, 1978). Other evidence for this view includes the finding that DNase II can cut the DNA of a 200 nucleotide pair nucleosome repeat unit into two equal halves without disruption of the particle (Altenburger et al., 1976), and electron microscopic observations that nucleosomes can open to half-nucleosomes at low ionic strength and revert to single units when the salt concentration is raised (Oudet et al., 1978a). A transition to symmetrically paired half nucleosomes has been postulated in models of transcription. Unfolding of the nucleosome would facilitate genetic readout without requiring displacement of the histones from the DNA template (Weintraub et al., 1976).

Reconstitution experiments using various combinations of the core histones and DNA have established that the arginine-rich histones H3 and H4 play a dominant role in nucleosome assembly. The H3–H4 tetramer interacts with the DNA strand to generate subnucleosomal particles with many of the properties of the native chromatin subunits, including DNA compaction and negative superhelicity (Bina-Stein and Simpson, 1977; Camerini-Otero et al., 1978; Oudet et al., 1978b), limited protease sensitivity (Sollner-Webb et al., 1976), characteristic nuclease digestion patterns (Camerini-Otero et al., 1976; Moss et al., 1977), and well-defined low-angle X-ray diffraction maxima (Moss et al., 1977; Bradbury et al., 1978). These effects cannot be achieved by mixing DNA with histones H2A and H2B, or with any combination of histones lacking H3 and H4. Such subnucleosomes prepared with H3–H4 tetramers can be distinguished from nucleosomes containing all four histone classes by high-resolution electron microscopy (Oudet et al., 1978b).

The fact that the constraints upon DNA conformation are due primarily to its interactions with histones H3 and H4 has particular significance for models of transcriptional control by modulation of histone–DNA interactions, because H3 and H4 are the primary nucleosomal targets for histone-modifying enzymes which weaken the electrostatic attraction between the basic proteins and the enveloping DNA strand (see Section II, C).

The topology of the DNA coil around the nucleosome is determined by periodic interactions with the eight histones of the core particle. These interactions are largely electrostatic and involve the positively charged amino-terminal regions of the histone chains and the negatively charged phosphate groups of the DNA double helix (Bradbury and Crane-Robinson, 1971; Weintraub and Van Lente, 1974; Baldwin et al., 1975; Li,

1975; Moss *et al.*, 1976a,b; Lilley and Tatchell, 1977; Whitlock and Simpson, 1977). Exhaustive digestion of chromatin with trypsin leads to the cleavage of only 20–30 amino acids from the NH_2-terminal regions of the nucleosomal histones (Weintraub and Van Lente, 1974). Tryptic treatment of nucleosome core particles releases constraints on the DNA strand, as judged by increased molecular ellipticity in the circular dichroism spectra and more rapid kinetics of DNA digestion by micrococcal nuclease and DNase I (Lilley and Tatchell, 1977; Whitlock and Simpson, 1977).

It is significant that the topology of the DNA coil in the nucleosome shows a pathway determined, at least in part, by symmetrical interactions with the histones of the core. These interactions have been studied by analyzing the fragments produced by nuclease digestion. Treatment of chromatin or isolated nucleosomes with DNase I reveals, after denaturation, a regular series of single-stranded DNA fragments differing in length by multiples of 10 nucleotides (Noll, 1974; Sollner-Webb and Felsenfeld, 1977; Whitlock, 1977; Lutter, 1978; Sollner-Webb *et al.*, 1978). The DNase I-susceptible sites in the nucleosome vary in their accessibility to nuclease attack, giving rise to a characteristic frequency pattern of DNA lengths in the digest. Sites at 30, 60–80, and 110 nucleotides from the 5′-ends of the 140 nucleotide pair DNA of the nucleosome core are relatively inaccessible to DNase I (Whitlock and Simpson, 1976). The excision of the NH_2-terminal ends of the histones by trypsin alters the DNase I cutting patterns; in particular, the regions between 20 and 35 nucleotides, and between 60 and 80 nucleotides from the 5′-termini become more accessible to nuclease attack (Whitlock and Simpson, 1977). Shielding of these regions by the NH_2-termini of histones H3 and H4 seems likely. The DNase I cuts are symmetrically located about a common axis, and the same is true for internucleosomal cuts by DNase II and staphylococcal nuclease (Sollner-Webb *et al.*, 1978).

The discrete and symmetric changes in the nuclease cutting patterns after removal of the NH_2-terminal regions of the histones implies that the basic arms of the eight core histones interact with the enveloping DNA strand at precise distances from the ends. The sequence of histone arrangement along DNA has been studied by a novel form of cross-linking. The DNA in the core particles was methylated and the methylated purines (7-methylguanine and 3-methyladenine) were partially removed (by incubation at 40°C at neutral pH). The aldehyde groups formed at the depurinated sites react with the ε-amino groups of the adjacent lysine residues of histones. This reaction labilizes the phosphodiester bonds and generates a series of DNA fragments attached at their 5′-ends to histone molecules. The fragments were separated according to size, and the DNA was removed prior to electrophoretic identification of the histones. This analysis showed that there are histone-free intervals in the core DNA at a regular distance of about $(10)n$ nucleotides from the 5'-termini, in agreement with the results of nuclease cutting experiments. The eight core histones occur in a symmetrical distribution along the DNA strand, and the two members of each histone class are bound similarly to the antiparallel strands of the DNA. Histone–DNA interaction is largely restricted to the wide groove of the DNA double helix (Mirzabekov and Melnikov, 1974), but 20 nucleotide segments at the 5'-ends of the core DNA do not appear to be attached to the core histones (Mirzabekov et al., 1978). In the linear map of histone arrangement along the 140 nucleotide pairs of DNA, the two H3 molecules have primary attachment sites at the 3'-ends and in the center, while both H4 molecules are located centrally. Two important features of the model are (1) that all the histone dimers formerly identified in solution and in cross-linked nucleosomes are adjacent in the model and (2) that the histone amino-termini cover a little less than one turn of the DNA helix. Thus, the incomplete helical clamps would allow the core particle to dissociate into histone octamer (or tetramers) and DNA. Such a dissociation to release constraints upon the DNA strand would be favored by conditions which weaken electrostatic interactions (such as high ionic strength) or by reactions which neutralize the positive charges on the histone lysine residues. This is effectively what occurs when histones are acetylated in situ, and this postsynthetic modification of histone primary structure provides a mechanism for the modulation of DNA–histone interactions in transcribing chromatin.

B. The Nucleosome and Transcription

A fundamental question in the analysis of chromatin structure is whether the histones of the nucleosome core constitute an obstacle to RNA polymerase readout of the associated DNA sequences. It has long been recognized that exogenous histones can inhibit RNA polymerase activity in chromatin (Huang and Bonner, 1962) and in isolated nuclei (Allfrey et al., 1963), but such observations do not translate readily to the effects of histone octamers on the DNA of the nucleosome. However, the observation that selective tryptic digestion of histones resulted in a 200–400% stimulation of RNA synthesis in thymus lymphocyte nuclei (Allfrey et al., 1963) fits the current view that excision of the NH_2-terminal regions of the histones would make the DNA more accessible to RNA polymerases.

The experimental evidence that nucleosomes act to retard passage of RNA polymerase is quite convincing. There is a reduction in the rate of

RNA chain elongation and in the number of RNA polymerase binding sites in chromatin as compared to deproteinized DNA (Cedar and Felsenfeld, 1973). This restriction of chromatin template function is retained in isolated nucleosomes. The rate of RNA synthesis from chromatin monomers is only 8–15% of that obtained with DNA extracted from the monomers, and the number of RNA polymerase binding sites on chromatin subunits is six to twenty times lower than that of the DNA extracted from those subunits (Bustin, 1978).

When purified core histones (H2A, H2B, H3, and H4) and bacteriophage T7 DNA were reconstituted to form a nucleoprotein complex, the histones inhibited both RNA chain initiation and elongation. At low ionic strengths, where electrostatic interactions would be high, RNA propagation was slowed and eventually halted by nucleosomes. At higher salt concentrations, where electrostatic interactions between histones and DNA would be substantially reduced, chain elongation rates approximated those on naked DNA (Williamson and Felsenfeld, 1978).

The parallelisms between strong histone–DNA binding and template restriction are clear. What is less clear is the degree to which the histone–DNA complex must be dissociated to permit passage of the RNA polymerase, and whether this must be accompanied by a change in the shape of the nucleosome, possibly by an unfolding about its dyad axes of symmetry (Weintraub et al., 1976).

Actively transcribed DNA sequences are known to be more accessible to endonucleases, such as DNase I which cut within the chromosomal subunit. Limited DNase I digestions of chromatin release a large fraction of the newly-synthesized RNA chains (Billing and Bonner, 1972), and nuclear subfractions enriched in transcriptionally active chromatin are more susceptible to DNase I attack than are transcriptionally inert fractions (Berkowitz and Doty, 1975). Of particular interest are observations showing that limited DNase I digestions selectively degrade those DNA sequences that are programmed for transcription in a given cell type. For example, a limited DNase I digestion of chick red cell nuclei removes the globin DNA sequences, but those sequences are not selectively degraded during DNase I treatment of brain or fibroblast nuclei (Weintraub and Groudine, 1976). Similarly, the ovalbumin genes are preferentially degraded during DNase I digestions of nuclei from the oviduct, but not from the liver; over 70% of the ovalbumin coding sequences are digested when only 10% of the oviduct nuclear DNA has been solubilized (Garel and Axel, 1976). Moreover, it was found that the nontranscribed ovalbumin sequences in the nuclei of red cells and fibroblasts are relatively resistant to DNase I digestion (Weintraub and Groudine, 1976).

DNase I sensitivity and transcriptional potential correlate well over a broad spectrum of cell types and in organisms as diverse as mammals, birds, fish, insects, and myxomycetes. In trout testis nuclei, for example, limited DNase I digestions preferentially deplete the chromatin of DNA sequences being transcribed into polyadenylated mRNAs (Levy-W. and Dixon, 1977), and in the slime mold, *Physarum polycephalum,* the actively transcribed ribosomal genes are more rapidly degraded by DNase I than is main brand DNA (Johnson *et al.,* 1978b). An important point is that the DNase I sensitivity of transcribable sequences does not appear to be related to the rates at which different genes are transcribed *in vivo:* over a range of 10–20 initiations/min for ovalbumin mRNA to only 5–10 initiations/day for some scarce mRNA populations (Garel *et al.,* 1977). It follows that the DNase I-sensitive conformation is probably not determined by the presence or density of RNA polymerases on the DNA template.

The persistent altered conformation of the potentially transcribable genes is also seen in integrated viral genomes. The transcribable sequences of adenovirus type 5, for example, are highly susceptible to DNase I degradation, while nontranscribed sequences are less susceptible. In adenovirus-transformed hamster cells there is a remarkable spatial correlation between the nuclease–sensitive portion of the integrated viral genome and the length of the transcription unit to within 2–3 nucleosomes at the 5'-end and between 2 and 15 nucleosomes at the 3'-end (Flint and Weintraub, 1977).

Such changes in the conformation of transcriptionally active chromatin are maintained in chromatin subfractions as small as the nucleosome itself, as judged by the selective degradation of the globin sequences in avian erythrocyte nucleosome populations (Weintraub and Groudine, 1976) and by the selective DNase I sensitivity of avian retrovirus genomes in chick fibroblast nucleosomes (Groudine *et al.,* 1978). This implies that nucleosomes are not all identical and that chromatin subunits containing transcribable DNA sequences differ in structure. Evidence that the structural differences involve postsynthetic modifications of the histones and the introduction of nonhistone proteins will be considered in Sections II,C and D and Section III.

In earlier discussions of the ultrastructure of chromatin in lampbrush "loops" and ribosomal transcription units (Section I,B), it was pointed out that nucleosomes are not visible in highly active regions and that the DNA template, studded with RNA polymerase molecules, exists in an almost fully extended configuration. The contrast in ultrastructure between transcribing and nontranscribing chromatin is particularly striking in ribosomal genes in which nontranscribed "spacer" sequences intervene between the coding sequences. In *Physarum polycephalum,* for example, a long, beaded central spacer separates the two rDNA transcription units which appear to be free of nucleosomes (Grainger and Ogle, 1978). As a consequence of their extended conformation, the coding regions of the ribosomal genes in *Physarum* are rapidly degraded by endonucleases, including staphylococcal nuclease (Johnson *et al.*, 1978a,b). However, cleavage of the coding sequences is not random; nucleoli treated with staphylococcal nuclease release a series of fragments of discrete lengths containing both transcribed and nontranscribed rDNA sequences. In short-term digestions, the coding regions of the rDNA are the first to be degraded, and the ribosomal coding sequences are cut into fragments about 140 nucleotide pairs in length. These fragments can be separated from the usual 11 S nucleosome monomers by gradient centrifugation, and a 5 S peak, enriched in sequences coding for 19 S and 26 S ribosomal RNA is recovered. Sequences from the central spacer are recovered in the 11 S nucleosomes and in higher oligomers. Electron microscopy of the 5 S peak reveals that the DNA is present in an extended fibrillar form with a DNA compaction ratio of only 1.2 to 1.3. The 11 S peak consists of typically 'beaded' nucleosomes (Johnson *et al.*, 1978b). The two types of monomeric chromatin subunits, both containing 140 nucleotide pairs of DNA, one extended and the other beaded, represent extremes in chromatin organization. Such fully extended chromatin subunits are probably present only during the most active periods of ribosomal RNA synthesis, and they are not seen in the transcriptionally inert microsclerotial form of *Physarum*. The extended DNA in the 5 S particles is not free DNA, but is associated with histones and nonhistone proteins. Histone H1 is absent, in accord with observations on the absence of histone H1 in the puffs of polytene chromosomes in *Drosophila* (Jamrich *et al.*, 1977). There is a marked deficiency of the two *Physarum* histones corresponding to mammalian H3 and H4 (Johnson *et al.*, 1978a). Since histone H3 and H4 are essential in the organization of the beaded state of nucleosomes (Camerini-Otero *et al.*, 1976, 1978; Moss *et al.*, 1977; Oudet *et al.*, 1978b), the extended state of the transcriptionally active rDNA sequences may indicate that the absence of H3 and H4 precludes any coiling of the rDNA into beaded structures. All four nucleosomal histones are present in the 11 S chromatin monomers of *Physarum* and these are beaded (Johnson *et al.*, 1978b). It has not yet been determined whether the release of histone H3 and H4 from the transcription unit is a transient phenomenon, perhaps facilitated by hyperacetylation of the lysine residues in the DNA-binding regions of the polypeptide chains.

The *Physarum* rDNA exemplifies the relationship between the extension of the DNA in the transcription unit and its susceptibility to attack by staphylococcal nuclease. A preferential degradation of the active ribosomal genes is also seen in *Xenopus* oocyte nuclei (Reeves, 1978a,b). Other examples include mRNA coding sequences, such as the ovalbumin gene that is cleaved faster than the globin gene during staphylococcal nuclease digestions of oviduct chromatin (Bellard *et al.*, 1978).

The enhanced susceptibility of actively transcribing DNA sequences to endonuclease digestion is frequently employed as a method of chromatin fractionation. Treatment of oviduct nuclei with staphylococcal nuclease for brief periods releases a set of nucleosomes that are five- to sixfold enriched in the ovalbumin genes, as compared to total nuclear DNA (Bloom and Anderson, 1978). Staphylococcal nuclease digestion of rat liver nuclei for 60 sec, followed by disruption in EGTA and differential centrifugation, yields a template-active chromatin fraction containing less than 10% of the total nuclear DNA but over 85% of the engaged RNA polymerase II molecules (Tata and Baker, 1978).

DNase II has also been employed for selective release of transcriptionally active chromatin. Rat liver nuclei, sheared with DNase II and fractionated on the basis of solubility in 2 mM MgCl$_2$, yield a soluble component enriched in specific subsets of DNA sequences which hybridize extensively to whole-cell RNA (Billing and Bonner, 1972; Gottesfeld *et al.*, 1974). A DNase II-sensitive fraction of chick reticulocyte chromatin contained less than 1% of the total chromatin DNA but over 50% of the nascent chromatin-bound RNA. A three- to fivefold enrichment of the globin genes was obtained in the nuclease-sensitive fraction of reticulocyte chromatin, but there was no enrichment of globin sequences in the corresponding fraction from liver chromatin (Hendrick *et al.*, 1977). When Friend erythroleukemia cells are treated with dimethyl sulfoxide, hemoglobin mRNA synthesis is induced (Ross *et al.*, 1972), and the globin sequences are recoverable in the DNase II-sensitive, Mg^{2+}-soluble fraction of the chromatin (Wallace *et al.*, 1977; Bonner *et al.*, 1978). While no enrichment for globin genes is observed in the Mg^{2+}-soluble fraction of uninduced cells, the globin sequences are significantly more abundant in the corresponding fraction from DMSO treated cells (Gottesfeld and Partington, 1977). The reverse distribution is seen for the ribosomal genes; transcription of rDNA is inhibited in DMSO-treated cells, and the DNase II-sensitive, Mg^{2+}-soluble fraction contains fewer ribosomal RNA coding sequences than observed in uninduced cells (Gottesfeld and Partington, 1977).

What is the reason for the enhanced staphylococcal nuclease and DNase II sensitivity of actively transcribed chromatin? Staphylococcal nuclease, like certain endogenous nucleases (Hewish and Burgoyne, 1973; Simpson and Whitlock, 1976), cuts the DNA linker strands between the nucleosomes to generate an ordered series of chromatin fragments (Rill and Van Holde, 1973; Noll, 1974; Shaw *et al.*, 1974; Axel, 1975; Sollner-Webb and Felsenfeld, 1975; Compton *et al.*, 1976; Shaw *et al.*, 1976; Woodcock and Frado, 1978). The greater accessibility of DNA in the linker regions is also evident in its high reactivity with cross-linking reagents such as trimethylpsoralen (Wiesehahn *et al.*, 1977; Cech and Pardue, 1977) and in its selective damage by chemical carcinogens (Metzger *et al.*, 1976, 1977; Ramanathan *et al.*, 1976; Jahn and Litman, 1977). As a consequence, limited digestions of carcinogen-treated chromatin with staphylococcal nuclease release a disproportionate amount of modified nucleotides. The nonrandom distribution of modified nucleotides after carcinogen administration is also reflected in a rapid rate of excision repair in the "linker" regions and a slow repair in the "core" DNA sequences (Bodell, 1977; Cleaver, 1977).

The question naturally arises as to whether the more accessible linker strands are likely to be enriched in those DNA sequences required for the control of transcription by the binding or activator or repressor molecules. This would imply a nonrandom distribution of nucleosomes on the DNA strand. Some studies designed to test this question have led to the opposite conclusion; e.g., nucleosomes are randomly distributed on SV40 viral chromatin (Polisky and McCarthy, 1975; Cremisi *et al.*, 1976) and on *Drosophila* chromatin (Prunell and Kornberg, 1978). It has also been observed that nucleosomes on linear SV40 minichromosomes can slowly migrate into a covalently attached but nucleosome-free viral DNA strand (Beard, 1978). However, more recent studies indicate that DNA sequence can direct the placement of histone 'cores' on SV40 chromatin (Wasylyk *et al.*, 1979), avian tRNA genes (Wittig and Wittig, 1979), and primate satellite DNAs (Brown *et al.*, 1979)

Spacer length between nucleosomes is a consideration in nuclease digestions. In *Physarum*, for example, the average repeat length of the DNA in chromatin subunits progressively diminishes during staphylococcal nuclease digestion, due to the preferential cutting of the longer linkers (Johnson *et al.*, 1976).

It follows that the rapid endonucleolytic cleavage of transcriptionally active chromatin is a direct consequence of a more extended and less shielded conformation of the DNA. Since the linker region between nucleosomes is associated with histone H1 (Baldwin *et al.*, 1975; Shaw *et al.*, 1976; Varshavsky *et al.*, 1976; Whitlock and Simpson, 1976b; Noll and Kornberg, 1977), a loss of H1 or modulation of its DNA cross-linking potential may occur at times of gene activation. In accord with this view are observations that the rate of release of nucleosomes during staphylococcal nuclease digestion is inversely proportional to H1 content (Sanders, 1978), and that phosphorylation of a specific site in histone H1 releases constraints on DNA in H1–DNA complexes (Adler *et al.*, 1972).

DNA sequences within the nucleosome are shielded by associations with the core histones, H2A, H2B, H3, and H4. Changes in the conformation of the 140 nucleotide pairs of DNA encircling the nucleosome are brought about by changes in the structure of the core histones and certain nonhistone proteins present in some of the nucleosomes, such as the HMG proteins (Section III).

C. Postsynthetic Modifications of Histones and DNA Template Accessibility

All of the histones of the nucleosome core, and histone H1, are subject to enzymatic modifications of structure which influence their interactions with DNA. For histones of the nucleosome core, the most prevalent modification is acetylation. The enzymatic basis of histone acetylation, evidence for the presence of acetyltransferases in chromatin, the structures and sites of modification of the histones, and correlations between histone acetylation and gene activation have been considered in recent reviews (Allfrey, 1977; Johnson and Allfrey, 1978). The present discussion will emphasize recent evidence that enzymatic acetylation of the histones alters the conformation of DNA in the nucleosome.

The organization of discrete lengths of DNA into nuclease-resistant "cores" depends mainly upon histone classes H3 and H4 (Camerini-Otero *et al.*, 1976; Sollner-Webb *et al.*, 1976). As described previously (Section II,A), the interaction between these histones and DNA primarily involves the positively-charged NH_2-terminal regions of the polypeptide chains. The lysine residues in these regions are specifically modified by enzymes which transfer acetyl groups from acetyl-coenzyme A to the ϵ-amino groups of lysine side chains, thus neutralizing their positive charge (Allfrey, 1964, 1970, 1977; Allfrey *et al.*, 1964; Gershey *et al.*, 1968). The reaction is reversible, in the sense that acetyl groups, once incorporated, can be removed by the action of histone deacetylases, without degradation of the histone chain (Pogo *et al.*, 1967, 1968; Inoue and Fujimoto, 1972; Vidali *et al.*, 1972; Edwards and Allfrey, 1973; Sanders *et al.*, 1973; Jackson *et al.*, 1975; Ruiz-Carrillo *et al.*, 1975, 1976; Boffa *et al.*, 1978; Sealy and Chalkley, 1978; Vidali *et al.*, 1978a,b).

Histones H3 and H4 each have multiple sites of acetylation. The modification of H3 *in vivo* affects lysine residues at positions 9, 14, 18, and 23 of the polypeptide chain (Candido and Dixon, 1972; Marzluff and McCarty, 1972; DeLange *et al.*, 1973; Hooper *et al.*, 1973, Brandt *et al.*, 1974; Dixon *et al.*, 1975). *In vitro*, there is an additional site of acetylation at lysine-4 (Thwaits *et al.*, 1976a). Histone H4 has a major site of acetylation at lysine-16 (DeLange *et al.*, 1969; Ogawa *et al.*, 1969) and minor sites at positions 5, 8, and 12 of the polypeptide chain (Sung and Dixon, 1970; Candido and Dixon, 1971; Dixon *et al.*, 1975; Thwaits *et al.*, 1976b). It is significant that all these sites of histone modification are limited to the NH_2-terminal regions of histones H3 and H4, and the same is true for acetylation of histones H2A and H2B (Dixon *et al.*, 1975).

The H3 sequence, 1–23, contains 5 lysine and 3 arginine residues, all of which would be positively charged at physiological pH values. Acetylation of the lysine residues at positions 9, 14, 18, and 23 would neutralize half of the positive charges in that region of the molecule. The H4 sequence, 1–20, also contains 5 lysine and 3 arginine residues, plus a histidine at position 18. Acetylation of lysine residues at positions 5, 8, 12, and 16 would also neutralize about half of the positive charge in the NH_2-terminal region of the molecule. Thus, tetraacetylation of the key nucleosomal histones, H3 and H4, would be expected to substantially weaken the electrostatic interactions between the histone "core" and the enveloping DNA strand. It follows that acetylation of histones H3 and H4 should release constraints on the DNA of the nucleosome core. Evidence in support of this view will now be presented.

In considering the role of histone acetylation in the control of DNA conformation, it is important to note that the potential for acetylation at multiple sites is not realized in every histone molecule; i.e., particular lysine residues may or may not be acetylated. Because this type of histone modification is a rapid and reversible process which can proceed in the absence of histone synthesis or degradation (Pogo *et al.*, 1967, 1968; Edwards and Allfrey, 1973; Sanders *et al.*, 1973; Jackson *et al.*, 1975), acetate "turnover" on stable histone molecules generates a changing population of histone subfractions. Consequently, preparations of purified histones, such as H3 and H4, are internally heterogeneous, each comprising a mixture of polypeptide chains of identical amino acid sequence, some of which are internally acetylated to different degrees while others are not acetylated at all. [Each of these subfractions may then differ with regard to other forms of substitution such as methylation and phosphorylation (Paik and Kim, 1975; Hohmann, 1978). Sequence heterogeneity in histone H3 adds additional complexity (Marzluff *et al.*, 1972; Pathy and Smith, 1975).]

Because of the charge neutralization associated with acetylation of the lysine ε-amino groups, histone subfractions which differ in their degree of acetylation migrate at different rates in acid–urea–polyacrylamide gels (Panyim and Chalkley, 1969) and in starch–urea–aluminum lactate gels (Sung and Dixon, 1970). The chromatographic separation of histone H4 subfractions of known ε-N-acetyllysine content made it possible to confirm the identity of the various histone brands as separated electrophoretically (Wangh et al., 1972). [In very high resolution gels it is possible to separate the diacetylated form of histone H4 from a form containing one acetyl and one phosphoryl group (Ruiz-Carrillo et al., 1975).] Using such methods it is now routine procedure to separate histone H4 into a series of five bands of decreasing mobility corresponding to the unacetylated form and derivatives containing 1, 2, 3, and 4-ϵ-N-acetyllysine residues. The degree of histone acetylation and the relative proportions of the various acetylated forms of H3 and H4 can be readily monitored by densitometry of stained histone gels. The results can be confirmed when necessary by labeling of the histones with radioactive acetate and measuring isotope distribution in each of the bands.

In most cells, the multiacetylated derivatives of histones H3 and H4 are present in small amounts relative to the nonacetylated and monoacetylated forms, and in cell types that are incapable of transcription [such as the mature sperm of Arbacia lixula (Wangh et al., 1972) or carp (Hooper et al., 1973)], all of the histone occurs in the nonacetylated form. The correlations between histone acetylation and transcription will be discussed in detail in Section II,D; the main point to be made here is that the degree of histone acetylation can be manipulated in living cells and monitored by electrophoretic analysis.

It has been observed that exposure of HeLa cells to 5 mM sodium butyrate leads to an accumulation of multiacetylated forms of histones H3 and H4 (Riggs et al., 1977). Similar effects have been reported for Friend erythroleukemia cells (Riggs et al., 1977; Candido et al., 1978) and for a wide variety of other vertebrate cell lines (Candido et al., 1978; Sealy and Chalkley, 1978a). The proportion of acetylated H4 in butyrate-treated HeLa cells increases progressively with time, and may exceed 80% of the total H4 present in about 20 hr (Riggs et al., 1977; Sealy and Chalkley, 1978a; Simpson, 1978; Vidali et al., 1978a). The effect is rapidly reversible, and most of the acetylated forms of histone H4 are converted back to the unmodified form within 15 min after removal of the butyrate (Vidali et al., 1978b). This is in accord with estimates that about 70% of the acetate incorporated into histones H3 and H4 during a short pulse of HTC cells with [^{3}H]acetate is removed with a half-life of about 3 min (Jackson et al., 1975).

The reason why exposure to sodium butyrate increases the level of histone acetylation has been clarified; 5 mM butyrate has been found to inhibit histone deacetylase activity in vitro (Boffa et al., 1978; Candido et al., 1978) and in vivo (Boffa et al., 1978; Candido et al., 1978; Sealy and Chalkley, 1978; Vidali et al., 1978a,b). Because there is no corresponding inhibition of acetyltransferase activity, the uptake of acetate into the histones continues while the removal of acetyl groups is effectively blocked (Candido *et al.*, 1978; Hagopian *et al.*, 1977; Vidali *et al.*, 1978). The resulting accumulation of the multiacetylated forms of histones H3 and H4 has important consequences for DNA conformation and template function.

It is known that the multiacetylated histones of butyrate-treated cells occur in nucleosome core particles (Simpson, 1978; Vidali *et al.*, 1978a). An alteration in the structure of the cores is indicated by a two- to tenfold increase in the rate of DNA degradation by DNase I (Nelson *et al.*, 1978a,b; Sealy and Chalkley, 1978b; Simpson, 1978; Vidali *et al.*, 1978a). Comparisons of the DNase I cutting patterns of highly acetylated and control core particles have shown that a site 60 nucleotides from the 5'-end of the 140 nucleotide pair DNA strand, which is resistant to DNase I attack in control cells, becomes susceptible to DNase I in core particles from butyrate-treated cells. Moreover, the 5'-terminal phosphates at the ends of the core DNA are removed two- to threefold faster during digestions of hyperacetylated core particles with staphyloccal nuclease than from core particles of control cells (Simpson, 1978). It follows that the weakening of histone–DNA interactions resulting from the acetylation of the lysine residues affects the accessibility of the DNA at 0, 60, and 140 nucleotides from the ends. The increased accessibility of DNA at 60 nucleotides from the 5'-ends is consistent with the view that histone H4, a major target of the acetylation reaction, interacts with the enveloping DNA strand in that region [as judged by histone–DNA cross-linking experiments (Mirzabekov *et al.*, 1978)]. The increased DNA accessibility at the 5'-ends is consistent with proposals that H3 and H4 interact with DNA on the side of the core particle which contains the entrance and exit sites for the DNA strand (Simpson, 1976; Bina-Stein and Simpson, 1977).

Similar conclusions about the modification of nucleosome core structure and increased DNase I sensitivity due to acetylation of the histones have been drawn from studies of chromatin modified by acetic anhydride (Wallace *et al.*, 1977) or acetyl adenylate (Shewmaker *et al.*, 1978). The physical changes in chromatin after chemical acetylation are more extreme than those noted as a consequence of enzymatic acetylation, probably because the chemical modification affects many lysine residues which are not normally substrates for the transacetylases. For example, acetic anhydride is known to acetylate lysine residues on histone H1 (Wong and Marushige, 1976), a protein class which is not subject to modification of its lysine residues *in vivo* (Candido and Dixon, 1972b). However, the chemical methods are likely to achieve a more complete acetylation of the NH_2-terminal regions of the nucleosomal histones than is achieved in butyrate-treated cells, and the results of chemical acetylation may be regarded as the limit of nucleosome destabilization which might be attained by enzymatic modification of the DNA-binding basic proteins. Evidence that the template function of chromatin is enhanced after treatment with acetic anhydride is discussed in Section II,D.

The results of acetylation, as seen in the chromatin or isolated nucleosomes of butyrate-treated cells, are largely due to modifications of histones H3 and H4, because acetylation levels of H2A and H2B are not as appreciably altered by butyrate (Candido *et al.*, 1978). The fact that acetylation of histone H4 might influence the conformation of DNA was suggested in earlier studies which compared the circular dichroism spectra of complexes containing DNA and chromatographically purified subfractions of H4 of differing ϵ-N-acetyllysine content. Acetylated H4 was found to be less effective than the nonacetylated form in producing conformational distortions of DNA (Adler *et al.*, 1974). Although the extrapolation of observations on DNA binding by a single histone class to the more complex situation of DNA binding by sets of histones in the nucleosome core is not entirely valid, the main point (that acetylation does influence DNA conformation) is now amply sustained by studies of the hyperacetylated chromatin of butyrate-treated cells.

The increased DNase I sensitivity of the nucleosomes in butyrate-treated cells has been assumed to have its origins in the reduced DNA-binding affinity of the acetylated histones. This conclusion is confirmed by studies showing that the multiacetylated forms of histones H3 and H4 are preferentially released during limited DNase I digestions of HeLa nuclei (Vidali *et al.*, 1978a) and HTC cell nuclei (Sealy and Chalkley, 1978b). In HeLa nuclei, about 32% of the acetylated histones are lost after removing only 11% of the DNA (Vidali *et al.* 1978a).

The increased accessibility of the DNA in the hyperacetylated chromatin of butyrate-treated cells is also evident in its reactivity with carcinogens. Both chlorozotocin and 1-(2-chloroethyl)-3-cyclohexyl-1-nitrosourea react preferentially with the extended (eu) chromatin fraction of the HeLa nucleus. Pretreatment of HeLa cells with 5 mM sodium butyrate was found to double the uptake of both nitrosoureas into chromatin and to raise the levels of alkylation of DNA and carbamolylation of associated chromosomal proteins (Tew *et al.*, 1978).

In accord with the view that hyperacetylation of histones should promote a greater extension of the DNA in chromatin, electron microscopy of butyrate-treated human bronchogenic carcinoma cells shows a progressive disperson of the heterochromatic clumps; all of the control cells had more than 10 clumps per nucelus, while 83% of the treated cells had 4 or less (Tralka *et al.*, 1979). It is significant that cells treated with butyrate responded with an ectopic production of chorionic gonadotropin and its α-subunit. Further evidence that hyperacetylation of the histones affects DNA template function will now be considered.

D. Histone Acetylation and Transcription

The view that histone acetylation plays a role in transcriptional control originated in studies of the effects of this modification on the capacity of the arginine-rich histones, H3 and H4, to inhibit RNA synthesis in cell-free systems. When histones H3 and H4, acetylated to varying degrees (by reaction with small amounts of acetic anhydride), were added to DNA in the presence of RNA polymerases from calf thymus nuclei, the degree of inhibition of RNA synthesis was decreased in proportion to the extent of histone acetylation (Allfrey *et al.*, 1964). Similarly, treatment of calf thymus chromatin with acetic anhydride markedly increased the rate of DNA-dependent RNA synthesis (as assayed with added *E. coli* RNA polymerase) (Marushige, 1976). The number of initiation sites for the *E. coli* enzyme was reported to be doubled when rat liver chromatin was exposed to 7 m*M* acetic anhydride (Oberhauser *et al.*, 1978). The increased availability of the DNA template to exogenous polymerases is in accord with the great increase in DNA accessibility to nucleases in chemically acetylated rat liver chromatin (Wallace *et al.*, 1977b).

Because the question of fidelity of histone acetylation by acetic anhydride clearly limits the direct chemical approach, the following discussion of histone modification will emphasize enzymatic conversions of histone subtypes and draw particular attention to the strong temporal and spatial correlations between acetylation and RNA synthesis *in vivo*.

Two observations on intranuclear sites of histone acetylation deserve special comment. In the mealy bug, *Planococcus citrii*, males preferentially utilize the maternal chromosome set and sequester the paternal chromosomes in a heterochromatic mass. The euchromatic, transcriptionally active maternal chromosomes incorporate about seven times more [³H]acetate than does the heterochromatic repressed paternal set (Berlowitz and Pallotta, 1972). In the ciliate, *Tetrahymena pyriformis*, transcription is largely limited to the macronucleus, while the micronucleus remains quiescent. Analyses of the histones is isolated macro- and micronuclei show that histone H4 is present in both, but macronuclei contain acetylated forms of H4, while micronuclei contain only the nonacetylated form (Gorovsky *et al.*, 1973).

The close association of acetylated histones with transcriptionally active chromatin has been confirmed by a variety of nuclear fractionation techniques. Sonication and differential centrifugation of calf thymus nuclei yields fractions which differ in transcriptional activity *in vivo* and *in vitro* (Frenster *et al.*, 1963). When such fractions were compared with regard to their content of radioactive histones after labeling with [¹⁴C]acetate, the histones of the transcriptionally active euchromatin fractions were four to six times more radioactive than those of the heterochromatic clumps (Allfrey, 1964). When chromatin is digested with DNase II and the solubilized products fractionated into Mg^{2+}-soluble and Mg^{2+}-insoluble fractions, the actively transcribing sequences are recovered in the Mg^{2+}-soluble fraction (Marushige and Bonner, 1971; Gottesfeld et al., 1974; Gottesfeld and Butler, 1977). When this procedure was applied to *Drosophila* cells after labeling with [³H]acetate, the histones of the template-active fraction had higher specific activities than those of the less active fraction (Levy-Wilson *et al.*, 1977). Similar fractionations of trout testis chromatin, combined with electrophoretic analysis of the modified histones, showed that the highly acetylated forms of histone H4 (di-, tri-, and tetraacetyl derivatives) were mainly associated with the template-active fraction (Davie and Candido, 1978). Another approach to chromatin fractionation is based on the rapid release of nucleosomes from transcribed chromatin during limited digestions with staphylococcal nuclease (Levy-W. and Dixon, 1978). The nucleosome fraction enriched in transcribed DNA sequences of trout testis was also shown to possess high levels of the multiacetylated forms of histone H4 (Levy-Wilson *et al.*, 1979).

The selective attack on transcribable DNA sequences by DNase I has been described in Section II,B. When avian erythrocyte nuclei are treated with DNase I under the conditions shown to preferentially degrade the globin genes (Weintraub and Groudine, 1976), there is a selective release of the acetylated forms of histones H3 and H4 (Allfrey *et al.*, 1977). Limited DNase I digestions of nuclei isolated from HeLa cells (Vidali *et al.*, 1978a) or HTC cells (Sealy and Chalkley, 1978b) preferentially release the multiacetylated forms of histone H4. The ratio of tetraacetylated H4 to nonacetylated H4 in the released histones is high (7 : 1) at early stages of digestion and progressively diminishes as digestion proceeds into the inactive DNA sequences (Sealy and Chalkley, 1978b).

Thus, several different experimental approaches confirm that multiacetylated forms of the histones are localized in the transcriptionally active regions of the chromatin of avian, mammalian, and insect cells. This disposition of modified histones does not, in itself, prove that histone acetylation is sufficient to induce the template-active state of the associated DNA sequences. Indeed, other evidence (to be discussed in Section III) establishes that certain nonhistone proteins play a key role in the control of DNA accessibility in the transcription unit. However, the modulation of nucleosome structure by acetylation of histones H3 and H4 does appear to influence transcription *in vivo* and *in vitro*.

The higher levels of histone acetylation seen in cultured cells after exposure to butyrate correlate with increased DNA accessibility to

DNase I, and one might expect a corresponding increase in the number of RNA polymerase binding sites and a facilitation of polymerase movement along a chromatin strand in which the histones have been so extensively modified. A recent report indicates that chromatin from HeLa cells cultured in the presence of butyrate is a more active template, showing an RNA elongation rate (as measured with *E. coli* RNA polymerase) about twice that of chromatin from HeLa cells not exposed to butyrate. The increase in template activity of hyperacetylated chromatin was roughly proportional to the extent of histone modification (Hagopian *et al.*, 1978). [It should be noted that even butyrate-treated cells have a low proportion of the tetraacetylated forms of histones H3 and H4. Thus, attempts to reconstitute nucleosomes from DNA and a random mixture of histones from butyrate-treated cells would be expected to yield only a small fraction of nucleosomes modified at all, or most, of the potential acetylation sites. This may account for the failure to detect significant differences in DNA template activity when nucleosomes reconstituted from SV40 DNA and histones from butyrate-treated HeLa cells were compared with nucleosomes reconstituted from SV40 DNA and histones from calf thymus lymphocytes (Mathis *et al.*, 1978). An equally important consideration is the failure to include in such reconstitution experiments those nonhistone proteins, such as HMG 14 and HMG 17, which are also acetylated and known to influence DNA accessibility and template function in native chromatin (See Section III).]

More compelling evidence that acetylation of nucleosomal proteins leads to changes in DNA template function is provided by numerous observations that cells cultured in the presence of butyrate are induced to synthesize "new" gene products. For example, the globin genes are activated in erythroleukemia cells exposed to 1 m*M* butyrate (Leder *et al.*, 1975). Tumorigenic Syrian hamster cells which have lost the ability to synthesize certain polypeptide species (of molecular weights 13,000 and 41,000) recover that ability when cultured in butyrate and lose it again when butyrate is removed from the medium (Leavitt and Moyzis, 1978). Butyrate induces the synthesis of human chorionic gonadotrophin and its α-subunit in nontrophoblastic human tumor cells (Chou *et al.*, 1977). In HeLa cells, low concentrations of butyrate induce high levels of alkaline phosphatase activity (Griffin *et al.*, 1974) and sialyltransferase activity (Fishman *et al.*, 1976). Other butyrate effects on HeLa cells include inductions of follicle-stimulating hormone (Ghosh and Cox, 1977), chorionic gonadotropin (Ghosh and Cox, 1976), and β-adrenergic receptors (Tallman *et al.*, 1977). In neuroblastoma cells, butyrate increases tyrosine hydroxylase (Waymire *et al.*, 1972; Prasad and Sinha, 1976), choline acetyltransferase, and adenylate cyclase activities (Prasad and Sinha,

1976). There are many reports of morphological and physiological changes in mammalian cells exposed to butyrate, most of which are reversible when the fatty acid is removed from the medium; for a review of this subject see Prasad and Sinha (1976). Of particular interest are reports that butyrate causes a loss of malignant characteristics in Syrian hamster tumor cells (Leavitt *et al.*, 1978) and inhibits DNA synthesis in neuroblastoma cells (Schneider, 1976), in HeLa cells (Hagopian *et al.*, 1977), and in mouse kidney cells transformed by murine sarcoma virus (Altenburg and Steiner, 1979). Nuclei from butyrate-treated HeLa cells fail to incorporate deoxyribonucleoside triphosphates into DNA (Hagopian *et al.*, 1977). Whether the suppression of DNA synthesis in butyrate-treated tumor cells is due to hyperacetylation of the histones and concomitant changes in nucleosome structure that might affect the binding and replicative function of the DNA polymerases remains to be determined. Cell growth and replication are possible at low butyrate concentrations, e.g., 2 mM (Leavitt and Moyzis, 1978), and 2 mM butyrate does not inhibit ongoing DNA synthesis in mouse splenic lymphocytes (Kyner *et al.*, 1976). But 2 mM butyrate prevents any increase in the rate of DNA synthesis following the exposure of lymphocytes to mitogens such as phytohemagglutinin (PHA) or concanavalin A (Con A) (Kyner *et al.*, 1976). Since mitogenic stimulation is known to increase the acetylation of lymphocyte histones (Pogo *et al.*, 1966), butyrate induction of still higher levels of acetylation may account for this effect. In any case, the evidence is highly suggestive that hyperacetylation of nucleosomal proteins may have very different consequences for DNA or RNA synthesis.

The response of lymphocytes to mitogens is known to involve changes in chromatin structure and composition. Some of these are particularly relevant to the mechanism of control of DNA conformation during a reprogramming of transcriptional activity. Human peripheral lymphocytes, which rarely divide in culture, are induced to reenter the growth cycle after exposure to PHA. Under the proper conditions, 70–80% of the T lymphocytes undergo a blastogenic transformation, as indicated by an increase in size, resumption of DNA synthesis, and eventual mitosis (Moorhead *et al.*, 1960). The synthesis of DNA and the "new" histones needed for cell division is a relatively late event in the transformation process; [^{14}C]thymidine incorporation, for example, is negligible for the first 24 hrs in culture (Pogo *et al.*, 1966). Changes in RNA and protein synthesis, on the other hand, can be detected shortly after the addition of PHA to the culture medium (Rubin and Cooper, 1965; Pogo *et al.*, 1966). A rapid reprogramming of transcription is indicated by changes in tRNA isoaccepting species within 4 hr (Griffin *et al.*, 1976). The lymphocyte response to PHA and other mitogens may be regarded as a triggering of chromosomal functions necessary for cell growth and division, and the system provides a useful paradigm for studies of changes in chromatin-associated proteins at the time of gene activation (Pogo et al., 1966; Kleinsmith et al., 1966; Levy et al., 1973; Johnson et al., 1974).

The kinetics of radioactive acetate uptake into histones after the addition of PHA are particularly suggestive. Within a few minutes, the "arginine-rich" histones, H3 and H4, show a major increase in their rates of [^{14}C]acetate incorporation (without any concomitant increase in the rate of histone synthesis). Pulse-labeling experiments with [^{14}C]uridine and isotopic acetate as precursors showed that the increase in acetylation of the histones precedes the increase in the rate of nuclear RNA synthesis (Pogo et al., 1966). An important point is that these reults, obtained by biochemical analyses of a mixed population of PHA-treated lymphocytes, could be confirmed at the cellular level by autoradiography of single cells; [^{14}C]acetate uptake was observed in the nucleus within 15 min after PHA stimulation, whereas enhanced RNA synthesis was detectable only after 3–6 hr in culture (Mukherjee and Cohen, 1968). Thus, it cannot be argued that histone acetylation happens in one set of cells and RNA synthesis in another.

The acetylation of the lysine residues in lymphocyte histones H3 and H4 would be expected to weaken constraints upon the associated DNA sequences (as discussed in Section II,C). That such changes follow PHA treatment of T lymphocytes is indicated by the changing reactivity of the chromatin toward the DNA-binding dye, acridine orange (AO). The amount of AO binding in the nuclei of PHA-treated cells increases rapidly over a time course which is very similar to that observed in the kinetic studies of histone acetylation. Of particular interest is the observation that a chemical acetylation of the proteins in unstimulated lymphocytes (using acetic anhydride) increases the binding of AO to DNA, while such acetylation of the PHA-treated cells does not lead to any further increase in their DNA dye-binding capacity (Killander and Rigler, 1965, 1969). The use of [^{3}H]actinomycin D as a probe for DNA accessibility led to similar conclusions about the increased reactivity of chromatin in PHA-stimulated cells (Darzynkiewicz et al., 1969). The results support the view that the enzymatic acetylation of the key nucleosomal histones, H3 and H4, can have profound effects on the accessibility and potential template activity of the DNA strand. Moreover, these changes in chromatin structure precede the increase in RNA synthetic capacity of the activated nucleus.

In considering the significance of histone acetylation, it is important to point out that this modification of the nucleosomal basic proteins is not, in itself, sufficient cause for the induction of RNA synthesis at previously repressed gene loci. For example, the blastogenic transformation of PHA-treated lymphocytes can be blocked by the addition of cortisol to the culture medium. No increase in RNA synthesis is observed, although basal rates of RNA synthesis are not significantly altered (Ono *et al.*, 1969). Under these conditions, an early increase in histone acetylation is still detectable, yet there is no obvious stimulation of transcription. The results suggest that changing the physical state of the chromatin is merely a prelude to other, more specific reactions which are needed to initiate RNA synthesis at particular gene loci. In this view, the acetylation of the nucleosomal proteins, H3 and H4 in particular, is part of the enzymatic mechanism for "releasing" the DNA template—the first step in a complex chain of events which must be set into motion in order to modify the patterns of transcription in the cells of higher organisms. Other changes in the chromatin are also involved; in T lymphocytes stimulated by mitogens, these other changes include increases in nuclear protein phosphorylation (Kleinsmith *et al.*, 1966; Johnson *et al.*, 1974), an extensive modification of nuclear nonhistone protein complement (Levy *et al.*, 1973; Johnson *et al.*, 1974), and a massive influx of proteins from the cytoplasm to the nucleus (Johnson *et al.*, 1974).

In the earlier discussion of positive control in eukaryotic transcription (Section I,A,2), evidence was presented for the activation of RNA synthesis during early embryonic development, during the differentiation of erythroid cells, and in the response of a variety of target tissues to stimulation by steroid and peptide hormones. In every case, the change in DNA template function can be correlated with antecedent changes in the acetylation of the histones. For example (a) the resumption of RNA synthesis between the blastula and gastrula stages in the *Arbacia* embryo is accompanied by increases in the proportions of the acetylated forms of histones H3 and H4 (Wangh *et al.*, 1972). There is a significant increase in the rate of acetate incorporation into the arginine-rich histones between those stages, suggesting that increased acetylation may be a preparative factor in the activation of new genes at the gastrula stage of sea urchin development (Burdick and Taylor, 1976). In chick embryonic muscle, the incorporation of [^{14}C]acetate into histone ϵ-N-acetyllysine residues is more than doubled during the period of activation of synthesis of the contractile proteins (Boffa and Vidali, 1971), while the activity of the histone deacetylases is suppressed during this inductive phase of muscle development (Boffa *et al.*, 1971).

(b) The acetylation of histones in the nuclei of erythroid cells provides further evidence for an involvement in transcriptional control. Hematopoietic stem cells differentiate into erythroblasts under the influence of the peptide hormone, erythropoietin. The spleen cells of polycythemic mice show an increase in RNA synthesis within 8 hrs after injection of the hormone, but a peak of histone acetylation occurs 4 hr earlier (Takaku *et al.*, 1969), as might be expected if acetylation is part of the mechanism for the restructuring of chromatin prior to gene activation. Conversely, the maturation of the nucleated avian erythrocyte involves a programmed series of nuclear and cytoplasmic changes that eventually lead to an almost complete suppression of RNA synthesis; e.g., the RNA synthetic capacity of isolated duck erythroblasts and early polychromatic erythrocytes is at least eight times higher than that observed in mature erythrocytes (Ruiz-Carrillo *et al.*, 1974, 1976). There is a parallel decline in the rate of histone acetylation and a substantial loss of the acetylated forms of histones H3 and H4 during red cell maturation (Wangh *et al.*, 1972; Ruiz-Carrillo *et al.*, 1974, 1976). Moreover, histone deacetylase activities are considerably higher in mature erythrocytes than in reticulocytes of the same species (Sanders *et al.*, 1973). An important fact is that histone acetylation generally takes place independently of the RNA polymerase reaction. Inhibitors of RNA synthesis such as rifamycin AF 013 and actinomycin D effectively block RNA synthesis in erythroid cells without a simultaneous inhibition of acetate incorporation into the histones, at least for a short time (Ruiz-Carrillo *et al.*, 1976). The absence of tight coupling is in accord with many observations showing that acetylation precedes a major increase in transcriptional activity.

(c) Many positive correlations have been noted between enhanced RNA synthesis and histone acetylation in hormone-sensitive tissues. The present discussion will emphasize this postsynthetic modification of nucleosomal proteins in systems previously cited for their pertinency to positive transcriptional control (Section I,A,2).

The administration of estrogens leads to rapid increases in the rates of RNA synthesis in the uterus (Mueller *et al.*, 1958; Teng and Hamilton, 1968). An enchancement of DNA template activity is evident in isolated uterine chromatin (Barker and Warren, 1966; Teng and Hamilton, 1968; Glasser *et al.*, 1972) and in isolated rat uterine nucleoli (Nicolette and Babler, 1974). An increase in [3H]acetate incorporation into uterine histones H3 and H4 is detectable within 2–5 min after injection of estradiol-17β. (There is no corresponding increase in acetylation of the liver histones in the hormone-treated animals.) In animals pretreated with nafoxidine (a potent antiestrogen which blocks the uterine response to estrogens) the administration of estradiol-17β does not stimulate histone acetylation in the uterus (Libby, 1972). Other estrogens, such as stilbestrol and estriol, also stimulate acetylation in the target tissue, but testosterone has no such effect (Libby, 1972). The mechanism of hormonal activation of histone acetylation within 2–5 min is not known, but there is evidence that uterine histone acetyltransferase activity may be directly stimulated by estradiol (Libby, 1968).

An analogous situation is seen in a simple organism, *Achlya ambisexualis*, which responds to the steroid sex hormone, antheridol, with an increase in the number of RNA polymerase initiation sites (Sutherland and Horgen, 1977) and an enhanced rate of ribosomal RNA synthesis (Timberlake, 1976). The acetylation of *Achlya* histones occurs prior to the increase in RNA synthetic capacity (Horgen and Ball, 1974).

Mineral corticoids, such as aldosterone, stimulate RNA synthesis in the kidney (Edelman and Fimognari, 1968), and increases are noted in the RNA polymerase activity of kidney and heart muscle nuclei following the administration of aldosterone to adrenalectomized rats (Liew *et al.*, 1972). A sharp increase in the acetylation of the kidney histones has been detected under these conditions (Libby, 1973; Liew *et al.*, 1973). The acetylation of histone H4 is increased nearly threefold within 5 min after a physiological dose of aldosterone (Libby, 1973). Other adrenocortical steroids with mineralocorticoid activity, such as progesterone, had no such effect. Spironolactone SC14266 (a compound with strong antimineralocorticoid activity) blocks the increase in acetylation of histone H4 when administered to animals 30 min before the injection of aldosterone (Libby, 1973). The organ specificity of the hormone response is indicated by the failure of aldosterone to stimulate histone acetylation in the liver (Libby, 1973; Liew *et al.*, 1973) and by the stimulation of histone acetylation and RNA synthesis in another target tissue, cardiac muscle (Liew *et al.*, 1972, 1973). In the response of the kidney to aldosterone, the increase in histone acetylation is a transient phenomenon in which the rate of acetylation of histone H4 returns to normal in about 20 min. Most of the increase in RNA synthesis occurs after the acetylation has peaked (Libby, 1973). The same is true for estradiol-17β stimulation of acetylation and DNA template function in the uterus (Libby, 1972).

The stimulatory effects of hydrocortisone on RNA synthesis in the liver (Feigelson *et al.*, 1962; Kenney and Kull, 1963) include a selective enhancement of synthesis of particular messenger RNA's, such as those for tryptophan oxygenase (Schutz *et al.*, 1975) and tyrosine aminotransferase (Nickol *et al.*, 1978). Studies of histone acetylation in adrenalectomized rats showed that the uptake of radioactive acetate into hepatic histones increases within 30 min after injection of hydrocortisone (Allfrey *et al.*, 1966). There is a corresponding increase in template activity of the liver chromatin of hormone-treated rats (Dahmus and Bonner, 1965). Comparisons of the kinetics of acetylation, RNA synthesis, and enzyme induction after hydrocortisone administration indicate that acetylation precedes the increase in RNA synthesis and the appearance of tyrosine aminotransferase (Graaff and von Holt, 1973).

In contrast to the stimulatory effects of glucocorticoids on acetylation and RNA synthesis in hepatocytes, lymphoid cells respond to glucocorticoids by a suppression of RNA polymerase activity (Fox and Gabourel, 1967; Makman *et al.*, 1970) and a decrease in RNA synthetic capacity (Kidson, 1965; Wagner, 1970; Darzynkiewicz and Andersson, 1971). Corticosteroids alter the structure of lymphocyte chromatin to restrict DNA accessibility, as judged by decreased binding of actinomycin D (Darzynkiewicz and Andersson, 1971) and AO (Alvarez and Truitt, 1977). Acridine orange binding to thymus lymphocyte chromatin is decreased within 15 min after the injection of dexamethasone. This decrease in DNA accessibility is accompanied by an increase in thermal stability, suggesting stronger interactions between DNA and associated proteins. This is the expected consequence of a decrease in histone acetylation, and tests of acetate incorporation into thymus lymphocyte nuclei have shown a rapid inhibition by β-methasone (Allfrey *et al.*, 1966). Thus, both negative and positive responses to steroid hormones, as monitored by transcription, involve coordinate changes in the acetylation of nucleosomal proteins.

Similar conclusions follow from studies of nuclear metabolism following the administration of polypeptide hormones. The stimulatory effects of erythropoietin on RNA synthesis and histone acetylation in spleen cells have already been described. Analogous effects are seen in tissues responding to insulin and gonadotropins. For example, insulin induces the synthesis of hepatic tyrosine aminotransferase by a mechanism requiring *de novo* RNA and protein synthesis (Wicks, 1969; Schimke and Doyle, 1970). This hormone increases the acetylation (and phosphorylation) of liver histones before the increase in RNA synthesis. The insulin-induced elevation of histone acetylation is even greater than that seen in hepatocytes responding to cortisol (Graaff and von Holt, 1973). Similarly, the injection of chorionic gonadotropin into prepubertal rats leads to a rapid stimulation of RNA synthesis in the ovary (Jungmann and Schweppe, 1972a). The effect is also evident in the enchanced RNA polymerase activity of ovarian nuclei isolated from the hormone-treated animals (Van Dyke and Katzman, 1968). The acetylation of histone H4 in the ovary is stimulated within 10 min after injection of chorionic gonadotropin (Jungmann and Schweppe, 1972b), again illustrating the close temporal correlation between this postsynthetic modification of the histone and activation of the chromatin for transcription.

Many other such correlations have been noted and reviewed elsewhere

(Allfrey, 1977; Johnson and Allfrey, 1978). They include a remarkable series of inductive responses to drugs as varied as phenobarbital, 3-methylcholanthrene, polyamines, and lysergic acid diethylamide. In each of the responding tissues (liver, cardiac muscle, and brain) there is a common pattern of an early stimulation of histone acetylation followed by an increase in RNA synthesis. Conversely, studies of the effects of the hepatotoxin, aflatoxin B_1, have shown that RNA synthesis is suppressed within 30 min (Pong and Wogan, 1970), and an extensive deacetylation of liver histones is evident in 15 min (Edwards and Allfrey, 1973). Upon subsequent recovery, the acetylation of histones H3 and H4 appears to precede the restoration of RNA synthetic capacity (Pong and Wogan, 1970; Edwards and Allfrey, 1973).

Increases in histone acetylation have been observed in a variety of cultured cells after transformation by oncogenic viruses, such as WI-38 fibroblasts transformed by SV40 virus (Krause and Stein, 1975) and human embryonic kidney cells infected with adenovirus 2 or adenovirus 12 (Ledinko, 1970). It is of particular interest that histones associated with the DNA of transforming viruses such as polyoma virus or SV40 virus are more highly acetylated than the corresponding histones of the host cells (Schaffhausen and Benjamin, 1976). A correlation between increased acetyl content of histones in viral chromatin and cell transformation is strongly suggested by the findings that nontransforming host-range mutants of polyoma virus fail to show a high level of histone acetylation (Schaffhausen and Benjamin, 1976) and that the acetylation of SV40 chromatin *in vitro* promotes the subsequent transformation and production of T antigen in cultured BALB/3T3 cells (Cohen *et al.*, 1979). The high levels of histone acetylation in SV40 chromatin are believed to result from a selective inhibition of the deacetylases associated with the viral chromatin (LaBella *et al.*, 1979).

Patterns of histone acetylation vary during the cell cycle in ways commensurate with changes in the compaction of the chromatin. The proportion of acetylated H4 molecules in synchronized CHO cells is highest during interphase and declines to a minimum during prophase and metaphase, the periods of maximal chromosomal condensation. A rapid increase in the proportion of acetylated H4 takes place as the chromosomes disperse at telophase (D'Anna *et al.*, 1977). Acetylation patterns during the S phase of dividing cells are complicated by the rapid modifications of newly synthesized histone molecules, some of which take place in the cytoplasm and are reversed within minutes in the nucleus (Ruiz-Carrillo *et al.*, 1975; Jackson *et al.*, 1976). Even more complex patterns of acetylation are seen during the mitotic and meiotic divisions of sperm precursor cells in the trout (Candido and Dixon, 1971, 1972b; Dixon *et al.*,

1975). In this organism, the histones of the spermatid are extensively modified by acetylation prior to their replacement by protamines. Similar observations have been made on histone acetylation during spermatid maturation in the rat (Grimes *et al.*, 1975). The changes observed are fully consistent with the view that extensive acetylation of the histone lysine residues weakens their interactions with DNA and thus facilitates their removal and eventual replacement by protamines and other sperm-specific basic proteins.

In summary, the reversible modification of histones by acetylation and deacetylation of clustered lysine residues in the DNA-binding domains provides an enzymatic mechanism for the control of nucleosome structure. The modifications of histones H3 and H4 are particularly significant for the release of local constraints upon the associated DNA strand. Hyperacetylation of these histones labilizes the nucleosome in ways that influence DNA accessibility to nucleases, such as DNase I, and would be expected to affect the binding and progression of RNA polymerases along the DNA template. This modification of nucleosome structure is nonspecific, insofar as it is rapidly induced by a wide range of diverse stimuli which differ in their modes of gene activation and in the duration of the effect. An increased acetylation of histones H3 and H4 generally precedes an increase in DNA-directed RNA synthesis, while deacetylation is associated with chromatin compaction and inactivity. In view of the recent discovery that ϵ-N-acetyllysine is present in the HMG proteins (see Section III) and subject to acetyl group turnover (Sterner *et al.*, 1978a), the enzymatic acetylation of chromosomal proteins is not restricted to histones but includes other DNA-binding proteins as well.

Finally, it should be noted that acetylation is only one of many post-synthetic modifications of DNA-associated proteins. Some other reactions relevant to the control of DNA conformation and template activity include phosphorylation, methylation, poly (ADP)-ribosylation, and isopeptide branching. These modifications and their consequences for chromatin structure will not be considered.

E. Histone Phosphorylation

With the discovery of histone phosphorylation in 1966 (Kleinsmith *et al.*, 1966a; Ord and Stocken, 1966) and the simultaneous observation that nuclear protein phosphorylation is significantly enhanced at early stages of gene activation in PHA-stimulated lymphocytes (Kleinsmith *et al.*, 1966b), this modification of chromosomal proteins has emerged as a major factor in the control of transcription and in the compaction of chromosomes at mitosis. Histone phosphorylation is a complex phenomenon that requires rigorous chemical methods for its analysis. The following discussion will focus on recent developments which have particular relevance to the structural problems faced in organizing higher levels of chromatin structure and in altering the template capacity of the nucleosome. For further information on the chemical and enzymatic basis of histone phosphorylation and the key role played by cyclic AMP in specific modifications of histone H1, the reader is referred to recent reviews (Johnson, 1977; Hohmann, 1978; Johnson and Allfrey, 1978).

All of the major histone classes are subject to phosphorylation by enzymatic mechanisms that may or may not be responsive to cyclic nucleotides, depending upon the histone, the site of modification within the polypeptide chain, and the nature of the kinase which is specific for that site. The complexity of the problem is particularly evident for histone H1, which contains many sites of phosphorylation, only one of which is clearly controlled by cyclic AMP. The complexity of H1 phosphorylation is compounded by the fact that this histone class consists of multiple subfractions which differ in amino acid composition (Kinkade and Cole, 1966a,b). Calf thymus H1 consists of four subtypes, each differing from the others, on the average, in 20% of the sequence (Rall and Cole, 1972). Spleen histones from the calf, rat, and chicken contain three, four, and five species-specific H1 subtypes, respectively, and the proportions of each subtype may vary in different tissues of the same organism (Kinkade, 1969). Moreover, the proportions of H1 subtypes vary during differentiation of the mammary gland (Hohmann and Cole, 1969, 1971) and testis (Kistler and Geroch, 1975).

Various kinases located in the cell nucleus phosphorylate specific sites in the histone polypeptide chains. The phosphorylation of histone H1 by cyclic AMP-dependent protein kinases modifies a single serine residue at position 37 (rabbit thymus H1) or position 38 (calf thymus H1) (Langan, 1971). It is noteworthy that serine-37 is a site of heterogeneity in rabbit H1, and some H1 molecules contain an alanine in that position. The substitution eliminates the phosphate acceptor capacity in that region of the polypeptide chain. This strongly suggests a divergent function for at least one H1 variant (Langan, 1971). Recent evidence that certain satellite DNA's in *Drosophila* are associated with particular histone variants in their phosphorylated forms (Blumenfeld *et al.,* 1978) supports the view that sequence diversity and phosphorylation of H1 subtypes are related to their DNA-binding and cross-linking functions.

Serine-37 is located near a cluster of basic amino acids in the amino-terminal region of the H1 chain, and phosphorylation of this site would be expected to reduce the net positive charge in this region of the molecule. It has been noted that the phosphorylation sites of many phosphoproteins appear to be adjacent to β-turns (Small *et al.*, 1977). Thus phosphorylation of H1 may energize a major conformational change, not only in the histone, but the H1–DNA complexes. Circular dichroic measurements of interactions between calf thymus H1 and DNA indicate that phosphorylation of serine-38 reduces the ability of the histone to induce characteristic conformational changes in the DNA (Adler *et al.*, 1971, 1972). Changes in DNA binding as a consequence of H1 phosphorylation have been detected by DNA affinity chromatography. A histone H1 subfraction of high DNA affinity was isolated and then phosphorylated at serine-38 by a cyclic AMP-dependent protein kinase from calf thymus nuclei. Rechromatography of the modified histone showed a significant decrease in its DNA-binding properties, as indicated by its elution from the DNA column at ionic strengths lower than those originally required for the displacement of the nonphosphorylated form (Fasy *et al.*, 1979).

There are indications that phosphorylation of H1 at the cyclic AMP-sensitive site may modify the template capacity of the associated DNA in chromatin. Chromatin reconstitution experiments with the phosphorylated and nonphosphorylated forms of H1 showed higher template activities with the phosphorylated form (Watson and Langan, 1973). Phosphorylation of calf thymus chromatin by a cyclic AMP-dependent pineal protein kinase increased DNA accessibility (as measured by binding of actinomycin D), and enhanced its template capacity in the RNA polymerase reaction (Fontana and Lovenberg, 1973).

The modification of histone H1 by cyclic AMP-dependent protein kinases also appears to have significance for transcription *in vivo*. Phosphorylation of serine-37 in rat liver H1 is stimulated by dibutyryl cyclic AMP, and by peptide hormones, such as glucagon, which activate adenyl cyclase (Langan, 1969a,b, 1971; Takeda and Ohga, 1973). Increased phosphorylation of serine-37 in rat liver H1 is stimulated by dibutyryl cyclic tration, suggesting that glucagon-mediated increases in cyclic AMP levels promote this modification of chromatin structure. An effect on transcription is suggested by studies of enzyme induction. The ability of various cyclic AMP analogues to stimulate histone H1 phosphorylation was compared with their capacity to induce tyrosine aminotransferase (TAT) in Reuber hepatoma cells. All analogues which induced the synthesis of the enzyme also enhanced phosphorylation of a specific residue in H1 (Wicks *et al.*, 1975). Similar conclusions were drawn from studies of the glucagon-mediated induction of TAT in rat liver. A rapid increase in intracellular cAMP levels and protein kinase activity preceded incorporation of [^{32}P]phosphate into a specific site on H1, and both phenomena were antecedent to TAT induction (Takeda and Ohga, 1973). The stimulation of hepatic protein kinase activity, as a cascade effect of increases in cAMP levels, is relevant to the induction of other liver enzymes, such as serine dehydratase (Jost et al., 1970; Jost and Sahib, 1971) and ornithine decarboxylase (Beck et al., 1972).

There are indications that the increased phosphorylation of histone H1 in hormone-stimulated cells may involve a translocation of cytoplasmic cAMP-dependent protein kinase(s) to the nucleus. Nuclei isolated from perfused rat liver after glucagon stimulation contain two to three times as much protein kinase activity as do nuclei from control livers. A corresponding decrease in cytoplasmic cAMP-dependent protein kinase activity of the glucagon-treated livers was also observed (Palmer et al., 1974). Intracellular redistributions of cAMP-dependent kinases have been reported in ovarian cells responding to gonadotropins (Jungmann et al., 1974, 1975), in the transsynaptic induction of tyrosine hydroxylase in the adrenal medulla (Costa et al., 1976), and in glial tumor cells responding to norepinephrine (Salem and deVellis, 1976). The induction of ornithine decarboxylase following the activation of cAMP-dependent protein kinases (Byus and Russell, 1976; Byus et al., 1976) has also been proposed to involve translocation of the kinase (Russell et al., 1976).

Studies of enzyme translocations between subcellular compartments are complicated by problems of enzyme redistribution and artifactual adsorption to subcellular organelles during homogenization of the tissue in aqueous media. Such artifacts have been observed in studies of the localization of cAMP-dependent protein kinase in cardiac muscle (Keely et al., 1975). However, the finding of elevated kinase activities in the nuclei of ovarian cells after stimulation by gonadotropins was confirmed by analyses of nuclei isolated in nonaqueous media, thus avoiding artifactual redistributions of soluble enzymes during isolation of the nuclei (Jungmann et al., 1975).

The increase in protein kinase activity in the hepatic nucleus responding to glucagon or dibutyryl cyclic AMP is probably due to an accumulation of the catalytic subunit of cAMP-dependent protein kinase (Castagna et al., 1975). Whether the cAMP-binding regulatory subunit of the enzyme is translocated to the nucleus at the same time remains to be clarified, but there are other indications that cAMP-binding proteins corresponding to the type I and type II regulatory subunits of cAMP-dependent protein kinase are present in hepatic nuclei (Friedman and Chambers, 1978; Neumann et al., 1978). Cyclic AMP activation of nuclear kinase activity has been repeatedly observed, and isolations in nonaqueous media have confirmed the localization of both the catalytic and regulatory subunits in lymphocyte nuclei (Johnson et al., 1975). Of particular interest are observations that repression of Walker 256 mammary carcinomas, as induced by dibutyryl cyclic AMP, leads to a threefold increase in nuclear cAMP binding and a 50% decrease in total cytoplasmic cAMP binding after 1 day of treatment (Cho-Chung *et al.*, 1977).

The finding that the catalytic subunit of cAMP-dependent protein kinase has DNA-binding properties which are not simply related to the basicity of the protein (Johnson *et al.*, 1975) raises a number of interesting questions about the distribution of the enzyme in active and inactive regions of the chromatin. On the basis of the many correlations between site-specific phosphorylation of H1 at serine-37 and gene activation for RNA synthesis, one might expect the kinase to be localized in actively transcribing regions. Kinases that phosphorylate nonhistone proteins have been shown to be associated with transcriptionally active fractions of chick oviduct chromatin (Keller *et al.*, 1975), but the intranuclear distribution of the cAMP-dependent protein kinase remains to be determined. The physiological significance of the modulation of chromatin structure by a cAMP-dependent phosphorylation of H1 is reaffirmed by the fact that an enzyme specific for the dephosphorylation of serine-37 has been detected and isolated (Meisler and Langan, 1969).

In considering mechanisms by which phosphorylation of H1 at serine-37 might influence transcription, a plausible model is one in which this modification facilitates displacement of H1 from the DNA sequences adjoining the nucleosome core. The displacement may be coupled to a substitution of H1 by proteins concerned with positive aspects of gene control. One reason for invoking a displacement of H1 is its apparent absence from actively transcribing regions of *Drosophila* chromosomes (Jamrich *et al.*, 1977). Certain proteins which appear to be preferentially localized in active chromatin, such as the HMG group (see Section III), have an H1-binding capability (Smerdon and Isenberg, (1976) and may take part in its detachment.

In assessing the functional consequences of histone H1 phosphorylation, it should be noted that the modification of serine-37 represents only a small proportion of total H1 phosphorylation. In rat liver the amount of H1 phosphorylated in response to cyclic AMP is only about 1% of the total lysine-rich histone. However, there are other cyclic AMP-independent mechanisms for modification of additional serine (and threonine) residues elsewhere in the polypeptide chain; e.g., a rat liver H1 kinase specifically phosphorylates serine-105 in a cAMP-independent fashion (Langan and Hohmann, 1975). In developing trout testis, H1 is phosphorylated at up to four different sites, sequences for which have been described (Dixon *et al.*, 1975). As many as six separate amino acid residues in the H1 sequence may be phosphorylated in HeLa cells during mitosis (Ajiro *et al.*, 1976). In addition to the common phosphorylation reactions involving esterification of serine and threonine hydroxyl groups, there is also evidence for a kinase-mediated formation of N—P bonds to produce phosphoryllysine in histone H1 (and phosphorylhistidine in H4) (Smith et al., 1974; Chen et al., 1977).

There have been numerous investigations of the relationship between H1 phosphorylation and the cell cycle (Balhorn et al., 1972, 1975; Louie and Dixon, 1972; Marks et al., 1973; Gurley et al., 1974, 1975; Hohmann et al., 1976). In many eukaryotic cells two to four sites are phosphorylated during the late G_1 and S phases, but as the cells enter mitosis a superphosphorylation event occurs in all of the H1 molecules (Gurley et al., 1975; Hohmann et al., 1976). Analyses of H1 tryptic peptides show more sites are phosphorylated in mitotic cells than in interphase cells (Lake, 1973). Threonine residues in both the NH_2-terminal and COOH-terminal portions of H1 are phosphorylated at mitosis, but not at other times in the cycle (Hohmann et al., 1976). The increased phosphorylation of H1 during mitosis correlates with a six- to tenfold increase in the activity of a growth-associated histone kinase (Lake and Salzman, 1972). The phosphate groups are removed rapidly as cells leave mitosis and enter G_1 (Gurley et al., 1974).

These results are consistent with a model in which phosphorylation of H1 at multiple sites directs condensation of the chromatin, presumably by favoring tight associations between H1 molecules to permit cross-linking of apposing DNA strands. Cooperativity in H1 interactions to favor the assembly of higher order structures in chromatin may depend on post-synthetic modifications of the polypeptide chains at the appropriate sites. This would be analogous to the modification of enzyme subunits to favor assembly of enzymes such as glycogen phosphorylase (Buc et al., 1973).

Correlations between H1 phosphorylation and the growth cycle are particularly striking in the case of Physarum polycephalum, an organism in which the natural synchrony of nuclear division permits a precise timing of events in the replication cycle. A peak of H1 phosphorylation was observed just before chromosome condensation and nuclear division (Bradbury et al., 1973, 1974), and it was proposed that specific phosphorylations of H1 by growth-associated phosphokinases constitute a trigger for the control of cell division (Bradbury et al., 1974). An exogenous growth-associated histone kinase was observed to enter growing Physarum plasmodia and to advance the timing of mitosis (Ingles et al., 1976). In vitro studies of H1–DNA interactions before and after modification of H1 by the growth-associated kinase showed that phosphorylated H1 cross-links DNA more effectively than does unphosphorylated H1 (Matthews and Bradbury, 1978). The results in Physarum, and in mammalian cells, are strongly supportive of the model in which superphosphorylation of H1 leads to compaction of chromatin.

A close association between H1 phosphorylation and chromatin compaction is also seen in *Drosophila virilis*, in which the phosphorylated forms of H1 are associated with satellite DNA's in the heterochromatin (Blumenfeld *et al.*, 1978).

The previous discussion has emphasized the various forms of modification of histone H1, but phosphorylation of all of the histones is known to occur (reviewed by Johnson and Allfrey, 1978). The significance of the postsynthetic phosphorylations of H2A, H2B, H3, and H4 remains enigmatic. H2A is phosphorylated throughout the cell cycle, while H3 is phosphorylated in an intense burst at metaphase (Gurley *et al.*, 1975). Phosphorylation of histone H4 is unusual in that it first occurs in the cytoplasm, leading to esterification of the NH_2-terminal acetylserine residue on the nascent histone chain. The phosphate group is removed shortly after the histone H4 enters the nucleus, but the acetyl group remains (Ruiz-Carrillo *et al.*, 1975).

In summary, the phosphorylation of histone H1 involves multiple sites and enzymes with different specificities and cofactor requirements. Phosphorylation of serine-37 by the cyclic AMP-dependent protein kinase generally correlates with gene activation for RNA synthesis, and it may constitute a mechanism for displacement of the histone. Phosphorylation of other sites by growth-associated H1 kinases generally correlates with compaction of the chromatin and cross-linking of DNA strands. All of the histones of the nucleosome core, including the key histones, H3 and H4, are modified to some extent by phosphorylation–dephosphorylation mechanisms, the significance of which is not yet understood.

Other modifications of histones and nucleosome-associated proteins include methylation, ADP-ribosylation, and isopeptide branching.

F. Histone Methylation

The methylation of histones is primarily a modification of the ϵ-amino groups of lysine residues in the core histones, H3 and H4; H1 is not methylated. [The enzymatic basis of protein methylation is reviewed by Paik and Kim (1975).] The substrate specificity of the methylating enzymes is seen in the precise methylation of lysine-20 in calf histone H4 (DeLange *et al.*, 1969) and lysines-9 and -27 of histone H3 in the pea (Pathy *et al.*, 1973) and trout (Honda *et al.*, 1975). In contrast to histone synthesis, which takes place on small polysomes in the cytoplasm, histone methylation is performed in the nucleus (Allfrey *et al.*, 1964). The reaction is cumulative, leading to the formation of ϵ-N-monomethyllysine, ϵ-N-dimethyllysine, and ϵ-N-trimethyllysine in histone H3, and to mono-

and dimethyllysine in histone H4. Unlike the postsynthetic modification of histones H3 and H4 by acetylation, methylation appears to be essentially irreversible (Borun et al., 1972).

While the function of histone methylation is not known, the timing of methylation during the cell cycle is suggestive. In regenerating rat liver, the peak of methylation occurs at a time when the rates of histone and DNA synthesis have already begun to decline, and it correlates with condensation of the chromatin in preparation for mitosis (Tidwell et al., 1968). In synchronized HeLa cells, maximum rates of methylation are observed in the G_2 phase and mitosis (Borun et al., 1972). Careful kinetic studies of histone methylation and synthesis in Ehrlich ascites tumor cells show that methylation lags behind synthesis, and that the rate and mechanism of methylation differ for histones H3 and H4 (Thomas et al., 1975).

The distributions of mono-, di-, and trimethyllysines in H3 and of mono- and dimethyllysine in H4 do not vary significantly in different tissues of the organism (Duerre and Charkabarty, 1975). This relative constancy and the irreversibility of the methylation reactions argue against a direct role of such modifications in the dynamic modulation of nucleosome structure. The presence of methyllysine residues at particular sites of H3 and H4 may signify that those regions are marked for recognition by other proteins that modulate nucleosome structure.

As noted earlier, methylation of histone H1 is not observed in vivo or in isolated cell nuclei. However, a distortion of the normal state of chromatin, induced by adding polycations to isolated rat liver nuclei, leads to aberrant methylation of H1 to produce ϵ-N-monomethyllysine (Byvoet et al., 1978). This is an interesting observation because it suggests that displacement of H1 from its sites of attachment to DNA, brought about by naturally occurring nuclear polycationic proteins, may affect the susceptibility of H1 to modification by kinases and ADP-ribosyltransferases.

Histones H3 and H4 normally contain methylated lysines, but they do not contain N^G, N^G-dimethylarginine. However, in animals treated with the alkylating carcinogen, 1,2-dimethylhydrazine, these histones are abnormally methylated to produce dimethylarginine. Conversely, nonhistone proteins, such as the HMGs and some proteins of nuclear ribonucleoprotein particles, normally contain N^G, N^G-dimethylarginine, but not methyllysine. In animals receiving the carcinogen, the nuclear nonhistone proteins contain both methyllysines and methylarginine (Allfrey, 1977b). The direct action of carcinogens on those amino acids (lysine and arginine) which are most likely to participate in electrostatic interactions with the DNA deserves comment. To the extent that these aberrant modifications of lysine and arginine alter the affinity of chromosomal proteins for DNA, they are likely to initiate irreversible alterations in structure and function of the chromatin.

G. Poly(ADP)-ribosylation of Chromosomal Proteins

Poly(ADP)-ribose, a macromolecule synthesized from NAD by a chromatin-associated enzyme, poly(ADP)-ribose polymerase, can be found in both histones and nonhistone proteins. The enzyme catalyzes the successive transfer of ADP-ribose subunits to particular sites in the polypeptide chains, but the nature of the linkage is not entirely clear (for reviews, see Smulson and Shall, 1976; Hayaishi and Ueda, 1977). Although the great majority of the observations on the transfer of ADP-ribose have been made on isolated nuclei, evidence has been obtained for the natural occurrence of poly(ADP)-ribosylated H1 in rat liver (Ueda *et al.*, 1975). Nuclear proteins modified *in vitro* by mono- or poly(ADP)-ribosylation can be purified by chromatography on dihydroxyboryl polyacrylamide bead columns (Okayama *et al.*, 1978) and then identified. The results of these purification methods concur with observations that ADP-ribosylation modifies histones H1 and H2B (Ord and Stocken, 1977) and also reveal the presence of ADP-ribosylated nonhistone proteins. Other studies have shown attachment of ADP-ribose to HMG proteins (Wong *et al.*, 1977; Giri *et al.*, 1978a) and to protein A24 (a complex of histone H2A and ubiquitin) (Okayama and Hayaishi, 1978).

The modification of histone H1 in isolated nuclei generates a dimeric complex containing two H1 molecules joined by a poly(ADP)-ribose polymer which is covalently linked to only one of the histones (Byrne *et al.*, 1978).

The distribution of the enzyme(s) catalyzing the transfer of ADP-ribose to histones has been studied by analysis of chromatin subunits prepared by staphylococcal nuclease digestion (Mullins *et al.*, 1977; Butt *et al.*, 1978). The poly(ADP)-ribose polymerase appears to bind to the internucleosomal linker region of chromatin (Butt *et al.*, 1978; Giri *et al.*, 1978b), but it is not uniformly distributed, as judged by the low enzyme activity in transcriptionally inert fractions of HeLa cell chromatin (Mullins *et al.*, 1977) and by an apparent periodicity in enzyme activity at intervals of 8–10 nucleosomes (Butt *et al.*, 1978). This periodicity is interesting because it agrees with evidence that the higher order repeat structure of chromatin may be built up of globular particles (''superbeads'') containing eight nucleosomes (Stratling *et al.*, 1978).

Studies of ADP-ribosylation during the cell cycle of synchronized HeLa cells shows that the amount of poly(ADP)-ribose increases from early S phase to a peak at mid-S, followed by a second, even larger increase at the

S–G$_2$ transition point (Kidwell and Mage, 1976). Poly(ADP)-ribosylation is enhanced when the DNA of permeabilized cells is damaged by DNase (Berger et al., 1978). These results are suggestive; ADP-ribosylation may be a signaling mechanism for indicating sites of DNA repair, or ligation of Okazaki fragments during normal DNA replication. The role of histone H1 and other modifiable chromosomal proteins in such phenomena remains to be clarified.

H. Histone Modification by Isopeptide Branching

Analyses of nucleolar acid-soluble proteins by two-dimensional gel electrophoresis revealed a protein, designated A24, which decreased in amount as nucleoli undergo hypertrophy during liver regeneration. After purification and characterization, protein A24 was found to contain a unique branched structure with two amino-termini and one carboxyl-terminus (Olson et al., 1976). Further characterization showed that A24 was the product of an isopeptide modification of histone H2A, in which an acidic sequence is joined through its carboxyl-terminus (glycine) to the ε-amino group of the lysine residue at position 119 of H2A (Goldknopf and Busch, 1977). When the sequence of the attached polypeptide was determined (Olson et al., 1976), a portion was found to be identical to the highly conserved, 74-residue polypeptide, ubiquitin, which is thought to be present in all living cells (Hunt and Dayhoff, 1977).

About 10% of the total H2A of cultured murine cells is present in the form of A24, and the latter is found in nucleosomes. This accounts for the observation that histone H2A is found in less than equimolar proportions with H2B, H3, and H4 in the chromatin subunits (Albright et al., 1979).

The modification of H2A by covalent attachment of ubiquitin seems to correlate with gene inactivation. Chromatin fractionations using the DNase II, Mg^{2+} procedure (Gottesfeld et al., 1974) showed that there was a reciprocal relationship between the content of A24 and free ubiquitin in the chromatin subfractions, with a diminished amount of A24 and an increased ubiquitin content in the transcriptionally active fractions (Goldknopf et al., 1978). Free ubiqutin has also been detected in trout testis chromatin, where it probably occurs in the linker regions between nucleosomes, as judged by its rapid release during limited digestions with staphylococcal nuclease (Watson et al., 1978).

The question of the reversibility of the ubiquitin modification of histone H2A remains to be analyzed. A protease associated with calf thymus chromatin has been found to act exclusively on H2A, cleaving the sequence between the valine residue at position 114 and leucine-115 (Eickbush et al., 1976). Cleavage at this position would remove the

COOH-terminal 15 amino acids together with any isopeptide branch at lysine-119. It is not known whether this proteolytic cleavage is followed *in vivo* by removal or repair of the remaining histone fragment. Because A24 disappears in mitosis with a corresponding increase in H2A content, it appears that ubiquitin is released. The isopeptide linkage between H2A and the ubiquitin-containing peptide reforms when cells enter the G_1 phase (Matsui *et al.*, 1979).

The enormous diversity of reactions modifying histones (acetylation, phosphorylation, methylation, ADP-ribosylation, and isopeptide branching) testifies to the need to modulate chromatin structure in a dynamic fashion and illustrates the enzymatic mechanisms employed for this purpose. Chromatin structure and function are also dependent upon DNA interactions with nonhistone proteins, and some of these will now be considered.

III. DNA-ASSOCIATED NONHISTONE PROTEINS

Current estimates of nuclear protein complexity place the number of different molecular species at well over 1000. Analyses of the proteins in HeLa cell chromatin by two-dimensional gel electrophoretic techniques has revealed more than 450 components, most of which are present in less than 10,000 copies per nucleus, and not detectable in the cytoplasm (Peterson and McConkey, 1976). There are numerous indications that nonhistone chromosomal proteins include DNA-binding components that determine the specificity of transcription in different cell types and mediate the genomic response to hormones, developmental stimuli, and carcinogens. Many aspects of their function are considered in recent reviews (Cameron and Jeter, 1974; Stein and Kleinsmith, 1975, 1978; Busch *et al.*, 1976; Stein *et al.*, 1977, 1978; Kleinsmith, 1978; Wang *et al.*, 1978; Allfrey and Boffa, 1979). The present discussion focuses on a limited subset of nuclear proteins which are receiving increasing attention because of their presence in nucleosomes and close association with transcriptionally active DNA sequences.

A. The High-Mobility Group Proteins

Extraction of nuclei or chromatin with 0.35 M NaCl yields a subset of nuclear proteins which can be fractionated by differential precipitations with trichloroacetic acid and chromatography on CM-cellulose. They have been designated high-mobility group (HMG) proteins because of their high electrophoretic mobility in low pH polyacrylamide gels (Goodwin and Johns, 1973; Goodwin *et al.*, 1973). There are four major

HMG proteins in calf thymus (HMG1, 2, 14, and 17) which appear to be distinct molecular species. All have molecular weights below 30,000 daltons. [A number of minor components originally classed as HMG's have been shown to arise as a result of proteolysis (Goodwin et al., 1978).] HMG proteins are widely distributed, occurring in mammalian cells (Goodwin and Johns, 1973), duck erythrocytes (Vidali et al., 1977), trout testis (Watson et al., 1977), Drosophila (Alfagame et al., 1976), and wheat germ and yeast (Spiker et al., 1978).

The major HMG proteins, HMG1 and HMG2, contain approximately 25% basic and 30% acidic amino acid residues. Peptide mapping and amino acid sequence studies show that the charge distribution within HMG molecules is not uniform and that there are major clusters of acidic amino acids within the polypeptide chains (Walker et al., 1976, 1978a,b). HMG1 contains an unusual sequence of 41 continuous aspartic and glutamic acid residues (Walker et al., 1978b). Partial sequences are known for HMG14 (Walker et al., 1978c) and the complete sequence of HMG17 from calf thymus has been determined (Walker et al., 1977). The latter protein is 89 amino acid residues long and has a molecular weight of 9247. The NH_2-terminal two-thirds of the molecule is highly basic, whereas the COOH-terminal region has an overall negative charge (Walker et al., 1977).

Isoelectric focusing studies show complex banding patterns for HMG1 and HMG2 (Goodwin et al., 1976). Much of this heterogeneity can probably be ascribed to postsynthetic modifications, since certain HMG's are now known to be acetylated (Sterner et al., 1978a), phosphorylated (Sun et al., 1979), methylated (Boffa et al., 1979), and ADP-ribosylated (Wong et al., 1977; Giri et al., 1978a).

Comparative sequence analyses of the HMG proteins show remarkable structural similarities, extending across phylogenetic boundaries (Watson et al., 1977, Rabbani et al., 1978; Sterner et al., 1978b). Antisera against HMG1 have been used in quantitative microcomplement fixation assays to estimate the sequence dissimilarity between HMG1 and other members of the HMG group. The maximum difference observed (between HMG1 and HMG14) was only 20%, and HMG1 and HMG2 differed by only 2% (Bustin et al., 1978). Comparisons of the individual HMG's from calf thymus, liver, spleen, and kidney gave no indications of tissue specificity (Rabbani et al., 1978); however, an unusual variant (HMG-E) was detected in erythrocytes, but not in thymocytes of the duck (Sterner et al., 1978b). The avian oviduct contains a large (95,000 dalton) HMG-like protein which is believed to be induced during the late, hormone-responsive stages of embryonic development (Teng et al., 1978).

The sequestration of positively and negatively charged regions within HMG sequences implies dual binding properties. HMG1 and HMG2 have been found to interact with different H1 subfractions with a high degree of specificity (Smerdon and Isenberg, 1976). There are also many studies of HMG binding to DNA (Goodwin *et al.*, 1975; Cary *et al.*, 1976; Abercrombie *et al.*, 1978; Javaherian and Amini, 1978; Javaherian *et al.*, 1978; Bidney and Reeck, 1978). Nuclear magnetic resonance spectra of complexes between HMG1 and DNA indicate that a lysine-rich portion of the molecule containing all the aromatic residues is bound to DNA, while the acidic region of the molecule remains free from DNA (Cary *et al.*, 1976). HMG17 interacts with DNA through a basic segment between residues 15 and 40 (Abercrombie *et al.*, 1978). Sequential chromatography of HMG's on double-stranded and single-stranded DNA columns has shown that HMG1 and HMG2 have a marked preference for single-stranded DNA (Bidney and Reeck, 1978). These HMG's were also shown to reduce the topological winding number of a circular DNA molecule, which strongly suggests that they have a DNA unwinding function (Javaherian *et al.*, 1978).

The total nuclear concentration of HMG's is low relative to that of the histones (about 3%), and their distribution within chromatin is of particular interest. The presence of HMG's in chromatin subunits prepared by staphylococcal nuclease digestion has been observed repeatedly (Goodwin *et al.*, 1977; Bakayev *et al.*, 1978; Levy-W. *et al.*, 1977, 1979; Mathew *et al.*, 1979; Neumann *et al.*, 1978). A particularly significant observation was the finding that HMG proteins are rapidly released during limited DNase I digestions of avian erythrocyte chromatin (Vidali *et al.*, 1977) under conditions previously shown to selectively degrade the globin genes (Weintraub and Groudine, 1976). Similar observations were made in trout testis cells (Levy-W. *et al.*, 1977) and in mouse Ehrlich ascites cells (Bakayev *et al.*, 1978). Mononucleosomes enriched in transcribed DNA sequences have been partially purified from trout testis cells and shown to contain H6 (equivalent to calf HMG17) (Levy-W. and Dixon, 1978; Levy-W. *et al.* 1979). A second HMG protein (HMG-T) appeared to be associated with the linker strands between the nucleosomes (Levy-W. *et al.*, 1977), where it is believed to make the associated DNA sequences more accessible to attack by staphylococcal nuclease (Levy-W. *et al.*, 1979). Nucleosomes containing H6 (HMG17) were lacking in histone H1, but contained all four core histones (Levy-W. *et al.*, 1979).

Further evidence for the participation of HMG proteins in controlling the conformation of transcriptionally active DNA sequences is provided by comparisons of the DNase I accessibility of the globin genes in erythrocyte nuclei before and after extraction in 0.35 M NaCl. In the extracted chromatin the globin genes are no longer preferentially degraded by DNase I. Restoration to the depleted chromatin of the 0.35 M extract or a subfraction enriched in HMG proteins led to renewed DNase I sensitivity of the globin sequences. The HMG proteins responsible for conferring the DNase I-sensitive structure of globin chromatin have electrophoretic and solubility properties similar to those of HMG14 and HMG17 (Weisbrod and Weintraub, 1979). It is significant that HMG's from other tissues, such as brain, can restore the DNase I-sensitive configuration to the globin genes in depleted erythrocyte chromatin, but erythrocyte HMG's are unable to confer DNase I sensitivity to the repressed globin genes in depleted brain chromatin (Weisbrod and Weintraub, 1979). This implies that HMG's alone are not sufficient to establish a template-active, DNase I-sensitive state in chromatin. They must interact with other chromosomal elements which confer specificity or regulate nucleosome structure.

The part played by postsynthetic modifications of the HMG proteins in histone or DNA binding or in interactions with recognition sites in chromatin remains to be investigated.

The presence of HMG-like proteins in active chromatin can be visualized in the polytene chromosomes of *Drosophila melanogaster,* using immunofluorescence microscopy. HMG-D1 was found to be localized at a limited number of specific loci, whereas histone H2B was uniformly distributed throughout the chromosomes (Alfagame *et al.,* 1976). An HMG-like protein with a molecular weight of 63,000, which is selectively released during limited DNase I digestions, was found to be associated with the puffing regions induced by heat-shock treatment (Mayfield *et al.,* 1978).

HMG proteins offer an interesting example of proteins with "address" sites in the nucleus. When HMG1 (coupled to rhodamine dye to permit its detection by fluorescence microscopy) is microinjected into the cytoplasm of HeLa cells, most of the fluorescent protein moves to the nucleus in a matter of hours (Stacey and Allfrey, 1979). Similarly, transfer of ^{125}I-labeled HMG1 to HeLa cells and chick fibroblasts by cell fusion techniques resulted in nuclear accumulation of the HMG protein. It was shown that the labeled HMG molecules can move between the nuclei of fused HeLa cells and between HeLa and chick fibroblast nuclei, but HMG's failed to accumulate in the transcriptionally repressed red cell nucleus of fused HeLa–chick erythrocyte heterokaryons (Rechsteiner and Kuehl, 1979). Whether this represents a deficiency (or shielding) of HMG-binding sites in the erythrocyte nucleus or the inability of the latter to introduce necessary postsynthetic modifications in the added HMG's is not known. The shuttling of HMG's between nuclei in heterokaryons is reminiscent of earlier observations on the dynamic equilibrium of the cytonucleoproteins of *Amoeba* (Goldstein, 1958, 1963, 1965; Prescott, 1963), and it is very likely that the proteins then observed to concentrate in the cell nuclei of *Amoeba* included members of the HMG class.

B. Nonhistone Protein Phosphorylation and Transcriptional Control

Many of the DNA-binding proteins in chromatin are phosphoproteins, subject to rapid "turnover" of their phosphate groups and selective in their interactions with DNA (Kleinsmith et al., 1970; Teng et al., 1970, 1971; Kleinsmith, 1978). Phosphorylation of the nonhistone proteins is an early event in lymphocyte activation by lectins and in the responses of a wide range of target cells to different steroid and peptide hormones. Changes in phosphorylation patterns are often detectable within minutes after stimulation, and different nuclear proteins are affected in widely divergent ways (Allfrey et al., 1973; Johnson et al., 1974). Much of the evidence relating nuclear protein phosphorylation to RNA synthesis has been reviewed recently (Kleinsmith, 1978). The aim of the present discussion is to draw attention to new findings which reaffirm the central role of phosphorylation mechanisms in the control of gene expression and suggest their involvement in malignant transformation.

First are the observations that the major eukaryotic RNA polymerases are themselves subject to phosphorylation. Five subunits of yeast RNA polymerase I, two subunits of polymerase II, and two subunits of polymerase III were found to incorporate [^{32}P]phosphate in vivo. The sites of phosphorylation of polymerase I were identified as phosphoserine and phosphothreonine (Bell et al., 1977). Phosphorylation of RNA polymerase I has also been observed in isolated rat liver nuclei incubated with [α-^{32}P]ATP and dibutyryl cyclic AMP. The cyclic AMP derivative was observed to stimulate nuclear protein phosphorylation and increase endogenous RNA polymerase activity in intact nuclei (Hirsch and Martelo, 1976). However, the effect of the cyclic nucleotide may well be indirect because the kinase which copurifies with RNA polymerase I was found to be cyclic AMP independent (Hirsch and Martelo, 1978).

RNA polymerase I is more extensively phosphorylated than the other two polymerases. All share a common 24,000 dalton subunit which is phosphorylated, but only polymerase I has a phosphorylated large (185,000 dalton) subunit (Bell et al., 1977). This suggests that phosphorylation is likely to be particularly significant in the regulation of ribosomal RNA synthesis. Attempts to demonstrate this by phosphorylating the isolated enzyme with a purified kinase and then measuring transcription in vitro have not been successful (Bell et al., 1977), but it is unlikely that the effects of phosphorylation would be detectable in the absence of the other chromosomal proteins which control DNA conformation and template function. Moreover, there are protein factors which stimulate the activity of RNA polymerase I (Goldberg et al., 1977; James et al., 1977), and they may also be subject to postsynthetic modifications. In analyzing such complex systems, it is advantageous to begin with intact nuclei or nucleoli and use reductive methods to separate and investigate their functions. Acid treatment of yeast nuclei or chromatin inactivates the endogenous RNA polymerases and permits an analysis of the specificity of added, purified polymerase I (Tekamp *et al.*, 1979). This should make it possible to determine whether phosphorylation of the polymerase or other chromosomal proteins affects the rate of ribosomal RNA chain initiation or elongation or the fidelity of rDNA transcription.

An additional observation relevant to the control of RNA synthesis by phosphorylation of the polymerases or by modification of the chromatin template is the finding that cyclic GMP (but not cyclic AMP) is distributed along the polytene chromosomes of *Drosophila* at the sites of RNA synthesis. Moreover, the intensity of cyclic GMP fluorescence was markedly enhanced at the specific puffs induced by heat-shock treatment (Spruill *et al.*, 1978). An increased concentration of RNA polymerase II at those sites has also been observed (Jamrich *et al.*, 1977).

The activation of RNA synthesis in target cells responding to steroid hormones depends upon interactions between the hormones and receptor proteins to form complexes which enter the nucleus. Formation of the glucocorticoid–receptor complex appears to require phosphorylation of the binding protein, because only the phosphorylated form of the molecule is capable of binding the hormone, and its binding capacity is lost after treatment with alkaline phosphatase (Nielson *et al.*, 1977).

In summary, phosphorylation mechanisms affect the conformation of the DNA template, the subunit structure of the RNA polymerases, and the function of regulatory proteins. Aberrations in protein phosphorylating mechanisms would be expected to have grave consequences for transcriptional control. The recent finding that the transforming gene of avian sarcoma virus encodes a protein kinase associated with a phosphoprotein (Levinson *et al.*, 1978) raises new questions about the role of nuclear protein phosphorylation in development of the malignant state.

IV. CONCLUSION

It is now taken as axiomatic that differential transcription provides a prime mechanism for the generation of diversity in cell structure and function. Individual cells suppress the transcription of most of their DNA sequences, while they selectively activate a relatively limited number of genes for the synthesis and assembly of the enzymes and structural proteins characteristic of the cell type and the species. In the course of embryogenesis and development different sets of genes are activated and repressed in response to programmed signals from the nucleus, the cytoplasm, and the organismic environment. The positive and negative control of gene function ultimately depends on factors that influence the binding and progression of RNA polymerases along the DNA template. The accessibility of the DNA strand to polymerases is restricted by its association with histones and certain nonhistone proteins in the nucleosome "cores" and in the "linker" regions between these chromatin subunits. Nucleosomes generally assume a tight, beaded configuration in inactive regions of the chromatin and are somehow elongated (perhaps by unfolding about an axis of symmetry) in rapidly transcribing regions. Higher orders of chromatin structure, such as the formation of "superbeads" in the 250 Å fibers and compaction of heterochromatin, depend on interactions of DNA with H1 histones and other structural proteins which restrict its template capacity even further.

Electron microscopic observations, enzymatic probes of DNA accessibility, and chromatin fractionation experiments make it clear that the chromatin subunit in "active" genes differs from the inactive nucleosome in its composition and physical state. It may be deficient in histones, such as H1, which have cross-linking properties. The key histones of the core particle, H3 and H4, may be extensively acetylated. Other core histones, such as H2A, may be less modified by isopeptide branching. The presence of nonhistone proteins, e.g., members of the HMG class, may influence nucleosome structure to facilitate access of polymerases to the DNA strand.

The system is dynamic; changes in the nucleosomal histones can occur in a matter of minutes in cells responding to steroid and peptide hormones, to mitogens, and to developmental stimuli. The delicate balance is upset by a variety of carcinogens and other drugs. Postsynthetic modifications affect the capacity of the histones to interact with DNA, either directly or indirectly. Acetylations of H3 and H4 release some of the constraints on the DNA strand encircling the nucleosome core. Phosphorylation of a single site (serine-38) in histone H1 may have a similar "relaxing" effect on DNA adjacent to the core, but higher levels of H1 phosphorylation seem to signify histone assembly into supercoils or "superbeads" and the formation of higher orders of chromatin structure necessary for the compaction of heterochromatin and aggregation of the metaphase chromosomes. These modifications of nucleosome structure are reversible. The net result of the competition between acetylases and deacetylases, kinases and phosphatases, and ADP-ribosylases and hydrolases is a statistical balance which is not uniform throughout the nucleus but varies in the transcribing and nontranscribing regions.

These variations in nucleosome structure reflect a cascade of enzymatic controls which are only beginning to be understood. The presence of the modifying enzymes, their regulatory subunits, activators, and inhibitors is only part of the overall control mechanism. Superimposed are postsynthetic modifications of the enzymes themselves. Phosphorylations of the kinases, acetylases, and deacetylases are equally dynamic and represent a major aspect of nuclear metabolism that is ultimately targeted to the control of the DNA template. A parallel regulatory system, directed at RNA polymerases I, II, and III, also appears to exist, and nonhistone proteins believed to play a role in transcription, including certain hormone receptors, are phosphorylated and dephosphorylated. Thus, the postsynthetic modification of chromosomal proteins has emerged as a major and indispensable attribute of transcriptional control. It is a promising new frontier in the study of the eukaryotic genome and its regulation.

REFERENCES

Abercrombie, B. D., Kneale, G. C., Crane-Robinson, C., Bradbury, E. M., Goodwin, G. H., Walker, J. M., and Johns, E. W. (1978). Studies on the conformational properties of the high mobility group chromosomal protein HMG17 and its interaction with DNA. *Eur. J. Biochem.* **84,** 173–177.

Adler, A. J., Schaffhausen, B., Langan, T. A., and Fasman, G. (1971). Altered conformational effects of phosphorylated lysine-rich histone (f-1) in f-1–deoxyribonucleic acid complexes. Circular dichroism and immunological studies. *Biochemistry* **10,** 909–913.

Adler, A. J., Langan, T. A., and Fasman, G. (1972). Complexes of deoxyribonucleic acid with lysine-rich (f-1) histone phosphorylated at two separate sites: Circular dichroism studies. *Arch. Biochem. Biophys.* **153,** 769–777.

Adler, A. J., Fasman, G. D., Wangh, L. J., and Allfrey, V. G. (1974). Altered conformational effects of naturally acetylated histone f2a1 (IV) in f2a1–deoxyribonucleic acid complexes. *J. Biol. Chem.* **249,** 2911–2914.

Ajiro, K., Borun, T. W., and Cohen, L. (1976). Differences in phosphorylation sites and levels among H1 histones during the HeLa cell cycle. *Fed. Proc., Fed. Am. Soc. Exp. Biol.* **35,** 1623 (abstract).

Ajiro, K., Zweidler, A., Borun, T., and Croce, C. (1978). Species-specific suppression of histone H1 and H2B production in human/mouse hybrids. *Proc. Natl. Acad. Sci. U.S.A.* **75,** 5599–5603.

Albright, S. C., Nelson, P. P., and Garrard, W. T. (1979). Histone molar ratios among different electrophoretic forms of mono- and dinucleosomes. *J. Biol. Chem.* **254,** 1065–1073.

Alfagame, C. R., Zweidler, A., Mahowald, A., and Cohen, L. H. (1974). Histones of *Drosophila* embryos. Electrophoretic isolation and structural studies. *J. Biol. Chem.* **249,** 3729–3736.

Alfagame, C. R., Rudkin, G. T., and Cohen, L. H. (1976). Locations of chromosomal proteins in polytene chromosomes. *Proc. Natl. Acad. Sci. U.S.A.* **73,** 2038–2042.

Allfrey, V. G. (1964). Structural modifications of histones and their possible role in the regulation of ribonucleic acid synthesis. *Proc. Can. Cancer Res. Conf.* **6,** 313–335.

Allfrey, V. G. (1970). Changes in chromosomal proteins at times of gene activation. *Fed. Proc., Fed. Am. Soc. Exp. Biol.* **29,** 1447–1460.

Allfrey, V. G. (1977a). Post-synthetic modifications of histone structure: A mechanism for the control of chromosome structure by the modulation of histone–DNA interaction. *In* "Chromatin and Chromosome Structure" (H. J. Li and R. A. Eckhardt, eds.), pp. 167–191. Academic Press, New York.

Allfrey, V. G. (1977b). Overview: Molecular changes associated with large bowel cancer and their potential as markers and chemotherapeutic agents. *Cancer (Philadelphia)* **40,** 2576–2579.

Allfrey, V. G., and Boffa, L. C. (1979). Modifications of nuclear protein structure and function during carcinogenesis. *In* "The Cell Nucleus (H. Busch, ed.), Vol. 7: Chromatin, Part D, pp. 521–562. Academic Press, New York.

Allfrey, V. G., Littau, V. C., and Mirsky, A. E. (1963). On the role of histones in regulating ribonucleic acid synthesis in the cell nucleus. *Proc. Natl. Acad. Sci. U.S.A.* **49,** 414–421.

Allfrey, V. G., Faulkner, R., and Mirsky, A. E. (1964). Acetylation and methylation of histones and their possible role in the regulation of RNA synthesis. *Proc. Natl. Acad. Sci. U.S.A.* **51,** 786–794.

Allfrey, V. G., Pogo, B. G. T., Pogo, A. O., Kleinsmith, L. J., and Mirsky, A. E. (1966). The metabolic behaviour of chromatin. *In* "Histones: Their Role in the Transfer of Genetic Information" (A. V. S. deReuck and J. Knight, eds.), pp. 42–62. Churchill, London.

Allfrey, V. G., Johnson, E. M., Karn, J., and Vidali, G. (1973). Phosphorylation of nuclear proteins at times of gene activation. *In* "Protein Phosphorylation in Control Mechanisms" (F. Huijing and E. Y. C. Lee, eds.), pp. 217–244. Academic Press, New York.

Allfrey, V. G., Vidali, G., Boffa, L. C., and Johnson, E. M. (1977). The role of DNA-associated proteins in the regulation of chromosome function. *In* "Birth Defects" (J. W. Littlefield and J. deGrouchy, eds.), pp. 41–65. Excerpta Medica, Amsterdam.

Altenburg, B. C., and Steiner, S. (1979). Cytochalasin B-induced multinucleation of murine sarcoma virus-transformed cells. *Exp. Cell Res. 118,* 31–37.

Altenburger, W., Horz, W., and Zachau, H. G. (1976). Nuclease cleavage of chromatin at 100 nucleotide intervals. *Nature (London)* **264,** 517–522.

Alvarez, M. R., and Truitt, A. J. (1977). Rapid nuclear cytochemical changes induced by dexamethasone in thymus lymphocytes of adrenalectomized rats. *Exp. Cell Res.* **106,** 105–110.

Appels, R., Tallroth, E., Appels, D. M., and Ringertz, N. R. (1975). Differential uptake of protein into the chick nuclei of HeLa × chick heterokaryons. *Exp. Cell Res.* **92,** 70–78.

Axel, R., (1975). Cleavage of DNA in nuclei and chromatin with staphylococcal nuclease. *Biochemistry* **14,** 2941–2945.

Axel, R., Feigelson, P., and Schutz, G. (1976). Analysis of the complexity and diversity of mRNA from chicken liver and oviduct. *Cell* **7,** 247–254.

Bakayev, V. V., Bakayeva, T. G., Schmatchenko, V. V., and Georgiev, G. (1978). Non-histone proteins in mononucleosomes and subnucleosomes. *Eur. J. Biochem.* **91,** 291–301.

Balwin, J. P., Boseley, P. G., Bradbury, E. M., and Ibel, K. (1975). The subunit structure of the eukaryotic chromosome. *Nature (London)* **253,** 245–249.

Balhorn, R., Chalkley, R., and Granner, D. (1972). Lysine-rich histone phosphorylation. A positive correlation with cell replication. *Biochemistry* **11,** 1094–1098.

Balhorn, R., Jackson, V., Granner, D., and Chalkley, R. (1975). Phosphorylation of the lysine-rich histones throughout the cell cycle. *Biochemistry* **14,** 2504–2511.

Bantle, J. A., and Hahn, W. E. (1976). Complexity and characterization of polyadenylated RNA in the mouse brain. *Cell* **8,** 139–150.

Barker, K. L., and Warren, J. C. (1966). Template activity of uterine chromatin: Control by estradiol. *Proc. Natl. Acad. Sci. U.S.A.* **56,** 1298–1302.

Barkley, M. D., Riggs, A. D., Jobe, A., and Bourgeois, S. (1975). Interaction of effecting ligands with *lac* repressor and repressor–operator complex. *Biochemistyr* **14,** 1700–1712.

Beard, P. (1978). Mobility of histones on the chromosomes of simian virus 40. *Cell* **15,** 955–967.

Beck, W. T., Bellantone, R. A., and Canellakis, E. S. (1972). The *in vivo* stimulation of rat liver ornithine decarboxylase activity by dibutyryl cyclic adenosine 3′,5′-monophosphate. *Biochem. Biophys. Res. Commun.* **48,** 1649–1655.

Bell, G. I., Valenzuela, P., and Rutter, W. J. (1977). Phosphorylation of yeast DNA-dependent RNA polymerases *in vivo* and *in vitro*. *J. Biol. Chem.* **252,** 3082–3091.

Bellard, M., Gannon, F., and Chambon, P. (1978). Nucleosome structure. III. The structure and transcriptional activity of the chromatin containing the ovalbumin and globin genes in chick oviduct nuclei. *Cold Spring Harbor Symp. Quant. Biol.* **42,** 779–791.

Berger, N. A., Kaichi, A. S., Steward, P. G., Klevecz, R. R., Forrest, G. L., and Gross, S. D. (1978). Synthesis of poly(adenosine diphosphate-ribose) in synchronized chinese hamster cells. *Exp. Cell Res.* **117,** 127–135.

Berkowitz, E. C., and Doty, P. (1975). Chemical and physical properties of fractionated chromatin. *Proc. Natl. Acad. Sci. U.S.A.* **72,** 3328–3332.

Berlowitz, L., and Pallotta, D. (1972). Acetylation of nuclear protein in the heterochromatin and euchromatin of mealy bugs. *Exp. Cell Res.* **71,** 45–48.

Bidney, D. L., And Reeck, G. R. (1978). Purification from cultured hepatoma cells of two nonhistone chromatin proteins with preferential affinity for single-stranded DNA: Apparent homology with calf thymus HMG proteins. *Biochem. Biophys. Res. Commun.* **85,** 1211–1218.

Billing, R. J., and Bonner, J. (1972). The structure of chromatin as revealed by DNase digestion studies. *Biochim. Biophys. Acta* **281,** 453–466.

Bina-Stein, M., and Simpson, R. T. (1977). Specific folding and contraction of DNA by histones H3 and H4. *Cell* **11,** 609–618.

Bloom, K. S., and Anderson, J. N. (1978). Fractionation of hen oviduct chromatin into transcriptionally active and inactive regions after selective micrococcal nuclease digestion. *Cell* **15,** 141–150.

Blumenfeld, M., Orf, J. W., Sina, B. J., Kreber, R. A., Callahan, M. A., Mullins, J. I., and Snyder, L. A. (1978). Correlation between phosphorylated H1 histones and satellite DNAs in *Drosophila virilis*. *Proc. Natl. Acad. Sci. U.S.A.* **75,** 866–870.

Bodell, W. J. (1977). Nonuniform distribution of DNA repair in chromatin after treatment with methylmethane sulfonate. *Nucleic Acids Res.* **4,** 2619–2628.

Boffa, L. C., and Vidali, G. (1971). Acid-extractable proteins from chick embryonic muscle nuclei. Characterization of histone and studies on histone acetylation. *Biochim. Biophys. Acta* **236,** 259–269.

Boffa, L. C., Gershey, E. L., and Vidali, G. (1971). Changes of the histone deacetylase activity during chick embryo muscle development. *Biochim. Biophys. Acta* **254,** 135–143.

Boffa, L. C., Vidali, G., and Allfrey, V. G. (1978). Suppression of histone acetylation in vivo and in vitro by Na butyrate. *J. Biol. Chem.* **253,** 3364–3366.

Boffa, L. C., Sterner, R., Vidali, G., and Allfrey, V. G. (1979). Presence of N^G, N^G-

dimethylarginine in proteins of the HMG class. *Biochem. Biophys. Res. Commun.*
89, 1322–1327.

Bolund, L., Ringertz, N. R., and Harris, H. (1969). Changes in the cytochemical properties
of erythrocyte nuclei reactivated by cell fusion. *J. Cell Sci.* **4**, 71–87.

Bonner, W. M., and Pollard, H. B. (1975). The presence of F3-F2a1 dimers and F1 oligomers
in chromatin. *Biochem. Biophys. Res. Commun.* **64**, 282–288.

Bonner, J. J., and Pardue, M. L. (1977). Ecdysone-stimulated RNA synthesis in salivary
glands of *Drosophila melanogaster:* Assay by *in situ* hybridization. *Cell* **12**, 219–225.

Bonner, J., Wallace, R. B., Sargent, T. D., Murphy, R. F., and Dube, S. K. (1978). The
expressed portion of eukaryotic chromatin. *Proc. Natl. Acad. Sci. U.S.A.* **42**, 851–855.

Borun, T. W., Pearson, D., and Paik, W. K. (1972). Studies of histone methylation during the
HeLa S-3 cell cycle. *J. Biol. Chem.* **247**, 4288–4298.

Boseley, P. G., Bradbury, E. M., Butler-Browne, G. S., Carpenter, B. G., and Stephens,,
R. M. (1976). Physical studies of chromatin. The recombination of histones with
DNA. *Eur. J. Biochem.* **62**, 21–31.

Bradbury, E. M., and Crane-Robinson, C. (1971). Physical and conformational studies of
histones and nucleohistones. *In* "Histones and Nucleohistones" (D. M. P. Phillips,
ed.), pp. 85–134. Plenum, New York.

Bradbury, E. M., Carpenter, B. G., and Rattle, H. W. E. (1973a). Magnetic resonance
studies of deoxyribonucleoproteins. *Nature (London)* **241**, 123–125.

Bradbury, E. M., Inglis, R. J., Matthews, H. R., and Sarner, N. (1973b). Phosphorylation of
very lysine-rich histones in *Physarum polycephalum.* Correlation with chromosome
condensation. *Eur. J. Biochem.* **33**, 131–139.

Bradbury, E. M., Inglis, R. J., and Matthews, H. R. (1974). Control of cell division by very
lysine-rich histone f1 phosphorylation. *Nature (London)* **247**, 257–261.

Bradbury, E. M., Moss, T., Hayashi, H., Hjelm, R. P., Suau, P., Stephens, R. M., Baldwin,
J. P., and Crane-Robinson, C. (1978). Nucleosomes, histone interactions and the role
of histones H3 and H4. *Cold Spring Harbor Symp. Quant. Biol.* **42**, 277–286.

Bradshaw, W. S., and Rutter, W. J. (1972). Multiple pancreatic lipases. Tissue distribution
and pattern of accumulation during embryological development. *Biochemistry* **11**,
1517–1528.

Bramwell, M. E. (1978). Detection of chick rRNA in the cytoplasm of heterokaryons containing reactivated red cell nuclei. *Exp. Cell Res.* **112**, 63–71.

Brandt, W. F., Strickland, W. N., Morgan, M., and von Holt, C. (1974). Comparison of the
N-terminal amino acid sequences of histone F3 from a mammal, a shark, an
echinoderm, a mollusc and a plant. *FEBS Lett.* **40**, 167–172.

Breathnach, R., Benoist, C., O'Hare, K., Gannon, F., and Chambon, P. (1978). Ovalbumin
gene: Evidence for a leader sequence in mRNA and DNA sequences at the exon–intron
boundaries. *Proc. Natl. Acad. Sci. U.S.A.* **75**, 4853–4857.

Briggs, R., and Cossens, G. (1966). Accumulation in the oocyte nucleus of a gene product
essential for embryonic development beyond gastrulation. *Proc. Natl. Acad. Sci.*
U.S.A. **55**, 1103–1108.

Brothers, A. J. (1976). Stable nuclear activation dependent on a protein synthesized during
oogenesis. *Nature (London)* **260**, 112–115.

Brown, D. D., and Gurdon, J. B. (1977). High-fidelity transcription of 5S DNA injected into
Xenopus oocytes. *Proc. Natl. Acad. Sci. U.S.A.* **74**, 2064–2068.

Brown, F. L., Musich, P. R., and Maio, J. J. (1979). The repetitive sequence structure of
component α DNA and its relationship to the nucleosomes of the African green
monkey. *J. Mol. Biol.* **131**, 777–799.

Buc, H., Buc, M. H., Garcia Blanco, F., Morange, M., and Winkler, H. (1973). Role of
5'-AMP and its analogs in the activation of muscle glycogen phosphorylase B. *In*

"Metabolic Interconversion of Enzymes" (E. M. Fischer, E. G. Krebs, H. Neurath, and E. R. Stadtman, eds.), pp. 21–31. Springer Verlag, Berlin and New York.

Burdick, C. J., and Taylor, B. A. (1976). Histone acetylation during early stages of sea urchin (*Arbacia punctulata*) development. *Exp. Cell Res.* **100**, 428–433.

Busch, H., Ballal, N. R., Olson, M. O. J., and Yeoman, L. C. (1976). Chromatin and its non-histone proteins. *In* "Methods in Cancer Research" (H. Busch, ed.), Vol. 11, pp. 44–121. Academic Press, New York.

Bustin, M. (1978). Binding of *E. coli* RNA polymerase to chromatin subunits. *Nucleic Acids Res.* **5**, 925–932.

Bustin, M., Goldblatt, D., and Sperling, R. (1976). Chromatin structure visualized by immunoelectron microscopy. *Cell* **7**, 297–304.

Bustin, M., Simpson, R. T., Sperling, R., and Goldblatt, D. (1977). Molecular homogeneity of the histone content of HeLa chromatin subunits. *Biochemistry* **16**, 5381–5385.

Bustin, M., Hopkins, R. B., and Isenberg, I. (1978). Immunological relatedness of high mobility group chromosomal proteins from calf thymus. *J. Biol. Chem.* **253**, 1694–1699.

Butt, T. R., Brothers, J. F., Giri, C. P., and Smulson, M. E. (1978). A nuclear-modifying enzyme is responsive to ordered chromatin structure. *Nucleic Acids Res.* **5**, 2775–2778.

Byrne, R. H., Stone, P. R., and Kidwell, W. R. (1978). Effect of polyamines and divalent cations on histone H1-poly(adenosine diphosphate-ribose) complex formation. *Exp. Cell Res.* **115**, 277–283.

Byus, C. V., and Russell, D. H. (1976). Possible regulation of ornithine decarboxylase activity in the adrenal medulla of the rat by a cyclic AMP-dependent mechanism. *Biochem. Pharmacol.* **25**, 1595–1600.

Byus, C. V., Wicks, W. D., and Russell, D. H. (1976). Induction of ornithine decarboxylase in Reuber H35 rat hepatoma cells. *J. Cyclic Nucleotide Res.* **2**, 241–250.

Byvoet, P., Baxter, C. S., and Sayre, D. F. (1978). Displacement and aberrant methylation *in vitro* of H1 histone in rat liver nuclei after half-saturation of chromatin with polycations. *Proc. Natl. Acad. Sci. U.S.A.* **75**, 5773–5777.

Callan, H. G. (1963). The nature of lampbrush chromosomes. *Int. Rev. Cytol.* **15**, 1–34.

Camerini-Otero, R. D., Sollner-Webb, B., and Felsenfeld, G. (1976). The organization of histones and DNA in chromatin: Evidence for an arginine-rich histone kernel. *Cell* **8**, 333–347.

Cameron, I. L., and Jeter, J. R., Jr., eds. (1974). "Acidic Proteins of the Nucleus." Academic Press, New York.

Campbell, G. R., Littau, V. C., Melera, P. W., Allfrey, V. G., and Johnson, E. M. (1979). Unique sequence arrangement of ribosomal genes in the palindromic rDNA molecule of *Physarum polycephalum*. *Nucleic Acids Res.* **6**, 1433–1447.

Candido, E. P. M., and Dixon, G. H. (1971). Sites of *in vivo* acetylation in trout testis histone IV. *J. Biol. Chem.* **246**, 3182–3188.

Candido, E. P. M., and Dixon, G. H. (1972a). Amino-terminal sequences and sites of *in vivo* acetylation of trout testis histones III and IV. *Proc. Natl. Acad. Sci. U.S.A.* **69**, 2015–2019.

Candido, E. P. M., and Dixon, G. H. (1972b). Trout testis cells. III. Acetylation of histones in different cell types from developing trout testis. *J. Biol. Chem.* **247**, 5506–5510.

Candido, E. P. M., Reeves, R., and Davie, J. R. (1978). Sodium butyrate inhibits histone deacetylation in cultured cells. *Cell* **14**, 105–113.

Cary, P. D., Crane-Robinson, C., Bradbury, E. M., Javaherian, K., Goodwin, G. H., and Johns, E. W. (1976). Conformational studies of two non-histone chromosomal proteins and their interactions with DNA. *Eur. J. Biochem.* **62**, 583–590.

Castagna, M., Palmer, W. K., and Walsh, D. A. (1975). Nuclear protein kinase activity in perfused rat liver stimulated with dibutyryl adenosine cyclic 3', 5'-monophosphate. *Eur. J. Biochem.* **55**, 193–199.

Catterall, J. F., O'Malley, B. W., Robertson, M. A., Staden, R., Tanaka, Y., and Brownlee, G. C. (1978). Nucleotide sequence homology at 12 intron–exon junctions in the chick ovalbumin gene. *Nature (London)* **275**, 510–513.

Cech, T., and Pardue, M. L. (1977). Cross-linking of DNA with trimethylpsoralen is a probe for chromatin structure. *Cell* **11**, 631–640.

Cedar, H., and Felsenfeld, G. (1973). Transcription of chromatin *in vitro*. *J. Mol. Biol.* **77**, 237–254.

Chalkley, R., and Hunter, C. (1975). Histone–histone propinquity by aldehyde fixation of chromatin. *Proc. Natl. Acad. Sci. U.S.A.* **72**, 1304–1308.

Chen, C. C., Bruegger, B. B., Kern, C. W., Lin, Y. C., Halpern, R. M., and Smith, R. A. (1977). Phosphorylation of nuclear proteins in rat regenerating liver. *Biochemistry* **16**, 4852–4855.

Chickaraishi, D. M., Deeb, S. S., and Sueoka, N. (1978). Sequence complexity of nuclear RNAs in rat tissues. *Cell* **13**, 111–120.

Cho-Chung, Y. S., Clair, T., and Porper, R. (1977). Cyclic AMP-binding proteins and protein kinase during regression of Walker 256 mammary carcinoma. *J. Biol. Chem.* **252**, 6342–6348.

Chou, J. Y., Robinson, J. C., and Wang, S.-S. (1977). Effects of sodium butyrate on synthesis of human chorionic gonadotropin in trophoblastic and non-trophoblastic tumors. *Nature (London)* **268**, 543–544.

Chung, S.-Y., Hill, W. E., and Doty, P. (1978). Characterization of the histone core complex. *Proc. Natl. Acad. Sci. U.S.A.* **75**, 1680–1684.

Cleaver, J. (1977). Nucleosome structure controls rates of excision repair in DNA of human cells. *Nature (London)* **270**, 451–453.

Clissold, P. M., Arnstein, H. R. V., and Chesterton, C. J. (1977). Quantitation of globin mRNA levels during erythroid development in the rabbit and discovery of a new β-related species in immature erythroblasts. *Cell* **11**, 353–361.

Cohen, B. N., Blue, W. T., and Wagner, T. E. (1979). Chemically-induced gene expression. Manipulation of the transforming ability of SV40 minichromatin by specific chemical acetylation of histones H3 and H4. Submitted for publication.

Compton, J. L., Bellard, M., and Chambon, P. (1976). Biochemical evidence of the variability in the DNA repeat length in the chromatin of higher eukaryotes. *Proc. Natl. Acad. Sci. U.S.A.* **73**, 4382–4386.

Costa, E., Kurosawa, A., and Guidotto, A. (1976). Activation and nuclear translocation of protein kinase during transsynaptic induction of tyrosine 3-monooxygenase. *Proc. Natl. Acad. Sci. U.S.A.* **73**, 1058–1062.

Cox, R. F. (1977). Estrogen withdrawal in chick oviduct. Selective loss of high abundance classes of polyadenylated messenger RNA. *Biochemistry* **16**, 3433–3443.

Cremisi, C., Pignath, P. F., and Yaniv, M. (1976). Random localization and the absence of movement of the nucleosomes on SV40 nucleoprotein complex isolated from infected cells. *Biochem. Biophys. Res. Commun.* **73**, 548–554.

Dahmus, M., and Bonner, J. (1965). Increased template activity of chromatin: A result of hydrocortisone administration. *Proc. Natl. Acad. Sci. U.S.A.,* **54**, 1370–1375

D'Anna, J. A., and Isenberg, I. (1974). A histone cross-complexing pattern. *Biochemistry* **13**, 4992–4997.

D'Anna, J. A., Tobey, R. A., Barham, S. S., and Gurley, L. R. (1977). A reduction in the degree of H4 acetylation during mitosis in Chinese hamster cells. *Biochem. Biophys. Res. Commun.* **77**, 187–194.

Darzynkiewicz, Z., and Andersson, J. (1971). Effect of prednisolone on thymus lymphocytes. I. Autoradiographic studies on ^{3}H-actinomycin D binding and ^{3}H-uridine incorporation. *Exp. Cell Res.* **67,** 39–48.

Darzynkiewicz, Z., Bolund, L., and Ringertz, N. R. (1969). Actinomycin binding of normal and PHA-stimulated lymphocytes. *Exp. Cell Res.* **55,** 120–123.

Davidson, E. H., Allfrey, V. G., and Mirsky, A. E. (1964). On the RNA synthesized during the lampbrush phase of amphibian oogenesis. *Proc. Natl. Acad. Sci. U.S.A.* **52,** 501–508.

Davie, J. R., and Candido, E. P. M. (1978). Acetylated histone H4 is preferentially associated with template-active chromatin. *Proc. Natl. Acad. Sci. U.S.A.* **75,** 3574–3578.

Deeley, R. G., Udell, D. S., Burns, A. T. H., Gordon, J. I., and Goldberger, R. F. (1977). Kinetics of avian vitellogenin messenger RNA induction. *J. Biol. Chem.* **252,** 7913–7915.

DeLange, R. J., Fambrough, D. M., Smith, E. L., and Bonner, J. (1969). Calf and pea histone IV: Presence of ϵ-N-acetyllysine. *J. Biol. Chem.* **244,** 319–334.

DeLange, R. J., Hooper, J. A., and Smith, E. L. (1973). Histone III. Sequence studies of the cyanogen bromide peptides: Complete amino acid sequence of calf thymus histone HIII. *J. Biol. Chem.* **248,** 3261–3274.

DeRobertis, E. M., and Gurdon, J. B. (1977). Gene activation in somatic nuclei after injection into amphibian oocytes. *Proc. Natl. Acad. Sci. U.S.A.* **74,** 2470–2474.

DeRobertis, E. M., Longthorne, R. F., and Gurdon, J. B. (1978). Intracellular migration of nuclear proteins in *Xenopus* oocytes. *Nature (London)* **272,** 254–256.

Dixon, G. H., Candido, E. P. M., Honda, B. M., Louie, A. J., MacLeod, A. R., and Sung, M. T. (1975). The biological roles of post-synthetic modification of basic nuclear proteins. *In* "The Structure and Function of Chromatin" (D. W. Fitzsimons and G. E. W. Wolstenholme, eds.), pp. 229–250. Associated Scientific Publishers, Amsterdam.

Duerre, J. A., and Chakrabarty, S. (1975). Methylated basic amino acid composition of histones from the various organs from the rat. *J. Biol. Chem.* **250,** 8457–8461.

Edelman, I. S., and Fimognari, G. M. (1968). On the biochemical mechanism of action of aldosterone. *Recent Prog. Horm. Res.* **24,** 1–37.

Edwards, G. S., and Allfrey, V. G. (1973). Aflatoxin B_1 and actinomycin D effects on histone acetylation and deacetylation in the liver. *Biochim. Biophys. Acta* **299,** 354–366.

Eickbush, T. H., Watson, D. K., and Moudrianakis, E. N. (1978). A chromatin-bound proteolytic activity with unique specificity for histone H2A. *Cell* **9,** 785–792.

Elgin, S. C. R., and Weintraub, H. (1975). Chromosomal proteins and chromatin structure. *Annu. Rev. Biochem.* **44,** 725–774.

Eron, L., and Block, R. (1971). Mechanism of initiation and repression of *in vitro* transcription of the *lac* operon in *Escherichia coli. Proc. Natl. Acad. Sci. U.S.A.* **68,** 1828–1832.

Fasman, G. D. (1977). Histone–DNA interactions: Circular dichroism studies. *In* "Chromatin and Chromosome Structure" (H. J. Li and R. A. Eckhardt, eds.), pp. 71–142. Academic Press, New York.

Fasy, T. M., Inoue, A., Johnson, E. M., and Allfrey, V. G. (1979). Phosphorylation of H1 and H5 histones by cyclic AMP-dependent protein kinase reduces DNA binding. *Biochim. Biophys. Acta* **564,** 322–334.

Feigelson, P., and Kurtz, D. T. (1978). Hormonal modulation of α 2u globulin mRNA: Sequence measurements using a specific cDNA probe. *Cold Spring Harbor Symp. Quant. Biol.* **42,** 659–663.

Feigelson, M., Gross, P. R., and Feigelson, P. (1962). Early effects of cortisone on nucleic acid and protein metabolism of rat liver. *Biochim. Biophys. Acta* **55,** 495–504.

Finch, J. T., and Klug, A. (1978). X-Ray and electron microscopic analyses of crystals of nucleosome cores. *Cold Spring Harbor Symp. Quant. Biol.* **42,** 1–9.

Finch, J. T., Lutter, L. C., Rhodes, D., Brown, R. S., Rushton, B., Levitt, M., and Klug, A. (1977). Structure of nucleosome core particles of chromatin. *Nature (London)* **269**, 29–36.

Fishman, P. H., Bradley, R. M., and Henneberry, P. C. (1976). Butyrate-induced glycolipid synthesis in HeLa cells: Properties of the induced sialyltransferase. *Arch. Biochem. Biophys.* **172**, 618–626.

Flint, S. J., and Weintraub, H. (1977). An altered subunit configuration associated with the actively transcribed DNA of integrated adenovirus genes. *Cell* **12**, 783–794.

Foe, V. E. (1978). Modulation of ribosomal RNA synthesis in *Oncopeltus fasciatus*. An electron microscopic study of the relationship between changes in chromatin structure and transcriptional activity. *Cold Spring Harbor Symp. Quant. Biol.* **42**, 732–739.

Foe, V. E., Wilkinson, L. E., and Laird, C. D. (1976). Comparative organization of active transcription units in *Oncopeltus fasciatus*. *Cell* **9**, 131–146.

Fontana, J. A., and Lovenberg, W. (1973). Pineal protein kinase: Effect of enzymic phosphorylation on actinomycin D binding by, and template activity of, chromatin. *Proc. Natl. Acad. Sci. U.S.A.* **70**, 755–758.

Ford, P. J., and Southern, E. M. (1973). Different sequences for 5 S RNA in kidney cells and ovaries of *Xenopus laevis*. *Nature (London), (New Biol.)* **241**, 7–12.

Fox, K., and Gabourel, J. (1967). Effect of cortisol on the RNA polymerase system of rat thymus. *J. Mol. Pharmacol.* **3**, 479–486.

Franke, W. W., Scheer, U., Spring, H., Trendelenburg, M. F., and Krohne, G. (1976). Morphology of transcriptional units of rDNA. *Exp. Cell Res.* **100**, 233–244.

Franke, W. W., Scheer, U., Trendelenburg, M. F., Zentgraf, H., and Spring, H. (1978). Morphology of transcriptionally active chromatin. *Cold Spring Harbor Symp. Quant. Biol.* **42**, 755–772.

Franklin, S. G., and Zweidler, A. (1977). Non-allelic variants of histones H2a, H2b and H3 in mammals. *Nature (London)* **266**, 272–275.

Frenster, J. H., Allfrey, V. G., and Mirsky, A. E. (1963). Repressed and active chromatin isolated from interphase lymphocytes. *Proc. Natl. Acad. Sci. U.S.A.* **50**, 1026–1032.

Friedman, D. L., and Chambers, D. A. (1978). Cyclic nucleotide-binding proteins detected by photoaffinity labeling in nucleus and cytoplasm of bovine liver. *Proc. Natl. Acad. Sci. U.S.A.* **75**, 5286–5290.

Galau, G. A., Britten, R. J., and Davidson, E. H. (1974). A measurement of the sequence complexity of polysomal messenger RNA in sea urchin embryos. *Cell* **2**, 9–20.

Galau, G. A., Klein, W. H., Davis, M. M., Wold, B. J., Britten, R. J., and Davidson, E. H. (1976). Structural gene sets active in embryos and adult tissues of the sea urchin. *Cell* **7**, 487–505.

Gall, J. G., and Callan, H. G. (1962). H³-Uridine incorporation in lampbrush chromosomes. *Proc. Natl. Acad. Sci. U.S.A.,* **48**, 562–570.

Garel, A., and Axel, R. (1976). Selective digestion of transcriptionally active ovalbumin genes from oviduct nuclei. *Proc. Natl. Acad. Sci. U.S.A.* **73**, 3966–3970.

Garel, A., Zolan, M., and Axel, R. (1977). Genes transcribed at diverse rates have a similar conformation in chromatin. *Proc. Natl. Acad. Sci. U.S.A.* **74**, 4867–4871.

Gaubatz, J., Hardison, R., Murphy, J., Eichner, M. E., and Chalkley, R. (1978). The role of H1 in the structure of chromatin. *Cold Spring Harbor Symp. Quant. Biol.* **42**, 265–271.

Germond, J. E., Hirt, B., Oudet, P., Gross-Bellard, M., and Chambon, P. (1975). Folding of the DNA double helix in chromatin-like structures from simian virus 40. *Proc. Natl. Acad. Sci. U.S.A.,* **72**, 1843–1847.

Gershey, E. L., Vidali, G., and Allfrey, V. G. (1968). Chemical studies of histone acetylation. The occurrence of ε-N-acetyllysine in the f2 al histone. *J. Biol. Chem.* **243**, 5018–5022.

Ghosh, N. K., and Cox, R. P. (1976). Production of human chorionic gonadotropin in HeLa cell cultures. *Nature* (*London*) **259**, 416–417.

Ghosh, N. K., and Cox, R. P. (1977). Induction of human follicle-stimulating hormone in HeLa cells by sodium butyrate. *Nature* (*London*) **267**, 435–437.

Gilbert, W., and Muller-Hill, B. (1966). Isolation of the *lac* repressor. *Proc. Natl. Acad. Sci. U.S.A.* **56**, 1891–1895.

Gilbert, W., and Müller-Hill, B. (1967). The *lac*-operator is DNA. *Proc. Natl. Acad. Sci. U.S.A.* **58**, 2415–2421.

Giri, C. P., West, M. H. P., and Smulson, M. E. (1978a). Nuclear protein modification and chromatin substructure. I. Differential poly(adenosine diphosphate)ribosylation of chromosomal proteins in nuclei versus isolated nucleosomes. *Biochemistry* **17**, 3495–3500.

Giri, C. P., West, M. H. P., Ramirez, M. L., and Smulson, M. E. (1978b). Nuclear protein modification and chromatin structure. 2. Internucleosomal localization of poly(adenosine diphosphate-ribose) polymerase. *Biochemistry* **17**, 3501–3504.

Glasser, S. R., Chytil, F. C., and Spelsberg, T. C. (1972). Early effects of estradiol-17β on the deoxyribonucleic acid dependent ribonucleic acid polymerases I and II of the rat uterus. *Biochem. J.* **130**, 947–957.

Goldberg, M. I., Perriard, J.-C., and Rutter, W. J. (1977). A protein cofactor that stimulates the activity of DNA-dependent RNA polymerase I on double-stranded DNA. *Biochemistry* **16**, 1648–1654.

Goldknopf, I. L., and Busch, H. (1977). Isopeptide linkage between nonhistone and histone 2A polypeptides of chromosomal conjugate–protein A24. *Proc. Natl. Acad. Sci. U.S.A.* **74**, 864–868.

Goldknopf, I. L., French, M. F., Daskal, Y., and Busch, H. (1978). A reciprocal relationship between contents of free ubiquitin and protein A24, its conjugate with histone 2A in chromatin fractions obtained by the DNase II, Mg^{++} procedure. *Biochem. Biophys. Res. Commun.* **84**, 786–793.

Goldstein, L. (1958). Localization of nuclear-specific protein as shown by transplantation experiments in *Amoeba proteus*. *Exp. Cell Res.* **15**, 635–637.

Goldstein, L. (1963). RNA and protein in nucleocytoplasmic interactions. *In* "Cell Growth and Cell Division" (R. J. C. Harris, ed.), pp. 129–149. Academic Press, New York.

Goldstein, L. (1965). Interchange of protein between nucleus and cytoplasm. *Int. Soc. Cell Biol.* **4**, 79–94.

Goldstein, L., Wise, G. E., Stephenson, C., and Ko, C. (1978). Small nuclear RNAs: An association with condensed chromatin in *Amoeba*. *J. Cell Sci.* **30**, 227–235.

Golomb, H. M., and Bahr, G. F. (1974). Electron microscopy of human interphase nuclei. Determination of total dry mass and DNA-packing ratio. *Chromosoma* (*Berlin*) **46**, 233–245.

Goodwin, G. H., and Johns, E. W. (1973). Isolation and characterization of two calf thymus chromatin non-histone proteins with high contents of acidic and basic amino acids. *Eur. J. Biochem.* **40**, 215–219.

Goodwin, G. H., Sanders, C., and Johns, E. W. (1973). A new group of chromatin-associated proteins with a high content of acidic and basic amino acids. *Eur. J. Biochem.* **38**, 14–19.

Goodwin, G. H., Nicolas, R. H., and Johns, E. W. (1976). Microheterogeneity in a nonhistone chromosomal protein. *FEBS Lett.* **64**, 412–414.

Goodwin, G. H., Nicolas, R. H., and Johns, E. W. (1977a). A quantitative analysis of histone H1 in rabbit thymus nuclei. *Biochem. J.* **167**, 485–488.

Goodwin, G. H., Rabbani, A., Nicolas, R. H., and Johns, E. W. (1977b). The isolation of the high mobility group non-histone chromosomal protein HMG14. *FEBS Lett.* **80,** 413–416.

Goodwin, G. H., Woodhead, L., and Johns, E. W. (1977c). The presence of high mobility group non-histone chromatin proteins in isolated nucleosomes. *FEBS Lett.* **73,** 85–88.

Goodwin, G. H., Walker, J. M., and Johns, E. W. (1978). Studies on the degradation of high mobility group non-histone chromosomal proteins. *Biochim. Biophys. Acta* **519,** 233–242.

Gorovsky, M. A., and Keevert, J. B. (1975). Absence of histone H1 in a mitotically dividing genetically inactive nucleus. *Proc. Natl. Acad. Sci. U.S.A.* **72,** 2672–2676.

Gorovsky, M. A., Pleger, G. L., Keevert, J. B., and Johmann, C. A. (1973). Studies on histone fraction f2a1 in macro- and micronuclei of *Tetrahymena pyriformis*. *J. Cell Biol.* **57,** 773–781.

Gottesfeld, J. M., and Butler, P. J. G. (1977). Structure of transcriptionally active chromatin subunits. *Nucleic Acids Res.* **4,** 3155–3173.

Gottesfeld, J. M., and Partington, G. A. (1977). Distribution of messenger RNA-coding sequences in fractionated chromatin. *Cell* **12,** 953–962.

Gottesfeld, J. M., Garrard, W. T., Bagi, G., Wilson, R. F., and Bonner, J. (1974). Partial purification of the template-active fraction of chromatin: A preliminary report. *Proc. Natl. Acad. Sci. U.S.A.* **71,** 2193–2197.

Gottesfeld, J. M., Murphy, R. F., and Bonner, J. (1975). Structure of transcriptionally-active chromatin. *Proc. Natl. Acad. Sci. U.S.A.* **72,** 4404–4408.

Graaff, G. deV., and von Holt, C. (1973). Enzymatic histone modification during the induction of tyrosine aminotransferase with insulin and hydrocortisone. *Biochim. Biophys. Acta* **299,** 480–484.

Grainger, R. M., and Ogle, R. C. (1978). Chromatin structure of the ribosomal RNA genes in *Physarum polycephalum*. *Chromosoma* (*Berlin*) **65,** 115–126.

Granboulan, N., and Granboulan, P. (1966). Cytochimie ultrastructurale du nucléole. II. Etude des sites de synthese du RNA dans le nucleole et le noyau. *Exp. Cell. Res.* **38,** 604–619.

Green, M. H., Buss, J., and Gariglio, P. (1975). Activation of nuclear RNA polymerase by Sarkosyl. *Eur. J. Biochem.* **53,** 217–225.

Griffin, M. H., Price, G. H., Bozzell, K. L., Cox, R. P., and Ghosh, N. K. (1974). A study of adenosine 3′,5′-cyclic monophosphate, sodium butyrate and cortisol as inducers of HeLa alkaline phosphatase. *Arch. Biochem. Biophys.* **164,** 619–623.

Griffin, G. D., Yang, W.-K., and Novelli, G. D. (1976). Transfer RNA species in human lymphocytes stimulated by mitogens and in leukemia cells. *Arch. Biochem. Biophys.* **176,** 187–196.

Griffith, J. D., and Christiansen, G. (1978). The multifunctional role of histone H1, probed with the SV40 minichromosome. *Cold Spring Harbor Symp. Quant. Biol.* **42,** 215–226.

Grimes, S. L., Jr., Chae, C. B., and Irvin, J. L. (1975). Acetylation of histones of rat testis. *Arch. Biochem. Biophys.* **168,** 425–435.

Gross, K. W., Lorena, J. M., Baglioni, C., and Gross, P. R. (1973). Cell-free translation of maternal messenger RNA from sea urchin eggs. *Proc. Natl. Acad. Sci. U.S.A.* **70,** 2614–2618.

Groudine, M., Das, S., Neiman, P., and Weintraub, H. (1978). Regulation of expression and chromosomal subunit conformation of avian retrovirus genomes. *Cell* **14,** 865–878.

Grunstein, M. (1978). Hatching in the sea urchin *Lytechinus pictis* is accompanied by a shift in histone H4 gene activity. *Proc. Natl. Acad. Sci. U.S.A.* **75,** 4135–4139.

Gurdon, J. B. (1974). "The Control of Gene Expression in Animal Development." Harvard Univ. Press, Cambridge, Massachusetts.

Gurley, L. R., Walters, R. A., and Tobey, R. A. (1974). Cell cycle specific changes in histone phosphorylation associated with cell proliferation and chromosome condensation. *J. Cell Biol.* **60**, 356–364.

Gurley, L. R., Walters, R. A., and Tobey, R. A. (1975). Sequential phosphorylation of histone subfractions in the chinese hamster cell cycle. *J. Biol. Chem.* **250**, 3936–3944.

Hagopian, H. K., Riggs, M. G., Swartz, L. A., and Ingram, V. M. (1977). Effect of *n*-butyrate on DNA synthesis in chick fibroblasts and HeLa cells. *Cell* **12**, 855–860.

Hagopian, H. K., Riggs, M. G., Newmann, J. R., Dobson, M. E., Owens, B. B., and Ingram, V. R. (1979). A model for differentiation: Modification of chromatin proteins in differentiating erythroid and non-erythroid cells. *In* "Cellular and Molecular Regulation of Hemoglobin Switching" (A. Nienhus and S. Stamatoyannopoulos, eds.), pp. 471–489. Grune & Stratton, New York.

Harding, J. D., and Rutter, W. J. (1978). Rat pancreatic amylase mRNA. Tissue specificity and accumulation during embryonic development. *J. Biol. Chem.* **253**, 8736–8740.

Harding, J. D., MacDonald, R. J., Przybla, A. E., Chirgwin, J. M., Pictet, R. L., and Rutter, W. J. (1977). Changes in the frequency of specific transcripts during development of the pancreas. *J. Biol. Chem.* **252**, 7391–7397.

Harris, H. (1965). Behaviour of differentiated nuclei in heterokaryons of animal cells from different species. *Nature (London)* **206**, 583–588.

Harris, H., and Cook, P. R. (1969). Synthesis of an enzyme determined by an erythrocyte nucleus in a hybrid cell. *J. Cell Sci.* **5**, 121–133.

Harris, S. E., Rosen, J. M., Means, A. R., and O'Malley, B. W. (1975). Use of a specific probe for ovalbumin messenger RNA to quantitate estrogen-induced gene transcripts. *Biochemistry* **14**, 2072–2081.

Harrison, P. R., Birnie, G. D., Hell, A., Humphries, S., Young, B. D., and Paul, J. (1974). Kinetic studies of gene frequency. I. Use of a DNA copy of reticulocyte 9 S RNA to estimate globin gene dosage in mouse tissues. *J. Mol. Biol.* **84**, 539–554.

Hastie, N. D., and Bishop, J. O. (1976). The expression of three abundance classes of messenger RNA in mouse tissues. *Cell* **9**, 761–774.

Hayaishi, O., and Ueda, K. (1977). Poly(ADP-ribose) and ADP-ribosylation of proteins. *Annu. Rev. Biochem.* **46**, 95–116.

Hendrick, D., Tolstoshev, P., and Randlett, D. (1977). Enrichment for the globin coding region in a chromatin fraction from chick reticulocytes by endonuclease digestion. *Gene* **2**, 147–158.

Hewish, D. R., and Burgoyne, L. A. (1973). Chromatin substructure: The digestion of chromatin DNA at regularly spaced sites by a nuclear deoxyribonuclease. *Biochem. Biophys. Res. Commun.* **52**, 504–510.

Hirsch, J., and Martelo, O. J. (1976). Phosphorylation of rat liver ribonucleic acid polymerase I by nuclear protein kinases. *J. Biol. Chem.* **251**, 5408–5413.

Hirsch, J., and Martelo, O. J. (1978). Purification and properties of a nuclear protein kinase associated with ribonucleic acid polymerase I. *Biochem. J.* **169**, 355–359.

Hjelm, R. P., Kneale, G. G., Suau, P., Baldwin, J. P., Bradbury, E. M., and Ibel, K. (1977). Small-angle neutron scattering studies of chromatin subunits. *Cell* **10**, 139–151.

Hohmann, P. (1978). The H1 class of histone and diversity in chromosomal structure. *In* "Subcellular Biochemistry" (D. B. Roodyn, ed.), Vol. 5, pp. 87–127, Plenum, New York.

Hohmann, P., and Cole, R. D. (1969). Hormonal effects on amino acid incorporation into lysine-rich histones. *Nature (London)* **223**, 1064–1066.

Hohmann, P., and Cole, R. D. (1971). Hormonal effects on amino acid incorporation into lysine-rich histones in the mouse mammary gland. *J. Mol. Biol.* **58**, 533–540.

Hohmann, P., Tobey, R. A., and Gurley, L. R. (1976). Phosphorylation of distinct regions of f1 histone. Relationship to the cell cycle. *J. Biol. Chem.* **251**, 3685–3692.

Honda, B. M., Dixon, G. H., and Candido, E. P. M. (1975). Sites of *in vivo* histone methylation in developing trout testis. *J. Biol. Chem.* **250**, 8681–8685.

Hooper, J. A., Smith, E. L., Summer, K. R., and Chalkley, R. (1973). Histone III. Amino acid sequence of histone III of the testes of the carp, *Letiobus bubalus*. *J. Biol. Chem.* **248**, 3275–3279.

Horgen, P. A., and Ball, S. F. (1974). Nuclear protein acetylation during hormone-induced sexual differentiation in *Achlya ambisexualis*. *Cytobios* **10**, 181–185.

Hough, B. R., Smith, M. J., Britten, R. J., and Davidson, E. H. (1975). Sequence complexity of heterogeneous nuclear RNA in sea urchin embryos. *Cell* **5**, 291–299.

Hough-Evans, B., Wold, B. J., Ernst, S. G., Britten, R. J., and Davidson, E. H. (1977). Appearance and persistence of maternal RNA sequences in sea urchin development. *Dev. Biol.* **60**, 258–277.

Hsu, L. S. L., Davies, J. A., and Chesterton, C. J. (1978). Mechanism of transcription repression during erythroid development in the rabbit. *Biochem. Soc. Trans. (London)* **6**, 1057–1060.

Huang, R. C. C., and Bonner, J. (1962). Histone, a suppressor of chromosomal RNA synthesis. *Proc. Natl. Acad. Sci. U.S.A.* **48**, 1216–1220.

Humphries, S., Windass, J., and Williamson, R. (1976). Mouse globin gene expression in erythroid and non-erythroid tissues. *Cell* **7**, 267–277.

Hunt, T., and Dayhoff, M. O. (1977). Amino-terminal sequence identity of ubiquitin and the nonhistone component of nuclear protein A24. *Biochem. Biophys. Res. Commun.* **74**, 650–655.

Hyde, J. E., and Walker, I. O. (1975). Covalent cross-linking of histones in chromatin. *FEBS Lett.* **50**, 150–154.

Iatrou, K., and Dixon, G. H. (1978). Protamine messenger RNA: Its life history during spermatogenesis in rainbow trout. *Fed. Proc., Fed. Am. Soc. Exp. Biol.* **37**, 2526–2533.

Ilyin, Y. V., Bayev, A. A., Zhuze, A. L., and Varshavsky, A. J. (1974). Histone–histone proximity in chromatin as seen by imidoester crosslinking. *Mol. Biol. Rep.* **1**, 343–348.

Inglis, R. J., Langan, T. R., Matthews, H. R., Hardie, D. G., and Bradbury, E. M. (1976). Advance of mitosis by histone phosphokinase. *Exp. Cell Res.* **97**, 418–425.

Inoue, A., and Fujimoto, D. (1972). Substrate specificity of histone deacetylase from calf thymus. *J. Biochem. (Tokyo)* **72**, 427–431.

Izawa, M., Allfrey, V. G., and Mirsky, A. E. (1963). The relationship between RNA synthesis and loop structure on lampbrush chromosomes. *Proc. Natl. Acad. Sci. U.S.A.* **49**, 544–551.

Jackson, V. (1978). Studies on histone organization in the nucleosome using formaldehyde as a reversible cross-linking agent. *Cell* **15**, 945–954.

Jackson, V., Shires, A., Chalkley, R., and Granner, D. (1975). Studies on highly metabolically active acetylation and phosphorylation of histones. *J. Biol. Chem.* **250**, 4856–4863.

Jackson, V., Shires, A., Tanphaichitr, N., and Chalkley, R. (1976). Modifications to histones immediately after synthesis. *J. Mol. Biol.* **104**, 471–483.

Jahn, C. L., and Litman, G. W. (1977). Distribution of covalently bound benzo(α)pyrene in chromatin. *Biochem. Biophys. Res. Commun.* **76**, 534–540.

James, G. T., Yeoman, L. C., Matsui, S.-I., Goldberg, A. H., and Busch, H. (1977). Isolation and characterization of nonhistone chromosomal protein C14 which stimulates RNA synthesis. *Biochemistry* **16**, 2384–2389.

Jamrich, M., Greenleaf, A. L., and Bautz, E. K. F. (1977). Localization of RNA polymerase in polytene chromosomes of *Drosophila melanogaster*. *Proc. Natl. Acad. Sci. U.S.A.* **74**, 2079–2083.

Javaherian, K., and Amini, S. (1978). Conformation study of calf thymus HMG 14 nonhistone protein. *Biochem. Biophys. Res. Commun.* **85**, 1385–1391.

Javaherian, K., Liu, L. F., and Wang, J. C. (1978). Nonhistone proteins HMG 1 and HMG 2 change the DNA helical structure. *Science* **199**, 1345–1346.

Jelinek, W. (1977). Specific nucleotide sequences in HeLa cell inverted repeat DNA: Enrichment of sequences found in double-stranded regions of heterogeneous nuclear RNA. *J. Mol. Biol.* **115**, 591–602.

Jelinek, W., and Leinwand, L. (1978). Low molecular weight RNAs hydrogen-bonded to nuclear and cytoplasmic poly(A)-terminated RNA from cultured hamster ovary cells. *Cell* **15**, 205–214.

Jelinek, W., Evans, R., Wilson, M., Salditt-Georgieff, M., and Darnell, J. (1978). Oligonucleotides in heterogeneous nuclear RNA: Similarity of inverted repeats and RNA from repetitious DNA sites. *Biochemistry* **17**, 2776–2783.

Joffe, J., Keene, M., and Weintraub, H. (1977). Histones H2A, H2B, H3 and H4 are present in equimolar amounts in chick erythroblasts. *Biochemistry* **16**, 1236–1238.

Johnson, E. M. (1977). Cyclic AMP-dependent protein kinase and its nuclear substrate proteins. *In* "Advances in Cyclic Nucleotide Research" (P. Greengard and G. A. Robison, eds.), Vol. 8, pp. 267–309. Raven, New York.

Johnson, E. M., and Allfrey, V. G. (1978). Post-synthetic modifications of histone primary structure. Phosphorylation and acetylation as related to chromatin conformation and function. *In* "Biochemical Actions of Hormones" (G. Litwack, ed.), Vol. 5, pp. 1–51. Academic Press, New York.

Johnson, E. M., Karn, J., and Allfrey, V. G. (1974). Early nuclear events in the induction of lymphocyte proliferation by mitogens. *J. Biol. Chem.* **249**, 4990–4999.

Johnson, E. M., Hadden, J. W., Inoue, A., and Allfrey, V. G. (1975). DNA-binding by cyclic adenosine 3′,5′-monophosphate dependent protein kinase from calf thymus nuclei. *Biochemistry* **14**, 3873–3884.

Johnson, E. M., Littau, V. C., Allfrey, V. G., Bradbury, E. M., and Matthews, H. R. (1976). The subunit structure of chromatin from *Physarum polycephalum*. *Nucleic Acids Res.* **3**, 3313–3329.

Johnson, E. M., Allfrey, V. G., Bradbury, E. M., and Matthews, H. R. (1978a). Altered nucleosome structure containing DNA sequences complementary to 19 S and 26 S ribosomal RNA in *Physarum polycephalum*. *Proc. Natl. Acad. Sci. U.S.A.* **75**, 1116–1120.

Johnson, E. M., Matthews, H. R., Littau, V. C., Lothstein, L., Bradbury, E. M., and Allfrey, V. G. (1978b). The structure of chromatin containing DNA complementary to 19 S and 26 S ribosomal RNA in active and inactive stages of *Physarum polycephalum*. *Arch. Biochem. Biophys.* **191**, 537–550.

Jost, H.-P., and Sahib, M. K. (1971). Role of cyclic adenosine 3′,5′-monophosphate on the induction of hepatic enzymes. II. Effect of N^6, O^2-dibutyryl cyclic adenosine 3′,5′-monophosphate on the kinetics of ribonucleic acid synthesis in purified rat liver nuclei. *J. Biol. Chem.* **246**, 1623–1629.

Jost, H.-P., Hsie, A., Hughes, S. D., and Ryan, K. (1970). Role of cyclic adenosine 3′,5′-monophosphate in the induction of hepatic enzymes. I. Kinetics of the induction of rat liver serine dehydratase by cyclic adenosine 3′,5′-monophosphate. *J. Biol. Chem.* **245**, 351–357.

Jungmann, R. A., and Schweppe, J. S. (1972a). Mechanism of action of gonadotropin. II. Control of ovarian nuclear ribonucleic acid polymerase activity and chromatin template capacity. *J. Biol. Chem.* **247**, 5543–5548.

Jungmann, R. A., and Schweppe, J. S. (1972b). Mechanism of action of gonadotropin. I. Evidence for gonadotropin-induced modifications of ovarian nuclear basic and acidic protein biosynthesis, phosphorylation and acetylation. *J. Biol. Chem.* **247**, 5535–5542.

Jungmann, R. A., Hiestand, P. C., and Schweppe, J. S. (1974). Mechanism of action of gonadotropin. IV. Cyclic AMP-dependent translocation of ovarian cyclic AMP-binding protein and protein kinase to nuclear acceptor sites. *Endocrinology* **94**, 168–183.

Jungmann, R. A., Lee, S. G., and DeAngelo, A. B. (1975). Translocation of cytoplasmic protein kinase and cyclic adenosine monophosphate-binding protein to intracellular acceptor sites. *In* "Advances in Cyclic Nucleotide Research" (P. Greengard and G. A. Robison, eds.), Vol. 5, pp. 281–306. Raven, New York.

Kao-Huang, Y., Revzin, A., Butler, A. P., O'Connor, P., Noble, D. W., and Von Hippel, P. H. (1977). Nonspecific DNA binding of genome regulating proteins as a biological control mechanism. Measurement of DNA-bound *Escherichia coli* lac repressor in vivo. *Proc. Natl. Acad. Sci. U.S.A.* **74**, 4228–4232.

Kaplan, B. B., Schachter, B. S., Osterburg, H. H., deVellis, J. S., and Finch, C. E. (1978). Sequence complexity of polyadenylated RNA obtained from rat brain regions and cultured rat cells of neural origin. *Biochemistry* **17**, 5516–5524.

Karrer, K. M., and Gall, J. G. (1976). The macromolecular ribosomal DNA of *Tetrahymena pyriformis* is a giant palindrome. *J. Mol. Biol.* **104**, 421–453.

Kedes, L. H. (1976). Histone messengers and histone genes. *Cell* **8**, 321–331.

Keely, S. L. Jr., Corbin, J. D., and Park, C. R. (1975). On the question of translocation of heart cyclic AMP-dependent protein kinase. *Proc. Natl. Acad. Sci. U.S.A.* **72**, 1501–1504.

Keller, R. K., Socher, S. H., Krall, J. F., Chandra, T., and O'Malley, B. W. (1975). Fractionation of chick oviduct chromatin. IV. Association of protein kinase with transcriptionally active chromatin. *Biochem. Biophys. Res. Commun.* **66**, 453–459.

Kemp, C. L. (1966). Electron microscope autoradiographic studies of RNA metabolism in *Trillium erectum* microspores. *Chromosoma (Berlin)* **19**, 137–148.

Kenney, F. T., and Kull, F. J. (1963). Hydrocortisone-stimulated synthesis of nuclear RNA in enzyme induction. *Proc. Natl. Acad. Sci. U.S.A.* **50**, 493–499.

Kidson, C. (1965). Kinetics of cortisol action on RNA synthesis. *Biochem. Biophys. Res. Commun.* **21**, 283–289.

Kidwell, W. R., and Mage, M. G. (1976). Changes in poly(adenosine diphosphate-ribose) and poly(adenosine diphosphate-ribose) polymerase in synchronous HeLa cells. *Biochemistry* **15**, 1213–1217.

Killander, D., and Rigler, R. (1965). Initial changes of deoxyribonucleoproteins and synthesis of nucleic acid in phytohemagglutinin-stimulated human leucocytes in vitro. *Exp. Cell Res.* **39**, 701–704.

Killander, D., and Rigler, R. (1969). Activation of deoxyribonucleoprotein in human leucocytes stimulated by phytohemagglutinin. 1. Kinetics of the binding of acridine orange to deoxyribonucleoprotein. *Exp. Cell Res.* **54**, 163–170.

Kincade, J. M. (1969). Qualitative species differences and quantitative tissue differences in the distribution of lysine-rich histones. *J. Biol. Chem.* **244**, 3375–3386.

Kincade, J. M., and Cole, R. D. (1966a). The resolution of four lysine-rich histones derived from calf thymus. *J. Biol. Chem.* **241**, 5790–5797.

Kincade, J. M., and Cole, R. D. (1966b). A structural comparison of different lysine-rich histones of calf thymus. *J. Biol. Chem.* **241**, 5798–5805.

Kistler, W. S., and Geroch, M. E. (1975). An unusual pattern of lysine-rich histone components is associated with spermatogenesis in rat testis. *Biochem. Biophys. Res. Commun.* **63,** 378–384.

Kleinsmith, L. J. (1978). Phosphorylation of nonhistone proteins. *In* "The Cell Nucleus" (H. Busch, ed.) Vol. 6; Chromatin, Part C, pp. 222–261. Academic Press, New York.

Kleinsmith, L. J., Allfrey, V. G., and Mirsky, A. E. (1966a). Phosphoprotein metabolism in isolated lymphocyte nuclei. *Proc. Natl. Acad. Sci. U.S.A.,* **55,** 1182–1189.

Kleinsmith, L. J., Allfrey, V. G., and Mirsky, A. E. (1966b). Phosphorylation of nuclear protein early in the course of gene activation in lymphocytes. *Science* **154,** 780–781.

Kleinsmith, L. J., Heidema, J., and Carroll, A. (1970). Specific binding of rat liver nuclear proteins to DNA. *Nature (London)* **226,** 1025–1028.

Kornberg, R. D. (1974). Chromatin structure: A repeating unit of histones and DNA. *Science* **184,** 868–871.

Kornberg, R. D. (1977). Structure of chromatin. *Annu. Rev. Biochem.* **46,** 931–954.

Kornberg, R. D., and Thomas, J. O. (1974). Chromatin structure: Oligomers of the histones. *Science* **184,** 865–868.

Krause, M. O., and Stein, G. S. (1975). Properties of the genome in normal and SV40-transformed WI38 human diploid fibroblasts. II. Metabolism and binding of histones. *Exp. Cell Res.* **92,** 175–190.

Kroeger, H. (1960). The induction of new puffing patterns by transplantation of salivary gland nuclei into egg cytoplasm of *Drosophila*. *Chromosoma (Berlin)* **11,** 129–145.

Kunkel, N. S., Hemminki, K., and Weinberg, E. S. (1978). Size of histone gene transcripts in different embryonic stages of the sea urchin, *Strongylocentrotus purpuratus*. *Biochemistry* **17,** 2591–2598.

Kurth, P. D., Moudrianakis, E. N., and Bustin, M. (1978). Histone localization in polytene chromosomes by immunofluorescence. *J. Cell Biol.* **78,** 910–918.

Kurtz, D. T., Sippel, A. E., Ansah-Yiadom, R., and Feigelson, P. (1976). Effects of sex hormones on the level of the messenger RNA for the rat hepatic protein α-2u globulin. *J. Biol. Chem.* **251,** 3594–3598.

Kwan, S. P., Wood, T. G., and Lingrel, J. B. (1977). Purification of a putative precursor of globin messenger RNA from mouse nucleated erythroid cells. *Proc. Natl. Acad. Sci. U.S.A.* **74,** 178–182.

Kyner, D., Zabos, P., Christman, J., and Acs, G. (1976). Effect of sodium butyrate on lymphocyte activation. *J. Exp. Med.* **144,** 1674–1678.

LaBella, F., Vidali, G., and Vesco, C. (1979). Histone acetylation in CV-1 cells infected with simian virus 40. *Virology* **96,** 564–575.

Lake, R. S. (1973). Further characterization of the f1 histone phosphokinase of metaphase-arrested animal cells. *J. Cell Biol.* **58,** 317–331.

Lake, R. S., and Salzmann, P. (1972). Occurrence and properties of a chromatin-associated f1 histone phosphokinase in mitotic Chinese hamster cells. *Biochemistry* **11,** 4817–4826.

Lakhotia, S. C., and Jacob, J. (1974). EM autoradiographic studies on polytene nuclei of *Drosophila melanogaster*. *Exp. Cell Res.* **86,** 253–263.

Langan, T. A. (1969a). Phosphorylation of liver histone following the administration of glucagon and insulin. *Proc. Natl. Acad. Sci. U.S.A.* **64,** 1276–1283.

Langan, T. A. (1969b). Action of adenosine 3′,5′-monophosphate dependent histone kinase *in vivo*. *J. Biol. Chem.* **244,** 5763–5765.

Langan, T. A. (1971). Cyclic AMP and histone phosphorylation. *Ann. N.Y. Acad. Sci.* **185,** 166–180.

Langan, T. A. (1973). Protein kinases and protein kinase substrates. *In* "Advances in Cyclic

Nucleotide Research (P. Greengard and G. A. Robison, eds.), Vol. 3, pp. 99–153. Raven, New York.

Langan, T. A., and Hohmann, P. (1975). Analysis of phosphorylation sites in lysine-rich H1 histone: An approach to the determination of structural chromosomal protein functions. *In* "Chromosomal Proteins and their Role in the Regulation of Gene Expression" (G. S. Stein and L. J. Kleinsmith, eds.), pp. 113–125. Academic Press, New York.

Langan, T. A., Rall, S. C., and Cole, R. D. (1971). Variation in primary structure at a phosphorylation site in lysine-rich histones. *J. Biol. Chem.* **246**, 1942–1944.

Langmore, J. P., and Wooley, J. C. (1975). Chromatin architecture: Investigation of a subunit of chromatin by dark-field electron microscopy. *Proc. Natl. Acad. Sci. U.S.A.* **72**, 2691–2695.

Leavitt, J., and Moyzis, R. (1978). Changes in gene expression accompanying neoplastic transformation of Syrian hamster cells. *J. Biol. Chem.* **253**, 2497–2500.

Leavitt, J., Barrett, J. C., Crawford, B. D., and Ts'o, P. O. P. (1978). Butyric acid suppression of the *in vitro* neoplastic state of Syrian hamster cells. *Nature (London)* **271**, 262–265.

Leder, A., Orkin, S., and Leder, P. (1975). Differentiation of erythroleukemic cells in the presence of inhibitors of DNA synthesis. *Science* **190**, 893–894.

Ledinko, N. (1970). Transient stimulation of deoxyribonucleic acid-dependent ribonucleic acid polymerase and histone acetylation in human embryonic kidney cultures infected with adenovirus 2 or 12: Apparent induction of host ribonucleic acid synthesis. *J. Virol.* **6**, 58–68.

Lee, D. C., McKnight, G. S., and Palmiter, R. D. (1978). The action of estrogen and progesterone on the expression of the transferrin gene. A comparison of the response in chick liver and oviduct. *J. Biol. Chem.* **253**, 3494–3503.

Levenson, R. G., and Marcu, K. B. (1976). On the existence of polyadenylated histone mRNA in *Xenopus laevis* oocytes. *Cell* **9**, 311–322.

Levinson, A. D., Oppermann, H., Levintow, L., Varmus, H. E., and Bishop, M. J. (1978). Evidence that the transforming gene of avian sarcoma virus encodes a protein kinase associated with a phosphoprotein. *Cell* **15**, 561–572.

Levy, R., Levy, S., Rosenberg, S. A., and Simpson, R. T. (1973). Selective stimulation of nonhistone chromatin protein synthesis in lymphoid cells by phytohemagglutinin. *Biochemistry* **12**, 224–228.

Levy-W., B., and Dixon, G. H. (1977). Renaturation kinetics of cDNA complementary to cytoplasmic polyadenylated RNA from rainbow trout testis. Accessibility of transcribed genes to pancreatic DNase. *Nucleic Acids Res.* **4**, 883–898.

Levy-W., B., and Dixon, G. H. (1978). Partial purification of transcriptionally active nucleosomes from trout testis cells. *Nucleic Acids Res.* **5**, 4155–4163.

Levy-W., B., Johnson, C. B., and McCarthy, B. J. (1976). Diversity of sequences in total and polyadenylated nuclear RNA from *Drosophila* cells. *Nucleic Acids Res.* **3**, 1777–1789.

Levy-W. B., Wong, N. C. W., and Dixon, G. H. (1977). Selective association of the trout-specific H6 protein with chromatin regions susceptible to DNase I and DNase II: Possible location of HMG-T in the spacer region between core nucleosomes. *Proc. Natl. Acad. Sci. U.S.A.* **74**, 2810–2814.

Levy-W., B., Wong, N. C. W., Watson, D. C., Peters, E. H., and Dixon, G. H. (1978). Structure and function of the low-salt extractable chromosomal proteins. Preferential association of trout testis proteins H6 and HMG-T with chromatin regions selectively sensitive to nucleases. *Cold Spring Harbor Symp. Quant. Biol.* **42**, 793–801.

Levy-W., B., Connor, W., and Dixon, G. H. (1979). A subset of trout testis nucleosomes enriched in transcribed DNA sequences contains high mobility group proteins as major structural components. *J. Biol. Chem.* **254**, 609–620.

Levy-Wilson, B., Gjerset, R. A., and McCarthy, B. J. (1977). Acetylation and phosphorylation of *Drosophila* histones. *Biochim. Biophys. Acta* **475**, 168–175.

Levy-Wilson, B., Watson, D. C., and Dixon, G. H. (1979). Multiacetylated forms of H4 are found in a putative transcriptionally competent chromatin fraction from calf thymus. *Nucleic Acids Res.* **6**, 259–273.

Lewis, P. N., Bradbury, E. M., and Crane-Robinson, C. (1975). Ionic-strength induced structure in histone H4 and its fragments. *Biochemistry* **14**, 3391–3397.

Li, H. J. (1975). A model for chromatin structure. *Nucleic Acids Res.* **2**, 1275–1284.

Li, H. J. (1977). Conformational studies on histones. *In* "Chromatin and Chromosome Structure" (H. J. Li and R. A. Eckhardt, eds.), pp. 1–36, Academic Press, New York.

Liao, S., Lin, A. H., and Barton, R. W. (1966). Selective stimulation of ribonucleic acid synthesis in prostatic nuclei by testosterone. *J. Biol. Chem.* **241**, 3869–3871.

Libby, P. R. (1968). Histone acetylation by cell-free preparations from rat uterus: *In vivo* stimulation by estradiol-17β. *Biochem. Biophys. Res. Commun.* **31**, 59–65.

Libby, P. R. (1972). Histone acetylation and hormone action. Early effects of estradiol-17β on histone acetylation in rat uterus. *Biochem. J.* **130**, 663–669.

Libby, P. R. (1973). Histone acetylation and hormone action. Early effects of aldosterone on histone acetylation in rat kidney. *Biochem. J.* **134**, 907–912.

Liew, C. C., Liu, D. K., and Gornall, A. D. (1972). Effects of aldosterone on RNA polymerase in rat heart and kidney nuclei. *Endocrinology* **90**, 488–495.

Liew, C. C., Suria, D., and Gornall, A. D. (1973). Effects of aldosterone on acetylation and phosphorylation of chromosomal proteins. *Endocrinology* **93**, 1025–1034.

Lilley, D. M. L., and Tatchell, K. (1977). Chromatin core particle unfolding induced by tryptic cleavage of histones. *Nucleic Acids Res.* **4**, 2039–2055.

Lin, S. Y., and Riggs, A. D. (1975). The general affinity of *lac* repressor for *E. coli* DNA. Implications for gene regulation in prokaryotes and eukaryotes. *Cell* **4**, 107–111.

Littau, V. C., Allfrey, V. G., Frenster, J. M., and Mirsky, A. E. (1964). Active and inactive regions of nuclear chromatin as revealed by electron microscope autoradiography. *Proc. Natl. Acad. Sci. U.S.A.* **52**, 93–100.

Littau, V. C., Burdick, C. J., Allfrey, V. G., and Mirsky, A. E. (1965). The role of histones in the maintenance of chromatin structure. *Proc. Natl. Acad. Sci. U.S.A.* **54**, 1204–1212.

Lohr, D., Corden, J., Tatchell, K., Kovacic, R. T., and Van Holde, K. (1977). Comparative subunit structure of HeLa, yeast and chicken erythrocyte chromatin. *Proc. Natl. Acad. Sci. U.S.A.* **74**, 79–83.

Louie, A. J., and Dixon, G. H. (1972). Trout testis cells. Synthesis and phosphorylation of histones and protamines in different cell types. *J. Biol. Chem.* **247**, 5498–5505.

Lutter, L. (1977). DNase I produces staggered cuts in the DNA of chromatin. *J. Mol. Biol.* **117**, 53–69.

McKenzie, S. L., Henikoff, S., and Meselson, M. (1975). Localization of RNA from heat-induced polysomes at puff sites in *Drosophila melanogaster*. *Proc. Natl. Acad. Sci. U.S.A.* **72**, 1117–1121.

McKnight, S. L., and Miller, O. L., Jr. (1976). Ultrastructural patterns of RNA synthesis during early embryogenesis of *Drosophila melanogaster*. *Cell* **8**, 305–319.

McKnight, S. L., Pennequin, P., and Schimke, R. T. (1975). Induction of ovalbumin mRNA sequences by estrogen and progesterone in chick oviduct as measured by hybridization to complementary DNA. *J. Biol. Chem.* **250**, 8105–8110.

McKnight, S. L., Bustin, M., and Miller, O. L., Jr. (1978). Electron microscopic analysis of chromosome metabolism in the *Drosophila melanogaster* embryo. *Cold Spring Harbor Symp. Quant. Biol.* **42**, 741–754.

Makman, M., Nakagawa, S., Dvorkin, D., and White, A. (1970). Inhibitory effects of cortisol and antibiotics on substrate entry and ribonucleic acid synthesis in rat thymocytes *in vitro*. *J. Biol. Chem.* **245**, 2556–2563.

Malcolm, D. B., and Sommerville, J. (1974). The structure of chromosome derived ribonucleoproteins in oocytes of *Triturus cristatus*. *Chromosoma (Berlin)* **48**, 137–158.

Mandel, J. L., Breathnach, R., Gerlinger, P., LeMeur, M., Gannon, F., and Chambon, P. (1978). Organization of coding and intervening sequences in the chicken ovalbumin split gene. *Cell* **14**, 641–653.

Marks, D. B., Paik, W. K., and Borun, T. W. (1973). The relationship of histone phosphorylation to deoxyribonucleic acid replication and mitosis during the HeLa S3 cell cycle. *J. Biol. Chem.* **248**, 5660–5667.

Marks, D. B., Kanefsky, T., Keller, B. J., and Marks, A. D. (1975). The presence of histone H1⁰ in human tissues. *Cancer Res.* **35**, 886–889.

Martinson, H. G., and McCarthy, B. J. (1975). Histone–histone association within chromatin: Cross-linking studies using tetranitromethane. *Biochemistry* **14**, 1073–1078.

Martinson, H. G., and McCarthy, B. J. (1976). Histone–histone interactions within chromatin. Preliminary characterization of presumptive H2B-H2A and H2B-H4 binding sites. *Biochemistry* **15**, 4126–4131.

Martinson, H. G., Shetlar, M. B.,and McCarthy, B. J. (1976). Histone–histone interactions within chromatin. Crosslinking studies using ultraviolet light. *Biochemistry* **15**, 2002–2007.

Marushige, K. (1976). Activation of chromatin by acetylation of histone side chains. *Proc. Natl. Acad. Sci. U.S.A.* **73**, 3937–3941.

Marushige, K., and Bonner, J. (1971). Fractionation of liver chromatin. *Proc. Natl. Acad. Sci. U.S.A.* **68**, 2941–2945.

Marzluff, W. F., Jr., and McCarty, K. S. (1972). Structural studies of calf thymus F3 histone. II. Occurrence of phosphoserine and ε-N-acetyllysine in thermolysin peptides. *Biochemistry* **11**, 2677–2681.

Marzluff, W. F., Jr., Sanders, L. A., Miller, D. M., and McCarty, K. S. (1972). Two chemically and metabolically distinct forms of calf thymus H3. *J. Biol. Chem.* **247**, 2026–2033.

Mathew, C. G. P., Goodwin, G. H., and Johns, E. W. (1979). Studies on the association of the high mobility group non-histone chromatin proteins with isolated nucleosomes. *Nucleic Acids Res.* **6**, 167–179.

Mathis, D. J., Oudet, P., Wasylyk, B., and Chambon, P. (1978). Effect of histone acetylation on structure and *in vitro* transcription of chromatin. *Nucleic Acids Res.* **5**, 3523–3547.

Matsui, S.-I., Seon, B. K., and Sandberg, A. A. (1979). Disappearance of a structural chromatin protein A24 in mitosis: Implications for molecular basis of chromatin condensation. *Proc. Natl. Acad. Sci. U.S.A.* **76**, 6386–6390.

Matthews, H. R., and Bradbury, E. M. (1978). The role of H1 histone phosphorylation in the cell cycle. Turbidity studies of H1–DNA interaction. *Exp. Cell Res.* **111**, 343–351.

Mayfield, J. E., Serunian, L. A., Silver, L. M., and Elgin, S. C. R. (1978). A protein released by DNase I digestion of *Drosophila* nuclei is preferentially associated with puffs. *Cell* **14**, 539–544.

Means, A. R., Comstock, J. P., Rosenfeld, G. C., and O'Malley, B. W. (1972). Ovalbumin messenger RNA of chick oviduct: Partial characterization, estrogen dependence, and translation *in vitro*. *Proc. Natl. Acad. Sci. U.S.A.* **69**, 1146–1150.

Meisler, M. H., and Langan, T. A. (1969). Characterization of a phosphatase specific for phosphorylated histones and protamines. *J. Biol. Chem.* **244**, 4961–4968.

Metzger, G., Wilhelm, F. X., and Wilhelm, M. L. (1976). Distribution along DNA of the bound carcinogen N-acetoxy-N-2-acetylaminofluorene in chromatin modified in vitro. *Chem. Biol. Interact.* **15**, 257–265.

Metzger, G., Wilhelm, F. X., and Wilhelm, M. L. (1977). Non-random binding of a chemical carcinogen to the DNA in chromatin. *Biochem. Biophys. Res. Commun.* **75**, 703–710.

Meyer, G. F., and Hennig, W. (1974). The nucleolus in primary spermatocytes of *Drosophila hydeii. Chromosoma (Berlin)* **46**, 121–144.

Michalski, C. J. (1978). Protein synthesis during hormone stimulation in the aquatic fungus, *Achlya. Biochem. Biophys. Res. Commun.* **84**, 417–427.

Miller, O. L., Jr. (1965). Fine structure of lampbrush chromosomes. *Natl. Cancer Inst. Monogr.* **18**, 79–99.

Miller, O. L., Jr., and Beatty, B. R. (1969). Visualization of nucleolar genes. *Science* **164**, 955–957.

Miller, O. L., Jr., and Bakken, A. H. (1972). Morphological studies of transcription. *Acta Endocrinol. Suppl.* **168**, 155–177.

Miller, O. L., Jr., and Hamkalo, B. A. (1972). Visualization of RNA synthesis on chromosomes. *Int. Rev. Cytol.* **33**, 1–23.

Miller, O. L., Jr., Beatty, B. R., and Hamkalo, B. A. (1972). Nuclear structure and function during amphibian oogenesis. *In* "Oogenesis" (J. D. Biggers and A. W. Schuetz, eds.), pp. 119–128. Univ. Park Press, Baltimore, Maryland.

Miller, O. J., Miller, D. A., Dev, V. G., Tantravahi, R., and Croce, C. (1976). Expression of human and suppression of mouse nucleolus organizer activity in mouse–human somatic cell hybrids. *Proc. Natl. Acad. Sci. U.S.A.* **73**, 4531–4535.

Minty, A. J., Birnie, G. D., and Paul, J. (1978). Gene expression in Friend leukemia cells following the induction of hemoglobin synthesis. *Exp. Cell Res.* **115**, 1–14.

Mirault, M.-E., Goldschmidt-Clermont, M., Moran, L., Arrigo, A. P., and Tissieres, A. (1978). The effect of heat-shock on gene expression in *Drosophila melanogaster. Cold Spring Harbor Symp. Quant. Biol.* **42**, 819–827.

Mirzabekov, A. D., and Melnikova, A. F. (1974). Location of chromatin proteins within the grooves of DNA by modification of chromatin with dimethyl sulfate. *Mol. Biol. Rep.* **1**, 385–390.

Mirzabekov, A. D., Shick, V. V., Belyavsky, A. V., and Bavykin, S. G. (1978). Primary organization of nucleosome core particle of chromatin: Sequence of histone arrangement along DNA. *Proc. Natl. Acad. Sci. U.S.A.* **75**, 4184–4188.

Molgaard, H. V., Matthews, H. R., and Bradbury, E. M. (1976). Organization of genes for ribosomal RNA in *Physarum polycephalum. Eur. J. Biochem.* **68**, 541–549.

Monahan, J. J., Harris, S. E., and O'Malley, B. W. (1976). Effect of estrogen on gene expression in the chick oviduct. Effect of estrogen on the sequence and population complexity of chick oviduct poly(A)-containing RNA. *J. Biol. Chem.* **251**, 3738–3748.

Moorhead, P. J., Nowell, P. C., Mellman, W., Battips, D. M., and Hungerford, D. A. (1960). Chromosome preparations of leukocytes cultured from human peripheral blood. *Exp. Cell Res.* **20**, 616–626.

Morris, N. R. (1976a). Nucleosome structure in *Aspergillus nidulans. Cell* **8**, 357–363.

Morris, N. R. (1976b). A comparison of the structure of chicken erythrocyte and chicken liver chromatin. *Cell* **9**, 627–632.

Moss, T., Cary, P. D., Crane-Robinson, C., and Bradbury, E. M. (1976a). Physical studies on the H3/H4 histone tetramer. *Biochemistry* **15**, 2261–2270.

Moss, T., Cary, P. D., Abercrombie, B. D., Crane-Robinson, C., and Bradbury, E. M.

(1976b). A pH-dependent interaction between histones H2A and H2B involving secondary and tertiary folding. *Eur. J. Biochem.* **71**, 337–346.

Moss, T., Stephens, R. M., Crane-Robinson, C., and Bradbury, E. M. (1977). A nucleosome-like structure containing DNA and the arginine-rich histones H3 and H4. *Nucleic Acids Res.* **4**, 2477–2485.

Mott, M. R., and Callan, H. G. (1975). An electron microscopic study of the lampbrush chromosomes of the newt, *Triturus cristatus. J. Cell Sci.* **17**, 241–261.

Mueller, G. C., Herranen, A. M., and Jervell, K. J. (1958). Studies on the mechanism of action of estrogens. *Recent Prog. Horm. Res.* **14**, 95–129.

Mukherjee, A. B., and Cohen, M. M. (1968). Histone acetylation: Cytologic evidence in human lymphocytes. *Exp. Cell Res.* **54**, 257–260.

Mullins, D. W., Giri, C. P., and Smulson, M. E. (1977). Poly(adenosine diphosphate-ribose) polymerase. The distribution of a chromosome-associated enzyme within the chromatin substructure. *Biochemistry* **16**, 506–513.

Murti, K. G., and Prescott, D. M. (1978). Electron microscopic visualization of transcribed genes in the nucleus of *Amoeba proteus. Exp. Cell Res.* **112**, 233–240.

Nelson, D. A., Perry, W. M., and Chalkley, R. (1978a). Sensitivity of regions of chromatin containing hyperacetylated histones to DNase I. *Biochem. Biophys. Res. Commun.* **82**, 356–363.

Nelson, D. A., Perry, W. M., Sealy, L., and Chalkley, R. (1978b). DNase I preferentially digests chromatin containing hyperacetylated histones. *Biochem. Biophys. Res. Commun.* **82**, 1346–1353.

Neumann, J. R., O'Meara, A. R., and Herrmann, R. L. (1978a). Cyclic AMP-dependent histone-specific nucleoplasmic protein kinase from rat liver. *Biochem. J.* **171**, 123–135.

Neumann, J., Whittaker,R., Blanchard, B., and Ingram, V. (1978b). Nucleosome-associated proteins and phosphoproteins of differentiating Friend erythroleukemia cells. *Nucleic Acids Res.* **5**, 1675–1687.

Newrock, K. M., Cohen, L. H., Hendricks, M. B., Donnelly, R. J., and Weinberg, E. S. (1978a). Stage-specific mRNAs coding for subtypes of H2A and H2B histones in the sea urchin embryo. *Cell* **14**, 327–336.

Newrock, K. M., Alfagame, C. R., Nardi, R. V., and Cohen, L. H. (1978b). Histone changes during chromatin remodeling in embryogenesis. *Cold Spring Harbor Symp. Quant. Biol.* **42**, 421–431.

Nickol, J. M., Lee, K.-L., and Kenney, F. T. (1978). Changes in hepatic levels of tyrosine aminotransferase messenger RNA during induction by hydrocortisone. *J. Biol. Chem.* **253**, 4009–4015.

Nicolette, J. A., and Babler, M. (1974). The role of protein in the estrogen-stimulated *in vitro* RNA synthesis of isolated rat uterine nuclei. *Arch. Biochem. Biophys.* **163**, 263–270.

Nielson, C. J., Sando, J. J., and Pratt, W. B. (1977). Evidence that dephosphorylation inactivates glucocorticoid receptors. *Proc. Natl. Acad. Sci. U.S.A.* **74**, 1398–1402.

Noll, M. (1974a). Subunit structure of chromatin. *Nature (London)* **251**, 249–251.

Noll, M. (1974b). Internal structure of the chromatin subunit. *Nucleic Acids Res.* **1**, 1573–1578.

Noll, M. (1976). Differences and similarities in chromatin structure of *Neurospora crassa* and higher eukaryotes. *Cell* **8**, 349–355.

Noll, M., and Kornberg, R. D. (1977). Action of micrococcal nuclease on chromatin and the location of histone H1. *J. Mol. Biol.* **109**, 393–404.

Nyberg, L. M., Hu, A.-L., Loor, R. M., and Wang, T. Y. (1976). Androgen-induced gene activation in the rat prostate. *Biochem. Biophys. Res. Commun.* **73**, 330–335.

Oberhauser, H., Csordas, A., Puschendorf, B., and Grunicke, H. (1978). Increase in initiation sites for chromatin directed RNA synthesis by acetylation of chromosomal proteins. *Biochem. Biophys. Res. Commun.* **84**, 110–116.

Ogata, R. T., and Gilbert, W. (1978). An amino-terminal fragment of *lac* repressor binds specifically to *lac* operator. *Proc. Natl. Acad. Sci. U.S.A.* **75**, 5851–5854.

Ogawa, Y., Quagliarotti, G., Jordan, J., Taylor, C. W., Starbuck, W. C., and Busch, H. (1969). Structural analysis of the glycine-rich, arginine-rich histone. III. Sequence of the amino-terminal half of the molecule containing the modified lysine residues and the total sequence. *J. Biol. Chem.* **244**, 4387–4392.

Okayama, H., and Hayaishi, O. (1978). ADP-Ribosylation of nuclear protein A24. *Biochem. Biophys. Res. Commun.* **84**, 755–762.

Okayama, H., Ueda, K.,and Hayaishi, O. (1978). Purification of ADP-ribosylated nuclear proteins by covalent chromatography on dihydroxyboryl polyacrylamide beads and their characterization. *Proc. Natl. Acad. Sci. U.S.A.* **75**, 1111–1115.

Old, R. W., Callan, H. G.,and Gross, K. W. (1977). Localization of histone gene transcripts in newt lampbrush chromosomes by in situ hybridization. *J. Cell Sci.* **27**, 57–79.

Olins, A. L., and Olins, D. E. (1974). Spheroid chromatin units (ν bodies). *Science* **183**, 330–332.

Olins, D. E., and Wright, E. B. (1973). Glutaraldehyde fixation of isolated eukaryotic nuclei. Evidence for histone–histone proximity. *J. Cell Biol.* **59**, 304–317.

Olson, M. O. J., Goldknopf, I. L., Guetzow, K. A., James, G. T., Hawkins, C., Mays-Rothberg, C. J., and Busch, H. (1976). The NH$_2$- and COOH-terminal amino acid sequence of nuclear protein A24. *J. Biol. Chem.* **251**, 5901–5903.

O'Malley, B. W., Tsai, M. J., Tsai, S. Y., and Towle, H. C. (1978). Regulation of gene expression in chick oviduct. *Cold Spring Harbor Symp. Quant. Biol.* **42**, 605–613.

Ono, T., Terayama, H., Takaku, F., and Nakao, K. (1969). Hydrocortisone effect upon the phytohemagglutinin-stimulated acetylation of histones in human lymphocytes. *Biochim. Biophys. Acta* **179**, 214–220.

Ord, M. G., and Stocken, L. A. (1966). Metabolic properties of histones from rat liver and thymus gland. *Biochem. J.* **98**, 888–897.

Ord, M. G., and Stocken, L. A. (1977). Adenosine diphosphate ribosylated histones. *Biochem. J.* **161**, 583–592.

Oudet, P., Gross-Bellard, M., and Chambon, P. (1975). Electron microscopic and biochemical evidence that chromatin structure is a repeating unit. *Cell* **4**, 281–300.

Oudet, P., Spadafora, C.,and Chambon, P. (1978a). Nucleosome structure. II. Structure of the SV40 minichromosome and electron microscopic evidence for reversible transitions of the nucleosome structure. *Cold Spring Harbor Symp. Quant. Biol.* **42**, 301–312.

Oudet, P., Germond, J. E., Sures, M., Gallwitz, D., Bellard, M., and Chambon, P. (1978b). Nucleosome structure. I. All four histones, H2A, H2B, H3 and H4, are required to form a nucleosome, but an H3-H4 subnucleosomal particle is formed with H3-H4 alone. *Cold Spring Harbor Symp. Quant. Biol.* **42**, 287–300.

Overton, G. C., and Weinberg, E. S. (1978). Length and sequence heterogeneity of the histone gene repeat unit of the sea urchin, *S. purpuratus*. *Cell* **14**, 247–257.

Packman, S., Aviv, H., Ross, J., and Leder, P. (1972). A comparison of globin genes in duck reticulocytes and liver cells. *Biochem. Biophys. Res. Commun.* **49**, 813–819.

Paik, W. K., and Kim, S. (1975). Protein methylation. Chemical, enzymological and biological significance. *In* "Advances in Enzymology" (A. Meister, ed.), Vol. 42, pp. 227–286. Wiley, New York.

Palmer, W. K., Castagna, M., and Walsh, D. A. (1974). Nuclear protein kinase activity in glucagon-stimulated perfused rat livers. *Biochem. J.* **143**, 469–471.

Palmiter, R. D. (1973). Rate of ovalbumin messenger ribonucleic acid synthesis in the oviduct of estrogen-primed chicks. *J. Biol. Chem.* **248**, 8260–8270.

Palmiter, R. D., and Carey, N. H. (1974). Rapid inactivation of ovalbumin messenger ribonucleic acid after acute withdrawal of estrogen. *Proc. Natl. Acad. Sci. U.S.A.* **71**, 2357–2361.

Palmiter, R. D., Mulvihill, E. R., McKnight, G. S., and Senear, A. W. (1978). Regulation of gene expression in the chick oviduct by steroid hormones. *Cold Spring Harbor Symp. Quant. Biol.* **42**, 639–647.

Panyim, S., and Chalkley, R. (1969a). High resolution acrylamide gel electrophoresis of histones. *Arch. Biochem. Biophys.* **130**, 337–346.

Panyim, S., and Chalkley, R. (1969b). A new histone found only in mammalian tissues with little cell division. *Biochem. Biophys. Res. Commun.* **37**, 1042–1049.

Pardon, J. F., Worcester, D. L., Wooley, J. C., Tatchell, K., Van Holde, K. E., and Richards, B. (1976). Low-angle neutron scattering from chromatin subunit particles. *Nucleic Acids Res.* **2**, 2163–2176.

Pathy, L., and Smith, E. L. (1975). Histone III. VI. Two forms of calf thymus histone III. *J. Biol. Chem.* **250**, 1919–1920.

Pathy, L., Smith, E. L., and Johnson, J. (1973). Histone III. The amino acid sequence of pea embryo histone III. *J. Biol. Chem.* **248**, 6834–6840.

Peterson, J. L., and McConkey, E. H. (1976). Non-histone chromosomal proteins from HeLa cells. *J. Biol. Chem.* **251**, 548–554.

Pogo, B. G. T., Allfrey, V. G., and Mirsky, A. E. (1966). RNA synthesis and histone acetylation during the course of gene activation in lymphocytes. *Proc. Natl. Acad. Sci. U.S.A.* **55**, 805–812.

Pogo, B. G. T., Allfrey, V. G., and Mirsky, A. E. (1967). The effect of phytohemagglutinin on ribonucleic acid synthesis and histone acetylation in equine leukocytes. *J. Cell Biol.* **35**, 477–482.

Pogo, B. G. T., Pogo, A. O., Allfrey, V. G., and Mirsky, A. E. (1968). Changing patterns of histone acetylation and RNA synthesis in regeneration of the liver. *Proc. Natl. Acad. Sci. U.S.A.* **59**, 1337–1344.

Polisky, B., and McCarthy, B. J. (1975). Location of histones on simian virus 40 DNA. *Proc. Natl. Acad. Sci. U.S.A.* **72**, 2895–2899.

Pong, S. S., and Loomis, W. F. (1973). Multiple nuclear ribonucleic acid polymerases during development of *Dictyostelium discoideum*. *J. Biol. Chem.* **248**, 3933–3939.

Pong, R. S., and Wogan, G. N. (1970). Time course and dose–response characteristics of aflatoxin B_1 effects on rat liver RNA polymerase and ultrastructure. *Cancer Res.* **30**, 294–304.

Prasad, K. N., and Sinha, P. K. (1976). Effect of sodium butyrate on mammalian cells in culture, A review. *In Vitro* **12**, 125–132.

Prescott, D. M. (1963). RNA and protein replacement in the nucleus during growth and division and the conservation of components in the chromosome. *In* "Cell Growth and Cell Division" (R. J. C. Harris, ed.), pp. 111–128. Academic Press, New York.

Prunell, A., and Kornberg, R. D. (1978). Relation of nucleosomes to DNA sequences. *Cold Spring Harbor Symp. Quant. Biol.* **42**, 103–108.

Pukkila, P. J. (1975). Identification of the lampbrush chromosome loops which transcribe 5 S ribosomal RNA in *Notopthalmus (Triturus) viridescens*. *Chromosoma (Berlin)* **53**, 71–89.

Rabbani, A., Goodwin, G. H., and Johns, E. W. (1978). Studies on the tissue specificity of the high mobility group non-histone chromosomal proteins from calf. *Biochem. J.* **173**, 497–505.

Ramanathan, R., Rajalakshmi, S., Sarma, D. S. R., and Farber, E. (1976). Nonrandom nature of *in vivo* methylation by dimethylnitrosamine and the subsequent removal of methylated products from rat liver chromatin DNA. *Cancer Res*. **36**, 2073–2079.

Ramirez, F., Gambino, R., Maniatis, G. M., Rifkind, R. A., Marks, P. A., and Bank, A. (1975). Changes in globin mRNA content during erythroid cell differentiation. *J. Biol. Chem*. **250**, 5054–5058.

Rechsteiner, M., and Kuehl, L. (1979). Microinjection of the nonhistone chromosomal protein HMG 1 into bovine fibroblasts and HeLa cells. *Cell* **16**, 901–908.

Reeves, R. (1978a). Structure of *Xenopus* ribosomal gene chromatin during changes in genomic transcription rates. *Cold Spring Harbor Symp. Quant. Biol*. **42**, 709–722.

Reeves, R. (1978b). Nucleosome structure of *Xenopus* oocyte amplified ribosomal genes. *Biochemistry* **17**, 4906–4916.

Renz, M., Nehls, P., and Hozier, J. (1977). Involvement of histone H1 in the organization of the chromosome fiber. *Proc. Natl. Acad. Sci. U.S.A*. **74**, 1879–1883.

Renz, M., Nehls, P., and Hozier, J. (1978). Histone H1 involvement in the structure of the chromosome fiber. *Cold Spring Harbor Symp. Quant. Biol*. **42**, 245–252.

Riggs, M. G., Whittaker, R. G., Newmann, J. R., and Ingram, V. M. (1977). *n*-Butyrate causes histone modification in HeLa and Friend erythroleukemia cells. *Nature (London)* **268**, 462–464.

Rill, R., and Van Holde, K. E. (1973). Properties of nuclease resistant fragments of calf thymus chromatin. *J. Biol. Chem*. **248**, 1080–1083.

Ross, J., Ikawa, Y., and Leder, P. (1972). Globin messenger RNA induction during erythroid differentiation of cultured leukemia cells. *Proc. Natl. Acad. Sci. U.S.A*. **69**, 3620–3623.

Rubin, A. D., and Cooper, H. L. (1965). Evolving patterns of RNA metabolism during transition from the resting state to active growth in lymphocytes stimulated by phytohemagglutinin. *Proc. Natl. Acad. Sci. U.S.A*. **54**, 469–476.

Ruiz-Carillo, A., Wangh, L. J., Littau, V. C., and Allfrey, V. G. (1974). Changes in histone acetyl content and in nuclear non-histone protein composition of avian erythroid cells at different stages of maturation. *J. Biol. Chem*. **249**, 7358–7368.

Ruiz-Carillo, A., Wangh, L. J., and Allfrey, V. G. (1975). Processing of newly synthesized histone molecules. Nascent H4 chains are reversibly phosphorylated and acetylated. *Science* **190**, 117–128.

Ruiz-Carrilo, A., Wangh, L. J., and Allfrey, V. G. (1976). Selective synthesis and modification of nuclear proteins during maturation of avian erythroid cells. *Arch. Biochem. Biophys*. **174**, 272–290.

Russell, D. H., Byus, C. V., and Manen, C. A. (1976). Proposed model of major sequential biochemical activity of a trophic response. *Life Sci*. **19**, 1297–1306.

Ryser, U., and Braun, R. (1974). The amount of DNA coding for rRNA during differentiation (spherulation) in *Physarum polycephalum*. *Biochim. Biophys. Acta* **361**, 33–36.

Sakai, M., Fujii-Kuriyama, Y., and Muramatsu, M. (1978). Number and frequency of protamine genes in rainbow trout testis. *Biochemistry* **17**, 5510–5515.

Salem, R., and deVellis, J. (1976). Protein kinase activity and cyclic AMP-dependent protein phosphorylation in subcellular fractions after norepinephrine treatment of glial cells. *Fed. Proc. Fed. Am. Soc. Exp. Biol*. **35**, 296 (Abstract).

Sanders, M. (1978). Fractionation of nucleosomes by salt elution from micrococcal nuclease-digested nuclei. *J. Cell Biol*. **79**, 97–109.

Sanders, L. A., Schechter, N. M., and McCarty, K. S. (1973). A comparative study of histone acetylation, histone deacetylation, and ribonucleic acid synthesis in avian reticulocytes and erythrocytes. *Biochemistry* **12**, 783–791.

Savage, M. J., Sala-Trepat, J. M., and Bonner, J. (1978). Measurement of the complexity and diversity of poly(adenylic acid)-containing messenger RNA from rat liver. *Biochemistry* **17**, 462–467.

Schaffhausen, B. S., and Benjamin, T. L. (1976). Deficiency in histone acetylation in nontransforming host range mutants of polyoma virus. *Proc. Natl. Acad. Sci. U.S.A.* **73**, 1092–1096.

Scheer, U. (1978). Changes of nucleosome frequency in nucleolar and non-nucleolar chromatin as a function of transcription: An electron microscopic study. *Cell* **13**, 535–549.

Scheer, U., Trendelenburg, M. F., and Franke, W. W. (1976a). Regulation of transcription of genes of ribosomal RNA during amphibian oogenesis. *J. Cell Biol.* **69**, 465–489.

Scheer, U., Franke, W. W., Trendelenburg, M. F., and Spring, H. (1976b). Classification of loops of lampbrush chromosomes according to the arrangement of transcriptional complexes. *J. Cell Sci.* **22**, 503–519.

Schimke, R. T., and Doyle, D. (1970). Control of enzyme levels in animal tissues. *Annu. Rev. Biochem.* **39**, 929–976.

Schneider, F. H. (1976). Effects of sodium butyrate on mouse neuroblastoma cells in culture. *Biochem. Pharmacol.* **25**, 2309–2317.

Schutz, G., Killewich, L., Chen, G., and Feigelson, P. (1975). Control of the mRNA for hepatic tyrosine oxygenase during hormonal and substrate induction. *Proc. Natl. Acad. Sci. U.S.A.* **72**, 1017–1020.

Sealy, L., and Chalkley, R. (1978a). The effect of sodium butyrate on histone modification. *Cell* **14**, 115–121.

Sealy, L., and Chalkley, R. (1978b). DNA associated with hyperacetylated histones is preferentially digested by DNase I. *Nucleic Acids Res.* **5**, 1863–1876.

Shaw, B. R., Corden, J. L., Sahasrabuddhe, C. G., and Van Holde, K. E. (1974). Chromatographic separation of chromatin subunits. *Biochem. Biophys. Res. Commun.* **61**, 1193–1198.

Shaw, B. R., Herman, T. M., Kovacic, R. T., Beaudreau, G. S., and Van Holde, K. E. (1976). Analysis of subunit organization in chicken erythrocyte chromatin. *Proc. Natl. Acad. Sci. U.S.A.* **73**, 505–509.

Shewmaker, C. K., Cohen, B. N., and Wagner, T. E. (1978). Chemically induced gene activation: Selective increase in DNase I susceptibility in chromatin acetylated with acetyl adenylate. *Biochem. Biophys. Res. Commun.* **84**, 342–349.

Simpson, R. T. (1976). Histone H3 and H4 interact with ends of nucleosome DNA. *Proc. Natl. Acad. Sci. U.S.A.* **73**, 4400–4404.

Simpson, R. T. (1978). Structure of chromatin containing extensively acetylated H3 and H4. *Cell* **13**, 691–699.

Simpson, R. T., and Whitlock, J. (1976). Chemical evidence that chromatin DNA exists in 160 base-pair beads interspersed with 40 base-pair bridges. *Nucleic Acids Res.* **3**, 117–127.

Simpson, R. T., Whitlock, J. P., Bina-Stein, M., and Stein, A. (1978). Histone–DNA interactions in chromatin core particles. *Cold Spring Harbor Symp. Quant. Biol.* **42**, 127–136.

Small, D., Chou, P. Y., and Fasman, G. (1977). Occurrence of phosphorylated residues in predicted β-turns: Implications for β-turn participation in control mechanisms. *Biochem. Biophys. Res. Commun.* **79**, 341–346.

Smerdon, M. J., and Isenberg, I. (1976). Interactions between subfractions of calf thymus H1 and non-histone chromosomal proteins HMG 1 and HMG 2. *Biochemistry* **15**, 4242–4247.

Smith, D. L., Chen, C. C., Bruegger, B. B., Holtz, S. L., Halpern, R. M., and Smith, R. A.

(1974). Characterization of histone kinases forming acid-labile histone phosphates in Walker 256 carcinoma cell nuclei. *Biochemistry* **13**, 3780–3785.

Smulson, M. E., and Shall, S. (1976). Poly (ADP-ribose). *Nature (London)* **263**, 14.

Sollner-Webb, B., and Felsenfeld, G. (1975). A comparison of the digestion of nuclei and chromatin by staphylococcal nuclease. *Biochemistry* **14**, 2915–2920.

Sollner-Webb, B., and Felsenfeld, G. (1977). Pancreatic DNase cleavage sites in nuclei. *Cell* **10**, 537–547.

Sollner-Webb, B., Camerini-Otero, R. D., and Felsenfeld, G. (1976). Chromatin structure as probed by nucleases and proteases: Evidence for the central role of histones H3 and H4. *Cell* **8**, 179–193.

Sollner-Webb, B., Melchior, W., and Felsenfeld, G. (1978). DNase I, DNase II and staphylococal nuclease cut at different, yet symmetrically-located sites in the nucleosome core. *Cell* **14**, 611–627.

Sommerville, J., and Malcolm, D. B. (1976). Transcription of genetic information in amphibian oocytes. *Chromosoma (Berlin)* **55**, 183–208.

Spiker, S., Mardian, J. K. W., and Isenberg, I. (1978). Chromosomal HMG proteins occur in three eukaryotic kingdoms. *Biochem. Biophys. Res. Commun.* **82**, 129–135.

Spradling, A., Pardue, M. L., and Penman, S. (1977). Messenger RNA in heat-shocked *Drosophila* cells. *J. Mol. Biol.* **109**, 559–587.

Spruill, W. A., Hurwitz, D. R., Lucchesi, J. C., and Steiner, A. L. (1978). Association of cyclic GMP with gene expression of polytene chromosomes of *Drosophila melanogaster*. *Proc. Natl. Acad. Sci. U.S.A.* **75**, 1480–1484.

Stacey, D., and Allfrey, V. G. (1979). Microinjection studies of protein transit across the nuclear envelope. Submitted for publication.

Stark, B. C., Kole, R., Bowman, E. J., and Altman, S. (1978). Ribonuclease P: An enzyme with an essential RNA component. *Proc. Natl. Acad. Sci. U.S.A.* **75**, 3718–3721.

Stein, G. S., and Kleinsmith, L. J., eds. (1975). "Chromosomal Proteins and Their Role in the Regulation of Gene Expression." Academic Press, New York.

Stein, G. S., Stein, J. L., Kleinsmith, L. J., Jansing, R. L., Park, W. D., and Thomson, J. A. (1977). Non-histone chromosomal proteins. Their role in the regulation of histone–gene expression. *Biochem. Soc. Symp.* **42**, 137–163.

Stein, G. S., Stein, J. L., and Thomson, J. A. (1978). Chromosomal proteins in transformed and neoplastic cells: A review. *Cancer Res.* **38**, 1181–1201.

Sterner, R., Vidali, G., Heinrikson, R. L., and Allfrey, V. G. (1978a). Post-synthetic modification of high mobility group proteins. Evidence that high mobility group proteins are acetylated. *J. Biol. Chem.* **253**, 7601–7604.

Sterner, R., Boffa, L. C., and Vidali, G. (1978b). Comparative structural analysis of high mobility group proteins from a variety of sources. Evidence for a high mobility group protein unique to avian erythrocyte nuclei. *J. Biol. Chem.* **253**, 3830–3836.

Stratling, W. H., Muller, U., and Zentgraf, H. (1978). Higher order repeat structure of chromatin is built up of globular particles containing eight nucleosomes. *Exp. Cell Res.* **117**, 301–311.

Strom, C. M., Moscona, M., and Dorfman, A. (1978). Amplification of DNA sequences during chicken cartilage and neural retina differentiation. *Proc. Natl. Acad. Sci. U.S.A.* **75**, 4451–4454.

Sun, I. Y-C., Johnson, E. M., and Allfrey, V. G. (1980). Affinity-purification of newly-phosphorylated protein molecules. *J. Biol. Chem.* (in press).

Sung, M. T., and Dixon, G. H. (1970). Modification of histones during spermiogenesis in trout: A molecular mechanism for altering histone binding to DNA. *Proc. Natl. Acad. Sci. U.S.A.* **67**, 1616–1623.

Sures, I., Lowry, J., and Kedes, L. H. (1978). The DNA sequences of sea urchin (*S. purpuratus*) H2A, H2B and H3 coding and spacer regions. *Cell* **15**, 1033–1044.

Sutherland, R. B., and Horgen, P. A. (1977). Effects of the sex hormone, antheridol, on the initiation of RNA synthesis in the simple eukaryote, *Achlya ambisexualis*. *J. Biol. Chem.* **252**, 8812–8820.

Takaku, F., Nakao, K., Ono, T., and Terayama, H. (1969). Changes in histone acetylation and RNA synthesis in the spleen of polycythemic mice after erythropoietin injection. *Biochim. Biophys. Acta* **195**, 396–400.

Takeda, M., and Ohga, Y. (1973). Adenosine 3′,5′-monophosphate and histone phosphorylation during enzyme induction by glucagon in rat liver. *J. Biochem.* (*Tokyo*) **73**, 621–629.

Tallman, J. F., Smith, C. C., and Henneberry, R. C. (1977) Induction of functional β-adrenergic receptors in HeLa cells. *Proc. Natl. Acad. Sci. U.S.A.* **74**, 873–877.

Tata, J. R. (1976). The expression of the vitellogenin gene. *Cell* **9**, 1–14.

Tata, J. R., and Baker, B. (1978). Enzymatic fractionation of nuclei: Polynucleosomes and RNA polymerase II as endogenous transcriptional complexes. *J. Mol. Biol.* **118**, 249–272.

Tekamp, P. A., Valenzuela, P., Maynard, T., Bell, G. I., and Rutter, W. J. (1979). Specific gene transcription in yeast nuclei and chromatin by added homologous RNA polymerases I and III. *J. Biol. Chem.* **254**, 955–963.

Teng, C. S., and Hamilton, T. H. (1968). The role of chromatin in estrogen action in the uterus. I. The control of template capacity and chemical composition and the binding of ³H-estradiol-17β. *Proc. Natl. Acad. Sci. U.S.A.* **60**, 1410–1417.

Teng, C. T., Teng, C. S., and Allfrey, V G. (1970). Species-specific interactions between nuclear phosphoproteins and DNA. *Biochem. Biophys. Res. Commun.* **41**, 690–696.

Teng, C. S., Teng, C. T., and Allfrey, V. G. (1971). Studies of nuclear acidic proteins. Evidence for their phosphorylation, tissue specificity, selective binding to deoxyribonucleic acid, and stimulatory effects on transcription. *J. Biol. Chem.* **246**, 3597–3609.

Teng, C. S., Gallagher, K., and Teng, C. T. (1978). Isolation of a high-molecular-weight high-mobility-type non-histone protein from hen oviduct. *Biochem. J.* **176**, 1003–1006.

Tew, K. D., Sudhaker, S., Schein, P. S., and Smulson, M. E. (1978). Binding of chlorozotocin and 1-(2-chloroethyl)-3-cyclohexyl-1-nitrosourea to chromatin and nucleosomal fractions of HeLa cells. *Cancer Res.* **38**, 3371–3378.

Thoma, F., and Koller, T. (1977). Influence of histone H1 on chromatin structure. *Cell* **12**, 101–107.

Thomas, J. O., and Furber, V. (1976). Yeast chromatin structure. *FEBS Lett.* **66**, 274–280.

Thomas, J. O., and Kornberg, R. D. (1975). An octamer of histones in chromatin and in solution. *Proc. Natl. Acad. Sci. U.S.A.* **72**, 2626–2630.

Thomas, J. O., and Thompson, R. J. (1977). Variation in chromatin structure in two cell types from the same tissue. A short DNA repeat length in cerebral cortex. *Cell* **10**, 633–640.

Thomas, G., Lange, H. W., and Hempel, K. (1975). Kinetics of histone methylation *in vivo* and its relation to the cell cycle in Ehrlich ascites tumor cells. *Eur. J. Biochem.* **51**, 609–615.

Thwaits, B. H., Brandt, W. F., and von Holt, C. (1976a). Sites of *in vitro* enzymatic acetylation of histone H3. *FEBS Lett.* **71**, 193–196.

Thwaits, B. H., Brandt, W. F., and von Holt, C. (1976b). Sites of *in vitro* enzymatic acetylation of histone H4. *FEBS Lett.* **71**, 197–200.

Tidwell, T., Allfrey, V. G., and Mirsky, A. E. (1968). The methylation of histones during regeneration of the liver. *J. Biol. Chem.* **243**, 707–715.

Timberlake, W. E. (1976). Alterations in RNA and protein synthesis associated with steroid hormone-induced sexual morphogenesis in the water mold *Achlya*. *Dev. Biol.* **51**, 202–214.

Todd, R. D., and Garrard, W. T. (1977). Two dimensional electrophoretic analysis of polynucleosomes. *J. Biol. Chem.* **252**, 4729–4738.

Tralka, T. S., Rosen, S. W., Weintraub, B. D., Leiblich, J. M., Engel, L. W., Wetzel, B. K., Kingsbury, E. W., and Rabson, A. S. (1979). Ultrastructural concomitants of sodium butyrate-enhanced ectopic hormone production of chorionic gonadotropin and its alpha subunit in human bronchogenic carcinoma cells. *J. Natl. Cancer Inst.* **62**, 45–61.

Trendelenburg, M. F., and Gurdon, J. B. (1978). Transcription of cloned *Xenopus* ribosomal genes visualized after injection into oocyte nuclei. *Nature (London)* **276**, 292–294.

Trendelenburg, M. F., Zentgraf, H., Franke, W. W., and Gurdon, J. B. (1978). Transcription patterns of amplified *Dytiscus* genes coding for ribosomal RNA after injection into *Xenopus* oocyte nuclei. *Proc. Natl. Acad. Sci. U.S.A.* **75**, 3791–3795.

Tsai, S. Y., Roop, D. R., Tsai, M. J., Stein, J. P., Means, A. R., and O'Malley, B. W. (1978). Effect of estrogen on gene expression in the chick oviduct. Regulation of the ovomucoid gene. *Biochemistry* **17**, 5773–5780.

Ueda, K., Omachi, A., Kawaichi, M., and Hayaishi, O. (1975). Natural occurrence of poly(ADP-ribosyl) histones in rat liver. *Proc. Natl. Acad. Sci. U.S.A.* **72**, 205–209.

Van Holde, K. E., Sahasrabuddhe, C. G., and Shaw, B. R. (1974). A model for particulate structure in chromatin. *Nucleic Acids Res.* **1**, 1579–1586.

Van Lente, F., Jackson, J. F., and Weintraub, H. (1975). Identification of specific crosslinked histones after treatment of chromatin with formaldehyde. *Cell* **5**, 45–50.

Varshavsky, A. J., Bakayev, V. V., and Georgiev, G. P. (1976). Heterogeneity of chromatin subunits *in vitro* and location of histone H1. *Nucleic Acids Res.* **3**, 477–492.

Vidali, G., Boffa, L. C., and Allfrey, V. G. (1972). Properties of an acidic histone-binding protein fraction from cell nuclei. Selective precipitation and deacetylation of histone F2A1 and F3. *J. Biol. Chem.* **247**, 7365–7373.

Vidali, G., Boffa, L. C., and Allfrey, V. G. (1977). Selective release of chromosomal proteins during limited DNase I digestion of avian erythrocyte chromatin. *Cell* **12**, 409–415.

Vidali, G., Boffa, L. C., Bradbury, E. M., and Allfrey, V. G. (1978a). Butyrate suppression of histone deacetylation leads to accumulation of multiacetylated forms of histones H3 and H4 and increased DNase I sensitivity of the associated DNA sequences. *Proc. Natl. Acad. Sci. U.S.A.* **75**, 2239–2243.

Vidali, G., Boffa, L. C., Mann, R. S., and Allfrey, V. G. (1978b). Reversible effects of Na butyrate on histone acetylation. *Biochem. Biophys. Res. Commun.* **82**, 223–227.

Vogt, V. M., and Braun, R. (1976). Structure of ribosomal DNA in *Physarum polycephalum*. *J. Mol. Biol.* **106**, 567–587.

Wagner, T. (1970). A trypsin-sensitive site for the action of hydrocortisone on calf thymus nuclei. *Biochem. Biophys. Res. Commun.* **38**, 890–893.

Walker, J. M., Shooter, K. V., Goodwin, G. H., and Johns, E. W. (1976). The isolation of two peptides from a non-histone chromosomal protein showing irregular charge distribution within the molecule. *Biochem. Biophys. Res. Commun.* **70**, 88–92.

Walker, J. M., Hastings, J. R. B., and Johns, E. W. (1977). The primary structure of a non-histone chromosomal protein. *Eur. J. Biochem.* **76**, 461–468.

Walker, J. M., Parker, B. M., and Johns, E. W. (1978a). Isolation and partial sequence of the cyanogen bromide peptides from calf thymus non-histone chromosomal protein HMG 1. *Int. J. Pep. Res.* **12**, 269–276.

Walker, J. M., Hastings, J. R. B., and Johns, E. W. (1978b). A novel continuous sequence of 41 aspartic and glutamic residues in a non-histone chromosomal protein. *Nature (London)* 271, 281–282.

Walker, J. M., Goodwin, G. H., and Johns, E. W. (1978c). Chromosomal proteins. The amino-terminal sequence of high-mobility-group non-histone chromosomal protein HMG 14, showing sequence homologies with two other chromosomal proteins. *Int. J. Pep. Res.* 11, 301–304.

Wallace, R. B., Dube, S. K., and Bonner, J. (1977a). Localization of the globin gene in the template active fraction of chromatin of Friend leukemia cells. *Science* 198, 1166–1168.

Wallace, R. B., Sargent, T. D., Murphy, R. F., and Bonner, J. (1977b). Physical properties of chemically acetylated rat liver chromatin. *Proc. Natl. Acad. Sci. U.S.A.* 74, 3244–3248.

Wang, T. Y., and Kostraba, N. C. (1978). Protein involved in positive and negative control of chromatin function. *In* "The Cell Nucleus" (H. Busch, ed.), Vol. 4: Chromatin, Part A, pp. 289–317. Academic Press, New York.

Wangh, L. J., Ruiz-Carrillo, A., and Allfrey, V. G. (1972). Separation and analysis of histone subfractions differing in their degree of acetylation. Some correlations with genetic activity in development. *Arch. Biochem. Biophys.* 150, 44–56.

Waslyk, B., Oudet, P., and Chambon, P. (1979). Preferential in vitro assembly of nucleosome cores on some AT-rich regions of SV40 DNA. *Nucleic Acids Res.* 7, 705–713.

Watson, G., and Langan, T. A. (1973). Effects of f1 histone and phosphorylated f1 histone on template activity of chromatin. *Fed. Proc. Fed. Am. Soc. Exp. Biol.* 32, 588 (Abstract).

Watson, D. C., Peters, E. H., and Dixon, G. H. (1977). The purification, characterization and partial sequence determination of a trout testis non-histone protein HMG-T. *Eur. J. Biochem.* 74, 53–60.

Watson, D. C., Levy-W., B., and Dixon, G. H. (1978). Free ubiquitin is a non-histone protein of trout testis chromatin. *Nature (London)* 276, 196–198.

Waymire, J. C., Weiner, N., and Prasad, N. K. (1972). Regulation of tyrosine hydroxylase activity in cultured mouse neuroblastoma cells. Elevation induced by analogs of adenosine 3′,5′-cyclic monophosphate. *Proc. Natl. Acad. Sci. U.S.A.* 69, 2241–2245.

Weintraub, H. (1978). The nucleosome repeat length increases during erythropoiesis in the chick. *Nucleic Acids Res.* 5, 1179–1188.

Weintraub, H., and Groudine, M. (1976). Chromosomal subunits in active genes have an altered conformation. *Science* 193, 848–856.

Weintraub, H., and Van Lente, F. (1974). Dissection of chromosome structure with trypsin and nucleases. *Proc. Natl. Acad. Sci. U.S.A.* 71, 4249–4253.

Weintraub, H., Palter, K., and Van Lente, F. (1975). Histones H2A, H2B, H3 and H4 form a tetrameric complex in solutions of high salt. *Cell* 6, 85–110.

Weintraub, H., Worcel, A., and Alberts, B. (1976). A model for chromatin based on two symmetrically paired half nucleosomes. *Cell* 9, 409–417.

Weisbrod, S., and Weintraub, H. (1979). Isolation of a subclass of nuclear proteins responsible for conferring a DNase I-sensitive structure on globin chromatin. *Proc. Natl. Acad. Sci. U.S.A.* 76, 630–634.

Whitlock, J. P., Jr. (1977). Staphylococcal nuclease and pancreatic DNase cleave the DNA within the chromatin core particle at different sites. *J. Biol. Chem.* 252, 7635–7639.

Whitlock, J. P., Jr., and Simpson, R. T. (1976a). Preparation and physical characterization of a homogeneous population of monomeric nucleosomes from HeLa cells. *Nucleic Acids Res.* 3, 2255–2266.

Whitlock, J. P., Jr., and Simpson, R. T. (1976b). Removal of histone H1 exposes a fifty base pair DNA fragment between nucleosomes. *Biochemistry* 15, 3307–3314.

Whitlock, J. P., Jr., and Simpson, R. T. (1977). Localization of the sites along nucleosome DNA which interact with NH$_2$-terminal histone regions. *J. Biol. Chem.* **252,** 6516–6520.

Wicks, W. D. (1969). Induction of hepatic enzymes by adenosine 3′,5′-monophosphate in organ culture. *J. Biol. Chem.* **244,** 3941–3950.

Wicks, W. D., Koontz, J., and Wagner, K. (1975). Possible participation of protein kinase in enzyme induction. *J. Cyclic Nucleotide Res.* **1,** 49–58.

Wiesehahn, G. P., Hyde, J. E., and Hearst, J. E. (1977). The photoaddition of trimethylpsoralen to *Drosophila melanogaster* nuclei: A probe for chromatin substructure. *Biochemistry* **16,** 925–932.

Wilce, P. A., Rossier, B. C., and Edelman, I. S. (1976a). Actions of aldosterone on rRNA and Na$^+$ transport in the toad bladder. *Biochemistry* **15,** 4286–4292.

Williamson, P., and Felsenfeld, G. (1978). Transcription of histone-covered T7 DNA by *Escherichia coli* RNA polymerase. *Biochemistry* **17,** 5695–5705.

Wittig, B., and Wittig, S. (1979). A phase relationship associates tRNA structural gene sequences with nucleosome cores. *Cell* **18,** 1173–1183.

Wong, T. K., and Marushige, K. (1976). Modification of histone binding in calf thymus chromatin and in the chromatin–protamine complex by acetic anhydride. *Biochemistry* **15,** 2041–2046.

Wong, N. C. W., Poirier, G. G., and Dixon, G. H. (1977). Adenosine diphosphoribosylation of certain basic chromosomal proteins in isolated trout testis nuclei. *J. Biochem.* **77,** 11–21.

Woodcock, C. L. F. (1973). Ultrastructure of inactive chromatin. *J. Cell Biol.* **59,** 368.

Woodcock, C. L. F., and Frado, L. L. Y. (1978). Ultrastructure of chromatin subunits during unfolding, histone depletion, and reconstitution. *Cold Spring Harbor Symp. Quant. Biol.* **42,** 43–55.

Woodcock, C. L. F., Safer, J. P., and Stanchfield, J. E. (1976a). Structural repeating units in chromatin. I. Evidence for their general occurrence. *Exp. Cell Res.* **97,** 101–110.

Woodcock, C. L. F., Frado, L. L. Y., Hatch, C. L., and Ricciardiello, L. (1976b). Fine structure of active ribosomal genes. *Chromosoma (Berlin)* **58,** 33–39.

Worcel, A., and Benyajati, C. (1977). Higher order coiling of DNA in chromatin. *Cell* **12,** 83–100.

Young, R. A., and Steitz, J. A. (1978). Complementary sequences 1700 nucleotides apart form a ribonuclease III cleavage site in *Escherichia coli* ribosomal precursor RNA. *Proc. Natl. Acad. Sci. U.S.A.* **75,** 3593–3597.

Zentgraf, H., Scheer, U., and Franke, W. W. (1975). Characterization and localization of the RNA synthesized in mature avian erythrocytes. *Exp. Cell Res.* **96,** 81–95.

Zhimulev, I. F., and Belyaeva, E. S. (1975). ^{2}H-uridine labelling patterns in the *Drosophila melanogaster* salivary gland chromosomes X, 2R, and 3L. *Chromosoma (Berlin)* **49,** 219–232.

9

Maturation Events Leading to Transfer RNA and Ribosomal RNA

Gail P. Mazzara, Guy Plunkett, III, and William H. McClain

I. INTRODUCTION

A. Classes of Cellular RNA

The genetic information which specifies the structure and function of an organism is encoded in the linear sequences of deoxyribonucleotides

439

which make up its deoxyribonucleic acid (DNA). Through a complex series of steps, this genetic information is transferred to a class of molecules composed of ribonucleotides, called ribonucleic acids (RNA's) (Fig. 1). The three types of RNA in cells are messenger RNA (mRNA), transfer RNA (tRNA), and ribosomal RNA (rRNA). All consist of a single polynucleotide strand, and each has a characteristic molecular weight and base composition (Table I). Each RNA type occurs in multiple molecular species: rRNA consists of three or four major species, tRNA includes about 60 different species, and mRNA may exist in hundreds or thousands

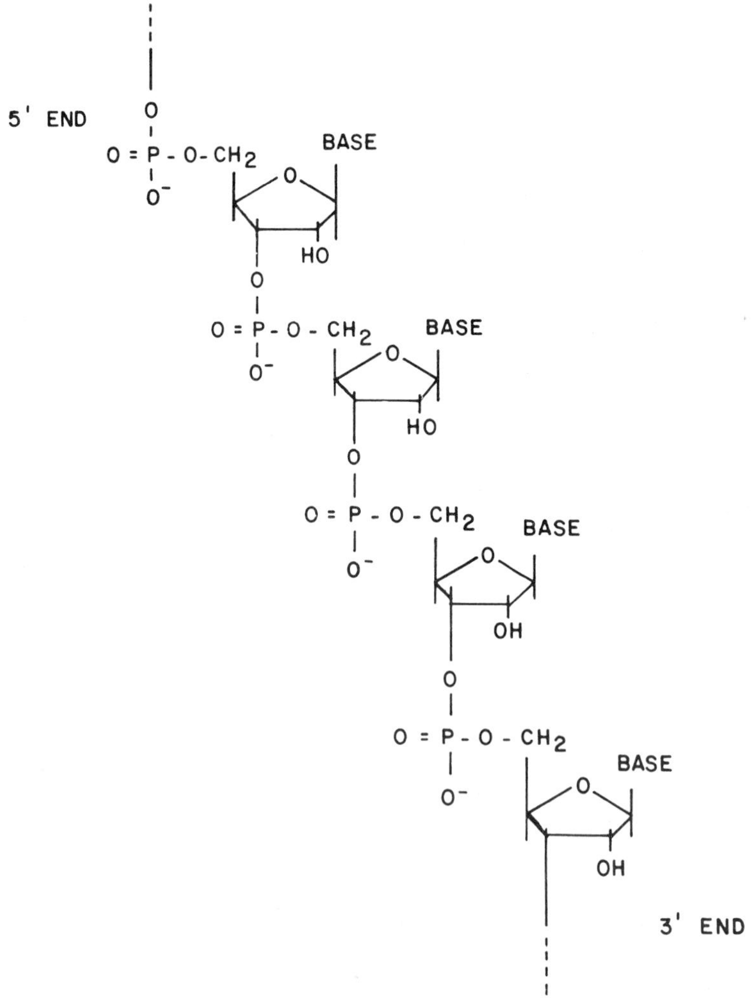

Fig. 1. The structure of RNA.

TABLE I

Characteristics of RNA's of *E. coli*

Type of RNA	Sedimentation coefficient	Molecular weight	Number of nucleotides	Number of molecules/cell	Percent of total cellular RNA
mRNA	6 S–25 S	$2.5 \times 10^{4-}$ 1.0×10^6	75–3000[a]	1.0–2.0×10^3	2–5
tRNA	4 S	2.3–3.0×10^4	75–90	2.3–2.8×10^5	10–15
rRNA	5 S	3.6×10^4	120	1.5–1.8×10^4 ⎫	
	16 S	5.5×10^5	1600	1.5–1.8×10^4 ⎬	80
	23 S	1.1×10^6	3300	1.5–1.8×10^4 ⎭	

[a] Monocistronic.

of species. (For a detailed classification of cellular RNA see Chapter 10 of this volume.)

Messenger RNA serves as a template for the sequential ordering of amino acids during protein synthesis. Triplets of nucleotides (codons) along the mRNA specify the linear amino acid sequence. Although mRNA composes only about 2 to 5% of the mass of cellular RNA, it consists of many species of various molecular weights and base sequences, since each of the proteins synthesized by cells is coded by a specific mRNA or segment of an mRNA molecule. Molecules of mRNA are generally metabolically unstable, with an average half-life of several minutes in prokaryotic organisms. However, in eukaryotes the half-life of mRNA species may be as long as 24 hours (Brawerman, 1974).

The other two classes of RNA are metabolically stable, with half-lives that may be expressed in terms of days. Although these do not serve as templates, they nevertheless play vital roles in protein synthesis. Transfer RNA's are small, about 25,000 daltons, and act as carriers of specific amino acids during protein synthesis. Each of the 20 amino acids has at least one corresponding tRNA species, and some have multiple tRNA species. These tRNA species differ in nucleotide sequence, but all share certain identifying features which are described in Section III,A.

About 80% of the mass of cellular RNA is ribosomal RNA, which is a structural component of the ribosomes on which protein synthesis occurs. The rRNA's fall into three size classes. All prokaryotic organisms contain three rRNA species which sediment at 23 S (1.1×10^6 daltons), 16 S (5.5×10^5 daltons), and 5 S (3.6×10^4 daltons). Eukaryotes contain two rRNA species which correspond to the prokaryotic 23 S and 16 S rRNA's, but which differ from them in molecular weight. In addition, they contain two small rRNA's, one of which is analogous to the prokaryotic 5 S rRNA species.

B. RNA Metabolism

Synthesis and degradation of RNA follow a cyclic pattern (Fig. 2). The initial step in RNA synthesis is transcription of RNA molecules upon DNA templates. In this process, nucleoside 5'-triphosphates pair with their Watson–Crick complements in the DNA. Chain growth occurs in the 5' to 3' direction; thus, the beginning nucleotide possesses a 5'-triphosphate group, and incoming residues are added as nucleoside 5'-monophosphates by the formation of a covalent sugar–phosphate bond, with the release of pyrophosphate. This reaction is catalyzed by the enzyme DNA-dependent RNA polymerase.

The initial product of transcription is an RNA molecule (precursor RNA) which must be processed to yield mature RNA. Because the precursor RNA contains extra nucleotide residues not present in the mature molecule, such processing includes one or more enzymatic cleavages of the precursor RNA to excise the nonfunctional RNA. Frequently, additional processing events include enzymatic modification of nucleotide residues (rRNA's and tRNA's) and addition of nucleotides to the terminal regions of the precursor RNA molecules (tRNA's). These molecular events, collectively referred to as *posttranscriptional processing,* are a prerequisite to the formation of RNA species in all cell types. The final event in the cycle of RNA metabolism is the degradation of RNA molecules, a process which releases nucleotides that eventually can serve as substrates for the synthesis of new RNA molecules. This cyclic pattern of RNA

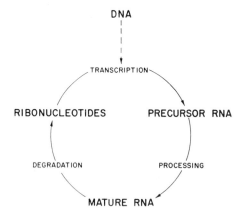

Fig. 2. The cyclic pattern of RNA metabolism.

metabolism is followed for all RNA's and is common to all known cell types.

This review will focus upon a limited aspect of RNA metabolism, namely, the posttranscriptional processes involved in the biosynthesis and degradation of tRNA and rRNA. The discovery that RNAs are synthesized via precursor RNA's that are larger than the mature species has raised several questions related to the process by which these molecules are converted into mature RNA species. Of special interest are the identities of the participating enzymes (processing enzymes) and the nucleotide sequences of the initial precursor RNA's and all precursor RNA intermediates generated in the production of mature RNA's. An appreciation of these details should provide a framework for asking how a precursor RNA is enzymatically recognized and handled so that its conversion to functional RNA is guaranteed. Perhaps the most intriguing question relates to the biological significance of this process: Why has nature chosen to utilize precursor RNA's in the production of RNA?

Similarly, in the study of RNA degradation, it is of interest to determine the enzymes involved and the manner in which the degradation process is controlled by the cell. Little is as yet known about this subject, and it will be treated only briefly in this chapter.

II. APPROACHES TO THE STUDY OF RNA PROCESSING

Complete elucidation of the biosynthetic steps involved in RNA processing requires identification of both the enzymes involved and the nucleotide sequences of the precursor RNA intermediates generated in the production of mature RNA's. Although the involvement of precursor RNA's in the synthesis of stable RNA's was first demonstrated in mammalian cells (Burdon *et al.*, 1967), characterization of these has been difficult because with mammalian cells it is difficult to obtain highly labeled, radiochemically pure preparations of an RNA. Thus, in most cases, sequence analysis of these precursor RNA's has been precluded.

Most of our knowledge of RNA processing is therefore derived from studies of prokaryotic organisms, whose relative genetic simplicity makes them extremely suitable for such studies. Precursor RNA's are biosynthetic intermediates normally present transiently and in minute amounts in the cell; thus, it is usually impractical to obtain individual species in amounts sufficient for their characterization. However, genetic manipulation of microorganisms has made it possible to produce strains which accumulate precursor RNA's as a result of mutations that block or delay precursor RNA processing by altering the processing enzymes or changing individual RNA genes. Those mutant organisms which accumulate precursor RNA's are especially useful for biochemical analysis.

Such biochemical analysis has included identification, isolation, and characterization of precursor RNA's. Most precursor RNA's have been isolated as ^{32}P-labeled RNA's following pulse-labeling; these ^{32}P-labeled precursor RNA's are most readily separated by polyacrylamide gel electrophoresis (Altman, 1971; Altman and Smith, 1971; Schedl and Primakoff, 1973; Sakano et al., 1974). Because these procedures often yield a variety of RNA's, identification of labeled RNA's as transiently existing precursor RNA's is necessary. Historically the precursor nature of an RNA species has been demonstrated by pulse-chase experiments, which show that the putative precursor RNA is converted to mature RNA-sized molecules in vivo (Burdon et al., 1967; Smillie and Burdon, 1970), or by competition hybridization, in which the presence of mature RNA sequences in the precursor RNA is demonstrated by its ability to compete with the mature RNA for hybridization to the parent DNA (Daniel et al., 1970). In prokaryotes, from which radiochemically pure preparations of both precursor RNA's and the mature RNA's derived from them can be obtained, direct demonstration that a precursor RNA contains the nucleotide sequence of a mature RNA can be accomplished by nucleotide sequence determination. This procedure, developed by Sanger and his colleagues (see Barrell, 1971) involves cleavage of ^{32}P-labeled RNA into fragments by enzymes specific for certain nucleotides. The fragments are fractionated by two-dimensional paper electrophoresis to yield a unique pattern of oligonucleotides called a fingerprint. The nucleotide sequences of these oligonucleotides are then determined by digesting them further with enzymes which have complementing specificities. Finally, the relative order of these oligonucleotides in the molecule is determined by subjecting it to limited ribonuclease digestion and analyzing the large fragments thus obtained to see which smaller oligonucleotides they contain. Such analysis reveals information about the structure of both the mature RNA and its precursor RNA intermediates, information which is necessary to a comprehensive understanding of the molecular principles involved in RNA processing. Complete nucleotide sequence analysis is currently practical only for small RNA's, such as tRNA's and their precursors; therefore, many studies of RNA processing have been devoted to analysis of tRNA biosynthesis.

Another technique which may prove useful in the study of RNA processing is nucleotide sequence analysis of the DNA which encodes RNA genes. Such analysis can provide valuable information about the structure of these RNA genes and possibly indicate the kinds of processing events required to produce mature RNA molecules. DNA sequence analysis has, for example, permitted the discovery of intervening sequences (e.g., Goodman *et al.*, 1977; Valenzuela *et al.*, 1978)—sequences present within the region encoding the mature RNA sequence but absent from the mature RNA (see Section III, C,4). DNA sequencing also allows one to determine unambiguously whether a particular nucleotide sequence is encoded in the DNA or is added to an RNA molecule by posttranscriptional processing, and may yield information about the nucleotide sequences which must be recognized by the processing enzymes. The technique is limited, however, since alone it cannot reveal which sequences are actually transcribed into precursor RNA.

III. TRANSFER RNA BIOSYNTHESIS

A. Structure of Transfer RNA and Precursor RNA's

Understanding of specific reactions involved in tRNA biosynthesis depends upon knowledge of tRNA structure. The transfer RNA's are required during protein synthesis for amino acid activation, as adaptors in mRNA-directed amino acid specification, and in binding the growing protein chains to ribosomes. They consist of about 60 different species, with at least one tRNA specific for each amino acid. Although these tRNA species differ in nucleotide sequence, their common function suggests that they must share common structural features. In fact, nucleotide sequence analysis of tRNA molecules from a variety of organisms has revealed that the different nucleotide sequences can be accommodated by a common secondary structure (Barrell and Clark, 1974).

Of approximately constant length (about 75 nucleotides), tRNA molecules are single stranded, with loops and folds involving short regions of base pairing (Cramer, 1971). A two-dimensional representation of tRNA consistent with known nucleotide sequences and giving maximum base pairing is known as the "cloverleaf pattern" (Fig. 3a). Each cloverleaf contains five sections. The acceptor stem is a hydrogen-bonded region with the unpaired sequence C-C-A to which the amino acid attaches located at the 3'-terminus of the tRNA chain. A second hydrogen-bonded stem terminates in the "T-ψ-C" loop, which may be involved in ribosome binding by hydrogen bonding between it and the 5 S RNA in the 50 S ribosomal subunit (Richter *et al.*, 1973). The variable region, which is located between the anticodon region and the "T-ψ-C" region, is of different sizes in different tRNA species, and may consist of a short loop or a base-paired region and loop. The hydrogen-bonded anticodon stem terminates in the anticodon loop, which contains the three nucleotides forming

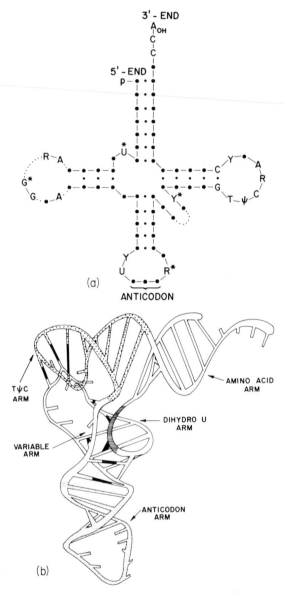

Fig. 3. (a) Generalized tRNA two-dimensional, or "cloverleaf," structure showing constant features. Large dots indicate bases, while the smaller dots represent the hydrogen bonds in base pairs. Y indicates a pyrimidine base, R indicates a purine base. An asterisk (*) denotes residues which may be modified. The dotted lines indicate regions of variable length (b) Schematic model of tRNA three-dimensional structure (yeast tRNA^Phe). The ribose phosphate backbone is drawn as a continuous cylinder with bars to indicate bases and

the anticodon; these nucleotides interact with the corresponding triplet (codon) in the mRNA. The fifth region, located between the acceptor stem and the anticodon region, is the "dihydro U" stem and loop, so named because it often contains the modified nucleotide dihydrouridine. An additional feature of tRNA structure is the presence of modified nucleotides (Fig. 4) which arise by enzymatic modification of the RNA (see Section III, B,4).

The three-dimensional structure of the cloverleaf has been investigated by X-ray diffraction studies of yeast tRNA[Phe], which showed that these molecules are arranged in an L shape, with the amino acid receptor at the end of one limb and the anticodon at the end of the other (Fig. 3b) (Robertus *et al.*, 1974). Most tRNA molecules are probably folded into a similar structure.

Studies of tRNA biosynthesis have revealed some structural features of precursor RNA's (Altman and Smith, 1971; Schedl *et al.*, 1975; Guthrie *et al.*, 1975) (Fig. 5). They may contain one, two, or more tRNA sequences, and are referred to, respectively, as monomeric, dimeric, or multimeric precursor RNA's. At the 5'-end, precursor RNA's contain additional nucleotides beyond the mature tRNA sequence; these extra residues are called the 5' leader region. At the 3'-end, additional nucleotides may extend beyond the C-C-A residues which correspond to the 3'-terminus of mature tRNA; however, in some precursor RNA's the C-C-A residues are absent, and precursor RNA-specific sequences occupy the 3'-termini of these molecules. During the maturation process, these sequences are removed prior to synthesis of the C-C-A residues at the 3'-termini of the precursor RNA's. In addition to extra nucleotides at 3'- and 5'-termini, precursor RNA's which contain more than one tRNA sequence also contain "spacer" nucleotides which separate the tRNA sequences.

As will be described in succeeding sections, several types of experimental results suggest that the individual tRNA sequences of precursor RNA's are folded into tRNA-like conformations (Chang and Smith, 1973; McClain and Seidman, 1975). This kind of conformational uniformity would provide a common basis by which the maturation enzymes can specifically recognize and handle a variety of precursor RNA's.

Knowledge of native tRNA conformation has been important in determining structural features necessary for the specificity of the processing enzymes. Those precursor RNA molecules whose structures are disturbed as a result of mutation are poorer substrates for those enzymes;

---

hydrogen-bonded base pairs. Hydrogen bonds involved in tertiary structure interactions are indicated by black bars. Some regions of the molecule are shaded, to aid in distinguishing the various structural features.

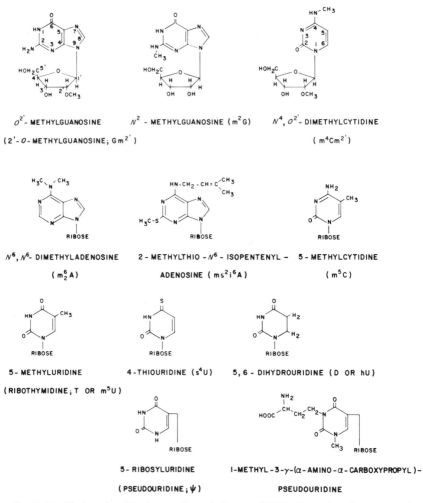

Fig. 4. Modified nucleosides in transfer and ribosomal RNA. Formulas of representative modified nucleosides are shown, including the numbering of the carbon atoms for the ribose moiety and both a purine and a pyrimidine base. In parentheses after the names of the compounds are abbreviations, and commonly used alternate names. In the abbreviations, symbols used to indicate modification of a standard base are in lower case letters immediately before the single capital letter indicating the standard nucleoside (e.g., m, methyl; i, isopentenyl; s, thio; ms, methylthio). Symbols used to indicate substituents on the ribose are placed immediately following the nucleoside symbol. Locants are indicated by superscripts; multipliers, when necessary, are indicated by subscripts.

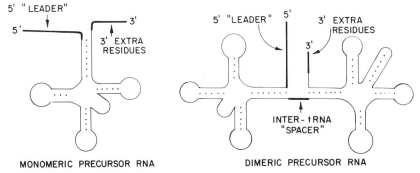

Fig. 5. Structural features of precursor RNA's. The RNA chain is represented by a continuous line, with dots indicating hydrogen bonds. Heavy lines indicate precursor-specific residues to be removed during processing.

identification of these structural deformations provides information about how processing enzymes recognize the precursor RNAs.

B. tRNA Synthesis in Prokaryotes

1. Arrangement of tRNA Genes

A basic question regarding transcription of tRNA genes into precursor RNAs is how the tRNA genes are arranged on the chromosome. Precise answers to this question currently can be obtained only for a few pro-karyotic organisms, such as *E. coli* or bacteriophage T4, in which tRNA genes can be located on the genome by genetic mapping.

In *E. coli,* tRNA genes are distributed around the chromosome (Bachmann *et al.,* 1976); however, several lines of evidence have suggested that many of the tRNA genes occur in small clusters which contain several closely spaced tRNA sequences (Brenner *et al.,* 1970). Transducing phages carrying a known tRNA gene often cotransduce other tRNA genes; this has been shown both by hybridization of tRNA's with phage DNA and by amplification of tRNA gene expression in infected cells (Russell *et al.,* 1970; Squires *et al.,* 1973). Further, hybridization of tRNA's with short DNA sequences obtained after sonication has indicated that DNA sequences of about 400 nucleotides may hybridize to two or three tRNA molecules (Brenner *et al.,* 1970; Doctor *et al.,* 1971).

In addition, precursor RNA molecules isolated from *E. coli* mutants defective in a processing enzyme, RNase P (see Section III, B,2), may contain the sequences of more than one tRNA, indicating that the genes for these tRNA's are not only clustered but are also cotranscribed (Schedl *et al.,* 1975). Among these are a group of three or four tRNA$_{\mathrm{III}}^{\mathrm{Gly}}$ genes

(Schedl *et al.*, 1975), the pair $tRNA_{II}^{Gly}$ and $tRNA^{Thr}$ (Carbon *et al.*, 1975) and a pair of $tRNA^{Tyr}$ genes (Ghysen and Celis, 1974).

Most recently, Ikemura and Ozeki (1977) have determined the chromosomal locations of *E. coli* genes specifying more than 20 different tRNA species, and their results confirm earlier observations that tRNA genes are clustered at several locations around the chromosome. These results give some indication of the maximum size of tRNA gene transcripts; however, it has not been possible to reach any general conclusions about which kinds of tRNA are grouped together. Those gene clusters which contain more than one copy of a single tRNA gene quite possibly arose by gene duplication; however, tRNA genes with similar sequences may occupy quite different positions, and genes within the same cluster may possess entirely different sequences.

Certain tRNA genes are found not within clusters of other tRNA genes, but instead within some or all of the approximately seven rRNA gene clusters of *E. coli* (Nomura *et al.*, 1977). The tRNA genes, including $tRNA_{1B}^{Ala}$, $tRNA_2^{Glu}$, and $tRNA_1^{Ile}$, are located between genes coding for the 16 S and 23 S rRNA's. Synthesis of these tRNA's must be coordinately controlled with the synthesis of rRNA, since the whole gene cluster is transcribed as a unit and then processed to yield 16 S and 23 S rRNA's together with the associated tRNA(s) (see Section IV). The reason for such coordinate regulation of biosynthesis of these particular RNA species is not known.

The genetic structure of the $tRNA_1^{Tyr}$ genes of *E. coli* has been analyzed in some detail. In *E. coli* there are two $tRNA_1^{Tyr}$ species, 1 and 2, which differ in sequence by two nucleotides (Goodman *et al.*, 1968, 1970). $tRNA_1^{Tyr}$ is specified by two tandem, identical genes separated by about 100 base pairs (Landy *et al.*, 1974). The $tRNA_2^{Tyr}$ gene is unlinked to these (Squires *et al.*, 1973). The $su_3{}^+$ $tRNA_1^{Tyr}$ is an amber suppressor which arises as a mutation in one of the two $tRNA_1^{Tyr}$ genes; this mutation changes the first residue of the tRNA anticodon from a modified G to a C, so that the tRNA translates the amber codon U-A-G (Goodman *et al.*, 1968, 1970). By selecting for transduction of the $su_3{}^+$ suppressor, $\phi80$ transducing phages carrying a segment of the *E. coli* genome including the two $tRNA_1^{Tyr}$ genes have been constructed (Andoh and Ozeki, 1968). These plaque-forming (p) "doublet" phages, $\phi80p$ ($su_3{}^+$) ($su_0{}^-$), thus carry one copy of the $su_3{}^+$ $tRNA_1^{Tyr}$ gene and one copy of the normal $tRNA_1^{Tyr}$ gene (designated as $su_0{}^-$). They segregate phages that have lost the $su_3{}^+$ gene but retain the $su_0{}^-$ gene, presumably by an unequal recombination event within the duplicated sequences (Russell *et al.*, 1970). The "singlet" phage are called $\phi80p$ ($su_0{}^-$). The $\phi80p$ ($su_3{}^+$) phage, carrying only the $su_3{}^+$ $tRNA^{Tyr}$ gene, can be similarly generated. The length of

DNA lost in this recombination event was originally determined, by DNA–DNA hybridization, to be about 100–150 base pairs (Miller *et al.*, 1971). A more precise measurement was derived from analysis of DNA fragments from the two transducing phages, produced by the action of restriction endonucleases (Landy *et al.*, 1974). Loss of one of the tRNA genes removes 200–260 base pairs; if the minimum size of a tRNA gene is 135–140 residues, the size of the transcribed precursor RNA (see Section III, B,2), then the region between the two tRNA^{Tyr} genes must be at most 100 base pairs. Since the two genes appear to be cotranscribed (Ghysen and Celis, 1974), the sequence between them is probably unique and does not encode a new promoter for RNA transcription.

Other regions of the tRNA$_1^{Tyr}$ genes have been analyzed more directly by nucleotide sequence analysis. The structures of mature tRNA$_1^{Tyr}$ (Goodman *et al.*, 1968) and of a precursor RNA to tRNA$_1^{Tyr}$ (Altman and Smith, 1971) have been established by RNA sequencing techniques. These will be discussed further in section III,B,2.

Khorana and his colleagues have determined the nucleotide sequence of the ϕ80p (su_3^+) DNA which precedes the 5'-end of the precursor RNA sequence (Seikiya *et al.*, 1976a,b). The sequence of 29 nucleotides immediately preceding the transcription starting point contains striking regions of twofold symmetry which allow the sequence to be represented as a pair of looped-out, hydrogen-bonded stems and loops, and which may be characteristic of RNA polymerase binding sites. This region is likely to be part of the promotor of the tRNA$_1^{Tyr}$ genes. Interestingly, the analogous promotor region of the gene for tRNA$_2^{Tyr}$ is identical to that of the tRNA$_1^{Tyr}$ genes, even though these genes are widely separated on the *E. coli* genome (Sekiya *et al.*, 1976a). Since about three times more tRNA$_2^{Tyr}$ is made than tRNA$_1^{Tyr}$ (Altman, 1978), one might hypothesize that regulation of the levels of these two tRNA^{Tyr} species may be influenced by differences in the folding of the chromosomal DNA in the regions of the tRNA^{Tyr} genes, which would make one of the two genes more accessible for transcription by RNA polymerase.

The nucleotide sequence of the portion of the tRNA$_1^{Tyr}$ gene which extends beyond the 3'-terminus of the mature tRNA molecule has also been studied. Khorana and his colleagues determined the sequence of 23 nucleotides beyond the 3'-terminus (Loewen and Khorana, 1973; Loewen *et al.*, 1974). More recently, Egan and Landy (1978) have used restriction fragments in the analysis of 540 base pairs beyond the DNA sequences coding for the mature tRNA$_1^{Tyr}$. This region is composed of a 178 base pair sequence, which is repeated 3.14 times without any intervening sequences between the basic repeating units. These repeating units are not entirely identical in sequence, as there are 14 sites at which one of the repeats differs from the others. The number and distribution of these altered residues suggested to Egan and Landy that in evolutionary terms the sequence reiteration is quite old. These researchers suggested that nonidentity in repeated sequences at one site may be relevant to *in vivo* transcription termination; a sequence in the second repeated unit may be the site for specific, ρ-dependent transcription termination. If this is true, then the *in vivo* size of the primary tRNA$_1^{Tyr}$ transcript can be predicted to be of about 640 nucleotides, 170 of which would comprise the tRNA$_1^{Tyr}$ sequences. Thus, sequence analysis of the genes coding for tRNA$_1^{Tyr}$ may prove useful in determining whether a precursor RNA to these tRNA's is a primary transcript. Such a transcript has not been found *in vivo*, although *in vitro* transcription of the tRNA^{Tyr} genes has produced a high molecular weight precursor RNA of the appropriate size (Daniel *et al.*, 1970); however, not all the experiments bearing on this question are in agreement (see Section III,B,5).

In bacteriophage T4, which codes for eight tRNA species, as well as two slightly larger stable RNA's of unknown function (McClain *et al.*, 1972), all ten low molecular weight RNA's map together in a region of less than 2500 base pairs (Wilson *et al.*, 1972; McClain *et al.*, 1972; Wilson and Abelson, 1972). The three precursor RNAs which have been identified thus far each contain the sequences of two tRNA's (Guthrie *et al.*, 1975). Because these precursor RNA's are not primary transcripts, they must have arisen by cleavage of a larger primary transcription product. However, it is not known whether the T4 tRNA's are transcribed as a single unit (see Section III,B,5).

2. Precursor RNA's in E. coli

a. E. coli tRNA$_1^{Tyr}$ Precursor RNA. The precursor to the suppressor tRNA su_3^+ tRNA$_1^{Tyr}$ from *E. coli* was the first to be sequenced. The precursor RNA was isolated from mutants unable to synthesize su_3^+ tRNA$_1^{Tyr}$; such mutants were identified by their defective or altered suppressor properties (Smith, 1975). The biosynthesis of these mutant tRNA's was followed in *E. coli* infected with the transducing phage $\phi80p$ (su_3^+); as described in Section III,B,1, this transducing phage carries a segment of the bacterial chromosome which contains the su_3^+ tRNA$_1^{Tyr}$ gene. In this system, the expression of the su_3^+ tRNA$_1^{Tyr}$ gene is greatly amplified because of the presence in the infected cell of 1000 or more su_3^+ tRNA$_1^{Tyr}$ genes that result from phage DNA replication (Smith *et al.*, 1966).

A number of mutants of su_3^+ tRNA$_1^{Tyr}$ defective in suppressor activity were isolated and characterized. Polyacrylamide gel electrophoresis of ^{32}P-labeled RNA synthesized after pulse labeling of these mutants revealed that many of them synthesized only a small amount of tRNA$_1^{Tyr}$

(Altman, 1971). Some of these mutants also exhibited transient accumulation of a larger RNA molecule, which was shown by nucleotide sequence analysis to contain the sequence of su_3^+ tRNA$_1^{Tyr}$ (Altman and Smith, 1971). This RNA was thus assumed to be a precursor to su_3^+ tRNA$_1^{Tyr}$.

The nucleotide sequence of the precursor RNA is shown in Fig. 6. It differs from the sequence of mature tRNA in two respects: it is longer, containing 41 additional residues at the 5′-terminus and 2–3 additional residues at the 3′-terminus, and it contains few modified nucleotides.

Sequence analysis of the accumulated precursor RNA's showed that many of the suppressor-negative mutations consist of single base substitutions at different positions in the tRNA sequences contained within the precursor RNA's (Smith, 1975). The expected effect of such sequence

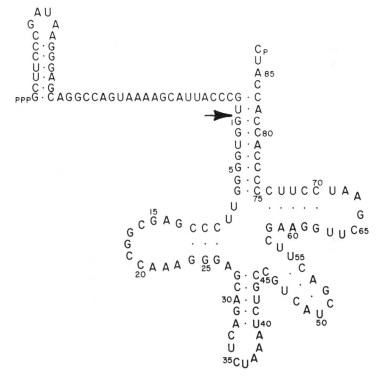

Fig. 6. The structure of the precursor to *E. coli* su_3^+ tRNA$_1^{Tyr}$. The dots represent hydrogen bonds. The arrow indicates the site of RNase P cleavage, which removes the 41-nucleotide 5′ "leader" sequence. The numbered residues comprise the mature tRNA sequence, which has a chain length of 85 nucleotides; note that the 3′-terminal residues in the precursor RNA, C-Up, are not part of the mature tRNA sequence. The precursor RNA contains few modified nucleotides; thus, modified nucleotides are not indicated in this figure. Internucleotide phosphate bonds are not represented.

changes is an alteration in tRNA conformation. These findings suggested a reason for the accumulation of precursor RNA's in these mutants: the sequence changes may deform the native conformation of precursor RNA's so that they cannot be efficiently cleaved by the processing enzymes. This idea has been supported by the isolation of revertants of suppressor defective mutants. These revertants acquire normal suppressor activity and normal tRNA biosynthesis as a result of a second mutation which apparently restores normal cloverleaf structure to the tRNA sequence within the precursor RNA (Smith *et al.*, 1970; Smith *et al.*, 1971). These results imply that the precursor RNA is folded into a tRNA-like structure and that this structure is important in its processing.

In partial digests of precursor RNA by ribonuclease T_1, the preferred splits within the tRNA sequence were the same as those in mature $tRNA_1^{Tyr}$ and were consistent with a cloverleaf secondary structure (Altman and Smith, 1971). More definite evidence for precursor RNA conformation has come from chemical modification experiments with the A15 mutant precursor RNA, which differs from su_3^+ $tRNA_1^{Tyr}$ by a G to an A change at residue 15 in the tRNA sequence (Chang and Smith, 1973) (Fig. 6). In these experiments, precursor RNA and tRNA were treated with methoxyamine, which reacts with cytidine, and with 1-cyclohexyl-3-[β-morpholinyl-(4)-ethyl]carbodiimide methotosylate, which reacts with guanosine, uridine, ribothymidine, and pseudouridine. Both reagents react extremely slowly with residues in hydrogen-bonded regions, and in mature tRNA only certain residues in the loops of the cloverleaf structure and the two C residues at the 3'-terminus are modified by these reagents. These restrictions reflect the accessibility of different residues in the tRNA structure; thus, chemical modification can be used to compare different conformations of the same primary sequence. Comparison of the patterns of modification in mutant A15 precursor RNA, mature A15 tRNA, and su_3^+ tRNA indicated that the conformation of the tRNA sequence contained within the precursor RNA is very similar to that of the mature tRNA.

The tRNA-like conformation of $tRNA^{Tyr}$ precursor RNA may be a common characteristic of many prokaryotic precursor RNA's. As another example, McClain and Seidman (1975) have proposed that in bacteriophage T4, the two tRNA sequences contained within dimeric precursor RNA's are also folded into tRNA-like structures (see Section III,B,3). The obvious implication is that precursor RNA's with different nucleotide sequences will possess similar conformations; this conformational uniformity would provide a basis by which the processing enzymes could recognize and handle many different precursor RNA's. The mutant $tRNA_1^{Tyr}$ precursor RNA's which fail to assume a tRNA-like structure are thus less efficient substrates for the processing enzymes, and therefore these mutant precursors transiently accumulate in the cell. The idea that mutant precursor RNA's are poorer substrates for processing enzymes has been supported by studies of *in vitro* cleavage of mutant tRNA$_1^{Tyr}$ precursor RNA's by a purified processing endonuclease, RNase P (Altman *et al.*, 1975). The initial rate of cleavage of precursor RNA by RNase P is three to seven times slower with mutant precursor RNA's (depending upon the mutant used) than with nonmutant substrate. These results reinforce the notion that single-site mutations which disrupt precursor RNA structure can drastically affect the interaction of processing enzymes with these substrates.

Further, in ^{32}P-labeling experiments, the kinetics of labeling of precursor RNA and tRNA suggest that a large proportion of mutant precursor RNA is degraded rather than processed to RNA (Altman and Smith, 1971). One hypothesis is that a defective precursor RNA assumes an incorrect structure whose stability is favored by the mutant base substitutions; such incorrect structures are substrates for attack by degrading nucleases (Altman and Smith, 1971; Smith, 1976). When the tRNA sequence within a precursor RNA assumes a tRNA-like conformation, the precursor RNA is no longer a substrate for these nucleases and can be correctly processed. This requirement for a specific conformation for conversion of precursor RNA to tRNA may thus reflect the operation of a cellular editing system that helps guarantee the overall fidelity of RNA transcription.

b. Precursor RNA's in *E. coli* Strains Defective in RNase P. The isolation and characterization of the su_3^+ tRNA^{Tyr} precursor RNA spurred the search for mutants in processing enzymes which participate in different steps of the tRNA^{Tyr} biosynthetic pathway. Schedl and Primakoff (1973) devised a selection method for isolating such mutants by identifying bacterial strains that, when infected with $\phi 80p$ (su_3^+) at high temperature, are unable to make functional su_3^+ tRNA^{Tyr} but are still able to transcribe and translate the β-galactosidase gene. Mutants of this kind would be expected to be defective in tRNA transcription, modification, or precursor RNA processing. One of the mutants thus isolated, strain A49, was temperature sensitive for the activity of the endoribonuclease RNase P, which cleaves the bond adjacent to the 5′-terminus of tRNA^{Tyr}. At the restrictive temperature, almost no mature tRNA^{Tyr} is synthesized in this mutant; instead, tRNA^{Tyr} precursor RNA accumulates. This precursor RNA is identical to the one described in the preceding section, except that it has several additional nucleotides at the 3′-terminus.

Synthesis of other mature tRNA species is also blocked in this mutant, and a variety of precursor RNA's accumulate (Fig. 7). Pulse labeling at 42°C followed by a chase at 30°C showed that these precursor RNA's behaved as intermediates in the biosynthesis of mature tRNA (Schedl *et*

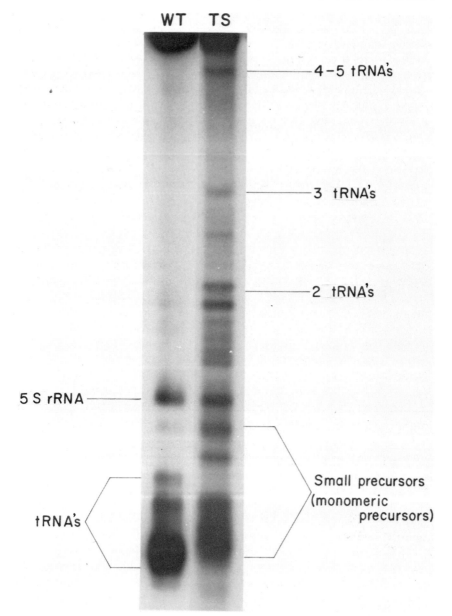

Fig. 7. Polyacrylamide gel profiles of ^{32}P-labeled RNA's isolated after pulse labeling of *E. coli*. WT, RNA's from wild-type cells; TS, RNA's from the RNase P mutant strain A49, labeled at the nonpermissive temperature. Accumulation of precursor RNA's, which are larger than mature tRNA's and have slower gel mobilities, is evident. The approximate sizes of the various A49 RNA's are indicated, in terms of numbers of tRNA molecules.

al., 1975). Nucleotide sequence analysis of the accumulated precursor RNA's showed that many of them lacked a 5'-terminal triphosphate residue; the presence of a nucleoside 5'-monophosphate at the 5'-termini of these precursor RNA's indicated that they are not primary transcripts but are instead intermediates in tRNA biosynthesis. One exception is the precursor RNA to tRNA$_3^{Gly}$, which contains three copies of the tRNA$_3^{Gly}$ sequence and has at the 5'-end a precursor RNA-specific sequence of about 60 nucleotides beginning pppA (Schedl *et al.*, 1975).

The precursor RNA's which accumulated in strain A49 ranged in size from 75 to about 600 nucleotides. The larger ones contained several tRNA sequences which were separated by inter-tRNA spacer nucleotides; in addition, the 3'- and 5'-ends of the precursor RNA's contained additional nucleotides not present in the mature tRNA sequences. However, most of the precursor RNA's were small, migrating faster than 5 S ribosomal RNA on polyacrylamide gels (see Fig. 7), and contained only one tRNA sequence; sometimes, the same tRNA sequence was also found in one or more larger precursor RNA's. For example, a precursor RNA of about 100 nucleotides contains the sequence of tRNA$_1^{Leu}$. This sequence is also found in a variety of larger precursor RNA's, the largest of which is about 500 nucleotides long and contains four or five copies of tRNA$_1^{Leu}$. One explanation for the presence of multiple precursor RNA's for a single tRNA species is that the smaller molecules are derived from the very large precursor RNA by endonucleolytic cleavage in the spacer regions separating tRNA sequences. Two types of evidence support this notion. Pulse-chase experiments in A49 cells at 42°C showed that the large tRNA$_1^{Leu}$ precursor RNA synthesized during pulse labeling was converted to the small precursor RNA's during the chase (Schedl *et al.*, 1975). The large precursor RNA could be similarly cleaved *in vitro* by a crude extract from A49 cells. Since strain A49 contains a mutation in the gene coding for RNase P, a second endonuclease is presumed to be responsible for splitting the large precursor RNA's. An endonuclease activity with the appropriate specificity, called RNase P$_2$ RNase O, has been isolated (see Section III,B,4).

RNase P mutant cells have been used to isolate new precursor RNA's for nucleotide sequence analysis. Among those sequenced is the dimeric precursor RNA containing tRNA$_2^{Gly}$ and tRNA^{Thr}, which begins with pG and is thus not a primary transcription product (Chang and Carbon, 1975). Like the tRNA^{Tyr} precursor RNA, it contains extra residues at both the 5'- and 3'-termini, in addition to a 6 nucleotide spacer region between the two tRNA sequences. Modified residues present in the mature tRNA's are also present in the precursor RNA, although some residues are incompletely modified. Both of the tRNA sequences in the precursor RNA contain the 3'-terminal C-C-A sequence; as will be seen subsequently, not all precursor RNA chains contain the C-C-A residues corresponding to the 3'-terminus of mature tRNA (see Section III,B,4).

3. Biosynthesis of Bacteriophage T4 tRNA's

a. **The tRNA's of T4.** The tRNA's of bacteriophage T4 have provided a useful system for the study of tRNA biosynthesis. The eight T4 tRNA genes are grouped together within a small region of the T4 genome (Wilson *et al.*, 1972; McClain *et al.*, 1972; Wilson and Abelson, 1972), and several of these are transcribed as precursor RNA's which contain more than one tRNA sequence (McClain *et al.*, 1972; Guthrie *et al.*, 1975). None of the T4 tRNA's are essential for bacteriophage growth (at least under laboratory conditions); therefore, the introduction into T4 tRNA genes of mutations that radically affect tRNA biosynthesis has been possible, since such mutations do not affect bacteriophage viability. In addition, T4 tRNA's can easily be isolated as radiochemically pure species: after infection of *E. coli* with T4, synthesis of host-specified RNA's immediately ceases, so that addition of [^{32}P]orthophosphate to T4-infected cells results in specific labeling of T4-specified nucleic acids. Low molecular weight, ^{32}P-labeled T4 RNA's can be completely resolved by two successive steps of polyacrylamide gel electrophoresis, a procedure which permitted the identification of eight T4 tRNA species, two larger stable RNA's, and several precursor RNA's (McClain *et al.*, 1972). Nucleotide sequence analysis revealed that each precursor RNA chain contained the sequences of two tRNA species (McClain *et al.*, 1972; Guthrie *et al.*, 1975). The processing of one of these, the precursor RNA to tRNA^{Pro} and tRNA^{Ser} (Pro-Ser precursor RNA) was extensively analyzed; this analysis defined a detailed biosynthetic pathway for the production of tRNA^{Pro} and tRNA^{Ser}.

b. **A Biosynthetic Pathway for tRNA Production.** As a first step in the study of tRNA^{Pro} and tRNA^{Ser} biosynthesis, the nucleotide sequences of the mature tRNA's and of the Pro-Ser precursor RNA were determined (Seidman *et al.*, 1975a; McClain *et al.*, 1975) (Fig. 8). The sequence of the Pro-Ser precursor RNA is characterized by the following features: it contains the sequences of two tRNA species arranged in the linear order 5'-tRNA^{Pro}-tRNA^{Ser}-3'; it contains a total of 13 precursor-specific nucleotides located at the 5'- and 3'-termini and in the region between the two tRNA sequences; it lacks the 5'-triphosphate group characteristic of an immediate transcription product, indicating that some processing had occurred prior to isolation of the molecule; it contains all of the modified nucleotides except the 2'-O-methylguanosine of tRNA^{Ser}; and it lacks the C-C-A sequences present at the 3'-termini of the mature tRNA molecules.

By inspection of the Pro-Ser precursor RNA sequence one can define the alterations required to convert it to mature tRNA^{Pro} and tRNA^{Ser}: a

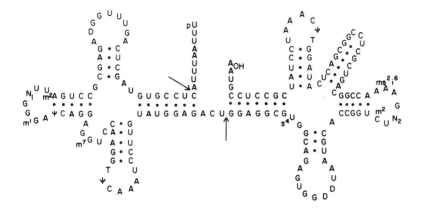

Precusor RNA

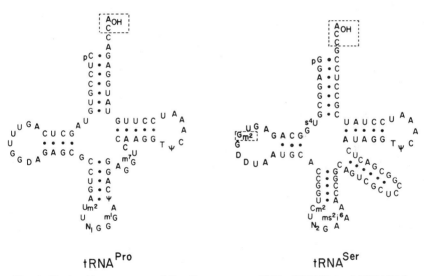

tRNA^Pro   tRNA^Ser

Fig. 8. Nucleotide sequences of Pro-Ser precursor RNA, tRNA^Pro, and tRNA^Ser from bacteriophage T4. The arrows indicate sites of RNase P cleavage. Broken boxes denote differences between the tRNA's and the precursor RNA. The internucleotide phosphate bonds are omitted for brevity.

pair of endonucleolytic cleavages, indicated by the arrows in Fig. 8, to generate two smaller precursor RNA's which contain the 5'-termini of the tRNAs; nucleolytic removal of residues from the 3' termini of these smaller precursor RNA's; synthesis of the 3' C-C-A sequence to the newly generated precursor RNA's; and formation of 2'-O-methylguanosine in tRNA^{Ser}.

Escherichia coli mutant strains defective in specific enzymes associated with RNA metabolism were used to identify the enzymes which catalyze these reactions and to elucidate the order of the steps in the biosynthetic pathway. By using a suppressor form of tRNA^{Ser} and examining suppressor activity in mutant strains, it was shown that tRNA^{Ser} function is unaffected in mutant cells lacking RNase I, RNase II, RNase III, or polynucleotide phosphorylase, implying that these enzymes do not participate in tRNA^{Ser} biosynthesis (McClain, 1977). This conclusion must be stated with some reservation, however, since residual enzyme activities in these mutants may be sufficient to sustain T4 tRNA^{Ser} production, and since enzyme activity lost by mutation may be replaced by an auxillary enzyme system. By contrast, functional tRNA^{Ser} is not produced in cells that lack RNase P, tRNA nucleotidyltransferase, or an exonuclease (BN exonuclease), indicating that these three enzymes may be involved in precursor RNA processing (Deutscher *et al.*, 1974; Seidman *et al.*, 1975a,b).

Additional investigation revealed how the three enzymes alter precursor RNA as well as the order in which they act. In cells that lack one of the processing enzymes, the appearance of mature tRNA is blocked, and a precursor RNA accumulates; normally, this precursor RNA is the substrate of the missing enzyme. The nucleotide sequence of the precursor RNA indicates the alterations it has undergone in previous steps. By sequencing the Pro-Ser precursor RNA's which accumulate in each of the three mutant *E. coli* strains, a biosynthetic pathway for the synthesis of tRNA^{Pro} and tRNA^{Ser} was established; this pathway is shown in Fig. 9 (Seidman *et al.*, 1975a,b; McClain and Seidman, 1975). Processing is initiated by the removal of U-A-A from the 3' terminus of the precursor RNA by BN exonuclease; this sequence is replaced by C-C-A through the action of tRNA nucleotidyltransferase. The requirement for tRNA nucleotidyltransferase for C-C-A formation is not common to the synthesis of all prokaryotic tRNAs, or even all T4 tRNAs; only four of the eight T4 tRNAs require this enzyme for biosynthesis (McClain *et al.*, 1978), and it apparently is not involved in the biosynthesis of *E. coli* tRNA's (Deutscher *et al.*, 1977) (see Section III,B,4).

The synthesis of C-C-A at the 3'-terminus of the precursor RNA is a prerequisite to its cleavage by RNase P, which is the next step in the biosynthetic pathway for Pro-Ser precursor RNA maturation. Thus, in

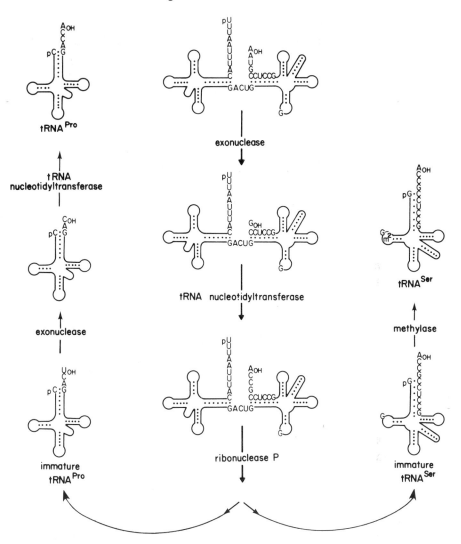

Fig. 9. Biosynthetic pathway for the production of bacteriophage T4 tRNA^{Pro} and tRNA^{Ser} from the Pro-Ser precursor RNA. Nucleotides in the regions acted upon by processing enzymes are indicated; the remainder of the RNA chain is indicated by the solid line. The dots represent hydrogen bonds.

cells that lack either tRNA nucleotidyltransferase or BN exonuclease, Pro-Ser precursor RNA accumulates. This presumably reflects the impor- tance of precursor RNA conformation for RNase P cleavage; if the tRNA sequences in the precursor RNA are in their final conformations, the

RNase P cleavage site will be near the 3'-terminus of the precursor RNA molecule. Thus RNase P can interact simultaneously with the 3'-terminus and the cleavage sites; possibly, RNase P must bind to the C-C-A sequence in order for cleavage to occur, although there is not enough data to evaluate this idea. Such binding would necessarily determine the order of initial steps shown in Fig. 9.

The sequential aspect of the biosynthetic pathway initially prevented analysis of the effect of mutations in BN exonuclease and tRNA nucelotidyltransferase on the final steps of tRNA^{Pro} biosynthesis, since the immature tRNA^{Pro} shown in Fig. 9 arises as a result of RNase P cleavage of Pro-Ser precursor RNA, and this cleavage does not occur in exonuclease- or tRNA nucleotidyltransferase-deficient cells. This difficulty was circumvented by analysis of tRNA biosynthesis in the related bacteriophage T2. Bacteriophage T2 produces a tRNA^{Pro} and a tRNA^{Ser} identical in sequence to those of T4; however, the T2 Pro-Ser precursor RNA differs in that it already contains the 3'-terminal C-C-A residues, which are formed in this phage during transcription, rather than by tRNA nucleotidyltransferase (Seidman *et al.,* 1975a). The T2 Pro-Ser precursor RNA is cleaved by RNase P in cells deficient in exonuclease or tRNA nucleotidyltransferase to yield tRNA^{Ser} and an immature form of tRNA^{Pro}. This immature tRNA^{Pro} species contained the 5'-terminus of tRNA^{Pro}, even though it lacked a complete 3'-terminal C-C-A sequence, indicating that the presence of this sequence is not necessary for RNase P cleavage at the 5'-terminus of immature tRNA^{Pro}.

Thus, RNase P cleaves Pro-Ser precursor RNA in two positions to generate two immature tRNA species that contain the 5'-termini of the mature tRNA's; the order of the cleavages has not been determined. The immature tRNA^{Ser} derived from the 3'-half of the precursor RNA lacks only the 2'-*O*-methylguanosine modification; formation of 2'-*O*-methylguanosine represents the terminal step in tRNA^{Ser} biosynthesis. The immature tRNA^{Pro} lacks part of the C-C-A residues; the combined actions of exonuclease and tRNA nucleotidyltransferase are required for the formation of mature tRNA^{Pro}.

Greater confidence in the validity of this pathway has come from a partial reconstitution of it *in vitro* using purified precursor RNA's and processing enzymes. BN exonuclease (Seidman *et al.,* 1975b), tRNA nucleotidyltransferase (Schmidt, 1975), and RNase P (Schmidt *et al.,* 1976) activities have been shown to be capable of catalyzing the reactions shown in the pathway. As expected, Pro-Ser precursor RNA terminating in C-C-A was a better substrate for RNase P *in vitro* than the corresponding RNA terminating in U-A-A. The Pro-Ser precursor RNA was cleaved twice *in vitro,* and the oligonucleotide pU-U-U-A-A-U-U-U-A was recovered as an additional product of this reaction. Since this product is not found in infected cells, it is presumed to be degraded nucleolytically once it is produced.

Because the dimeric precursor RNA's isolated from T4 are not primary transcripts, other processing steps must precede the production of these precursor RNA's. One enzyme likely to be involved in this early processing is RNase III (McClain, 1979) (See section III,B,4,a).

c. Specificity of Processing Enzymes. One of the striking features of the biosynthetic pathway which generates tRNA^{Pro} and tRNA^{Ser} is the highly specific nature of the enzymatic reactions involved. Although the molecular mechanisms underlying enzyme–RNA interactions which give rise to such specificity cannot yet be precisely defined, appreciation of the grosser features associated with this interaction has been achieved partly through the study of T4 strains unable to synthesize tRNA^{Ser} because of a mutation in the T4 genome. These mutants were isolated by using similar techniques to those described for finding su_3^+ tRNA^{Tyr} mutations: tRNA^{Ser} was converted to a suppressor tRNA, and its function was assayed by its ability to suppress an amber mutation in an essential T4 gene (McClain *et al.*, 1973). These selection procedures uncovered two classes of suppressor negative mutants. As expected, many mutants had base substitutions within the tRNA^{Ser} gene; however, a second class of mutants mapped outside the tRNA gene cluster and were shown to exert a pleiotropic effect on T4 tRNA synthesis (Wilson and Abelson, 1972; McClain *et al.*, 1973). In these mutants tRNA^{Ser}, tRNA^{Pro}, and tRNA^{Ile} are not synthesized at all, and the yield of other T4 tRNAs is reduced. Thus, the gene defined by this mutation, now called *mb* gene, specifies a component essential for the synthesis of some T4 tRNA's; possibly, the *mb* gene product is involved in transcription or processing of these tRNA's.

Many of the suppressor negative mutants contained base substitutions in tRNA^{Ser} which resulted in accumulation of Pro-Ser precursor RNA at the expense of mature tRNA species (McClain *et al.*, 1973). In some cases, accumulation is not extensive, apparently because nucleolytic degradation of precursor RNA is also occurring.

Certain nucleotide substitutions blocked only the RNase P reaction (McClain *et al.*, 1975). In these mutants the U-A-A is replaced by C-C-A at the 3'-terminus, although this reaction is not always complete. Since these mutations occur in the amino acid stem of tRNA^{Ser}, these results indicate that the integrity of this stem is not necessary for the actions of exonuclease and tRNA nucleotidyltransferase. This is most obvious in the case of a mutation which deletes the terminal base pair of this stem.

About 95% of the nucleotide substitution mutants isolated were defective in the first step of the biosynthetic pathway, removal of U-A-A res-

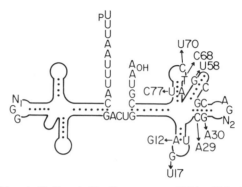

Fig. 10. Nucleotide substitutions in Pro-Ser precursor RNA which affect the first step of the biosynthetic pathway, removal of U-A-A residues. Note that these nucleotide substitutions, which block production of both tRNA's, are found only in the portion of the precursor RNA containing the sequence of tRNA^{Ser}. Any of the individual nucleotide substitutions shown is sufficient to block removal of the U-A-A residues.

idues (W. H. McClain, unpublished results). These mutations are shown in Fig. 10. Possibly, these precursor RNA's are not recognized by exonuclease, or, if they are recognized, the action of the enzyme is not limited to U-A-A residues and continued degradation of the molecules occurs.

Certainly, not all of the residues pictured in Fig. 10 participate directly in the exonuclease–precursor RNA interactions, since sufficient specificity and strength in an enzyme–RNA complex might be attained through the formation of only a few bonds (Schimmel, 1977). More likely, nucleotide substitutions shown in Fig. 10 affect the conformation of the precursor RNA so that it can no longer successfully interact with the processing enzymes. This conclusion is substantiated by the finding that, as was found for su_3^+ tRNA^{Tyr} precursor RNA, adverse effects of nucleotide substitutions in the stem of the cloverleaf structure can be reversed by a second substitution opposite the first one to restore normal base pairing (W. H. McClain, unpublished results).

Nucleotide substitutions that block synthesis of tRNA^{Ser} have thus far been found in only one part of the precursor RNA molecule, that containing the sequence of tRNA^{Ser}. This result implies that the nucleotide sequence of tRNA^{Pro} is unimportant for the reactions leading to tRNA^{Ser}. Thus, processing enzymes that catalyze the first three steps shown in Fig. 9 primarily interact with the residues of tRNA^{Ser}, and for these reactions to occur, tRNA^{Ser} must be in its normal conformation.

The importance of conformation is further demonstrated by the fact that mutant Pro-Ser precursor RNAs lack many modified nucleotides, but these deficiencies are confined to tRNA^{Ser} residues (McClain and Seidman, 1975). Thus, there seem to be no significant interactions between the residues of the two tRNA sequences, for such interactions should have been reflected by a reduction of nucleotide modification in tRNA^{Pro}. It is likely, then, that the precursor RNA contains two independent entities in the form of tRNA sequences folded in their respective three-dimensional conformations. Analogous results were obtained in the study of a second precursor RNA from T4, that which contains the sequences of tRNA^{Gln} and tRNA^{Leu} (McClain and Seidman, 1975).

The proposal that normal precursor RNA conformation is defined by the conformations of its member species implies that processing enzymes recognize the common feature shared by these molecules, namely, their three-dimensional, tRNA-like structures. This type of recognition presumably allows these enzymes to act on a great number of precursor RNA chains.

4. Enzymes Involved in Processing of Precursor RNA's

Determination of the structures of several precursor RNA's has permitted identification of the specific enzymatic reactions required for their conversion to mature tRNA's. These include removal of extra nucleotide sequences in the precursor RNA and modification of specific bases in the tRNA sequence. In *E. coli,* several of the enzymes involved in these steps have been identified. The availability of mutant precursor RNA's which cannot be completely processed by these enzymes has also made possible an appreciation of some of the grosser features associated with the specificity and mode of action of these processing enzymes.

Much of the study of conversion of precursor RNA to mature tRNA has relied on analyses of processing reactions *in vitro*. Although these *in vitro* analyses are valuable, they must be approached critically, for they may not always accurately reflect the *in vivo* processing events. For example, if processing in the cell is carried out by a specific enzyme complex, such a complex might be difficult to reconstruct *in vitro,* especially from soluble enzyme fractions. Similarly, if processing takes place during transcription, the precursor RNA's isolated from mutant cells and used as substrates for *in vitro* processing may actually have no counterpart in wild-type cells—their accumulation in mutant cells may be an artifact of the mutation. Also, the *in vitro* level of enzyme activity may represent only a gross qualitative estimate of enzyme levels within the cell. Further, enzymes which perform quite similar reactions *in vitro* may have entirely different functions *in vivo* as a result of compartmentalization or localization of these enzymes in the cell. *In vitro* studies must be approached with these limitations in mind.

   a. Processing Endoribonucleases. Only one enzyme, RNase P, has been shown to perform an obligatory function in precursor RNA processing. Robertson and co-workers (Robertson *et al.*, 1972) purified the RNase P activity from *E. coli* and showed that it specifically cleaves tRNA$_1^{Tyr}$ precursor RNA to generate the mature 5'-terminus of tRNA$_1^{Tyr}$. The enzyme has also been shown to cleave a variety of other precursor RNA's of known sequence *in vitro*, and fingerprint analysis of the cleaved products has demonstrated that RNase P cleaves the precursor RNA molecule to generate the correct 5'-terminus of the tRNA contained therein (Robertson *et al.*, 1972; Chang and Carbon, 1975; Guthrie, 1975). Further, the properties of the RNase P-defective mutants, which accumulate a wide variety of precursor RNA's, have shown that the enzyme is involved in the processing of virtually all *E. coli* tRNA's (Schedl *et al.*, 1975). It is also required for the biosynthesis of bacteriophage T4 tRNA's (Seidman *et al.*, 1975a) (see Section III,B,3) and the tRNA's specified by phage BF23, a relative of T5 (Sakano *et al.*, 1974).

Examination of the nucleotide sequences in precursor RNAs which are cleaved by RNase P shows that enzyme specificity is not determined by the nucleotide sequence immediately surrounding the cleavage site. In addition, the correct nucleotide sequence of the cleavage site is not in itself sufficient for RNase P action. Schmidt *et al.* (1976) isolated the two oligonucleotides from the precursor to T4 tRNA[Pro] and tRNA[Ser] which span the RNase P cleavage sites, and tested these as possible substrates for RNase P. Under conditions where 40% of the precursor RNA was cleaved to mature tRNAs by RNase P, no cleavage of the oligonucleotides was detected. These results demonstrate that precursor RNA secondary and/or tertiary structure is required for RNase P cleavage. This conclusion is supported by the finding that mutations at numerous locations in precursor RNAs, distinct from the site of RNase P cleavage, affect processing by this enzyme (Altman and Smith, 1971; McClain *et al.*, 1975). Since the effect of these mutations is to alter the conformation of the precursor RNA's, often by disrupting the cloverleaf structure of the tRNA's contained therein, RNase P presumably recognizes the tRNA conformations of these precursor RNA's. A tRNA-like structure for precursor RNA's, has, in fact, been supported by several types of experimental results (Chang and Smith, 1973; Seidman and McClain, 1975). The ability to recognize a common precursor RNA structure would enable RNase P to cleave a variety of precursor RNA's of different nucleotide sequences.

The structural features associated with accurate RNase P cleavage have not been precisely defined. On the 5'-side of the cleavage site, as few as nine nucleotides are sufficient for cleavage of the tRNA$_1^{Tyr}$ precursor RNA (Altman *et al.*, 1975). However, the rate at which precursor RNA containing only nine additional residues at the 5'-end was cleaved by RNase P was not reported; knowledge of this rate compared to the rate of cleavage of intact precursor RNA is important for the evaluation of the relevance of *in vitro* data to reactions which actually occur in the cell.

The nucleotide sequence at the 3'-terminus of precursor RNA's is involved in recognition of the substrate by RNase P, presumably because of its spatial proximity to the cleavage site. In the case of a bacteriophage T4 tRNA precursor in which the 3' C-C-A sequence is added posttranscriptionally, the presence of a complete C-C-A sequence at the 3'-terminus is required both *in vivo* and *in vitro* for efficient cleavage of the precursor by RNase P (Seidman *et al.*, 1975a, b; McClain and Seidman, 1975). On the other hand, removal of 3'-residues from the *E. coli* $tRNA_1^{Tyr}$ precursor decreases the efficiency of *in vitro* cleavage of the precursor by RNase P. Altman *et al.* (1975) have reported that chemical or enzymatic removal of three or four nucleotides from the 3' end of that precursor, eliminating part or all of the transcribed C-C-Ap, reduces the rate of cleavage by RNase P to about 50% of that found with intact precursor RNA. Thus in contrast to the findings with the T4 precursor RNA's, RNase P cleavage of the $tRNA_1^{Tyr}$ precursor RNA does not require substrate terminating at the 3' C-C-A sequence; indeed *in vivo* cleavage may occur before trimming of the 3'-terminus of the precursor RNA is completed.

Although these studies give some indication of the structural features of precursor RNA's which are recognized by RNase P, exactly how this recognition occurs is still unknown. Robertson *et al.* (1972) suggested that RNase P may be associated with some nucleic acid; more recently, Stark *et al.* (1978) have presented evidence that RNase P requires an associated (non-transfer) RNA for its function. They showed that highly purified RNase P preparations contain discrete RNA species; furthermore, treatment of the enzyme with micrococcal nuclease or RNase A abolished its activity. Based on these findings, Stark and co-workers concluded that the interaction of RNase P with its substrates depends upon the presence of RNA in the enzyme complex. Whether the RNA is required for stabilization of an RNA–protein complex or if it is actually involved in substrate recognition is not known.

A second endonucleolytic activity may be responsible for the partial cleavage of precursor RNA's which occurs in RNase P mutant cells (see Section III,B,2). Many small precursor RNA's found in RNase P mutants are not primary transcripts, but are instead cleavage products of larger precursor RNA's which contain several tRNA species (Schedl *et al.*, 1975). Thus, another endonuclease may split within the spacer regions of these multimeric precursor RNA's to generate smaller precursor RNA's with extra sequences at the 3'- and 5'-termini. An endonucleolytic activity which has these properties *in vitro*, called RNase P_2 (Schedl *et al.*, 1975) or RNase O (Sakano and Shimura, 1975), has been partially purified. *In vitro*, this enzyme is capable of cleaving a number of dimeric and multimeric precursor RNA's to give molecules containing one or more tRNA sequences.

Schedl *et al.* (1975) have studied the actions of RNase P and RNase P_2 *in vitro*, using the large precursor RNA's which accumulate in RNase P mutant cells as substrates. Although these precursor RNA's could be cleaved by both endonucleases, their conversion to products containing only one tRNA sequence could only be achieved by a combination of RNase P and RNase P_2. These results suggested that cleavage of precursor RNA's occurs sequentially, such that cleavage by one endonuclease exposes a site for the second enzyme. However, *in vivo* some processing probably occurs during transcription; in this case, any restrictions on the action of enzymes imposed by the conformation of an intact precursor RNA *in vitro* may have no counterpart *in vivo*.

Studies of RNase P_2 action *in vitro* are subject to additional limitations. The characterization of RNase P_2 and RNase O is not complete; it is not known whether the two activities represent the same enzyme, or whether they differ from another *E. coli* ribonuclease, RNase III, which shares some properties with these enzymes (Dunn, 1976). Their involvement in the processing of precursor RNA's can be rigorously demonstrated only by the isolation of conditionally lethal mutants affecting these functions.

RNase III has itself been shown to participate in some instances of precursor RNA processing. In RNase III-defective *E. coli* strains, T4 tRNA^{Gln} is not produced, although the other T4 tRNAs, including tRNA^{Leu} (which is normally processed from a dimeric precursor RNA containing tRNA^{Gln} and tRNA^{Leu}) are present (McClain, 1979). Although an accumulated precursor RNA containing tRNA^{Gln} has not yet been isolated from these strains, the absence of mature tRNA^{Gln} indicates that RNase III is involved in its biosynthesis. If, as is suggested by *in vitro* transcription experiments (Goldfarb *et al.*, 1978), all tRNA genes are cotranscribed, tRNA^{Gln} would be the first tRNA to be transcribed, as it maps to the 5'-side of all other T4 tRNAs. Perhaps RNase III is involved in cleaving off a large leader region at the 5'-end of this primary transcript; when this region is not endonucleolytically removed, the smaller precursor to tRNA^{Gln} is not formed. Evaluation of this hypothesis awaits isolation and characterization of a precursor RNA from RNase III-mutant cells which contains the sequence of tRNA^{Gln}.

Yet another endoribonucleolytic activity, called RNase P_4, has been partially purified by Bikoff and Gefter (1975). This enzyme, shown to be distinct from RNase P_2 and RNase III, is believed to be involved in cleavage of the primary transcript from the $tRNA_1^{Tyr}$ gene to yield a precursor RNA similar to that found in RNase P mutant cells.

b. Exoribonucleases. Precursor RNA's often contain, at the 3'-termini, extra nucleotide residues either following the C-C-A sequence or replacing it. In the former case, the additional residues are removed, leaving the mature C-C-A sequence; in the latter case, the extra residues are removed to allow tRNA nucleotidyltransferase to synthesize the mature C-C-A sequence. In both instances, the responsible exonucleases must remove only precursor RNA-specific residues, leaving the remainder of the tRNA molecule intact.

Crude extracts of *E. coli* are capable of removing nucleotides from the 3'-termini of precursor RNA's *in vitro*. However, several *E. coli* exoribonucleases are able to perform this action, and attempts to identify the enzymes responsible for processing of the 3'-termini of precursor RNA's *in vivo* have yielded conflicting results. Schedl *et al.* (1975) discovered an activity, RNase P_3, which copurified with RNase II, a 3' → 5' exonuclease; although this enzyme can apparently generate the mature 3'-terminus of $tRNA_1^{Tyr}$ *in vitro,* its role in the *in vivo* synthesis of tRNA has never been established, and is seriously doubted. Sakano and Shimura (1975) described an exonuclease, designated RNase Q, capable of removing extra nucleotides from the 3'-termini of precursor RNAs; its relationship to RNase II is not known. An apparently distinct activity, identified by Bikoff *et al.* (1975), can perform this same exonucleolytic function. Nevertheless, although these *E. coli* exonucleases are capable of 3'-end processing *in vitro,* not all of them may perform that function *in vivo*.

One enzyme known to participate in precursor RNA processing *in vivo* is the BN exonuclease. In *E. coli* mutant strains which lack this enzyme, some T4 precursor RNA's can not be processed to mature tRNA's because extra nucleotides at the 3'-termini are not removed from them (Seidman and McClain, 1975; Seidman *et al.,* 1975b). BN exonuclease is required for the removal of extra nucleotides from the 3'-ends of precursors to T4 $tRNA^{Pro}$, $tRNA^{Ser}$, and $tRNA^{Ile}$ before the C-C-A terminus can be synthesized by tRNA nucleotidyltransferase. Schmidt and McClain (1978) have partially purified the BN exonuclease, and have shown it to be capable of removing, *in vitro,* 3'-residues from precursors to $tRNA^{Pro}$, $tRNA^{Ser}$, and $tRNA^{Ile}$. Since mutant cells which lack this enzyme are viable, with growth properties indistinguishable from wild-type *E. coli,* the BN enzyme may not function in the maturation of *E. coli* tRNAs. Furthermore, residues extraneous to the C-C-A sequence in certain bacteriophage precursor RNA's, namely, the precursor to T4 $tRNA^{Thr}$ (W. H. McClain, unpublished results) and T2 $tRNA^{Ser}$ (Seidman *et al.,* 1975b), are removed in the strain which lacks BN exonuclease. These observations point to the existence of at least two 3'-ribonucleases for the maturation of tRNA precursors, although there is a possibility that strain BN contains residual 3'-ribonuclease activity sufficient to support the normal synthesis of some tRNA's. Nevertheless, the possible existence of two 3'-ribonucleases makes logical sense; of the two putative 3'-exonucleases, BN enzyme would normally function to remove from precursor RNAs nucleotides that occupy the position of the C-C-A sequence, while a second 3'-exonuclease would remove residues extraneous to the C-C-A sequence.

Recently, Ghosh and Deutscher (1978) have reported the purification of an exonuclease, RNase D, which seems likely to be involved in the removal of nucleotides extraneous to the C-C-A sequence of precursor RNAs. Using synthetic precursor RNAs as substrates they purified RNase D and showed that it is highly specific for precursor RNA's containing extra nucleotides at the 3'-terminus, and has low activity on intact tRNA. The enzyme can also remove extra nucleotides from the 3'-end of *E. coli* tRNA$_1^{Tyr}$ precursor RNA, but only after prior cleavage with RNase P. Thse findings suggest that the *in vivo* order of processing is RNase P cleavage followed by exonuclease action at the 3'-terminus. Although preliminary results indicate that RNase D generates the mature 3'-terminus of tRNA$_1^{Tyr}$, further experiments are required to determine whether the enzyme can quantitatively and specifically remove the extra 3'-residues from a natural precursor RNA.

RNase D thus possesses those properties which are expected of a 3'-processing exonuclease which removes extra residues, but stops at the C-C-A sequence. How might such specificity be achieved? Ghosh and Deutscher (1978) showed that RNase D acts upon tRNA molecules with altered conformations. Perhaps a precursor RNA is recognized as an altered tRNA. Alternatively, the C-C-A sequence itself may confer resistance to the nuclease, since removal of these nucleotides from a precursor RNA will render it susceptible to degradation by RNase D.

An additional role for 3'-exonucleases in tRNA biosynthesis is degradation. Mutant precursor RNA's and the fragments released by RNase P cleavage of precursor RNA's are degraded to mononucleotides, and 3'-exonucleases can perform this function.

c. **tRNA Nucleotidyltransferase.** All tRNA's contain the 3'-terminal sequence C-C-A, which is required for the acceptor activity of mature tRNA. The biosynthetic origin of this sequence has been in question, since there exists an enzyme, tRNA nucleotidyltransferase, capable of adding the C-C-A sequence to those tRNA molecules from which it is absent (Deutscher, 1973). The existence of this enzyme makes it unnecessary for the C-C-A sequence to be encoded in the DNA, yet the C-C-A sequence is present in many bacterial precursor RNA's (Altman and Smith, 1971; Chang and Carbon, 1975). Thus, the question was raised of what role tRNA nucleotidyltransferase plays in precursor RNA processing.

The involvement of this enzyme in tRNA biosynthesis has been studied in *E. coli* mutants with reduced tRNA nucleotidyltransferase activity (*cca* mutants) (Deutscher *et al.,* 1974, 1975). The enzyme was shown to be required for the biosynthesis of some bacteriophage T4 tRNA's; in *cca* mutants, the production of some T4 tRNA's is blocked, and precursor RNA's lacking the 3'-terminal C-C-A residues accumulate (see Section III,B,3). However, there is no evidence for a requirement for tRNA nucleotidyltransferase in *E. coli* tRNA biosynthesis. Measurement of the activities of six suppressor tRNA's in the presence of the *cca* mutation showed no decrease in these activities compared to wild type (Morse and Deutscher, 1975). In addition, aminoacylation studies of tRNA's produced in *cca* mutant strains indicated that only a limited number of these tRNA's contained defective 3'-termini (Deutscher *et al.,* 1975). Additional experiments showed that these defective termini arose by degradation of the tRNA chains subsequent to biosynthesis (Deutscher *et al.,* 1977). In the presence of chloramphenicol, an antibiotic which inhibits exonucleolytic degradation at the 3'-terminus of tRNA's while permitting continued tRNA synthesis, the proportion of defective tRNA molecules rapidly declined. Thus, in tRNA nucleotidyltransferase-deficient strains, tRNA's with intact C-C-A termini are synthesized. This was most directly demonstrated for tRNA[Cys] (Mazzara and McClain, 1977). Nucleotide sequence analysis indicated that newly synthesized tRNA[Cys] isolated after a short pulse labeling of *cca* mutant cells contained a larger fraction of molecules terminating C-C-A than did tRNA[Cys] which had been present in the cells for several generations; this result suggested that tRNA[Cys] is subject to nucleolytic degradation at the 3'-terminus during its lifetime in the cell. By contrast, the tRNA[Cys] molecules synthesized after pulse labeling in the presence of chloramphenicol all possessed an intact C-C-A sequence. Thus, under conditions where nucleolytic degradation of 3'-termini of tRNA's is inhibited, tRNA[Cys] was shown to be synthesized with a mature 3' C-C-A terminus, even when the level of tRNA nucleotidyltransferase in the cell was reduced by the presence of the *cca* mutation. These combined results strongly suggest that the primary function of tRNA nucleotidyltransferase in *E. coli* is repair of defective tRNA species which arise as a result of nucleolytic degradation of 3'-termini and not biosynthesis of the C-C-A sequence during precursor RNA maturation. This conclusion must be stated with reservation, since it was based on studies using a mutant strain which contains 2% residual tRNA nucleotidyltransferase activity. It is possible that the enzyme does participate in tRNA biosynthesis, but residual enzyme activity is sufficient to perform this function in the mutant cells. Resolution of this question might be provided by the isolation and study of mutant strains which contain no residual enzyme activity.

d. Modification Enzymes. After transcription of a tRNA sequence, specific nucleotides within the sequence are modified by rearrangements and substitutions of additional groups (for reviews, see Nishimura, 1972; McCloskey and Nishimura, 1977). These modified nucleotides are located in specific regions of the tRNA molecule as arranged in the cloverleaf structure (see Fig. 3a). These regions include the first position of the anticodon; the residue next to the 3′-side of the anticodon; the "T-ψ-C" loop, which contains ribothymidine (T) and pseudouridine (ψ); and several other positions in the tRNA molecule. Modification of nucleotides is accomplished by a number of enzymes, several of which have been isolated and purified.

The enzymes which catalyze the transfer of an intact methyl group (CH_3^- (from S-adenosyl-L-methionine to a C, N, or O atom of a purine or pyrimidine base or of ribose are called methylases (Borek and Srinivasan, 1966). The tRNA methylases can be placed in four groups. Uracil tRNA methylase and cytosine tRNA methylase alkylate the C-5 position of the pyrimidine ring to form ribothymine and 5-methylcytosine, respectively (see Fig. 4). The adenine tRNA methylases give rise to 2-methyladenosine, 6-methyladenosine, and 6-dimethyladenosine. Two guanine tRNA methylases form 1-methylguanosine, and a third is specific for the synthesis of 7-methylguanosine (Hurwitz et al., 1964).

Another type of nucleotide modification is the formation of sulfur-containing nucleotides, such as 4-thiouridine. A tRNA sulfur transferase which catalyzes transfer of sulfur from L-cysteine to tRNA uracil in vitro has been isolated from E. coli (Söll, 1971). Another enzyme, Δ^2-isopentenylpyrophosphate tRNA transferase, which catalyzes the synthesis of 6-(Δ^2-isopentenyl)adenosine, has also been isolated (Söll, 1971).

Pseudouridine formation is catalyzed by at least two different enzymes. One of these, pseudouridine synthetase I, has been partially characterized and is specific for pseudouridine formation in the anticodon region of tRNA's (Cortese et al., 1974). Pseudouridine synthetase II is responsible for pseudouridine synthesis in the T-ψ-C sequence common to most tRNA's (Schaefer et al., 1973). The properties of a partially purified preparation of this enzyme have been studied using tRNA's from cells grown in the presence of 2-thiouracil, which are approximately 50% deficient in pseudouridine in the T-ψ-C region (Kwong et al., 1977).

Although several modification enzymes have been isolated, little is known about how they recognize the specific nucleotides which they modify. However, studies of tRNA methylase specificity have provided some information about this process. Using pure species of tRNA as methyl acceptors *in vitro,* Baguley *et al.* (1970) and Kuchino and Nishimura (1970) showed that tRNA methylases recognize specific regions of the cloverleaf structures of different tRNA's as well as specific nucleotide sequences. The importance of tRNA conformation for recognition by tRNA methylases has also been shown by using tRNA fragments as methyl acceptors (Kuchino *et al.,* 1971). *Escherichia coli* tRNA$_f^{Met}$ was cleaved in the dihydrouridine loop to yield two fragments. The larger of these, which contained three-quarters of the original tRNA molecule, could be methylated to a slight extent to yield 1-methyladenylic acid. When the two fragments of the tRNA molecule were combined to form a complex, the extent of methylation was restored to that observed with intact tRNA$_f^{Met}$. These results showed that tRNA conformation is essential for recognition by tRNA methylases. Further studies, described below, indicate that correct conformation is probably important for recognition by other modification enzymes as well.

One may ask when during precursor RNA processing does nucleotide modification occur. Munns and Sims (1975) studied posttranscriptional methylation in an attempt to determine the amount of time required for tRNA's to acquire their full complement of methylated constituents, and whether formation of specific classes of methylated nucleotides occur at different stages of precursor RNA processing. Their results indicated that methyl groups incorporated into base moieties of tRNA (1- and 7-methylguanosine, 2- and N^6-methyladenosine, and 5-methyluridine) occurred at early or intermediate times during precursor RNA processing, while those incorporated into the ribose moieties (2'-O-methylribose) occurred late in precursor RNA processing. These methylated nucleotides are located in specific regions of the tRNA molecule as arranged in the cloverleaf structure; thus, selected regions within tRNA molecules may become methylated at specific times during precursor RNA processing.

Although the experiments described above suggest that nucleotide modifications occur at specific times during precursor RNA processing, they do not reveal the nature of the substrates recognized by the modification enzymes. However, several *E. coli* precursor RNA's, including the su_3^+ tRNA$_1^{Tyr}$ precursor RNA (Altman and Smith, 1971) and the dimeric precursor RNA to tRNA$_{II}^{Gly}$ and tRNA^{Thr} (Chang and Carbon, 1975), have been shown to contain modified nucleotides, although they are usually not present in molar yield. Thus, at least some nucleotide modifications can occur at the precursor RNA level in tRNA biosynthesis. The extent of *in vivo* modification of precursor RNA is probably related to its half-life in the cell; it may be related as well to the ability of uncleaved precursor RNA to serve as a substrate for the various modification enzymes. It has been reported that certain modified nucleotides normally found in mature su_3^+ tRNA$_1^{Tyr}$ cannot be formed enzymatically *in vitro* unless the 5' extra segment of the precursor RNA has been removed by RNase P (Schaefer *et al.*, 1973). In particular, stoichiometric yields of ribothymidine and both pseudouridines were found only with cleaved precursor RNA as substrate: the conversion of U40 to pseudouridine (see Fig. 6) was not observed with intact precursor RNA as substrate, and the conversion of U63 to ribothymidine, of U64 to pseudouridine, and of A38 to isopentenyladenosine apparently proceeded more efficiently on cleaved substrate. If these results are confirmed, they will provide direct evidence that either the extra fragment sterically interferes with the binding of some modification enzymes to the tRNA segment or that the tRNA segment has a somewhat different structure when it is part of the precursor RNA molecule. The chemical modification studies used to investigate precursor RNA conformation (see Section III,B,2) provide no evidence for the latter suggestion, however.

Several lines of evidence suggest, then, that nucleotide modification is not a prerequisite to cleavage of a precursor RNA. As just described, cleavage of the precursor RNA to su_3^+ tRNA by RNase P *in vitro* proceeds in the absence of complete nucleotide modification. Further, tRNA molecules which are undermodified can be isolated from a variety of sources, including wild-type cells (Gefter and Russell, 1969), cells grown under particular restrictive conditions (Kitchingham and Fournier, 1975; Kwong *et al.*, 1977), or cells which contain mutations in modification enzymes (Björk and Neidhardt, 1975; Colby *et al.*, 1976; Lawther and Hatfield, 1977). These findings indicate that cleavage of the precursor RNA's to produce these tRNA's does not require the presence of modified nucleotides. Further, those modification enzymes which have been studied *in vitro* can use these undermodified tRNA molecules as substrates, so there is no evidence thus far for modifications that occur exclusively on the precursor RNA.

Unlike *E. coli* precursor RNA's, the precursor RNA's of bacteriophage T4 are almost completely modified: in the tRNA^{Pro}-tRNA^{Ser} and the tRNA^{Gln}-tRNA^{Leu} dimeric precursor RNA's, all modifications are present with the exception of 2'-O-methylguanosine, which is entirely absent (Guthrie *et al.*, 1973). The more complete modification of T4 precursor RNA's may reflect the fact that they are processed more slowly than *E. coli* precursor RNAs, and thus have half-lives *in vivo* at least ten times as long as those of *E. coli* precursor RNA's (Altman, 1971; Guthrie *et al.*, 1973). This is compatible with the idea that the extent of modification is governed by the relative rate of reaction of modifying enzymes with the precursor RNA's. The absence of $2'$-O-methylguanosine in these precursor RNA's is consistent with the results of Munns and Sims (1975) who reported that methylation of the ribose moieties in tRNA occurs relatively late in precursor RNA processing. Certain base-substitution mutants in the T4 precursor RNA's are undermodified, presumably because the conformation of the tRNA sequence within the precursor RNA is altered (McClain and Seidman, 1975). These results reaffirm the idea that a tRNA-like conformation in precursor RNA is important for nucleotide modification.

The function of nucleotide modifications in tRNA has been recently reviewed (McCloskey and Nishimura, 1977). The modified nucleotides located adjacent to the $3'$-side of the anticodon are probably involved in codon recognition of the tRNA, possibly by facilitating formation of precise codon–anticodon base pairs by stabilizing the three-dimensional structure of the anticodon loop (Nishimura, 1972). Several researchers have demonstrated that these modified nucleotides are essential for the amino acid transfer function of tRNA's: Gefter and Russell (1969), for example, showed that unmodified su_3^+ tRNA$_1^{Tyr}$ which lacks 2-methylthio-N^6-isopentenyl adenosine (ms^{2}i^{6}A) adjacent to the anticodon is defective in its ability to bind ribosomes *in vitro*, although aminoacylation is not affected.

Modified nucleotides found in the first position of the anticodon are probably involved in specific codon recognition (Nishimura, 1972). Colby *et al.* (1976) demonstrated that the absence of a modified nucleotide in the first position of the anticodon of a T4 *ochre* suppressor tRNA restricts the suppressor function of this tRNA at the level of tRNA–mRNA–ribosome interaction. Their findings imply that the modified nucleotide is important for recognition of the *ochre* codon.

The G-T-ψ-C sequence, common to almost all tRNA's, apparently interacts with the complimentary sequence G-A-A-C in 5 S ribosomal RNA (Richter *et al.*, 1973). The T and ψ modifications are presumed to facilitate this interaction, since the oligonucleotide T-ψ-C-G binds to 5 S rRNA with an association constant almost three times that for the unmodified nucleotide (Erdmann *et al.*, 1973). However, the T modification, at least, is not required for tRNA function, since mutants unable to synthesize T are similar to wild type in most respects, although they do not compete well with wild-type cells in mixed cultures (Björk and Isaksson, 1970; Björk and Neidhardt, 1975). In several cases, the G-T-ψ-C sequence is replaced by a different sequence in the tRNA. For example, initiator tRNA's of eukaryotic cells contain the sequence G-A-U-C in place of G-T-ψ-C (Simsek *et al.*, 1974; Piper and Clark, 1974). Among prokaryotes, the extreme thermophile *Thermus thermophilus* HB8 contains 5-methyl-2-thiouridine (m^5s^2U) in place of T in this region; the m^5s^2U modification is believed to be important for the capacity of the tRNA to synthesize protein at high temperature (Watanabe *et al.*, 1974).

The requirement for modified nucleotides for tRNA activity in functions other than protein synthesis has been demonstrated in *Salmonella typhimurium* (Chang *et al.*, 1971; Singer *et al.*, 1972) and *E. coli* (Bruni *et al.*, 1977; Lawther and Hatfield, 1977). Mutant strains of these organisms defective in pseudouridine synthetase I lack pseudouridine in the anticodon stem of tRNA^{His}; these mutants are also derepressed for histidine biosynthesis.

5. In Vitro Transcription of tRNA Genes

While *in vivo* and *in vitro* studies of the processing of precursor RNA's which accumulate in mutant cells have been valuable in the analysis of tRNA structure and the maturation process, it has not yet been established that any of the precursor RNA's analyzed thus far corresponds to a primary gene product. In order to gain a complete understanding of the reactions involved in RNA processing, it will be necessary to identify the uncleaved, unmodified primary transcripts and the enzymatic reactions which convert them to mature tRNA's. To achieve this goal, a number of laboratories have been studying the transcription of tRNA genes *in vitro*.

Several researchers have reported successful *in vitro* transcription of *E. coli* tRNA genes. In the experiments of Daniel and co-workers, the tRNA$_1^{Tyr}$ genes were transcribed from the DNA of the $\phi80p(su_3^+)(su_0^-)$ transducing phage by purified RNA polymerase to yield a large RNA molecule, which sediments at about 9 S (Daniel *et al.*, 1970; Grimberg and Daniel, 1974). (The mature tRNA sediments at 4 S.) Incubation of this large RNA transcript with a crude extract from *E. coli* cells resulted in the generation of several intermediate-sized RNA molecules as well as mature tRNA$_1^{Tyr}$ (Daniel *et al.*, 1975). Most of these RNA's were characterized only by their mobilities on polyacrylamide gels following electrophoresis; only the mature tRNA$_1^{Tyr}$ and one slightly larger RNA product were subjected to nucleotide sequence analysis. Nevertheless, the finding that incubation of these longer RNA's with crude cell extracts results in the production of tRNA^{Tyr} implies that they represent precursor RNA's, and, further, that maturation of the precursor RNA's is a stepwise process involving the formation of intermediate-sized precursor RNA's. In additional studies, it was shown that these precursor RNA's could be recognized by specific modification enzymes, and that tRNA$_1^{Tyr}$ produced in this system possessed amino acid acceptor activity (Zeevi and Daniel, 1976). These researchers reported similar results for the *in vitro* transcription of the $tRNA_2^{Tyr}$, $tRNA_2^{Gly}$, $tRNA_3^{Thr}$ gene cluster, which is carried by the λ h80t bacteriophage. This gene cluster is apparently transcribed as a single multimeric precursor RNA which is processed in a stepwise fashion to yield mature tRNA's that can be partially modified *in vitro* and that possess amino acid acceptor activity. These researchers also noted that unmodified tRNA produced *in vitro* could also be aminoacylated, suggesting that nucleotide modification is not required for precursor RNA maturation or for tRNA aminoacylation.

Zubay *et al.* (1971) used DNA from $\phi80p(su_3^+)$ in a cell-free transcription system in which su_3^+ $tRNA_1^{Tyr}$ was synthesized. In its nucleotide sequence and its capacity to suppress as amber mutation in the gene for β-galactosidase, the su_3^+ $tRNA_1^{Tyr}$ synthesized in this *in vitro* system was identical with natural su_3^+ $tRNA_1^{Tyr}$ and had complete biochemical activity. However, because crude cell extracts were used in this system, the mechanism of tRNA transcription and processing was not investigated. Bikoff and co-workers (Bikoff and Gefter, 1975; Bikoff *et al.*, 1975) investigated this *in vitro* system further by purifying and characterizing four enzyme activities required for the synthesis of su_3^+ $tRNA_1^{Tyr}$. These were (1) RNA polymerase; (2) "fraction V" or RNase P_4, an endonuclease which presumably cleaves the primary transcript to yield a precursor RNA similar to that found in RNase P-deficient cells; (3) RNase P, which endonucleolytically removes extra nucleotides from the 5'-end of precursor RNA; and (4) RNase P_3, which specifically catalyzes removal of 3' extra nucleotides from the precursor RNA. This last enzyme is distinct from RNase II, the exonuclease presumed by Schedl and co-workers to be responsible for 3' maturation of precursor RNA's. Although this research identified the enzymatic components required for $tRNA^{Tyr}$ synthesis, they did not directly identify the primary transcript of the $tRNA^{Tyr}$ gene, except to observe that a large (about 175 nucleotides) RNA molecule was synthesized by RNA polymerase and converted to a smaller precursor RNA through the action of "fraction V" nuclease.

In an attempt to determine the identity of the primary transcript, Fournier *et al.* (1977) used *E. coli* RNA polymerase for the transcription of $\phi80p(su_3^+)$, then treated the resulting precursor RNA with a partially enriched preparation of processing ribonucleases. Fractionation of the transcripts by electrophoresis in polyacrylamide gels containing formamide, which prevents aggregation of RNA molecules, revealed that only one species of precursor RNA of about 180 nucleotides was synthesized. This precursor RNA could be converted to functional su_3^+ $tRNA_1^{Tyr}$, and it was suggested that this precursor RNA could represent the primary transcript of the $tRNA^{Tyr}$ gene. However, it is possible that the RNA polymerase preparation used by these workers contained a maturation activity which cleaved a larger primary transcript to produce the precursor RNA identified in these studies.

Attempts have also been made to transcribe the tRNA genes of bacteriophage T4 *in vitro*, in order to establish whether these genes, which are clustered in a small region of the T4 genome, are cotranscribed. In early experiments, Nierlich *et al.* (1973) and Lamfrom *et al.* (1973) reported the *in vitro* transcription of T4 DNA to yield RNA's corresponding to some of the T4 tRNA's produced *in vivo*. However, the primary transcription product was not characterized.

More recently, Goldfarb *et al.* (1978) used *E. coli* RNA polymerase to transcribe T4 DNA *in vitro*. By using as templates DNA from T4 mutants containing deletions in the tRNA gene region, they were able to identify a polycistronic precursor of T4 tRNA's among the primary transcription products. This precursor RNA could be processed into mature sized tRNA molecules. These results suggest that T4 tRNA genes are cotranscribed, although as with any *in vitro* analysis, it is difficult to prove that the findings actually correspond to the *in vivo* events (see Section III,B,4).

Thus, the study of tRNA synthesis *in vitro* has begun to make progress toward the goal of establishing a completely defined cell-free system capable of producing fully mature, functional tRNA. Although it is clear that mature *E. coli* tRNA can now be reliably synthesized *in vitro*, characterization of tRNA gene transcripts and identification of the processing enzymes which catalyze the early steps in precursor RNA maturation remain to be achieved.

C. tRNA Synthesis in Eukaryotes

1. Arrangement of tRNA Genes

Escherichia coli has about 60 different tRNA genes, and possesses one or a few copies of each of these genes (Schweizer *et al.*, 1969); by contrast, eukaryotes have a much higher number of tRNA genes: in yeast, 320–400 (Schweizer *et al.*, 1969); in *Drosophila*, 750 (Ritossa *et al.*, 1966); in HeLa cells, 1300 per haploid genome (Hatlen and Attardi, 1971); and in *Xenopus*, about 8000 (Clarkson *et al.*, 1973a). Despite the large number of tRNA genes in eukaryotes, the chromatographic profiles of tRNA's from these organisms show little more complexity than *E. coli* tRNA. Since chromatographic analysis resolves tRNA species of different nucleotide sequences, one interpretation of these results is that a relatively small number of individual tRNA genes are present in multiple copies, rather than many tRNA genes of different sequence present in one or a few copies.

The location and arrangement of eukaryotic tRNA genes has been studied in several organisms. Molecular hybridization experiments with purified tRNA and DNA from animal cells confirmed that, with the exception of mitochondrial tRNA's, tRNA molecules arise as a result of transcription from nuclear DNA (Burdon, 1975). Unlike the genes for 18 S and 28 S rRNA's (the eukaryotic counterparts of 16 S and 23 S rRNA's of prokaryotes), they are not located in the nucleolus (Ritossa *et al.*, 1966; Brown and Weber, 1968).

In HeLa cells (Aloni *et al.*, 1971) and in *Drosophila* (Grigliatti *et al.*, 1973) the tRNA genes are distributed among chromosomes of all size ranges, and in yeast the eight tRNA^{Tyr} genes are unlinked (Hawthorne and Mortimer, 1968; Olson *et al.*, 1977). Recently, however, *in vitro* transcription and processing of the yeast genes for tRNA$_3^{Arg}$ and tRNA^{Asp}, which are cloned into the plasmid vector pBR322, has demonstrated that these tRNA's are cotranscribed *in vitro* to yield a dimeric precursor RNA, which is subsequently processed to yield the mature tRNA's. The presence of this dimeric precursor RNA has not yet been demonstrated *in vivo* (Ogden *et al.*, 1979).

In *Xenopus*, the majority of tRNA genes are clustered within DNA segments which are about ten times the size of one tRNA sequence (Clarkson *et al.*, 1973b). It is not known whether each tRNA gene is separated from its neighbors by "spacer" DNA sequences or whether small clusters of adjacent tRNA genes exist. It is also not yet known how these tRNA sequences are transcribed. The only tRNA precursor RNAs isolated from eukaryotes in sufficient purity and amount for sequencing contain only one tRNA sequence, but these are probably not the primary transcripts of tRNA genes.

2. Precursor RNA's in Eukaryotes

Precursor tRNA's have been identified in mammalian cells (Burdon *et al.*, 1967; Burdon and Clason, 1969; Bernhardt and Darnell, 1969), in silkworm silk glands (Garber *et al.*, 1978), and in yeast (Blatt and Feldmann, 1973; Hopper *et al.*, 1978; Knapp *et al.*, 1978). They are generally smaller than prokaryotic precursor RNA's, containing only one tRNA sequence per precursor RNA, and probably represent intermediate steps in the biosynthesis and processing of tRNA.

Precursor RNA's were first identified by pulse labeling of HeLa or Krebs ascites tumor cells, which revealed a metabolically unstable and heterogeneous RNA fraction which migrated on polyacrylamide gels or gel filtration columns between 5 S ribosomal RNA and 4 S (transfer) RNA (Burdon *et al.*, 1967; Bernhardt and Darnell, 1969).

As the cells were exposed to longer labeling times, the proportion of radioactivity in this intermediate decreased concomitant with an increase in radioactivity in the tRNA region. These results suggested a possible precursor relationship between the rapidly labeled low molecular weight RNA and the tRNA. Support for this idea has come from experiments in which pulse labeling was followed by a chase in the presence of actinomycin D, a chemical which prevents further transcription; in these experiments, the heterogeneous low molecular weight RNA was shown to be converted to RNA species the size of tRNA (Burdon *et al.*, 1967; Bernhardt and Darnell, 1969; Choe and Taylor, 1972). Analysis of this precursor RNA after treatment of it with formamide or urea indicated that it differs from tRNA by chain length, rather than conformation; calculation from gel mobilities of RNA's so treated suggested that precursor RNA's were about 20 nucleotides longer than tRNA's (Burdon and Clason, 1969).

The characteristics of these precursor RNA's have not been examined in detail because of the difficulty in obtaining radiochemically pure preparations of individual species. In most cases, it has not been directly shown, for example, by hybridization competition studies, that they contain tRNA sequences. (One exception to this is in *Bombyx mori;* this will be discussed in detail below.) However, preliminary analysis has indicated that these precursor RNA's contain about 20 extra residues located at both the 5'- and 3'-ends of the molecule (Smillie and Burdon, 1970; Burdon, 1975). They do not have nucleoside 5'-triphosphate residues, and so do not represent primary transcripts of the tRNA genes. Digestion of precursor RNA with snake venom phosphodiesterase, a 3'-exonuclease, proceeds readily with the immediate release of uridylate residues; thus, these molecules do not possess the sequence C-C-A at the extreme 3'-terminus, although they may contain this trinucleotide some distance from the terminus (Burdon, 1975). A final characteristic of these precursor RNA's is that they contain both methylated bases and pseudouridine in amounts less than that found in mature tRNA (Burdon, 1975).

Although little is known about the structure of eukaryotic precursor RNA's, some attempts have been made to characterize the processing steps and enzyme activities involved in their conversion to mature tRNA species. The half-life of precursor RNA's in Chinese hamster ovary cells is about 12–15 min, while that of the mature tRNA's is about 100 hr (Choe and Taylor, 1972). *In vivo* conversion of these precursor RNA's to tRNA does not require methylation or pseudouridine formation, so the cleavage and modification events may overlap in time; however, modification is completed only after processing to tRNA-sized molecules has occurred (Burdon, 1971). *In vivo* studies have indicated that specific types of modifications occur at particular times during the processing of precursor RNA's (Munns and Sims, 1975).

In mammalian cells, the final processing steps must take place in the cytoplasm, since precursor RNA's and tRNA's can be isolated from the cytoplasm but not from the nuclei (Burdon *et al.*, 1967). *In vivo* labeling shows that methylation of tRNA occurs in the cytoplasm (Muramatsu and Fujisawa, 1968). In addition, several enzymes involved in tRNA processing, including tRNA methylases (Burdon *et al.*, 1967), tRNA nucleotidyltransferase (Mukerji and Deutscher, 1972), and the enzyme converting precursor RNA to 4 S RNA (Moshowitz, 1970), are all found in the cytoplasm.

Some progress has been made in the identification of an endonucleolytic activity from eukaryotic cells which may be responsible for precursor RNA cleavage. Since specificity of tRNA processing enzymes appears to depend upon tRNA conformation, it seemed likely that this endonuclease might recognize prokaryotic precursor RNA's which are presumably similar in structure to those of eukaryotes. Therefore, attempts to identify specific nucleases in eukaryotic cell extracts relied upon the use of the tRNA$_1^{Tyr}$ precursor RNA as a substrate to search for such processing enzymes.

Using this approach, Koseki *et al.* (1976) identified enzyme activities in human KB cells that cleave tRNA^{Tyr} precursor RNA. One of these removes some of the extra 3'-terminal nucleotides and is probably an exonuclease. The other activity possesses an RNase P-like specificity; like RNase P, it specifically cleaves tRNA^{Tyr} precursor RNA, removing the 41 extra nucleotides at the 5'-end of the RNA to produce the 5'-terminal sequence of mature tRNA^{Tyr}. Other *E. coli* precursor RNA's, as well as KB cell tRNA precursor RNA's, are also cleaved by this enzyme; however, it is not active on low molecular weight RNA's with long half-lives *in vivo*, such as 4 S RNA, 5 S RNA, or certain low molecular weight RNA's from KB cells. Thus, this endonuclease resembles RNase P in its substrate specificity. Nevertheless, the significance of this enzyme activity for RNA processing is not clear. It has not been possible to show that it actually creates the correct 5'-terminus of homologous eukaryotic tRNA's, since no individual precursor RNA from KB cells has been isolated in a radiochemically pure form. In addition, given the lack of human cell lines with lesions in tRNA biosynthesis, it is difficult to establish that the enzyme is essential in the *in vivo* synthesis of human tRNA, or that it is the only enzyme in human cells which could carry out this function.

3. Precursor RNA's in Bombyx mori

A promising system for characterizing eukaryotic tRNA biosynthesis is the silk glands of *Bombyx mori*, where the synthesis of a few tRNA species is greatly amplified (Garber *et al.*, 1978; Chen and Siddiqui, 1973). During the larval period, just before this specialized tissue secretes silk fibroin for the cocoon, 42% of total tRNA is specific for aminoacylation with glycine, while about 22% represents alanine and 14% serine. Thus, isolation of precursor RNA's from this organism is facilitated by the presence in a significant amount of only a small number of precursor RNAs. Evidence for the presence of such precursor RNA's has come from pulse chase experiments, which showed that a heterogeneous 4.5 S RNA species can be chased *in vivo* into 4 S (tRNA-sized) molecules (Chen and Siddiqui, 1975; Tsutsumi *et al.*, 1976). No evidence was found in these gels for larger precursor RNA's, which could contain more than one tRNA sequence. *In vitro* incubation of these 4.5 S RNA's with subcellular protein extracts from *B. mori* shifts the electrophoretic mobilities of these RNA's to that of tRNA. The 4.5 S RNAs also contain modified nucleotides, and can compete with tRNA in DNA–RNA hybridization experiments.

The 4.5 S RNA's can be separated by two-dimensional polyacrylamide gel electrophoresis into many distinct species, some of which are radiochemically pure by the standards of RNA fingerprinting (Garber *et al.*, 1978). Among these, the individual precursor RNA's to $tRNA_{2a}^{Ala}$, $tRNA_{2b}^{Ala}$, $tRNA_1^{Gly}$, and $tRNA_2^{Gly}$ were identified. Fingerprint analysis of these precursor RNA's showed that each contained a single tRNA sequence along with precursor-specific sequences at the 3'- and 5'-ends. It has not been determined whether these particular precursor RNA's contain nucleoside 5'-triphosphates of the 5'-termini; however, unfractionated precursor RNA mixtures contain pppA and pppG, indicating that at least a fraction of these are primary transcription products.

Both $tRNA^{Ala}$ precursor RNA's contained nucleotide modifications characteristic of the mature tRNA species. Molar yields of all the modified nucleotides were not reported, but at least some residues were completely modified. Modification levels in the $tRNA^{Gly}$ precursor RNA's were not determined.

These precursor RNA's can be cleaved *in vitro* by cell extracts to yield tRNA molecules. They are also substrates for a partially purified, RNase P-like activity isolated from *B. mori* silk glands, although proof that cleavage occurs at the 5'-termini of mature tRNA sequences awaits total sequence analysis of individual precursor RNA's and their cleavage products.

4. Nucleotide Sequence of tRNA^Tyr Genes in Yeast

One possible way to study tRNA biosynthesis, given an inability to isolate radiochemically pure precursor RNA species, is to determine the nucleotide sequence of the DNA encoding a particular tRNA species. Contemporary techniques for cloning a specific DNA fragment, together with improved procedures for sequencing DNA, have made such an approach feasible for some tRNA genes. This approach is somewhat limited, as it does not reveal how much of the gene is transcribed into precursor RNA; nevertheless, it may uncover important information about tRNA gene structure.

DNA sequencing was used for the analysis of tRNA^Tyr genes in yeast. These genes comprise a eukaryotic system that is unusually amenable to genetic and biochemical analysis. There are eight unlinked genetic loci, each of which can mutate to a tyrosine-inserting nonsense suppressor (Hawthorne and Leupold, 1974; Piper *et al.*, 1977); correspondingly, eight different restriction fragments produced by digestion of yeast DNA with *Eco*R1 can hybridize to tRNA^Tyr. Since only one tRNA^Tyr sequence was detected in wild-type yeast cells (Madison and Kung, 1967), the eight genes are presumed to be identical or nearly identical in the region which encodes the mature tRNA sequence.

Goodman *et al.* (1977) identified the yeast *Eco*R1 restriction fragment which encodes the tRNA^Tyr suppressor locus *SUP4*, and cloned this fragment from wild-type (*sup4*^+) and suppressor (*SUP4-o*) yeast strains. The DNA sequences of these genes, and of two other genetically uncharacterized tRNA^Tyr genes which had also been cloned, were determined. Strikingly, the sequence analysis revealed that perfect sequence homology exists within the DNA encoding the mature tRNA^Tyr sequences, but the flanking sequences are almost totally dissimilar. If 5'-leader regions of yeast tRNA genes are transcribed and subsequently removed by a single enzyme, as they are for prokaryotic tRNA's, then processing at the 5'-terminus of such hypothetical precursor RNA's must depend very little, if at all, on the nucleotide sequence of the 5'-leader region.

None of the tRNA^Tyr genes contain a coding triplet for the 3'-terminal C-C-A residues. This result implies that one step in the processing of yeast tRNA^Tyr is the posttranscriptional addition of the C-C-A sequence to precursor RNA.

An unusual feature of these DNA sequences is that they differ from that of tRNA^Tyr by virtue of a 14 base-pair tract which occurs just to the 3'-side of the anticodon triplet. Such extra DNA sequences, called "intervening sequences," have been found in several other eukaryotic genes, including genes coding for mouse and rabbit globin, ovalbumin, and immunoglobin (see Darnell, 1978). Four alleles of yeast tRNA^{Phe} genes which have been sequenced also contain an intervening 18 or 19 base-pair sequence at a similar position in the tRNA sequence (Valenzuela *et al.*, 1978). The presence of an intervening sequence within the tRNA^{Tyr} and tRNA^{Phe} genes indicates that the biosynthetic pathways for these tRNAs must involve elimination of the intervening sequence. Various possibilities for removal of this sequence include DNA splicing, RNA polymerase "jumping," or cleavage and ligation of precursor RNA. Recent investigations have provided evidence for the latter mechanism. Hopper *et al.* (1978) demonstrated that the yeast mutant *ts*-136, which has a lesion in the *rna*1 gene, accumulates a small number of precursors to tRNA's at the nonpermissive temperature. Some of these precursor RNA's contain modified nucleosides characteristic of mature tRNA's, and several of them were shown to hybridize to yeast tRNA genes carried on *E. coli* plasmids. Knapp *et al.* (1978) and O'Farrell *et al.* (1978) characterized two of these precursor RNA's, the precursors to tRNA^{Tyr} and tRNA^{Phe}. They demonstrated, by nucleotide sequence analysis, that these precursor RNA's contain the intervening sequences, but do not possess additional nucleotides at the 5'-ends. The 3'-ends contain the mature C-C-A terminus, which is not encoded by the genes for these tRNAs and thus has been added post-transcriptionally. Modified nucleotides are also present in the precursor RNA's, although modification is incomplete.

An enzyme activity from wild-type yeast capable of excising the intervening sequence from these precursor RNA's *in vitro* was also identified. These results demonstrated that the intervening sequence within the tRNA gene is transcribed and subsequently removed during processing of the precursor RNA. Removal of the intervening sequence is presumably a late step in the processing pathway, since the mature 3'- and 5'-termini of the precursor RNA's are already present in the molecules analyzed by Knapp *et al.* (1978) and O'Farrell *et al.* (1978). Evidence indicates that the intervening sequences in at least some mRNA's are also transcribed and subsequently excised from the RNA molecule (Darnell, 1978).

Analysis of the transcription and processing of cloned yeast tRNA genes using intact *Xenopus* oocytes (DeRobertis and Olson, 1979) or extracts from *Xenopus* oocytes (Ogden *et al.*, 1979) has confirmed and extended knowledge of the process by which these tRNA's are synthesized. In the case of tRNA^{Tyr}, for example, biosynthesis begins with the transcription of a precursor RNA which contains extra sequences at the 5'-end, lacks the 3' C-C-A sequence, and possesses an intervening sequence. In an ordered series of steps, this primary transcript is processed: the 5'-leader sequence is removed in at least two steps, the 3' C-C-A sequence is added, and several base modifications occur to yield a precursor RNA with mature 3'- and 5'-termini which still contains the intervening sequence. The final major processing step is the removal of the intervening sequence and the religation of the RNA (De Robertis and Olson, 1979).

O'Farrell *et al.* (1978) also demonstrated that the precursor RNA cannot be aminoacylated *in vitro*. This result may reflect the non-tRNA-like conformation of the precursor RNA. Analysis of the structure of the precursor RNA indicated that, although similar to that of mature tRNA, the anticodon region in the precursor RNA is rearranged to accommodate residues of the intervening sequence in a base-paired structure. Thus far, no one has reported attempts to aminocylate precursor RNA's from prokaryotic cells; these molecules might be substrates for aminoacyl-tRNA synthetases because they are similar in conformation to mature tRNA.

Interestingly, only a few precursor RNA's accumulate in *ts*-136, and those precursor RNA's which have been analyzed all contain intervening sequences. By contrast, yeast tRNA$_3^{Arg}$ and tRNA^{Asp} genes do not contain intervening sequences, and their corresponding precursor RNA's do not accumulate in *ts*-136 (Knapp *et al.*, 1978). This information suggests that the *rna*1 gene function is required directly or indirectly in the removal of intervening sequences. It has been shown, however, that strain *ts*-136 contains the enzyme(s) required for the removal of intervening sequences. The physiological reason that only a fraction of precursor RNAs contain intervening sequences is not known, nor is the function of the intervening sequence in the production of biologically active tRNA.

IV. RIBOSOMAL RNA BIOSYNTHESIS

A. Description of Ribosomes and rRNA

The most abundant RNA species in all cells are the ribosomal RNA (rRNA) molecules, which comprise about 80% of the bulk cellular RNA. These RNA species and their associated proteins constitute the ribosome, the hub of the cellular protein synthesizing machinery. The production of rRNA is therefore integral to the regulation of cell growth and to the differentiation and maintenance of all tissues. In all organisms, the production of the mature, functional rRNA's involves a group of post-transcriptional alterations of the molecules. These alterations include various modifications and substitutions upon certain nucleotide residues, as well as cleavage events which reduce the chain lengths of precursors to those of the mature forms.

The ribosomes of prokaryotes and eukaryotes (and of eukaryotic chloroplasts and mitochondria) are quite similar, but numerous structural, and probably functional, differences are also apparent. All ribosomes are roughly ellipsoidal complexes consisting of two ribonucleoprotein subunits; they are customarily characterized by sedimentation coefficients (in Svedberg units), which in part reflect the relative sizes of the particles. Bacterial cells, typified by *E. coli,* contain about 2×10^4 ribosomes which sediment at 70 S ($\sim2.5 \times 10^6$ daltons), with large and small subunits of 50 S and 30 S, respectively. A mammalian cell, on the other hand, contains about 5×10^6 ribosomes; these sediment at 80 S ($\sim4 \times 10^6$ daltons) and consist of 60 S and 40 S subunits. The ribosomes of chloroplasts are very similar to those of bacteria, but mitochondrial ribosomes are apparently unlike those of both bacteria and eukaryotes: reported values include 73 S for fungal mitochondria (consisting of 50 S and 38 S subunits) and 55 S for mammalian mitochondria (40 S and 30 S subunits).

The rRNA components of ribosomal subunits can also be classified by approximate sedimentation velocities or by molecular weight as determined by electrophoretic mobility in polyacrylamide gels. Bacterial ribosomes contain three distinct rRNA species: 16 S rRNA in the 30 S subunit and 23 S and 5 S rRNAs in the 50 S subunit. The molecular weights of these molecules are 1.1×10^6 (23 S), 0.55×10^6 (16 S), and 0.04×10^6 (5 S), which correspond to polynucleotide chain lengths of about 3300, 1600, and 120 residues, respectively. Similar molecular species are found in all prokaryotes examined, as well as in chloroplasts. In some organisms, however, there are indications that the large (23 S) rRNA does not exist in ribosomes as a single polynucleotide chain, but instead as two smaller molecules. The RNA is apparently cleaved during or after ribosomal maturation, and the cleavage products remain noncovalently associated. The utility to the organisms of cleaving the molecule remains unclear. Similar cleavage of the large rRNA has been reported for numerous eukaryotes.

Among eukaryotes, the small (40 S) ribosomal subunit usually contains an "18 S" rRNA component (0.75×10^6 daltons, about 2000 nucleotides), but the principal RNA in the large (60 S) subunit varies in size. In higher plants, ferns, algae, fungi, and some protozoa the large rRNA is about "25 S" ($\sim1.3 \times 10^6$ daltons), but in metazoa (where it is referred to as "28 S" rRNA) this component has apparently evolved with each major step in animal evolution, and ranges from $\sim1.4 \times 10^6$ daltons in sea urchins to $\sim1.75 \times 10^6$ daltons (about 5500 nucleotides) in mammals. In addition to this 25–28 S rRNA, the large subunits of eukaryotic ribosomes contain two smaller RNA components: a 5.8 S species (about 160 nucleotides) which is intimately associated with 28 S rRNA, and a 5 S species (about 120 nucleotides). Mitochondrial rRNA's are not as well characterized, but in the case of mammalian mitochondria 12 S (0.36 × 10^6 daltons) and 21 S (0.56 × 10^6 daltons) rRNA's have been reported for the small and large subunits, respectively; they apparently lack 5 S rRNA.

An unusual feature of ribosomal RNA, both in prokaryotes and in eukaryotes, is its content of nucleoside modifications, especially methyl groups. These moieties are added to the ribosomal RNA after its transcription, during the maturation process (see below, Sections IV,B,4 and IV,C,3,c).

B. Ribosomal RNA Synthesis in Prokaryotes

1. Structure of Prokaryotic rRNA's

Many questions regarding the structures of the rRNA molecules and the processing steps involved in their synthesis and maturation will only be understood when the complete nucleotide sequences of the molecules are available. This is not a trivial task in the cases of 16 S and 23 S rRNA's, given their large sizes, but recent advances in DNA sequencing techniques make the task less formidable. Several laboratories have undertaken the determination of the primary structures of these molecules.

At this time, most of the completely known nucleotide sequences of prokaryotic rRNA's are those of the 5 S rRNA molecules of a variety of bacterial species, including *Escherichia coli* (Brownlee *et al.*, 1968), *Pseudomonas fluorescens* (DuBuy and Weissman, 1971), three *Bacillus* species (Pribula *et al.*, 1974; Marotta *et al.*, 1976), and two species of blue-green algae (cyanobacteria) (Corry *et al.*, 1974a, 1974b). The sequences of the 5 S rRNA's from this phylogenetically diverse group of bacteria differ substantially, but certain common features are conserved, including the occurrence of complementary sequences at the 5'- and 3'-ends, structurally capable of forming an antiparallel, hydrogen-bonded "stalk." They also contain the sequence CGAAC, which at least in principle could interact by hydrogen-bonding with the GTψCG sequence common to almost all tRNA molecules. Fox and Woese (1975) have proposed a model for the secondary structure of 5 S rRNA which is quite consistent with the known sequences, as well as being in agreement with most of the evidence obtained by physical and chemical methods. (The sequence of 5 S rRNA from *Bacillus subtilis* is shown in this arrangement in Fig. 11.) However, it must be noted that the secondary and tertiary structures of any of the rRNA molecules in solution may be different from their conformations when packaged within the ribosome in close proximity to potentially interacting proteins and the other rRNA molecules.

Due to their large sizes, the nucleotide sequences of 16 S and 23 S rRNA molecules have been difficult to determine. However, several investigators are in the process of sequencing those rRNA species from *E. coli*. Nucleotide sequences around methylated sites in 23 S and 16 S rRNA's of *E. coli* (Fellner and Sanger, 1968; Fellner, 1969) indicated that methylation is specific, occurring at only a few sites. Using [methyl-^{14}C]-labeled RNA's, unique fragments could be derived from each rRNA following treatment with ribonuclease T_1, suggesting that each molecule is homogeneous, at least at methylation sites. By using the sequencing techniques of Sanger with ^{32}P-labeled 16 S rRNA, the nucleotide sequence of about 95% of that molecule, distributed over several large parts of it, has been extensively studied (Ehresmann *et al.*, 1975). Noller and his associates (Brosius *et al.*, 1978) have recently determined the complete sequence of a 16 S rRNA gene from *E. coli* carried by a specialized transducing derivative of λ.

The available sequence data on ribosomal RNA's indicate several positions in the RNA chains which are variable. The extent of heterogeneity is in all cases confined to a small number of sites and most likely represents more or less independent evolution of multiple rRNA genes (see Section IV,B,2). For example, *E. coli* strain MRE600 produces two forms of 5 S rRNA, differing only at the thirteenth nucleotide residue (numbering from the 5'-terminus of the molecule), with one form having a G at that position and the other having a U instead (Brownlee *et al.*, 1968). It is unknown whether these minor structural variations within the rRNA populations of a given organism have any functional significance.

2. Organization of Bacterial Genes for rRNA

a. **Number of Genes and Linkage.** The bacterial genome contains several copies of the genes coding for each rRNA; all copies of a gene are nearly identical, with small differences, probably due to point mutations arising during independent evolution, accounting for sequence heterogeneities noted above. This redundancy of genetic information is unusual; it may have arisen to supply a large quantity of rRNA to fast-growing cells, but it is also possible that such redundancy reflects functional differences among the various gene copies. In contrast, ribosomal proteins are each represented by only one gene per chromosome. Because there is "translational amplification" of the ribosomal protein genes, each mRNA can direct the synthesis of many copies of a protein, while rRNA gene products must be utilized directly.

After extracting DNA from stationary phase cultures of *E. coli*, which should contain only one copy of the genome in each cell, Spadari and Ritossa (1970) found that the amount of 16 S and 23 S rRNA's which can hybridize to the DNA corresponds to between 5.7 and 7.4 genes. This suggests that there are about 6 or 7 genes for each rRNA in the genome. By using fragments of DNA of different lengths, they showed that there is a high extent of clustering of the rRNA genes. In *B. subtilis* the proportion of DNA saturated by hybridization with rRNA corresponds to about 10 genes for each rRNA. DNA–RNA hybridization saturation studies with the individual rRNA species have shown that the relative number of genes coding for each of the rRNA's is the same (Smith *et al.*, 1968; Pace and Pace, 1971).

In addition to being present in equal numbers, the genes for the 3 species of rRNA's have been shown to be closely linked. Colli *et al.* (1971) studied the arrangement of the 5 S, 16 S, and 23 S rRNA genes of *B. subtilis* by means of Cs_2SO_4–$HgCl_2$ density gradient centrifugation of rRNA–DNA hybrids. Single-stranded DNA binds mercuric ions more effectively than the DNA in DNA–RNA hybrids, and this differential binding permits separation of DNA hybridized to 16 S or 23 S rRNA from the remaining single strands. By using alkaline hydrolysis to degrade the RNA moiety of hybrids recovered from the gradient, DNA fragments hybridized with one rRNA species can be tested for their ability to hybridize with other rRNA species. When the DNA fragments used for hybridization are approximately 2 to 4×10^6 MW in size, fragments which had first bound 16 S rRNA always bind both 23 S and 5 S rRNA's also; fragments which had first bound 23 S rRNA can also bind both 16 S and 5 S rRNA's. But when the single-stranded DNA is degraded to fragments of MW about 1×10^6 (about the size of 23 S rRNA) prior to the initial hybridization, only about half of the fragments isolated by hybridization to 16 S can subsequently bind 23 S rRNA, and many fragments which had first bound 23 S rRNA cannot bind 16 S rRNA; this suggests that the 16 S and 23 S rRNA genes occur in pairs. In addition, the ability of 1×10^6 MW DNA fragments to bind 5 S rRNA is strongly related to their ability to bind 23 S rRNA rather than 16 S rRNA. A gene order of 16 S–23 S–5 S is consistent with these results.

b. Order of Transcription. In eukaryotes (see Section IV,C,2), coordinate synthesis of 18 S, 28 S, and 5.8 S rRNA molecules is guaranteed by the fact that the genes specifying each of these rRNA components are all part of the same "transcriptional unit." A transcriptional unit is a segment of DNA (containing one or many genes) bounded by transcription initiation and termination sites, which is read without interruption by an RNA polymerase molecule beginning RNA synthesis at the initiation site (Pace, 1973). The term "operon" is sometimes used interchangeably with "transcriptional unit," but it is actually appropriate only for those transcriptional units which contain a demonstrable operator locus (i.e., a repressor-sensitive site). The rRNA transcription units have not been shown to contain such loci and are therefore not operons in the strictest sense.

The first indication that prokaryotic rRNA genes were also part of such a transcription unit came from the work of Bleyman *et al.* (1969), who examined the sensitivity of the synthesis of rRNA's to actinomycin D. This antibiotic creates more or less random blocks to transcription, so that the sensitivity of a transcription unit to actinomycin D should depend on its length, and genes which are further away from the point of transcription initiation should have a greater risk of a block preventing transcription through them. In *B. subtilis,* the synthesis of 23 S rRNA is about twice as sensitive to actinomycin D as is 16 S rRNA synthesis, while 5 S rRNA synthesis was as sensitive as that of 23 S rRNA. It was proposed that the 5 S rRNA gene was part of a transcriptional unit with the 23 S gene, but that the 16 S rRNA gene was transcribed independently (although they noted that the data did not completely disallow individual transcription units of the order 16 S, 23 S, 5 S). The antibiotic rifampicin provides a better tool for transcriptional mapping because it specifically inhibits the initiation of transcription without influencing the completion of nascent RNA chains. Hence longer transcription units can continue to be transcribed for greater lengths of time after the addition of rifampicin to growing bacterial cells, because they possess a greater number of RNA polymerase molecules engaged in transcription. On the average, if there are "n" polymerases transcribing 16 S rRNA, there should be "$2n$" polymerases transcribing the 23 S rRNA gene (given the relative sizes of the RNA's), so that twice as much 23 S rRNA should be produced compared with 16 S rRNA after the addition of rifampicin. But if the two genes are cotranscribed, with the gene for 16 S rRNA preceding that for 23 S rRNA, then 23 S rRNA will be synthesized by all the polymerase molecules located on the 16 S rRNA gene as well as those located on the 23 S rRNA gene. A situation in which the 23 S rRNA gene precedes that for 16 S rRNA generates analogous predictions. By adding [^{32}P]orthophosphate together with rifampicin and measuring the label's relative incorporation into 16 S and 23 S rRNA molecules made in the presence of the antibiotic, Doolittle and Pace (1971) were able to show that the most likely arrangement of genes is as a single unit in the order 16 S–23 S–5 S, with transcription beginning at 16 S and proceeding through 23 S and 5 S without reinitiation.

This result is corroborated by the observations of other investigators. Kossman *et al.* (1971) starved a "stringent" strain of *E. coli* for a required amino acid (such strains respond to the deprivation of a required amino acid by halting RNA synthesis), and then restored the amino acid to the bacteria, resulting in the synchronous initiation of rRNA synthesis in the culture. Following this synchronized rRNA synthesis by means of

[^{3}H]uridine incorporation, they showed that the first RNA polymerase molecules to initiate rRNA chains synthesize 16 S rRNA (as identified by RNA–DNA hybridization), and only later do sequences of 23 S rRNA appear: based on the time of appearance, the polymerase which transcribes 23 S rRNA genes must first synthesize another large RNA sequence about the same size as 16 S rRNA. The data do not exclude the possibility that this latter RNA is some unknown molecule and not 16 S rRNA itself (if so, it would have to be rapidly degraded after its synthesis, because no such RNA in the required abundance is observed); however, the close physical linkage of the 16 S and 23 S rRNA genes argues against such an arrangement. Finally, as measured both by incorporation of [^{3}H]uridine (Bremer and Berry, 1971) and by direct electron microscopic observation of "ribosomal DNA" and the associated nascent rRNA chains (obtained by lysing osmotically fragile cells) (Hamkalo and Miller, 1973), the duration of 16 S and 23 S rRNA synthesis in *E. coli* after the addition of rifampicin is compatible with their respective genes being arranged in compound transcriptional units in the order 16 S–23 S.

    c. Map Location: rRNA Transcription Units and Associated tRNA Genes. Determination of the locations of the rRNA genes within the bacterial chromosome has been difficult, because the rRNA genes cannot be easily identified genetically (e.g., there are no known bacterial mutants in the rRNA genes analogous to the suppressor mutants of tRNA genes). Early studies examined saturation levels for rRNA hybridized to DNA prepared from *E. coli* strains which were merodiploid for a small region of the chromosome. An increase in saturation levels was taken as an indication of the presence of rRNA genes on the F' factor carried by the strain in question (Yu *et al.,* 1970; Gorelic, 1970; Birnbaum and Kaplan, 1971; Unger *et al.,* 1972). A difficulty of this approach is that since there are about seven genes for each rRNA, an increase in the number of gene copies by only one in the merodiploid might have been difficult to detect.

    Exploiting the natural sequence heterogeneity within the 5 S rRNA's of *E. coli* strains K12 and B, Jarry and Rosset (1973a,b) undertook to map 5 S rRNA genes. The absence of any detectable phenotype made precise mapping difficult; for identification of the type of 5 S rRNA molecules, it was necessary to isolate labeled 5 S rRNA from recombinants or transductants and to estimate the mole percentage of certain characteristic nucleotide sequences after fingerprinting the products of ribonuclease T$_1$ digestion. They were able to map four of the seven 5 S rRNA genes with some precision; later analyses have shown that 16 S and 23 S rRNA genes map in the same locations.

    More recent mapping studies have utilized segments of the *E. coli* chromosome carried by transducing phages and plasmids. Several specialized transducing phages, derivatives of λ and $\phi80$, have been isolated which carry complete or partial rRNA transcription units ("operons") from *E. coli;* these are analogous to the $\phi80p(su_3^+)$ transducing phage utilized in the analysis of tRNA^{Tyr} (see Section III,B,2). Use has also been made of plasmids from the collection constructed by Clarke and Carbon (1976), which consist of fragments of mechanically sheared *E. coli* DNA inserted into the ColE1 plasmid by *in vitro* recombinant DNA ("genetic engineering") techniques (for a discussion of such techniques see Chapter 11 of this volume). The phages and plasmids carrying rDNA (ribosomal DNA is the DNA encoding the rRNA sequences) were analyzed by means of RNA–DNA hybridization techniques and heteroduplex mapping. In this second technique, preparations of two related DNA molecules, which differ in some sequences (two plasmids carrying different insertions, for example), are denatured and then renatured together. Among the resultant molecules are heteroduplexes, or hybrids, in which a single strand of one parental molecule has paired with the complementary single strand of the other parent; they are comprised of duplex regions, which identify sequences present in both parental molecules, and single-strand regions of nonhomology, which represent sequences unique to one of the two parental molecules (e.g., deletion or insertion "loops" and substitution "bubbles"). Such structural features can be recognized and physically mapped when the renatured preparation is spread for electron microscopy (Davis *et al.,* 1971). These studies confirmed that *E. coli* contains seven rRNA "operons," each containing a promotor and genes coding for 16 S, 23 S, and 5 S rRNA's. The results are summarized in Table II; for a more detailed description of these mapping studies see reviews by Nomura (1976) and Nomura *et al.* (1977).

In the course of the studies described above, it was discovered that tRNA genes are located in the spacer regions between the genes for 16 S and 23 S rRNA's (Wu and Davidson, 1975; Lund *et al.,* 1976). On the basis of heteroduplex analyses of rRNA transducing phages and hybrid plasmids carrying rRNA "operons," the rRNA transcription units of *E. coli* can be divided into two classes with respect to spacer regions (Kenerley *et al.,* 1977). Hybridization analyses with bulk tRNA's, charged with individual ^{3}H-amino acids, confirmed the existence of two different spacer tRNA gene arrangements between the genes for 16 S and 23 S rRNA. Among the rRNA operons, three had spacer regions encoding both tRNA$_{1B}^{Ala}$ and tRNA$_1^{Ile}$, while the other four had spacer genes for tRNA$_2^{Glu}$ only (Morgan *et al.,* 1977). It should be noted that such hybridization experiments alone cannot distinguish isoaccepting species of tRNA's. However, tRNA's hybridized to DNA fragments containing the spacer region and parts of the 16 S and 23 S rRNA genes (generated by *in vitro*

TABLE II

Ribosomal RNA Transcription Units in *E. coli* K12[a]

Transcription unit designation	Chromosomal location	Known genes in transcription unit					
		16 S	tRNA(s)	23 S	5 S	tRNA(s)	
rrnA	85 min	*rrsA*	(*tilA, talA*)	*rrlA*	*rrfA*		
rrnB	88 min	*rrsB*	*tgtB*	*rrlB*	*rrfB*		
rrnC	83 min	*rrsC*	*tgtC*	*rrlC*	*rrfC*	*ttrC*	*tasC*
rrnD	71 min	*rrsD*	(*tilD, talD*)	*rrlD*	*rrfD*		
rrnE	89 min	*rrsE*	*tgtE*	*rrlE*	*rrfE*		
rrnF	74 min	*rrsF*	unknown	*rrlF*	*rrfF*		
Group I	Unmapped	*rrs*	(*til, tal*)	*rrl*	*rrf*	*tas*	
Group VI	Unmapped	*rrs*	*tgt*	*rrl*	*rrf*		

[a] Each rRNA transcription unit (designated *rrn*) contains genes encoding the indicated RNAs: *rrs* (16 S rRNA), *rrl* (23 S rRNA), *rrf* (5 S rRNA), *tgt* (tRNA$_2^{Glu}$), *til* (tRNA$_1^{Ile}$), *tal* (tRNA$_{1B}^{Ala}$), *tas* (tRNA$_1^{Asp}$), or *ttr* (tRNA^{Trp}). Group I and group VI designate rRNA transcription units which have been isolated but not mapped; they have been shown to have chromosomal locations different from *rrnA* through *rrnE*, but one may prove to be identical to *rrnF*. The gene for tRNA$_2^{Glu}$ was previously designated *gltT*, that for tRNA^{Trp} was called *trpT*, and the two tRNA^{Asp} genes were called *aspT* and *aspU*; the designations used in this table reflect the genes' location within rRNA transcription units, and permit a distinction between redundant genes. The relative order of the spacer tRNA$_1^{Ile}$ and tRNA$_{1B}^{Ala}$ genes has been determined for at least one (as yet unidentified) transcription unit, as *rrs til tal rrl rrf*. The distal tRNA^{Thr} is probably associated with the *rrnD* transcription unit (E. A. Morgan, personal communication). (Adapted from Nomura *et al.*, 1977.)

restriction endonuclease treatment, and identified by heteroduplex analyses and hybridization with both 16 S and 23 S rRNA's) can be eluted and subjected to fingerprint analysis. Such analysis shows that the particular tRNA species encoded by spacer genes are the three noted above; no other spacer tRNA's have been identified to date.

Additional tRNA genes have been found to be associated with some rRNA transcription units. Heteroduplex analyses of restriction fragments, and RNA–DNA hybridization analyses, indicate that these tRNA's ("distal tRNA's") are not encoded by the spacer region, but by genes at the promotor-distal ends of the operons in question (i.e., to the 3'-side of the 5 S gene). Experiments with various hybrid plasmids, some lacking regions of the rRNA "operons," demonstrated that the expression of the distal tRNA genes required the rRNA promotor, and rifampicin run-out experiments (i.e., determination of the relative duration of synthesis after the addition of rifampicin) indicated that the distal tRNA genes were further from the promoter than the 5 S rRNA genes (Morgan *et al.*, 1978). These results, and the work of Lund and Dahlberg (1977), strongly suggest that these distal tRNA genes are cotranscribed with genes for ribosomal RNA's and spacer tRNA's. Transfer RNA's identified as distal tRNA's are tRNA$_1^{Asp}$, tRNA^{Trp}, and a tRNA^{Thr} species. Not all rRNA operons seem to have distal tRNA genes, and there is no apparent correlation between kinds of spacer tRNA genes and kinds of distal tRNA genes (e.g., tRNA$_1^{Asp}$ genes are associated with both types of spacers). Included in Table II are the tRNA genes known to be associated with rRNA operons.

The basic structure of an rRNA transcription unit (in *E. coli*) can be represented as: promotor–16 S rRNA gene–spacer tRNA gene(s)–23 S rRNA gene–5 S rRNA gene [distal tRNA gene(s)]. Hence the post-transcriptional processing of rRNA's, as discussed in the following sections, may overlap with that of tRNA's. The utility to the organism of cotranscribing genes for rRNA's and these selected tRNA's, thereby producing these RNA species in a coordinate fashion, is unknown. It is also unknown whether a similar situation exists outside of *E. coli*.

3. Posttranscriptional Processing

a. **Hypothetical Compound Transcript.** In eukaryotic cells the transcriptional units encoding the rRNA molecules are composed of one gene each for 18 S, 28 S, and 5.8 S rRNA's. The immediate product of transcription is a large molecule (45 S in mammalian cells) containing the sequences of all three rRNA's, plus excess material which is removed and discarded during maturation events (see Section IV,C). As described above, the rDNA of bacteria is also arranged in compound transcription units composed of one gene each for 16 S, 23 S, and 5 S rRNA's [and tRNA(s)]; any RNA polymerase molecule initiating transcription of the 16 S gene presumably continues on to produce one molecule each of the 23 S rRNA, 5 S rRNA, and the associated tRNA(s). But unlike the situation in eukaryotes, no large precursor RNA has yet been isolated from normally growing bacterial cells. Instead, the precursors identified thus far for 16 S and 23 S rRNA's are only slightly larger than the respective mature molecules. These precursors are designated p16 and p23 rRNA's according to the nomenclature of Hecht and Woese (1968), and the mature RNA species are m16 and m23 rRNA's; precursor and mature 5 S rRNA's are designated p5 and m5, respectively.

Two reasons have been suggested for the failure to detect a compound rRNA transcript in bacteria. The first is that, as the RNA polymerase molecules pass from the 16 S gene into the 23 S gene (or the spacer region), they release the completed p16 molecules before beginning synthesis of p23. Alternatively, an endonuclease might cleave the p16 molecule from the nascent RNA chain soon after the polymerase completes transcribing that part of the transcription unit. The existing evidence strongly supports the second model.

Pettijohn *et al*. (1970) have isolated intact DNA complexes from *E. coli;* such "nucleoids" retain the RNA polymerase molecules and nascent RNA strands which are associated with the DNA at the time of cell disruption. These RNA polymerase molecules, upon addition of the four nucleoside triphosphates, complete the transcription of the transcription unit with which they are associated, releasing the completed RNA chains. Synthesis occurs as an extension of nascent RNA chains in the isolated complex, and new chains are not initiated. When the RNA's completed *in vitro* are analyzed by polyacrylamide gel electrophoresis, two major species are resolved. One, designated "large p23," is somewhat larger than the previously mentioned precursor to 23 S rRNA; the other, with an electrophoretic mobility equivalent to a 30 S RNA molecule, is termed p30. The p30 RNA species is about the size of a molecule with the combined molecular lengths of p16 and the "large p23" molecule. The *in vitro* completion yields no p16-sized RNA. DNA–RNA competition hybridization experiments showed that the "large p23" material contained m23 rRNA sequences, while the p30 RNA contained both m16 and m23 rRNA sequences; p30 is apparently a transcript of much of the rRNA transcription unit. The presence of 5 S RNA and tRNA sequences in these molecules was not examined. If nascent RNA is pulse labeled *in vivo* immediately prior to isolation of nucleoids, and subsequently completed *in vitro* with unlabeled substrates, the label appears in both classes of released molecules; however, when p30 RNA's were examined for distribution of the pulse label, it was predominantly in m16 sequences. Since 16 S rRNA sequences are found only in p30 RNA, which is completed *in vitro*, and these m16 sequences are initiated *in vivo*, the m16 sequences must be in the 5' portion of the p30 molecule. Few p30-associated 23 S sequences are labeled *in vivo*, so most of the p30 RNA must be the product of polymerase molecules associated with the 16 S gene at the time of nucleoid isolation, which *in vitro* continue to read through the 23 S gene, and presumably the 5 S and tRNA genes.

One interpretation of the *in vitro* production of a compound transcript of the rDNA is that the release of p16 from the RNA polymerase as it moves into the 23 S gene is effected by a specific endonuclease, which is soluble and is removed from the nucleoids during their preparation. (Alternatively, a specific conformation or organization of a processing complex could be lost or altered during nucleoid preparation.) Thus all the 16 S sequences completed *in vitro* are part of p30 RNA because there is no endonuclease present to cleave the p16 from the nascent chains. The absence of this enzyme or an analogous one may also explain the observation that the *in vitro* generated "p23" is larger than the p23 rRNA appearing *in vivo*. The *in vitro* "p23," and p30 as well, may contain the 5 S sequences—and perhaps spacer tRNA sequences—in addition to the p23 sequence. *In vivo* most of the p23 rRNA molecules are probably cleaved from the nascent RNA chain soon after the polymerase enters the 5 S gene, although the labeling kinetics of 5 S rRNA indicate that some may pass through a p23–p5 intermediate (Pace, 1973). This interpretation predicts that a mutant defective in this endonuclease could be isolated—if a specific functional handle on one of the rRNA's is available (and this function is encoded by only one gene).

b. Immediate Precursors to 16 S and 23 S rRNA's. The kinetics of accumulation of stable RNA species (rRNA's and tRNA's) in *E. coli* reveals that all such RNA's are formed from precursors by posttranscriptional processes (Pace *et al.*, 1970). Because the processing reactions in prokaryotes follow the transcription event much more rapidly than in eukaryotes, precursor forms of the ribosomal RNA's are less readily observed. However, when the RNA of growing cells is pulse labeled, the labeled RNA components of ribonucleoprotein precursors of ribosomes are found to be somewhat different from the mature 16 and 23 S rRNA's (Osawa *et al.*, 1969; Nierhaus *et al.*, 1973). These immature rRNA molecules, when compared to their mature counterparts, sediment more rapidly and have lower electrophoretic mobilities on polyacrylamide gels. This suggests that they represent longer polynucleotide chains, and that the maturation of the precursors to mature rRNA's involved a size reduction (Hecht and Woese, 1968). That these larger RNA molecules (p23 and p16) are, in fact, precursors to mature 23 S and 16 S rRNA's (m23 and m16) was demonstrated by analysis of the kinetics of radioisotope incorporation into p16 and p23 relative to m16 and m23, and by the observation that maturation—the accumulation of m16 and m23 and the concomitant loss of p16 and p23—occurs even when further RNA synthesis is blocked by the addition of actinomycin D to the cells (Adesnik and Levinthal, 1969). In addition, a precursor species of 5 S rRNA (p5) was observed, and kinetic analysis of its appearance showed that it is not derived from p23 or p16. Although most of the investigations of prokaryotic rRNA processing have involved *E. coli* or *B. subtilis*, similar precursor molecules have been observed in a number of other prokaryotes (Pace, 1973), implying that the processing of rRNA is essentially the same in many prokaryotes.

The precursor forms of rRNA's are detectable not only in pulse-labeling experiments, they accumulate under any conditions which prohibit protein synthesis without affecting RNA production. Antibiotic inhibitors of protein synthesis such as chloramphenicol (Adesnik and Levinthal, 1969) prevent rRNA maturation, as does amino acid starvation of auxotrophic strains carrying the *rel* mutation. [The ability of cells to respond to amino acid deprivation, by greatly reducing stable RNA synthesis, is governed at least in part by the *rel* locus. Wild-type (*rel*$^+$) strains are designated as "stringent," whereas *rel* strains, in which RNA synthesis continues unabated after removal of a required amino acid, are termed "relaxed."] This is not due to an artificial situation brought on by starvation of a mutant which has lost the ability to regulate RNA synthesis; the same accumulation of precursors occurs in *rel*$^+$ strains when starved for a required amino acid (Chang and Irr, 1973), although RNA synthesis occurs at a much reduced rate in such situations. Among other conditions which interfere with the production of functional proteins and prevent the maturation of p16 and p23 in *E. coli* are the addition of fluorouracil to cultures, treatment of cells with toluene or with cobalt chloride, and potassium depletion (see Pace, 1973). The prevention of rRNA maturation by inhibition of protein synthesis could be interpreted as implying that the processing enzymes turn over rapidly, so that in the absence of their continued production they are unavailable for cleavage of the precursors. However, the precursor rRNA's also accumulate under nonpermissive conditions in *E. coli* mutants which are conditionally defective in ribosomal assembly [*sad* (subunit assembly defective) mutants] (Guthrie *et al.*, 1969; Nashimoto and Nomura, 1970). These mutants are incapable of assembling ribosomal subunits at restrictive temperatures, but protein synthesis is not immediately affected. One interpretation of these results is that prevention of rRNA maturation by inhibiting protein synthesis is a consequence not of stopping processing enzyme synthesis, but rather of stopping ribosomal protein synthesis, thereby preventing the assembly of ribosomal subunits. According to this interpretation, processing enzymes apparently do not act upon naked p16 and p23, but upon specific rRNA–ribosomal protein aggregates.

In *E. coli* and *B. subtilis* the p16 and p23 molecules are both about 100–200 nucleotides longer than their mature counterparts. Several groups of researchers have isolated the p16 rRNA from *E. coli*, either from pulse-labeled cells, chloramphenicol-treated cells, or ribosome assembly mutants, and have compared the ribonuclease T$_1$ fingerprints of these molecules with that of m16 (Brownlee and Cartwright, 1971; Hayes *et al.*, 1971; Lowry and Dahlberg, 1971; Sogin *et al.*, 1971). The excess sequences appear to be associated with both the 5'- and 3'-termini of the precursor, and no 5'-nucleotide triphosphates were observed, implying that p16 is itself processed from a larger primary transcript. Tang and Guha (1975) reported the existence of a 5'-triphosphate terminus (pppA-) on p16, but they isolated the RNA from spheroplasts, in which they noted that rRNA processing was very inefficient—p16 and p23 accumulated, but no m16 or m23 were observed; their p16 may represent a precursor to the p16 observed by other investigators. Because of their large sizes, p23 and p16 have not yet been completely sequenced, and investigators are, therefore, somewhat limited in relating these species to the maturation process.

c. **Enzymes Involved in Processing p16 S and p23 S rRNA's.** There is relatively little detailed information regarding the maturation nuclease(s) involved in m16 and m23 formation. Corte *et al.* (1971) and Yuki (1971) observed that a temperature-sensitive (*ts*) mutant of *E. coli* which produced a thermolabile RNase II was incapable of producing mature ribosomes or mature rRNA's at the restrictive temperature. *In vitro* treatment with cell-free extracts (or with ostensibly purified RNase II) from the mutant promoted the conversion of purified p16 RNA to material with the gel mobility of m16, but no conversion was observed if the extract was heated. The interpretation was that RNase II cleaved p16 to m16, and its inactivation led to a block in rRNA maturation. However, observations by Weatherford *et al.* (1972) contradict this conclusion. They showed that *ts*+ transductants of the mutant still produced the thermolabile RNase II, but were able to grow normally at elevated temperatures, carrying out normal rRNA synthesis and maturation. In studies involving direct enzyme assays, Dean and Sykes (1974) found no correlation between the level of RNase II activity and the accumulation of p16 RNA.

Using one of the *ts*+ strains with a thermolabile RNase II, however, Apirion and his colleagues (Meyhack *et al.,* 1974; Apirion, 1975) have reported the partial purification of an activity which they designate RNase M (maturase). *In vitro,* RNase M converts the p16 rRNA in a ribosomal precursor particle to a product with the electrophoretic mobility of m16 rRNA, but has no affect on p23-containing precursor particles. Initial purification steps yielded material with high RNase II activity, but heating the preparation reduced RNase II activity to 0.4% relative to the unheated preparation, without decreasing the RNase M activity—verifying the genetic analyses which indicated that RNase II was not involved in p16 maturation. Fractionation on a Sephadex G-150 column separated the RNase M activity from the remaining RNase II activity and from contaminating RNase III activity. The strain from which RNase M was isolated is deficient in RNase I and polynucleotide phosphorylase (PNPase), so these activities can also be ruled out as possible identities of RNase M. Hayes and Vasseur (1976) have also reported the partial purification of an enzyme, possibly the same as RNase M, which converts p16 in precursor particles to a product with the electrophoretic mobility of m16; fingerprint analysis revealed that while the product had the 3'-terminus of m16, its 5'-terminus was not that of the mature molecule. Thus two enzymes are implicated in the maturation of p16: RNase M to cleave extra residues from the 3'-terminus and another, unknown, enzyme to process the 5'-terminus. The 3' maturation activity (RNase M) has the substrate specificity for ribonucleoprotein particles which *in vivo* results had predicted—treatment of free p16 rRNA leads to extensive degradation without significant formation of m16 rRNA (Hayes and Vasseur, 1976). The activity necessary for 5' processing has been detected in crude extracts only, in which the large size of p16 rRNA renders it extremely sensitive to nonspecific degradation (see Hayes and Vasseur, 1976; also Lund and Dahlberg, 1977). Aside from this *in vitro* specificity, it should be noted that RNase M and the other activities have not been shown to have any physiological significance for rRNA production. There also exists the possibility that there is more than one enzyme capable of carrying out a given reaction; the existence of such auxiliary enzyme systems is a limitation in any attempt to assign enzyme functions.

d. Precursor to 5 S rRNA and Processing of p5 S rRNA. As would be expected from discussions above, the involvement of posttranscriptional cleavage steps in the maturation of rRNA is not limited to 16 S and 23 S rRNA's: the mature 5 S rRNA components of all prokaryotes so far examined are also derived from larger precursors (Pace, 1973). Since 5 S rRNA is small in size (120 nucleotides in *E. coli,* 116 in *B. subtilis*), the details of its maturation are more amenable to experimental determination; its size also makes it less susceptible to nonspecific degradation, thereby facilitating the process of maturation enzyme isolation and purification.

The p5 rRNA of *E. coli* is identifiable in pulse-labeled cells and accumulates in the absence of protein synthesis (Forget and Jordan, 1970; Feunteun *et al.,* 1972); upon restoration of protein synthesis, p5 is converted to m5 rRNA. As isolated, p5 rRNA is a mixed population of molecules that are only one, two, or three nucleotides larger than the mature molecule, and all the additional residues are located at the 5'-terminus. The 5'-terminal T_1 oligonucleotides are $p5_I$, pAUUUG; $p5_{II}$, pUUUUg; $p5_{III}$, pUUG; and m5, pUG. Kinetic studies indicate that during maturation, the extra nucleotides are removed in a stepwise fashion until the final m5 rRNA length is achieved, the above sequences reflecting intermediates (Feunteun *et al.,* 1972). Since similar p5 RNA's have been detected in other organisms, there is the possibility that the maturation scheme found in *E. coli* may be fairly common among prokaryotes (Pace, 1973). Unfortunately the scheme is so simple (probably sequential exonucleolytic cleavages) that it is not useful as a model system for understanding the processing events involved in the production of the high molecular weight rRNA's.

However, in *Bacillus* species a more complex situation exists, bearing

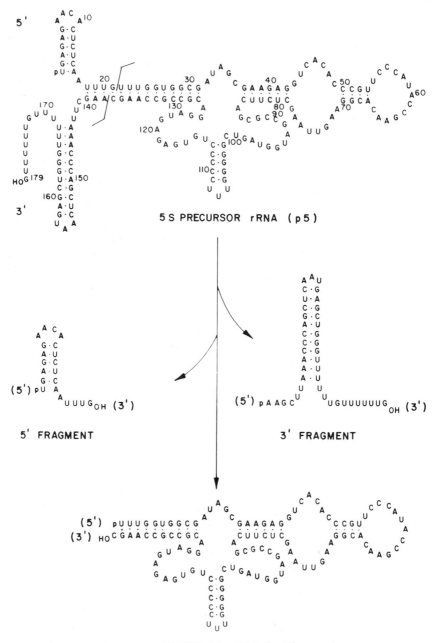

5 S PRECURSOR rRNA (p5)

5' FRAGMENT

3' FRAGMENT

MATURE 5S rRNA (m5)

more resemblance to the scheme of p16 maturation observed in *E. coli*. In pulse-labeling experiments, or in the absence of protein synthesis, three precursors of 5 S RNA have been identified in *B. subtilis* by Pace and his co-workers (Pace *et al.*, 1973; N. R. Pace, unpublished results). The largest of these, $p5_C$, is about 240 nucleotides in length; $p5_A$ contains 179 nucleotides (Sogin *et al.*, 1976), and $p5_B$ is composed of about 150 nucleotides. The relatedness of $p5_A$, $p5_B$, $p5_C$, and m5 rRNA is indicated by fingerprint analyses of the products of RNase T_1 and RNase A digestions (Pace *et al.*, 1973; Pace and Sogin, 1975; N. R. Pace, unpublished results). All three precursors have excess sequences (as in p16 of *E. coli*) associated with both their 5'- and 3'-termini, relative to m5. Kinetic analyses indicate that the three species, accumulated in the absence of protein synthesis, are rapidly and independently converted to m5 upon restoration of protein synthesis; no precursor species is an intermediate between other precursors and the mature molecule. The complete nucleotide sequences of m5 (Marotta *et al.*, 1976) and $p5_A$ (Sogin *et al.*, 1976) have been determined (see Fig. 11), and the other precursor species have also been examined. Because the sequences of these three precursor molecules are different, they are apparently the products of different genes. The 3'-termini of $p5_A$ and $p5_C$ can be written with helical regions followed by $U_{5-6}Pu_{OH}$, a feature indicative of transcription termination (similar pyrimidine-rich sequences, terminating in a purine, are associated with termini of other *in vivo* and *in vitro* RNA species) (see Bertrand *et al.*, 1975; Küpper *et al.*, 1978). On the other hand, the 3'-terminus of $p5_B$ consists of only 10–12 nucleotides beyond the m5 sequence, and lacks the $U_{5-6}Pu_{OH}$; this suggests the possibility that other genes in the $p5_B$-containing rRNA transcription units lie adjacent to the 3'-termini of the 5 S genes—perhaps distal tRNA genes, analogous to those in *E. coli*.

In the absence of protein synthesis, some p5 is slowly converted to m5, suggesting that the nuclease(s) responsible could act upon the free RNA. Encouraged by that observation, Pace and his co-workers have isolated and extensively purified a specific endonuclease responsible for the maturation of p5 (Pace and Sogin, 1975; Sogin *et al.*, 1977), employing $p5_A$ and $p5_B$ as substrates. Termed RNase M5 (maturation of 5 S), it cleaves the 5' and 3' precursor-specific segments from the p5 molecules. Most of their

Fig. 11. Structure of 5 S precursor RNA (p5), and the reaction catalyzed by RNase M5 of *B. subtilis*. Complementary bases in the mature 5 S rRNA (m5) sequence, and in the mature component of the precursor sequence, are paired to form the secondary structure suggested by Fox and Woese (1975) for prokaryotic 5 S rRNA's. Hydrogen bonds are denoted by dots; internucleotide phosphate bonds are omitted. Arrows indicate the points of scission by RNase M5 within p5 RNA to generate m5 rRNA and the 3'- and 5'-fragments. (Adapted from Sogin *et al.*, 1976, and Schroeder *et al.*, 1977.)

work has focused on $p5_A$, from which 21- and 42-nucleotide long fragments are cleaved from the 5'- and 3'-termini, respectively, to yield m5 RNA. The basic reaction is shown in Fig. 11. The same nuclease excises both the 5'- and 3'-terminal segments, and the time course of the *in vitro* accumulation of the products, mature 5 S rRNA (m5), the 3'-terminal fragment, and the 5'-terminal fragment, suggests that both segments are excised from $p5_A$ without intermediary release of the substrate (Sogin *et al.*, 1977). RNase M5 consists of two components, α and β; their specific roles are unknown, but both are required for the cleavage event.

Some work has been done on identifying the features of p5 with which RNase M5 interacts, using as substrates reconstructed precursor molecules lacking one or the other precursor-specific segments (Meyhack *et al.*, 1977). Molecules lacking the 3'-segment are still excellent substrates (as determined by excision of the 5'-segment), but those lacking the 5'-segment are 90% diminished in susceptibility of the 3'-segment to RNase M5 action. Further experiments (N. R. Pace, personal communication) utilizing partially synthetic substrates constructed using RNA ligase indicate that the entire 5'-segment is not necessary; the covalent addition of U_3G to the 5'-end of molecules lacking the 5' precursor-specific segment is sufficient to restore full substrate capacity. Furthermore, the addition of C_6G instead of U_3G as a synthetic "precursor-specific" segment largely restores susceptibility of the substrate to RNase M5. Finally, the covalent addition of U_3G to the 5'-end of *mature* 5 S rRNA was found to generate an excellent substrate for RNase M5. Thus it seems that the major substrate features required by the enzyme lie within the sequence and/or conformation of the mature component of the molecule.

e. RNase III and Cleavage of the Primary Transcript. In normal bacterial cells the nascent rRNA transcript, either free or associated with ribosomal proteins, is apparently cut during transcription to produce the immediate rRNA precursors; hence, although the rRNA genes are organized as a compound transcription unit, no *in vivo* analogue of the *in vitro* p30 S compound transcript (Pettijohn *et al.*, 1970; see Section IV,B,3,a) is observed. However, in RNase III-deficient mutants of *E. coli* a large (~2×10^6 daltons) RNA species which is the expected size for such a primary transcript transiently exists (Dunn and Studier, 1973; Nikolaev *et al.*, 1973; Nikolaev *et al.*, 1974). [Immediate precursors and mature forms of 23 S, 16 S, and 5 S rRNA's are also produced, as are additional novel transient RNA species designated 25 S (about 3600 to 3900 nucleotides) and 18 S (about 2100 bases).] Hybridization competition experiments revealed that this molecule, designated 30 S RNA, does contain 23 S and 16 S rRNA sequences. Patterns of label incorporation into various RNA species were interpreted as indicating that 30 S RNA is the initial rRNA precursor in these strains; 30 S rRNA was shown to have a 5'-triphosphate (pppA-), supporting its identity as an initial transcription product (Ginsburg and Steitz, 1975; Hayes et al., 1975). Upon in vitro treatment with purified RNase III, 30 S RNA is converted to molecules the size of p16 and p23 rRNA's plus several low molecular weight RNA fragments (Dunn and Studier, 1973; Nikolaev et al., 1973; Ginsburg and Steitz, 1975). In the presence of chloramphenicol, RNase III-deficient strains accumulate 30 S RNA in bulk quantities (Schlessinger et al., 1974; Ginsburg and Steitz, 1975), in addition to p23 and p16 rRNA's (Hayes et al., 1975), permitting the isolation of sufficient 30 S RNA to further investigate its apparent in vitro processing by RNase III. This in vitro processing is competitively inhibited by added double-stranded RNA (Dunn and Studier, 1973) and is totally inhibited by the intercalating agent ethidium bromide (Nikolaev et al., 1975). These observations are consistent with the idea that the recognition and cleavage of 30 S RNA by RNase III involves some sort of double-stranded secondary structure in the RNA molecule. In light of this, the observation that the 16 S and 23 S genes are preceded and followed by short self-complementary sequences ("inverted repeats"), which permit formation of single-stranded loops on short double-stranded stems (Wu and Davidson, 1975), is interesting. Presumably nascent 30 S RNA transcripts could also form such secondary structures, the stems of which could be attacked by RNase III.

The products of the in vitro cleavvage of 30 S RNA have been examined by RNase T_1 fingerprint analyses, as have the in vivo immediate precursor molecules (p16 and p23) observed in RNase III-deficient cells during pulse-chase experiments and in the presence of chloramphenicol (Ginsburg and Steitz, 1975; Hayes et al., 1975). The p16-sized molecule from in vitro cleavage of 30 S RNA is essentially the same as the in vivo p16, and their fingerprints resemble those published for p16 from chloramphenicol-treated wild-type cells and from ts ribosome assembly mutants (see Lowry and Dahlberg, 1971, and Section IV,B,3,b); i.e., they contain about eight oligonucleotides absent from fingerprints of m16, and lack the 3'- and 5'-terminal oligonucleotides of the mature molecule. The p23 from RNase III-deficient cells is somewhat larger than that from wild-type cells; it contains all of the m23 oligonucleotides except the 3', and possibly the 5', terminal oligonucleotide(s), and contains additional sequences not present in m23. The p23-sized product of RNase III-treated 30 S RNA was apparently not examined; 25 S RNA, which also accumulates transiently in RNase III-deficient strains, was extensively treated with RNase III to yield a product similar, but not identical, to m23 rRNA (Hayes et al., 1975). Among the smaller RNase III-generated products of 30 S RNA, a fragment about 300 nucleotides long contained the 5 S rRNA sequences, although it lacked the 5'-terminus of m5 (Ginsburg and Steitz, 1975); this fragment is much larger than the p5 observed *in vivo*. Using crude cell-free (S-30) extracts to process 30 S *in vitro*, Lund and Dahlberg (1977) demonstrated that 30 S RNA also contains spacer and distal tRNA's.

Thus, in normal cells, RNase III is the endonuclease involved in specifically cleaving the elongating nascent rRNA transcript to produce the immediate monocistronic precursors to mature rRNA species. In the RNase III-deficient mutants certain novel RNA molecules are observed, including the uncleaved rRNA transcript, 30 S RNA (which contains 16 S rRNA, 23 S rRNA, 5 S r RNA, spacer tRNA, and distal tRNA sequences). These mutant strains are fully viable and are able to produce the mature rRNA's despite the absence of any detectable RNase III activity (presumably by means of auxiliary RNases). On the basis of labeling patterns, rifampicin chase experiments, and kinetic studies, Nikolaev and his colleagues have suggested that the 30 S RNA is the *in vivo* precursor to the monocistronic precursor species (Nikolaev *et al.*, 1973; also Dunn and Studier, 1973). They suggested that the 30 S RNA is slowly processed by residual RNase III activity (or a second enzyme system) and that chloramphenicol treatment resulted in bulk accumulation of 30 S RNA by directly or indirectly inhibiting that residual activity. Schlessinger *et al.* (1974) reported that if RNase III-deficient cells were labeled with [³H]uridine in the presence of chloramphenicol, and then washed and resuspended in fresh drug-free medium containing no label, the label would be lost from 30 S RNA and would appear (or increase) in the 16 S and 23 S fractions. Unfortunately, their data as presented to not permit the stoichiometry of the supposed processing to be determined; i.e., from their figure, it is unclear whether the label in 30 S RNA is sufficient to account for the label in m23 and m16 at the later time. The ''rifampicin-chase'' experiments of Nikolaev *et al.* (1973) (pulse labelings in which further incorporation of label was blocked by a subsequent rifampicin treatment, so that the fate of labeled RNA's could be followed) are also difficult to interpret, since no data on the amount of label in the 23 S or 16 S rRNAs are given at early times following the addition of the drug; the lack of data is attributed to ''the apparent presence of immature rRNA precursors and mRNA with . . . mobilities'' (Nikolaev *et al.*, 1973) similar to those of the mature rRNA's.

The above experiments were carried out using the RNase III-deficient strain AB301-105 (initially designated AB105), isolated by Kindler *et al.* (1973). Its isolation from the immediate parental strain A19 (wild type for RNase III activity) involved nitrosoguanidine mutagenesis, which is known to induce multiple closely linked mutations due to its mode of action at the growing fork of replicating DNA (Drake and Baltz, 1976). Apirion and co-workers have extensively investigated AB301-105 and various RNase III-deficient strains derived from it. They have shown that AB301-105 differs from A19 by at least eight mutations (Apirion and Watson, 1974), including a mutation which affects RNA metabolism by somehow altering the stability of various RNA molecules (Apirion and Watson, 1975a,b). However, the observed qualitative effects upon rRNA metabolism [i.e., the appearance of 30 S, 25 S, "p23" (large p23), and 18 S RNA's] are correlated with the RNase III mutant locus, designated *rnc*-105 (Apirion *et al.*, 1976a,b), as demonstrated with otherwise isogenic pairs of *rnc*+ and *rnc*-105 strains. They have examined the kinetics of labeling of the various rRNA species produced in RNase III-deficient cells, using "cleaned up" derivatives of AB301-105 in which the *rnc*-105 allele has been separated (by transduction, reversion, or other genetic techniques) from some or all of the other mutations affecting growth and/or RNA metabolism (Apirion, 1975; Gegenheimer and Apirion, 1975). Their analyses indicate that, contrary to the hypothesis of Nikolaev and his colleagues, 30 S is not a major precursor to the 16 S and 23 S rRNA's synthesized by these strains; they concluded that the 16 S and 23 S rRNA's are cut from the elongating nascent transcript, possibly by an alternative pathway not involving RNase III.

This claim has not yet been confirmed by other workers, but the conclusion that processing of at least some rRNA precursors can occur during transcription even in RNase III-deficient cells is supported by observations of Hofmann and Miller (1977). By gently lysing cells and preparing their contents for transmission electron microscopy, they were able to visualize individual rRNA cistrons in an *rnc* strain "caught in the act" of transcription. The pattern of nascent rRNA fibrils was a combination of uncleaved transcripts which reached the length of 30 S RNA, and of transcripts which had been cleaved during transcription, apparently at a site between the 16 S and 23 S sequences, and were significantly shorter than the fibrils which contained 30 S RNA.

Studies of the residual activity in RNase III-deficient cells which is active against double-stranded RNA do not support the contention that it is RNase III; its activity profiles with respect to pH and temperature differ from those of RNase III from *rnc*+ cells (Apirion *et al.*, 1976a,b). It has apparently never been assayed for specific cleavage of 30 S RNA; if it were the enzyme responsible for cutting 30 S RNA, as suggested by Nikolaev and his co-workers, such a specificity might be expected *in vitro*. While there is no argument with the observation that RNase III can specifically cleave 30 S RNA *in vitro*, there is no evidence that it does or can cut the complete transcript *in vivo*. In the case of RNase P cleavage of a tRNA precursor, Schmidt and McClain (1978) have shown that the *in vitro* action of that enzyme does not fully reflect the *in vivo* events in the maturation pathway. It may be that in the normal metabolic state, transcription and precursor RNA processing are coupled in an organized fashion, contributing to a stepwise specificity of RNA processing. If a nascent elongating molecule is not cleaved, it may be released as the 30 S RNA— possibly a "dead end" molecule which is not processed, but rather eventually degraded.

4. Processing Pathway for rRNA's: A Model

Given the organization of rRNA genes within transcriptional units, the data on precursor RNA's and processing intermediates, and what little is known about the enzymes involved, a model scheme for the processing of the primary rRNA transcript can be developed (at least for *E. coli*). Apirion and his colleagues (Gegenheimer *et al.*, 1977; Apirion and Lassar, 1978) have proposed such a processing "map," based in part on further kinetic analyses of rRNA labeling in wild-type and mutant cells; Fig. 12 is adapted from their model. The processing of precursors of rRNA is a complex set of events involving a fairly large number of enzymatic cleav-

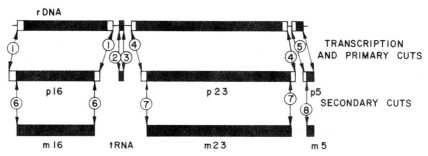

Fig. 12. Proposed processing pathway for prokaryotic ribosomal RNA (processing in wild-type strains). The first line shows the transcriptional sequence of the rRNA genes. Thick solid bars represent mature RNA's; thick open bars represent precursor sequences removed in secondary processing; and thin lines represent sections of the primary transcript discarded during primary processing. Each "cut" (which may involve several actual cleavage events) is given a separate number, as described in the text. Spacer tRNA's are collectively represented by a single tRNA species; distal tRNA's are not represented, but are presumably excised by similar events. Cuts 1 and 4, ribonuclease III. Cuts 2 and 3, ribonuclease(s) involved in tRNA excision and processing; may generate mature tRNA's, or precursors to be further processed. Cut 5, "ribonuclease E"; its existence is discussed in the text. Cut 6, "ribonuclease M16," a generic term for the enzymatic system that matures p16 rRNA; the two cuts are probably the result of separate enzymes, as discussed in the text. Cut 7, "ribonuclease M23," a generic term for the enzymatic system that matures p23 rRNA. Cut 8, "ribonuclease M5," by analogy with *B. subtilis;* in *E. coli* it may be a less specific exonuclease. (Adapted from Gegenheimer *et al.*, 1977.)

ages. These cleavages can be classed as primary or secondary processing events: primary processing events occur during transcription, cleaving the elongating nascent rRNA transcript to generate the immediate precursors; the processing of the immediate precursors to yield mature rRNA species comprises the secondary processing events.

The normal steps in processing the nascent transcripts are proposed to be as follows (see Fig. 12). When RNA polymerase finishes transcribing the 16 S sequences, RNase III cleavage (cut 1) releases p16 RNA and a 200 base fragment containing the 5'-terminus. After the polymerase moves through the tRNA and 23 S sequences and into the region preceding 5 S, a pair of cleavages (cut 4) release p23 and the spacer region. Precursors to the spacer tRNA's are excised either before or after release of the spacer region. At least two endonucleolytic cuts (cuts 2 and 3) are required to release the tRNA sequences from the spacer region; additional cuts would be necessary in spacers with more than one tRNA. Finally, p5 sequences are released (cut 5) from the remaining fragment of the transcript, which may contain distal tRNA's as well. The enzyme involved in this last cut is apparently not RNase III, since the time course of appearance and maturation of p5 RNA, and its size, seems identical in *rnc*$^+$ and *rnc* cells. The putative enzyme responsible for this cut between the 23 S and 5 S sequences has been designated "RNase E" by Apirion and his co-workers; they have reported the isolation of a conditional lethal mutant which apparently affects this enzyme, such that the cleavage does not occur at nonpermissive temperatures (Apirion and Lassar, 1978). Other (secondary) processing events yield m16, m23, and m5 rRNA's.

In RNase III-deficient (*rnc*) mutants, cuts 1 and 4 are not made; instead, the transcript is apparently processed by the remaining primary and secondary processing enzymes or others. The enzymes involved in tRNA processing that introduce cuts 2 and 3 (see Section III,B,4) might separate 16 S from 23 S sequences by cutting in the spacer region. RNase P apparently participates in processing of the rRNA transcripts to yield spacer tRNA's, because in *rnc*$^+$ cells with a thermosensitive RNase P, precursors to all known spacer tRNA's accumulate at restrictive temperatures (T. Ikemura, personal communication to Nomura, cited in Nomura *et al.*, 1977). The *in vitro* data of Lund and Dahlberg (1977) support the conclusion that rRNA processing can continue using the remaining enzymes, although their studies involved *in vitro* cleavage of 30 S RNA, as opposed to nascent elongating transcripts. They reported that a cell-free extract prepared from AB301-105 could be used to "process" 30 S RNA, yielding most of the small RNA's seen when wild-type cells are used to prepare the extracts, including p5 rRNA and the spacer tRNA's. (They do not report data on larger RNA's resulting from treatment with the RNase III-deficient extracts, i.e., whether or not species resembling 18 S, "p23 S", or 25 S are generated.)

The primary processing of rRNA in RNase III-deficient strains results in molecules which differ in size from those of wild-type strains (e.g., "p23" versus p23). Nonetheless, in mutant strains some of these molecules are trimmed to generate mature rRNA's in functional ribosomes. It thus seems that the sites of secondary processing cleavages can be recognized and properly cut in molecules which contain sequences in addition to those of the normal precursor. On the other hand, the processing of the nascent rRNA transcript may proceed by an ordered sequence of cleavages, such that a growing transcript from which 16 S sequences are not removed is destined to become a 30 S molecule. Perhaps those sequences somehow prevent cut 5 from occurring.

It should be emphasized that this model is only a working hypothesis and probably simplifies the situation. Some of the enzymes involved are only postulated, and there may be unknown auxiliary pathways. While the 30 S RNA evidently arises when none of the processing cleavages are made, the origin of the other novel RNA species observed in *rnc* mutants is unclear. In at least some genetic backgrounds the RNA's in the 23–25 S region of the gel profile and those in the 16–18 S region appear to be heterogeneous in size and do not clearly resolve into separate bands (see Apirion, 1975; Ginsburg and Steitz, 1975; Gegenheimer *et al.*, 1977). Lund and Dahlberg (1977) showed that RNA's in both regions, from AB301-105, yielded spacer tRNA's upon treatment *in vitro* with cell-free extracts; perhaps this reflects incomplete conversion of the larger to smaller size molecules in the absence of RNase III. It is not known whether these larger RNA's (e.g., 25 S) are also processed *in vivo,* or whether they represent dead ends in terms of processing. Apparently contradicting results among various investigators, with respect to these aspects of rRNA processing, are difficult to resolve; this results from differences in strains used, methods of RNA preparation and separation, and method of identification of the various RNA species.

5. Posttranscriptional Nucleotide Modifications

In addition to cleavage events, posttranscriptional metabolism of rRNA's includes the modification of a few specific nucleotides in the 16 S and 23 S rRNA's; the 5 S rRNA lacks modified nucleotides completely (Brownlee *et al.*, 1968). All the modified nucleosides known to occur in bacterial rRNA's, except one [5-ribosyluridine or pseudouridine (ψ)], are methyl substituted. The nucleosides are methylated by methyl groups transferred from S-adenosylmethionine to the polynucleotide acceptors after transcription. The majority of detailed studies of methylated nucleosides in rRNA from prokaryotes have focused on *E. coli,* in which the methyl-substituted nucleosides are of two general types: those containing methylated bases and those possessing a methyl substitution at the 2'-*O*-position on the ribose moiety of the nucleoside. About 1% of the total rRNA nucleotides are methylated in *E. coli,* the majority (~90%) of those being base substituted (Dubin and Günlap, 1967; Fellner and Sanger, 1968; Nichols and Lane, 1966). Figure 4 shows the structures of some methylated nucleosides.

The methylations are associated with particular sequences (Fellner and Sanger, 1968; Sogin *et al.,* 1972). Most methylated nucleotides and the sequences in their immediate vicinity (e.g., the ribonuclease T_1-generated oligonucleotides which contain the modified residues) are conserved in the 16 S rRNA's of phylogenetically diverse prokaryotes which share relatively little sequence similarity otherwise (Sogin *et al.,* 1972). The conservation of these highly specific methylations presumably indicates some function, although the nature of that function is unknown; at least some methylations seem inconsequential with respect to cell growth under laboratory conditions, in that their presence or absence has no apparent effect on cell growth.

The methylation of *E. coli* 16 S rRNA seems to occur in two stages: the first during the formation and lifetime of p16, and the second apparently concomitantly with the size transition of p16 to m16 (or immediately preceding or following that transition). When pulse-labeled p16 RNA from growing cells, or the p16 RNA which accumulates in cells treated with chloramphenicol or in *rel* strains starved for an essential amino acid (other than methionine, the methyl donor), is examined, it contains only about 20–50% of the normal m16 complement of methyl groups (Sypherd, 1968; Dubin and Günlap, 1967). In addition, the methylated nucleosides in such molecules are not representative of the total present in m16 RNA; only specific sequences are found to be methylated (Sogin *et al.,* 1971, 1972; Hayes *et al.,* 1971), and those sequences may be up to 80–100% methylated relative to the same sequences in m16. These methylations are carried out in the absence of protein synthesis, so the methylating enzymes responsible must be able to recognize the free polynucleotide chain. However, Dahlberg and his co-workers (Lowry and Dahlberg, 1971; Dahlberg *et al.,* 1975) report little if any methylation in p16 from chloramphenicol-treated cells or ribosomal assembly defective mutants at nonpermissive temperatures, or in the p16 portion of the 30 S molecules which accumulates in RNase III-deficient strains. Specifically they find no indication of the methylated oligonucleotide $m^4Cm^{2'}$-C-m_2C-C-G, which Sogin *et al.* (1971, 1972) observed in both *E. coli* and *B. megaterium* p16 RNA's at about 50% the m16 level. The reason for this discrepancy may lie in the particular strains used by the different groups or in differences in procedures (e.g., method of preparing p16 RNA).

The second stage of methylation, during which the remaining methylated nucleosides characteristic of m16 rRNA are formed, occurs only during active protein synthesis; restoration of a required amino acid to starved *rel* cells apparently allows completion of methylation of at least some of the undermethylated rRNA which accumulates during starvation (Sypherd, 1968). Presumably the enzymes responsible for these final methyl substitutions require the ribonucleoprotein precursors of the ribosomal subunits as their substrates.

The modification of the two large rRNA's is not coordinate, and 23 S RNA sequences may be methylated in reactions accompanying transcription. Unlike p16 RNA, the p23 RNA from chloramphenicol-treated cells or amino acid-starved *rel* cells contains most of the m23 complement of methylations, with only the ribose $2'-O$-positions being specifically undermethylated (Dubin and Günlap, 1967; Dahlberg *et al.*, 1975). In RNase III-deficient cells, the 30 S RNA contains all or most of the 23 S methylations, but little if any of the 16 S methylations (Dahlberg *et al.*, 1975).

There have been some studies of specific rRNA methylases, and several of these enzymes have been partially purified and characterized *in vitro*. Gordon and Bowman (1964) reported that submethylated rRNA, in the form of ribosomal precursor particles from methionine-starved *rel* cells, is capable of accepting methyl groups *in vitro*. They found that S-adenosylmethionine was the immediate methyl donor, but made no attempt to isolate the enzymes involved.

Sipe *et al.* (1972) have extensively purified the enzyme from *E. coli* responsible for N^6-methyladenine (m^6A) formation in rRNA. The enzyme utilizes S-adenosylmethionine as a donor and methylates a limited number of adenine moieties in methyl-deficient rRNA (isolated from methionine-starved *rel met* cells) from *E. coli*, but not in mature rRNA from the same strain. Although the adenine residues methylated *in vitro* have not been rigorously shown to be within the same sequences as in mature rRNA, the methylase appears from their result to be quite specific in its action.

Mutants of *E. coli* which are resistant to the antibiotic kasugamycin (which acts on the 30 S ribosomal subunit) do not have the sequence m_2^6A-m_2^6A-C-C-U-G in their 16 S rRNA. This sequence somehow participates in binding the drug, and its absence confers the resistant phenotype (Helser *et al.*, 1971). The methylase responsible for m_2^6A formation has been partially purified by Helser *et al.* (1972); it is absent from kasugamycin resistant strains, which define a genetic locus, *ksgA*, associated with the methylase. *In vitro*, the enzyme uses S-adenosylmethionine as a methyl donor to specifically dimethylate the appropriate two adjacent adenine residues in 16 S rRNA from kasugamycin-resistant strains of *E. coli*. The 16 S rRNA from normal (sensitive) cells is not an acceptor; it already possesses the dimethylated sequence. Unlike the methylase described by Sipe *et al.* (1972), the dimethylase functions only with rRNA–ribosomal protein complexes; purified rRNA is not an effective substrate. Thammana and Held (1974) have studied this requirement further *in vitro*, using 16 S rRNA from *ksgA* (resistant) cells, in combination with various ribosomal proteins, as a substrate for the methylase. They were able to identify specific 30 S subunit ribosomal proteins required for the dimethylation, and others which inhibited the reaction. By comparison with the *in vitro* pathway of ribosome assembly, they ascertained that the methylation occurs at an intermediate stage of assembly and is inhibited at late stages.

A similar situation exists in strains of *Staphylococcus aureus* which are induced to erythromycin resistance and in constitutively-resistant derivatives of them. In this case the resistance is correlated with an N^6,N^6-dimethyladenine (m_2^6A) residue in the 23 S rRNA of resistant cells; it is the sensitive (wild-type) cells which lack the modified nucleotide. The amount of m_2^6A in the 16 S rRNA remains the same in sensitive and resistant cells, indicating that separate methylases may be involved in the formation of m_2^6A in 16 S and 23 S rRNA's (Lai and Weisblum, 1971; Lai *et al.*, 1973).

Björk and his colleagues have developed a screening procedure for mutants with defects in RNA methylation, in which they extract the RNA from mutagenized cells and assay it for methyl acceptor activity *in vitro*, using a wild-type cell-free extract as a source of methylases (Björk and Isaksson, 1970; Björk and Kjellin-Stråby, 1978). Among the mutants isolated by this technique was a strain carrying the mutation *rrmA*, the 23 S rRNA of which completely lacks m^1G, although the m^1G levels in tRNA's are unaffected (16 S rRNA does not contain this particular modification even in normal cells). Isaksson (1973a) partially purified a rRNA(m^1G)-methylase from wild-type cells, and showed it to be missing from *rrmA* mutants. The methylase does not require a ribonucleoprotein complex as a substrate and methylates free 23 S rRNA from *rrmA* cells, but not 16 S rRNA; it can also methylate the 23 S rRNA in 50 S subunits and complete 70 S ribosomes, presumably because the site it recognizes and acts on is accessible even in the intact ribosome (Isaksson, 1973b). A rRNA(m^2G)-methylase was also purified, and mutants apparently defective in the formation of m^6A and m^5C residues in 23 S rRNA have been isolated.

The enzymes involved in methylating the $2'$-O-positions on ribose moieties of polynucleotides have also proved to be amenable to *in vitro* study. Nichols and Lane (1968), using crude extracts of *E. coli*, demonstrated the transfer of methyl groups from S-adenosylmethionine to the precursor rRNA components from methionine-starved *rel* cells to form $O^{2'}$-methylguanosine ($Gm^{2'}$) and $O^{2'}$-methylcytidine ($Cm^{2'}$). However, this system was not able to generate $N^4,O^{2'}$-dimethylcytidine ($m^4Cm^{2'}$) or $O^{2'}$-methyluridine ($Um^{2'}$). The enzymes responsible for these latter modifications may require rRNA–ribosomal protein aggregates from a later stage in ribosomal assembly, or the enzymatic activities may be lost during the preparation of the cell-free extracts.

There is very little information available regarding the formation of 5-ribosyluridine (pseudouridine) in rRNA; it is apparently not formed by the same enzyme as the ψ of tRNA anticodons. Chloramphenicol inhibits its formation in the rRNA of *E. coli* (Dubin and Günlap, 1967; Dahlberg *et al.*, 1975), implying pseudouridine formation occurs posttranscriptionally. In chloramphenicol-treated cells, 16 S rRNA is especially undermodified with respect to pseudouridine, and the responsible enzyme may require ribonucleoproteins as substrates.

C. Ribosomal RNA Synthesis in Eukaryotes

1. Gene Dosage and Organization

In eukaryotic organisms the "28 S" (25–28 S) and "18 S" (17–18 S) rRNA genes are present in clusters of hundreds to thousands of copies. (The terms 28 S and 18 S are used here to denote the general classes of rRNA; actual values vary for different organisms, and will be specified when necessary for discussion. In this general sense, 28 S will usually include the associated 5.8 S rRNA species, described below.) The rDNA in these clusters consists of multiple copies of repeating units, joined end-to-end to generate tandemly reiterated arrays. Each repeating unit consists of three basic elements: rRNA sequences, transcribed spacer sequences (sequences found in the precursor RNA's, but not in the mature species, including "leader" sequences), and nontranscribed spacer sequences. The nontranscribed spacer DNA presumably contains the control sequences—the signals for the initiation and termination of transcription—for the synthesis of the transcribed RNA's. The physical structures of rDNA repeating units from a variety of eukaryotes have been examined, using such techniques as DNA–RNA hybridization and restriction mapping. Construction of a restriction map requires a defined piece of DNA, such as a large restriction fragment containing the rDNA sequences. This is separately digested by various restriction endonucleases to generate sets of restriction fragments; these resultant fragments can then be ordered by comparison with the products of partial digestions and by further digestion with other nucleases of different site specificity.

These studies reveal that in some eukaryotes the nontranscribed regions are homogeneous in size, and in such an organism all repeating units may be similar, if not identical; in other organisms the nontranscribed sequences are heterogeneous in length. However, in all eukaryotes the transcribed sequences in any given organism are homogeneous with respect to length; it is not known whether they are also homogeneous with respect to sequence, or if they display sequence heterogeneity analogous to that observed in prokaryotes. An exception to this generality occurs in *Drosophila melanogaster,* in which some of the repeating units contain large intervening sequences dividing the 28 S rDNA into two blocks (see Glover and Hogness, 1977); it is not known, however, if these repeating units are physiologically active. Except for *Drosophila,* and some mitochondrial and chloroplast genomes (Bernardi, 1978), rDNA regions comprise transcriptional units of the general form: spacer–18 S rDNA–spacer–28 S rDNA. There is no evidence for the presence of tRNA genes analogous to those observed in *E. coli* rRNA transcription units.

2. Compound Precursor rRNA's (18 S, 28 S, and 5.8 S rRNA's)

It was in eukaryotic cells that precursor RNA's were first observed. Scherrer and Darnell (1962) observed a relatively short-lived high molecular weight RNA in HeLa cells pulse-labeled with [^{3}H]- or [^{14}C]uridine. In the sedimentation pattern of total RNA from such cells, a peak was observed emerging over a background of heterogeneous RNA; with increasing pulse length this 45 S peak (corresponding to a molecule of about 4.5×10^6 daltons) became more prominent, and then label began to appear in a 32 S component and in 18 S rRNA. After a still longer pulse, radioactive 28 S rRNA was observed. While this observation implied a precursor–product relationship between the 45 S RNA and 18 S and 28 S rRNA's within the cell, such relationships are difficult to establish on kinetic grounds alone; one complication arises from cellular pools of radioactive metabolites that are not easily "chased." The use of drug treatments allows investigators to label presumptive precursors, stop further labeling, and observe the fate of the labeled species. Brief treatment of HeLa cells with a low dose of actinomycin D, to block further transcription, reduces [^{3}H]uridine uptake by about 50%; nucleolar RNA synthesis (observed by radioautography) and 45 S RNA synthesis (assayed by zonal centrifugation) are almost completely stopped. In cells treated with actinomycin D after a short labeling, the label in the 45 S RNA quickly disappears; coincident with its disappearance, radioactivity appears in 32 S and 18 S RNA, and after a longer time label in the 32–28 S region gradually shifts to predominantly 28 S (Darnell, 1968).

a. **Intracellular Location of Precursors and Processing Intermediates.** When cellular components are fractionated and assayed as to RNA content individually, the results confirm the cytological data (radioautography, etc.), indicating that the nucleolus is the exclusive site of 45 S RNA synthesis. Sedimentation analysis of nuclear RNA reveals a peak at 45 S and a larger, polydisperse peak at about 30 S, containing the 32 S and 28 S RNA species. In pulse-chase experiments, as label disappears from 45 S RNA it simultaneously appears in the nuclear 32 S peak and the cytoplasmic 18 S peak; there is no appreciable 18 S rRNA in the nuclear fraction. The 32 S material undergoes a transition to 28 S, which then enters the cytoplasm (Penman, 1966). If the nuclear component of HeLa cells is further fractionated, a particulate fraction can be obtained which consists largely of organelles that resemble the nucleoli of intact cells, as determined by phase-contrast and electron microscopy (Penman *et al.*, 1966; Holtzman *et al.*, 1966). All precursors to cytoplasmic rRNA are associated with this fraction, while the surrounding nucleoplasm apparently contains 28 S rRNA, and 18 S rRNA is found exclusively in the cytoplasm. While the actual site of the transformation of 32 S and 28 S RNA is not indicated by the data, it is apparent that the event is relatively slow and occurs in the nuclear fraction. On the other hand, 18 S rRNA is apparently generated from 45 S RNA in the nucleolus and immediately transported to the cytoplasm.

RNA–DNA hybridization competition experiments (Jeanteur and Attardi, 1969; Quagliarotti *et al.*, 1970) indicate that nucleolar 45 S RNA contains one 28 S rRNA sequence and one 18 S rRNA sequence, comprising about 35–43% and 13–17%, respectively, of the molecule; the remaining 50–40% of the length is composed of non-rRNA sequences (transcribed spacers). The 32 S RNA is about 70% 28 S rRNA sequences and does not compete at all with 18 S rRNA. Examination of the methylation patterns of HeLa cell rRNA and its precursors yielded equivalent results. The fractionation patterns of alkali-resistant material (due to 2'-*O* methylations on ribose moieties) from 28 S and 18 S rRNA's are reproducible and distinct from one another. That of 45 S RNA is the equivalent of an equimolar mixture of 28 S and 18 S rRNA's, while the pattern of 32 S RNA is the same as 28 S rRNA alone (Wagner *et al.*, 1967).

When the RNA's are fractionated by polyacrylamide gel electrophoresis instead of sucrose gradient centrifugation, the increased resolution reveals several molecules intermediate in size between 45 S RNA and the mature rRNA's, with estimated molecular weights equivalent to 41 S, 36 S, 32 S, 24 S, and 20 S (Weinberg and Penman, 1970; Weinberg *et al.*, 1967). As characterized by methylation patterns, 41 S is a precursor to 28 S and 18 S rRNAs, 32 S is a precursor to 28 S rRNA, and 20 S is a precursor to 18 S rRNA. The 36 S and 24 S components were in much lower yield and were not further characterized; it was suggested that they might be generated when the order of cleavage steps was altered. These precursor–product relationships have been confirmed by fingerprint analyses of [^{14}C]methyl labeled RNA's (Maden et al., 1972).

In lower eukaryotes the initial precursor is smaller (36–38 S; 2.6 to 2.8 × 10^6 daltons), and close to 80% of the sequence is conserved during processing (i.e., consists of mature rRNA sequences); the spacer segments are shortened, and the 28 S rRNA is also smaller, relative to mammalian rRNA. Precursors and intermediates analogous to those in mammalian cells have been observed in a variety of eukaryotes, including amphibia (Landesman and Gross, 1969), insects (Edström and Daneholt, 1967; Greenberg, 1969; Dalgarno et al., 1972), plants (Rogers et al., 1970; Leaver and Key, 1970), a protozoan (Kumar, 1967), and yeast (Taber and Vincent, 1969; Retèl and Planta, 1970; Udem and Warner, 1972).

b. Structure of Precursor rRNA's. The structure of the immediate transcript, and hence the transcription unit, has been investigated for several organisms by Wellauer and Dawid (1973, 1975), using secondary structure mapping. In this technique, RNA is spread on grids for electron microscopy under partially denaturing conditions, resulting in a pattern of self-complementary loops (stabilized by a relatively high G · C content) which is highly reproducible for any given molecule. By comparing mature rRNA's, precursors, and intermediate molecules prepared in this manner, the arrangement of conserved and nonconserved (rRNA and spacer) regions can be determined, as well as information regarding the maturation pathway. All the precursor RNA's examined showed analogous arrangements of the sequences: 28 S–transcribed spacer–18 S–transcribed spacer.

Attempts to determine the 5' → 3' polarity of the rRNA transcriptional units have yielded conflicting results, but the bulk of the evidence favors the arrangement placing the 28 S rRNA sequence near the 3'-OH end of the precursor, as indicated in Fig. 13. Dawid and Wellauer (1976) investigated the polarity using well-characterized DNases (the 3'-specific exonuclease III of E. coli and the 5'-specific exonuclease of bacteriophage λ) to partially digest identifiable restriction fragments of Xenopus laevis rDNA. The resulting double-stranded rDNA fragments with single-stranded tails were separately assayed for hybridization with 18 S and 28 S rRNA, and the results were interpreted in terms of polarity of the coding strand of rDNA. These data were consistent with the location of the 28 S rRNA sequences being distal to the promotor (and hence at the 3'-OH end of the precursor); a promotor-proximal location for the 28 S rRNA cistron would have yielded exactly opposite hybridization results and is therefore ruled out. This conclusion is also supported by experiments based on visualization of growing transcripts on rRNA genes that were cleaved with a restriction endonuclease prior to spreading for electron microscopy (Reeder *et al.*, 1976) and by restriction mapping and partial DNA sequencing of an rDNA repeating unit from yeast (Bell *et al.*, 1977).

In all eukaryotic organisms, the processing of rRNA (with the exception of 5 S rRNA, see Section IV,C,4 below) occurs almost entirely in the nucleolus, an organelle that forms at the chromosomal site containing the rRNA gene cluster; the number of nucleoli per nucleus depends on the extent of dispersion of the rRNA transcription units (i.e., upon the number of "nucleolus-organizing regions"). It is possible to isolate and purify nucleoli, so they can be used as a relatively pure source of rRNA precursors and maturation products, as well as a starting point in the identification of enzymes involved in the processing events.

3. Processing Pathway

a. **Cleavage Events at Specific Sites.** While the details of rRNA processing differ among various eukaryotes due to differences in the sizes of the initial transcript and the mature rRNA species, the basic features of the processing pathways are remarkably similar (Wellauer and Dawid, 1975; Winicov and Perry, 1975; Winicov, 1976). Most can be described in terms of four principal events involving cleavages at or near the sites numbered 1 through 4 in Fig. 13. The first cleavage (site 1) removes the 5'-terminal nonconserved segment corresponding to a leader sequence. While this terminal segment is presumably degraded, it is transiently observed in HeLa cells as a 24 S component. (This identification is supported by secondary structure patterns, kinetics of labeling, and composition.) In some primitive eukaryotes the initial cleavage might occur prior to completion of transcription, as normally occurs in bacterial cells. Variability in the timing of this cleavage could account for the discrepancies in the size of the initial rRNA precursor found in yeast (compare Udem and Warner, 1972; Retèl and Planta, 1970; Taber and Vincent, 1969).

The temporal order of cleavages at sites 2 and 3 varies among different organisms. In mouse L cells and *Xenopus laevis,* among others, the intermediate produced by cleavage 1 is cleaved at the 3'-end of the 18 S sequence (site 2) to generate 18 S rRNA and another intermediate (36 S, or 34 S in *Xenopus*) which is rapidly processed by removal of part of the transcribed spacer (cleavage at site 3) to yield 32 S RNA (30 S in *Xenopus*). In other organisms, typified by HeLa cells, the cleavage at site 3 occurs first, generating two smaller intermediates: a 20 S component, rapidly processed to 18 S rRNA (by site 2 cleavage), and a 32 S compo-

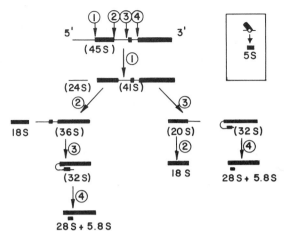

Fig. 13. Proposed processing pathway for eukaryotic ribosomal RNA. The first line shows the primary transcript of the rRNA genes, with numbered arrows indicating the approximate sites of the cleavage involved in the four principle processing events, as discussed in the text. Thick bars represent the mature rRNA sequences; thin lines, transcribed spacer sequences discarded during processing. The sizes (sedimentation values, in Svedberg units) of the primary transcript and the intermediate precursor molecules from mammalian cells are indicated in parentheses; the corresponding molecules from lower eukaryotes are smaller. The bend in the 32 S molecule is to indicate the noncovalent association which occurs between the 28 S and 5.8 S sequences. The temporal order of cleavages at sites 2 and 3 varies among different species, and even in the same species under different conditions. On the left is the order typified by mouse L cells, with site 2 cleavage preceding that at site 3. The opposite order, typified by HeLa cells, is shown on the right. Inset: As described in the text, 5 S rRNA is a primary transcription product. The thin line represents the DNA encoding the 5 S sequence; the thick bar represents the nascent 5 S rRNA transcript, with its associated RNA polymerase molecule (open circle). (Adapted from Perry, 1976.)

nent analogous to that observed in L cells (Wellauer and Dawid, 1975). In some cells, both orders can occur, the predominant pathway apparently determined by factors influencing the conformational state of the precursor RNA. In BHK-21 (Syrian hamster) cells, 45 S processing normally occurs as in HeLa cells, with the site 3 cleavage preceding that at site 2. A mutant cell line, *ts* 422E, exists which is temperature-sensitive for the production of 28 S rRNA, with the main defect in the conversion of 32 S RNA to 28 S rRNA (cleavage 4) (Toniolo *et al.,* 1973). The mutation also affects the temporal order of rRNA processing; at the nonpermissive temperature the L cell pathway predominates, as evidenced by the lack of 20 S RNA and the presence of the 36 S component (Winicov and Perry, 1975; Winicov, 1976). The existence of both alternate pathways in some cells argues for a commonality of cleavage sites.

The final trimming (cleavage 4) is often the rate-limiting step, leading to a substantial accumulation of the proximal intermediate (32 S RNA in mammals). This relative slowness may be due to the fact that at least two cleavage events are involved. In addition to the trimming of the "28 S" rRNA there is an event affecting a segment that is located between sites 3 and 4 (Spiers and Birnstiel, 1974; Maden and Robertson, 1974; Bell *et al.,* 1977). This segment, about 150–160 nucleotides long, is the 5.8 S rRNA component (previously designated 7 S, and lRNA). It is covalently linked to the product of the site 3 cleavage, and all larger (previous) precursors, and after site 4 cleavage it remains stably, noncovalently associated with the 28 S rRNA (Pene *et al.,* 1968; Udem and Warner, 1972; Maden and Robertson, 1974). It is difficult to determine whether these final cleavages occur simultaneously, or whether site 4 cleavage generates substrate for yet another cleavage event. A small fraction ($\sim$10%) of yeast 5.8 S rRNA is extended by 6 to 7 nucleotides at the 5'-terminus; these extended forms are stable and are apparently not precursor forms of the molecule (Rubin, 1973, 1974). Heterogeneity at the 5'-terminus has also been reported for other 5.8 S rRNA's (Nazar *et al.,* 1975; Nazar and Roy, 1978). These data suggest the possibility of an occasional variation in the exact location of the cleavage. A structurally related molecule about 250 nucleotides long, shown by sequence analysis to contain the 5.8 S rRNA sequence plus additional residues, has been isolated from pulse-labeled yeast protoplasts (Trapman *et al.,* 1975). It is noncovalently bound to a larger RNA in ribosomal precursor particles and may represent the immediate precursor to 5.8 S rRNA.

Due to its relatively simple structure, 5.8 S rRNA is a potential model for posttranscriptional processing of RNA in eukaryotes. Pace's group has begun a study of the 28 S–5.8 S complex and have concluded that it is not a simple hydrogen-bonded duplex; the association is apparently maintained by both secondary and tertiary structural interactions. They have isolated the "junction complex" from mouse rRNA as a ribonuclease-resistant complex (Pace *et al.,* 1977). It includes the 3'-terminal 42–43 nucleotides of the 5.8 S rRNA, but it is proposed that only the 3'-terminal 20–21 nucleotides are associated by hydrogen bonding with the 28 S rRNA. The remaining residues of the 5.8 S component of the junction complex are envisaged as forming an intramolecular hairpin which stacks upon, and hence stabilizes, the RNA double helix formed by the 28 S–5.8 S association. Whether this complex plays a role in processing of the rRNA precursor (such as by providing a specific site for an enzyme) or is involved in the assembly or function of the mature ribosome is unknown.

The 5.8 S rRNA is released from "28 S" rRNA under conditions which denature the complex; such conditions include heating during extraction or treatment with 8 *M* urea or 66% dimethyl sulfoxide (Pene *et al.,* 1968).

When the 26 S rRNA of most, but not all, insects is heated in this manner, it dissociates into the 5.8 S molecule and two molecules about the size of the 18 S rRNA (Shine and Dalgarno, 1973). Jordan (1975) demonstrated that in pulse-labeled *Drosophila* cells there is an intact 26 S rRNA which is further processed by a central cleavage prior to being transported to the cytoplasm. Jordan (1974) has also described a "2 S" RNA component noncovalently associated with 26 S rRNA and derived from it in a late step of processing. This 30-nucleotide long RNA is unique to *Drosophila* and is generated at the same time as 5.8 S rRNA; the "5.8 S" rRNA of *Drosophila* is only about 120–130 nucleotides long, and apparently in this organism the "5.8 S" and "2 S" RNA's together correspond to the 5.8 S molecules of other eukaryotes (Jordan *et al.*, 1976). The function of these additional cuts in the 5.8 S and 26 S RNA's is unknown.

All of the above processing events are more or less polar, proceeding from one end of the molecule to the other in an orderly sequence, rather than by multiple attack upon all possible cleavage sites. The 18 S RNA can apparently be cleaved from prematurely terminated transcripts (Siev *et al.*, 1969; Van den Bos *et al.*, 1971; Hackett and Sauerbier, 1975), but 5.8 S RNA apparently cannot be generated until after the cleavage at site 3 (Fig. 13) has occurred. [Wellauer and Dawid (1973) observed molecules that might have been generated by an initial cleavage at site 4, as well as molecules apparently generated by a cleavage event within the sequence of 28 S RNA; however, these were all in very low yield, and presumably arose during aberrant processing or during the extraction procedure.] Similar polarity of processing has been observed for multimeric tRNA precursors (see Section III,B,4), and in prokaryotic rRNA processing (see Section IV,B,4). Possibly this results from a sequential removal of steric constraints, with consequent exposure of new cleavage sites.

b. Enzymes Involved in Cleavage Events. The enzymes involved in rRNA processing are largely unknown, despite considerable effort in this area. Theoretical considerations, based upon the arrangement of the rRNA sequences in precursors and intermediates in the processing pathways, indicate the involvement of an endonuclease to produce 5'-phosphate and 3'-hydroxyl termini; exonucleolytic trimming may also be involved. End group analyses of 18 S rRNA's reveal almost universal 5'-pU and 3'-A_{OH} termini; possibly these evolutionarily conserved residues contribute to a specificity of cleavage at sites 1 and 2. On the other hand, nucleotide sequences contributing to specificity, if they exist, do not have to be at the cleavage site.

Toniolo *et al.* (1973) showed, by obtaining revertants capable of growth at the nonpermissive temperature, that the temperature-sensitive defect in 28 S rRNA production exhibited by the *ts* 422E line of BHK cells is due to a single mutation. Hence this system offers a potential for genetic manipulation of the processing events. However, the nature of the mutation has apparently not been investigated further, and it is unknown whether it involves an alteration in a specific nuclease or in some other component of the rRNA maturation system.

Since pre-rRNA processing nucleases would have to reside in the nucleolus, one approach to determining the enzymes involved has been to search in isolated nucleoli or nucleolar extracts for nuclease activities which are capable of cleaving pre-rRNA in a manner resembling the *in vivo* event. Partially purified intact nucleoli apparently have the capacity to carry on *in vitro* the initial stages of processing of rRNA precursors. When prelabeled nucleoli are purified and incubated under appropriate conditions to reduce nonspecific degradation, a slow "processing" of 45 S RNA is observed. The initial products appear similar in size and methyl group content to the physiological 41 S and 32 S RNA intermediates (Vesco and Penman, 1968). Exogenous 45 S RNA added to the nucleoli is similarly degraded to smaller molecules, while exogenous 32 S, 28 S, or 18 S RNA undergoes comparatively little degradation (Liau *et al.,* 1968). However, upon prolonged incubation the process becomes aberrant, and the products are further degraded. The presence of active nucleases which nonspecifically degrade the pre-rRNA components is a problem which plagues the few *in vitro* studies carried out thus far.

Attempts have been made to resolve the *in vitro* systems further, using precursor ribonucleoprotein particles or free 45 S RNA as substrates for partially purified enzyme preparations. Niessing and Sekeris (1970) reported the conversion of rapidly labeled high molecular weight RNA to 30 S material upon incubation with a partially purified nuclease present in nuclear ribonucleoprotein particles, but the nuclease was not further characterized. An exoribonuclease has been partially purified from isolated nucleoli (Lazarus and Sporn, 1967; Kelley and Perry, 1971); it has an apparent preference, on the basis of studies with defined substrates, for degrading single-stranded random-coil RNA's, as opposed to RNA's with a higher degree of secondary structure (Lazarus and Sporn, 1967), and methyl-deficient segments are apparently hydrolyzed at higher rates than methylated regions (Perry and Kelley, 1972). The exonuclease may play a subsidiary role in processing, involving a limited trimming of the products of endonucleolytic cleavage, or it may serve as a scavenger of the excised spacer segments.

An endoribonuclease capable of apparent rRNA processing in cell-free systems has been isolated and studied by several groups (Mirault and Scherrer, 1972; Winicov and Perry, 1974). The nuclease is capable of converting preribosomal particles containing 45 S RNA to particles with RNA's the sizes of 41 S, 32 S, 28 S, and 20 S nucleolar RNA's (as determined by polyacrylamide gel electrophoresis). Treatment of free 45 S RNA with a similar nuclease (Prestayko et al., 1972) results in the degradation of the RNA into fragments of about 10–16 S, while 28 S or 18 S RNA undergo less degradation. While the reaction products resemble the physiological intermediates in size, there is no data presented on methyl enrichment, end group analysis, or fingerprinting, and the size resemblance may be fortuitous. The enzyme has been characterized (Prestayko et al., 1973; Winicov and Perry, 1975), and its products, possessing 3'-phosphates, are not those expected on theoretical grounds. If it is involved in pre-rRNA processing, additional trimming must occur to generate the termini seen in mature molecules. The endonuclease is C-specific, and its mechanism of action involves a 2', 3'-cyclic phosphodiester intermediate; hence its activity is blocked by 2'-O-methylation of the substrate—it cleaves poly(AC) but not poly(ACm$^{2'}$).

Several other possible candidates for processing enzymes have been isolated (see Winicov and Perry, 1975, for a list of ribonucleases ascribed to the nucleus and/or nucleolus). However, there is a lack of conclusive evidence implicating the enzymes in rRNA processing; many of them were isolated using physiologically irrelevant test substrates, such as homopolynucleotides, and have not been tested using nucleolar RNA species.

Another approach, patterned after in vitro results with E. coli RNA's and RNase III (see Section IV,B,3,e above), has been to use purified E. coli RNase III to examine 45 S (mammalian) or 35 S (yeast) pre-rRNA's for regions susceptible to this double-strand-specific nuclease. Such regions do exist, and cleavage by RNase III is inhibited under conditions in which the conformation or stability of double-helical regions is altered (Nikolaev et al., 1975). Some agents which alter the properties of double-helical regions also inhibit in vivo processing (Snyder et al., 1971). The finding of a double-strand-specific endoribonuclease in eukaryotic cells (Robertson and Mathews, 1973) makes this an attractive possibility; however, it should be noted that in secondary structure mapping studies employing electron microscopy, none of the cleavage sites involved in pre-rRNA processing were located within a possible loop (Wellauer and Dawid, 1975).

c. Noncleavage Processing Events. Prior to cleavage, the eukaryotic pre-rRNA transcript undergoes at least two alterations, both of which are likely to affect subsequent processing: (1) the methylation of rRNA sequences and (2) the combination with proteins in characteristic ribonucleoprotein particles.

Modifications of nucleotides in eukaryotic rRNA's occur predominantly on the initial transcript, and appear to be confined to those rRNA sequences that comprise the final products (Wagner *et al.*, 1967; Maden *et al.*, 1972). Most of the methyl groups are on the ribose portion of nucleotide residues, and the rest are on the bases (Brown and Attardi, 1965; Lane and Tamaoki, 1969). In HeLa cells the 28 S rRNA contains 65 $2'$-O-ribose methylations (plus one in the 5.8 S rRNA) and five base methylations, all added at the 45 S stage. With the 18 S rRNA, 40 $2'$-O-ribose methyls are added at 45 S, but six base methylations occur at later stages, including four in the sequence G-A-A-C to yield G-m$_2^6$A-m$_2^6$A-C. At least one "late" 18 S base methylation occurs after the 18 S rRNA enters the cytoplasm as part of the 40 S ribosomal subunit (Maden *et al.*, 1972; Maden and Salim, 1974). The same general pattern of methylation is found in yeast, including "late" methylation of a 17 S rRNA sequence to generate two adjacent dimethyladenine residues (Klootwijk *et al.*, 1972). The late modification of the small subunit rRNA (16 S rRNA) in general is a characteristic of prokaryotic rRNA processing as well, including the formation of the same G-m$_2^6$A-m$_2^6$A-C . . . sequence (except in *E. coli ksgA* mutants; see Section IV,B,5 above). The significance, if any, of this evolutionarily conserved highly methylated sequence is unknown and, at least in *E. coli,* is dispensible.

Unlike prokaryotic rRNA's, the 18 S rRNA of some eukaryotes contains a "hypermodified" nucleotide. In Chinese hamster cells, the 18 S rRNA contains a single such residue per molecule, tentatively identified as 1-methyl-3-γ-(α-amino-α-carboxypropyl)pseudouridine (see Fig. 4) (Saponara and Enger, 1974). A similar or identical nucleotide has been identified in yeast 17 S rRNA and is apparently also present in 18 S rRNA's of chick fibroblasts, *Xenopus*, and *Drosophila* (Maden *et al.*, 1975); possibly this nucleotide is present in all eukaryotes.

Methylation seems required for proper cleavage of rRNA (Vaughn *et al.*, 1967), although how it is involved is not clear. The phosphodiester bonds on the $3'$-side of ribose-methylated nucleotides are totally resistant to cleavage by nucleases that produce $2'$, $3'$-cyclic intermediates; perhaps methylations are involved in protecting exposed bonds from detrimental cleavages.

Both ribosomal proteins and extra (additional) nonribosomal proteins are found associated with the initial rRNA precursor molecule in HeLa cell nucleoli, and the protein–RNA ratio of preribosomal particles decreases over the course of maturation (Kumar and Warner, 1972). The "extra" proteins are specific and are apparently reutilized, while the ribomal proteins enter the cytoplasm as part of ribosomal subunits. These nucleolar proteins may play a role in processing, perhaps by protecting the RNA against "wrong" cleavages. However, until such a role is demonstrated, the possibility cannot be dismissed that their association with rRNA may be an artifact of the isolation procedure. Preribosomal particles are digested to a limited extent by partially purified nucleolar RNases under conditions in which free RNA is almost totally degraded. In cells in which protein synthesis has been inhibited, as by addition of cycloheximide, pre-rRNA processing is incomplete (Craig and Perry, 1970; Willems *et al.*, 1969).

4. Processing of Eukaryotic 5 S rRNA

In lower eukaryotes, exemplified by yeast, the 5 S RNA genes are physically linked to the 35 S pre-rRNA sequences (and hence to the genes for 18 S, 5.8 S, and 28 S rRNA's) and interspersed with them in an alternating order (Rubin and Sulston, 1973; Kaback *et al.*, 1976). This arrangement has been confirmed by recent physical mapping studies and partial DNA sequencing of cloned yeast rDNA repeating units (Valenzuela *et al.*, 1977; Bell *et al.*, 1977), which indicate, however, that the 5 S RNA is transcribed from the opposite strand of the DNA relative to the 35 S pre-rRNA. In higher eukaryotes, the 5 S genes are completely unlinked with the other rRNA sequences (Brown and Weber, 1968; Aloni *et al.*, 1971), although they themselves may be arranged in a single continuous cluster in some organisms (Artavanis-Tsakonas *et al.*, 1977) in a manner analogous to the rDNA repeating unit structure. While the arrangement in yeast results in approximately equal numbers of 5 S and 35 S rRNA coding regions (about 150 coding regions each, per haploid genome), in the higher eukaryotes the 5 S sequences are present in large excess over the other rRNA genes—perhaps a 40-fold or greater excess in the case of *Xenopus laevis* (Brown and Weber, 1968).

Regardless of linkage to the other rRNA genes, 5 S RNA synthesis in all eukaryotes is independent of the other rRNA's (Perry and Kelley, 1968; Udem and Warner, 1972; Wegnez *et al.*, 1972); in at least some cases the 5 S sequences are transcribed by an entirely different RNA polymerase than the other rRNA sequences (Parker and Roeder, 1977). The utility to the cell of this distinction between 5 S rRNA, on the one hand, and the 18 S, 5.8 S, and 28 S rRNA's, on the other, is a mystery. These differences in synthesis and gene number between the 5 S rRNA and the other rRNA species suggest that the 5 S RNA may have another function, as yet unknown, in addition to its role as a ribosomal component.

Under normal metabolic conditions, no precursor to 5 S RNA is observed in eukaryotes, even in short pulse labelings. Kinetic analyses indicate that 5 S RNA appears labeled almost immediately upon the exposure of cells to labeled ribonucleosides. The presence of nucleoside di- and triphosphates at its 5′-terminus (Hatlen *et al.*, 1969) suggests that 5 S RNA is the initial transcript of the 5 S genes. A possible precursor to 5 S RNA was recently detected in heat-shocked *Drosophila* tissue culture cells (Jacq *et al.*, 1977). This 135-nucleotide long molecule differs from 5 S RNA in that it contains an additional 15–16 nucleotides beyond the 3′-terminus of the mature 5 S molecule (no additional residues occur at the 5′-terminus). This additional 3′-terminal sequence ends in a stretch of U residues, suggestive of a polymerase termination signal; a similar stretch is present in the sequence of the yeast 5 S gene, but it is not known if these are transcribed (Valenzuela *et al.*, 1977). In addition to the sequence data, the identity of this molecule as a precursor to 5 S RNA was indicated by pulse-chase temperature shift experiments, in which labeled "p5 S" could be chased into 5 S RNA upon recovery of the cells from the heat shock (Jacq *et al.*, 1977). The cellular location of this apparent processing was not determined; 5 S RNA is not synthesized in the nucleous, but it must eventually become associated with the preribosomal ribonucleoprotein particles found there.

The nucleotide sequences of several eukaryotic 5 S RNA's have been determined (e.g., Forget and Weissman, 1969; Hatlen *et al.*, 1969; Jordan *et al.*, 1974), and they can be compared with the sequences of 5.8 S rRNA's and prokaryotic 5 S rRNA's. Interestingly, although both of the small eukaryotic RNA's show some homology with the prokaryotic 5 S RNA, the 5.8 S RNA of yeast exhibits greater similarities to *E. coli* 5 S RNA than to yeast 5 S RNA (Erdmann, 1976). The 5.8 S molecules contain several G-A-A-Py sequences; as in the case of the G-A-A-C sequence of prokaryotic 5 S RNA's, these sequences are complementary to the G-T-ψ-C sequence common to almost all tRNA's involved in polypeptide elongation. On the other hand, eukaryotic 5 S RNA's contain the sequence Py-G-A-U, which is complementary to the A-U-C-G sequence of eukaryotic initiation tRNA's. Finally, in reconstitution experiments (noted in Erdmann, 1976) eukaryotic 5.8 S rRNA was found to interact specifically with *E. coli* 5 S RNA-binding proteins, while eukaryotic 5 S RNA does not interact with any *E. coli* 50 S ribosomal proteins.

Eukaryotic ribosomes contain two small rRNA components (5 S and 5.8 S), while prokaryotic ribosomes contain only one (5 S rRNA). Although the eukaryotic 5 S molecule is the same size (about 120 nucleotides) as the prokaryotic 5 S species, they differ in both sequences and mode of synthesis; as described above, it is the 5.8 S rRNA which in both respects more closely resembles the small prokaryotic rRNA. Thus, it seems likely that the 5.8 S rRNA species, and not the 5 S species, is the eukaryotic functional homologue of the prokaryotic 5 S rRNA.

V. DEGRADATION OF RNA

In addition to the processing events described in preceding sections, which are involved in the maturation of RNA species, the posttranscriptional metabolism of RNA includes the degradation of various RNA molecules. This degradative metabolism apparently serves to rid cells of those RNA's rendered useless to, and/or no longer required by, the metabolic machinery. It may also provide a control point for the regulation of the amounts of RNA's which accumulate in cells, in response to environmental and/or developmental signals (Bacon and Rosenberg, 1967; Cooper, 1973). In prokaryotes, the RNA's which normally undergo degradation include most, or all, mRNA's (Pato and Meyenberg, 1970), and precursor-specific fragments trimmed from rRNA and tRNA precursors during processing. Precursors not destined to become mature RNA's (mutant tRNA precursors, for example) are also subject to degradation. Although the so-called stable RNA's, including mature rRNA and tRNA, are subject to little degradation during conditions of steady state growth, they are slowly broken down and degraded when cells are under nutritional stress (e.g., starvation). In eukaryotes, all cellular RNA is eventually degraded, and the same thing is probably true for prokaryotes.

There is relatively little information available regarding the means by which cells carry out these degradative processes. The identities of the enzymes involved are for the most part unknown, as is the basis for the selectivity of the process. There must exist some mechanism for distinguishing the stable RNA's from the unstable molecules, and for selecting molecules destined for degradation. It is not known whether RNA molecules have predestined life spans during which they "wear out" and only then become subject to degradation, or whether they are selected at random from the total intracellular population.

Apirion and his colleagues (Apirion, 1975; Kaplan and Apirion, 1975a) have investigated the enzymology of starvation-induced RNA degradation in *E. coli* by analyzing the rate of degradation of total RNA in mutants deficient in various nucleases. Most of their work involved cells starved for carbon, but similar results were obtained under conditions of nitrogen, phosphorus, or amino acid starvation (Kaplan and Apirion, 1975b). Their results implicated RNase II with a major responsible activity in the degradative process, with PNPase possibly playing a minor role. However, when individual RNA species were examined instead of total RNA degradation, 16 S and 23 S rRNA's were observed to disappear even in cells in which the level of both exonucleases was reduced considerably. In such cells, small RNA molecules, about 4 S in size, were observed. It was concluded that rRNA degradation is initiated by endonucleolytic cleavages to produce these small fragments, which are subsequently degraded to mononucleotides by the exonucleases. (It should be noted that no homology between rRNA and the 4 S molecules has been demonstrated.) The identity of the endonuclease is unknown; on the basis of further mutant studies, it is neither RNase I nor RNase III.

Parenthetically, it must be noted that there exist several potential pitfalls in determining the enzymology of a process on the basis of mutant studies alone; these must be considered when interpreting the results of any such studies. The presence or absence of enzyme activity as measured *in vitro* may not accurately reflect the *in vivo* level of activity. In addition, auxiliary enzyme systems may exist which can assume the role of a relevant enzyme inactivated by mutation. Finally, mutant strains may contain other, unknown mutations in addition to the characterized ones, and these may contribute to the observed results.

The breakdown of stable RNA in *E. coli* during starvation is a relatively slow process, brought about by preexisting enzymes; no protein synthesis is involved (Apirion, 1975). Apparently, the rRNA in polysomes and monosomes is not accessible to the nucleolytic activities, and only free ribosomal subunits are subject to RNA degradation (the ribosomal proteins are not degraded—they seem to dissociate from the cleaved rRNA and become associated with the cell membrane, possibly to be reused later). Hence the stability of rRNA in *E. coli* may be at least in part due to protection while in the form of intact ribosomes, and factors controlling the equilibrium between ribosomes and free subunits may play an important role in rRNA stability. The process of stable RNA degradation is adaptive (i.e., bacteria adapt to starvation conditions) and is correlated with cell viability and the capacity to recover from starvation.

Ohnishi and Schlessinger (1972) have described a mutant which very rapidly degrades essentially all (80–100%) stable RNA when RNA synthesis is stopped in cultures at 42°C; under the same conditions the wild-type parental strain shows only 10–30% RNA breakdown. The increased degradation is specific for stable RNA's (mRNA breaks down at the normal rate in the mutant cells) and is apparently not an adaptive process: only 10% of the mutant cells (as opposed to 50% of wild-type cells) are capable of forming colonies at 30°C after the culture had been incubated for 1 hr at 42°C in the presence of rifampicin or actinomycin D. The rapid and specific degradation of stable RNA (predominantly rRNA) is manifested only in the absence of RNA synthesis (and therefore does not involve the production of mRNA encoding a new nuclease); rifampacin-resistant derivatives of the mutant are similar to wild type when treated with rifampicin at 42°C, but exhibit the rapid degradation if actinomycin D is used.

If certain inhibitors of ribosome function (including chloramphenicol, streptomycin, erythromycin, or neomycin) are added to cultures of the mutant at 42°C simultaneously with the rifampicin or actinomycin D, the degradation of stable RNA is inhibited—with dose–response curves essentially the same as those for the inhibition of protein synthesis. The stable RNA of streptomycin-resistant derivatives of the mutant is not "protected" by streptomycin, but can still be "protected" by chloramphenicol; the inhibition of stable RNA breakdown by these antibiotics thus seems to be directly related to their action upon ribosomes. The "protection" of stable RNA is less effective if the antibiotic is added at later times; if 20 min elapse between the beginning of rifampicin treatment and the addition of chloramphenicol, there is no inhibition of degradation. The "protection" is attributed to the stabilization of polysomes, oligoribosomes, and/or monosomes (i.e., to an inhibition of ribosomal subunit dissociation). It thus seems that, as with starvation degradation, the free ribosomal subunit is the initial substrate of rapid rRNA degradation.

The rapid breakdown of stable RNA has been examined genetically by Ohnishi and his colleagues, who showed that two genes were involved (see Ohnishi *et al.*, 1977). Both genes have been mapped, and it is interesting that while *srnA* (*srn,* stable RNA) is a chromosomal gene, mapping at 9.4 min on the *E. coli* genome, *srnB* is encoded by an F plasmid gene. The Srn⁻ phenotype requires both the chromosomal *srnA* mutation and the *srnB*⁺ gene of the F plasmid (which may be present either in the F⁺, F′, or Hfr state); F⁻ *srnA* strains are phenotypically Srn⁺, and F plasmid derivatives which lack the *srnB*⁺ region (defined by deletion mapping) cannot cause the Srn⁻ phenotype in *srnA* strains. Although defined genetically, the Srn⁻ phenotype is not understood in terms of mechanism; the functions of the *srnA* and *srnB* genes are unknown, and there is apparently no correlation between this phenotype and the known ribonucleases which have been defined genetically.

While genetic analyses may yield information as to the mode of degradation, they are unlikely to provide much insight as to the specificity of the process—the mechanism of distinguishing stable from unstable RNA's. The studies described above would seem to implicate the close association with ribosomal proteins, and the association of ribosomal subunits with each other, as factors contributing to the stability of rRNA. But what of tRNA's; are they stabilized by some structural feature which distinguishes them from unstable RNA's? Work with mutant tRNA's indicates that structural alterations of these molecules influence their stability. Abelson *et al.* (1970) observed that, compared to su_0 (wild-type) and su_3^+ strains of *E. coli*, su_3^- mutants showed a reduced synthesis of $tRNA_1^{Tyr}$; they suggested that the mutant tRNA's, or their precursors, were being degraded. Nomura (1974) studied the temperature-sensitive mutant $su3^{ts-6}$, in which the mutant $tRNA_1^{Tyr}$ is irreversibly inactivated at high temperatures. He observed a selective decay of the mutant $tRNA_1^{Tyr}$ *in vivo* at the nonpermissive temperature and proposed that *E. coli* possesses a means of scavenging deformed tRNA molecules.

Ghosh and Deutscher (1978) have partially purified an *E. coli* nuclease which may be involved in such an intracellular scavenging mechanism for damaged tRNA's and other inactive RNA molecules. The nuclease, RNase D, is active against heat-denatured tRNA, and against tRNA from which 3'-terminal residues or the 5'-terminal phosphate have been removed (by snake venom phosphodiesterase or alkaline phosphatase, respectively); the ultimate products of the degradation are 5'-monophosphates. RNase D acts poorly on intact tRNA, and is inactive with rRNA, poly(A), poly(U), or poly(A) · poly(U). On the basis of its substrate specificity and ionic requirements, the RNase D activity is distinguishable from RNase I, RNase II, RNase III, and PNPase; it seems to be a 3'-exonuclease, but could conceivably be an endonuclease contaminated with a nonspecific exonuclease.

Schroeder *et al.* (1977) have also investigated intracellular scavenging mechanisms, studying the fate of the products from *B. subtilis* RNase M5 action upon 5 S rRNA precursor. As described previously (see Section IV,B,3,d and Fig. 11), RNase M5 cleaves the p5 precursor molecule to generate the mature 5S rRNA (m5) and two precursor-specific fragments. In growing cells these precursor-specific fragments do not accumulate, whereas m5 rRNA appears to be stable indefinitely, even in the prolonged presence of chloramphenicol (implying that the stability of m5 rRNA is not due solely to protection by ribosomal proteins). When p5 precursor RNA is treated *in vitro*, with cell-free extracts providing RNase M5 activity, the resultant m5 rRNA is stable, while the precursor-specific fragments are rapidly destroyed. However, the fragments are indefinitely stable if purified RNase M5, free of extraneous nuclease activity, is used to treat the precursor. The fragment scavenging activity is, therefore, separate from the RNase M5 (processing) activity; no scavenging nuclease was purified, but the results of *in vitro* studies with cell-free extracts seem to implicate a 3'-exonucleolytic process, releasing 5'-nucleoside monophosphates. Thus the scavenging of the precursor-specific fragments resembles the degradation of denatured tRNA's by RNase D from *E. coli,* and may involve an analogous enzyme. The availability for comparative purposes of defined RNA molecules which are stable (m5 rRNA), and unstable (precursor fragments) in growing cells makes this system an attractive one for investigating the specificity of degradative reactions.

REFERENCES

Abelson, J. N., Gefter, M. L., Barnett, L., Landy, A., Russell, R. L., and Smith, J. D. (1970). Mutant tyrosine transfer ribonucleic acids. *J. Mol. Biol.* **47**, 15–28.

Adesnik, M., and Levinthal, C. (1969). Synthesis and maturation of ribosomal RNA in *Escherichia coli. J. Mol. Biol.* **46**, 281–303.

Aloni, Y., Hatlen, L. E., and Attardi, G. (1971). Studies of fractional HeLa cell metaphase chromosomes. II. Chromosomal distribution of sites for transfer RNA and 5 S RNA. *J. Mol. Biol.* **56**, 555–563.

Altman, S. (1971). Isolation of tyrosine tRNA precursor molecules. *Nature (London), New Biol.* **229**, 19–20.

Altman, S. (1978). Transfer RNA biosynthesis. *In* "Biochemistry of Nucleic Acids II" (B. F. C. Clark, ed.), pp. 19–44. International Review of Biochemistry, Volume 17. Univ. Park Press, Baltimore, Maryland.

Altman, S., and Smith, J. D. (1971). Tyrosine tRNA precursor molecule polynucleotide sequence. *Nature (London), New Biol.* **233**, 35–39.

Altman, S., Bothwell, A. L. M., and Stark, B. C. (1975). Processing of *E. coli* tRNA^{Tyr} precursor RNA *in vitro. Brookhaven Symp. Biol.* **26**, 12–25.

Andoh, T., and Ozeki, H. (1968). Suppressor gene *su3*$^+$ of *E. coli,* a structural gene for tyrosine tRNA. *Proc. Nat. Acad. Sci. U.S.A.* **59**, 792–799.

Apirion, D. (1975). The fate of mRNA and rRNA in *Escherichia coli. Brookhaven Symp. Biol.* **26**, 286–306.

Apirion, D., and Lassar, A. B. (1978). A conditional lethal mutant of *Escherichia coli* which affects the processing of ribosomal RNA. *J. Biol. Chem.* **253**, 1738–1742.

Apirion, D., and Watson, N. (1974). Analysis of an *Escherichia coli* strain carrying physiologically compensating mutations one of which causes an altered ribonuclease III. *Mol. Gen. Genet.* **132**, 89–104.

Apirion, D., and Watson, N. (1975a). Unaltered stability of newly synthesized RNA in strains of *Escherichia coli* missing a ribonuclease specific for double-stranded RNA. *Mol. Gen. Genet.* **136**, 317–326.

Apirion, D., and Watson, N. (1975b). Mapping and characterization of a mutation in *Escherichia coli* that reduces the level of ribonuclease III specific for double-stranded ribonucleic acid. *J. Bacteriol.* **124**, 317–324.

Apirion, D., Neil, J., and Watson, N. (1976a). Consequences of losing ribonuclease III on the *Escherichia coli* cell. *Mol. Gen. Genet.* **144** 185–190.

Apirion, D., Neil, J., and Watson, N. (1976b). Revertants from RNase III negative strains of *Escherichia coli. Mol. Gen. Genet.* **149**, 201–210.

Artavanis-Tsakonas, S., Schedl, P., Tschudi, C., Pirrotta, V., Steward, R., and Gehring, W. (1977). The 5 S genes of *Drosophila melanogaster. Cell* **12**, 1057–1067.

Bachmann, B. J., Low, K. B., and Taylor, A. L. (1976). Recalibrated linkage map of *Escherichia coli* K12. *Bacteriol. Rev.* **40**, 116–167.

Bacon, K., and Rosenberg, E. (1967). Ribonucleic acid synthesis during morphogenesis in *Myxococcus xanthus. J. Bacteriol.* **94**, 1883–1889.

Baguley, B. C., Wehrli, W., and Staehelin, M. (1970). *In vitro* methylation of yeast serine transfer ribonucleic acid. *Biochemistry* **9**, 1645–1649.

Barrell, B. G. (1971). Fractionation and sequence analysis of radioactive nucleotides. *In* "Procedures in Nucleic Acid Research" (G. L. Cantoni and D. R. Davies, eds.), pp. 751–799. Harper, New York.

Barrell, B. G., and Clark, B. F. C. (1974). "Handbook of Nucleic Acid Sequences." Joynson-Bruvvers, Oxford.

Bell, G. I., DeGennaro, L. J., Gelfand, D. H., Bishop, R. J., Valenzuela, P., and Rutter, W. J. (1977). Ribosomal RNA genes of *Saccharomyces cerevisiae*. I. Physical map of the repeating unit and location of the regions coding for 5 S, 5.8 S, 18 S, and 25 S ribosomal RNAs. *J. Biol. Chem.* **252**, 8118–8125.

Bernardi, G. (1978). Intervening sequences in the mitochondrial genome. *Nature (London)* **276**, 558–559.

Bernhardt, D., and Darnell, J. E. (1969). tRNA synthesis in HeLa cells: A precursor to tRNA and the effects of methionine starvation on tRNA synthesis. *J. Mol. Biol.* **42**, 43–56.

Bertrand, K., Korn, L., Lee, F., Platt, T., Squires, C. L., Squires, C., and Yanofsky, C. (1975). New features of the regulation of the tryptophan operon. *Science* **189**, 22–26.

Bikoff, E. K., and Gefter, M. L. (1975). *In vitro* synthesis of transfer RNA. I. Purification of required components. *J. Biol. Chem.* **250**, 6240–6247.

Bikoff, E. K., LaRue, B. F., and Gefter, M. L. (1975). *In vitro* synthesis of transfer RNA. II. Identification of required enzymatic activities. *J. Biol. Chem.* **250**, 6248–6255.

Birnbaum, L., and Kaplan, S. (1971). Localization of the ribosomal RNA genes in *Escherichia coli*. *Proc. Natl. Acad. Sci. U.S.A.* **68**, 925–929.

Björk, G. R., and Isaksson, L. A. (1970). Isolation of mutants of *Escherichia coli* lacking 5-methyluracil in transfer ribonucleic acid or 1-methylguanosine in ribosomal RNA. *J. Mol. Biol.* **51**, 83–100.

Björk, G. R., and Kjellin-Stråby, K. (1978). General screening procedure for RNA modificationless mutants: Isolation of *Escherichia coli* strains with specific defects in RNA methylation. *J. Bacteriol.* **133**, 499–507.

Björk, G. R., and Neidhardt, F. C. (1975). Physiological and biochemical studies on the function of 5-methyluridine in the transfer ribonucleic acid of *Escherichia coli*. *J. Bacteriol.* **124**, 99–111.

Blatt, B., and Feldmann, H. (1973). Characterization of precursors to tRNA in yeast. *FEBS Lett.* **37**, 129–133.

Bleyman, M., Kondo, M., Hecht, N., and Woese, C. (1969). Transcriptional mapping: Functional organization of the ribosomal and transfer ribonucleic acid cistrons in the *Bacillus subtilis* genome. *J. Bacteriol.* **99**, 535–543.

Borek, E., and Srinivasan, P. R. (1966). The methylation of nucleic acids. *Annu. Rev. Biochem.* **35**, 275–298.

Brawerman, G. (1974). Eukaryotic messenger RNA. *Annu. Rev. Biochem.* **43**, 621–642.

Bremer, H., and Berry, L. (1971). Co-transcription of 16 S and 23 S ribosomal RNA in *Escherichia coli*. *Nature (London), New Biol.* **234**, 81–83.

Brenner, D. J., Fournier, M. Y., and Doctor, B. P. (1970). Isolation and partial characterization of the transfer ribonucleic acid cistrons from *Escherichia coli*. *Nature (London)* **227**, 448–451.

Brosius, J., Palmer, M. L., Kennedy, P. J., and Noller, H. F. (1978). Complete nucleotide sequence of a 16 S ribosomal RNA gene from *Escherichia coli*. *Proc. Natl. Acad. Sci. U.S.A.* **75**, 4801–4805.

Brown, G., and Attardi, G. (1965). Methylation of nucleic acids in HeLa cells. *Biochem. Biophys. Res. Commun.* **20**, 298–302.

Brown, D., and Weber, C. (1968). Gene linkage by RNA–DNA hybridization. I. Unique DNA sequences homologous to 4 S RNA, 5 S RNA and ribosomal RNA. *J. Mol. Biol.* **34**, 661–680.

Brown, D. D., Wesnik, P. C., and Jordan, E. (1971). Purification and some characteristics of 5 S DNA from *Xenopus laevis*. *Proc. Natl. Acad. Sci. U.S.A.* **68**, 3175–3179.

Brownlee, G., and Cartwright, E. (1971). Sequence studies on precursor 16 S ribosomal RNA of *Escherichia coli*. *Nature (London), New Biol.* **232**, 50–52.

Brownlee, G. G., Sanger, F., and Barrell, B. G. (1968). The sequence of 5 S ribosomal ribonucleic acid. *J. Mol. Biol.* **34**, 379–412.

Bruni, C. B., Calantuoni, V., Shordone, L., Cortese, R., and Blasi, F. (1977). Biochemical and regulatory properties of *Escherichia coli* K12 *hisT* mutants. *J. Bacteriol.* **130**, 4–10.

Burdon, R. H. (1971). Ribonucleic acid maturation in animal cells. *Prog. Nucleic Acid Res. Mol. Biol.* **11**, 33–79.

Burdon, R. H. (1975). Processing of tRNA precursors in higher organisms. *Brookhaven Symp. Biol.* **26**, 138–153.

Burdon, R. H., and Clason, A. E. (1969). Intracellular location and molecular characteristics of tumour cell transfer RNA precursors. *J. Mol. Biol.* **39**, 113–124.

Burdon, R. H., Martin, B. T., and Lal, B. M. (1967). Synthesis of low molecular weight ribonucleic acid in tumour cells. *J. Mol. Biol.* **28**, 357–371.

Carbon, J., Chang, S., and Kirk, L. L. (1975). Clustered tRNA genes in *Escherichia coli*: Transcription and processing. *Brookhaven Symp. Biol.* **26**, 26–36.

Chang, S., and Carbon, J. (1975). The nucleotide sequence of a precursor to the glycine- and threonine-specific transfer ribonucleic acids of *Escherichia coli*. *J. Biol. Chem.* **250**, 5542–5555.

Chang, B., and Irr, J. (1973). Maturation of ribosomal RNA in stringent and relaxed bacteria. *Nature (London), New Biol.* **243**, 35–37.

Chang, S. E., and Smith, J. D. (1973). Structural studies on a tyrosine tRNA precursor. *Nature (London), New Biol.* **246**, 165–168.

Chang, G. W., Roth, J. R., and Ames, B. N. (1971). Histidine regulation in *Salmonella typhimurium*. VIII. Mutations of the *hisT* gene. *J. Bacteriol.* **108**, 410–414.

Chen, G. S., and Siddiqui, M. A. Q. (1973). Biosynthesis of transfer RNA: *In vitro* conversion of transfer RNA precursors from *Bombyx mori* to 4 S RNA by *Escherichia coli* enzymes. *Proc. Natl. Acad. Sci. U.S.A.* **70**, 2610–2613.

Chen, G. S., and Siddiqui, M. A. Q. (1975). Biosynthesis of transfer RNA: Isolation and characterization of precursors to transfer RNA in the posterior silk gland of *Bombyx mori*. *J. Mol. Biol.* **96**, 153–170.

Choe, B. K., and Taylor, M. W. (1972). Kinetics of synthesis and characterization of transfer RNA precursors in mammalian cells. *Biochim. Biophys. Acta* **272**, 275–287.

Clarke, L., and Carbon, J. (1976). A colony bank containing synthetic Col E1 plasmids representative of the entire *E. coli* genome. *Cell* **9**, 91–99.

Clarkson, S. G., Birnsteil, M. L., and Serra, V. (1973a). Reiterated transfer RNA genes of *Xenopus laevis*. *J. Mol. Biol.* **79**, 391–410.

Clarkson, S. G., Birnsteil, M. L., and Purdom, I. F. (1973b). Clustering of transfer RNA genes of *Xenopus laevis*. *J. Mol. Biol.* **79**, 411–429.

Colby, D., Schedl, P., and Guthrie, C. (1976). A functional requirement for modification of the wobble nucleotide in the anticodon of a T4 suppressor tRNA. *Cell* **9**, 449–463.

Colli, W., Smith, I., and Oishi, M. (1971). Physical linkage between 5 S, 16 S and 23 S ribosomal RNA genes in *Bacillus subtilis*. *J. Mol. Biol.* **56**, 117–127.

Cooper, H. L. (1973). Degradation of 28 S RNA late in ribosomal RNA maturation in nongrowing lymphocytes and its reversal after growth stimulation. *J. Cell Biol.* **59**, 250–254.

Corry, M., Payne, P., and Dyer, T. (1974a). The nucleotide sequence of 5 S rRNA from the blue-green alga *Anacystis nidulans*. *FEBS Lett.* **46**, 63–66.

Corry, M., Payne, P., and Dyer, T. (1974b). A sequence analysis of 5 S rRNA from the blue-green alga *Oscillatoria tenuis* and a comparison of blue-green alga 5 S rRNA with those of bacterial and eukaryotic origins. *FEBS Lett.* **46**, 67–70.

Corte, G., Schlessinger, D., Longo, D., and Venkov, P. (1971). Transformation of 17 S to 16 S ribosomal RNA using ribonuclease II of *Escherichia coli*. *J. Mol. Biol.* **60**, 325–338.

Cortese, R., Kammen, H. O., Spengler, S. J., and Ames, B. N. (1974). Biosynthesis of pseudouridine in transfer ribonucleic acid. *J. Biol. Chem.* **249**, 1103–1108.

Craig, N. C., and Perry, R. P. (1970). Aberrant intranucleolar maturation of ribosomal precursors in the absence of protein synthesis. *J. Cell Biol.* **45**, 554–564.

Cramer, F. (1971). Three-dimensional structure of tRNA. *Prog. Nucleic Acid Res. Mol. Biol.* **11**, 391–421.

Dahlberg, J. E., Nikolaev, N., and Schlessinger, D. (1975). Post-transcriptional modification of nucleotides in *E. coli* ribosomal RNAs. *Brookhaven Symp. Biol.* **26**, 194–200.

Dalgarno, L., Hosking, D. M., and Shen, C. H. (1972). Steps in the biosynthesis of ribosomal RNA in cultured *Aedes aegypti* cells. *Eur. J. Biochem.* **24**, 498–506.

Daniel, V., Sarid, S., Beckmann, J. S., and Littauer, U. Z. (1970). *In vitro* transcription of a transfer RNA gene. *Proc. Natl. Acad. Sci. U.S.A.* **66**, 1260–1266.

Daniel, V., Grimberg, J. I., and Zeevi, M. (1975). *In vitro* synthesis of tRNA precursors and their conversion to mature size tRNA. *Nature (London)* **257**, 193–197.

Darnell, J. E. (1968). Ribonucleic acids from animal cells. *Bacteriol. Rev.* **32**, 262–290.

Darnell, J. E. (1978). Implications of RNA · RNA splicing in evolution of eukaryotic cells. *Science* **202**, 1257–1260.

Davis, R. W., Simon, M., and Davidson, N. (1971). Electron microscope heteroduplex method for mapping regions of base sequence homology in nucleic acids. *In* "Methods in Enzymology: Vol. 21, Nucleic Acids, Part D" (L. Grossman and K. Moldave, eds.), pp. 413–428. Academic Press, New York.

Dawid, I. B., and Wellauer, P. K. (1976). A reinvestigation of 5′→3′ polarity in 40 S ribosomal RNA precursor of *Xenopus laevis*. *Cell* **8**, 443–448.

Dean, J., and Sykes, J. (1974). The role of ribonuclease II in the maturation of precursor 16 S ribosomal ribonucleic acid in *Escherichia coli*. *Biochem. J.* **140**, 443–450.

DeRobertis, E. M., and Olson, M. V. (1979). Transcription and processing of cloned yeast tRNA genes microinjected into frog oocytes. *Nature (London)* **278**, 137–143.

Deutscher, M. P. (1973). Synthesis and functions of the -C-C-A terminus of transfer RNA. *Prog. Nucleic Acid Res. Mol. Biol.* **13**, 51–92.

Deutscher, M. P., Foulds, J., and McClain, W. H. (1974). Transfer ribonucleic acid nucleotidyltransferase plays an essential role in the normal growth of *Escherichia coli* and in the biosynthesis of some bacteriophage T4 transfer ribonucleic acids. *J. Biol. Chem.* **249**, 6696–6699.

Deutscher, M. P., Foulds, J., Morse, J. W., and Hilderman, R. H. (1975). Synthesis of the CCA terminus of transfer RNA. *Brookhaven Symp. Biol.* **26**, 124–137.

Deutscher, M. P., Lin, J. J.-C., and Evans, J. A. (1977). tRNA metabolism in *Escherichia coli* cells deficient in tRNA nucleotidyltransferase. *J. Mol. Biol.* **117**, 1081–1094.

Doctor, B. P., Brenner, D. J., Fanning, G. R., Faulkner, A. G., Fournier, M. J., Miller, W. L., Peterkofsky, A., and Sodd, M. A. (1971). Further characterization of purified *E. coli* tRNA cistrons. *Fed. Proc. Fed. Am. Soc. Exp. Biol.* **30**, 1218 (Abstr.).

Doolittle, W., and Pace, N. (1971). Transcriptional organization of the ribosomal RNA cistrons in *Escherichia coli*. *Proc. Natl. Acad. Sci. U.S.A.* **68**, 1786–1790.

Drake, J., and Baltz, R. (1976). The biochemistry of mutagenesis. *Annu. Rev. Biochem.* **45**, 11–37.

DuBuy, B., and Weissman, S. (1971). Nucleotide sequence of *Pseudomonas fluorescens* 5 S ribonucleic acid. *J. Biol. Chem.* **246**, 747–761.

Dubin, D. T., and Günlap, A. (1967). Minor nucleotide composition of ribosomal precursor, and ribosomal ribonucleic acid in *Escherichia coli*. *Biochim. Biophys. Acta* **134**, 106–123.

Dunn, J. (1976). RNase III cleavage of single stranded RNA. The effect of ionic strength on the fidelity of cleavage. *J. Biol. Chem.* **251**, 3807–3814.

Dunn, J., and Studier, F. W. (1973). T7 early RNAs and *Escherichia coli* ribosomal RNAs are cut from large precursor RNAs *in vivo* by ribonuclease III. *Proc. Natl. Acad. Sci. U.S.A.* **70**, 3296–3300.

Edström, J.-E., and Daneholt, B. (1967). Sedimentation properties of the newly synthesized RNA from isolated nuclear components of *Chironomus tentans* salivary gland cells. *J. Mol. Biol.* **28**, 331–343.

Egan, J., and Landy, A. (1978). Structural analysis of the tRNA$_1^{Tyr}$ gene of *Escherichia coli:* A 178 base pair sequence that is repeated 3.14 times. *J. Biol. Chem.* **253**, 3607–3622.

Ehresmann, C., Stiegler, P., Mackie, G., Zimmermann, R., Ebel, J., and Fellner, P. (1975). Primary sequence of the 16 S ribosomal RNA of *Escherichia coli*. *Nucleic Acids Res.* **2**, 265–278.

Erdmann, V. A. (1976). Structure and function of 5 S and 5.8 S RNA. *Prog. Nucleic Acid Res. Mol. Biol.* **18**, 45–90.

Erdmann, V. A., Sprinzl, M., and Pongs, O. (1973). The involvement of 5 S RNA in the binding of tRNA to ribosomes. *Biochem. Biophys. Res. Commun.* **54**, 942–948.

Fellner, P. (1969). Nucleotide sequences from specific areas of the 16 S and 23 S rRNAs of *E. coli*. *Eur. J. Biochem.* **11**, 12–27.

Fellner, P., and Sanger, F. (1968). Sequence analysis of specific areas of the 16 S and 23 S ribosomal RNAs. *Nature (London)* **219**, 236–238.

Feunteun, J., Jordan, B., and Monier, R. (1972). Study of the maturation of 5 S RNA precursors in *Escherichia coli*. *J. Mol. Biol.* **70**, 465–474.

Forget, B. G., and Jordan, B. (1970). 5 S RNA synthesized by *Escherichia coli* in the presence of chloramphenicol: Different 5′-terminal sequences. *Science* **167**, 382–384.

Forget, B., and Weissman, S. (1969). The nucleotide sequence of ribosomal 5 S ribonucleic acid from KB cells. *J. Biol. Chem.* **244**, 3148–3165.

Fournier, M. J., Webb, E., and Tang, S. (1977). *In vitro* biosynthesis of functional *Escherichia coli* su_3^+ tyrosine transfer RNA. *Biochemistry* **16**, 3608–3616.

Fox, G., and Woese, C. (1975). 5 S RNA secondary structure. *Nature (London)* **256**, 505–507.

Garber, R. L., Siddiqui, M. A. Q., and Altman, S. (1978). Identification of precursor molecules to individual tRNA species from *Bombyx mori*. *Proc. Natl. Acad. Sci. U.S.A.* **75**, 635–639.

Gefter, M. L., and Russell, R. L. (1969). A role of modifications in tyrosine transfer RNA: A modified base affecting ribosome binding. *J. Mol. Biol.* **39**, 145–157.

Gegenheimer, P., and Apirion, D. (1975). *Escherichia coli* ribosomal ribonucleic acids are not cut from an intact precursor molecule. *J. Biol. Chem.* **250**, 2407–2409.

Gegenheimer, P., Watson, N., and Apirion, D. (1977). Multiple pathways for primary processing of ribosomal RNA in *Escherichia coli*. *J. Biol. Chem.* **252**, 3064–3073.

Ghosh, R. K., and Deutscher, M. P. (1978). Identification of an *Escherichia coli* nuclease acting on structurally altered transfer RNA molecules. *J. Biol. Chem.* **253**, 997–1000.

Ghysen, A., and Celis, J. E. (1974). Joint transcription of two tRNA$_1^{Tyr}$ genes from *Escherichia coli*. *Nature (London)* **249**, 418–421.

Ginsburg, D., and Steitz, J. (1975). The 30 S ribosomal precursor RNA from *Escherichia coli:* A primary transcript containing 23 S, 16 S, and 5 S sequences. *J. Biol. Chem.* **250,** 5647–5654.

Glover, D. M., and Hogness, D. S. (1977). A novel arrangement of the 18 S and 28 S sequences in a repeating unit of *Drosophila melanogaster* rDNA. *Cell* **10,** 167–176.

Grimberg, J. I., and Daniel, V. (1974). *In vitro* transcription of three adjacent *E. coli* transfer RNA genes. *Nature (London)* **250,** 320–323.

Guthrie, C. (1975). The nucleotide sequence of the dimeric precursor to glutamine and leucine transfer RNAs coded by bacteriophage T4. *J. Mol. Biol.* **95,** 529–547.

Guthrie, C., Nashimoto, H., and Nomura, M. (1969). Studies on the assembly of ribosomes *in vitro. Cold Spring Harbor Symp. Quant. Biol.* **34,** 69–75.

Guthrie, C., Seidman, J. G., Altman, S., Barrell, B. G., Smith, J. D., and McClain, W. H. (1973). Identification of tRNA precursor molecules made by phage T4. *Nature (London), New Biol.* **246,** 6–11.

Guthrie, C., Seidman, J. G., Comer, M. M., Bock, R. M., Schmidt, F. J., Barrell, B. G., and McClain, W. H. (1975). The biology of bacteriophage T4 transfer RNAs. *Brookhaven Symp. Biol.* **26,** 106–123.

Hackett, P. B., and Sauerbier, W. (1975). The transcriptional organization of the ribosomal RNA genes in mouse L cells. *J. Mol. Biol.* **91,** 235–256.

Hamkalo, B. A., and Miller, O. L., Jr. (1973). Electron microscopy of genetic activity. *Annu. Rev. Biochem.* **41,** 379–397.

Hatlen, L., and Attardi, G. (1971). Proportion of the HeLa cell genome complementary to transfer RNA and 5 S RNA. *J. Mol. Biol.* **56,** 535–553.

Hatlen, L. E., Amaldi, F., and Attardi, G. (1969). Oligonucleotide patterns after pancreatic ribonuclease digestion and the 3′ and 5′ termini of 5 S ribonucleic acid from HeLa cells. *Biochemistry* **8,** 4989–5005.

Hawthorne, D. C., and Leupold, U. (1974). Suppressor mutations in yeast. *Curr. Top. Microbiol. Immunol.* **64,** 1–47.

Hawthorne, D. C., and Mortimer, R. K. (1968). Genetic mapping of nonsense suppressors in yeast. *Genetics* **60,** 735–742.

Goldfarb, A., Seaman, E., and Daniel, V. (1978). *In vitro* transcription and isolation of a polycistronic RNA product of the T4 tRNA operon. *Nature (London)* **273,** 562–564.

Goodman, H. M., Abelson, J., Landy, A., Brenner, S., and Smith, J. D. (1968). Amber suppression: A nucleotide change in the anticodon of a tyrosine transfer RNA. *Nature (London)* **217,** 1019–1024.

Goodman, H. M., Abelson, J. N., Landy, A., Zadrazil, S., and Smith, J. D. (1970). The nucleotide sequences of tyrosine transfer RNAs of *Escherichia coli*. Sequences of the amber suppressor su_{III} transfer RNA, the wild-type su^-_{III} transfer RNA and tyrosine transfer RNAs species I and II. *Eur. J. Biochem.* **13,** 461–483.

Goodman, H. M., Olson, M. V., and Hall, B. D. (1977). Nucleotide sequence of a mutant eukaryotic gene: The yeast tyrosine-inserting ochre suppressor *SUP4-o. Proc. Natl. Acad. Sci. U.S.A.* **74,** 5453–5457.

Gordon, J., and Bowman, H. G. (1964). Studies on microbial RNA. II. Transfer of methyl groups from methionine to the RNA of a ribonucleoprotein particle. *J. Mol. Biol.* **9,** 638–653.

Gorelic, L. (1970). Chromosomal location of ribosomal RNA cistrons in *Escherichia coli. Mol. Gen. Genet.* **106,** 323–327.

Greenberg, J. R. (1969). Synthesis and properties of ribosomal RNA in *Drosophila. J. Mol. Biol.* **46,** 85–98.

Grigliatti, T. A., White, B. N., Tener, G. M., Kaufman, T. C., Holden, J. J., and Suzuki,

D. T. (1973). Studies on the transfer RNA genes of *Drosophila*. *Cold Spring Harbor Symp. Quant. Biol.* **38**, 461–474.

Hayes, F., and Vasseur, M. (1976). Processing of the 17 S *Escherichia coli* precursor RNA in the 27 S pre-ribosomal particle. *Eur. J. Biochem.* **61**, 433–442.

Hayes, F., Hayes, D., Fellner, P., and Ehresmann, C. (1971). Additional nucleotide sequences in precursor 16 S ribosomal RNA from *Escherichia coli*. *Nature (London), New Biol.* **232**, 54–55.

Hayes, F., Vasseur, M., Nikolaev, N., Schlessinger, D., Sri Wada, J., Krol, A., and Branlant, C. (1975). Structure of a 30 S pre-ribosomal RNA of *E. coli*. *FEBS Lett.* **56**, 85–91.

Hecht, N. B., and Woese, C. R. (1968). Separation of bacterial ribosomal ribonucleic acid from its macromolecular precursors by polyacrylamide gel electrophoresis. *J. Bacteriol.* **95**, 986–990.

Helser, T. L., Davies, J. E., and Dahlberg, J. E. (1971). Change in methylation of 16 S ribosomal RNA associated with mutation to kasugamycin resistance in *Escherichia coli*. *Nature (London), New Biol.* **233**, 12–14.

Helser, T. L., Davies, J. E., and Dahlberg, J. E. (1972). Mechanism of kasugamycin resistance in *Escherichia coli*. *Nature (London), New Biol.* **235**, 6–9.

Hofmann, S., and Miller, O. L., Jr. (1977). Visualization of ribosomal ribonucleic acid synthesis in a ribonuclease III-deficient strain of *Escherichia coli*. *J. Bacteriol.* **132**, 718–722.

Holtzman, E., Smith, I., and Penman, S. (1966). Electron microscopic studies of detergent-treated HeLa cell nuclei. *J. Mol. Biol.* **17**, 131–135.

Hopper, A. K., Banks, F., and Evangelidis, V. (1978). A yeast mutant which accumulates precursor tRNAs. *Cell* **14**, 211–219.

Hurwitz, J., Gold, M., and Anders, M. (1964). The enzymatic methylation of ribonucleic acid and deoxyribonucleic acid. III. Purification of soluble ribonucleic acid-methylating enzymes. *J. Biol. Chem.* **239**, 3462–3473.

Ikemura, T., and Ozeki, H. (1977). Gross map location of *Escherichia coli* transfer RNA genes. *J. Mol. Biol.* **117**, 419–446.

Isaksson, L. A. (1973a). Partial purification of ribosomal RNA(m^1G)- and rRNA(m^2G)-methylases from *Escherichia coli* and demonstration of some proteins affecting their apparant activity. *Biochim. Biophys. Acta* **312**, 122–133.

Isaksson, L. A. (1973b). Formation *in vitro* of 1-methylguanine in 23 S RNA from *Escherichia coli:* The effects of spermidine and two proteins. *Biochim. Biophys. Acta* **312**, 134–146.

Jacq, B., Jourdan, R., and Jordan, B. R. (1977). Structure and processing of precursor 5 S RNA in *Drosophila melanogaster*. *J. Mol. Biol.* **117**, 785–795.

Jarry, B., and Rosset, R. (1973a). Localization of some 5 S RNA cistrons on *Escherichia coli* chromosome. *Mol. Gen. Genet.* **121**, 151–162.

Jarry, B., and Rosset, R. (1973b). Further mapping of 5 S RNA cistrons in *Escherichia coli*. *Mol. Gen. Genet.* **126**, 29–35.

Jeanteur, P., and Attardi, G. (1969). Relationship between HeLa cell ribosomal RNA and its precursors studied by high resolution RNA–DNA hybridization. *J. Mol. Biol.* **45**, 305–324.

Jordan, B. R. (1974). '2 S' RNA, a new ribosomal RNA component in cultured *Drosophila* cells. *FEBS Lett.* **44**, 39–42.

Jordan, B. R. (1975). Demonstration of intact 26 S ribosomal RNA molecules in *Drosophila* cells. *J. Mol. Biol.* **98**, 277–280.

Jordan, B. R., Galling, G., and Jourdan, R. (1974). Sequence and conformation of 5 S RNA from *Chlorella* cytoplasmic ribosomes: Comparison with other 5 S RNA molecules. *J. Mol. Biol.* **87**, 205–225.

Jordan, B. R., Jourdan, R., and Jacq, B. (1976). Late steps in the maturation of *Drosophila* 26 S ribosomal RNA: Generation of 5.8 S and 2 S RNAs by cleavages occurring in the cytoplasm. *J. Mol. Biol.* **101**, 85–105.

Kaback, D. B., Halvorson, H. O., and Rubin, G. M. (1976). Location and magnification of 5 S RNA genes in *Saccharomyces cerevisiae*. *J. Mol. Biol.* **107**, 385–390.

Kaplan, R., and Apirion, D. (1975a). The fate of ribosomes in *Escherichia coli* cells starved of a carbon source. *J. Biol. Chem.* **250**, 1854–1863.

Kaplan, R., and Apirion, D. (1975b). Decay of ribosomal ribonucleic acid in *Escherichia coli* cells starved for various nutrients. *J. Biol. Chem.* **250**, 3174–3178.

Kelley, D. E., and Perry, R. P. (1971). The production of ribosomal RNA from high molecular weight precursors. II. Demonstration of an exonuclease in isolated nucleoli. *Biochim. Biophys. Acta* **238**, 357–362.

Kenerley, M. E., Morgan, E. A., Post, L., Lindahl, L., and Nomura, M. (1977). Characterization of hybrid plasmids carrying individual ribosomal ribonucleic acid transcription units of *Escherichia coli*. *J. Bacteriol.* **132**, 931–949.

Kindler, P., Keil, T. U., and Hofschneider, P. H. (1973). Isolation and characterization of a ribonuclease III deficient mutant of *Escherichia coli*. *Mol. Gen. Genet.* **126**, 53–69.

Kitchingham, G. R., and Fournier, M. J. (1975). Inhibition of post-transcriptional modification of *E. coli* tRNA. *Brookhaven Symp. Biol.* **26**, 44–52.

Klootwijk, J., Van den Bos, R. C., and Planta, R. J. (1972). Secondary methylation of yeast ribosomal RNA. *FEBS Lett.* **27**, 102–106.

Knapp, G., Beckmann, J. S., Johnson, P. F., Fuhrman, S. A., and Abelson, J. (1978). Transcription and processing of intervening sequences in yeast tRNA genes. *Cell* **14**, 221–236.

Koseki, A., Bothwell, A. L. M., and Altman, S. (1976). Identification of a ribonuclease P-like activity from human KB cells. *Cell* **9**, 101–116.

Kossman, C., Stamato, T., and Pettijohn, D. (1971). Tandem synthesis of the 16 S and 23 S ribosomal RNA sequences of *Escherichia coli*. *Nature (London), New Biol.* **234**, 102–104.

Kuchino, Y., and Nishimura, S. (1970). Nucleotide sequence specificities of guanylate residue-specific tRNA methylases from rat liver. *Biochem. Biophys. Res. Commun.* **40**, 306–313.

Kuchino, Y., Seno, T., and Nishimura, S. (1971). Fragmented *E. coli* methionine tRNA$_f$ as methyl acceptor for rat liver tRNA methylase: Alteration of the site of methylation by the conformational change of tRNA structure resulting from fragmentation. *Biochem. Biophys. Res. Commun.* **43**, 476–483.

Kumar, A. (1967). Patterns of ribosomal RNA synthesis in *Tetrahymena*. *J. Cell Biol.* **35**, 74A.

Kumar, A., and Warner, J. R. (1972). Characterization of ribosomal precursor particles from HeLa cell nucleoli. *J. Mol. Biol.* **63**, 233–246.

Küpper, H., Sekiya, T., Rosenberg, M., Egan, J., and Landy, A. (1978). A ρ-dependent termination site in the gene coding for tyrosine tRNA su_3 of *Escherichia coli*. *Nature (London)* **272**, 423–428.

Kwong, L. K., Moore, V. G., and Kaiser, I. I. (1977). Pseudouridine-deficient transfer RNAs from *Escherichia coli* B and their use as substrates for pseudouridine synthetase. *J. Biol. Chem.* **252**, 6310–6315.

Lai, C. J., and Weisblum, B. (1971). Altered methylation of ribosomal RNA in an erythromycin-resistant strain of *Staphylococcus aureus*. *Proc. Natl. Acad. Sci. U.S.A.* **68**, 856–860.

Lai, C. J., Weisblum, B., Fahnestock, S., and Nomura, M. (1973). Alteration of 23 S ribosomal RNA and erythromycin-induced resistance to lincomycin and spiramycin in *Staphylococcus aureus*. *J. Mol. Biol.* **74**, 67–72.

Lamfrom, H., Sarabhai, A., Nierlich, D. P., and Abelson, J. (1973). Synthesis of tRNA in cell-free extracts. *Nature (London), New Biol.* **246**, 11–12.

Landesman, R., and Gross, P. R. (1969). Patterns of macromolecule synthesis during development of *Xenopus laevis*. II. Identification of the 40 S precursor to ribosomal RNA. *Dev. Biol.* **19**, 244–260.

Landy, A., Foeller, C., and Ross, W. (1974). DNA fragments carrying genes for tRNA$_1^{Tyr}$. *Nature (London)* **249**, 738–742.

Lane, B. G., and Tamaoki, T. (1969). Methylated bases and sugars in 16 S and 28 S RNA from L cells. *Biochim. Biophys. Acta* **179**, 332–340.

Lawther, R. P., and Hatfield, G. W. (1977). Biochemical characterization of an *Escherichia coli hisT* strain. *J. Bacteriol.* **130**, 552–557.

Lazarus, H., and Sporn, M. (1967). Purification and properties of a nuclear exoribonuclease from Ehrlich ascites tumor cells. *Proc. Natl. Acad. Sci. U.S.A.* **57**, 1386–1393.

Leaver, C. J., and Key, J. L. (1970). Ribosomal synthesis in plants. *J. Mol. Biol.* **49**, 671–680.

Liau, M. C., Craig, N. C., and Perry, R. P. (1968). The production of ribosomal RNA from high molecular weight precursors. I. Factors which influence the ability of isolated nucleoli to process 45 S RNA. *Biochim. Biophys. Acta* **169**, 196–205.

Loewen, P. C., and Khorana, H. G. (1973). Studies on polynucleotides. XXII. The dodecanucleotide sequence adjoining the C-C-A end of the tyrosine transfer ribonucleic acid gene. *J. Biol. Chem.* **248**, 3489–3499.

Loewen, P. C., Sekiya, T., and Khorana, H. G. (1974). The nucleotide sequence adjoining the CCA end of an *Escherichia coli* tyrosine transfer ribonucleic acid gene. *J. Biol. Chem.* **249**, 217–226.

Lowry, C., and Dahlberg, J. E. (1971). Structural differences between the 16 S ribosomal RNA of *Escherichia coli* and its precursor. *Nature (London), New Biol.* **232**, 52–54.

Lund, E., and Dahlberg, J. E. (1977). Spacer transfer RNAs in ribosomal RNA transcripts of *E. coli:* Processing of 30 S ribosomal RNA *in vitro*. *Cell* **11**, 247–262.

Lund, E., Dahlberg, J. E., Lindahl, L., Jaskunas, S. R., Dennis, P., and Nomura, M. (1976). Transfer RNA genes between 16 S and 23 S rRNA genes in rRNA transcription units of *E. coli*. *Cell* **7**, 165–177.

McClain, W. H. (1977). Seven terminal steps in a biosynthesic pathway leading from DNA to transfer RNA. *Acc. Chem. Res.* **10**, 418–425.

McClain, W. H. (1979). A role for ribonuclease III in synthesis of bacteriophage T4 transfer RNAs. *Biochem. Biophys. Res. Commun.* **86**, 718–724.

McClain, W. H., and Seidman, J. G. (1975). Genetic perturbations that reveal tertiary conformation of tRNA precursor molecules. *Nature (London)* **257**, 106–110.

McClain, W. H., Guthrie, C., and Barrell, B. G. (1972). Eight transfer RNAs induced by infection of *Escherichia coli* with bacteriophage T4. *Proc. Natl. Acad. Sci. U.S.A.* **69**, 3703–3707.

McClain, W. H., Guthrie, C., and Barrell, B. G. (1973). The *psu*$_1^+$ amber suppressor gene of bacteriophage T4: Identification of its amino acid and transfer RNA. *J. Mol. Biol.* **81**, 157–171.

McClain, W. H., Barrell, B. G., and Seidman, J. G. (1975). Nucleotide alterations in bacteriophage T4 serine transfer RNA that affect the conversion of precursor RNA into transfer RNA. *J. Mol. Biol.* **99**, 717–732.

McClain, W. H., Seidman, J. G., and Schmidt, F. J. (1978). Evolution of the biosynthesis of 3'-terminal C-C-A residues in T-even bacteriophage transfer RNAs. *J. Mol. Biol.* **119**, 519–536.

McCloskey, J. A., and Nishimura, S. (1977). Modified nucleosides in transfer RNA. *Acc. Chem. Res.* **10**, 403–410.

Maden, B. E. H., and Robertson, J. S. (1974). Demonstration of the "5.8 S" ribosomal sequence in HeLa cell ribosomal precursor RNA. *J. Mol. Biol.* **87**, 227–235.

Maden, B. E. H., and Salim, M. (1974). The methylated nucleotide sequences in HeLa cell ribosomal RNA and its precursors. *J. Mol. Biol.* **88**, 133–164.

Maden, B. E. H., Salim, M., and Summers, D. F. (1972). Maturation pathway for ribosomal RNA in the HeLa cell nucleolus. *Nature (London), New Biol.* **237**, 5–9.

Maden, B. E. H., Forbes, J., DeJonge, P., and Klootwijk, J. (1975). Presence of a hypermodified nucleotide in HeLa cell 18 S and *Saccharomyces carlsbergensis* 17 S ribosomal RNAs. *FEBS Lett.* **59**, 60–63.

Madison, J. T., and Kung, H. K. (1967). Large oligonucleotides isolated from yeast tyrosine transfer ribonucleic acid after partial digestion with ribonuclease T_1. *J. Biol. Chem.* **242**, 1324–1330.

Marotta, C., Varricchio, F., Smith, I., Weissman, S., Sogin, M., and Pace, N. (1976). The primary structure of *Bacillus subtilis* and *Bacillus stearothermophilus* 5 S ribonucleic acids. *J. Biol. Chem.* **251**, 3122–3127.

Mazzara, G. P., and McClain, W. H. (1977). Cysteine transfer RNA of *Escherichia coli*: Nucleotide sequence and unusual metabolic properties of the 3' C-C-A terminus. *J. Mol. Biol.* **117**, 1061–1079.

Meyhack, B., Meyhack, I., and Apirion, D. (1974). Processing of precursor particles containing 17 S rRNA in a cell free system. *FEBS Lett.* **49**, 215–219.

Meyhack, B., Pace, B., and Pace, N. R. (1977). Involvement of precursor-specific segments in the *in vitro* maturation of *Bacillus subtilis* precursor 5 S ribosomal RNA. *Biochemistry* **16**, 5009–5015.

Miller, R. C., Jr., Besmer, P., Khorana, H. G., Fiandt, M., and Szybalski, W. (1971). Studies on polynucleotides. XCVII. Opposing orientations and location of the su_{III}^+ gene in the transducing coliphages $\phi80psu_{III}^+$ and $\phi80dsu_{III}^+$ su_{III}^-. *J. Mol. Biol.* **56**, 363–368.

Mirault, M. E., and Scherrer, K. (1972). *In vitro* processing of HeLa cell preribosomes by a nucleolar endoribonuclease. *FEBS Lett.* **20**, 233–238.

Morgan, E. A., Ikemura, T., and Nomura, M. (1977). Identification of spacer tRNA genes in individual ribosomal RNA transcription units of *Escherichia coli*. *Proc. Natl. Acad. Sci. U.S.A.* **74**, 2710–2714.

Morgan, E. A., Ikemura, T., Lindahl, L., Fallon, A. M., and Nomura, M. (1978). Some rRNA operons in *E. coli* have tRNA genes at their distal ends. *Cell* **13**, 335–344.

Morse, J. W., and Deutscher, M. P. (1975). Apparent noninvolvement of transfer RNA nucleotidyltransferase in the biosynthesis of *Escherichia coli* suppressor transfer RNAs. *J. Mol. Biol.* **95**, 141–144.

Moshowitz, D. B. (1970). Transfer RNA synthesis in HeLa cells. II. Formation of tRNA from a precursor *in vitro* and formation of pseudouridine. *J. Mol. Biol.* **50**, 143–151.

Mukerji, S. K., and Deutscher, M. P. (1972). Reactions at the 3' terminus of transfer ribonucleic acid. V. Subcellular localization and evidence for a mitochondrial transfer ribonucleic acid nucleotidyltransferase. *J. Biol. Chem.* **247**, 481–488.

Munns, T. W., and Sims, H. F. (1975). Methylation and processing of transfer ribonucleic acid in mammalian and bacterial cells. *J. Biol. Chem.* **250**, 2143–2149.

Muramatsu, M., and Fujisawa, F. (1968). Methylation of ribosomal RNA precursor and tRNA in rat liver. *Biochim. Biophys. Acta* **157**, 476–492.

Nashimoto, H., and Nomura, M. (1970). Structure and function of bacterial ribosomes. XI. Dependence of 50 S subunit assembly on simultaneous assembly of 30 S subunits. *Proc. Natl. Acad. Sci. U.S.A.* **67**, 1440–1447.

Nazar, R. N., and Roy, K. L. (1978). Nucleotide sequence of rainbow trout (*Salmo gairdneri*) ribosomal 5.8 S ribonucleic acid. *J. Biol. Chem.* **252**, 395–399.

Nazar, R. N., Sitz, T. O., and Busch, H. (1975). Structural analysis of mammalian ribosomal ribonucleic acid and its precursors. Nucleotide sequence of ribosomal 5.8 S ribonucleic acid. *J. Biol. Chem.* **250**, 8591–8597.

Nichols, J. L., and Lane, B. G. (1966). N^4-Methyl-2′-O-methylcytidine and other methyl-substituted nucleoside constituents of *Escherichia coli* ribosomal and soluble RNA. *Biochim. Biophys. Acta* **119**, 649–651.

Nichols, J. L., and Lane, B. G. (1968). The *in vitro* $O^{2'}$-methylation of RNA in the ribonucleoprotein precursor particles from a *relaxed* mutant of *Escherichia coli*. *Can. J. Biochem.* **46**, 109–115.

Nierhaus, K., Bordasch, K., and Homann, H. (1973). Ribosomal proteins. XLIII. *In vivo* assembly of *Escherichia coli* ribosomal proteins. *J. Mol. Biol.* **74**, 587–597.

Nierlich, D. P., Lamfrom, H., Sarabhai, A., and Abelson, J. (1973). Transfer RNA synthesis *in vitro*. *Proc. Natl. Acad. Sci. U.S.A.* **70**, 179–182.

Niessing, J., and Sekeris, C. E. (1970). Cleavage of high-molecular-weight DNA-like RNA by a nuclease present in 30 S ribonucleoprotein particles of rat liver nuclei. *Biochim. Biophys. Acta* **209**, 484–492.

Nikolaev, N., Silengo, L., and Schlessinger, D. (1973). Synthesis of a large precursor to ribosomal RNA in a mutant of *Escherichia coli*. *Proc. Natl. Acad. Sci. U.S.A.* **70**, 3361–3365.

Nikolaev, N., Schlessinger, D., and Wellauer, P. (1974). 30 S pre-ribosomal RNA of *Escherichia coli* and products of cleavage by ribonuclease III: Length and molecular weight. *J. Mol. Biol.* **86**, 741–747.

Nikolaev, N., Birge, C. H., Gotoh, S., Glazier, K., and Schlessinger, D. (1975). Primary processing of high molecular weight preribosomal RNA in *Escherichia coli* and HeLa cells. *Brookhaven Symp. Biol.* **26**, 175–193.

Nishimura, S. (1972). Minor components in transfer RNA: Their characterization, localization, and function. *Prog. Nucleic Acid Res. Mol. Biol.* **12**, 49–85.

Nomura, Y. (1974). Biological evidence for possibility of the intracellular mechanism scavenging 'deformed tRNA molecules.' *FEBS Lett.* **45**, 223–227.

Nomura, M. (1976). Organization of bacterial genes for ribosomal components: Studies utilizing novel approaches. *Cell* **9**, 634–644.

Nomura, M., Morgan, E. A., and Jaskunas, S. R. (1977). Genetics of bacterial ribosomes. *Annu. Rev. Genet.* **11**, 297–347.

O'Farrell, P. Z., Cordell, B., Valenzuela, P., Rutter, W. J., and Goodman, H. M. (1978). Structure and processing of yeast precursor tRNAs containing intervening sequences. *Nature (London)* **274**, 438–445.

Ogden, R. C., Beckmann, J. S., Abelson, J., Kang, H. S., Söll, D., and Schmidt, O. (1979). *In vitro* transcription and processing of a yeast tRNA gene containing an intervening sequence. *Cell* **17**, 399–406.

Ohnishi, Y., and Schlessinger, D. (1972). Total breakdown of ribosomal and transfer RNA in a mutant of *Escherichia coli*. *Nature (London), New Biol.* **238**, 228–231.

Ohnishi, Y., Iguma, H., Ono, T., Nagaishi, H., and Clark, A. J. (1977). Genetic mapping of the F plasmid gene that promotes degradation of stable ribonucleic acid in *Escherichia coli*. *J. Bacteriol.* **132**, 784–789.

Olson, M. V., Montgomery, D. L., Hopper, A. K., Page, G. S., Horodyski, F., and Hall, B. D. (1977). Molecular characterization of the tyrosine tRNA genes of yeast. *Nature (London)* **267**, 639–641.

Osawa, S., Otaka, E., Itoh, T., and Fukui, T. (1969). Biosynthesis of ribosomal subunit in *Escherichia coli*. *J. Mol. Biol.* **40**, 321–351.

Pace, N. (1973). Structure and synthesis of the ribosomal ribonucleic acid of prokaryotes. *Bacteriol. Rev.* **37**, 562–603.

Pace, B., and Pace, N. (1971). Gene dosage for 5 S ribosomal ribonucleic acid in *Escherichia coli* and *Bacillus megaterium*. *J. Bacteriol.* **105**, 142–149.

Pace, N. R., and Sogin, M. L. (1975). *In vitro* maturation of precursors of 5 S ribosomal RNA from *Bacillus subtilis*. *Brookhaven Symp. Biol.* **26**, 224–239.

Pace, B., Peterson, R. L., and Pace, N. R. (1970). Formation of all stable RNA species in *Escherichia coli* by posttranscriptional modification. *Proc. Natl. Acad. Sci. U.S.A.* **65**, 1097–1104.

Pace, N. R., Pato, M. L., McKibbin, J., and Radcliffe, C. W. (1973). Precursors of 5 S ribosomal RNA in *Bacillus subtilis*. *J. Mol. Biol.* **75**, 619–631.

Pace, N. R., Walker, T. A., and Schroeder, E. (1977). The structure of the 5.8 S RNA component of the 5.8 S–28 S ribosomal RNA junction complex. *Biochemistry* **16**, 5321–5328.

Parker, C. S., and Roeder, R. G. (1977). Selective and accurate transcription of the *Xenopus laevis* 5 S RNA genes in isolated chromatin by purified RNA polymerase III. *Proc. Natl. Acad. Sci. U.S.A.* **74**, 44–48.

Pato, M. L., and Meyenburg, K. von (1970). Residual RNA synthesis in *Escherichia coli* after inhibition of initiation of transcription by rifampicin. *Cold Spring Harbor Symp. Quant. Biol.* **35**, 497–504.

Pene, J., Knight, E., Jr., and Darnell, J. E., Jr. (1968). Characterization of a new low molecular weight RNA in HeLa cell ribosomes. *J. Mol. Biol.* **33**, 609–623.

Penman, S. (1966). RNA metabolism in the HeLa cell nucleus. *J. Mol. Biol.* **17**, 117–130.

Penman, S., Smith, I., and Holtzman, E. (1966). Ribosomal RNA synthesis and processing in a particulate site in the HeLa cell nucleus. *Science* **154**, 786–789.

Perry, R. (1976). Processing of RNA. *Annu. Rev. Biochem.* **45**, 605–629.

Perry, R. P., and Kelley, D. E. (1968). Persistent synthesis of 5 S RNA when production of 28 S and 18 S ribosomal RNA is inhibited by low doses of actinomycin D. *J. Cell. Physiol.* **72**, 235–246.

Perry, R. P., and Kelley, D. E. (1972). The production of ribosomal RNA from high molecular weight precursors. III. Hydrolysis of pre-ribosomal and ribosomal RNA by a 3′-OH specific exoribonuclease. *J. Mol. Biol.* **70**, 265–279.

Pettijohn, D. E., Stonington, O. G., and Kossman, C. R. (1970). Chain termination of ribosomal RNA synthesis *in vitro*. *Nature (London)* **228**, 235–239.

Piper, P. W., and Clark, B. F. C. (1974). Primary structure of a mouse myeloma cell initiator transfer RNA. *Nature (London)* **247**, 516–518.

Piper, P. W., Wasserstein, M., Engbaek, F., Kaltoft, K., Celis, J. E., Zeuthen, J., Leibman, S., and Sherman, F. (1977). Nonsense suppressors of *Saccharomyces cerevisiae* can be generated by mutation of the tyrosine tRNA anticodon. *Nature (London)* **262**, 757–761.

Prestayko, A. W., Lewis, B. C., and Busch, H. (1972). Endoribonuclease activity associated with nucleolar ribonucleoprotein particles from Novikoff hepatoma. *Biochim. Biophys. Acta* **269**, 90–103.

Prestayko, A. W., Lewis, B. C., and Busch, H. (1973). Purification and properties of a nucleolar endoribonuclease from Novikoff hepatoma. *Biochim. Biophys. Acta* **319**, 323–335.

Pribula, C., Fox, G., Woese, C., Sogin, M., and Pace, N. (1974). Nucleotide sequence of *Bacillus megaterium* 5 S RNA. *FEBS Lett.* **44**, 322–323.

Quagliarotti, G., Hidvegi, E., Wikman, J., and Busch, H. (1970). Structural analysis of nuclear precursors of ribosomal ribonucleic acid. Comparative hybridizations of nucleolar and ribosomal ribonucleic acid with nucleolar deoxyribonucleic acid. *J. Biol. Chem.* **245**, 1962–1969.

Reeder, R. H., Higashinakagawa, T., and Miller, O., Jr. (1976). The $5' \rightarrow 3'$ polarity of the *Xenopus* ribosomal RNA precursor molecule. *Cell* **8**, 449–454.

Retèl, J., and Planta, R. J. (1970). Non-conservative processing of ribosomal precursor RNA in yeast. *Biochim. Biophys. Acta.* **199**, 286–288.

Richter, D., Erdmann, V. A., and Sprinzl, M. (1973). Specific recognition of GTΨC loop (loop IV) of tRNA by 50 S ribosomal subunits from *E. coli. Nature (London), New Biol.* **246**, 132–135.

Ritossa, F. M., Atwood, K. C., and Speigelman, S. (1966). On the redundancy of DNA complementary to amino acid transfer RNA and its absence from the nucleolar organizer region of *Drosophila melanogaster. Genetics* **54**, 663–676.

Robertson, H., and Mathews, M. (1973). Double-stranded RNA as an inhibitor of protein synthesis and as a substrate for a nuclease in extracts of Krebs II ascites cells. *Proc. Natl. Acad. Sci. U.S.A.* **70**, 225–229.

Robertson, H. D., Altman, S., and Smith, J. D. (1972). Purification and properties of a specific *Escherichia coli* ribonuclease which cleaves tyrosine transfer ribonucleic acid precursor. *J. Biol. Chem.* **247**, 5243–5251.

Robertus, J. D., Lander, J. E., Finch, J. T., Rhodes, D., Brown, R. S., Clark, B. F. C., and Klug, A. (1974). Structure of yeast phenylalanine tRNA at 3 Å resolution. *Nature (London)* **250**, 546–551.

Rogers, M. E., Loening, U. E., and Fraser, R. S. S. (1970). Ribosomal RNA precursors in plants. *J. Mol. Biol.* **49**, 681–692.

Rubin, G. M. (1973). The nucleotide sequence of *Saccharomyces cerevisiae* 5.8 S ribosomal ribonucleic acid. *J. Biol. Chem.* **248**, 3860–3875.

Rubin, G. M. (1974). Three forms of the 5.8 S ribosomal RNA species in *Saccharomyces cerevisiae. Eur. J. Biochem.* **41**, 197–202.

Rubin, G. M., and Sulston, J. E. (1973). Physical linkage of the 5 S cistrons to the 18 S and 28 S ribosomal RNA cistrons in *Saccharomyces cerevisiae. J. Mol. Biol.* **79**, 521–530.

Russell, R. L., Abelson, J. N., Landy, A., Gefter, M. L., Brenner, S., and Smith, J. D. (1970). Duplicate genes for tyrosine transfer RNA in *Escherichia coli. J. Mol. Biol.* **47**, 1–13.

Sakano, H., and Shimura, Y. (1975). Sequential processing of precursor tRNA molecules in *Escherichia coli. Proc. Natl. Acad. Sci. U.S.A.* **72**, 3369–3373.

Sakano, H., Yamada, S., Ikemura, T., Shimura, Y., and Ozeki, H. (1974). Temperature sensitive mutants of *Escherichia coli* for tRNA synthesis. *Nucleic Acid Res.* **1**, 355–371.

Saponara, A. G., and Enger, M. D. (1974). The isolation from ribonucleic acid of substituted uridines containing α-aminobutyrate moieties derived from methionine. *Biochim. Biophys. Acta* **349**, 61–77.

Schaefer, K. P., Altman, S., and Söll, D. (1973). Nucleotide modification *in vitro* of the precursor of transfer RNA^{Tyr}. *Proc. Natl. Acad. Sci. U.S.A.* **70**, 3626–3630.

Schedl, P., and Primakoff, P. (1973). Mutants of *Escherichia coli* thermosensitive for the synthesis of transfer RNA. *Proc. Natl. Acad. Sci. U.S.A.* **70**, 2091–2095.

Schedl, P., Primakoff, P., and Roberts, J. (1975). Processing of *E. coli* tRNA precursors. *Brookhaven Symp. Biol.* **26**, 53–76.

Scherrer, K., and Darnell, J. E. (1962). Sedimentation characteristics of rapidly labeled RNA from HeLa cells. *Biochem. Biophys. Res. Commun.* **7**, 486–490.

Schimmel, P. R. (1977). Approaches to understanding the mechanism of specific protein-transfer RNA interactions. *Acc. Chem. Res.* **10**, 411–418.

Schlessinger, D., Ono, M., Nikolaev, N., and Silengo, L. (1974). Accumulation of 30 S pre-ribosomal ribonucleic acid in an *Escherichia coli* mutant treated with chloramphenicol. *Biochemistry* **13**, 4268–4271.

Schmidt, F. J. (1975). A novel function of *Escherichia coli* transfer RNA nucleotidyltransferase. *J. Biol. Chem.* **250**, 8399–8403.

Schmidt, F. J., and McClain, W. H. (1978). Transfer RNA biosynthesis: Alternate orders of ribonuclease P cleavage occur *in vitro* but not *in vivo*. *J. Biol. Chem.* **253**, 4730–4738.

Schmidt, F. J., Seidman, J. G., and Bock, R. M. (1976). Transfer ribonucleic acid biosynthesis. Substrate specificity of ribonuclease P. *J. Biol. Chem.* **251**, 2440–2445.

Schroeder, E., McKibbin, J., Sogin, M. L., and Pace, N. R. (1977). The mode of degradation of precursor-specific RNA fragments by *Bacillus subtilis*. *J. Bacteriol.* **130**, 1000–1009.

Schweizer, E., MacKechnie, C., and Halvorson, H. O. (1969). The redundancy of ribosomal and transfer RNA genes in *Saccharomyces cerevisiae*. *J. Mol. Biol.* **40**, 261–277.

Seidman, J. G., and McClain, W. H. (1975). Three steps in conversion of large precursor RNA into serine and proline transfer RNAs. *Proc. Natl. Acad. Sci. U.S.A.* **72**, 1491–1495.

Seidman, J. G., Barrell, B. G., and McClain, W. H. (1975a). Five steps in the conversion of a large precursor RNA into bacteriophage proline and serine transfer RNAs. *J. Mol. Biol.* **99**, 733–760.

Seidman, J. G., Schmidt, F. J., Foss, K., and McClain, W. H. (1975b). A mutant of *Escherichia coli* defective in removing 3′ terminal nucleotides from some transfer RNA precursor molecules. *Cell* **5**, 389–400.

Sekiya, T., Contreras, R., Küpper, H., Landy, A., and Khorana, H. G. (1976a). *Escherichia coli* tyrosine transfer ribonucleic acid genes: Nucleotide sequence of their promotors and of the regions adjoining C–C–A ends. *J. Biol. Chem.* **251**, 5124–5140.

Sekiya, T., Gait, M. J., Noris, K., Ramamoorthy, B., and Khorana, H. G. (1976b). The nucleotide sequence in the promotor region of the gene for an *E. coli* tyrosine transfer ribonucleic acid. *J. Biol. Chem.* **251**, 4481–4489.

Shine, J., and Dalgarno, L. (1973). Occurrence of heat-dissociable ribosomal RNA in insects: The presence of three polynucleotide chains in 26 S RNA from cultured *Aedes aegypti* cells. *J. Mol. Biol.* **75**, 57–72.

Siev, M., Weinberg, R., and Penman, S. (1969). The selective interruption of nucleolar RNA synthesis in HeLa cells by cordycepin. *J. Cell Biol.* **41**, 510–520.

Simsek, M., RajBhandary, U. L., Boisnard, M., and Petrissant, G. (1974). Nucleotide sequence of rabbit liver and sheep mammary gland cytoplasmic initiator transfer RNAs. *Nature (London)* **247**, 518–520.

Singer, C. E., Smith, G. R., Cortese, R., and Ames, B. N. (1972). Mutant tRNA[His] ineffective in repression and lacking two pseudouridine modifications. *Nature (London), New Biol.* **238**, 72–74.

Sipe, J. E., Anderson, W. M., Jr., Remy, C. N., and Love, S. H. (1972). Characterization of

S-adenosylmethionine:ribosomal ribonucleic acid–adenine (N^6-)methyltransferase of *Escherichia coli* strain B. *J. Bacteriol.* **110**, 81–91.

Smillie, E. J., and Burdon, R. H. (1970). Enzymic conversion of tRNA precursor to 4 S RNA *in vitro*. *Biochim. Biophys. Acta* **213**, 248–250.

Smith, J. D. (1975). Mutants which allow accumulation of tRNA^{Tyr} precursor molecules. *Brookhaven Symp. Biol.* **26**, 1–11.

Smith, J. D. (1976). Transcription and processing of transfer RNA precursors. *Prog. Nucleic Acid Res. Mol. Biol.* **16**, 25–73.

Smith, J. D., Abelson, J. N., Clark, B. F. C., Goodman, H. M., and Brenner, S. (1966). Studies on amber suppressor tRNA. *Cold Spring Harbor Symp. Quant. Biol.* **31**, 479–485.

Smith, I., Dubnau, D., Morell, P., and Marmur, J. (1968). Chromosomal location of DNA base sequences complementary to transfer RNA and to 5 S, 16 S, and 23 S ribosomal RNA in *Bacillus subtilis*. *J. Mol. Biol.* **33**, 123–140.

Smith, J. D., Anderson, K., Cashmore, A., Hooper, M. L., and Russell, R. (1970). Studies on the structure and synthesis of *Escherichia coli* tyrosine transfer RNA. *Cold Spring Harbor Symp. Quant. Biol.* **35**, 21–27.

Smith, J. D., Barnett, L., Brenner, S., and Russell, R. L. (1971). More mutant tyrosine transfer ribonucleic acids. *J. Mol. Biol.* **54**, 1–14.

Snyder, A. L., Kann, H. E., Jr., and Kohn, K. W. (1971). Inhibition of the processing of ribosomal precursor RNA by intercalating agents. *J. Mol. Biol.* **58**, 555–565.

Sogin, M., Pace, B., Pace, N. R., and Woese, C. (1971). Primary structural relationship of p16 to m16 ribosomal RNA. *Nature (London), New Biol.* **232**, 48–49.

Sogin, M. L., Pechman, K. J., Zublen, L., Lewis, B. J., and Woese, C. R. (1972). Observations on the post-transcriptionally modified nucleotides in the 16 S ribosomal ribonucleic acid. *J. Bacteriol.* **112**, 13–16.

Sogin, M. L., Pace, N. R., Rosenberg, M., and Weissman, S. M. (1976). Nucleotide sequence of a 5 S ribosomal RNA precursor from *Bacillus subtilis*. *J. Biol. Chem.* **251**, 3480–3488.

Sogin, M. L., Pace, B., and Pace, N. R. (1977). Partial purification and properties of a ribosomal RNA maturation endonuclease from *Bacillus subtilis*. *J. Biol. Chem.* **252**, 1350–1357.

Söll, D. (1971). Enzymatic modification of transfer RNA. *Science* **173**, 293–299.

Spadari, S., and Ritossa, F. (1970). Clustered genes for ribosomal RNAs in *E. coli*. *J. Mol. Biol.* **53**, 357–368.

Spiers, J., and Birnstiel, M. (1974). Arrangement of the 5.8 S RNA cistrons in the genome of *Xenopus laevis*. *J. Mol. Biol.* **87**, 237–256.

Squires, C., Konrad, B., Kirschbaum, J., and Carbon, J. (1973). Three adjacent transfer RNA genes in *Escherichia coli*. *Proc. Natl. Acad. Sci. U.S.A.* **70**, 438–441.

Stark, B. C., Kole, R., Bowman, E. J., and Altman, S. (1978). Ribonuclease P: An enzyme with an essential RNA component. *Proc. Natl. Acad. Sci. U.S.A.* **75**, 3717–3721.

Sypherd, P. S. (1968). Ribosome development and the methylation of ribosomal ribonucleic acid. *J. Bacteriol.* **94**, 1844–1850.

Taber, R. L., Jr., and Vincent, W. S. (1969). The synthesis and processing of ribosomal RNA precursor molecules in yeast. *Biochim. Biophys. Acta* **186**, 317–325.

Tang, P. W., and Guha, A. (1975). Determination of 5'-triphosphate terminus of 16 S ribosomal RNA precursor. *Nature (London)* **257**, 157–158.

Thammana, P., and Held, W. A. (1974). Methylation of 16 S RNA during ribosome assembly *in vitro*. *Nature (London)* **251**, 682–686.

Toniolo, D., Meiss, H., and Basilico, C. (1973). A temperature sensitive mutation affecting 28 S ribosomal RNA production in mammalian cells. *Proc. Natl. Acad. Sci. U.S.A* **70**, 1273–1277.

Trapman, J., De Jonge, P., and Planta, R. J. (1975). On the biosynthesis of 5.8 S ribosomal RNA in yeast. *FEBS Lett.* **57**, 26–30.

Tsutsumi, K., Majima, R., and Shimura, K. (1976). The biosynthesis of transfer RNA in insects. II. Isolation of transfer RNA precursors from the posterior silk gland of *Bombyx mori. J. Biochem. (Tokyo)* **80**, 1039–1045.

Udem, S. A., and Warner, J. R. (1972). Ribosomal RNA synthesis in *Saccharomyces cerevisiae. J. Mol. Biol.* **65**, 227–242.

Unger, M., Birnbaum, L., and Kaplan, S. (1972). Location of the ribosomal RNA cistrons of *Escherichia coli:* A second site. *Mol. Gen. Genet.* **119**, 377–380.

Valenzuela, P., Bell, G. I., Venegas, A., Sewell, E. T., Masiarz, F. R., DeGennaro, L. J., Weinberg, F., and Rutter, W. J. (1977). Ribosomal RNA genes of *Saccharomyces cerevisiae.* II. Physical map and nucleotide sequence of the 5 S ribosomal RNA gene and adjacent intergenic regions. *J. Biol. Chem.* **252**, 8126–8135.

Valenzuela, P., Venegas, A., Weinberg, F., Bishop, R., and Rutter, W. J. (1978). Structure of yeast phenylalanine-tRNA genes: An intervening DNA segment within the region coding for the tRNA. *Proc. Natl. Acad. Sci. U.S.A.* **75**, 190–194.

Van den Bos, R. C., Retèl, J., and Planta, R. J. (1971). The size and location of the ribosomal RNA segments in ribosomal RNA precursor RNA of yeast. *Biochim. Biophys. Acta.* **232**, 494–508.

Vaughn, M. H., Jr., Soeiro, R., Warner, J. R., and Darnell, J. E., Jr. (1967). The effects of methionine deprivation on ribosome synthesis in HeLa cells. *Proc. Natl. Acad. Sci. U.S.A.* **58**, 1527–1534.

Vesco, C., and Penman, S. (1968). The fractionation of nuclei and the integrity of purified nucleoli in HeLa cells. *Biochim. Biophys. Acta.* **169**, 188–195.

Wagner, E. K., Penman, S., and Ingram, V. M. (1967). Methylation patterns of HeLa cell ribosomal RNA and its nucleolar precursors. *J. Mol. Biol.* **29**, 371–387.

Watanabe, K., Oshima, T., Saneyoshi, M., and Nishimura, S. (1974). Replacement of ribothymidine by 5-methyl-2-thiouridine in sequence GTΨC in tRNA of an extreme thermophile. *FEBS Lett.* **43**, 59–63.

Weatherford, S. C., Rosen, L., Gorelic, L., and Apirion, D. (1972). *Escherichia coli* strains with thermolabile ribonuclease II activity. *J. Biol. Chem.* **247**, 5404–5408.

Wegnez, M., Monier, R., and Denis, H. (1972). Sequence heterogeneity of 5 S RNA in *Xenopus laevis. FEBS Lett.* **25**, 13–20.

Weinberg, R. A., and Penman, S. (1970). Processing of 45 S nucleolar RNA. *J. Mol. Biol.* **47**, 169–178.

Weinberg, R. A., Loening, U., Willems, M., and Penman, S. (1967). Acrylamide gel electrophoresis of HeLa cell nucleolar RNA. *Proc. Natl. Acad. Sci. U.S.A.* **58**, 1088–1095.

Wellauer, P. K., and Dawid, I. B. (1973). Secondary structure maps of RNA: Processing of HeLa ribosomal RNA. *Proc. Natl. Acad. Sci. U.S.A.* **70**, 2827–2837.

Wellauer, P. K., and Dawid, I. B. (1975). Structure and processing of ribosomal RNA: A comparative electron microscopic study in three animals. *Brookhaven Symp. Biol.* **26**, 214–223.

Willems, M., Penman, M., and Penman, S. (1969). The regulation of RNA synthesis and processing in the nucleolus during inhibition of protein synthesis. *J. Cell Biol.* **41**, 177–187.

Wilson, J. H., and Abelson, J. N. (1972). Bacteriophage T4 transfer RNA. II. Mutants of T4 defective in the formation of functional suppressor transfer RNA. *J. Mol. Biol.* **69**, 57–73.

Wilson, J. H., Kim, J. S., and Abelson, J. N. (1972). Bacteriophage T4 transfer RNA. III. Clustering of the genes for T4 transfer RNAs. *J. Mol. Biol.* **71**, 547–556.

Winicov, I. (1976). Alternate temporal order in ribosomal RNA maturation. *J. Mol. Biol.* **100**, 141–155.

Winicov, I., and Perry, R. P. (1974). Characterization of a nucleolar endonuclease possibly involved in ribosomal ribonucleic acid maturation. *Biochemistry* **13**, 2908–2914.

Winicov, I., and Perry, R. P. (1975). Enzymological aspects of processing of mammalian rRNA. *Brookhaven Symp. Biol.* **26**, 201–213.

Wu, M., and Davidson, N. (1975). Use of gene 32 protein staining of single-strand polynucleotides for gene mapping by electron microscopy; application to the $\phi80d_3ilvsu^+7$ system. *Proc. Natl. Acad. Sci. U.S.A.* **72**, 4506–4510.

Yu, M., Vermeulen, C., and Atwood, K. (1970). Location of the genes for 16 S and 23 S ribosomal RNA in the genetic map of *Escherichia coli*. *Proc. Natl. Acad. Sci. U.S.A.* **67**, 27–31.

Yuki, A. (1971). Apparant maturation of *Escherichia coli* ribosomal RNA *in vitro*. *J. Mol. Biol.* **56**, 435–439.

Zeevi, M., and Daniel, V. (1976). Aminoacylation and nucleoside modification of *in vitro* synthesized transfer RNA. *Nature (London)* **260**, 72–74.

Zubay, G., Cheong, L., and Gefter, M. (1971). DNA-directed cell-free synthesis of biologically active transfer RNA: su_{III}^+ tyrosyl-tRNA. *Proc. Natl. Acad. Sci. U.S.A.* **68**, 2195–2197.

10

The Processing of hnRNA and Its Relation to mRNA

Robert Williamson

I. INTRODUCTION

There is a great deal of DNA in most eukaryotic cell nuclei; in humans, there is enough to code for about ten million polypeptide chains each containing 150 amino acids. On the other hand, no single differentiated cell appears to contain more than a few thousand different polypeptides, and some cell types have only a small number of functional proteins, which are characteristic of the tissue of which they are part. The obvious disparity between the large number of gene sequences and the small number of genes expressed as proteins led to interest in the regulation of transcrip-

CELL BIOLOGY, VOL. 3

Copyright © 1980 by Academic Press, Inc.

tion of DNA into nuclear RNA and the processing of nuclear RNA to cytoplasmic messenger RNA (mRNA). This chapter will focus on mammalian cells, since these have been studied most extensively; however, the regulation of transcription seems similar for plants, lower eukaryotes, and other animals, although the complexity of the nuclear DNA increases with ascension of the evolutionary ladder.

Earlier reviews of this subject will give the reader background information and a wealth of detail; in particular, Darnell *et al.* (1973) have sketched in the theoretical framework of the problem of transcriptional and translational controls, and Perry (1976) has given detailed experimental data. Lewin (1975a,b) has placed the subject in the general context of molecular biology. Robertson and Dickson (1975) published a prescient review of nuclear RNA processing before many were thinking about the subject at all. However, following the advances based on recombinant DNA technology allowing direct analysis of both the genome and primary transcripts, a general reassessment of the processing of nuclear RNA transcripts into functional mRNA is appropriate.

II. LABELING OF hnRNA

When the roles of DNA and mRNA were understood, attempts were made to isolate nuclear RNA and to demonstrate that it is the precursor of cytoplasmic mRNA of polysomes. Evidence for such a similar relationship was initially sought in bacterial cells. When a labeled RNA precursor such as [^{3}H]- or [^{14}C]uridine is given in a short pulse to bacterial cells, RNA is rapidly labeled, turns over quickly, becomes associated with ribosomes (but is nonribosomal), and has a base composition resembling that of bulk DNA (Lipmann, 1963). It proved to be remarkably difficult, however, to make sense of data from similar experiments for cultured mammalian cells. Nuclear RNA is indeed rapidly labeled and has a base composition different from ribosomal RNA and resembles cellular DNA, but it is transferred to the cytoplasm in a low amount. Harris, who pointed this out quite vigorously for a number of years, was even led to question the validity of the concept of mRNA in eukaryotic cells (Harris, 1962).

After a short pulse of [^{3}H]uridine, labeled nuclear RNA, isolated by a variety of techniques, sediments through sucrose gradients or runs on acrylamide gels as a broad peak from approximately 10 S to 60 S (MW 0.25×10^6 to 10×10^6). This profile (the so-called bell-shaped curve) is distinctive and invariant; this rapidly labeled nuclear RNA is generally referred to as hnRNA (heterogeneous nuclear RNA). HeLa cells and nucleated avian reticulocytes gives similar profiles, even though the reticulocytes synthesize hemoglobin and little else, whereas HeLa cells make many different proteins. These profiles are particularly distinct with pulse times of less than 10 min, or in the presence of actinomycin D at a level that inhibits primarily ribosomal RNA synthesis, as a result of which no ribosomal precursor (which sediments at 45 S) is visible. Greater than 90% of the RNA labeled during the first 10 min is broken down in the nucleus and does not exit to form polysome-associated messenger RNA in the cytoplasm; in differentiated cells, the figure is nearer to 98% (Brandhorst and McConkey, 1974).

III. THE SEQUENCE COMPLEXITIES OF NUCLEAR AND CYTOPLASMIC RNA's

If post-transcriptional events influence gene expression and account for the low proportion of nuclear RNA that is processed into cytoplasmic RNA, it follows that the number of different sequences transcribed in the nucleus must be greater than the number found in the cytoplasm. Radioactive isotope incorporation is a poor measure of the number of different genes undergoing transcription, since rapid synthesis (and breakdown) of a few sequences cannot be distinguished from synthesis of many different sequences, each at a slower rate. However, DNA–RNA hybridization can be used to determine the number of sequences represented in nuclear and cytoplasmic RNA's. Since the proportion of the genome that hybridizes to ribosomal RNA's is known, and is very small compared to the proportion of the genome that hybridizes to total nuclear or cytoplasmic RNA from most sources, hybridization of total RNA can be used to give an accurate estimate of how much of the DNA is being transcribed. The number of sequences, expressed as total number of nucleotide sequences or as a proportion of the genome, is known as the sequence complexity of a nucleic acid. The sequence complexity of a population of RNA's is the sum of the different sequences represented, and the complexities of nuclear and cytoplasmic RNA's for a tissue or organism can be compared by hybridizing each to DNA separately or by measuring their hybridization of DNA in competition with each other.

In at least six different eukaryotic cell types, the complexities of populations of nuclear RNA's is five to ten times greater than those of cytoplasmic polyadenylated mRNA. This is true for sea urchin gastrulae (Bantle and Hahn, 1976), *Xenopus* liver (Ryffel, 1976), *Drosophila* cells in culture (Levy and McCarthy, 1976), HeLa cells (Herman *et al.,* 1976), and a variety of rat tissues (Chikaraishi *et al.*, 1978). Chikaraishi *et al.* used iodinated single-copy DNA and followed annealing to total nuclear RNA from rat brain, liver, kidney, spleen, and thymus. They assumed that transcription was asymmetric and used great excess of nuclear RNA to ensure that even those sequences present in few copies hybridized to DNA.

Under these conditions, approximately 31% of the unique sequences in the genome are represented as transcripts in brain nuclear RNA (5.9×10^8 nucleotides) and 9% in the case of thymus nuclear RNA. Mixtures of RNA from the two tissues were hybridized, and they found that many nuclear RNA sequences homologous to single-copy DNA are held in common between the tissues studied and form a set of mutually inclusive sequences. There are between 0.1 and 0.3 copies per cell of the "low frequency" class of nuclear RNA's in brain, liver, and kidney; that is, approximately one cell in five has one copy of a given low frequency sequence. Previous results with cDNA transcribed from poly(A)-containing nuclear RNA gave similar, but slightly lower, values; it is not known whether they were lower because of the presence of non-polyadenylated species of RNA, or incomplete transcription by RNA-dependent DNA polymerase, or a methodological problem in the estimation of the kinetics of reannealing.

DNA sequences may be present in multiplicities ranging from single copies to several hundred thousand copies each per single haploid genome; most messenger RNA's are transcribed from DNA sequences which are only present in single copies. (It should be noted that "single" here does not exclude two, three, or even four copies, as is indeed the case for α- and γ-globin genes in humans; however, there are not as many copies of these genes as there are for the genes specifying histone mRNA's or ribosomal RNA's.) The unique and repeated DNA sequences are interspersed with each other in the genome. For several species, several hundred base pairs of reiterated sequence alternate with a thousand or so base pairs of unique sequence for long stretches of the genome (Davidson and Britten, 1973).

Most mRNA's, however, are transcripts of single-copy DNA and do not contain sequences transcribed from repetitious DNA (Campo and Bishop, 1974). In contrast, hnRNA molecules from a number of sources, including rat ascites cells and sea urchin gastrulae, contain interspersed unique and repetitive sequences in host high molecular weight transcripts, resembling the interspersion found in DNA in overall distribution (Smith *et al.*, 1974; Holmes and Bonner, 1974). Thus, assuming hnRNA is a precursor of mRNA, processing must remove repetitious sequences from the nuclear precursors.

IV. PRECURSORS FOR SPECIFIC mRNA's

The experiments described above with complex mixtures of mRNA populations provided strong evidence that mRNA molecules are transcribed as parts of larger molecules. However, experiments with nuclear transcripts of single, defined genes (e.g., the gene for β-globin) were, until recently, much less conclusive. Such experiments require both a means of detecting the specific sequence and a method for estimating accurately the size of molecules containing that sequence. Translation assays were used to detect specific sequences in nuclear RNA (e.g., Williamson *et al.,* 1973), but these were not satisfactory; they can be invalidated by contamination with small amounts of cytoplasmic mRNA, are difficult to quantitate, and presuppose that the translation system can process large nuclear RNA sequences to give active messenger RNA.

Because of this, hybridization to sequence-specific probes is the technique of choice, particularly since this also allows detection of very small amounts of RNA in a quantitative manner. Accurate sizing of nuclear RNA species also poses difficulties. Aggregation can now be eliminated using denaturing agents such as formamide (Holmes and Bonner, 1973); however, it is very difficult to ensure that nicking of large species by adventitious ribonucleases does not occur, and it is impossible to eliminate the possibility of very rapid processing of a primary transcript to a shorter sequence, perhaps even while transcription is proceeding, as occurs in bacteria. Although Melli and Pemberton (1972) introduced the concept of mRNA-specific probes for the detection of nuclear RNA transcripts, the first report of the use of cDNA for this purpose was from Imaizumi *et al.* (1973), who studied the nuclear RNA of duck erythroblasts containing globin mRNA sequences. High molecular weight nuclear RNA hybridized to cDNA for β-globin showed significant (but low) levels of mRNA sequences for globin. These were found in mostly the smaller hnRNA molecules because (a) processing was occurring, (b) the primary transcripts were smaller than most hnRNA, or (c) there was significant contamination with cytoplasmic mRNA for β-globin. A similar analysis using formamide gels from the same laboratory (Spohr *et al.,* 1974, 1976) gave similar results with more accuracy; once again, hybridization to RNA molecules of greater than 28 S occurred, but, under fully denaturing conditions, these larger RNA molecules were an extremely small fraction of the total, and most of the RNA with β-globin sequences comigrated with cytoplasmic globin mRNA.

A number of similar studies have been reported. McNaughton *et al.* (1974) found that the bulk of primary transcripts in duck erythroblast nuclear RNA to be approximately three times the size of the mRNA, but could find no evidence in denaturing gradients for globin transcripts larger than cytoplasmic mRNA. Lanyon *et al.* (1975) found significant levels of globin mRNA in nuclear RNA greater than 30 S from human fetal liver and bone marrow, but McKnight and Schimke (1974) failed to find ovalbumin transcripts larger than ovalbumin mRNA in nuclear RNA from estrogen-stimulated chick oviduct.

Such experiments studied only the steady-state distribution of nuclear RNA sequences and, therefore, could not identify the kinetic relationships between nuclear transcripts and cytoplasmic mRNA. Ideal kinetic analysis requires the isolation of rapidly labeled nuclear RNA specific for a single gene transcript. Lizardi (1976) used column chromatography on a matrix to which the sequence GG$\hat{\text{C}}$GCU was bound as a specific ligand for the isolation of fibroin mRNA from *Bombyx*. After injection of radioactive label, at all times greater than 6 min the size of the labeled fibroin RNA corresponded exactly to that of the cytoplasmic mRNA—but there was a suggestion of the shortest pulse time of a precursor 50% larger than the mRNA.

Other similar experiments, relying on chromatographic isolation of globin-specific nuclear RNA sequences, have been reported during the past few years. Ross (1976) pulse labeled with [³H]uridine primary culture mouse fetal liver cells, which are erythropoietic, and isolated total cellular RNA, which he then fractionated on denaturing formamide–sucrose gradients. Portions of each fraction were hybridized both to highly labeled cDNA to allow detection of mature globin mRNA (unlabeled by the pulse) and to excess unlabeled globin cDNA to detect the pulse-labeled globin gene transcripts. The mature mRNA, as expected, sedimented at 9 S–10 S, but the pulse-labeled globin gene transcript gave a broad band from 10 S to 20 S, with a peak at 14 S. In pulse-chase experiments the 14 S rapidly labeled peak shifted to the 10 S region stoichiometrically, supporting the notion that the longer sequences are indeed precursors to globin mRNA.

Rather than analyzing total cell RNA, Curtis and Weissman (1976) attempted to isolate globin-specific nuclear RNA using affinity chromatography. Erythroleukemia cells were induced with dimethyl sulfoxide (DMSO) to differentiate into erythroblasts and pulse labeled with ³²P. Then globin-specific RNA was isolated by hybridizing to globin cDNA elongated with poly(dC) and chromatographically retained with poly(I)–Sephadex. Two ³²P-labeled peaks, one at 16 S and the other comigrating with mRNA at 9 S to 10 S, were obtained. The processing time found with shorter pulses is less than 20 min. These experiments demonstrated that for globin, a high molecular weight nuclear precursor RNA approximately 2.5 times the size of the mRNA has been found, but no very high molecular weight globin-specific hnRNA was detected.

V. STRUCTURAL GENE SEQUENCES AND INTRONS

The nature of the primary transcript for β-globin mRNA in mammals is now clear. The structural gene for β-globin in mice, rabbits, and humans (and presumably all mammals) contains two inserted noncoding sequences: one several hundred base pairs in length and the other somewhat shorter; both are contained in the region of hnRNA that specifies the mRNA (see Chapter 11). These have been called "introns," "gene inserts," "intervening sequences," etc., and do not occur in the polysomal mRNA (Jeffreys and Flavell, 1977). When rabbit β-globin hnRNA is hybridized to messenger-derived cDNA or to plasmid DNA containing DNA sequences derived from globin mRNA, two loops of RNA that cannot pair with the DNA are seen, indicating that the introns are present in the hnRNA but are processed out to give active mRNA (Tilghman et al., 1978).

This has been confirmed by treating the hybrid between β-globin hnRNA and β-globin cDNA with two enzymes: ribonuclease A, which cleaves single-strand regions of RNA much more rapidly than RNA annealed in a hybrid with DNA, and ribonuclease H, which is specific for hybrids between RNA and DNA (Smith and Lingrel, 1978; Smith et al., 1978). The cleavage products are precisely those expected if introns are transcribed and present in hnRNA and absent from mRNA. Thus, after treatment with RNase A, a single hnRNA–cDNA hybrid remains prior to denaturation. After denaturation, three RNA fragments derived from hnRNA, of the sizes of the RNA regions bounding the introns, are seen.

Since some of these experiments were performed using cloned globin cDNA from plasmid recombinants, no nuclear RNA impurities are represented in the cDNA, and all the observed hybridization is due to globin sequences. As a result of this kind of analysis, J. B. Lingrel (personal communication) has suggested that the 27 S "precursor" found by Bastos and Aviv (1977) using cDNA is not a globin mRNA precursor, but is a reticulocyte mRNA for another protein present in low amounts.

The structure of the chick ovalbumin gene also has been studied intensively by similar techniques; it is present in a single copy in the haploid genome and can be activated by treatment of the chicken with estrogen. Seven different introns have been found to be scattered in the structural gene for ovalbumin, mostly toward the 5′-end of the coding sequence. These contain collectively approximately four times as many base pairs as the sequence that codes for mRNA. The genomic DNA containing both the structural gene and the introns has been cloned in plasmids and is now largely sequenced. Each intron–extron junction appears to be derived from a primeval sequence TCAGGT; all of the junction sequences are similar to such a master sequence, which has presumably diverged from one original sequence (Catterall *et al.*, 1978). No other or longer homologies are found among the introns; they all are of different lengths, and no palindromes are apparent.

Early experiments argued against a high molecular weight precursor for ovalbumin mRNA (McKnight and Schimke, 1974). However, it is now apparent that the primary transcript is, in fact, at least approximately 8.0 kb in length; McKnight and Schimke, who designed their experiments with a smaller precursor in mind, probably lost this large molecule in the phenol–water interface during phenol extraction. This large precursor is slightly longer than the complete structural gene including introns (7.0 kb). A second ovalbumin mRNA precursor, at 6.0 kb, is also seen. There is preliminary evidence for stepwise processing where first the RNA corresponding to one intron, and then another, and so on, is removed, and the remaining RNA is ligated. This implies a mechanism in which a processing site dependent on tertiary structure is not uncovered until a prior processing step occurs.

The introns present in globin and ovalbumin structural genes have now been studied sufficiently to allow certain generalizations to be made. (Several other genes also have been studied in this regard, and it now seems likely that these generalizations will apply for most eukaryotic genes.) All intron sequences are unique; they are not repeated elsewhere in the genome, except that similarities are found in the corresponding introns of related genes. Thus the introns for the mouse major β-globin and minor β-globin genes have related sequences, particularly near to the coding sequences to either side. In each example studied, the intron are transcribed, but there is no evidence that processing occurs outside the nucleus. Since neither palindromic sequences nor regions of extensive base pairing are found, it may be that nuclear proteins interact with specific sequences to allow processing; in fact, nuclear proteins include cleavage enzymes which recognize specific sequences (Williamson, 1973). Such processing also requires enzymes with ligase activity. A careful study of the molecular mechanism of processing of an hnRNA molecule to mRNA is now possible.

Several structural genes, the corresponding mRNA's, and the nuclear precursor RNA's are now being isolated and studied; conalbumin, ovomucoid, lysozyme, thyroglobulin, tubulin (Cleveland *et al.*, 1978), immunoglobulins (Schibler *et al.*, 1978), albumin (Strair *et al.*, 1978), and of course the families of globin genes in a variety of species (Lawn *et al.*, 1978). Each gene contains transcribed introns, and the hnRNA isolated by very careful techniques is the size predicted for the transcription of the final polysomal mRNA sequence plus introns and little else. At most, a further few hundred nucleotides in the several thousand could be near the 5'- or 3'-end of the polysomal mRNA. Therefore, we may now state that the start signal for transcription is almost certainly just to the 5'- end of the DNA sequence corresponding to the start of the mRNA, i.e., a "leader" sequence is present before the transcription initiation codon. Since it is known that the distance between some of these genes and their nearest neighbor is approximately 5 kb (see Chapter 11), it is apparent that most of the intergene sequence is not transcribed (Flavell *et al.*, 1978; Lanyon *et al.*, 1975).

VI. HISTONE GENE TRANSCRIPTION

Although the number of structural genes that have been studied is still small, it seems likely that the arrangement of sequences outlined above is typical for a gene present in nonrepetitive DNA and transcribed to give a polyadenylated mRNA. It may also be true for immunoglobulin mRNA, although in this case a somatic rearrangement of DNA takes place prior to the transcription of the structural gene plus intron to bring the requisite V and C regions together. However, one set of structural genes differs from the other genes in most regards considered above: the histone genes are present in multiple copies (several hundred in sea urchins, several dozen in man) and the polysomal mRNA is not polyadenylated (Kedes, 1979). The genes for the five separate histones are present in a repeat unit in the genome, and contain no introns; at least during cleavage stages, when histone biosynthesis is very rapid, there is no evidence for any transcript larger than the polysomal mRNA for a single histone, i.e., no polycistronic mRNA is observed. However, at later stages such as mesenchyme blastulae, larger transcripts, which may be transcribed from a different set of histone genes, can be identified; these may be a true polycistronic message, containing several different mRNA's, rather than a hnRNA of high molecular weight that is the precursor to a single mRNA (Kunkel *et al.*, 1978; Kunkel and Weinberg, 1978).

VII. POLYADENYLATION AND METHYLATION

At one time it was suggested that polyadenylation served as a marker to distinguish sequences that had to be processed before transfer from the nucleus to the cytoplasm from sequences that would be entirely degraded within the nucleus (Darnell *et al.*, 1973). However, three results argue against such an hypothesis. First, the sequence complexity of nuclear polyadenylated RNA is some five to ten times that of cytoplasmic polyadenylated mRNA (Getz *et al.*, 1975). Second, in nonerythroid tissues, where globin RNA sequences remain in the nucleus and are not processed, they are nevertheless for the most part polyadenylated (Humphries *et al.*, 1976). Finally, Bastos and Aviv (1977) have suggested that the largest globin nuclear precursors are not polyadenylated, while the 15 S precursor is, and that polyadenylation occurs in the middle of a time sequence involving several processing steps. However, since the nature of the 27 S hnRNA molecule is now in dispute, this result must await further clarification.

Some polysomal RNA's that are rapidly labeled with radioactive uridine and have other characteristics of mRNA are not polyadenylated; this class includes histone mRNA's, but also many other sequences. These include previously polyadenylated mRNA's that are aging and therefore have lost poly(A), but also a set of non-cross-hybridizing sequences that are unrelated to poly(A)-containing mRNA's (Milkarek *et al.*, 1974; Nemer *et al.*, 1974). None of these sequences has been identified as coding for particular proteins, and nothing is known of their nuclear transcription.

The addition of poly(A) to the 3'-terminus of the RNA transcript occurs by end addition in the nucleus; the 15 S β-globin precursor is fully polyadenylated. This suggests that polyadenylation occurs in a step unrelated to processing of introns and export of the nuclear mRNA to the polysomes.

Similarly, it is known that polysomal mRNA's are "capped" with a methylated sequence of the general type 7-Me-G$^{5'}$ppp$^{5'}$N-Me-p- at their 5'-terminus (Perry and Kelley, 1974; Salditt-Georgieff *et al.*, 1976). These methylated caps are present in nuclear RNA and apparently are introduced after transcription approximately at the same time as the addition of the poly(A) sequence to the 3'-terminus. In mixtures of mRNA's, as from mouse L cells, and in mixtures of hnRNA's, other methylated bases are found, but for the one defined mRNA studied, that for β-globin, the only methylated sequence is the "cap." This is added in a definite sequence of enzymatic steps; in addition, the methyl groups turn over slowly in the cytoplasm. For several years the fact that both the methylated "cap" and the poly(A) sequence were found simultaneously in hnRNA's was mystifying; now that it is known that processing involves the excision of introns, the simultaneous presence of both terminal markers in a precursor is no longer difficult to understand. Like polyadenylation, it now seems probable that methylation of hnRNA is not related to processing, as had been suspected earlier.

VIII. NUCLEAR RIBONUCLEOPROTEIN

In the cell, all nucleic acid molecules occur as complexes with protein, and these associations can be instrumental in defining nuclease sensitivity and thus processing sites (Williamson, 1973; Robertson et al., 1977). Nuclear ribonucleoprotein complexes (hnRNP) can be isolated as a heterodisperse population sedimenting at up to 400 S, and can be converted into 30 S "units" by RNase (Samarina et al., 1968), suggesting a "beaded" structure similar to that of chromatin or polysomes. Although a number of proteins of molecular weight approximately 40,000 predominate, it is now clear that the total protein population of hnRNP isolated from various tissues is both highly heterogeneous and tissue specific (Pederson, 1974). This may still reflect a specific interaction of only a few proteins with a single transcript; however, to date it has not been possible to isolate an mRNA-specific hnRNP. Data supporting this suggestion of one or a few proteins bound to a particular type of transcript (Pederson, 1974) have been provided by Scott and Somerville (1974), who demonstrated that antibodies to a single structural nuclear protein from *Triturus* oocytes attach specifically only to a few loops of the oocyte lampbrush chromosome, each thought to represent a single transcription unit.

Chromatin-bound hnRNP contains proteins which turn over less rapidly than those associated with "nucleoplasmic" hnRNP (Augenlicht and Lipkin, 1976). Oligo(dT)–cellulose can be used to isolate a class of polyadenylated hnRNP (Kumar and Pederson, 1975), but the separation is not complete and is difficult to evaluate. hnRNP isolated from HeLa cells is held in the cell in a base-paired secondary structure which is similar to the secondary structure found for isolated hnRNA; these double-stranded sequences may well represent functional sites for regulatory proteins involved in mRNA processing (Calvet and Pederson, 1977). Robertson and Dickson (1975) have pointed out analogies for such determinants in prokaryotic systems, and added the novel suggestion that RNA sequences excised during hnRNA processing might serve as primers for RNA transcription at other sites in the genome. Other roles in regulation are, of course, also possible; however, DNA sequences coding for introns are known not to occur at a high multiplicity throughout the genome, which makes general control functions such as those proposed by Britten and Davidson (1969) unlikely.

IX. ADDITIONAL GENETIC ASPECTS

Further data on processing of hnRNA as a genetically determined step come from a study of the hereditary disease β-thalassaemia described by

Comi *et al.* (1977). For the patient studied, no β-globin mRNA sequences are found in the cytoplasm of peripheral red blood cells, but high levels are present in nuclei of bone marrow cells, indicating that a genetic lesion prevents accurate processing. Such a lesion could be enzymatic or could involve a sequence change or deletion at the putative recognition sites at the intron boundaries.

It is very difficult to examine transcription and processing in the intact cell. However, Laird *et al.* (1976) spread nuclei from cultured cells of *Drosophila* and examined the lengths of families of RNA transcript from what were assumed to be structural genes. Long transcripts were seen, but most interesting was the fact that all transcript sizes extrapolated back to a unique origin of transcription. Even now, this is the clearest data available that the start of mRNA precursor transcription occurs at a single site on the DNA in eukaryotes.

The regulation of gene expression may thus be seen as an overlapping hierarchy of controls, as first postulated in Scherrer's cascade hypothesis (Scherrer, 1973). The most basic levels of control, which might include specification of euchromatin and heterochromatin, are the least specific, and may change only with cell division. Other levels, of increasing flexibility, would include fine control of transcription, polyadenylation, methylation, processing of introns and of RNA sequences which lie distal to the 3'- or 5'-ends of the structural genes, and association of hnRNA or nuclear mRNA with specific proteins involved in transport and polysomal association. Further levels of control may exist at the translational and post-translational levels, both for mRNA half-life and for initiation, elongation, and termination stages of translation.

Implicit in such a scheme is a possible economy of regulatory molecules, which operate through class specificity rather than molecular species specificity. Thus, translational specificity may allow the synthesis of a class of primary transcripts, only some of which are processed by specific regulatory molecules to give functional mRNA's.

X. CONCLUSIONS

In conclusion, most mRNA transcripts are from "nonrepetitive" DNA, while nuclear hnRNA contains both repetitive and "nonrepetitive" sequence transcripts. For the 50 to 75% of the total DNA sequences that are "nonrepetitive" (i.e., present in ten copies or less), the majority is transcribed at least at some stage of development and for some type of cell. Of the sequences transcribed for any given cell, only 2 to 20% are processed from nucleus to cytoplasm. Both 5'-methyl group addition ("capping") and 3′-polyadenylation are early nuclear events, occurring before processing is complete.

The primary transcript for structural genes includes RNA transcribed from introns, but appears to contain no long sequences to either end of the structural gene sequence. If "flanking" sequences are transcribed, they are processed from the nuclear RNA while transcription is in progress. Since introns vary from one-fifth or less the size of the coding sequence (α-globin) to three times the size of the coding sequence (ovalbumin), they do not fully explain the difference in hnRNA and mRNA complexities.

The signal for processing of transcripts of introns from hnRNA is not a palindrome; it may be a primary sequence or a secondary structure. Neither polyadenylation nor methylation appears to play a direct part in processing. The nuclear half-life of several mammalian hnRNA's for specific mRNA's is of the order of 5 min.

Many questions, however, remain. Studies of gene organization and transcription in the case of several differentiation-specific proteins have shown the existence of introns (e.g., globin, immunoglobin, fibroin, ovalbumin). It is not certain whether introns occur and are transcribed generally throughout the genome. It is also not known whether the very low levels of nuclear transcript seen for many coding genes represent a normal transcription and processing event leading to translation or an abortive event required for transcriptional control. Neither the nature of the initiation signals for transcription in eukaryotes nor the role of the introns are understood. The entire process acts on RNA's as nuclear nucleoproteins, yet this is ill-characterized and should be the proper subject for the study of the enzymology of processing.

In case the above appears too daunting, it should be emphasized that for the first time, with the development of recombinant DNA methodology, the techniques match the problem. This explains the enormous increase in interest in nuclear RNA at the present time.

ACKNOWLEDGMENTS

I wish to thank my colleagues for their help in studying these problems over some years, and in particular Dr. Chris Coleclough, whose thesis was often used in preparing this review. This work was supported by grants from the Medical Research Council.

REFERENCES

Augenlicht, L. H., and Lipkin, M. (1976). Appearance of rapidly labeled, high molecular weight RNA in nuclear ribonucleoprotein. *J. Biol. Chem.* **251**, 2592.

Bantle, J. A., and Hahn, W. E. (1976). Complexity and characterization of polyadenylated RNA in the mouse brain. *Cell* **8**, 139–150.

Bastos, R. N., and Aviv, H. (1977). Globin RNA precursor molecules: Biosynthesis and processing in erythroid cell. *Cell* **11**, 641–650.

Brandhorst, B. P., and McConkey, E. H. (1974). Stability of nuclear RNA in mammalian cells. *J. Mol. Biol.* **85**, 451.

Britten, R. J., and Davidson, E. H. (1969). Gene regulation for higher cells: A theory. *Science* **165**, 349.

Calvet, J. P., and Pederson, T. (1977). Secondary structure of heterogeneous nuclear RNA: Two classes of double-stranded RNA in native ribonucleoprotein. *Proc. Natl. Acad. Sci. U.S.A.* **74**, 3705–3709.

Campo, S., and Bishop, J. O. (1974). Two classes of messenger RNA in cultured rat cells: Repetitive sequence transcripts and unique sequence transcripts. *J. Mol. Biol.* **90**, 649.

Catterall, J. F., O'Malley, B. W., Robertson, M. A., Staden, R., Tanaka, Y., and Brownlee, G. G. (1978). Nucleotide sequence homology at 12 intron–extron junction in the chick ovalbumin gene. *Nature (London)* **275**, 510–513.

Chikaraishi, D. M., Deeb, S. S., and Sueoka, N. (1978). Sequence complexity of nuclear RNAs in adult rat tissues. *Cell* **13**, 111–120.

Cleveland, D. W., Kirschner, M. W., and Cowan, N. J. (1978). Isolation of separate mRNAs for α- and β-tubulin and characterization of the corresponding *in vitro* translation products. *Cell* **15**, 1021–1033.

Comi, P., Giglioni, G., Barbarano, L., Ottolenghi, S., Williamson, R., Novakova, M., and Masera, G. (1977). Transcriptional and post-transcriptional defects in β⁰-thalassaemia. *Eur. J. Biochem.* **79**, 617–622.

Curtis, P. J., and Weissmann, C. (1976). Purification of globin messenger RNA from dimethylsulfoxide-induced Friend cells and detection of a putative globin messenger RNA precursor. *J. Mol. Biol.* **106**, 1061–1076.

Darnell, J. E., Jelinek, W., and Molloy, G. (1973). Biogenesis of mRNA: Genetic regulation in mammalian cells. *Science* **181**, 1215.

Davidson, E. H., and Britten, R. J. (1973). Organisation, transcription and regulation in the animal genome. *Q. Rev. Biol.* **48**, 565–613.

Flavell, R. A., Kooter, J. M., DeBoer, E., Little, P. F. R., and Williamson, R. (1978). Analysis of the β-δ globin gene loci in normal and Hb Lepore DNA: Direct determination of gene linkage and intergene distance. *Cell* **15**, 25–41.

Getz, M. J., Birnie, G. D., Young, B. D., MacPhail, E., and Paul, J. (1975). A kinetic estimation of base sequence complexity of nuclear poly(A)-containing RNA in mouse Friend cells. *Cell* **4**, 121–129.

Harris, H. (1962). The labile nuclear ribonucleic acid of animal cells and its relevance to the messenger-ribonucleic acid hypothesis. *Biochem. J.* **84**, 60P–61P.

Herman, R. C., Williams, J. G., and Penman, S. (1976). Message and non-message sequences adjacent to poly(A) in steady state heterogeneous nuclear RNA of HeLa cells. *Cell* **7**, 429–437.

Holmes, D. S., and Bonner, J. (1973). Preparation, molecular weight, base composition and secondary structure of giant nuclear ribonucleic acid. *Biochemistry* **12**, 2330.

Holmes, D. S., and Bonner, J. (1974). Interspersion of repetitive and single-copy sequences in nuclear ribonucleic acid of high molecular weight. *Proc. Natl. Acad. Sci. U.S.A.* **71**, 1108–1112.

Humphries, S., Windass, J. D., and Williamson, R. (1976). Mouse globin gene expression in erythroid and nonerythroid tissues. *Cell* **7**, 267–277.

Imaizumi, T., Diggelmann, H., and Scherrer, K. (1973). Demonstration of globin messenger sequences in giant nuclear precursors of messenger RNA of avian Erythroblasts. *Proc. Natl. Acad. Sci. U.S.A.* **70**, 1122–1126.

Jeffreys, A. J., and Flavell, R. A. (1977). The rabbit β-globin gene contains a large insert in the coding sequence. *Cell* **12**, 1097–1108.

Kedes, L. H. (1979). Histone genes and histone messengers. *Annu. Rev. Biochem.* (in press).

Kumar, A., and Pederson, T. (1975). Comparison of proteins bound to heterogeneous nuclear RNA and messenger RNA in HeLa cells. *J. Mol. Biol.* **96**, 353.

Kunkel, N. S., and Weinberg, E. S. (1978). Histone gene transcripts in the cleavage and mesenchyme blastula embryo of the sea urchin, *S. purpuratus. Cell* **14**, 313–326.

Kunkel, N. S., Hemminki, K., and Weinberg, E. S. (1978). Size of histone gene transcripts in different embryonic stages of sea urchin. *Biochemistry* **17**, 2591–2597.

Laird, C. D., Foe, B. E., and Konson, M. E. (1976). Comparative organization of active transcription units in *Oncopeltus sasciatus. Cell* **9**, 131–146.

Lanyon, W. G., Ottolenghi, S., and Williamson, R. (1975). Human globin gene expression and linkage in bone marrow and fetal liver. *Proc. Natl. Acad. Sci. U.S.A.* **72**, 258–262.

Lawn, R. M., Fritsch, E. F., Parker, R. C., Blanke, G., and Maniatis, T. (1978). The isolation and characterisation of linked α- and β-globin genes from a cloned library of human DNA. *Cell* **15**, 1157–1174.

Levy, W. B., and McCarthy, B. J. (1976). Relationship between nuclear and cytoplasmic RNA in *Drosophilia* cells. *Biochemistry* **15**, 2415.

Lewin, B. (1975a). Units of transcription and translation: The heterogeneous nuclear RNA and messenger RNA. *Cell* **4**, 11–20.

Lewin, B. (1975b). Units of transcription and translation: Sequence components of heterogeneous nuclear RNA and messenger RNA. *Cell* **4**, 77–93.

Lipmann, F. (1963). Messenger ribonucleic acid. *Prog. Nucl. Acid. Res.* **1**, 135–161.

Lizardi, P. M. (1976). The size of pulse-labeled fibroin messenger RNA. *Cell* **7**, 239–245.

McKnight, G. S., and Schimke, R. T. (1974). Ovalbumin messenger RNA: Evidence that the initial product of transcription is the same size as polysomal ovalbumin messenger. *Proc. Natl. Acad. Sci. U.S.A.* **71**, 4327–4331.

McNaughton, M., Freeman, K. B., and Bishop, J. O. (1974). A precursor to haemoglobin mRNA in nuclei of immature duck red blood cells. *Cell* **1**, 117–125.

Melli, M., and Pemberton, R. E. (1972). New method of studying the precursor–product relationship between high molecular weight RNA and messenger RNA. *Nature (London), New Biol.* **236**, 172–174.

Milkarek, C., Price, R., and Penman, S. (1974). Metabolism of a poly(A) minus mRNA fraction in HeLa cells. *Cell* **3**, 1–10.

Nemer, M., Graham, M., and Dubroff, L. M. (1974). Co-existence of nonhistone messenger RNA species lacking and containing polyadenylic acid in sea urchin embryos. *J. Mol. Biol.* **89**, 435.

Pederson, T. (1974). Proteins associated with heterogeneous nuclear RNA in eukaryotic cells. *J. Mol. Biol.* **83**, 163.

Perry, R. P. (1976). Processing of RNA. *Annu. Rev. Biochem.* **45**, 605–629.

Perry, R. P., and Kelley, D. E. (1974). Existence of methylated messenger RNA in mouse cells. *Cell* **1**, 37–42.

Robertson, H. D., and Dickson, E. (1975). RNA processing and the control of gene expression. *Brookhaven Symp. Biol.* **26**, 240–266.

Robertson, H. D., Dickson, E., and Jelinek, W. (1977). Determination of nucleotide sequences from double-stranded regions of HeLa cell nuclear RNA. *J. Mol. Biol.* **115**, 571.

Ross, J. (1976). A precursor of globin messenger RNA. *J. Mol. Biol.* **106**, 403.

Ryffel, G. U. (1976). Comparison of cytoplasmic and nuclear poly(A) containing RNA sequences in *Xenopus* liver cells. *Eur. J. Biochem.* **62**, 417–423.

Salditt-Georgieff, M., Jelinek, W., Darnell, J. E., Furuichi, Y., Morgan, M., and Shatkin, A. (1976). Methyl labeling of HeLa cell hnRNA: A comparison with mRNA. *Cell* 7, 227–237.

Samarina, O. P., Lukanidin, E. M., Molnar, J., and Georgiev, G. P. (1968). Structural organisation of nuclear complexes containing DNA-like RNA. *J. Mol. Biol.* **33**, 251.

Scherrer, K. (1973). Messenger RNA in eukaryotic cells: The life history of duck globin messenger RNA. *Acta Endocrinol. (Copenhagen), Suppl.* No. 180, 95–129.

Schibler, U., Marcu, K. B., and Perry, R. P. (1978). The synthesis and processing of the messenger RNAs specifying heavy and light chain immunoglobulins in MPC-11 cells. *Cell* **15**, 1495–1510.

Scott, S. E. M., and Sommerville, J. (1974). Location of nuclear proteins on the chromosomes of newt oocytes. *Nature (London)* **250**, 680–682.

Smith, K., and Lingrel, J. B. (1978). Sequence organisation of the β-globin mRNA precursor. *Nucleic Acid Res.* **5**, 3295–3301.

Smith, K., Rosteck, P., Jr., and Lingrel, J. B. (1978). The location of the globin mRNA sequence within its 16 S precursor. *Nucleic Acid Res.* **5**, 105–115.

Smith, M. J., Hough, B. R., Chamberlin, M. E., and Davidson, E. H. (1974). Repetitive and nonrepetitive sequence in sea urchin heterogeneous nuclear RNA. *J. Mol. Biol.* **85**, 103–126.

Spohr, G., Imaizumi, T., and Scherrer, K. (1974). Synthesis and processing of nuclear precursor–messenger RNA in avian erythroblasts and HeLa cells. *Proc. Natl. Acad. Sci. U.S.A.* **71**, 5009–5013.

Spohr, G., Dettori, G., and Manzari, V. (1976). Globin mRNA sequences in polyadenylated and nonpolyadenylated nuclear precursor messenger RNA from avian erythroblasts. *Cell* **8**, 505–512.

Strair, R. K., Yap, S. H., Nadal-Ginard, B., and Sharfritz, D. A. (1978). Identification of a high molecular weight presumptive precursor to albumin mRNA in the nucleus of rat liver and hepatoma cell line $H_4A_zC_2$. *J. Biol. Chem.* **253**, 1328–1331.

Tilghman, S. M., Tiemeier, D. C., Seidman, J. G., Matija Peterlin, B., Sullivan, M., Maizel, J. V., and Leder, P. (1978). Intervening sequence of DNA identified in the structural portion of a mouse β-globin gene. *Proc. Natl. Acad. Sci. U.S.A.* **75**, 725–729.

Williamson, R. (1973). The protein moieties of animal messenger ribonucleoprotein. *FEBS Lett.* **37**, 1–6.

Williamson, R., Drewienkiewicz, C. E., and Paul, J. (1973). A direct proof of the presence of globin mRNA sequences in the high molecular weight RNA (>35S) from mouse embryonic liver. *Nature (London), New Biol.* **241**, 66–68.

11

Recombinant DNA Procedures in the Study of Eukaryotic Genes

Tom Maniatis

CELL BIOLOGY, VOL. 3

I. INTRODUCTION

The purpose of this chapter is to review recent technical advances in recombinant DNA research and to describe briefly how these techniques are being applied to the study of eukaryotic genes. In the 5 years since the first recombinant DNA experiments, significant advances have been made in the development of new and more sophisticated techniques. The application of these techniques to the study of eukaryotic genes has provided exciting and unexpected findings, the most notable being that many genes are interrupted by noncoding DNA sequences (intervening sequences or introns), including *Drosophila* ribosomal genes (Glover and Hogness, 1977; White and Hogness, 1977), the rabbit β-globin gene (Jeffreys and Flavell, 1977b), mouse β-globin genes (Leder *et al.*, 1977b; Tilghman *et al.*, 1977; Tiemeier *et al.*, 1978), the chicken ovalbumin gene (Breathnach *et al.*, 1977; Doel *et al.*, 1977; Weinstock *et al.*, 1978; Lai *et al.*, 1978), mouse immunoglobulin genes (Brack and Tonegawa, 1977; Tonegawa *et al.*, 1978), yeast tRNA genes (Goodman *et al.*, 1977; Valenzuela *et al.*, 1978), and human δ- and β-globin genes (Flavell *et al.*, 1978; Mears *et al.*, 1978; Lawn *et al.*, 1978).

At the same time, studies with mammalian tumor viruses revealed that only partial colinearity exists between mRNA sequences and the viral DNA template (Berget *et al.*, 1977; Chow *et al.*, 1977; Klessig, 1977; Aloni *et al.*, 1977; Hsu and Ford, 1977; Celma *et al.*, 1977). That is, 5′-terminal sequences of adenovirus and SV40 mRNA's are complementary to viral DNA sequences that are remote from the DNA which is transcribed into sequences coding for structural proteins. In addition, both adenovirus and SV40 early genes contain noncoding DNA within their coding sequences (Berk and Sharp, 1978a,b; C. S. Lai *et al.*, 1978). The use of recombinant DNA techniques will undoubtedly play a central role in studies directed toward the understanding of the function of the noncoding regions within eukaryotic viral and cellular genes.

The primary impact of recombinant DNA research on the study of eukaryotic gene expression thus far has been at the level of understanding gene structure and sequence organization. In the case of immunoglobulins this approach has led to profound conceptual advances in the understanding of the mechanism for generating antibody diversity. With the continuing development of procedures for introducing cloned genes into mammalian cells in culture (see Section VIII) and *in vitro* systems for studying selective gene activation, recombinant DNA research will provide an approach to eukaryotic molecular genetics comparable to that used so effectively to study prokaryotic gene regulation. There seems little doubt that this approach will ultimately lead to major advances in the understanding of eukaryotic genetics and the mechanism of cellular differentiation.

During the past year, two reviews on recombinant DNA research were published (Sinsheimer, 1977; Glover, 1977). In addition, the proceedings from three symposia devoted either fully or partially to recombinant DNA research have been published (Nierlich et al., 1970, Beers and Bassett, 1977; Scott and Werner, 1970). Also, one issue of Science (Volume 196, 1977) was devoted in part to the description of recombinant DNA techniques. Recently, the first volume of a series entitled "Genetic Engineering" edited by J. K. Setlow and A. Hollaender was published. Finally, the United States National Institutes of Health publishes the Recombinant DNA Technical Bulletin on a regular basis, which includes a wide range of topics from the development of new host–vector systems, to reports of scientific meetings, to revisions in regulations which apply to recombinant DNA research. An overview of the application of recombinant DNA procedures to the study of eukaryotic genes up to June 1977 can be found in the Proceedings of the Cold Spring Harbor Symposium of Quantitative Biology on chromatin (Volume 42, 1978). A more recent collection of papers covering the same topic will appear in the proceedings of the 1979 ICN-UCLA Conference on Eukaryotic Gene Regulation (R. Axel et al. 1979).

II. METHODS FOR IN VITRO RECOMBINATION

A. Ligation

1. Cohesive Ends

A large number of restriction endonucleases generate DNA fragments with cohesive ends by making staggered single-stranded cuts [see Roberts (1976) for a review]. A number of these endonucleases can be used for in vitro recombination [see Table 2 in Sinsheimer (1977) for summary]. Glover (1977) presents protocols for ligating cohesive ends using either E. coli or bacteriophage T4 DNA ligase. Purification of these enzymes has been facilitated by transferring their respective genes to bacteriophage λ (Panasenko et al., 1977; N. Murray, personal communication). For example, the gene for E. coli DNA ligase has been transferred to a bacteriophage λ DNA vector which is capable of forming a stable lysogen. Temperature induction of this lysogen results in a 500-fold overproduction of E. coli ligase (Panasenko et al., 1977); similarly, the T4 ligase gene has been transferred to bacteriophage λ (N. Murray, personal communication).

2. Blunt Ends

T4 ligase reseals single-strand nicks in DNA and covalently joins complementary cohesive ends. In addition, the enzyme can be used to join covalently two DNA molecules with blunt or flush ends with 5′-phosphate and 3′-OH groups (Sgaramella, 1972). Sugino *et al.* (1977) compared the cohesive and blunt end ligation activities of T4 DNA ligase. Using the ligation of a blunt-ended synthetic decamer DNA duplex as an assay, they calculated an apparent K_m of 50 μM enzyme for the blunt end activity. This K_m is nearly 100 times higher than that measured for the nicking sealing (cohesive end joining) activity of the enzyme. In addition to a high concentration of enzyme, the blunt end activity requires a high concentration of DNA ends (greater than 1 μM (Sgaramella, 1972). Sugino *et al.* (1977) found that the blunt end activity of T4 DNA ligase could be enhanced significantly by the addition of T4 RNA ligase. The mechanism of this enhancement is not known.

The blunt end ligation activity of T4 ligase makes it possible to join virtually any restriction fragment to a cloning vector. All known blunt ended restriction fragments contain 5′-phosphate and 3′-hydroxyl termini, the substrate for DNA ligase, and it is possible to convert staggered ends of vector DNA to blunt ends by "filling in" the ends using DNA polymerase I or reverse transcriptase. An attractive feature of blunt end cloning procedures is that the original staggered end cleavage site can be regenerated if the appropriate pair of blunt end and enzymatically repaired cohesive ends are joined by ligation (see Fig. 1, for example). A number of different pairs of restriction endonucleases can be employed to regenerate restriction enzyme recognition sequences (Backman *et al.*, 1976).

Another important application of the blunt end ligation activity of T4 DNA ligase has been made possible by the development of synthetic duplex DNA "linkers" (Scheller *et al.*, 1977a; Bahl *et al.*, 1977; Bahl *et al.*, 1978). As shown in Fig. 2 three different synthetic decamer linkers have been synthesized by Scheller *et al.* (1977a). One of these molecules is cleaved by *Eco*RI, one by *Hpa*II or *Bam*HI, and one by *Hind*III. With these linkers virtually any blunt-ended DNA fragment can be adapted for cohesive end ligation, and the cloned DNA fragment can be recovered from the vectors by digestion with the appropriate restriction enzyme. Recently two new types of linkers have been developed for cloning (Bahl *et al.*, 1978). First, a preformed *Bam*HI adapter has been synthesized which does not require cleavage with a restriction enzyme for creating cohesive ends. Second, a *Bam*HI/*Eco*RI conversion adapter has been made which can be used to convert one cohesive end to another.

Synthetic DNA linkers have been used to clone a synthetic lactose

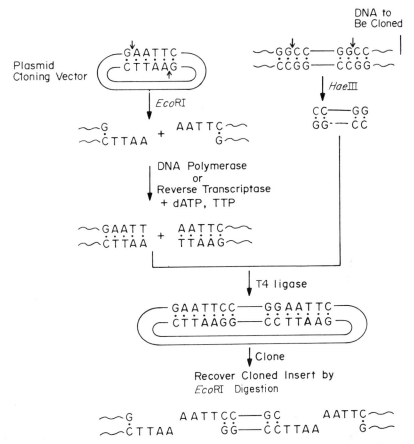

Fig. 1. Schematic diagram illustrating the use of blunt end ligation in gene cloning. The plasmid vector is prepared for ligation by digesting with *Eco*RI and filling in the cohesive ends using the polymerizing activity of *E. coli* DNA polymerase I or reverse transcriptase. The DNA to be cloned is prepared by digesting with the enzyme *Hae*III which produces blunt ends with 5′-terminal cytosine residues. When the two molecules are joined by blunt end ligation the *Eco*RI site in the plasmid is reconstituted, allowing the inserted fragment to be recovered subsequent to cloning. A number of different combinations of staggered and blunt end restriction endonuclease activities can be used to reconstitute the original recognition site. (From Backman *et al.*, 1976.)

operator (Heyneker *et al.*, 1976), repetitive sequences from sea urchin DNA (Scheller *et al.*, 1977b), and synthetic duplex cDNA's (Seeburg *et al.*, 1977). The method has also been used to clone large, random eukaryotic DNA fragments on bacteriophage λ (Maniatis *et al.*, 1978). The most serious technical difficulty with linker-adapted cloning is producing blunt ends on the DNA fragment to be cloned. For example, only a small frac-

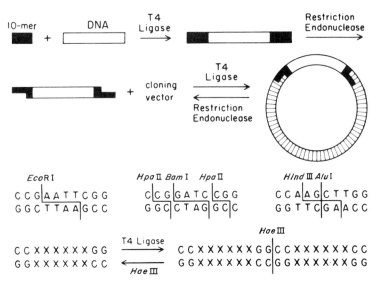

Fig. 2. Schematic diagram showing the use of synthetic linkers in gene cloning. Top: A decameric linker bearing a restriction enzyme cleavage site is joined by T4 DNA ligase to both ends of the DNA to be cloned, and cohesive ends are then produced by treatment with restriction endonuclease. This "sticky ended" DNA can then be incorporated into a vector that has been opened with the same restriction endonuclease. Bottom: Three different chemically synthesized decameric linkers. After blunt end ligation a site recognized by *Hae*III is produced. (Reproduced from Scheller *et al.*, 1977a).

tion of the fragments which are treated with S1 nuclease during their preparation appear to be blunt ended (Scheller *et al.*, 1977b). Attempts to adjust the S1 nuclease reaction conditions (for example, low temperature and high salt) have not had a significant effect on the efficiency of blunt end ligation (Scheller *et al.*, 1977b; Maniatis *et al.*, 1978). Recently this difficulty has been minimized by treating S1-digested complex DNA with *E. coli* DNA polymerase I in the presence of deoxynucleotide triphosphates prior to blunt end ligation (Seeburg *et al.*, 1977). *E. coli* DNA polymerase I will fill in DNA fragments which contain a 5′-terminal single-stranded DNA overhang, and the 3′ → 5′-exonuclease activity of the polymerase will remove 3′ protruding single-stranded DNA sequences (Kornberg, 1969). Seeburg *et al.* (1977) report an enhanced efficiency of linker addition to synthetic duplex cDNA using this procedure.

B. dA-dT, dG-dC Joining

The dA-dT joining procedure of Jackson *et al.* (1972) and Lobban and Kaiser (1973) was first used by Wensink *et al.* (1974) to introduce recombinant DNA into *E. coli*. The procedure [which is described in detail by

Glover (1977)] involves the removal of a small number of bases at the 5'-termini of duplex DNA with λ exonuclease to expose single-stranded 3'-OH termini, and the enzymatic addition of 50–150 residues of dATP or TTP with terminal transferase. By adding dATP to the vector DNA and TTP to the DNA to be cloned, the two molecules can be joined by hydrogen bonding between the homopolymeric tails to form circular hybrid molecules capable of propagating in *E. coli*. Roychoudhury *et al.* (1976) found that the λ exonuclease step could be eliminated by carrying out the terminal transferase reaction in the presence of cobalt rather than magnesium. The disadvantage of this procedure is that the terminal transferase will also add nucleotides to nicks in the DNA with 3'-OH termini. It is, therefore, essential to use DNase-free terminal transferase and intact duplex DNA when using the cobalt reaction conditions. Brutlag *et al.* (1977) have optimized conditions for the terminal transferase reaction. They found that polypurines could be added most efficiently in the presence of high ionic strength cacodylate buffer and cobalt, while the polypyrimidine reaction proceeds most efficiently at low ionic strength in the presence of magnesium. Neither of these conditions requires λ exonuclease treatment of the DNA prior to the terminal transferase reaction.

Eukaryotic DNA inserted into plasmids via the dA-dT joining procedure can be excised and recovered in either one of two ways. The first method involves S1 nuclease digestion of the hybrid plasmid DNA under moderately denaturing conditions which preferentially melt the dA-dT linkers (Hofstetter *et al.* (1976). The efficiency of this reaction depends on the length of the dA-dT linkers. Inserts with short linkers are difficult to excise. Increasing the stringency of the denaturing conditions leads to substantial degradation of the insert and contamination of the insert DNA with plasmid DNA. An alternative excision procedure involves the conversion of plasmid DNA to linear duplex molecules using a restriction enzyme which does not cleave within the inserted DNA (Goff and Berg, 1978). The linear DNA is denatured and briefly renatured to form "snapback" structures between the dA-dT segment on each strand. The vector DNA is then removed by digestion with *E. coli* exonuclease VII, which digests single-stranded DNA in both the 5' to 3' and 3' to 5' directions. The dA-dT region is then melted and the structure is reannealed to form a complete duplex of the excised DNA.

Rougeon *et al.* (1975) have used a dG-dC joining procedure to clone double-stranded rabbit β-globin cDNA. The first step in their procedure is to fill in the *Eco*RI cohesive ends of the plasmid vector and to add poly(dG) to the resulting blunt ends using terminal transferase. Poly(dC) is then added to the DNA to be inserted into the plasmid. Annealing of the dG-dC sequences and transformation should regenerate an *Eco*RI site at each end of the insert. Unfortunately, the *Eco*RI site was rarely reinstated in practice, presumably because of the destruction of the *Eco*RI site by nucleases which contaminated the enzymes used in the joining procedure. It is also possible to reinstate the *Pst*I recognition site by using a dG-dC joining procedure (W. Rowenkamp and R. A. Firtel, personal communication); in this case the plasmid vector is cleaved with *Pst*I 5' (CTGCA↓G) 3' and then tailed with dG. Thus, by tailing the DNA to be inserted with dC, the *Pst*I site is reconstructed during bacterial transformation. As many as 85% of the inserts cloned in this manner can be recovered by digestion with *Pst*I.

III. CLONING VECTORS

A. Plasmids

Sinsheimer (1977) has tabulated most of the plasmid vectors presently in use. Recently, a new "multipurpose" plasmid designated pBR322 has been constructed (Bolivar *et al.*, 1977). This plasmid is small (4+ kb), amplifiable, carries five unique restriction sites into which DNA can be inserted, carries genes for resistance to the antibiotics ampicillin and tetracycline, and is nontransposable. The entire nucleotide sequence of the plasmid has been determined (Sutcliff, 1978a,b). Insertion of DNA into the *Hind*III, *Bam*HI, or *Sal*I restriction sites inactivates the tetracycline-resistance gene, whereas joining DNA to the *Pst*I site inactivates the ampicillin-resistance gene in some cases [see Villa-Komaroff *et al.* (1978) for exceptions]. Insertion of DNA fragments into the *Eco*RI site does not inactivate either drug-resistance gene. This plasmid as well as four others has been certified for use in the bacterial host χ1776 as an EK2 host vector system by the National Institutes of Health (see the *Recombinant DNA Technical Bulletin* for current listing).

One difficulty encountered in the use of plasmid DNA vectors is self-joining to form transforming, nonrecombinant plasmids. The presence of a large number of such transformants significantly increases the amount of labor required to identify clones which carry the inserts of interest. This problem has been dealt with in a number of different ways. (1) One of the primary advantages of the dA-dT or dG-dC joining procedure is that the tailed plasmid DNA is incapable of annealing to form a circular molecule capable of transformation. Thus, most of the transformants obtained with this procedure carry the desired DNA insert. (2) As mentioned above, the plasmid pBR322 carries the resistance genes for both ampicillin and tetracycline. By digesting the DNA with any one of the single cutting enzymes, *Bam*HI, *Hind*III, or *Sal*I, and inserting a foreign DNA fragment, the tetracycline-resistance gene can be selectively inactivated. Those cells transformed by plasmid DNA which have undergone self-ligation can be eliminated from the population by growing transformed cells in the presence of tetracycline and cycloserine. Tetracycline-resistant cells which carry the nonrecombinant plasmid continue to grow and are killed by the incorporation of cycloserine. Because tetracycline is a bacteriostatic drug, the tetracycline-sensitive cells carrying hybrid plasmid DNA can be recovered after removing the tetracycline and cycloserine and growing in the presence of ampicillin. Greater than 90% of the transformants which survive the selection carry inserted DNA in the *Eco*RI site (Rodriguez *et al.*, 1976). (3) T4 ligase requires 5'-terminal phosphate groups for activity. Plasmid DNA which is cleaved with a restriction enzyme and treated with alkaline phosphatase to remove 5'-phosphates can circularize only if the two ends are bridged by a foreign DNA fragment containing 5'-phosphate ends. Thus, alkaline phosphatase treatment of singly cut plasmid DNA prevents the formation of nonrecombinant plasmid DNA (Shine *et al.*, 1977). (4) If pBR322 DNA (Tet[Resistant], Amp[Resistant]) is linearized by digestion with *Eco*RI plus either *Hind*III, *Bam*HI, or *Sal*I, only those molecules which carry one *Eco*RI cohesive end, and one *Bam*HI, *Hind*III, or *Sal*I cohesive end can ligate to the plasmid DNA to form a circular molecule which is Tet[Sensitive], Amp[Resistant]. Similarly, a double digest with *Pst*I and *Eco*RI will yield a vector DNA which is Tet[Resistant], Amp[Sensitive] and can only accomodate *Pst*I/*Eco*RI fragments. The double-digested plasmid vector cannot circularize by intramolecular ligation.

B. Bacteriophage λ

Bacteriophage λ has proved to be an important tool in recombinant DNA research. The relatively detailed understanding of the genetics and biology of the virus (Hershey, 1971) has facilitated the construction of a number of different EK1 and EK2 cloning vectors (Thomas *et al.*, 1974; Murray and Murray, 1974; Rambach and Tiollais, 1974; Murray, 1977; Leder *et al.*, 1977a; Blattner *et al.*, 1977; Williams and Blattner, 1979). The features of all the λ-EK2 vectors presently available are discussed by Skalka (1978). Greater than one-third of the λ chromosome is dispensable for lytic growth and is present as a continuous stretch of DNA in the middle of the genome. Relatively large fragments of DNA (>20 kb) can therefore be cloned and propagated. An additional advantage of using bacteriophage λ cloning vectors is the availability of *in vitro* packaging systems which provide an exceedingly efficient means of introducing recombinant λ DNA into bacterial hosts (Hohn and Murray, 1977; Sternberg *et al.*, 1977). Typically, the efficiency of transfection using intact λ DNA is 10^5–10^6 plaque-forming units (pfu)/μg (Mandel and Higa, 1970). *In vitro* packaging systems are 10–100 times more efficient than this (Hohn and Murray, 1977; Sternberg *et al.*, 1977; Maniatis *et al.*, 1978; Blattner *et al.*, 1978). In this procedure recombinant DNA is mixed with extracts of two different lysogenic bacteria which contain partially assembled λ virion proteins. This DNA is packaged into a phage head by *in vitro* complementation of the two extracts to give an infectious particle carrying the recombinant DNA. Sternberg *et al.* (1977) constructed defective λ lysogens which can be used in recombinant DNA experiments which require EK2-level biological containment (Enquist and Sternberg, 1978). The use of *in vitro* packaging procedures in recombinant DNA experiments is reviewed in detail by Enquist and Sternberg (1978).

Another advantage of using phage λ vectors is that *in vitro* recombinants can be readily screened for inserts of interest by *in situ* plaque hybridization (Benton and Davis, 1977) (see Section IV).

C. Bacteriophage M13

The single-stranded, filamentous bacteriophages M13 and fd have recently been developed as cloning vectors (Messing *et al.*, 1977). The single-stranded DNA of these phages is converted into a circular double-stranded replicative form upon infection, is amplified to 300 copies per cell, and packaged as a single-stranded DNA. Thus, both double- and single-stranded cloned DNA can be obtained from infected cells [see Ray (1971) for a review of the biology of filamentous bacteriphage]. Because of the filamentous shape of the phage, there are no stringent size constraints imposed by the packaging mechanism, as is the case in bacteriophage λ. For example, fd phage have been constructed which carry three copies (15 kb) of an *E. coli* transposon (Herrmann *et al.*, 1978). The recombinant DNA is nearly three times as large as fd DNA, and still grows to high titers (10^{11}/ml). One of the most significant advantages of the cloning system is that single-strand DNA templates which are necessary for the rapid DNA sequencing technique of Sanger *et al.* (1977) can be easily obtained. Since the fragment of interest can be cloned in either orientation both strands of the duplex DNA can be obtained in single-stranded form. In addition, hybridization experiments requiring separated strands of cloned DNA can be accomplished without resorting to laborious strand-separation procedures.

The identification of a nonessential sequence in M13 DNA that can be utilized for *in vitro* recombination was accomplished by inserting a 880 bp *Hind*III restriction fragment carrying the lactose operon into one of the ten *Bsu*I restriction sites present in the phage DNA and showing that the recombinant is viable (Messing *et al.*, 1977). Recently an M13 histidine transducing phage has been constructed which carries a single *Eco*RI site in the nonessential region and can therefore be used as an *Eco*RI vector (Barnes, 1978). Similarly, a portion of the *E. coli* lactose operon which contains the lactose operator–promotor region and the coding sequence for the first 145 amino acids of the β-galactosidase gene has been joined to M13 (Messing and Gronenborn, 1978). Expression of this DNA results in the production of a small peptide which complements the function of the α-region of the β-galactosidase gene and, therefore, results in the formation of blue plaques in the appropriate host and under suitable plating conditions. An *Eco*RI site was introduced into the β-galactosidase coding sequence in such a way that the insertion of a foreign DNA molecule into this site can be detected by the loss of α-complementation. Thus, phage which carry an insert produce a white plaque, while the plaques formed by nonrecombinant phage are blue. A *Hind*III vector which behaves in the same way has recently been constructed (J. Messing, personal communication). In order to use filamentous phage in recombinant DNA experiments it is necessary to use a host bacterium carrying an F factor deficient in the ability to transfer (tra⁻) to other bacterium. Messing has constructed a tra⁻ strain which is an approved EK1 host and is suitable for detecting α-complementation by the M13 strains described above.

D. Cosmids

The packaging of DNA into bacteriophage λ heads *in vivo* and *in vitro* requires at least one cohesive end site (Cos λ) and approximately 35–50 kb of DNA, in addition to the λ head proteins and the λ A gene product (see Umene *et al.*, 1978 for discussion). Hybrid plasmids consisting of colicin EI DNA plus a Cos λ site can be efficiently packaged into phage λ heads *in vivo* (Umene *et al.*, 1978). Such plasmids, which have been termed "cosmids" (Collins and Hohn, 1978), can also be packaged *in vitro* and efficiently introduced into *E. coli* cells (Collins and Hohn, 1978; Collins and Bruning, 1978). Thus a 4–5 kb plasmid containing the Cos λ site and a drug-resistance marker could be joined to as much as 45 kb of DNA and injected into bacteria via the *in vitro* packaging system of λ. This exceeds by a factor of two the size capacity of currently available bacteriophage λ EK2 vectors. The feasibility of this technique has been demonstrated recently by the cloning of large segments of chicken chromosomal DNA containing three different ovalbumin-like genes (Royal *et al.*, 1979). This system cannot be used for EK2-level experiments in the United States at the present time.

IV. METHODS FOR SCREENING RECOMBINANT CLONES FOR
SPECIFIC DNA SEQUENCES

Most eukaryotic genes isolated to date were obtained by cloning partially purified or total genomic DNA fragments and screening the resulting clones using labeled hybridization probes. This approach is practical because of the development of rapid screening methods for both recombinant plasmids (Grunstein and Hogness, 1975) and bacteriophage λ (Benton and Davis, 1977) vector systems. The colony hybridization procedure of Grunstein and Hogness (1975) consists of transferring colonies from a plate onto a nitrocellulose filter lying on top of an agar plate. The colonies are grown on the filter and lysed in alkali. After neutralization, the DNA from the lysed cells binds to the filter and can be hybridized to labeled probe. Hybridization is detected by autoradiography, and the appropriate colonies are picked by aligning a replica plate with the autoradiogram. The most time-consuming steps in this procedure are picking individual colonies onto filters and preparing and cataloging stabs for each colony.

Benton and Davis (1977) found that a single plaque of recombinant phage contains enough phage DNA to be detected by hybridization to labeled probe and that the DNA could be transferred directly from the plaque to a nitrocellulose filter. Thus, the most time-consuming step of transferring individual plaques for regrowth on filters was eliminated. The Benton–Davis procedure consists of plating a large number of phage (5000–10,000) on a single petri dish, blotting the phage and DNA onto a nitrocellulose filter, and then denaturing and fixing the DNA onto the filter as described above. A significant amount of the phage remains on the plate so that after hybridization and autoradiography of the nitrocellulose filter, plaques from the region of the plate which hybridized can be recovered and replated. Thus, this procedure eliminates the need for picking individual plaques and preparing replica plates. Recently, Cami and Kourilsky (1978) developed rapid procedures for screening colonies of bacteria containing recombinant plasmids. In one procedure defective lysogenic bacteria which carry a temperature-sensitive λ repressor are used as hosts in plasmid cloning experiments. The recombinant clones are grown at 32°C overnight, and then shifted to 42°C for a few hours. The temperature shift induces the lysogen which results in the production of lytic λ proteins followed by cell lysis. The DNA from the lysed colonies which contain the hybrid plasmid DNA is blotted onto a nitrocellulose filter and hybridized as described in the Benton–Davis procedure. Fortunately, after the temperature induction every colony contains large numbers of viable bacteria which can be recovered and rescreened. Thus, as with the plaque hybridization procedure, replica plating before screening is unnecessary. With this procedure as many as 25,000 colonies can be screened on a single plate. In another procedure recombinant plasmids in normal nonlysogenic hosts are screened by growing the bacterial colonies on one plate, replicating them onto a nitrocellulose filter, then incubating the filter on plates containing ampicillin which promotes cell lysis. The procedure works, but the hybridization signals were more than tenfold weaker than those obtained using the lysogenic strains as hosts (Cami and Kourilsky, 1978).

V. ISOLATION OF SINGLE-COPY STRUCTURAL GENE SEQUENCES

A. cDNA Cloning

With the development of procedures for synthesizing and cloning double-stranded DNA copies of poly(A)-containing mRNA's, it has been possible to obtain pure mRNA sequences in relatively large amounts. The feasibility of this approach to gene isolation was established using rabbit globin mRNA (Rougeon et al., 1975; Maniatis et al., 1976; Rabbits, 1976; Higuchi et al., 1976) and later applied to numerous other mRNA sequences [see Efstratiadis and Villa-Komaroff (1979), for a review of cDNA cloning procedures and a complete listing of published cDNA clones]. The procedure most often used to synthesize double-stranded cDNA involves reverse transcription of the mRNA under conditions favoring the production of cDNA copies equal to the template length; synthesis of a second DNA strand from a hairpin loop at 3'-end of the cDNA using E. coli DNA polymerase I or reverse transcriptase; and cleavage of the loop joining the two DNA strands with S1 nuclease. The synthetic duplex DNA is inserted into a bacterial plasmid or λ DNA cloning vector in one of several ways: (1) dA-dT joining to an EcoRI cleavage site (Maniatis et al., 1976; Higuchi et al., 1976), (2) dG-dC joining to a PstI cleavage site (W. Rowenkamp and R. A. Firtel, personal communication; Villa-Komaroff et al., 1978), (3) blunt end ligation to synthetic DNA linker molecules, followed by cleavage of the linkers with the appropriate restriction enzyme to generate cohesive ends which are then ligated to the vector DNA (Seeburg, 1977).

In one procedure of cDNA cloning, a mRNA–cDNA complex is tailed directly with poly(dA) and then inserted into the dT-tailed *Eco*RI site of a plasmid (Wood and Lee, 1976). This procedure has not been used extensively because of the relatively low efficiency of RNA–DNA duplex cloning, but has been useful in the cloning of certain purified mRNA's (Zain *et al.*, 1979).

Cloned cDNA has provided an important tool for studying the structure of eukaryotic genes. For example, the nucleotide sequence of a number of mRNA's has been derived from cDNA clones (Efstratiadis *et al.*, 1977; Browne *et al.*, 1977; Liu *et al.*, 1977; Seeburg *et al.*, 1977; Shine *et al.*, 1977; Ullrich *et al.*, 1977; McReynolds *et al.*, 1978). cDNA clones have also been used to map restriction sites surrounding eukaryotic genes in genomic DNA (Jeffreys and Flavell, 1977a,b; Breathnach *et al.*, 1977; Lai *et al.*, 1978; Weinstock *et al.*, 1978), using a method which was first used to map restriction sites surrounding integrated SV40 DNA (Botchan *et al.*, 1976). The method consists of digesting total genomic DNA with one or more restriction enzymes, separating the products by agarose gel electrophoresis, transferring (blotting) the fractionated DNA from the gel onto a nitrocellulose filter (Southern, 1975), hybridizing the filter to a cDNA plasmid which has been ^{32}P-labeled *in vitro* by nick translation (Maniatis *et al.*, 1975; Rigby *et al.*, 1977), and visualizing the hybridization by autoradiography. I will refer to this procedure as "genomic blotting."

B. Isolation of Genes by Transformation and Selection

Several different yeast genes have been isolated using biochemical selection procedures (Struhl *et al.*, 1976; Struhl and Davis, 1977; Ratzkin and Carbon, 1977). This is accomplished by joining random fragments of yeast DNA to bacteriophage or plasmid DNA vectors, introducing the hybrid DNA molecules into an *E. coli* host carrying an appropriate auxotrophic mutation, and growing the transformed cell under selective conditions. Survivors carry a hybrid DNA bearing a yeast gene which can be expressed in *E. coli*. For example, hybrid DNA molecules have been isolated that complement mutants of *E. coli* lacking the enzyme activity necessary for histidine biosynthesis (Struhl and Davis, 1977; Ratzkin and Carbon, 1977). The complementing activities were shown to be derived from the yeast DNA by both biochemical and genetic experiments.

C. Isolation of Genes by Partial Purification and Cloning

One approach to cloning specific segments of eukaryotic genomes is to partially purify restriction fragments of genomic DNA that carry the gene of interest, join them to an appropriate cloning vector, and screen the resulting recombinants with specific hybridization probes. The partial purification step is included to reduce the number of clones which must be screened. For example, digestion of total mouse DNA with *Eco*RI produces approximately one million different fragments with a number average size of 2300 base pairs (Botchan *et al.*, 1973). With a genome size of 3×10^9 base pairs it would be necessary to screen approximately six million plaques to achieve a 99% probability of obtaining any given sequence from the genome (Clarke and Carbon, 1976). Two different gene enrichment schemes have been successfully applied to the problem of cloning single copy genes from mammalian DNA (Tonegawa *et al.*, 1977a,b; Tilghman *et al.*, 1977). First, Tonegawa *et al.* (1977a,b) used a two-step purification procedure to achieve a 350-fold enrichment of a genomic mouse DNA fragment bearing an immunoglobulin variable region gene. The first step in the procedure was to digest mouse DNA with *Eco*RI and fractionate the products by preparative agarose gel electrophoresis. The gel was sliced, the DNA was eluted, and a portion was hybridized to ^{125}I-labeled λ-chain immunoglobulin mRNA to identify the fractions from the gel which contain immunoglobulin gene sequence. The second step in the procedure involved preparative cesium chloride sedimentation of specific mRNA–DNA hybrids (R loop structures). The fractions from the first step of the enrichment procedure were mixed with purified mRNA under conditions in which RNA–DNA duplexes were formed without complete denaturation of the duplex DNA [R loop formation (White and Hogness, 1977)]. If the correct conditions are chosen, the RNA will hybridize specifically to the DNA fragment of interest to produce a molecule which contains a region of RNA–DNA duplex which can be separated from the duplex DNA when centrifuged to equilibrium in a cesium chloride gradient. The effectiveness of this enrichment step depends on the relative size of the mRNA probe and the *Eco*RI fragment bearing the gene. If the ratio is very small, little if any, enrichment would be expected.

A second gene enrichment procedure involves the use of reverse-phase chromatography combined with preparative agarose gel electrophoresis (Tilghman *et al.*, 1977; Edgell *et al.*, 1979). Large amounts of restriction enzyme-cleaved DNA is fractionated by reverse-phase chromatography (Landy *et al.*, 1976; Hardies and Wells, 1976), and the fractions containing the sequence of interest are identified by hybridization of portions of each fraction to a labeled probe. An enrichment of approximately ten- to twentyfold can be achieved by this procedure. A second "dimension" of purification is then accomplished by preparative agarose gel electrophoresis of the appropriate fractions from the reverse-phase column using a specially constructed preparative agarose gel electrophoresiselectroelution device (Edgell *et al.*, 1979). An overall enrichment of approximately 500-fold can be achieved by these two purification steps (Tilghman *et al.*, 1977).

The partially purified immunoglobulin (Tonegawa *et al.*, 1977a,b) and globin (Tilghman *et al.*, 1977) genes were joined to the bacteriophage λ EK2 vector λgtWES · λB (Leder *et al.*, 1977b), the recombinant DNA introduced into *E. coli* by transfection (Mandel and Higa, 1970) and the recombinant phage screen by *in situ* plaque hybridization procedures (Benton and Davis, 1977). A number of different genes have been isolated using this procedure, including mouse β-globin genes (Tilghman, 1977), fragments of the chicken ovalbumin gene (Garapin *et al.*, 1978b; Woo *et al.*, 1978), and κ-chain mouse immunoglobulin (Seidman *et al.*, 1978). The major disadvantage of the procedure is that genes containing *Eco*RI sites must be cloned in pieces (Garapin *et al.*, 1978b; Woo *et al.*, 1978).

D. Direct Isolation of Structural Genes from Cloned Genomic DNA

With the development of efficient *in situ* plaque hybridization procedures it was possible to isolate genes without a preenrichment step. A limited *Eco*RI digest of unfractionated genomic DNA is cloned in bacteriophage λ vectors using *in vitro* packaging procedures, and the recombinants are screened directly for the gene of interest (Garapin *et al.*, 1978a; Blattner *et al.*, 1978). This procedure is more straightforward than gene enrichment, but does not solve the problem of cloning genes containing *Eco*RI sites in pieces.

Recently, a procedure has been established which does not require partial purification of the gene prior to cloning, and the gene is isolated intact (Maniatis *et al.*, 1978). If random fragments of genomic DNA are cloned and the number of recombinants is large enough for complete sequence representation, in principle, any gene can be isolated by screening the collection or "library" with the appropriate hybridization probe. The feasibility of constructing libraries from the DNA of organisms with small genomes such as *Drosophila* or yeast has been previously demonstrated (Wensink *et al.*, 1974; Glover *et al.*, 1975; Carbon *et al.*, 1977). However, construction of libraries from larger genomes was not attempted because of technical limitations, especially those related to screening large numbers of clones. The availability of EK2 certified bacteriophage cloning vectors (Leder *et al.*, 1977a; Blattner *et al.*, 1977; Williams and Blattner, 1979), the development of a rapid *in situ* plaque hybridization procedure (Benton and Davis, 1977), and the development of more efficient procedures for introducing λ DNA into *E. coli* (*in vitro* packaging, Sternberg *et al.*, 1977; Hohn and Murray, 1977) have made possible the construction and screening of DNA libraries of complex genomes (Maniatis *et al.*, 1978). The strategy used to construct libraries of eukaryotic DNA is outlined in Fig. 3. In brief, high molecular weight DNA is fragmented either by

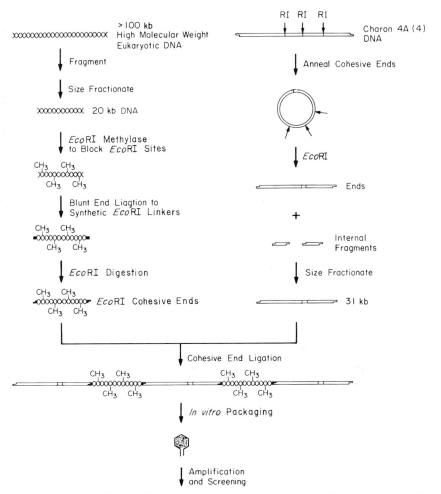

Fig. 3. Schematic diagram illustrating procedure for constructing libraries of eukaryotic DNA. High molecular weight eukaryotic DNA is fragmented, size fractionated, and reacted with *Eco*RI methylase to render the *Eco*RI recognition sites resistant to cleavage. It is then possible to join synthetic *Eco*RI linkers to the DNA and cohesive ends subsequently produced by digestion with *Eco*RI. The vector is prepared by annealing the cohesive ends and digesting with *Eco*RI to remove the internal fragments carrying genes nonessential to phage viability. The annealed end fragments are then ligated to the eukaryotic DNA at high concentrations to generate concatemeric hybrid DNA molecules which are packaged into viable recombinant phage *in vitro*. (Reproduced from Maniatis *et al.*, 1978).

shearing followed by S1 nuclease treatment or by a nonlimit endonuclease digestion with restriction enzymes generating molecules with blunt ends. Molecules of approximately 20 kb are selected by preparative sucrose gradient centrifugation and rendered EcoRI-resistant by treatment with EcoRI methylase. Synthetic DNA linkers bearing EcoRI recognition sites are covalently attached by blunt end ligation using T4 ligase. Cohesive ends are then generated by digestion with EcoRI. The Charon 4A strain of bacteriophage λ (Blattner et al., 1977) was used as a cloning vector. As shown in Fig. 3, λ Charon 4A DNA is cleaved three times by EcoRI to produce two internal fragments and two end fragments. The internal fragments carry genes which are not essential to phage viability and, therefore, can be removed and replaced with eukaryotic DNA. The internal fragments are separated from the end fragments by annealing the cohesive ends of the molecule, digesting the circular DNA with EcoRI, and fractionating the products on a sucrose gradient. The 31 kb end fragments are readily separated from the 7 and 8 kb internal fragments. Recombinant phage molecules can then be obtained by joining the 31 kb end fragments of the vector to the 20 kb eukaryotic DNA by EcoRI cohesive end ligation.

Because of the enormous complexity of the mammalian genome, a very large number of independent phage recombinants is required if the collection of cloned sequences is to contain all of the sequences present in the genome. For example, the calculated number of independently derived recombinant phage necessary to achieve a 99% probability of having any given DNA sequence in a mammalian library (for 20 kb DNA inserts) is approximately 7×10^5. It is possible to obtain the necessary number of phage recombinants using the in vitro packaging procedures described above. Once a large number of viable phage recombinants is obtained, they can be amplified 10^6-fold by low-density growth in bacteria on agar plates with no apparent loss in sequence complexity. The recombinant phage then constitute a "permanent" library of genomic DNA fragments which can be repeatedly screened for sequences of interest. Cloned libraries of Drosophila, silkmoth, and rabbit DNA (Maniatis et al., 1978), chicken DNA (Dodgson et al., 1979), and human DNA (Lawn et al., 1978), have been constructed using the linker procedure described in Fig. 3. At the same time, libraries of high molecular weight DNA fragments produced by nonlimit EcoRI digestion have been constructed using sea urchin (D. Anderson and E. Davidson, personal communication), Drosophila (R. Robinson, N. D. Hershey, and N. Davidson, personal communication), rat (T. Sargent, B. Wallace, and J. Bonner, personal communication), and mouse (M. Davis and L. Hood, personal communication) DNA. All of these libraries have been successfully screened for specific gene sequences.

The primary advantage of this rapid method of gene isolation is that all of the members of a family of evolutionarily or developmentally related genes can be isolated in a single step by screening a library with mixed probe. Furthermore, isolation of a set of overlapping clones, all of which contain a given gene permits study of sequences extending many kilobases from the gene in the 5' and 3' directions. Moreover, even more distant regions along the chromosome can be obtained by rescreening the libraries using terminal fragments of the initially selected clones as hybridization probes. Thus, it is possible to study the organization of closely linked genes using this procedure of gene isolation.

VI. ISOLATION AND CHARACTERIZATION OF SPECIFIC EUKARYOTIC GENES

In the following section I will review the application of recombinant DNA procedures to the study of four different types of eukaryotic gene systems: histone, ovalbumin, immunoglobulin, and globin. I chose these systems because they are the most thoroughly studied at the present time and because they all provide excellent examples of the impact of recombinant DNA procedures on the study of eukaryotic genes. By necessity, the discussion in each case is not comprehensive. However, I refer to reviews which include a more complete discussion of the biology of each system.

A. Histone Genes

Histone gene expression is regulated during the cell cycle, closely coupled to DNA replication (see Kedes, 1976, 1979, for discussion, and Robbins and Borun, 1967). In addition, different sets of histone genes are expressed during different stages of development (Cohen *et al.*, 1975; Arceci *et al.*, 1976; Weinberg *et al.*, 1977). For example, analysis of developmental stage-specific histone mRNA's indicates that a different set of histone genes is expressed during different stages of sea urchin embryogenesis (Newrock *et al.*, 1978; Kunkel and Weinberg, 1978; M. Grunstein, personal communication). Sea urchin histone genes have been cloned using plasmid (Kedes *et al.*, 1975) and bacteriophage λ DNA (Clarkson *et al.*, 1976) cloning vectors [see Kedes (1976) for review]. The unique base composition and organization of the sea urchin histone genes was exploited to achieve substantial purification of the DNA prior to cloning. Characterization of the cloned DNA revealed that the gene order and direction of transcription of the five histone genes is $5' \rightarrow H4 \rightarrow H2B \rightarrow H3 \rightarrow H2A \rightarrow H1 \rightarrow 3'$ (Schaffner *et al.*, 1976; Cohn *et al.*,

1976; Gross *et al.*, 1976). The G + C-rich genes are separated from each other by A + T-rich spacer sequences. The nucleotide sequence of a large portion of the gene and spacer regions of a single cluster has been determined (Sures *et al.*, 1976; Birnstiel *et al.*, 1977; Schaffner *et al.*, 1978). These data indicate that none of the histone genes in *Drosophila* contain noncoding intervening sequences.

Drosophila melanogaster histone genes were isolated by screening a collection of hybrid plasmids with sea urchin histone mRNA (Lifton *et al.*, 1977). *In situ* hybridization experiments indicated that the histone plasmid sequences are located in region 39DE of *D. melanogaster* salivary gland polytene chromosomes. As shown in Fig. 4 the *Drosophila* histone genes, in contrast to those of the sea urchin, are not transcribed from the same DNA strand. This argues strongly against the possibility that the histone gene cluster is copied as a single polycistronic transcript that is cleaved to yield five mature histone mRNA's. Nucleotide sequence analysis of the histone gene cluster indicates that none of the histone genes in *Drosophila* contain noncoding intervening sequences. Some heterogeneity in the organization of the histone gene cluster has been detected in *Drosophila*. A fraction of the histone gene clusters contain an insertion between *H1* and *H3* genes (Lifton *et al.*, 1977).

B. Immunoglobulin Genes

Studies on the structure and organization of cloned immunoglobulin genes have provided important insights into the molecular basis of antibody diversity (Tonegawa *et al.*, 1977b). Evidence for rearrangement of

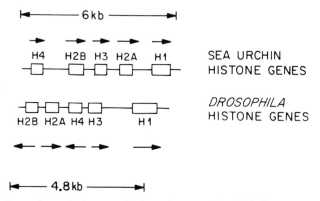

Fig. 4. Comparison of the organization of sea urchin (Schaffner *et al.*, 1978) and *Drosophila* (Lifton *et al.*, 1977) histone genes. The arrows indicate the direction of transcription of each gene.

immunoglobulin λ_{II} light chain gene sequences during differentiation of lymphocytes was obtained by digesting mouse embryonic or myeloma DNA with the restriction enzyme BamHI, fractionating the DNA according to size by agarose gel electrophoresis, recovering the DNA from the gel, and hybridizing to constant or variable region-specific probes. The hybridization profiles of the two DNA's were different, suggesting that the developmental rearrangement model for the generation of antibody diversity is correct (Hozumi and Tonegawa, 1977). However, alternative interpretations of the data could not be rigorously excluded. In order to obtain more definitive evidence, the λ_{II} immunoglobulin light chain variable gene was cloned from embryonic (Tonegawa et al., 1977a) or plasmocytoma (Brack and Tonegawa, 1977) DNA. When constant or variable region mRNA probes were hybridized to the cloned embryo DNA, only variable region sequences were found, confirming the conclusion that the variable and constant region sequences are not contiguous in germ line DNA. Definitive evidence for this conclusion was obtained by determining the nucleotide sequence of the cloned gene (Tonegawa et al., 1978). In addition, comparison of the nucleotide sequence with the amino acid sequence data revealed the presence of a 93 base pair intervening sequence located between the sequence for hydrophobic leader and the rest of the gene sequence. When another light chain immunoglobulin gene was isolated from myeloma DNA, both constant and variable region sequences were found on the cloned DNA fragment, but they were separated by 1250 base pairs of noncoding intervening sequences (Brack and Tonegawa, 1977). Thus, it seems that both DNA translocation and RNA splicing are involved in immunoglobulin gene expression.

The cloning experiments with λ light chain genes favors the somatic mutation model for anitbody diversity, since only one or a few variable sequences are present in embryonic DNA. In contrast, recent cloning experiments with κ light chain variable regions favor the germ line hypothesis for antibody diversity (Seidman et al., 1978). When mouse myeloma DNA is digested with EcoRI and fractionated by RPC5 chromatography followed by agarose gel electrophoresis and the DNA is probed with the variable region κ chain probe, six different EcoRI fragments can be detected. Two of these fragments were cloned and characterized (Seidman et al., 1978). Only the variable region sequence was present in the two clones, indicating that these particular variable region sequences are separated from the constant region in myeloma DNA. Nucleotide sequence analysis of the variable region genes carried out on the two different clones indicates that the two genes have similar but not identical sequences. Thus, it appears that two EcoRI fragments carry different members of a κ variable region subgroup. Assuming that all

*Eco*RI fragments studied in hybridization experiments carry different κ chain variable region sequences, 25 to 30 κ gene subgroups identified by amino acid sequence data studies could account for 125–150 distinct variable region genes. There could be as many as 100 subgroups, suggesting that at least some of the observed antibody diversity can be accounted for without proposing a somatic mutation mechanism.

Insight into the mechanism of immunoglobulin gene rearrangement was provided by DNA sequence analysis of the regions where the variable and constant regions of both the λ and κ immunoglobulin light chains are joined (Bernard *et al.*, 1978; Sakano *et al.*, 1979a). In embryonic mouse DNA a short sequence called *J* is located 1.2 kb upstream from the λI type light chain constant region coding sequence. The *J* sequence is homologous to a sequence near the junction of the constant and variable region in immunoglobulin λ light chain mRNA (Brack *et al.*, 1978). This and the fact that the constant and variable region sequences are separated by 1.2 kb in myeloma DNA suggests that *V-C* joining takes place by site-specific recombination between the 3' end of the *V* gene sequence and the 5' end of the *J* sequence. This suggestion was confirmed by DNA sequence analysis of the *V-C* junction region of the λ light chain constant and variable gene sequences in embryonic and myeloma DNA (Brack *et al.*, 1978). No obvious sequence homologies are detected at the two recombination sites in embryonic DNA.

In the κ light chain system the secreted immunoglobulin chain is also encoded in three separate DNA segments (V_κ, J_κ, and C_κ) and joining takes place between embryonic V_κ and J_κ segments. Nucleotide sequence analysis of the *J* cluster and the 3' end of embryonic V_κ DNA led to the identification of a sequence 5' to each *J* region which could form an inverted stem structure with a sequence *J'* to the embryonic V_κ genes. This information and genomic blotting experiments with total embryo and myeloma DNA led Sakano *et al.* (1979a) to propose that site-specific somatic recombination between V_κ and J_κ may result in the excision of an entire DNA segment between the *V* gene and the *J* DNA segment (see Fig. 5). Production of a mature mRNA is then accomplished by transcribing the entire *V, J*, intron, *C* regions following by splicing out the intron and extraneous J region sequences. Evidence for transcription and processing of a large nuclear RNA precursor is provided by Gilmore-Hebert *et al.* (1978).

Analysis of the structure of cloned immunoglobulin heavy chains (Sakano *et al.*, 1979b; Early *et al*, 1979) has suggested at least one possible function of introns: to allow new proteins to be formed from parts of old ones by recombination within intron sequences (see Gilbert, 1978, and Blake, 1979a,b for a discussion of this point). The essential observation is

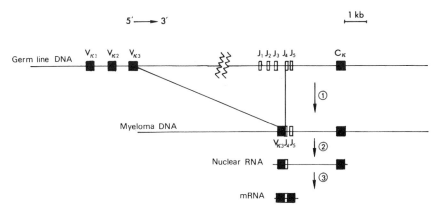

Fig. 5. Somatic recombination model for joining κ light chain variable and constant region coding sequences during lymphocyte development. In germ line DNA different variable region coding sequences ($V_{\kappa 1}$, $V_{\kappa 2}$, . . . , V_{κ_n}) are separated by an unknown number of base pairs from a cluster of different J region coding sequences (J_1, J_2, . . . , J_n) and a single constant region gene (C_κ). A site-specific somatic recombination event occurs between the 3'-end of one of the V_κ sequences and the 5'-end of one of the J region sequences to produce a fused $V_\kappa J_\kappa$ segment separated from the constant region gene by a noncoding intervening sequence (intron) which includes the $J5$ region (1). The region between V and J is thought to be deleted as a result of the recombination event. The entire V to C region is transcribed into RNA (2) and the intron and $J5$ sequence are removed by splicing to form a mature immunoglobulin mRNA (3). (See Sakano *et al.*, 1979a, for discussion.)

that the DNA sequences encoding the three constant region functional domains in γ (Sakano *et al.*, 1979b) and α (Early *et al.*, 1979) heavy chain immunoglobulins are separated by introns.

C. The Chicken Ovalbumin Gene

The chicken ovalbumin gene is one of the best studied examples of a gene whose activation is hormone dependent (Shrader and O'Malley, 1976). Nearly full-length double-stranded DNA copies of ovalbumin mRNA have been synthesized (Monahan *et al.*, 1977) and cloned in bacterial plasmids (McReynolds *et al.*, 1977; Humphries *et al.*, 1977). Recently, the complete ovalbumin mRNA sequence was determined using one of these cDNA clones (McReynolds *et al.*, 1978). cDNA plasmids have also been used as hybridization probes to study the position of restriction enzyme cleavage sites within and surrounding the ovalbumin gene in cellular DNA (Breathnach *et al.*, 1977; Doel *et al.*, 1977; Weinstock *et al.*, 1978; Lai, C. J. *et al.*, 1978a). Comparison of the restriction endonuclease cleavage map of cloned duplex DNA with that of the gene in cellular DNA indicated that at least two noncoding intervening sequences are present in the ovalbumin gene.

The structural characterization of the ovalbumin gene was accomplished by analyzing individual *Eco* RI framents obtained by gene enrichment cloning procedures (Garapin *et al.*, 1978a,b; Mandel *et al.*, 1978; Woo *et al.*, 1978; Dugaiczyk *et al.*, 1978) and by the isolation of the complete gene from a bacteriophage λ library of chicken DNA (Gannon *et al.*, 1979; Dugaiczyk *et al.*, 1979). The structural studies included restriction enzyme mapping experiments, DNA sequence analysis, and R loop studies. The salient features of the structure are as follows (See Fig. 6): (1) The ovalbumin mRNA sequence of 1859 nucleotides is interrupted by seven noncoding intervening sequences which range in size from approximately 300 to 1600 bp (see Fig. 5). The minimal size of the transcriptional unit is 7.7 kb. (2) Unlike globin genes, the ovalbumin gene contains a leader sequence of 45 bp which is separated from the rest of the 5′ untranslated mRNA sequence by approximately 1500 bp of DNA. (3) A nuclear RNA transcript of approximately 7800 nucleotides can be detected in hormonally induced chick oviduct, indicating that the entire gene is transcribed and the intervening sequences removed by "splicing" to form mature cytoplasmic mRNA. (4) All of the coding–noncoding junctions have similar sequences which fit the general sequence 5′-TCAGGTA-3′ at the 5′ end of the intervening sequence and 5′-TXCAGG-3′ at the 3′ end of the same intervening sequence. The dinucleotides GT and AG are found at the 5′- and 3′-ends, respectively, of all the intervening sequences.

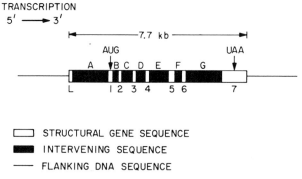

Fig. 6. The arrangement of introns in the chicken ovalbumin gene. The DNA encoding ovalbumin mRNA sequence (1859 bp) is interrupted by seven intervening sequences with the following sizes: A–G: 1729, 492, 303, 812, 206, 55, 890. The first 45 untranslated nucleotides of the mRNA (L) are separated from the rest of the mRNA molecule by the A intron. Thus, the gene has a "leader" sequence similar to that observed in mammalian tumor viruses. The arrow indicates the direction of transcription of the gene. The complete gene structure was determined by Gannon *et al.*, (1979) and Dugaiczyk *et al.*, (1979).

Recently, cosmid cloning procedures have been used to isolate two ovalbumin-like genes which are closely linked to the ovalbumin gene (Royal *et al.*, 1979). These two genes designated X and Y are located on the 5' side of the ovalbumin gene, they are transcribed from the same DNA strand and both are under estrogen control. The two ovalbumin-like genes share weak but significant sequence homology with the ovalbumin gene and they contain 8 and 7 introns, respectively. Thus the two genes have an organization strikingly similar to the ovalbumin gene. The characterization of this set of closely linked hormonally controlled genes provides an excellent system for studying the mechanism of coordinate gene activation.

D. Globin Genes

Vetebrate globin genes provide a well-characterized model system for the study of a small multigene family whose members are differentially expressed during both the adult and embryonic red cell development (see Bunn *et al.*, 1977 and Nienhuis and Benz, 1977, for review). In addition, the expression of globin genes is mediated directly or indirectly by the hormone erythropoietin (Graber and Krantz, 1978). Rabbit globin cDNA's were the first globin sequences to be cloned and characterized (Rougeon *et al.*, 1975; Maniatis *et al.*, 1976; Higuchi *et al.*, 1976; Rabbits, 1976). The nucleotide sequence of rabbit β- (Efstratiadis *et al.*, 1977) and α-globin (Heindell *et al.*, 1978) mRNA's were determined using cloned cDNA. One clone PβG1 (Maniatis *et al.*, 1976) was used to map the position of various restriction enzyme cleavage sites surrounding the β-globin gene in rabbit DNA (Jeffreys and Flavell, 1977a). Inconsistencies between the restriction map of PβG1 and the one derived for the β-globin gene in cellular DNA led to the detection of a large noncoding intervening sequence in a rabbit β-globin gene (Jeffreys and Flavell, 1977b).

In the past two years, the isolation of globin genes from the DNA of several different organisms, including mouse (Tilghman *et al.*, 1977), rabbit (Maniatis *et al.*, 1978; Van Den Berg *et al.*, 1978), man (Lawn *et al.*, 1978; Smithies *et al.*, 1978), and chicken (Dodgson *et al.*, 1979) has been achieved. The two adult mouse β-globin genes (β major and β minor) which are found in adult reticulocytes in the ratio of 0.8 to 0.2 were isolated by gene enrichment and cloning procedures (Tilghman *et al.*, 1977). Electron microscopic analysis of R loop structures and nucleotide sequence analysis identified a large noncoding intervening sequence in the cloned β-globin gene located between codons for amino acids 104 and 105 in both of the mouse β-globin genes. A second smaller intervening sequence was detected near the 5'-end of the gene. Electron microscopic analysis of heteroduplexes between the two mouse β-globin genes indicated that only the sequences immediately adjacent to the coding sequences and those at the junction between the coding region and the large intervening sequence exhibit homology (Tiemeier, 1978). The possible significance of this observation is discussed in relationship to evolutionary stabilization of the two genes by restricting the target size for recombination during meiosis (Tiemeier *et al.*, 1978).

The discovery of one or more intervening sequences in globin genes presents the problem of how the intervening sequences are removed before mature globin mRNA appears in the cytoplasm (Jeffreys and Flavell, 1977b; Tilghman *et al.*, 1978). The possibility that intervening sequences are removed at the DNA level in differentiated erythroid cells was ruled out by the observation that the intervening sequence is present in the DNA of all tissues examined, including nucleated erythroid cells (Jeffreys and Flavell, 1977b). The intervening sequence is, in fact, transcribed. This was shown by hybridizing purified 15 S mouse β-globin nuclear RNA precursor to genomic clones of the mouse β-globin gene (Tilghman *et al.*, 1978). The presence of a perfect RNA–DNA duplex definitively shows that the noncoding sequences are transcribed. This suggests that the intervening sequences are removed by splicing of internal segments of nuclear RNA to form the mature mRNA.

Recently the adult rabbit β-globin gene has been isolated using gene enrichment procedures (Van Den Berg *et al.*, 1978) and by direct isolation from a library of cloned rabbit DNA (Maniatis *et al.*, 1978). Characterization of the cloned β-globin genes locates the 600 base pair noncoding intervening sequence detected in genomic blotting experiments between the codons for amino acids 104 and 105, exactly the same position as the large intervening sequence within the mouse β-globin genes (Tilghman *et al.*, 1978). In addition, the second intervening sequence of 126 base pairs, which was not detected in genomic blotting experiments (Jeffreys and Flavell, 1977b), was located between the codons for amino acids 30 and 31, again in exactly the same position as the smaller intervening sequence found in the mouse β-globin genes. In the process of characterizing clones of large rabbit DNA fragments bearing the β-globin genes, a second closely linked β-related globin gene was detected (Maniatis *et al.*, 1978; Lacy *et al.*, 1978). This gene is located to 9 kb in the 5' direction from the β-globin gene and is transcribed from the same DNA strand. Two more β-related rabbit genes were detected by plaque hybridization to a human γ-globin cDNA plasmid (pHW151, Wilson *et al.*, 1978). These two genes are separated from each other by approximately 8 kb and are located approximately 6 kb in the 5' direction from the β-related gene adjacent to the adult β-globin gene. Thus, over 40 kb of contiguous rabbit DNA sequences have been isolated and shown to contain four different, closely linked rabbit β-related globin genes, two of which are expressed in the embryo and two which are expressed in adult erythroid cells (R. Hardison, E. Lacy, E. Butler and T. Maniatis, unpublished results).

Physical linkage between non-α-globin genes has also been detected in human DNA using gene isolation procedures (Lawn et al., 1978). A library of human embryonic DNA was constructed and screened with a human β-globin cDNA plasmid (pHW102, Wilson et al., 1978). Two clones were isolated which carry both the adult δ- and β-globin genes. The two genes are separated by approximately 5.4 kb of DNA, and their orientation with respect to the direction of transcription is 5' δβ 3' as predicted by genetic data. Both adult δ- and β-globin genes contain a large noncoding intervening sequence (950 and 900 base pairs, respectively) located between the codons for amino acids 104 and 105, the same position as the large intervening sequence in the rabbit and mouse β-globin genes. Although the location of the large intervening sequence within the coding regions of the two human globin genes is identical, the two noncoding sequences bear little sequence homology. A second, smaller intervening sequence similar to that found in other mammalian β-globin genes was detected near the 5'-end of the human β-globin gene. The linkage arrangement and identification of noncoding intervening sequences in these two genes was independently established using genomic blotting procedures (Flavell et al., 1978; Mears et al., 1978). (See Fig. 7 for a summary of this information.) A similar analysis was carried out for the two human fetal globin genes designated Gγ and Aγ (Little et al., 1979; Fritsch et al., 1979; Tuan et al., 1979; Bernards et al., 1979). The two genes are closely linked, separated by only 3.5 kb of DNA. Both genes contain one large noncoding intervening sequence of approximately 850 base pairs and as with all other β-like globin genes studied, a smaller intervening sequence nearer to the 5'-end of the two genes (Smithies et al., 1978; O. Smithies, personal communication). Physical linkage between the fetal and adult β-like genes was demonstrated by making use of a well-characterized deletion which extends beyond the 5'-end of the γ-globin gene (Fritsch et al., 1979; Tuan et al., 1979) and by studying partial digestion products of total human genomic DNA (Bernards et al., 1979). These analyses reveal that the Aγ globin gene is located approximately 16 kb to the 5' side of the δ gene as shown in the map of Fig. 7. This, in combination with previous studies (Little et al., 1979) indicated that all four genes are transcribed from the same DNA strand in the direction 5'-Gγ-Aγ-δ-β-3'. Recently, an embryonic β-like gene, ε, has been cloned and characterized (Proudfoot and Baralle, 1979). By isolating a series of overlapping clones extending beyond the 5'-end of the Gγ-globin gene it was possible to show that the ε-globin gene isolated

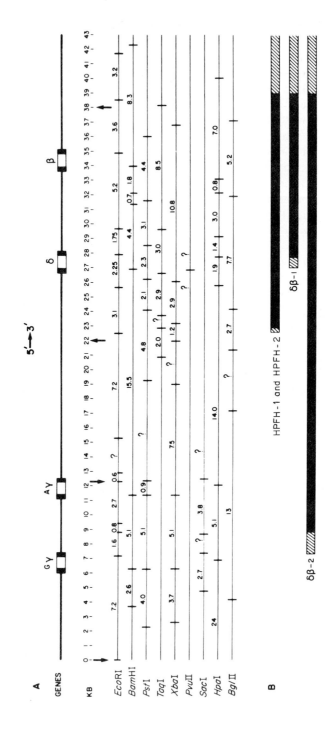

by Proudfoot and Baralle lies 15 kb to the 5' side of the $^G\gamma$-globin gene (E. Fritsch, R. M. Lawn, and T. Maniatis, manuscript in preparation). Thus, over 70 kb of human chromosomal DNA containing five differentially expressed β-like globin genes has been cloned and characterized. These clones will provide valuable tools in the study of the developmental stage-specific expression of this small gene family.

The existence of well-defined functional deficiencies in human hemoglobin expression (thalassemias) provides the opportunity for studying relationships between globin gene organization and function (Weatherall and Clegg, 1972). A number of laboratories have used the genomic blotting procedure to study the organization of human globin genes in the DNA of normal and thalassemic individuals (Mears *et al.*, 1978; Flavell *et al.*, 1978; Orkin *et al.*, 1978, 1979b). The availability of cloned genomic DNA with its associated noncoding sequences will extend the utility of genomic blotting procedures by making it possible to detect structural changes in sequences not accessible to cDNA probes. This was recently demonstrated by the analysis of deletions which affect the expression of fetal globin genes in man (Fritsch *et al.*, 1979; Tuan *et al.*, 1979). A hybridization probe which lies well outside the structural gene sequences of the δ- and β-globin genes was used to map large deletions in the β-like globin gene cluster. Because different deletions have different effects on fetal globin gene expression in adults, a comparison of the end points of many such

---

Fig. 7. A map of restriction endonuclease cleavage sites within and surrounding the human γ-, δ-, and β-globin genes. (A) The arrangement of fetal and adult β-like globin genes within a 43 kilobase (kb) segment of human DNA is presented. The direction of transcription of the four linked genes is left to right (5' → 3'). The solid boxes represent the locations of mRNA coding regions. The open boxes represent the large noncoding intervening sequence in each gene. This intervening sequence is located between codons 104 and 105 in the δ, β, and $^A\gamma$ genes. Sizes of the $^G\gamma$-, $^A\gamma$-, δ-, and β-globin intervening sequences are approximately 850, 850, 950, and 900 bp, respectively. A smaller, intervening sequence identified in the β and $^A\gamma$ genes and presumably present in the $^G\gamma$ and δ genes is not shown. For each of the restriction endonucleases indicated on the left, the relative locations of cleavage sites are marked by a vertical line. The sizes for the restriction enzyme fragments are given in Kilobases. Question marks indicate that the presence of additional cleavage sites has not been determined. The locations of cleavage sites within the region delineated by the upward pointing arrows (↑) were determined from an analysis of the clone HβG1 which contains the linked δ- and β-globin genes. Some cleavage sites within the region delineated by the downward pointing arrows (↓) were determined by an analysis of a clone HγG1 which contains the $^G\gamma$- and $^A\gamma$-globin genes. Additional cleavage sites within and surrounding the γ-, δ-, and β-globin genes were mapped by blot hybridization experiments. (B) The regions of the β-like globin gene locus deleted in the DNA's isolated from individuals with hereditary persistence of hemoglobin (HPFH) or $\delta\beta$-thalassemia are indicated by solid boxes. The precise locations of the endpoints of the deletions are within the regions specified by the hatched boxes. The locations of the rightward ends of the deletions are not known.

deletions might reveal which regions within the gene cluster play a role in the regulation of the switch from fetal to adult globin gene expression. Although the correlation between the mapping data and the clinical observations is complicated and not enough cases have been examined for a general pattern to emerge, it is clear that deletions which lie as much as 12 kb from the fetal γ-globin genes can have a significant effect on their differential expression during development (see Fritsch *et al.*, 1979, for discussion). These observations suggest the interesting possibility that the gene cluster as a whole functions as a regulatory unit rather than each gene being independently regulated.

A number of laboratories are in the process of isolating β-globin genes from the DNA of individual homozygous for abnormalities in β-globin gene expression, such as β^0 or β^+ thalassemia. Comparison of the structure of these genes with the normal β-globin gene should provide important information regarding the molecular basis of these genetic diseases.

VII. IDENTIFICATION OF REGULATORY SEQUENCES IN CLONED EUKARYOTIC DNA

As indicated in the examples cited above, the effective use of recombinant DNA procedures in elucidating the structure of eukaryotic genes is firmly established. Although these studies may provide important insights into the mechanisms of genetic regulation, definitive identification of regulatory sequences, such as transcriptional initiation and termination sites, binding sites for regulatory proteins, and RNA processing sites, will require the development of functional *in vitro* or *in vivo* assays for gene expression. One of the most promising directions in this regard is to alter cloned DNA *in vitro* (*in vitro* mutagenesis, Carbon *et al.*, 1975; Domingo *et al.*, 1976; Shortle and Nathans, 1978; Shortle *et al.*, 1979; Weissman *et al.*, 1979) and then introduce the gene into an environment in which it is normally expressed (for example, in a well-defined *in vitro* system or the appropriate differentiated cell type.

A. *In Vitro* Systems

Excellent *in vitro* transcription systems have been developed for genes which are transcribed by the class III type eukaryotic RNA polymerase. This includes the genes for *Xenopus* 5 S RNA and transfer RNA, and the 5.5 S RNA encoded by adenovirus. For example, transcripts which are indistinguishable from the *in vivo* gene products can be obtained by injecting cloned 5 S DNA (Brown and Gurdon, 1978; Parker *et al.*, 1978) or tDNA (Kressman *et al.*, 1978a,b) into the oocyte nucleus of *Xenopus laevis*. Similarly, faithful transcripts of the adenovirus genes coding for 5.5 S RNA can be produced in a cell-free system from human KB cells (Wu *et al.*, 1977; Wu, 1978). Purified viral DNA is added directly to the extract, which contains a cytoplasmic, postmitochondrial supernatant. Over 90% of the cellular RNA polymerase III activity is found in this supernatant fraction.

Genes normally transcribed by RNA polymerase II also have been microinjected into *Xenopus* oocytes and the products of transcription and translation analyzed (Mertz and Gurdon, 1977; DeRobertis and Mertz, 1977; Kressman *et al.*, 1978b). When SV40 DNA is microinjected, SV40-specific RNA transcripts can be detected, and some of the RNA is translated to produce the viral capsid proteins VP1 and VP3. Similarly, histone-like proteins can be observed when cloned *Drosophila* histone genes are injected into oocyte nuclei. The synthesis of both types of proteins are α-amanitin sensitive, suggesting that the injected genes are copied by RNA polymerase II. The fidelity of transcription of the injected DNA cannot be determined in the experiment, since it is possible that only a small fraction of the transcripts produced are translated into identifiable proteins. Thus, the feasibility of using this system for identifying and studying eukaryotic promoters has not been rigorously established. In order for the oocyte injection system to be useful in this regard, the faithful initiation and asymmetric transcription of injected genes must be demonstrated. This has not been possible using cloned sea urchin histone DNA (Kressman *et al.*, 1978b) or mouse globin DNA (J. Gurdon, personal communication).

Faithful *in vitro* transcription of cloned DNA "reconstituted" with chromosomal proteins has been demonstrated using purified RNA polymerase III and 5 S DNA (Parker *et al.*, 1978). Similar experiments involving RNA polymerase II have not been reported. [See Chambon (1978) for a discussion of difficulties encountered in experiments involving *in vitro* transcription of chromatin.]

A cell-free system which removes the intervening sequences from yeast tRNA precursors and religates the ends to form mature tRNA has recently been described (Knapp *et al.*, 1978). Activities which remove the intervening sequences from vertebrate structural gene sequences are being searched for in several laboratories, but to date no successful experiments have been reported. Such systems could, of course, be important in elucidating the mechanism by which intervening sequences are processed.

Several eukaryotic host–vector systems are presently being developed which may provide the opportunity to use such an approach.

B. Yeast as a Eukaryotic Host–Vector Cloning System

Yeast host–vector cloning systems are being developed in a number of laboratories to provide a system in which to study regulatory sequences in cloned eukaryotic DNA. The most significant advance in this direction is the finding that it is possible to transform auxotrophic yeast cells stably to a wild-type phenotype using hybrid DNA molecules isolated by biochemical selection in *E. coli* (Hinnen *et al.*, 1978; Ilgen *et al.*, 1979). Spheroplasts of a *leu*2⁻ yeast strain are mixed with bacterial plasmid DNA carrying the yeast *leu*2⁺ gene in the presence of calcium and polyethylene glycol, and plated on a medium deficient in leucine. Approximately 1 out of 10^7 cells survive, whereas no survivors are observed when the cells are plated without the addition of DNA. Direct proof that the survivors were transformed by the hybrid DNA was obtained by showing that the plasmid vector DNA is integrated into the chromosomes of the transformed cells and by genetically mapping *leu*2⁺ marker to several different places in the yeast genome (Hinnen *et al.*, 1978).

Much higher efficiencies of transformation (up to 3×10^{-3}) can be achieved using specially designated yeast cloning vectors which are thought to contain one or more yeast replication origins (Beggs, 1978; R. Davis, personal communication; see also K. Nasmyth, 1978). For example, one vector consists of a yeast 2 μm circle of DNA, the cloned yeast *leu*2 gene and the bacterial plasmid PMB59 (Beggs, 1978). These vectors can exist as unintegrated replicons with 50–100 copies per cell. In theory any gene can be attached to the vector and its expression studied in yeast under nonselective conditions. For example, a gene could be cloned in *E. coli,* subjected to *in vitro* mutagenesis, and then placed back into the yeast cell to study the effect of the mutation.

C. SV40 DNA as a Mammalian Cloning Vector

This small mammalian DNA tumor virus is being developed as a cloning vector in several laboratories (Jackson *et al.*, 1972; Goff and Berg, 1976; Ganem *et al.*, 1976; Hamer *et al.*, 1977; Upcroft *et al.*, 1978; W. Schaffner and M. Botchan, personal communication). The virus exhibits two distinct modes of interaction with cells in culture. When a permissive cell is infected, the virus grows lytically, producing several thousand infectious particles per cell. Infection of a nonpermissive host can result in the integration of the viral DNA into the host chromosome and stable transformation of the growth properties of the cell. Thus, a suitably constituted SV40-cellular DNA hybrid could be propagated in mammalian cells either as an extrachromosomal replicon or as a defective integrated virus. The relatively simple genetic map of SV40 allows easy *in vitro* manipulation of the viral chromosome. The gene for SV40 T antigen (the protein necessary for SV40 DNA replication and for cellular transformation) is located on one region of the viral chromosome separated from the genes encoding viral capsid proteins. Using restriction endonucleases it is possible to remove the DNA coding for capsid proteins and replace it with foreign DNA. The hybrid DNA can be used to transform nonpermissive cells, or, if the insert is small enough, the recombinant can be propagated lytically in permissive cells in the presence of a complementing helper virus which provides viral capsid proteins.

The feasibility of using SV40 as a viral vector system was demonstrated by joining SV40 DNA to bacterial (Hamer *et al.*, 1977; Upcroft *et al.*, 1978) or bacteriophage (Ganem *et al.*, 1976; Goff and Berg, 1976) DNA and propagating the hybrid DNA in either the integrated or lytic state. If permissive monkey cells are transfected with hybrid molecules in the absence of a complementing helper virus a persistent infection develops and unintegrated, replicating viral DNA molecules can be detected in the cells (Upcroft *et al.*, 1978).

Recently, a double-stranded cDNA copy of rabbit β-globin mRNA has been joined to SV40 DNA in such a way that the sequence is efficiently transcribed during lytic growth in the presence of a helper virus (Mulligan et al., 1979). Large amounts of globin mRNA and protein can be detected in the cytoplasm of cells infected with the hybrid virus, indicating that the hybrid nuclear transcript is processed and the message efficiently translated *in vivo*. Attempts are being made to insert β-globin genes isolated from cellular DNA into SV40 vectors with the hope of obtaining globin gene transcription originating from a promotor on the globin DNA. If these experiments are successful it will be possible to study the function of sequences within and surrounding β-globin genes.

D. The Direct Transfer of Eukaryotic Genes by Biochemical Transformation

The transfer of single copy genes from one mammalian cell to another has been accomplished in tissue culture by transformation with intact metaphase chromosomes (McBride and Ozer, 1973; Willicke and Ruddel, 1975) or purified cellular DNA (Wigler *et al.*, 1978). The latter procedure was developed using the thymidine kinase (*tk*) gene of herpes simplex virus (HSV) as a model system (Bacchetti and Graham, 1977; Wigler *et al.*, 1977; Maitland and McDougall, 1977; Pellicer *et al.*, 1978) and then extended to the *tk* genes of several different vertebrate organisms (Wigler *et al.*, 1978).

The tk$^+$ or tk$^-$ phenotype can be selected for in tissue culture by grow-ing cells under conditions in which the synthesis of thymidine kinase is either necessary for survival or results in cell death. When DNA is sus-pended in CaCl$_2$ and phosphate buffer, a calcium phosphate–DNA precipi-tate forms which is taken up by cells in culture (Graham and van der Eb, 1973). Using intact HSV DNA (Bacchetti and Graham, 1977) or purified restriction fragments from the viral DNA (Wigler et al., 1977; Maitland and McDougall, 1977), it was possible to effect the stable transformation of the tk$^-$ cells in culture using the calcium phosphate precipitation proce-dure. The efficiency of transformation varies between 10^{-5} and 10^{-6}/μg viral DNA, well above the natural reversion frequency of the tk$^-$ cells. The possibility that the altered phenotype of the cells was due to reversion rather than transformation was definitively ruled out by showing that the thymidine kinase activity in the transformed cells is identical to that of the HSV enzyme (which differs from that of the wild-type counterpart of the recipient cell) by several different biochemical and serological criteria. The physical state of the HSV *tk* gene in the transformed cells was exam-ined using hybridization procedures (Pellicer et al., 1978). These studies showed that one copy of the gene is stably integrated per diploid chromosomal complement of all transformants examined, and that the site of integration is random.

The integrated HSV *tk* gene can be transferred from the *tk* transformant to a second tk$^-$ cell using unfractionated cellular DNA (Wigler et al., 1978; Wigler et al., 1979a). This was made possible by improvements in the calcium phosphate transformation procedure. Efficiencies as high as 1 transformant per 10^6 cells per 40 pg of purified HSV *tk* gene can now be obtained (Wigler et al., 1978). Considering that the integrated viral gene represents only 1 ppm of the cellular DNA, it should be possible to obtain 1 transformant per 10^6 cells using 30–40 μg of total mammalian genomic DNA. In practice it is possible to obtain 10 transformants per 10^6 cells per 20 μg of DNA, a frequency 20-fold higher than expected. The secondary transformants express HSV-specific tk activity and carry the viral *tk* gene in an integrated state. Having optimized conditions for the transfer of the integrated HSV *tk* gene, it was then possible to attempt the transfer of in-digenous *tk* genes from a tk$^+$ to a tk$^-$ cell using total, unfractionated cellular DNA (Wigler et al., 1978). This was accomplished using DNA isolated from human, calf, chicken, or hamster cells. The efficiency of transfer was comparable to that obtained with the integrated HSV *tk* gene (7 colonies/10^6 cells/20 μg DNA). The thymidine kinase in the transformed cells was shown to be donor derived by examining the isoelectric point of the tk enzyme in transformed cells (Wigler et al., 1978).

Recently, the feasibility of introducing any eukaryotic gene into mammalian cells in culture was demonstrated using the rabbit β-globin gene and the mouse tk$^-$ L cell system (Wigler et al., 1979a,b; Mantei et al., 1979). Wigler et al. (1979a,b) accomplished this using a "cotransformation procedure" in which a small amount of the HSV tk gene and a vast excess of cloned rabbit β-globin DNA with tk$^-$ L cells are added to L cells using the calcium phosphate procedure and selecting the tk$^+$ transformants. When the transformants are scored for rabbit β-globin gene sequences by genomic blotting procedures, six out of eight were found to contain the gene. The number of copies of rabbit β-globin gene cell ranged from 1 to 20. Mantei et al. (1979) introduced the rabbit β-globin gene into mouse L cells by covalently joining a plasmid containing the genomic copy of β-globin gene with a plasmid containing the HSV tk gene. When the tk$^-$ L cells were transformed with the hybrid molecules, 19 out of 21 transformants carried the rabbit β-globin gene and from 3 to 20 copies were present per cell. In both the cotransformation and the covalent joining procedures the rabbit β-globin gene is transcribed into RNA. However, many more copies of the mRNA are found when the covalent joining procedure is used. For example, in one mouse L cell line containing 20 copies of the rabbit β-globin gene introduced by cotransformation only 5 copies of rabbit β-globin RNA can be detected (Wold et al., 1979). In contrast, a cell line which was prepared using the tk–globin hybrid molecules also contains 20 copies of the rabbit globin gene, but over 2000 copies of rabbit β-globin RNA can be detected (Mantei et al., 1979). In general, the mouse cells in which the tk and globin genes are physically linked produce considerably more rabbit globin RNA than those in which the globin and tk genes were introduced as separate molecules. Although too few cell lines have been examined to warrant any definite conclusions, it is possible that close proximity of the tk and globin genes can enhance globin gene expression. This could result from read through from a promotor on the tk gene fragment or merely from being adjacent to an active gene. There is also a qualitative difference in the RNA produced in the two types of cells containing rabbit globin genes. In both cases cytoplasmic poly(A)-containing rabbit globin mRNA is synthesized, and the small and large introns are accurately excised. However, in the case of the tk-globin hybrid transformation experiments the rabbit globin RNA contains the complete 5'-end of rabbit globin mRNA, while the RNA in cells prepared by cotransformation are missing the first 48 nucleotides at the 5'-end of the rabbit globin RNA. Unfortunately, since only one cell line was analyzed, it is not possible to determine whether this is a general property of the cotransformation procedure. Attempts are presently being made in a number of laboratories to introduce cloned globin genes into mouse erythroleukemia cells. If this is successful and if the heterologous globin genes can be induced by dimethyl sulfoxide it will be possible to use the system to study the mechanism of globin-specific gene expression *in vivo* using cloned DNA.

REFERENCES

Aloni, Y., Dhar, R., Laub, O., Horowitz, M., and Khoury, G. (1977). Novel mechanisms for RNA maturation: The leader sequences of SV40 mRNA are not transcribed adjacent to the coding sequences. *Proc. Natl. Acad. Sci. U.S.A.* **74**, 3686–3690.

Arceci, R. J., Senger, D. R., and Gross, P. R. (1976). The programmed switch of lysine-rich histone synthesis at gastrulation. *Cell* **9**, 171–178.

Axel, R., Maniatis, T., Fox, C. F., eds. (1979). *In Eukaryotic Gene Regulation*. Academic Press, New York.

Anonymous (1978). News and views. *Nature (London)* **274**, 741.

Bacchetti, S., and Graham, F. L. (1977). Transfer of the gene for thymidine kinase to thymidine kinase deficient human cells by purified herpes simplex viral DNA. *Proc. Natl. Acad. Sci. U.S.A.* **74**, 1590–1594.

Backman, K., Ptashne, M., and Gilbert, W. (1976). Construction of plasmids carrying the C_1 gene of bacteriophage λ. *Proc. Natl. Acad. Sci. U.S.A.* **73**, 4174–4178.

Bahl, C. P., Marians, K. J., Wu, R., Stawinski, S., and Narang, A. (1977). A general method for inserting specific DNA sequences into cloning vehicles. *Gene* **1**, 81–92.

Bahl, C. P., Wu, R., Brousseau, R., Sood, A. K., Hsiung, H., and Narang, S. A. (1978). Chemical synthesis of versatile adaptors for molecular cloning. *Biochem. Biophys. Res. Commun.* **81**, 695–703.

Barnes, W. M. (1978). Construction of an M13 histidine transducing phage: A single-stranded single RI site cloning vehicle. *Gene* **5**, 127–139.

Beers, R. F., and Basset, E. G., eds. (1977). "Recombinant Molecules: Impact on Science and Society." Raven, New York.

Beggs, J. D. (1978). Transformation of yeast by a replicating hybrid plasmid. *Nature (London)* **275**, 104–108.

Benton, W. D., and Davis, R. W. (1977). Screening λgt recombinant clones by hybridization to single plaques *in situ*. *Science* **196**, 180–183.

Berget, S. M., Moore, C., and Sharp, P. A. (1977). Spliced segments at the 5′ terminus of adenovirus 2 late mRNA. *Proc. Natl. Acad. Sci. U.S.A.* **74**, 3171–3175.

Berk, A. J., and Sharp, P. A. (1978a). Spliced early mRNAs of simian virus 40. *Proc. Natl. Acad. Sci. U.S.A.* **75**, 274–278.

Berk, A. J., and Sharp, P. A. (1978b). Structure of the adenovirus 2 early mRNAs. *Cell* **14**, 695–711.

Bernard, O., Hozumi, N., and Tonegawa, S. (1978). Sequences of mouse immunoglobulin light chain genes before and after somatic changes. *Cell* **15**, 1133–1144.

Bernards, R., Little, P.F.R., Annison, G., Williamson, R., and Flavell, R. A. (1979). Structure of the human $^G\gamma$-$^A\gamma$-δ-β globin gene locus. *Proc. Natl. Acad. Sci. U.S.A.* (in press)

Birnstiel, M. L., Schaffner, W., and Smith, H. O. (1977). DNA sequences coding for the H2b histone of *Psammechinus milaris*. *Nature (London)* **266**, 603–607.

Blake, C. C. F. (1979). "News and Views" *Nature (London)* **273**, 267; **277**, 598.

Blattner, F. R., Williams, B. G., Blechl, A. E., Denniston-Thompson, K., Faber, H. E., Furlong, L. A., Grunwald, D. J., Kiefer, D. O., Moore, D. D., Schumm, H. W.,

Sheldon, E. L., and Smithies, O. (1977). Charon phages: Safer derivatives of bacteriophage λ DNA cloning. *Science* **196**, 161–169.

Blattner, F. R., Blechl, A. E., Denniston-Thompson, K., Faber, H. E., Richards, J. E., Slightom, J. L., Tucker, P. W., and Smithies, O. (1978). Cloning of human fetal γ and mouse α-type globin DNA: Preparation and screening of shotgun collections. *Science* **202**, 1279–1283.

Bolivar, F., Rodriguez, R. L., Greene, P. J., Betlach, M. C., Heyneker, H. L., and Boyer, H. W. (1977). Construction and characterization of new cloning vehicles. II. A multipurpose cloning system. *Gene* **2**, 95–113.

Botchan, M., McKenna, G., and Sharp, P. (1973). Cleavage of mouse DNA by a restriction enzyme as a clue to the arrangement of genes. *Cold Spring Harbor Symp. Quant. Biol.* **38**, 383–395.

Botchan, M., Topp, W., and Sambrook, J. (1976). The arrangement of simian virus 40 sequences in the DNA of transformed cells. *Cell* **9**, 269–281.

Brack, C., and Tonegawa, S. (1977). Variable and constant parts of the immunoglobulin light chain gene of a mouse myeloma cell are 1250 nontranslated bases apart. *Proc. Natl. Acad. Sci. U.S.A.* **74**, 5652–5757.

Brack, C., Hirama, M., Lenhard-Schuller, R., and Tonegawa, S. (1978). A complete immunoglobulin gene is created by somatic recombination. *Cell* **15**, 1–14.

Breathnach, R., Mandel, J. L., and Chambon, P. (1977). Ovalbumin gene is split in chicken DNA. *Nature (London)* **270**, 314–319.

Brown, D. D., and Gurdon, J. D. (1978). High fidelity transcription of 5S DNA injected in *Xenopus* oocytes. *Proc. Natl. Acad. Sci. U.S.A.* **75**, 2064–2068.

Browne, J. K., Paddock, G. V., Liu, A., Clarke, P., Heindell, H. C., and Salser, W. (1977). Nucleotide sequences from the rabbit β-globin gene inserted into *E. coli* plasmids. *Science* **195**, 389–391.

Brutlag, D., Fry, K., Nelson, T., and Hung, P. (1977). Synthesis of hybrid bacterial plasmids containing highly repeated satellite DNA. *Cell* **10**, 509–519.

Bunn, F. H., Forget, B. G., and Ranney, H. M. (1977). "Human Hemoglobins." Saunders, Philadelphia.

Cami, B., and Kourilsky, P. (1978). Screening of cloned recombinant DNA in bacteria by *in situ* colony hybridization. *Nucleic Acids Res.* **5**, 2381–2390.

Carbon, J., Shenk, T., and Berg, P. (1975). Biochemical procedure for production of small deletions in SV40 DNA. *Proc. Natl. Acad. Sci. U.S.A.* **72**, 1392–1396.

Carbon, J., Clarke, L., Ilgen, C., and Ratzkin, B. (1977). *In* "Recombinant Molecules: Impact or Science and Society" (R. F. Beers and E. G. Bassat, eds.), pp. 335–378. Raven, New York.

Celma, M. L., Dhar, R., and Weissman, S. M. (1977). Comparison of the nucleotide sequence of the mRNA for the major structural protein of SV40 with the DNA sequence encoding the amino acids of the protein. *Nucleic Acids Res.* **4**, 2549–2559.

Chambon, P. (1978). The molecular biology of the eukaryotic genome is coming of age. *Cold Spring Harbor Symp. Quant. Biol.* **42**, 1209–1234.

Chow, L. T., Gelinas, R. E., Broker, T. R., and Roberts, R. J. (1977). An amazing sequence arrangement of the 5′ ends of adenovirus 2 mRNA. *Cell* **12**, 1–8.

Clarke, L., and Carbon, J. (1976). A colony bank containing synthetic colE1 hybrid plasmids representative of the entire *E. coli* genome. *Cell* **9**, 91–99.

Clarkson, S. G., Smith, H. O., Schaffner, W., Gross, K. W., and Birnstiel, M. (1976). Integration of eukaryotic genes for 5 S RNA and histone proteins into a phage lambda receptor: Stage-specific switches in histone synthesis during embryogenesis of the sea urchin. *Nucleic Acids Res.* **3**, 2617–2632.

Cohen, L. H., Newrock, K. N., and Zweidler, A. (1975). Stage-specific switches in histone synthesis during embryogenesis of the sea urchin. *Science* **190**, 994–997.

Cohn, R. H., Lowry, J. C., and Kedes, L. H. (1976). Histone genes of the sea urchin (*S. purpuratus*) cloned in *E. coli;* order, polarity, and strandedness of the five histone coding and spacer regions. *Cell* **9**, 147–161.

Collins, J., and Bruning, H. J. (1978). Plasmids usable as gene cloning vectors in an *in vitro* packaging by Coliphage "cosmids." *Gene* **4**, 85–107.

Collins, J., and Hohn, B. (1978). Cosmids: A type of plasmid gene-cloning vector that is packageable *in vitro* in bacteriophage λ heads. *Proc. Natl. Acad. Sci.* **75**, 4242–4246.

DeRobertis, E. M., and Mertz, J. E. (1977). Coupled transcription–translation of DNA injected into *Xenopus* oocytes. *Cell* **12**, 175–182.

Dodgson, J. B., Strommer, J., and Engel, J. D. (1979). The isolation of the chicken β-globin gene and a linked embryonic β-like globin gene from a chicken DNA recombinant library. *Cell* **17**, 879–887.

Doel, M. T., Houghton, M., Cook, E. A., and Carey, N. H. (1977). The presence of ovalbumin mRNA coding sequences in multiple restriction fragments of chicken DNA. *Nucleic Acids Res.* **4**, 3701–3713.

Domingo, E., Flavell, R. A., and Weissman, C. (1976). *In vitro* site-directed mutagenesis: Generation and properties of an infectious extracistronic mutant of bacteriophage Qβ. *Gene* **1**, 3–25.

Dugaiczyk, A., Woo, S. L. C., Lai, E. C., Mace, M. L., McReynolds, L., and O'Malley, B. W. (1978). The natural ovalbumin gene contains seven intervening sequences. *Nature (London)* **274**, 328–333.

Dugaiczyk, A., Woo, S. L. C., Colbert, D. A., Lai, E. C., Mace, M. L. Jr., and O'Malley, B. W. (1978). The ovalbumin gene: Cloning and molecular organization of the entire natural gene. *Proc. Natl. Acad. Sci.* **76**, 2253–2257.

Early, P. W., Davis, M. M., Kaback, D. B., Davidson, N., and Hood, L. (1979). Immunoglobulin heavy chain gene organization in mice: Analysis of a myeloma genomic clone containing variable and α constant regions. *Proc. Natl. Acad. Sci.* **76**, 857–861.

Edgell, M. H., Weaver, S., Haigwood, N., and Hutchison, C. A., III. (1979). Gene enrichment. *In* "Genetic Engineering" (J. K. Setlow and A. Hollaender, eds.), Vol. 1, pp. 37–49. Plenum, New York.

Efstratiadis, A., and Villa-Komaroff, L. (1979). Cloning of double-stranded DNA. *In* "Genetic Engineering" (J. Setlow and A. Hollaender, eds.), Vol. I. pp. 15–36. Plenum, New York.

Efstratiadis, A., Kafatos, F. C., and Maniatis, T. (1977). The primary structure of rabbit β-globin mRNA as determined from cloned DNA. *Cell* **10**, 571–585.

Enquist, L., and Sternberg, N. (1980). *In vitro* packaging of λ D*am* vectors and their use in cloning DNA fragments in recombinant DNA. *In* "Methods in Enzymology" (R. Wu, ed.) vol. 68, New York.

Flavell, R. A., Kooter, J. M., DeBoer, E., Little, P. F. R., and Williamson, R. (1978). Analysis of the human βδ-globin gene loci in normal and hemoglobin Lepore DNA: Direct determination of gene linkage and intergene distance. *Cell* **15**, 25–41.

Fritsch, E. F., Lawn, R. M., and Maniatis, T. (1979) Characterization of deletions which affect the expression of fetal globin genes in man. *Nature (London)* **279**, 598–603.

Ganem, D., Nussbaum, P. L., Davoli, D., and Fareed, G. C. (1976). Propagation of a segment of bacteriophage λ DNA in monkey cells after covalent linkage to a defective SV40 genome. *Cell* **7**, 349–359.

Gannon, F., O'Hare, K., Perrin, F., LePennec, J. P., Benoist, C., Cochet, M., Breathnach, R., Royal, A., Garapin, D., Cami, B., and Chambon, P. (1979). Organization and sequences at the 5' end of a cloned complete ovalbumin gene. *Nature (London)* **278**, 428–433.

Garapin, A. C., Cami, B., Roskam, W., Kourilsky, P., LePennec, J. P., Perrin, F., Gerlinger, P., Cochet, M., and Chambon, P. (1978a). Electron microscopy and restriction enzyme mapping reveal additional intervening sequences in the chicken ovalbumin split gene. *Cell* **14**, 629–638.

Garapin, A. C., LePennec, J. P., Roskam, W., Perrin, F., Cami, B., Krust, A., Breathnach, R., Chambon, P., and Kourilsky, P. (1978b). Isolation by molecular cloning of a fragment of the split ovalbumin gene. *Nature (London)* **273**, 349–354.

Gilbert, W. (1978). "News and Views" *Nature (London)* **273**, 267.

Gilmore-Hebert, M., Hercules, K., Komaromy, M., and Wall, R. (1978). Variable and constant regions are separated in the 10 kb transcription unit coding for immunoglobulin κ light chains. *Proc. Natl. Acad. Sci.* **75**, 6044–6048.

Glover, D. M. (1977). The construction and cloning of hybrid DNA molecules. *In* "New Techniques in Biophysics and Cell Biology" (R. H. Pain and B. J. Smith, eds.), Vol. 3, pp. 125–145. Wiley (Interscience), New York.

Glover, D. M., and Hogness, D. S. (1977). A novel arrangement of the 18S and 28S sequences in a repeating unit of *D. melanogaster* rDNA. *Cell* **10**, 167–176.

Glover, D. M., White, R. L., Finnegan, D. J., and Hogness, D. S. (1975). Characterization of six cloned DNAs from *D. melanogaster* including one that contains the genes for rRNA. *Cell* **5**, 149–157.

Goff, S., and Berg, P. (1976). Construction of hybrid viruses containing SV40 phage DNA segments and their propagation in cultured monkey cells. *Cell* **9**, 695–705.

Goff, S., and Berg, P. (1978). The excision of DNA segments introduced into cloning vectors by the poly(dA·dT) joining method. *Proc. Natl. Acad. Sci. U.S.A.* **75**, 1763–1767.

Goodman, H. M., Olson, M. J., and Hall, B. D. (1977). Nucleotide sequence of a mutant eukaryotic gene: The yeast tyrosine ochre suppressor sup4-0. *Proc. Natl. Acad. Sci. U.S.A.* **74**, 5453–5457.

Graber, S. E., and Krantz, M. D. (1978). Erythropoietin and the control of red cell production. *Annu. Rev. Med.* **29**, 51–66.

Graham, F. L., and van der Eb, A. J. (1973). A new technique for the assay of infectivity of human adenovirus 5 DNA. *Virology* **52**, 456–467.

Gross, K., Schaffner, W., Telford, J., and Birnstiel, M. (1976). Molecular analysis of the histone gene cluster of *Psammechinus miliaris*. III. Polarity and asymmetry of the histone coding sequences. *Cell* **8**, 479–484.

Grunstein, M., and Hogness, D. (1975). Colony hybridization: A method for the isolation of cloned DNAs that contain a specific gene. *Proc. Natl. Acad. Sci. U.S.A.* **72**, 3961–3965.

Hamer, D., Davoli, D., Thomas, C. A., and Fareed, G. C. (1977). SV40 carrying an *E. coli* suppressor gene. *J. Mol. Biol.* **112**, 155–182.

Hardies, S. C., and Wells, R. D. (1976). Preparative fractionation of DNA restriction fragments by reverse phase column chromatography. *Proc. Natl. Acad. Sci. U.S.A.* **73**, 3117–3121.

Heindell, H. C., Liu, A. Y., Paddock, G. V., Studnicka, G. M., and Salser, W. (1978). The primary sequence of rabbit α-globin mRNA. *Cell* **15**, 43–54.

Herrmann, R., Neugebauer, K., Zentgraf, H., and Schaller, H. (1978). Transposition of a DNA sequence determining kanamycin resistance into the single-stranded genome of bacteriophage fd. *Mol. Gen. Genet.* **159**, 171–178.

Hershey, A. D. (1971). "The Bacteriophage λ." Cold Spring Harbor Laboratory, Cold Spring Harbor, New York.

Heyneker, H. L., Shine, J. M., Goodman, H. M., Boyer, H., Rosenberg, J., Dickerson, R. E., Narang, S. A., Itakura, K., Lin, S., and Riggs, A. D. (1976). Synthetic lac operator DNA is functional *in vivo*. *Nature (London)* **263**, 748–752.

Higuchi, R., Paddock, G. V., Wall, R., and Salser, W. (1976). A general method for cloning eukaryotic structural gene sequences. *Proc. Natl. Acad. Sci. U.S.A.* **73**, 3146–3150.

Hinnen, A., Hicks, J. B., and Fink, G. R. (1978). Transformation of yeast. *Proc. Natl. Acad. Sci. U.S.A.* **75**, 1929–1933.

Hofstetter, H., Schambock, A., Van den Berg, J., and Weissman, C. (1976). Specific excision of the inserted DNA segment from hybrid plasmids constructed by the poly(dT) method. *Biochim. Biophys. Acta* **454**, 587–591.

Hohn, B., and Murray, K. (1977). Packaging recombinant DNA molecules into bacteriophage particles *in vitro*. *Proc. Natl. Acad. Sci. U.S.A.* **74**, 3259–3263.

Hozumi, N., and Tonegawa, S. (1976). Evidence for somatic rearrangements of immunoglobulin genes coding for variable and constant regions. *Proc. Natl. Acad. Sci. U.S.A.* **73**, 3628–3632.

Hsu, M. T., and Ford, J. (1977). Sequence arrangement of the 5' ends of SV40 16 S and 19 S mRNAs. *Proc. Natl. Acad. Sci. U.S.A.* **74**, 4982–4985.

Humphries, P., Cochet, M., Krost, A., Gerlinger, P., Kourilsky, P., and Chambon, P. (1977). Molecular cloning of extensive sequences of the *in vitro* synthesized chicken ovalbumin structural gene. *Nucleic Acids Res.* **4**, 2389–2406.

Ilgen, C., Farabaugh, P. J., Hinnen, A., Walsh, J. M., and Fink, G. R. (1979). Transformation of yeast. *In* "Genetic Engineering" (J. Setlow and Hollaender, A., eds.). Vol. 1, pp. 117–132. Plenum, New York.

Jackson, D. A., Symons, R. M., and Berg, P. (1972). A biochemical method for inserting new genetic information into SV40 DNA: Circular SV40 DNA molecules containing λ phage genes in the galactose operon of *E. coli*. *Proc. Natl. Acad. Sci. U.S.A.* **69**, 2904–2909.

Jeffreys, A. J., and Flavell, R. A. (1977a). A physical map of the DNA regions flanking the rabbit β-globin gene. *Cell* **12**, 429–439.

Jeffreys, A. J., and Flavell, R. A. (1977b). The rabbit β-globin gene contains a large insert in the coding sequence. *Cell* **12**, 1097–1108.

Kedes, L. H. (1976). Histone messengers and histone genes. *Cell* **8**, 321–333.

Kedes, L. H. (1979). Histone genes and histone messengers. *Annu. Rev. of Biochem.* (in press).

Kedes, L. H., Chang, A. C. Y., Houseman, D., and Cohen, L. H. (1975). Isolation of histone genes from unfractionated sea urchin DNA by subculture cloning in *E. coli*. *Nature (London)* **255**, 533–538.

Klessig, D. F. (1977). Two adenovirus mRNAs have a common 5' terminal leader sequence encoded at least 10 kb upstream from their main coding regions. *Cell* **12**, 9–21.

Knapp, G., Beckmann, J. S., Johnson, P. F., Fuhrman, S. A., and Abelson, J. (1978). Transcription and processing of intervening sequences in yeast tRNA genes. *Cell* **14**, 221–236.

Kornberg, A. (1969). Active center of DNA polymerase. *Science* **163**, 1410–1418.

Kressman, A., Clarkson, S. G., Pirrotta, V., and Birnstiel, M. L. (1978a). Transcription of cloned tRNA gene fragments and subfragments injected into the oocyte nucleus of *Xenopus laevis*. *Proc. Natl. Acad. Sci. U.S.A.* **75**, 1176–1180.

Kressman, A., Clarkson, S. G., Telford, J. L., and Birnstiel, M. L. (1978b). Transcription of *Xenopus* tDNA and sea urchin histone DNA injected into the *Xenopus* oocyte nucleus. *Cold Spring Harbor Symp. Quant. Biol.* **42**, 1077–1082.

Kunkel, N. S., and Weinberg, E. S. (1978). Histone gene transcripts in the cleavage and mesenchyme blastula embryo of the sea urchin *S. purpuratus*. *Cell* **14**, 313–326.

Lacy, E., Lawn, R. M., Fritsch, E., Hardison, R. C., Parker, R. C., and Maniatis, T. (1978). Isolation and characterization of mammalian globin genes. *In* "Cellular and Molecular Regulation of Hemoglobin Switching" (G. Stomatoyannopoulos, and A. Nienhuis, eds.), pp. 501–519. Grune and Stratton, New York.

Lai, C. J., Dhar, R., and Khoury, G. (1978a). Mapping the spliced and unspliced late lytic SV40 mRNAs. *Cell* **14**, 971–982.

Lai, E. C., Woo, S. L. C., Dugaiczyk, A., Catterall, J. F., and O'Malley, B. W. (1978b). The ovalbumin gene: Structural sequences in native chicken DNA are not contiguous. *Proc. Natl. Acad. Sci. U.S.A.* **75**, 2205–2209.

Landy, A., Foeller, C., Reszelbach, R., and Dudock, B. (1976). Preparative fractionation of DNA restriction fragments by high pressure column chromatography on RPC5. *Nucleic Acids Res.* **3**, 2575–2592.

Lawn, R. M., Fritsch, E. F., Parker, R. C., Blake, G., and Maniatis, T. (1978). The isolation and characterization of linked δ- and β-globin genes from a cloned library of human DNA. *Cell* **15**, 1157–1174.

Leder, P., Tiemeier, D., and Enquist, L. (1977a). EK2 derivatives of bacteriophage λ useful in the cloning of DNA from higher organisms: The λgtWES systems. *Science* **196**, 175–177.

Leder, P., Tilghman, S. M., Tiemeier, D. C., Polsky, F. I., Seidman, J. G., Edgell, M. H., Enquist, L. W., Leder, A., and Norman, B. (1977b). The cloning of mouse globin and surrounding gene sequences in bacteriophage λ. *Cold Spring Harbor Symp. Quant. Biol.* **42**, 915–920.

Lifton, R. P., Goldberg, M. L., Karp, R. W., and Hogness, D. (1977). The organization of the histone genes in *D. melanogaster:* Functional and evolutionary implications. *Cold Spring Harbor Symp. Quant. Biol.* **42**, 1047–1051.

Lin, A. Y., Paddock, G. V., Heindell, H. C., and Salser, W. (1977). Nucleotide sequences from a rabbit alpha globin gene inserted into a chimeric plasmid. *Science* **196**, 192–194.

Little, P. F. R., Flavell, R. A., Kooter, J. M., Annison, G., and Williamson, R. (1979). Structure of the human fetal globin gene locus. *Nature (London)* **278**, 227–231.

Lobban, P. E., and Kaiser, A. D. (1973). Enzymatic end to end joining of DNA molecules. *J. Mol. Biol.* **78**, 453–471.

McBride, O. W., and Ozer, H. L. (1973). Transfer of genetic information by purified metaphase chromosomes. *Proc. Natl. Acad. Sci. U.S.A.* **70**, 1258–1262.

McReynolds, L. A., Catterall, J. F., and O'Malley, B. W. (1977). The ovalbumin gene: Cloning of a complete DS-cDNA in a bacterial plasmid. *Gene* **2**, 217–231.

McReynolds, L., O'Malley, B. W., Nisbet, A. D., Fothergill, J. E., Givol, D., Fields, S., Robertson, M., and Brownlee, G. C. (1978). Sequence of chicken ovalbumin mRNA. *Nature (London)* **273**, 723–728.

Maitland, N. J., and McDougall, J. K. (1977). Biochemical transformation of mouse cells by fragments of herpes simplex virus DNA. *Cell* **11**, 233–241.

Mandel, J. L., Breathnach, R., Gerlinger, P., Le Meur, M., Gannon, F., and Chambon, P. (1978). Organization of coding and intervening sequences in the chicken ovalbumin split gene. *Cell* **14**, 641–653.

Mandel, M., and Higa, A. (1970). Calcium-dependent bacteriophage DNA transfection. *J. Mol. Biol.* **53**, 159–162.

Maniatis, T., Jeffrey, A., and Kleid, D. (1975). Nucleotide sequence of the rightward operator of phage λ. *Proc. Natl. Acad. Sci. U.S.A.* **72**, 1184–1188.

Maniatis, T., Sim, G. K., Efstratiadis, A., and Kafatos, F. C. (1976). Amplification and characterization of a β-globin gene synthesized *in vitro*. *Cell* **8**, 163–182.

Maniatis, T., Hardison, R. C., Lacy, E., Lauer, J., O'Connell, C., Quon, D., Sim, G. K., and Efstratiadis, A. (1978). The isolation of structural genes from libraries of eukaryotic DNA *Cell* **15**, 687–701.

Mantei, N., van Ooyen, A., van den Berg, J., Beggs, J. D., Boll, W., Weaver, R. F., and Weissman, C. (1979). Synthesis of rabbit β-globin-specific RNA in mouse L cells and yeast transformed with cloned rabbit chromosomal β-globin DNA. *In* "Eukaryotic Gene Regulation" (R. Axel, T. Maniatis, and C. F. Fox, eds.). Academic Press, New York.

Mears, J. G., Ramirez, F., Leibowitz, D., and Bank, A. (1978). Organization of human δ- and β-globin genes in cellular DNA and the presence of intragenic inserts. *Cell* **15**, 15–23.

Mertz, J. E., and Gurdon, J. B. (1977). Purified DNAs are transcribed after microinjection into *Xenopus* oocytes. *Proc. Natl. Acad. Sci. U.S.A.* **74**, 1502–1506.

Messing, J., Gronenborn, B., Müller-Hill, B., and Hofschneider, P. H. (1977). Filamentous coli phage M13 as a cloning vehicle: Insertion of a *Hin*dII fragment of the lac regulatory region in M13 replicative form *in vitro*. *Proc. Natl. Acad. Sci. U.S.A.* **74**, 3642–3646.

Messing, J., and Gronenborn, B. (1978). The filamentous phage M13 as a carrier DNA for operon fusions *in vitro*. *In* "The Single Stranded DNA Phages" (D. Denhardt, D. Dressler, and D. Ray, eds.), pp. 449–453. Cold Spring Harbor Laboratory, Cold Spring Harbor, New York.

Mulligan R. C., Howard, B. H., and Berg, P. (1979). Synthesis of rabbit β-globin in cultured monkey kidney cells following infection with a SV40 β-globin recombinant clone. *Nature (London)* **277**, 108–114.

Murray, K. (1977). Applications of bacteriophage λ in recombinant DNA research. *In* "Molecular Cloning of Recombinant DNA" (W. A. Scott and A. Werner, eds.), Vol. 13, pp. 133–151. Academic Press, New York.

Murray, N. E., and Murray, K. (1974). Manipulation of restriction targets in phage λ to form receptor chromosomes for DNA fragments. *Nature (London)* **251**, 476–481.

Nasmyth, K. (1978). News and Views. *Nature (London)* **274**, 741.

Newrock, K. M., Cohen, L. H., Hendricks, M. B., Donnelly, R. J., and Weinberg, E. S. (1978). Stage-specific mRNAs coding for subtypes of H2a and H2b histones in the sea urchin embryo. *Cell* **14**, 327–336.

Nienhuis, A. W., and Benz, E. J. (1977). Regulation of hemoglobin synthesis during development of the red cell. *New England J. Med.* **297**, 1318–1328, 1371–1381, 1430–1436.

Nierlich, D. P., Rutter, W. J., and Fox, C. F., eds. (1976). Molecular Mechanisms in the Control of Gene Expression, Vol. 5. Academic Press, New York.

Orkin, S. H., Alter, B. P., Altay, C., Mahoney, M. J., Lazarus, H., Hobbins, J. C., and Nathans, D. G. (1978). Application of endonuclease mapping to the analysis and prenatal diagnosis of thalassemias caused by globin gene deletion. *N. Engl. J. Med.* **299**, 166–172.

Orkin, S. H., Old, J. M., Weatherall, D. J., and Nathan, D. G. (1979a). Partial deletion of β-globin gene DNA in certain patients with β°-thalassemia. *Proc. Natl. Acad. Sci.* **76**, 2400–2404.

Orkin, S. H., Old, J., Lazarus, H., Altay, C., Gurgey, A., Weatherall, D. J., and Nathan, D. G. (1979b). The molecular basis of α-thalassemias: Frequent occurrence of dysfunctional α loci among non-Asians with Hb H disease. *Cell* **17**, 33–42.

Panasenko, S. M., Cameron, J. R., Davis, R. M., and Lehman, I. R. (1977). Five hundred-fold overproduction of DNA ligase after induction of a hybrid λ lysogen constructed *in vitro*. *Science* **196**, 188–189.

Parker, C. S., Jaehning, J. A., and Roeder, R. G. (1978). Faithful gene transcription by eukaryotic RNA polymerases in reconstructed systems. *Cold Spring Harbor Symp. Quant. Biol.* **42,** 577–582.

Pellicer, A., Wigler, M., Axel, R., and Silverstein, S. (1978). The transfer and stable integration of HSV thymidine kinase gene into mouse cells. *Cell* **14,** 133–141.

Proudfoot, N., and Baralle, F. E. (1979). Molecular cloning of human ε-globin gene. *Proc. Natl. Acad. Sci.* **76,** 5435–5439.

Rabbits, T. H. (1976). Bacterial cloning of plasmids carrying copies of rabbit β-globin mRNA. *Nature (London)* **260,** 221–225.

Rambach, A., and Tiollais, P. (1974). Bacteriophage λ having *Eco* RI endonuclease sites only in the nonessential region of the genome. *Proc. Natl. Acad. Sci. U.S.A.* **71,** 3927–3930.

Ratzkin, B., and Carbon, J. (1977). Functional expression of cloned yeast DNA in *Escherichia coli*. *Proc. Natl. Acad. Sci. U.S.A.* **74,** 487–491.

Ray, D. S. (1977). Replication of filamentous bacteriophages. *In* "Comprehensive Virology" Frankel-Convat, H., (eds.), Vol. 17, pp. 105–178. Plenum, New York.

Rigby, P. W. J., Dieckmann, M., Rhodes, C., and Berg, P. (1977). Labeling DNA to high specific activity *in vitro* by nick translation with DNA polymerase I. *J. Mol. Biol.* **113,** 237–251.

Robbins, E., and Borun, T. W. (1967). The cytoplasmic synthesis of histones in HeLa cells and its temporal relationship to DNA replication. *Proc. Natl. Acad. Sci. U.S.A.* **57,** 409–416.

Roberts, R. J. (1976). Restriction endonucleases. *Crit. Rev. Biochem.* **3,** 123–164.

Roderiguez, R. L., Bolivar, F., Goodman, H. M., Boyer, H. W., and Betlach, M. (1976). Construction and characterization of cloning vehicles. *In* "Molecular Mechanisms in the Control of Gene Expression" (D. P. Nierlich, W. J. Rutter, and C. F. Fox, eds.), pp. 471–478. Academic Press, New York.

Roop, D. R., Nordstrom, J. L., Tsai, S. Y., Tsai, M. J., and O'Malley, B. M. (1978) Transcription of structural and intervening sequences in the ovalbumin gene and identification of potential ovalbumin mRNA precursors. *Cell* **15,** 671–685.

Rougeon, F., Kourilsky, P., and Mach, B. (1975). Insertion of a rabbit β-globin gene sequence into an *E. coli* plasmid. *Nucleic Acids Res.* **2,** 2365–2378.

Royal, A., Garapin, A., Cami, B., Perrin, F., Mandel, J., LeMeur, M., Bregegegre, Gannon, F., LePennec, J. P., Chambon, P., and Kourilsky, P. (1979). The ovalbumin gene region: Common features in the organization of three genes expressed in chicken oviduct under hormonal control. *Nature (London)* **279,** 125–132.

Roychoudhury, R., Jay, E., and Wu, R. (1976). Terminal labeling and addition of homopolymer tracks to duplex DNA fragments by terminal deoxynucleotidyl transferase. *Nucleic Acids Res.* **3,** 863–877.

Sakano, H., Hüppi, K., Heinrich, G., and Tonegawa, S. (1979a). Sequences at somatic recombination sites of immunoglobulin light chain genes: Their implications for the mode of recombination, antibody diversity and evolution. *Nature (London)* **280,** 288–294.

Sakano, H., Rogers, J. H., Hüppi, K., Brack, C., Traunecker, A., Maki, R., Wall, R., and Tonegawa, S. (1979b). Domains and the hinge region of an immunoglobulin heavy chain are encoded in separate DNA segments. *Nature (London)* **277,** 627–633.

Sanger, F., Nicklen, S., and Coulson, A. R. (1977). DNA sequencing with chain terminating inhibitors. *Proc. Natl. Acad. Sci. U.S.A.* **74,** 5463–5467.

Schaffner, W., Gross, K., Telford, J., and Birnstiel, M. (1976). Molecular analysis of the histone gene cluster of *Psammechinus miliaris*: II. The arrangement of the five histone coding and spacer sequences. *Cell* **8,** 471–478.

Schaffner, W., Kunz, G., Daetwyler, H., Telford, J., Smith, H. O., and Birnstiel, M. L.

(1978). Genes and spacers of cloned sea urchin histone DNA analyzed by sequencing. *Cell* **14**, 655–671.

Scheller, R. H., Dickerson, R. E., Boyer, A. W., Riggs, A. D., and Itakura, K. (1977a). Chemical synthesis of restriction enzyme recognition sites useful for cloning. *Science* **196**, 177–180.

Scheller, R. H., Thomas, T. L., Lee, A. S., Klein, W. H., Niles, W. D., Britten, R. J., and Davidson, E. H. (1977b). Clones of individual repetitive sequences from sea urchin DNA constructed with synthetic Eco RI sites. *Science* **196**, 197–200.

Scott, W. A., and Werner, R. (1977). "Molecular Cloning of Recombinant DNA," Vol. 13. Academic Press, New York.

Seeburg, P. H., Shine, J., Martial, J. A., Baxter, J. D., and Goodman, H. M. (1977). Nucleotide sequence and amplification in bacteria of structural genes for rat growth hormone. *Nature* (*London*) **270**, 486–494.

Seidman, J. G., Leder, A., Edgell, M. H., Polsky, F., Tilghman, S. M., Tiemeier, D. C., and Leder, P. (1978). Multiple-related immunoglobulin variable region genes identified by cloning and sequence analysis. *Proc. Natl. Acad. Sci. U.S.A.* **75**, 3881–3885.

Sgaramella, V. (1972). Enzymatic oligomerization of bacteriophage P22 DNA and of linear SV40 DNA. *Proc. Natl. Acad. Sci. U.S.A.* **69**, 3389–3393.

Shine, J., Seeberg, P. H., Martial, J. A., Baxter, J. D., and Goodman, H. M. (1977). Construction and analysis of recombinant DNA for human chorionic somatomammotropin. *Nature* (*London*) **270**, 494–499.

Shortle, D., and Nathans, D. (1978). Local mutagenesis: A method for generating viral mutants with base substitutions in preselected regions of the viral genome. *Proc. Natl. Acad. Sci. U.S.A.* **75**, 2170–2174.

Shortle, D., Pipas, J., Lazarowitz, S., Di Maio, D., and Nathans, D. (1979). Constructed mutants of simian virus 40. *In* "Genetic Engineering" (J. K. Setlow, and A. Hollaender, eds.). Vol. 1, pp. 73–92. Plenum, New York.

Shrader, W., and O'Malley, B. W. (1976). *In* "Receptors and Hormone Action" (B. W. O'Malley and L. Birnbaumer, eds.), pp. 189–224. Academic Press, New York.

Sinsheimer, R. L. (1977). Recombinant DNA. *Annu. Rev. Biochem.* **46**, 415–438.

Skalka, A. (1978). Current status of coli phage λ EK2 vectors. *Gene* **3**, 29–38.

Smithies, O., Blechl, A. E., Denniston-Thompson, K., Newell, N., Richards, J. E., Slightom, J. L., Tucker, D. W., and Blattner, F. R. (1978). Cloning human fetal γ-globin and mouse α-type globin DNA: Characterization and partial sequencing. *Science* **202**, 1284–2189.

Sternberg, N., Tiemeier, D., and Enquist, L. (1977). *In vitro* packaging of a λ *Dam* vector containing Eco RI DNA fragments of *E. coli* in phage P1. *Gene* **1**, 255–280.

Struhl, K., and Davis, R. W. (1977). Production of a functional eukaryotic enzyme in *Escherichia coli:* Cloning and expression of the yeast structural gene for imidazoleglycerol phosphate dehydratase (His 3). *Proc. Natl. Acad. Sci. U.S.A.* **74**, 5255–5259.

Struhl, K., Cameron, J. R., and Davis, R. W. (1976). Functional genetic expression of eukaryotic DNA in *E. coli. Proc. Natl. Acad. Sci. U.S.A.* **73**, 1471–1475.

Sugino, A., Goodman, H. M., Heyneker, H. L., Shine, J., Boyer, H. W., and Cozzarelli, N. R. (1977). Interaction of bacteriophage T4 RNA and DNA ligases in the joining of duplex DNA at base paired ends. *J. Biol. Chem.* **252**, 3987–3994.

Sures, I., Maxam, A., Cohn, R. H., and Kedes, L. H. (1976). Identification and localization of the histone H2a and H3 genes by sequence analysis of the sea urchin (*S. purpuratus*) DNA cloned in *E. coli. Cell* **9**, 495–502.

Sutcliff, J. G. (1978a). Nucleotide sequence of the ampicillin resistance gene of *E. coli* plasmid pBR322. *Proc. Natl. Acad. Sci. U.S.A.* **75**, 3737–3741.

Sutcliff, J. G. (1978b). pBR322 restriction map derived from the DNA sequence: Accurate DNA size markers up to 4361 nucleotide pairs long. *Nucleic Acids Res.* **5**, 2721–2728.

Thomas, M., Cameron, J. R., and Davis, R. W. (1974). Viable molecular hybrids of bacteriophage λ and eukaryotic DNA. *Proc. Natl. Acad. Sci. U.S.A.* **71**, 4579–4583.

Tiemeier, D. C., Tilghman, S. M., Polsky, F. I., Seidman, J. G., Leder, A., Edgell, M. H., and Leder, P. (1978). A comparison of two cloned mouse β-globin genes and their surrounding and intervening sequences. *Cell* **14**, 237–245.

Tilghman, S. M., Tiemeier, D. C., Seidman, J. G., Peterlin, B. M., Sullivan, M., Maizel, J. V., and Leder, P. (1977). Intervening sequence of DNA identified in the structural portion of a mouse β-globin gene. *Proc. Natl. Acad. Sci. U.S.A.* **75**, 725–729.

Tilghman, S. M., Curtis, P. J., Tiemeier, D. C., Leder, P., and Weissman, C. (1978). The intervening sequence of a mouse β-globin gene is transcribed within the 15 S β-globin mRNA precursor. *Proc. Natl. Acad. Sci. U.S.A.* **75**, 1309–1313.

Tonegawa, S., Brack, C., Hozumi, N., and Schuller, R. (1977a). Cloning of an immunoglobulin variable region gene from mouse embryo. *Proc. Natl. Acad. Sci. U.S.A.* **74**, 3518–3522.

Tonegawa, S., Brack, C., Hozumi, N., and Pirrotta, V. (1977b). Organization of immunoglobulin genes. *Cold Spring Harbor Symp. Quant. Biol.* **42**, 921–931.

Tonegawa, S., Maxam, A. M., Tizard, R., Bernard, O., and Gilbert, W. (1978). Sequence of a mouse germ line gene for variable regions of an immunoglobulin light chain. *Proc. Natl. Acad. Sci. U.S.A.* **75**, 1485–1489.

Tuan, D., Biro, A. P., de Riel, J. K., Lazarus, H., and Forget, B. (1979). Restriction endonuclease mapping of the human γ-globin gene loci. *Nucleic Acids Res.* **6**, 2519–2544.

Ullrich, A., Shine, J., Chirgwin, J., Pialet, R., Tischer, E., Rutter, W. J., and Goodman, H. M. (1977). Rat insulin genes: Construction of plasmids containing the coding sequences. *Science* **196**, 1313–1317.

Umene, K., Shimada, K., and Takagi, Y. (1978). Packaging of col EI DNA having λ phage cohesive ends. *Mol. Gen. Genet.* **159**, 39–45.

Upcroft, P., Skolnik, H., Upcroft, J. A., Solomon, D., Khoury, G., Hamer, D. H., and Fareed, G. C. (1978). Transduction of a bacterial gene into mammalian cells. *Proc. Natl. Acad. Sci. U.S.A.* **75**, 2117–2121.

Valenzuela, P., Venegas, A., Weinberg, F., Bishop, R., and Rutter, W. J. (1978). Structure of yeast phenylalanine tRNA genes: An intervening DNA segment within the region coding for the tRNA. *Proc. Natl. Acad. Sci. U.S.A.* **75**, 190–194.

Van den Berg, J., Van Ooyen, A., Mantei, N., Schamböck, A., Grosveld, G., Flavell, R., and Weissman, C. (1978). Comparison of cloned rabbit and mouse β-globin genes: Strong evolutionary divergence of two homologous pairs of introns. *Nature (London)* **276**, 37–44.

Villa-Komaroff, L., Efstratiadis, A., Broome, S., Lomedico, P., Tizard, R., Naber, S. P., Chick, W. L., and Gilbert, W. (1978). A bacterial clone synthesizing proinsulin. *Proc. Natl. Acad. Sci. U.S.A.* **75**, 3727–3731.

Weatherall, D. J., and Clegg, J. B. (1972). "The Thalassemia Syndromes," 2nd ed. Blackwell, Oxford.

Weinberg, E. S., Overton, G. C., Hendricks, M. B., Newrock, K. M., and Cohen, L. H. (1977). Histone gene heterogeneity in the sea urchin *Strongylocentrotus purpuratus*. *Cold Spring Harbor Symp. Quant. Biol.* **42**, 1093–1100.

Weissman, C., Nagota, S., Taniguchi, T., Weber, H., and Meyer, F. (1979). The use of site directed mutagenesis in revised genetics. *In* "Genetic Engineering" (J. Setlow and A. Hollaender, eds.), Vol. 1, pp. 133–150. Plenum, New York.

Weinstock, R., Sweet, R., Weiss, M., Cedar, H., and Axel, R. (1978). Intragenic DNA spacers interrupt the ovalbumin gene. *Proc. Natl. Acad. Sci. U.S.A.* **75**, 1299–1303.

Wensink, P. C., Finnegan, D. H., Donelson, J. E., and Hogness, D. S. (1974). A system from mapping DNA sequences in the chromosomes of *Drosophila melanogaster*. *Cell* **3**, 315–325.

White, R. L., and Hogness, D. S. (1977). R-loop mapping of the 18 S and 28 S sequences in the long and short repeating units of *D. melanogaster* rDNA. *Cell* **10**, 177–192.

Wigler, M., Silverstein, S., Lee, L. S., Pellicer, A., Cheng, Y. C., and Axel, R. (1977). Transfer of purified herpes virus thymidine kinase gene to cultured mouse cells. *Cell* **11**, 223–232.

Wigler, M., Pellicer, A., Silverstein, S., and Axel, R. (1978). Biochemical transfer of single copy eukaryotic genes using total cellular DNA as a donor. *Cell* **14**, 725–731.

Wigler, M., Pellicer, A., Axel, R., and Silverstein, S. (1979a). Transformation of mammalian cells. *In* "Genetic Engineering" (J. Setlow and A. Hollaender, eds.), Vol. 1, pp. 51–72. Plenum, New York.

Wigler, M., Sweet, R., Sim, G. K., Wold, B., Pellicer, A., Lacy, E., Maniatis, T., Silverstein, S., and Axel, R. (1979b). Transformation of mammalian cells with prokaryotic and eukaryotic genes. *In* "Eukaryotic Gene Regulation" (R. Axel, T. Maniatis, and C. F. Fox, eds.), Academic Press, New York.

Wigler, M., Sweet, R., Sim, G. K., Wold, B., Pellicer, A., Lacy, E., Maniatis, T., Silverstein, S., and Axel, R. (1979c). Transformation of mammalian cells with genes from procaryotes and eucaryotes. *Cell* **16**, 777–785.

Williams, B. G., and Blattner, F. R. (1979). Construction and characterization of the hybrid bacteriophage λ charon vectors for DNA cloning. *J. Virol.* **29**, 555–575.

Willicke, K., and Ruddle, F. H. (1975). Transfer of the human gene for hypoxanthine-guanine phosphoribosyltransferase via isolated human metaphase chromosomes into mouse L-cells. *Proc. Natl. Acad. Sci. U.S.A.* **72**, 1792–1796.

Wilson, J. T., Wilson, L. B., de Riel, J. K., Villa-Komaroff, L., Efstratiadis, A., Forget, B. G., and Weissman, S. M. (1978). Insertion of synthetic copies of human globin genes into bacterial plasmids. *Nucleic Acids Res.* **5**, 563–581.

Wold, B., Wigler, M., Lacy, E., Maniatis, T., Silverstein, S., and Axel, R. (1979). Expression of an adult rabbit β-globin gene stably inserted into the genome of mouse L cells. *Proc. Natl. Acad. Sci. U.S.A.* **76**, 5684–5688.

Woo, S. L. C., Dugaiczyk, A., Tsai, M. J., Lai, E. C., Catteral, J. F., and O'Malley, B. W. (1978). The ovalbumin gene: Cloning of the natural gene. *Proc. Natl. Acad. Sci. U.S.A.* **75**, 3688–3692.

Wood, K. O., and Lee, J. C. (1976). Integration of synthetic globin genes into an *E. coli* plasmid. *Nucleic Acids Res.* **3**, 1961–1971.

Wu, G. J., Luciw, P., Mitra, S., Zubay, G., and Ginsberg, H. S. (1977). *In* "Nucleic Acid-Protein Recognition" (H. Vogel, ed.), pp. 171–186. Academic Press, New York.

Wu, G. R. (1978). Adenovirus DNA directed transcription of 5.5 S RNA *in vitro*. *Proc. Natl. Acad. Sci. U.S.A.* **75**, 2175–2179.

Zain, S., Sambrook, J., Roberts, R. J., Keller, W., Fried, M., and Dunn, A. (1979). Nucleotide sequence analysis of the leader segments in a cloned copy of adenovirus 2 fiber mRNA. *Cell* **16**, 851–861.

12

Basic Characteristics of Different Classes of Cellular RNA's: A Directory

Yong C. Choi and Tae-Suk Ro-Choi

609

I. INTRODUCTION

There are a number of RNA classes in a single eukaryotic cell. For example, the approximate number of RNA species can be calculated to be 10^4–10^5 with an assumption that RNA's with average length of 10^3 bases are transcribed from approximately 10% of the genome (10^9 base pairs) of mammalian cells. Table I shows the distribution of various RNA species in a single HeLa cell. A majority of RNA's are present in the cytoplasm and are associated with the machinery of protein synthesis (10^6 ribosomes and 10^8 tRNA's). The next most abundant RNA's are the low molecular weight RNA's (LMW nRNA's or snRNA's), which are associated with nuclear substructures. The abundance of mRNA's which code for the various cell proteins is not high, but can be estimated to be 10^4–10^5 copies in a single cell.

In order to classify an individual RNA species, it is necessary to establish a unified system of nomenclature. The common usages are summarized as follows.

1. *Localization:* nuclear RNA (nRNA) (nucleoplasmic RNA and nucleolar RNA), cytoplasmic RNA (ctRNA), mitochondrial RNA (mtRNA)
2. *Function:* transfer RNA (tRNA), messenger RNA (mRNA), ribosomal RNA (rRNA), gene regulator RNA (primer RNA), structural RNA (RNP-associated RNA)

TABLE I

Intracellular Distribution of Various RNA's of HeLa Cells[a]

RNA	Percentage	Copies in nucleus	Copies in cytoplasm
hnRNA	1	1×10^4	—
45 S pre-rRNA	1	1×10^4	—
32 S pre-rRNA	3	4×10^4	—
28 S rRNA	53	7×10^4	5×10^6
18 S rRNA	24	6×10^4	5×10^6
5.8 S rRNA	1	6×10^4	5×10^6
5S rRNA	1	6×10^5	5×10^6
tRNA	12	2×10^5	1×10^8
snRNA	0.5	2×10^6	—
mRNA	3	—	1×10^5
Nonpolysomal polydisperse	<1	—	1×10^4

[a] The designation of snRNA is referred to nucleus-specific LMW nRNA's including 4.5 S nRNA$_{I–III}$, 5 S nRNA$_{III}$, U-1 nRNA, U-2 nRNA, and U-3 nRNA. Data from Darnell (1968) and Weinberg and Penman (1968).

3. *Metabolic properties:* stable RNA (long half-life), unstable RNA (short half-life), precursor RNA
4. *Size:* low molecular weight RNA (LMW RNA), intermediate molecular weight RNA (IMW RNA), high molecular weight RNA (HMW RNA)
5. *Composition characteristics:* GC-rich RNA, AU-rich RNA, poly(A) + RNA, 5'-cap RNA
6. *Abundance:* most abundant RNA ($>10^4$ copies), highly abundant RNA (10^3 copies), abundant RNA (10^2 copies), scarce RNA (1–10 copies)

A. Structure

RNA is a polyribonucleotide which contains a 2'-OH on the ribose moiety and a directional linkage consisting of a unit building block (base–ribose–phosphate) with a backbone of 3' → 5' phosphodiester bonds. Because of the bulky 2'-OH moiety, RNA has a secondary and tertiary structure different from DNA. These remarkable differences have been demonstrated by sequence analysis and model building (Holley *et al.,* 1965; Fier *et al.,* 1976) and X-ray crystallography (Kim *et al.,* 1974a; Robertus *et al.,* 1974). Examples include the following: (1) Possible secondary structures of RNA are represented by the cloverleaf and flower models. (2) There is no long stretch of bihelical structure in RNA and no clear demarcation between major and minor grooves. An RNA helix looks more like a band wrapped around an imaginary cylinder. (3) The tertiary structures are maintained by nonclassical hydrogen bondings (base-pair triple) and tight base stackings.

In addition, cellular RNA's have special structures introduced by posttranscriptional modifications (Perry, 1976; see also Chapter 11, this volume). It has been firmly established that most mature RNA's are derived from precursor RNA's which have sequences complementary to DNA sequences of chromosomes and which are subsequently modified by maturation processes. Posttranscriptional modifications of primary structure of RNA's include noncomplementary structures such as poly(A) and a 5'-cap.

B. Populations of RNA's

When gene expression is modified by a number of conditions (drugs, hormones, viruses, cell cycle, differentiation, development, and others), qualitative and quantitative changes occur in RNA metabolism. The metabolic changes are reflected in different levels, including availability of active genes by "turn on" and "turn off" mechanisms, rate of gene readout by RNA polymerases, rate of RNA processing, and turnover of mature RNA's. Accordingly, the sizes of RNA populations are a "balance sheet" resulting from (1) the synthetic rate of transcriptional products linked to the number of active genes, (2) the rate of maturation, (3) the functional life time of RNA's, and (4) their rate of degradation.

1. Compartmentalization and Association of RNA's with Subcellular Components

All the eukaryotic RNA's (except tRNA's) exist in the form of ribonucleoprotein complexes (RNP's) that are associated with subcellular components. Nuclear RNP's include pre-mRNA–RNP's associated with nucleoplasm, pre-rRNA–RNP's associated with the nucleolus, and LMW nRNA–RNP's possibly associated with the "nuclear skeleton."

Cytoplasmic RNP's include nonpolysomal mRNA–RNP's, and membrane-bound and nonmembrane-bound polysomes, which contain tRNA's, 5 S rRNA, 5.8 S rRNA, 18 S rRNA, 28 S rRNA, and mRNA's. Mitochondrial RNP's include mitochondrial polysomes and others. The best-defined RNP's are polysomes, which are involved in protein synthesis. The least-defined RNP's are LMW nRNA-RNP's.

2. Expression of Genes

The number of cellular RNA species is a function of the number of active genes in a cell. Expression of each gene is modulated by many mechanisms. For example, the genetic mechanisms of abnormal hemoglobin disease (thalassemia) show defects in many levels of gene regulation, including gene structures, transcription, and translation (Forget, 1978). The mechanisms of gene expression are currently a topic of intensive investigation.

C. Evolution of RNA Structures

To establish a unified picture of the structure–function relationships of RNA's, extensive studies of structural homology have been made to determine functional and phylogenic significance of RNA sequences in terms of evolution. Although a unified picture of genes and RNA's can be obtained by comparative studies of gene organization (DNA sequence), it is noted that the approach by RNA sequence method shows an additional special feature, namely, posttranscriptional modifications that cannot be determined by DNA sequence methods. The structural comparisons among prokaryotes and eukaryotes show two major trends. The first is concerned with a continuous trend which indicates conserved sequences through evolution. This is exemplified in the general structure of tRNA's (universal structure and invariant sequences), homology of 5 S rRNA and 5.8 S rRNA, and conserved sequences of 18 S rRNA and 28 S rRNA. The second is the discontinuous trend which indicates diverse sequences specific for an organism. This is exemplified in variant sequences of tRNA's and nonconserved sequences of 18 S rRNA and 28 S rRNA.

This chapter deals with basic characteristics of cellular RNA's. Since this subject is rapidly expanding, the galaxies of information are impossible to include in detail. This chapter is limited to the current trends in the study of the structure–function aspects of RNA's and is presented in the form of a miniature directory.

II. NUCLEAR RNA's

Nuclear RNA's can be classified as shown below:

1. *Compartments:* nucleoplasmic RNA's, nucleolar RNAs, and RNA's associated with the nuclear skeleton
2. *Physical discreteness and size:* low molecular weight nuclear RNA's, intermediate molecular weight nuclear RNA's, high molecular weight nuclear RNA's, and heterogeneous nuclear RNA's

Although earlier work showed that nuclear RNA's consisted of RNA's with rapid and slow rates of turnover, the systematic study of nuclear RNA's began with the first successful demonstration of undegraded nRNA's fractionated by sucrose density gradient centrifugation (Scherrer and Darnell, 1962). Subsequently, the improved methods of fractionation of nuclei (Busch and Smetana, 1970) and of fractionation of nuclear RNA's (Weinberg, 1973) led to the definition of the classes of RNA's present in nuclei and their significance in nuclear metabolism. The detailed characterization of nuclear RNA's has revealed (1) the phenomena of hnRNA processing and pre-rRNA processing, (2) the existence of LMW nRNA's, and (3) the primary structures of various nRNA's. Table II shows the discrete nRNA species separated by polyacrylamide gel electrophoresis, using Novikoff hepatoma cells as an example (Reddy *et al.,* 1974b; Ro-Choi *et al.,* 1973). The nomenclature is somewhat arbitrary because no unified system is available. However, a number of different RNA classes clearly are present in the nucleus.

A. Low Molecular Weight Nuclear RNA's (LMW nRNA's)

LMW nRNA's were initially recognized as 4–7 S nRNA's which can be fractionated as a slowly sedimenting RNA class in sucrose density gra-

TABLE II

Discrete Nuclear RNA Species (Novikoff Ascites Cells) Fractionated by Polyacrylamide Gel Electrophoresis[a]

LMW nRNA		IMW nRNA 8–18 S		HMW nRNA	
4–7 S	Location[b]	Band	Location[b]	>18 S	Location[b]
4 S	Nu/No	1 ⎫		18 S	Nu/No
(tRNA)		⎪ ⎬ Nu			
4.2 S	Nu	4 ⎭			
				20 S	No
4.5 S_I	Nu	A_1 ⎫			
4.5 S_{II}	Nu	⎪ ⎬ Nu		23 S	No
4.5 S_{III}	Nu	A_4 ⎭			
				28-S	Nu/No
5 S_I	Nu	5 ⎫			
5 S_{II}	Nu/No	⎪ ⎬ Nu			
5 S_{III}	Nu	1 0 ⎭		32 S	No
A (5.4 S)	Nu	11 ⎫		36 S	No
B (5.4 S)	Nu	⎪ ⎬ Nu/No			
C		15 ⎭		42 S	No
5.8 S	Nu/No	16 ⎫		45 S	No
		⎪ ⎬ Nu			
U-1a	Nu	18 ⎭			
U-1b	Nu				
		19 ⎫			
U-2	Nu	⎪ ⎬ Nu/No			
		22 ⎭			
U-3a	No				
U-3b	No	23 ⎫			
U-3c	No	⎪ ⎬ Nu/No			
		28 ⎭			
		29 ⎫			
		⎪ ⎪			
		$C_1 — C_8$ ⎬ Nu/No			
		⎪ ⎪			
		36 ⎭			

[a] LMW nRNA data from Busch and Smetana (1970), Prestayko *et al.* (1971), Reddy *et al.* (1974), and Ro-Choi *et al.* (1970, 1973, 1974); IMW nRNA data from Savage *et al.* (1974); HMW nRNA data from Busch and Smetana (1970) and Ro-Choi *et al.* (1973).

[b] Nu, nucleus; No, nucleolus; Nu/No, nucleus and nucleolus.

dients. As soon as the techniques of polyacrylamide gel electrophoresis were applied to study LMW nRNA's, a number of discrete RNA species were recognized (Galibert *et al.*, 1967; Peacock and Dingman, 1967). Subsequently, systematic studies were undertaken that established that (1) LMW nRNA's are not artifactual products derived from degradation of HMW nRNA's during the isolation of nRNA's and (2) LMW nRNA's are the inherent components of nuclear substructures (Ro-Choi and Busch, 1974; Hellung-Larsen, 1977).

1. Nomenclature

There are several systems of nomenclature for LMW nRNA species. Table III lists the different systems used mostly for LMW nRNA species derived from vertebrates.

TABLE III

Different System of Nomenclature for LMW nRNA's[a]

System I[b]	System II	System III
K		
L		8 S
M		
N		
A		U-3
B		
C	H, VII	U-2
D*		U-1c
	G, VI	
D		U-1b
		U-1a
E	F, V	
		5.8 S
F		A, B, C
		5 S RNA$_I$
G	E	
		5 S RNA$_{II}$
G'	D, IV	5 S RNA$_{III}$
H$_1$		4.5 S RNA$_I$
H$_2$	C, III	4.5 S RNA$_{II}$
H$_3$		4.5 S RNA$_{III}$
I	B, II	4 S tRNA

[a] System I, Weinberg and Penman (1968); system II, Zapisek *et al.* (1969); system III, Ro-Choi *et al.* (1970).

[b] D and D* are conformational isomers.

2. Classification of LMW nRNA's

LMW nRNA's can be classified by their location in subcellular compartments and by known metabolic characteristics.

a. Nucleolar LMW nRNA's. Nucleolar LMW nRNA's include 4 S nRNAs (tRNA's), 5 S rRNA, 5.8 S rRNA, U-3 nRNA's (U-3a, U-3b, and U-3c), and 8 S nRNA.

b. Etranucleolar or Nucleoplasmic LMW nRNA's. Extranucleolar or nucleoplasmic LMW nRNA's include 4 S nRNA's (tRNA's), 4.5 S RNA$_I$, 4.5 S RNA$_{II}$ and 4.5 S RNA$_{III}$, 5 S rRNA (5 S RNA$_I$ and 5 S RNA$_{II}$), 5 S nRNA$_{III}$, 5.8 S rRNA, U-1 nRNA's (U-1a and U-1b), and U-2 nRNA

c. LMW nRNA's which are precursors for cytoplasmic RNA's: pre-tRNA's, 5 S rRNA, and 5.8 S rRNA

d. Shuttling and Nonshuttling LMW nRNA's. This classification of low molecular weight RNA's was proposed by Goldstein (1975) on the basis of experiments with nuclear transplantation in amoebae. Nuclei containing ^{3}H-labeled, stable, LMW RNA's were transplanted into the cells which contained no radioactive RNA molecules. Those species of RNA which translocated from transplanted "hot" nuclei through the cytoplasm to the unlabeled host cell nucleus were named shuttling RNA, and those which were not taken up by the host nucleus or those that remained in the transplanted "hot" nucleus were referred to as nonshuttling RNA.

Amoebae were found to possess snRNA's that have been designated as 3.5 S, 4 S, 5 S, 5.5 S, U-1, U-2, U-3, and 8 S RNA. Of these, shuttling RNA's are 5 S nRNA, U-1 nRNA, U-3 nRNA, and 8 S nRNA and nonshuttling RNA's are 3.5 S nRNA, 5.5 S rRNA, and U-2 nRNA (Goldstein and Ko, 1974).

e. Characteristic Structures of the 5'-End of LMW nRNA's. LMW nRNA's can also be classified by characteristic structures of the 5'-end: (1.) 5'-Capped LMW nRNA's—5 S nRNA$_{III}$, U-1 nRNA, U-2 nRNA, and U-3 nRNA. It is clear that the 5'-cap of LMW nRNA [m$_3^{2,2,7}$ GpppAmpN(m)p--] is somewhat different from that of hnRNA and mRNA [m^{7}Gppp(m)N(m)pN(m)p--]. (2) Noncapped LMW nRNA with phosphoryl moieties (mono- or triphosphoryl): tRNA's, 4.5 S nRNA$_I$, 4.5 S RNA$_{III}$, 5 S rRNA, and 5.8 S RNA

The chemistry of the two different forms (capping bases) of the 5'-cap structure has been demonstrated first by the determination of novel 5' → 5'-pyrophosphodiester linkage at the 5'-end of LMW nRNA's, which contain m$_3^{2,2,7}$G (Ro-Choi et al., 1974), and by the subsequent suggestion of similar backbone linkage at the 5'-end of mRNA's which contain m^{7}G (Rottman et al., 1974). Soon after the first description, a number of extensive studies were made to establish various types of 5'-cap structure (Furuichi et al., 1975; Adams and Cory, 1975; Busch, 1976; Shatkin,

1976). The term 5'-cap was coined by Perry and Kelley (1976). It has been observed so far that the capping molecule is the 7-methyl derivative of guanosine.

3. Homology and Evolutionary Trends

Extensive comparative studies have been made to examine the population profiles of LMW nRNA's among different eukaryotic cells, including vertebrates and invertebrates (Hellung-Larsen, 1977). Although these studies were based on the electrophoretic mobility without examination of structural characteristics, two important findings were obtained by comparison among frog, lizard, chicken, rabbit, mouse, rat, hamster, monkey, and man. (1) The number of LMW nRNA's is very similar among vertebrates. An exception is shown in nucleated red blood cells (frog and chicken), which lack nucleolus-specific U-3 RNA. (2) The sizes of U-1 nRNA, U-2 nRNA, U-3 nRNA, 8 S nRNA, and 5.8 S rRNA differ among mammalian and nonmammalian cells. These differences are related to the evolutionary diversity of LMW rRNA's, specific for a given organism.

Furthermore, results of comparative studies of LMW nRNA's of invertebrates, unicellular eukaryotes, and prokaryotes (*Mycoplasma, Saccharomyces, Physarum, Tetrahymena, Hylotropus, Tenebrio, Calliphora, Paracentrotus,* and *Psammechinus*) show that (1) the number of LMW nRNA's is different and (2) the common LMW nRNA's are 5 S rRNA and 5.8 S rRNA. An exception is amoeba, which shows similarity to vertebrates. At the present, no available structural information shows an evolutionary trend.

4. Structural Characteristics of LMW nRNA's

Although earlier studies described the nucleotide composition of LMW nRNA's from various organisms (Weinberg and Penman, 1968; Larsen *et al.,* 1969), the most extensive characterization of LMW nRNA's has been made for rat hepatoma cells (Ro-Choi *et al.,* 1974; Busch *et al.,* 1975). Table IV shows some structural characteristics of LMW nRNA's from Novikoff ascites cells.

a. 4 S RNA (tRNA's). The 4 S RNA mainly consists of tRNA's, a considerable amount of which is aminoacylated. Nucleolar 4 S RNA's have about 20–40% and nuclear 4 S RNA's have about 50–70% of the aminoacyl acceptor activity that cytoplasmic tRNA's have. Detailed examination of nuclear isoaccepting tRNA[Leu] and tRNA[Val] showed little difference from the cytoplasmic counterparts (Ritter and Busch, 1972). Determination of modified nucleotides indicated lower levels of dihydrouridylate and N^2-methylguanylate in nucleolar 4 S RNA in comparison to nuclear and cytoplasmic 4 S RNA (Reddy et al., 1972).

TABLE IV

Structural Characteristics of LMW nRNA's of Novikoff Hepatoma[a]

	4 S RNA	4.5 S RNA			5 S RNA			U1 RNA			
		I	II	III	I	II	III	5.8 S	U-1b,c	U-2 RNA	U-3 RNA
Localization											
Nucleolar	+	?	?	?	+	?	−	+	−	−	+
Nucleoplasmic	+	+	+	+	+	+	+	+	+	+	−
Cytoplasmic	+	−	−	−	?	+	−	+	−	−	?
Structural characteristics											
Chain length	80 ± 5	96	100	93	121	121	130	158	171	196	240 ± 10
5'-End	pGp,pAp, pCp,pUp	(pp)pGp	−	pAp	(pp)pGp	(pp)pGp	"Cap" $pm_3^{2,2,7}G$ | p | p pAmpUmpAp	pCp pGp	"Cap" $pm_3^{2,2,7}G$ | p | p pAmpUmpAp	"Cap" $pm_3^{2,2,7}G$ | p | p pAmpUmpCp	"Cap" $pm_3^{2,2,7}G$ | p | p pAmpA(m)pAp
3'-End	A_{OH}	U_{OH}	U_{OH}	Um	U_{OH}	U_{OH}	U_{OH} A_{OH}	U_{OH}	U_{OH} G_{OH}	C_{OH} A_{OH}	A_{OH} U_{OH} G_{OH}
Modified nucleotides	m^1A, m^3U hU, m_2^2G, m^2G m^1G, m^3C, m^5C ψ, I, m^7G AcC etc.		?	Am-Ap Gm-Ap 2Gm-Gp m^2G $m^6A, 3\psi$			Um-Up Gm-Cp 2ψ	Um-Gp Gm-Cp ψ	Am-Cp	Gm-Gm-Cp Gm-Gp Gm-Ap m^6Am-Gp Cm-ψp, Um-Ap, Cm-Up, 13ψ	G_{OH} ψ

b. 4.5 S nRNA$_I$. The primary structure has been determined (Ro-Choi *et al.*, 1972). The 4.5 S nRNA$_I$ has 96–97 nucleotides with no modified nucleotides. The 5'-region of 4.5 S nRNA$_I$ is rich in purines, whereas the 3'-region is rich in pyrimidines. The presence at the 5'-end of a triphosphoryl moiety suggests that 4.5 S nRNA$_I$ may be a primary transcription product. The possibility that this RNA may be a pre-tRNA was ruled out by sequence determination.

c. 4.5 S nRNA$_{II}$. The 4.5 S nRNA$_{II}$ has 90–100 nucleotides with no modifications. Nucleotide composition shows equal distribution of four major nucleotides (Reddy *et al.*, 1972), and its 3'-terminal has been found to be 80% U. The complete structure of this RNA is not known.

d. 4.5 S nRNA$_{III}$. Partial structural analysis of this RNA has been made (El-Khatib *et al.*, 1970) and found to be characterized by a 2'O-methylated uridine at the 3'-end and by m^{6}A (Reddy *et al.*, 1972).

e. 5 S rRNA (5 S nRNA$_I$ and 5 S nRNA$_{II}$). The 5 S rRNA contains two different conformation isomers and show no primary structure difference from ribosomal 5 S rRNA (Ro-Choi *et al.*, 1971). No modified nucleotides were found. 5 S rRNA is associated with pre-rRNA RNP.

f. 5 S nRNA$_{III}$. The 5 S nRNA$_{III}$ is characterized by the presence of 2'-O-methylated nucleotides and a type II 5'-cap (m$_3^{2,2,7}$GpppAmpUmpAp) (Ro-Choi *et al.*, 1971, 1978). These structural features are the best criteria for differentiating between 5 S nRNA$_{III}$ and 5 S rRNA, which show similar electrophoretic mobilities at neutral pH.

g. 5.8 S rRNA. The primary sequence of this RNA has been determined (Nazar *et al.*, 1975). It is associated with 28 S pre-rRNA of the nucleolus (Prestayko *et al.*, 1970) and is derived from 32 S pre-rRNA (Nazar *et al.*, 1975).

h. U-1 nRNA. The primary structure has been determined (Reddy *et al.*, 1974b). U-1 nRNA from Novikoff hepatoma contains a type II 5'-cap (m$_3^{2,2,7}$GpppAmpUmpAp) (Busch *et al.*, 1975; Ro-Choi and Henning, 1977). U-1 nRNA of mouse myeloma cell contains the same 5'-cap structure (Cory and Adams, 1975). Although U-1 nRNA contains the 5'-cap structure and AUG sequence (initiation codon), it does not code for a protein and, in fact, inhibits the translation of mRNA *in vitro* (Rao *et al.*, 1977).

i. U-2 nRNA. The primary structure of Novikoff hepatoma U-2 RNA has been determined (Shibata *et al.*, 1975; Ro-Choi *et al.*, 1975). U-2 RNA is the most unusual nRNA species because the 5'-end portion of 69 residues contains a type II 5'-cap (m$_3^{2,2,7}$GpppAmpUmpCp), an alkali-resistant trinucleotide (Gm-Gm-C), six alkali-resistant dinucleotides, and thirteen pseudouridylate residues. U-2 nRNA of mouse contains a similar 5'-cap structure (Cory and Adams, 1975).

j. U-3 nRNA. U-3 nRNA is characterized by the presence of type I ($m_3^{2,2,7}$GpppAmpAp) and II ($m_3^{2,2,7}$GpppAmpAmpAp) 5'-caps (Ro-Choi *et al.*, 1978). U-3 nRNA is specific for the nucleolus and shows microheterogeneity (U-3a, U-3b, and U-3c) (Prestayko *et al.*, 1971).

k. Function of LMW nRNA's. The function of LMW nRNA's has not been defined experimentally, although there are hypotheses (Ro-Choi and Busch, 1974). LMW nRNA's exist in the form of RNP's (Prestayko *et al.*, 1970; Rein, 1971; Enger *et al.*, 1974; Raj *et al.*, 1975; Sekeris and Niessen, 1975) and are associated with the nuclear skeleton (Herman *et al.*, 1976). The chromosomal association of some LMW nRNA's has been clearly demonstrated by Goldstein *et al.* (1977). Also, LMW nRNA's, such as U-1 nRNA, U-2 nRNA, and U-3 nRNA, can be found in the cytoplasm (Herman *et al.*, 1976; Hellung-Larsen, 1977).

B. Intermediate Molecular Weight Nuclear RNA's (IMW nRNA's)

Intermediate molecular weight nRNA's, were initially referred to as 8–18 S RNA's (10 S, 14 S, 18 S nRNA's) because they sediment on sucrose gradients between 4–7 S RNA's and 28–45 S RNA's. Refined fractionation by polyacrylamide gel electrophoresis shows at least 36 discrete bands (Table II) (Savage *et al.*, 1974). Examination of nucleotide composition showed that most of the discrete RNA species were GC-rich RNA. The new species of RNA's described by Benecke and Penman (1977) are included in this class. The function of these discrete IMW nRNA's is not clearly defined.

C. Precursor Ribosomal RNA's (Pre-rRNA's)

High molecular weight nRNA's include the discrete class of nucleolar pre-rRNA's and the heterogeneous class of hnRNA's. These classes of RNA's also are referred to as the rapidly sedimenting RNA's. hnRNA's are discussed in the following section (Section II, D). Nucleolar pre-rRNA's, especially of mammalian cells, were initially found to be 18 S (20 S), 23 S, 28 S, 32 S (or 35 S), and 45 S pre-rRNA's by sucrose density gradient centrifugation (Steele and Busch, 1966). Higher resolution polyacrylamide gel electrophoresis showed the presence of discrete pre-rRNA species, 20 S, 24 S, 28 S, 32 S, 36 S, 41 S, and 45 S (Weinberg and Penman, 1970).

Early studies of metabolic characterization of nucleolar pre-rRNA's (Perry, 1967; Darnell, 1968; Busch and Smetana, 1970) and recent studies of structural characterization of nucleolar pre-rRNA's (Hadjiolov and Nikolaev, 1976; Cox, 1977) have firmly established the scheme of nucleolar processing of 45 S pre-rRNA. Recent studies have characterized rDNA and have given the general structure of rDNA and the primary transcriptional unit (Rungger and Crippa, 1977).

1. Size of the Primary Transcriptional Product

The primary pre-rRNA is the largest pre-rRNA, the precursor of the two HMW rRNA's (18 S and 28 S). Theoretically, the primary pre-rRNA should be the transcriptional product which contains the complete readout of the rDNA, including a 5'-triphosphoryl moiety. However, so far only the 40 S pre-rRNA of *Xenopus* oocytes has been observed to contain the 5'-triphosphoryl moiety (Reeder *et al.*, 1977). Accordingly, the primary pre-rRNA is referred to as a form derived from the primary transcriptional product. Recently, more improved kinetic techniques have been used to make the observations discussed below:

a. Kinetically Unstable Pre-rRNA ("Short-Lived" Pre-rRNA). The size of pre-rRNA detected by short pulse labeling techniques in mouse (47 S) (Tiollais *et al.*, 1971), *Tetrahymena* (42 S) (Kumar, 1967), and yeast (2.8 megadaltons) (Gierson *et al.*, 1971) is larger than that detected by long labeling techniques. The size of pre-rRNA synthesized in isolated nuclei is larger than that synthesized in frog egg cells (Caston and Jones, 1970).

b. "Run-Through Readout" Pre-rRNA. In addition to pre-rRNA (45 S), larger pre-rRNA (65–85 S) has been demonstrated, suggesting a continuous readout of more than one unit of rDNA (Hidvegi *et al.*, 1971).

c. Facultative Pre-rRNA. Pre-rRNA's of many plants show two independent pre-rRNA's characterized by different maturation kinetics (Rungger and Cripps, 1977).

Extensive comparative studies indicate that the size of primary pre-rRNA (30–45 S) increases from prokaryotes to eukaryotes. Table V shows the relationship of primary prerRNA to rDNA.

2. Processing Scheme of the Primary Transcriptional Product

Discrete nucleolar pre-rRNA's can be related to nucleolar processing. One of the best-defined schemes of nucleolar processing was initially developed in mammalian cells. The scheme was constructed by identification of intermediate pre-rRNA's (41 S, 36 S, 32 S, 28 S–5.8 S complex, 24 S, 20 S) (Weinberg and Penman, 1970; Wellauer and Dawid, 1973; Prestayko *et al.*, 1971) and of their intermediate pre-rRNA–RNP's (Warner and Soeiro, 1967; Liau and Perry, 1969) and by structural analysis (Egawa *et al.*, 1971; Nazar *et al.*, 1975; Maden *et al.*, 1977).

Recently, a number of organisms have been examined to clarify different schemes of processing of the primary transcriptional products (Hadjiolov and Nikolaev, 1976). Furthermore, different rDNA's have been

TABLE V

Size of Primary Pre-rRNA's in Relation to rDNA

Species	Unit of rDNA ($\times 10^{-3}$ base pairs)	Reference[a]	S value	Pre-rRNA ($\times 10^{-6}$ daltons)	Reference[a]
E. coli	5.5	a	30	2.1	j
Yeast	9.3	b	35	2.5	k
Pea					
Seedlings	—		33	2.4,2.3	l
Root tips	—			2.3 ± 0.1	m
			36	—	n
Tetrahymena	13.6	c			o
			42	—	p
Drosophila	13.7–18.3 (insertion)	d	38	2.85	q
Xenopus	10.7–16.7 (spacer heterogeneity)	e	40	2.5–2.6	r
Chick	27.0	f	45	3.9	q
Mouse	44.0	g	45	4.1	q
	38.0–39.6 (spacer heterogeneity)	h	—	—	
Rabbit	28.6	i	45	4.2	s
Rat	—	—	45	4.2	t
Human	31.7	h	45	4.4	u

[a] Key to references: a, Deonier *et al.* (1973); b, Nath and Bollon (1977); c, Karrer and Gall (1976); d, Pellgrini *et al.* (1977); e, Wellauer *et al.* (1976); f, McClements and Skalka (1977); g, Cory and Adams (1977); h, Arnheim and Southern (1977); i, Southern (1975); j, Nikolaev *et al.* (1974); k, Udem and Warner (1972); l, Gierson *et al.* (1971); m, Rogers *et al.* (1970); n, Kumar (1970); o, Prescott *et al.* (1971); p, Pousada *et al.* (1975); q, Perry *et al.* (1970; r, Loening *et al.* (1969); s, Schibler *et al.* (1975); t, Quagliarotti *et al.* (1970); u, McConkey and Hopkins (1969).

mapped to define the processing schemes (Rungger and Crippa, 1977). It is now possible to visualize the following general scheme of processing:

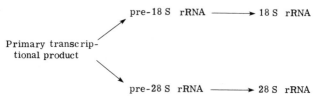

Table VI shows a list of various schemes of rRNA processing.

TABLE VI
Processing Schemes of Pre-rRNA's

Species	Primary pre-rRNA		Trimmed intermediate pre-rRNA		Split intermediates				Mature rRNA's		Reference[a]
	S value	(megadaltons)	S value	(megadaltons)	S value	(megadaltons)	S value	(megadaltons)	S value	(megadaltons)	
E. coli	30 S	(2.1)			17.5 S	(0.65)	p16 S	(0.60)	16 S	(0.55)	a
									$tRNA_2^{Glu}$		b
									$tRNA_1^{Ileu}$		
									$tRNA_{1B}^{Ala}$		
Yeast	35 S	(2.5)			25 S	(1.17)	p23 S	(1.1)	23 S	(1.05)	c
							20 S	(0.9)	5 S		
							29 S	(1.6)	18 S	(0.7)	d
									26 S	(1.3)	
									5.8 S		
Drosophila	38 S	(3.3)			20 S	(.1)			18 S	(0.7)	e
					30 S	(2.0)			26 S	(1.5)	
									5.8 S		
									2 S		
Xenopus	40 S	(2.74)	38 S	(2.38)	18 S	(0.67)	30 S	(1.65)	18 S	(0.67)	f
					34 S	(1.81)			28 S	(1.56)	g
									5.8 S		
Mouse	45 S	(4.7)	41 S	(3.3)	18 S	(0.68)	32 S	(2.19)	18 S	(0.68)	h
					36 S	(2.66)			28 S	(1.56)	
									5.8 S		
Rat	45 S	(4.2)			23 S	(1.37)			18 S	(0.68)	i
					35 S	(2.87)			28 S	(1.7)	j
									5.8 S		k
Human	45 S	(4.7)	41 S	(3.3)			20 S	(1.04)	18 S	(0.68)	l
							32 S	(2.27)	28 S	(1.76)	m
									5.8 S		n

[a] Key to references: a, Nikolaev et al. (1975); b, Lund and Dahlberg (1977); c, Warner (1974); d, Perry et al. (1970); e, Jordan et al. (1974); f, Loening et al. (1969); g, Speirs and Birnstiel (1974); h, Wellauer et al. (1974); i, Quagliarotti et al. (1970); j, Egawa et al. (1971); k, Nazar et al. (1975); l, Weinberg and Penman (1970); m, Wellauer and Dawid (1973); n, Maden and Robertson (1974).

3. Structure of the Primary Transcriptional Product

Structural characterization of pre-rRNA's of vertebrates reveals the following salient features: (1) Polarity of pre-rRNA examined by electron microscopy (Dawid and Wellauer, 1976) and by restriction mapping of rDNA (Reeder *et al.*, 1976) shows that 18 S rRNA is proximal to the 5'-end and that 28 S rRNA is proximal to the 3'-end. (2) Examination of the 5'-end of pre-rRNA shows no 5'-triphosphoryl group (Choi and Busch, 1970; Slack and Loening, 1974), except in the case of *Xenopus* (Reeder *et al.*, 1977). (3) Homochromatographic mapping of T1 RNase-derived oligonucleotides shows some minor heterogeneity that cannot be detected by other methods (Choi *et al.*, 1976a,b). (4) Pre-rRNA is divided into sequences of ribosomal RNA (18 S, 28 S, and 5.8 S rRNA) and a sequence of transcribed spacer regions (Inagaki and Busch, 1972; Maden and Robertson, 1974; Nazar *et al.*, 1975). (5) Posttranscriptional modification is detected only in the nonspacer regions of rRNA's (Maden *et al.*, 1977).

a. **Ribosomal Sequences.** Comparison of sequences from 45 S pre-rRNA with those of 18 S rRNA and 28 S rRNA (and 5.8 S rRNA) shows regions of complete homology. Further characterization of modified nucleotides of pre-rRNA and rRNA's shows stoichiometry for their modified nucleotides and provides evidence of conservation of ribosomal sequences through processing (Table VII) (Maden *et al.*, 1977).

More detailed studies have shown that posttranscriptional modification occurs at two stages: one is at the pre-rRNA level in the nucleus (early modification), and the other is in the cytoplasm (late modification). However, possible splicing mechanisms have not been ruled out for processing. There is evidence for a DNA insertion in the rDNA sequence for 26 S rRNA of *Drosophila* (Pellegrini *et al.*, 1977).

b. **Transcribed Spacer Sequences.** Studies of T1 RNase-derived oligonucleotides of 45 S pre-rRNA show clear evidence of two specific types of long oligonucleotide spacers (Inagaki and Busch, 1972; Choi *et al.*, 1976a,b). One is C-rich, and the other is U-rich, with the latter being different from those found in hnRNA's (Bubroff, 1977).

c. **Evolution of Pre-rRNA.** The evidence supporting pre-rRNA evolution (Cox, 1977), briefly summarized, is the following: (1) diversity of size differences accompanied by size increase of rRNA's and of transcribed spacers, (2) diversity of nucleotide composition of rRNA's and differences in modified nucleotides (Table VII), (3) extensive structural homology, including sequences containing modified nucleotides (Table VII) (Khan *et al.*, 1978), and (4) conservation of the order of rRNA's (18 S rRNA and 28 S rRNA) in the precursor. However, there is variability in the spacers between 18 S rRNA and 28 S rRNA. In prokaryotes the spacers between

TABLE VII

Methylated Nucleotides of Pre-rRNA's and rRNA's[a]

Modified nucleotides	Yeast				Man			
	37 S	26 S	17 S	5.8 S	45 S	28 S	18 S	5.8 S
Early methylation								
m^1Ap	2	2			1	1		
m^6Ap					1	1		
m^5Cp	1	1						
mCp					2	2		
Am-Ap	5	1			6	2	4	
Am-Cp	4	3	1		6	4	2	
Am-Gp	4	3	1		10	8	2	
Am-Up	3	1	2		9.2	4.7	4.5	
Cm-Ap	1	1	0		4	3	1	
Cm-Cp	4	2	2		6.5	4	2.5	
Cm-Gp	1		1		5	3	2	
Cm-Up	3	3			5	4	1	
Gm-Ap	3	2	1		3.5	3	0.5	
Gm-Cp	1	1			3	1	1	1
Gm-Gp	6	4	2		16	10	6	
Gm-Up	3	1	2		6	5	1	
Um-Ap	2	1	1		3	1	2	
Um-Cp	1	1	1		5	2	3	
Um-Gp	4	3	1		5.9	2.2	3.5	0.2
Um-Up	2	2			2		2	
ψm-Gp					1	1		
Am-Am-Up	1	1						
Am-Cm-Gp	1	1						
Am-Gm-Cp	1	1						
Um-Gm-Up					1	1		
Um-Gm-ψp					1	1		
Am-Gm-Cm-Ap					1	1		
Late methylation								
m^6Ap							1	
m^7Gp			1				1	
m^3Up		2						
$m^1ac^3\psi p$-Cp			1				1	
m_2^6A-m_2^6A-Cp			1				1	
Um-Gm-ψp	1							

[a] Data from Brand *et al.* (1977), Khan *et al.* (1978), Maden and Salim (1974), and Maden *et al.* (1977).

16 S rRNA and 23 S rRNA contain tRNA's (Lund and Dahlberg, 1977). In eukaryotes, the spacers contain 5.8 S rRNA (Speirs and Birnstiel, 1974).

4. Pre-28 S rRNA

Pre-28 S rRNA is referred to as the immediate precursor of nucleolar 28 S rRNA (28 S pre-rRNA), which in mammals is 32 S pre-rRNA. The functional significance of 32 S pre-rRNA was first described by Pene *et al.* (1968). 32 S pre-rRNA was found to be the immediate precursor of both 28 S rRNA and 5.8 S rRNA, which exists as a complex with 28 S rRNA through hydrogen bonding. Subsequently, the precursor–product relationship was clarified by structural examination of 32 S pre-rRNA (Maden and Robertson, 1974; Nazar *et al.*, 1975). The position of 5.8 S rRNA is proximal to the 5'-end of the 32 S pre-rRNA (Speirs and Birnstiel, 1974).

5. Pre-18 S rRNA

In mammalian cells, 20 S pre-rRNA is the immediate precursor of 18 S rRNA, as was first described by Weinberg and Penman (1970). Subsequently, the structural examination of "23 S" pre-rRNA supported the precursor–product relationship (Egawa *et al.*, 1971).

D. Heterogeneous Nuclear RNA's (hnRNA's)

Heterogeneous nuclear RNA's (8–100 S) are a class of nucleoplasmic RNA's composed of approximately 10^4–10^5 species. The term "heterogeneous nuclear RNA" was originally designated by Warner *et al.* (1966). They are more labile than other RNA's, as readily shown by labeling kinetics. hnRNA is AU-rich and similar to cellular DNA, whereas that of pre-rRNA's is GC-rich. Earlier work (Georgiev, 1967; Darnell, 1968) and subsequent progress (Molloy and Puckett, 1976) indicated that mRNA's are derived from hnRNA. Hybridization competition of hnRNA's with cytoplasmic mRNA's reveals (1) unique mRNA sequences in hnRNA's which are found in the cytoplasm, and (2) non-mRNA sequences (reiterated sequences) which are confined to the nucleus and degraded.

Recently, as the structural characterization of mRNA's has progressed, new experimental approaches have been developed to characterize hnRNA's (Darnell, 1978). The remarkable advances are (1) isolation of pure mRNAs, (2) synthesis of cDNA from pure mRNA, (3) characterization of posttranscriptional modifications, (4) visualization of nucleic acid hybrids, (5) utilization of restriction enzymes for isolation of specific genes, (6) development of DNA recombinant technology, and (7) DNA sequencing methods. Accordingly, the study of hnRNA's emphasizes the integration of two different approaches: RNA molecular biology and DNA molecular biology.

Most recently, it has become possible to isolate, amplify, and characterize a number of genes coding for the specific mRNA's for globin, histones, insulin, and many others. Furthermore, it has become also feasible to identify specific pre-mRNA in hnRNA's (Darnell, 1978).

1. Size of Primary Transcriptional Product

The size of hnRNA's shows a wide range, from 8 S to 100 S (10^2–10^5 bases). The significance of the enormous size difference between giant hnRNA's (10^4–10^5 bases) and mRNA's (10^3 bases) has been difficult to understand, since eukaryotic mRNA's (not viral mRNA's) are commonly believed to be monocistronic. The giant hnRNA's were even thought at one time to be physical aggregates of smaller RNA's. However, a number of studies and conditions that prevent aggregation have shown the reality of giant hnRNA by (1) direct electron microscopic visualization, (2) rapidly labeling techniques utilizing very short exposures to radioactive precursors, and (3) identification of "insert" sequences in mRNA genes.

Although 5'-cap structures and poly(A) at the 3'-end appear to be conserved, the question of enormous size differences between mRNA's and hnRNA's has been settled by evidence that hnRNA's are processed by special mechanisms, including posttranscriptional splicing mechanisms (Berget et al., 1977; Chow et al., 1977; Gelina and Roberts, 1977; Klessig, 1977). Table VIII shows a list of hnRNA's identified as pre-mRNA's. However, there is no clear evidence to support the idea that the largest pre-mRNA's are the primary transcription products which contain no structural modification.

TABLE VIII

Processing of hnRNA's and Discrete Pre-mRNA's Found in hnRNA's

mRNA	hnRNA's	mRNA	Reference[a]
Histone	37–46 S → 28–35 S → (?20 S) → 9 S		a
Light chain of immunoglobulin	40 S → 24 S → 13 S → 13 S		b
Globulin	15 S → 10 S		c, d
Ovalbumin	>7800 bases → six intermediate species → 2050 bases		e
			f
Silk fibroin	45–65 S → 40–65 S		g
			h
Balbiani ring	75 S → 75 S		i

[a] Key to references: a, Melli et al. (1977);   b, Gilmore-Herbert and Wall (1978); c, Bastos and Aviv (1977);   d, Tilghman et al. (1978);   e, Mandel et al. (1978); f, Roop et al. (1978);   g, Suzuki and Brown (1972);   h, Lizardi (1976);   i, Daneholt (1975).

2. Posttranscriptional Modification

a. Chain Elongation. *i. Poly(A) synthesis (polyadenylation) at the 3'-end of hnRNA.* Soon after the first demonstration of poly(A) in mRNA (Kates, 1970; Lim and Canellakis, 1970), hnRNA also was found to contain poly(A) (Darnell *et al.*, 1971; Edmonds *et al.*, 1971; Lee *et al.*, 1971). The synthesis of poly(A) takes place after the termination of transcription and is catalyzed by poly(A) polymerase (Edmonds and Winters, 1975). The chain length of hnRNA poly(A) is usually 200–250 and is considerably longer than the poly(A) of cytoplasmic mRNA. The initial evidence for posttranscriptional synthesis of poly(A) and its significance was obtained by (1) the findings that nuclear DNA lacks a poly(dT) stretch complementary to poly(A) (Birnboim *et al.*, 1973); (2) the metabolic inhibition of poly(A) synthesis by cordycepin (3'-deoxyadenosine) (Adesnik *et al.*, 1972) and the resultant inhibition of transport of mRNA into cytoplasm, and (3) the structural determination of poly(A) at the 3'-end of mRNA and hnRNA (Molloy and Darnell, 1973). Some mRNA's contain no poly(A), for example, some histone and protamine mRNA's (see Table IX). The importance of polyadenylation was recognized as one of the mRNA maturation steps. Without polyadenylation, the subsequent steps leading to transport of pre-mRNA into cytoplasm are blocked (Molloy and Puckett, 1976; Perry, 1976).

ii. 5'-Cap synthesis. Modification of the 5'-end of hnRNA, an enzyme-catalyzed process, is referred to as 5'-capping (Moss *et al.*, 1976; Shatkin, 1976). The steps of 5'-capping are (1) the formation of guanosine $5' \rightarrow 5'$-triphosphodiester internucleotide linkage $GpppNp^1Np^2-$, where N^1 is predominantly a purine and N^2 is either a purine or pyrimidine, (2) N^7 methylation to form $m^7GpppNp^1Np^2-$ and a cap O structure, and (3) 2'-O-methylation of the ultimate nucleotide, to form the cap I structure, $m^7GpppN^1mpNp^2-$.

5'-Capping has been observed to precede poly(A) synthesis, but is not universal, since some mRNA's, including histone mRNA's, contain 5'-caps but no poly(A). 5'-Caps have been suggested to promote an initiation of the translation of mRNAs (Shatkin, 1976).

b. Base Methylation. Besides the 5'-capping, additional methylation of hnRNA's occurs at internal sites. N^6 Methylation of adenine is the only modification found so far (Desrosier *et al.*, 1974; Wei *et al.*, 1977). The number varies among many hnRNA's, including the absence of any methylation in some instances.

c. Chain Shortening by Selective Enzymatic Cleavage and Ligation (Splicing). The splicing mechanism has not been identified. Williamson

(1977) and Gilbert (1978) have proposed that hnRNA consists of "intronic" sequences (inserts) that are cleaved and discarded selectively during processing and "extronic" (or expressed) sequences that are separated from intronic sequences and ligated selectively in processing. Splicing has been visualized by electron microscopic examination of mRNA–DNA gene hybrids. RNA–DNA hybrids show DNA R loops that result from DNA segments lacking complementary RNA. The DNA sequences of R loops are referred to as inserts. The absence of RNA sequences complementary to inserts can be interpreted in two ways: one is a transcription jumping mechanism by which RNA polymerase skips the insertion sites, and a splicing mechanism which involves selective excision of the noncoding sequences of the transcript that includes insertion and coding sequences, followed by ligation of the coding sequences. At present, the latter mechanism is favored.

3. Structure of hnRNA's

According to the proposal of Gilbert (1978), mRNA is composed of integrated extronic sequences containing a 5' noncoding region, an internal coding region, and a 3' noncoding region. The precursor hnRNA contains additional sequences (intronic sequences) which may have enormous significance. For example, 15 S β-globin pre-mRNA contains two intronic sequences (Tilghman et al., 1978), and high molecular weight ovalbumin pre-mRNA (>7800 nucleotides) contains seven intron sequences (Mandel et al., 1978; Roop et al., 1978). The structural evidence of intron–extron junctions was obtained in ovalbumin gene (Catterall et al., 1978). Structural characteristics of hnRNA's are the following.

1. Features based on posttranscriptional modification: (1) poly(A) + and poly(A) − at the 3'-end; (2) 5'-cap(+) and 5'-cap(−) at the 5'-end; and (3) $m^6A(+)$ and $m^6A(−)$ at the internal structure.
2. A high percentage (80% or higher) of hnRNA's contains repetitive sequences, whereas mRNA's contain a much lower proportion of repetitive sequence (Holmes and Bonner, 1974; Molloy et al., 1974; Smith et al., 1974).
3. Secondary structures characterized as dsRNA (double-stranded RNA) resistant to endonuclease digestion (Jelinek and Darnell, 1972; Kronenberg and Humphrey, 1972; Ryskov et al., 1973). Partial characterization showed these nuclease-resistant sequences to be derived from repetition of inverted sequences of DNA (Jelinek, 1977).
4. The presence of U-rich sequences (Molloy et al., 1974) and

A-rich sequences (Nakazato *et al.*, 1974) that differ from that of other RNA's. The frequency of U-rich sequences is higher in hnRNA's than in mRNA's.

III. CYTOPLASMIC RNA's

Cytoplasmic RNA's are those contained in the cytoplasm, but exclude those of cytoplasmic organelles such as mitochondria (chloroplasts) and centrioles. A majority (more than 80%) of cytoplasmic RNA's are present in polysomes involved in protein synthesis. The polysomes are divided into two classes: (1) membrane-bound and (2) non-membrane-bound. Minor RNA species include nonpolysomal mRNA (Brawerman, 1974; Heywood and Kennedy, 1976) and LMW RNA's.

A. Messenger RNA's (mRNA's)

The term "messenger" was derived from genetic studies of genetic control of protein synthesis (Jacob and Monod, 1961). Some major steps in the development of our understanding of mRNA are the following.

1. The first detection of *in vivo* mRNA by Volkin and Astrachan (1956).
2. The first detection of *in vivo* mRNA associated with ribosomes (Brenner *et al.*, 1961).
3. The demonstration by Nirenberg and Matthaei (1961) that *in vitro* protein synthesis depends on the presence of an RNA template and the definition of the triplet genetic codes by Brimacombe *et al.* (1965) and Söll *et al.* (1965).
4. The demonstration by Warner *et al.* (1963) and Gierer (1963) that polysomes consist of an RNA molecule attached to several ribosomes, and the ability of polysomal RNA to direct *in vitro* peptide synthesis (Lockard and Lingrel, 1969).
5. The characterization of posttranscriptional modifications of mRNA: poly(A) by Kates (1970) and Lim and Canellakis (1970) and the 5'-cap by Rottman *et al.* (1974), Cory and Adams (1975), and Furiuchi *et al.* (1975) and many others in 1975 (Shatkin, 1976).
6. The purification of several different mRNA's (Brawerman, 1974).

Table IX shows a list of selected mRNA's that have been purified and defined for messenger activity. Some of the characteristics of purified mRNA's are given below.

1. Size of mRNA's

mRNA's show a wide variation in size from 6.5 S for protamine mRNA's to 75 S for Balbiani ring mRNA. The size ratio of mRNA to its corresponding pre-mRNA also shows a wide range of variation (Table VIII).

2. mRNA Structure and Its Functional Significance

a. 5'-Caps. Three types of 5'-caps have been observed:

5'-cap	Type 0	$m^7GpppN^1pNp^2$----
5'-cap	Type I	$m^7GpppN^1mpNp^2$----
5'-cap	Type II	$m^7Gppp(m)N^1mpNmp$----

Type 0 was observed in plant viral RNA (Dasgupta *et al.*, 1975) and in yeast mRNA's (Sripati *et al.*, 1976). Type I and type II are the most frequent 5'-caps in mRNA's. These different types of 5'-cap are attributed to nuclear events (early modification) for Type 0 and type I, and cytoplasmic events (late modification) for the conversion of type I to type II (Perry *et al.*, 1976; Rottman *et al.*, 1976).

The origin of N^1, as shown above, is attributed not only to the 5'-end (pppAp--- and pppGp---) of the primary transcriptional products, but also to the 5'-end (ppNp-- and pNp--) derived from the internal structure of hnRNA's (Shibler and Perry, 1976). A physical model of the 5'-caps suggests that m^7Gp intercalates between N^1 and N^2, and that the 2'-*O*-methyl moiety plays an important role in the spacings of the three stacked bases (Kim and Sarma, 1978).

The function of the 5'-cap is not completely understood because some mRNA's which are highly efficient translation templates have no 5'-caps (Hewlitt *et al.*, 1976; Nomoto *et al.*, 1976; Frisby *et al.*, 1976). However, other evidence suggests that the stereochemical structure (cap conformation) of m^7Gppp is important for promoting initiation of protein synthesis, since (1) the degradation of the 5'-cap reduces messenger activity (Rao *et al.*, 1975; Muthukrishnan *et al.*, 1975) and (2) m^7Gpp is a better competitive inhibitor of messenger activity than m^7Gp (Adams *et al.*, 1978).

b. 5'-Untranslated Region (Leader Sequence). This sequence includes sequences from the 5'-end of the RNA to the initiation codon (AUG or GUG). It has been proposed that for the initiation of protein synthesis the leader forms a transient hydrogen bonded complex with the 3'-end of 18 S rRNA (Shine and Delgarno, 1974; Steitz and Jakes, 1975). Extensive structural comparisons showed that AUG is the only initiation codon of eukaryotic mRNA's and that CUUPyUG is often found close to the 5'-cap structure, suggesting a possible recognition site for an initiation factor (Baralle and Brownlee, 1978). The chain length of leader sequence varies considerably (approximately 10–100 nucleotides).

TABLE IX

Structural Characteristics of Messenger RNA's

Source of mRNA	Size[a]	5'-Cap	m^6A	poly(A)	Sequence determined	Reference[b]
Protamine						
Trout testes	6.5 S			20 bases		a
1	275 + 20 Bases			20 bases	Partial	b
2	250 + 20 Bases			20 bases		
3	235 + 20 Bases			20 bases		
4	215 + 20 Bases			20 bases		
Histones						
Sea urchin	9 S					
H_4	400 bases			—		c
H_2A	500 bases			—		d
H_2B	530 bases	$m^7GpppAmN$		—		
H_3	620 bases			—		
H_1	700 bases			—		
HeLa cells						
H_4	8.6 S			—		e
H_2A	9.2 S	$m^7Gppp\left(\begin{matrix}m^6A\\Am\\Gm\end{matrix}\right)\left(\begin{matrix}Um\\Cm\\Am\end{matrix}\right)$		—		f
H_2B				—		
H_3	10.7 S			—		
H_1	12–13 S			—		g
Globins						
Rabbit						
α-Chain	9S	$m^7Gpppm^6AmC(m)$		30–40 Bases	Partial	
β-Chain	590 + (30–40) bases	$m^7Gpppm^6AmC(m)$		30–40 Bases	Complete	h, i
Human						
α-Chain		$m^7GpppAC$		+	Partial	j
β-Chain		$m^7GpppAC$		+	Near complete	k, l

	Size					Completeness	Ref.[b]
Immunoglobulin							
Mouse myeloma							
L-Chain	13 S	+	+	200 Bases		Partial	m
H-Chain	17 S	+	+	150–200 Bases		Partial	n
Ovalbumin							
Chick	16 S		+	20–140 Bases		Near total	o
							p
							q
Milk proteins							
Guinea pig							
Caseins	5.3×10^5 daltons			90 Bases			r
α-Lactalbumin	4.6×10^5 daltons			90 Bases			
	3.2×10^5 daltons			90 Bases			
Caseins							
Rat	12 S			38 Bases			s
	15 S			42 Bases			
Procollagen							
Chick	22 S			+			t
Myosin							
Chick	26 S			+			u
Thyroglobulin							
Beef	33 S			+			v
	2.5×10^6 daltons						
Silk fibrion							
Bombyx mori	45 S–65 S			100 Bases		Partial	w
Balbiani ring							
Chironymus tentans	75 S			+			x

[a] Size is expressed by S (sedimentation rate), daltons, and bases.

[b] Key to references: a, Iatrou and Dixon (1977); b, Davies et al. (1977); c, Grunstein and Schedl (1976); d, Surrey and Nemer (1976); e, Borun et al. (1977); f, Moss et al. (1977); g, Lockard and RajBhandary (1976); h, Baralle (1977); i, Efstratiadis et al. (1977); j, Proudfoot and Longley (1976); k, Proudfoot (1977); l, Marotta et al. (1977); m, Milstein et al. (1974); n, Cowan et al. (1976); o, Woo et al. (1975); p, McKeynolds et al. (1978); q, Kuebbing and Liarokas (1978); r, Craig et al. (1976); s, Rosen (1976); t, Harwood et al. (1974); u, Heywood and Kennedy (1976); v, Vassart et al. (1975); w, Suzuki and Brown (1972); x, Daneholt (1975).

c. Coding Region. This region consists of the sequence of the genetic message that specifies the amino acid sequence of a protein. Theoretically, this sequence can be of randomly selected frequencies of the 64 triplet codons. However, complete sequence determination of mRNA's shows selective usage of certain triplets where degeneracy of the code allows for a choice (Fiers *et al.*, 1976; Efstratiadis *et al.*, 1977; Marotta *et al.*, 1977). The basis for triplet codon selection is undetermined, but may simply be related to the frequency of anticodons available in the tRNA population of a particular cell (Marotta *et al.*, 1977).

d. 3'-Untranslated Region. This region includes sequences from the termination codon (UAA, UGA, and UAG) to the start of poly(A) on the 3'-end 'of the molecule. The length of this region varies considerably among different mRNA's (approximately 100–700 nucleotides) (McReynolds *et al.*, 1978), and its significance has not been determined. However, it may be significant that the sequence AUAAA has been conserved in several mRNA's (Proudfoot and Brownlee, 1976).

e. Poly(A). The length (under 200) of poly(A) in mRNA's is less than that in hnRNA's (200–250 nucleotides). Examination of poly(A)+ and poly(A)− mRNA indicates that approximately 30% of the total mRNA's are poly(A)− mRNA's (Greenberg, 1976). Although the function of poly(A) remains unclear, it has been suggested that the half-life of poly(A) is related to that of mRNA, and, accordingly, the chain length of poly(A) may be related to the stability of mRNA.

3. Comparative Significance of Posttranscriptional Modification

The 5'-cap and poly(A) generally are not observed in prokaryotic cells. However, poly(A) has been detected when *E. coli* is grown under special conditions (Srinivasan *et al.*, 1975).

a. Methylation. Yeast mRNA's contain type 0 and type I 5'-caps (Scripati *et al.*, 1976; DeKloet and Andrean, 1976), but do not contain m^6A. Wheat germ mRNA's have type I and type II 5'-caps (Saini and Lane, 1977), but also lack m^6A.

b. Poly(A). Histone mRNA's in *Xenopus laevis* oocyte exist as both poly(A)+ and poly(A)− mRNA's (Levenson and Marcu, 1976). Protamine mRNA's in developing trout testes also are both poly(A)+ and poly(A)− mRNA's (Iatrou and Dixon, 1977). Histone mRNA's in higher animals are poly(A)− (Adesnik and Darnell, 1972; Greenberg and Perry, 1972).

B. Transfer RNA's (tRNA's)

The best-characterized RNA's are the tRNA's, and the primary nucleotide sequence and the three-dimensional structure of many tRNA's have been rigorously studied (Rich and RajBhandary, 1976; Goddard, 1977). Holley *et al.* (1965) made the first successful sequence determination of yeast tRNA and proposed the cloverleaf model as its secondary structure. Since then the nucleotide sequences of more than 100 different species of tRNA's have been studied (Sprinzl *et al.*, 1977). More recently, the three-dimensional structure has been determined by X-ray crystallography (Kim *et al.*, 1974a; Robertus *et al.*, 1974).

The detailed relationship between the structure and function of tRNA's has been clarified, and now the more refined absolute structure of tRNA (better than 2.5 Å resolution) needs to be elucidated to achieve further understanding of tRNA function (Kim, 1976; Rich, 1977).

1. Localization

tRNA's show two general associations with cellular ultrastructures. One is their association with the ribosomes of the cytoplasm and mitochondria, and the other is their localization elsewhere. For example, they are found in nucleoplasm and the nucleolus, which do not contain functional ribosomes. As predicted from the number of possible triplet codons, a number of kinds of isoaccepting tRNA's exist for each of the amino acids, and their population densities are dynamically influenced by factors such as metabolic states, cell types, and differentiation (Littauer and Inouye, 1973).

2. The Proposed Generalized Structure of tRNA's

The chain length of tRNA's ranges from 74 (mini tRNA) (Keith *et al.*, 1970) to 96 (maxi tRNA's) (Yamada and Ishikura, 1973). Despite this variation, all the tRNA's can be fitted into a generalized three-dimensional cloverleaf structure (Kim *et al.*, 1974b). Based on this generalization, the secondary structure can be divided into the following domains (Fig. 1):

1. The stem "a" (the acceptor stem) contains the 5'-end and the 3'-end of tRNA and is usually made up of seven base pairs.
2. The D arm (dihydrouridine arm) is comprised of stem "b" with three to four base pairs and the loop "I" (D loop) of variable chain length.
3. The anticodon arm includes the stem "c" with five base pairs and the anticodon loop "II" with seven nucleotides.
4. The extra arm or loop is variable in chain length (4–21). In some cases, this arm includes the stem "d" of variable chain length and the loop "III" of three to four nucleotides or only a loop of three to five nucleotides.
5. The TψC arm consists of the stem "e" of five base pairs and the loop "IV" of seven nucleotides.

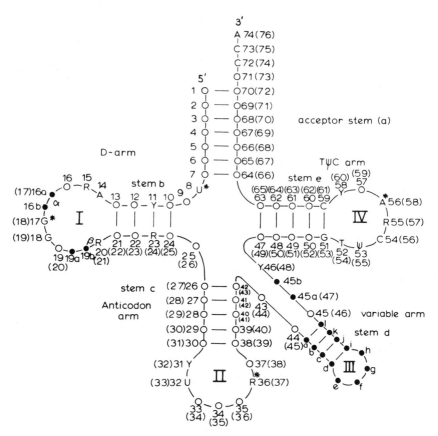

Fig. 1. The generalized "cloverleaf" secondary structure of transfer RNA. Open circles represent residues always present in tRNA's and are numbered consecutively from the 5'-end. Filled circles, labeled a, b, etc., represent residues occasionally present in tRNA and are restricted to the variable arm and regions, α and β of the dihydrouridine (D) loop. Stems and loops are systematically labeled a, b, c, d, e and I, II, III, IV, respectively, and the more commonly used trivial nomenclature is also shown. The specific numbering for yeast tRNA[Phe] is shown in parenthesis where this differs from the generalized numbering. Invariant and semi-invariant (R = purine, Y = pyrimidine, * = sometimes modified) residues are shown as is the secondary base–base hydrogen bonding in the helical stems. The modification at G17 is a ribose 2'-O-methylation. From Goddard (1977). By permission of *Prog. Biophys. Mol. Biol.*, Pergamon Press and Dr. J. P. Goddard.

The classification of the above domains is extremely useful in helping to visualize (1) the accommodation of variable chain lengths of tRNA within the three regions of the "cloverleaf," two regions (α and β) of the D loop, and the variable arm; (2) the sequences of invariance, semi-invariance, and variance; and (3) the location and structure of modified nucleotides. Accordingly, the diverse primary sequences of tRNA's are fitted into the general model of unified structural features with evolutionarily conserved structures and varying structures.

The three-dimensional structure of yeast tRNA's has been determined (Kim *et al.*, 1974a; Robertus, 1974). The overall conformation of the tRNA has been called both "T-shaped" (Robertus *et al.*, 1974) and "L-shaped" (Kim *et al.*, 1974a), but most frequently the latter. It should be noted, however, that this difference in nomenclature is derived from different perspectives of the tRNA structure, so neither term is completely descriptive. These general observations are based on extensive structural analysis: (1) Extensive base stacking in the stems provides the stabilization force for the molecule. (2) The normal double helical structure is in the A conformation. (3) Various tertiary hydrogen bondings (non-Watson–Crick) are observed among bases and backbones in loops. (4) The functional design is shown in (1) availability of the anticodon loop to interact with mRNA, (2) flexible conformation of the CCA_{OH} end for peptide formation, and (3) possible conformational opening between the D and TψC loops.

3. Function of Modified Nucleotides in tRNA

More than 50 modified nucleotides have been identified in various tRNA's. Their location is not random, and their significance has been extensively studied (McCloskey and Nishimura, 1977). For example, it has been proposed that methyl groups can form transient $—CH_2—$ bonds to ribosomes to be functional (Feldman, 1977). Their functions may be briefly summarized as follows.

Modified nucleotides, located adjacent to the anticodon, show some regularities. Hydrophobically modified nucleosides (such as N^6-isopentenyladenosine, i^6A) are found for codons of mRNA starting with U. Hydrophilically modified nucleosides (such as N-[N-(9-β-D-ribofuranosylpurin-6-yl)carbamoyl] threonine, t^6A) are found for codons of mRNA starting with A. They are involved in anticodon–codon interactions.

Modified nucleotides, located in the first position of the anticodon, are often hypermodified and are involved in the anticodon–codon interactions.

Modified nucleotides, located in other positions, show functional roles in (1) binding of the arm TψC of tRNA to the ribosome, (2) stabilization of tRNA conformation, (3) enhancement of resistance to RNases, and (4) the recognition of aminoacyl-tRNA synthetase.

Modified nucleotides may have a role in the regulation of cell function: (1) Pseudouridylate in tRNA[His] is important in the regulatory expression of the histidine operon (Singer *et al.*, 1972). (2) The Q nucleotide

[7-(4,5-cis-dihydroxy-1-cyclopenten-3-ylaminomethyl)-7-deazaguanylate] of tRNA is important in cell differentiation of *Drosophila* (White *et al.*, 1973). (3) Q* nucleotide (a mixture of β-D-mannosyl Q and β-D-galactosyl Q) of tRNA is important in membrane function (Okada *et al.*, 1977).

4. Synthesis of tRNA's

Mature tRNA's are derived from pre-tRNA's. It has been suggested that a splicing mechanism is involved in processing the pre-tRNA into a mature tRNA (Valenzuela *et al.*, 1978).

5. tRNA Function

The tRNA function (Goddard, 1977) includes the following: (a) Translation of mRNA into protein by polysomal protein synthesis. The steps are well defined as (1) initiation by initiator tRNA; (2) elongation, including formation of the elongation factor Tu-GTP-aminoacyl-tRNA, binding of aminoacyl-tRNA to the A site of the ribosome, codon–anticodon interaction, peptide bond formation and translocation to the P site of the ribosome, and termination; (b) direct transfer of amino acids from aminoacyl-tRNA's to specific acceptor molecules; (c) regulation, by aminoacyl-tRNA's, of amino acid biosynthesis; and (d) tRNA's as primers for DNA synthesis in RNA tumor viruses.

C. High Molecular Weight Ribosomal RNA's (HMW rRNA's)

The term "ribosomal RNA" was first used to describe the RNA contained in the ribosome (Roberts, 1958). The systematic study of rRNA's began with bacterial ribosome subunits by Kurland (1960) and with eukaryotic ribosomal subunits by Hall and Doty (1959). It was observed that 16 S rRNA in bacteria and 18 S rRNA in eukaryotes is associated with the small ribosomal subunit (30–40 S) and that 23 S rRNA and 28 S rRNA from bacteria and eukaryotes, respectively, are associated with the large ribosomal subunits (50–60 S). Subsequently, Rosset and Monier (1963) described the presence of 5 S rRNA in the bacterial 50 S subunit, and Zylber and Penman (1969) described a similar rRNA associated with the eukaryotic 60 S subunit. Furthermore, an additional small rRNA (5.8 S rRNA) was found in the 60 S subunit by Forget and Weissman (1967) and Pene *et al.* (1968).

In order to understand the role of ribosomes in protein synthesis (ribosome cycle), a number of approaches have been taken to clarify the role of rRNA structure in relation to ribosome function. The bacterial ribosome is the best characterized and its characteristics have been defined by (1) the *in vitro* reconstitution of functional ribosome (Nomura and Held, 1974); (2)

the elucidation of the sequence of 5 S rRNA and its role in ribosome function (Brownlee *et al.*, 1968; Monier, 1974); (3) the determination of the structure of 16 S rRNA and 23 S rRNA (Fellner, 1974; Branlant *et al.*, 1975; Ehresmann *et al.*, 1975); and (4) the investigation of RNA–protein interactions (Zimmermann, 1974). Table X shows a list of well-defined properties of subunits.

1. Comparison of High Molecular Weight rRNA's

Extensive studies of rRNA's of many organisms show both diversity and conservation of rRNA structures. Loening (1968) demonstrated a distinct evolutionary trend in the size of high molecular weight rRNA's and classified the different types of rRNA's associated with ribosomes (70 S for prokaryotes and 80 S for eukaryotes).

1. The prokaryotic rRNA's consist of "16 S" rRNA (0.56×10^6 daltons) and "23 S" rRNA (1.1×10^6 daltons) found in bacteria, actinomycetes, blue-green algae, and chloroplasts of higher plants.
2. rRNA's in the higher plants, ferns, algae, fungi, and some protozoa consist of "18 S" rRNA (0.7×10^6 daltons) and "25 S" rRNA (1.3×10^6 daltons).
3. The metazoan rRNA's consist of "18 S" rRNA (0.7×10^6 daltons) and "28 S" rRNA (1.4×10^6 to 1.75×10^6 daltons), where "28 S" rRNA shows a size range of 1.4×10^6 daltons in sea urchins to 1.75×10^6 daltons in mammals.

However, peculiar rRNA's are found which cannot be fitted into the above classification. Amoeba and *Euglena* contain "18 S" rRNA (0.85 to 0.89×10^6 daltons) and "25 S" rRNA (1.53×10^6 daltons). Dinoflagellates contain eukaryotic "18 S" rRNA (0.7×10^6 daltons) and prokaryotic "23 S" rRNA (1.23×10^6 daltons) (Gressel *et al.*, 1975).

Table XI shows certain characteristics of HMW rRNA's illustrating both diversity and conservation of these molecules. It is apparent that "23–28 S" rRNA's are more diverse in size and nucleotide composition than "16–18 S" rRNA's. Ribosomal proteins also show variation in number: 21 proteins in 30 S subunit and 34 proteins in 50 S subunit versus 30 proteins in 40 S subunit and 40 proteins in 60 S subunit (Wool and Stöffler, 1974).

2. Structure of High Molecular Weight rRNA's

In contrast to prokaryotic rRNA's, of which a nearly complete sequence (95%) of 16 S rRNA (Ehresmann *et al.*, 1975) and partial sequence of 23 S rRNA (Branlant *et al.*, 1975) are known, only limited sequence

TABLE X

Bacterial Ribosome (*E. coli*)

Charac-teristic	Small 30 S subunit	Refer-ence[a]	Large 50 S subunit	Refer-ence[a]
General properties	0.9×10^6 daltons $55 \times 220 \times 220$ Å^3 67% RNA 33% Protein (21 S proteins)	a	1.55×10^6 daltons $115 \times 230 \times 230$ Å^3 71% RNA 29% Proteins (34 L proteins)	a
	Center of mass (RNA and proteins) coincided	b	Center of mass (RNA and proteins) displaced	d
Structure of RNA	16 S rRNA 0.56×10^6 daltons 1550 nucleotides 10 methylated nucleotides	c	23 S rRNA 1.1×10^6 daltons 3000–3500 Nucleotides 14 Methylated nucleotides	e
			5 S rRNA 120 Nucleotides No modified nucleotide	f

Function of RNA	
16 S rRNA	
1. Binding sites for 13 S proteins	g
3'-End portion	
2. Binding sites for initiation factors	h
3. Binding sites for S proteins important to initiation	i
4. Binding sites for initiation codon (5'-end of mRNA)	j
5'-End portion	
5. Site for decoding	k
23 S rRNA	
1. Binding sites for 10 L proteins	l
2. Part of (or proximal to) the peptidyltransferase center (P site)	
5'-end	
3. Binding site for 3'-end of 16 S rRNA	m
5 S rRNA	
4. Binding site for TψC of tRNA (A site)	n

[a] Key to references: a, van Holde and Hill (1974); b, Moore et al. (1975); c, Fellner (1974); d, Sturhmann et al. (1976); e, Bralant et al. (1975); f, Brownlee et al. (1968); g, Hochheppel et al. (1976); h, Kenner (1973), Hawley et al. (1974), Bollen et al. (1975), Dahlberg and Dahlberg (1975); i, Held et al. (1973), van Duin and Knippenberg (1974), Szer and Leffler (1974), Dahlberg (1974); j, Shine and Dalgarno (1974), Steitz and Jakes (1975), Shine and Dalgarno (1975); k, Wagner et al. (1976); l, Bispink and Matthaei (1973), Greenwell et al. (1974), Girshovich et al. (1974), Barta et al. (1975), Yukioka et al. (1975), Breitmeyer and Noller (1976), Sonnenberg et al. (1976); m, Bralant et al. (1976); n, Dube (1973), Erdmann et al. (1973), Richter et al. (1973), Erdman (1976).

641

information of eukaryotic rRNA's is currently available. Nevertheless, extensive studies of the partial sequences of both 18 S rRNA and 28 S rRNA's have provided important structural information.

a. "16 S" and "18 S" rRNA. *i. 3'-end.* The 3'-end portion of "16–18 S" rRNA has a general structure characterized by complementarity to the initiation codon of mRNA and the presence of modified nucleotides $m_2^6Am_2^6AC$. Shine and Dalgarno (1974) and Steitz and Jakes (1975) proposed that the 3'-end sequence is important in the initiation of translation (Shine and Dalgarno, 1975). The region of complementarity normally found in "16 S" rRNA is not present in "18 S" rRNA, and its potential for hydrogen bonding is different.

ii. 5'-end. The 5'-end of "18 S" rRNA from a number of vertebrates has been observed to show structural homology (Table XI). The function of the 5'-end portion of "18 S" rRNA is probably similar to that of "16 S" rRNA, as suggested by Wagner *et al.* (1976). It was shown that the 5'-end portion (450 nucleotides) is involved in the decoding site of 30 S subunit.

iii. Internal sequence. Partial sequence studies of oligonucleotides produced by complete T1 RNase digestion or complete RNase A digestion of rRNA's show remarkable similarities among rat, mouse, hamster, and man (Fuke *et al.*, 1976). In addition to conserved sequences among vertebrates, specific differences were found which may reflect species-specific diversity. The complete catalog of T1 RNase-derived oligonucleotides of rat "18 S" rRNA's shows the characteristic oligonucleotides of "18 S" rRNA, which are useful to define conserved sequence, nonconserved sequence, early modification products, and late modification products (Choi and Busch, 1978).

b. "28 S" rRNA. "28 S" rRNA exists as a hybrid with 5.8 S rRNA (and, in some cases, with an additional 2 S rRNA) in the 60 S subunit. Because of this property, "28 S" rRNA has different physical states when prepared under various conditions. The temperature of RNA extraction is used to designate two different states of "28 S" rRNA. Cold "28 S" rRNA refers to the hybrid form with 5.8 S rRNA and hot "28 S" rRNA refers to only "28 S" rRNA because the high temperature (60°–65°C) used in the extraction dissociates 5.8 S rRNA from "28 S" rRNA (Pene *et al.*, 1968). Jordan *et al.* (1976) reported that cold 26 S rRNA of *Drosophila* also contains 5.8 S rRNA and 2 S rRNA, and that hot 26 S rRNA is often found to contain a hidden breakage. Further studies showed that 26 S rRNA is a continuous polyribonucleotide.

TABLE XI

Structural Characteristics of 16–18 S and 23–28 S rRNA's

Species	MW 10^6 daltons	Nucleotide composition[a]				Homology with yeast[b] (%)	5'-End	3'-End	Reference[c]
		C	A	U	G				
16–18 S rRNA									
E. coli	0.56	22.3	24.2	21.3	32.1		pAAAUUG	AUCACCUCCUUA$_{OH}$	a
Yeast	0.72	19.1	26.6	28.1	26.1	100	pUUG	UUA$_{OH}$	b, c
Pea	0.71	20.1	23.7	25.2	31.1	63			
Tetrahymena	0.69	15.8	31.9	31.7	20.5	48			
Drosophila	0.73	20.3	28.8	27.4	23.5		pUAC(C,U)G	AUCAUUA$_{OH}$	d
Xenopus	0.70	24.1	24.1	22.9	28.9		pUAC(C,U)G	UUA$_{OH}$	e, f
Chick	0.70	26.6	20.7	21.2	31.5				e
Mouse	0.70	24.7	23.5	21.9	29.9	42	pUACCUG	AUCAUUA$_{OH}$	g
Rabbit	0.70	28.8	20.5	20.0	30.7			AUCAUUA$_{OH}$	h
Rat	0.70	27.8	22.4	19.6	30.2		pUACCUG	AUCAUUA$_{OH}$	i, j
Human	0.70	27.6	20.0	21.6	30.8		pUAC(C,U)G	AUCAUUA$_{OH}$	g, i
23–28 S rRNA									
E. coli	1.07	21.0	25.5	21.0	32.5		pGGU	CUUAACCCUU$_{OH}$	k
Yeast	1.30	19.2	26.4	26.0	28.4	100	pU(N$_4$)G	NpGpU$_{OH}$	b, c
Pea	1.30	22.6	23.6	21.6	31.2	50			
Tetrahymena	1.30	16.1	34.7	32.1	16.8	38			
Drosophila	1.40	19.4	30.8	27.1	22.5		pU(N)$_{2-3}$G	NpCGA$_{OH}$	m
Xenopus	1.54	27.9	19.7	17.4	35.0				e
Chick	1.58	30.0	17.5	16.7	35.8		pCG		e, m
Mouse	1.71	27.8	19.5	18.2	34.5	40	pCG	U$_{OH}$	e
Rabbit	1.72	31.7	16.9	15.5	35.9			GUUUGU$_{OH}$	h
Rat	1.75	29.8	18.3	19.0	32.9				
Human	1.75	32.2	15.9	16.8	35.8		pCG		e

[a] Data of nucleotide composition from Lava-Sanchez *et al.* (1972).

[b] Data of homology from Martin *et al.* (1970).

[c] Key to references: a, Ehresman *et al.* (1975); b, Sugiura and Takanami (1967); c, Oliver and Lane (1970); d, Dalgarno and Shine (1973); e, Sakuma *et al.* (1976); f, Slack and Loening (1974); g, Eladari and Galibert (1976); h, Hunt (1970); i, Fuke *et al.* (1976); j, Choi and Busch (1978); k, Bralant *et al.* (1975); l, Shine *et al.* (1974); m, Khan and Maden (1976a).

Partial characterization of "28 S" rRNA has been made, and the most characteristic structure of mammalian "28 S" rRNA is the presence of a high G + C core consisting of approximately 600 nucleotides (G + C = 81%) (Kanamaru *et al.*, 1972). The primary sequence of one such core fragment (from a total of 14 fragments) was determined by Kanamaru *et al.* (1974).

3. Secondary Structure

a. "18 S" rRNA. The secondary structure was studied by spectral analyses (Cox, 1970; Cox *et al.*, 1976a,c). The "18 S" rRNA of *Drosophila* and protozoa is AU-rich and contains AU-rich cores of secondary structure. The "18 S" rRNA of *Xenopus* and rabbit is GC-rich and contains GC-rich cores of secondary structure. Approximately 65% of the residues are found to form base pairs, and 20% of the residues are found to form base stackings.

b. "28 S" rRNA. The secondary structure was studied by spectral analyses (Cox *et al.*, 1973, 1976b). The "28 S" rRNA of *Drosophila* and protozoa is AU-rich and contains AU-rich cores of secondary structure. Approximately 65% of the residues are found to form base pairs and 20% are found to form base stackings.

4. Modified Nucleotides

A number of modified nucleotides are found in "18 S" rRNA and "28 S" rRNA. In contrast to bacterial rRNA's containing predominantly base-modified nucleotides, most modified nucleotides in eukaryotic rRNA's are 2'-O methylated forms of the ribose moeity and base rearrangement products (e.g., pseudouridylate). Base methylation occurs, but are less frequent than in prokaryotes. Least frequent are hypermodified nucleotides such as 1-methyl-3-aminocarboxypropyl pseudouridylate (Maden *et al.*, 1975: Choi *et al.*, 1978). Extensive structural homologies at modification sites occur among vertebrate rRNA's (Khan *et al.*, 1978). The early (nuclear event) and late (cytoplasmic event) modifications are well defined in yeast and mammalian cells (Brand *et al.*, 1977; Khan *et al.*, 1978). Table VII shows the numbers and types of methylated nucleotides formed in the primary pre-rRNA and mature rRNA's of yeast and HeLa cells as early and late methylation products. A complete catalog of oligonucleotides containing all the modified nucleotides was obtained for "18 S" rRNA of rat (Choi and Busch, 1978).

D. Low Molecular Weight rRNA's (LMW rRNA's)

1. 5 S rRNA's

a. **Structure.** The 5 S rRNA contains approximately 120 nucleotides. Since the first sequence determination of bacterial 5 S rRNA (Brownlee *et al.*, 1968) and of eukaryotic 5 S rRNA (Forget and Weissman, 1969), a number of different species of 5 S rRNA have been sequenced (Erdmann, 1978).

Bacterial 5 S rRNA's are not identical in primary sequence, but are functionally equivalent since the cross or hybrid reconstitution of various 50 S subunits showed little loss of activity (Wrede and Erdmann, 1973; Bellemare *et al.*, 1973). Eukaryotic 5 S rRNA's also have been sequenced for a wide range of eukaryotic species, including yeast, green algae, flowering plants, insects, reptiles, amphibia, birds, and mammals (Hori, 1975). The structural homology between yeast and mammals is approximately 57% and that between yeast and flowering plants is very similar (Payne and Dyer, 1976).

b. **Function.** The comparative primary sequences show special features of 5 S rRNA's (Richter *et al.*, 1973; Simsek *et al.*, 1973). These are characterized by structures complementary to tRNAs as shown in the tabulation below, where Y = pyrimidine and R = purine.

Prokaryotic 5 S rRNA	Y-G-A-A-C (positions ca. 43–47)
Complementary sequence of tRNA	G-T-ψ-C-R
5 S rRNA's of green algae and flowering plants	R-G-A-A-C (positions ca. 43–47)
Complementary sequence of tRNA	G-T-ψ-C-
5 S rRNA's of yeast, reptiles, amphibia, birds, and mammals	Y-G-A-U-C
Complementary sequence of tRNA (initiation)	G-A-U-C-R

It is suggested that eukaryotic 5 S rRNA may be involved at the site of initiation tRNA binding to the ribosome (Nishikawa and Takemura, 1974); i.e., eukaryotic 5 S rRNA may form part of the P site, while prokaryotic 5 S rRNA forms the A site.

2. 5.8 S rRNA

The 5.8 S rRNA is approximately 150 nucleotides in length and contains few modified nucleotides (1–2 ψ and two 2'-O methylated nucleotides).

Since the first determination of yeast 5.8 S rRNA (Rubin, 1973), other eukaryotic 5.8 S rRNA's have been sequenced (Erdmann, 1978). The structural homology between yeast and HeLa cell 5.8 S rRNA is approximately 73% (Nazar *et al.*, 1975). A special feature of 5.8 S rRNA is the presence of two sequences complementary to the G-T-ψ-C sequence of tRNA. The sequence A-G-A-A-C is located at approximately positions 41–45, and G-A-A-C is at approximately positions 103–107. The 5.8 S rRNA has been proposed to interact with peptidyl-tRNA during the formation of peptide bond (Nishikawa and Takemura, 1974). The presence of these two G-A-A-C sequences poses an interesting question whether they are involved in the binding of aminoacyl-tRNA and in translocation.

IV. MITOCHONDRIAL RNA's

Mitochondrial RNA's are the RNA's confined to the mitochondrion, a semiautonomous organelle which contains a unique system of replication, transcription, and translation (Borst, 1972; Mahler and Raff, 1975; Saccone and Quagliariello, 1975). Two different approaches have been used to study mitochondrial RNA's. One is the direct characterization of the physicochemical structure and biological function of mitochondrial RNA's. The other is the characterization of mitochondrial DNA, leading to an understanding of the mitochondrial gene expression. The latter approach is exemplified by gene mappings by use of mutants, by electron microscopy following RNA–DNA hybridization methods, and by biochemical techniques using restriction enzyme and sequence analysis methods.

These approaches have permitted determination of the size, number, position, and order of genes encoded in the two strands (H and L strands) of mitochondrial genomes (Kroon and Saccone, 1976). One of the most important findings is that the mitochondrial genome (10^4–10^5 base pairs, i.e., less than 1% of nuclear genomes) can encode fewer than 100 genes. Approximately 70 genes (72,000 base pairs) are estimated to exist in yeast mtDNA (Rabinowitz *et al.*, 1976). The total sequence of HeLa mtDNA (15,000 base pairs) will be known in the near future, and the structural information will be enormously valuable for understanding mitochondrial gene expression. Classification is as follows:

1. *Transcriptional origin*—RNA's coded by mitochondrial DNA and RNA's (imported from the extramitochondrial cytoplasm) coded by nuclear DNA
2. *Functional RNA's*—rRNA's, tRNA's, and mRNA's [poly(A)+ and poly(A)−]

3. *Precursor RNA's*—pre-rRNA's and pre-mRNA's. (Little information about transcriptional units or primary transcriptional products is available

Mitochondrial RNA's show certain unusual characteristics. Extensive comparative studies of intramitochondrial RNA's, extramitochondrial RNA's, and prokaryotic RNA's show the mitochondrial RNA's to be in part similar to prokaryotic RNA's and in part similar to eukaryotic RNA's. These dual characteristics are observed in size, posttranscriptional modifications, and sequences and are interpreted as reflecting diversity caused by two opposing trends in evolution. For example, the trends are expressed in the function and structure of ribosomes: in size and sensitivity to translation inhibitors they show great similarity with bacterial ribosomes, whereas their rRNA sequences show similarity with eukaryotic ribosomes (Saccone and Quagliariello, 1975). Some of the peculiar diversities are discussed later.

A. Mitochondrial Transfer RNA's (mt-tRNA's)

the populations of mt-tRNA's have been best defined by gene mapping studies. All the yeast mt-tRNA's, including isoacceptors, are probably coded by mt-DNA (Rabinowitz *et al.*, 1976). However, HeLa cell mt-tRNA's appear to consist of mixed populations, some nuclear coded tRNA's (16 species, excluding those for asparagine and glutamine) (Attardi *et al.*, 1976). Similar tRNA origins were reported for *Tetrahymena* (Chiu *et al.*, 1975).

Prokaryotic features of mt-tRNA's include the presence of tRNA[f-Met] found in yeast, *Neurospora,* and rat (Suyama, 1976b; Heckman *et al.*, 1978). Twenty-seven species of mt-tRNA's (*Locusta migratoria*) are observed by two-dimensional polyacrylamide gel electrophoresis (Feldman and Kleinow, 1976). Studies of nucleotide composition and modified nucleotides show differences from cytoplasmic tRNA's (Chia *et al.*, 1976; Feldman and Kleinow, 1976; Schneller *et al.*, 1976).

B. Mitochondrial Messenger RNA's (mt-mRNA's)

The existence of two subclasses of mitochondrial-coded mRNA's was established by examinations of poly(A)+ and poly(A)− RNA's using transcription map techniques (Battey and Clayton, 1978). Poly(A)+ mRNA's are found in yeast (Hendler *et al.*, 1975), fungi (Rosen and Erdman, 1976), and animals (Hirsch *et al.*, 1974; Ojala and Attardi, 1974; Aujame *et al.*, 1976). The size of the poly(A) tail varies from 20–30 nucleotides for yeast, to 50–70 nucleotides for animals, i.e., shorter than the cytoplasmic counterpart. However, the class of mt-mRNA's may include immature mRNA's which are precursors of mt-mRNA's (Battey and Clayton, 1978).

HeLa mt-mRNA's are among the best characterized (Attardi *et al.*, 1976). Electrophoretic fractionation of poly(A)+ RNA's showed eighteen discrete components: one 7 S RNA, four smaller than 12S rRNA $(3.5 \times 10^5$ daltons), three between 12 S rRNA and 16 S rRNA $(5.4 \times 10^5$ daltons), and ten larger than 16 S rRNA. Some of the latter may represent precursors of smaller components (Battey and Clayton, 1978). In addition, electrophoretic fractionation of poly(A)− RNA's showed twelve discrete components: one smaller than 12 S rRNA, four between 12 S rRNA and 16 S rRNA, and seven larger than 16 S rRNA. Similar discrete poly(A)+ RNA's were observed in other systems: five components in mouse (Battey and Clayton, 1978), eleven components in *Drosophilia* (Spralding *et al.*, 1977), and eight to ten components in yeast (Rabinowitz *et al.*, 1976). Posttranscriptional modification, such as 5′-caps and m⁶A, have not been studied in these RNA's.

C. Mitochondrial pre-rRNA's (mt-pre-rRNA's)

Unlike the nuclear primary pre-rRNA's universally found in eukaryotic cells, the primary mitochondrial pre-rRNA's are difficult to define. Although this difficulty is mainly due to isolation methods, it seems likely that there are inherent differences in the transcriptional processes of mitochondrial genome.

One *Neurospora* mt-pre-rRNA has been well characterized (Kuriyama and Luck, 1974; Lambowitz *et al.*, 1975). This primary mt-pre-rRNA has been shown to be 32 S, and is subsequently processed to P19 S mt-rRNA and P25 S mt-tRNA; both finally mature into 19 S mt-rRNA and 25 S mt-rRNA. In addition, posttranscriptional methylation was observed to be necessary for efficient processing of the precursor. The pathway is illustrated in the scheme below.

The primary mt-pre-rRNA of other systems has not been clearly identified. There is tentative acceptance of the idea that a single primary pre-rRNA exists in the mitochondrial transcription system. Nevertheless, two types of pre-rRNA's may be envisioned on the basis of arrangements of two mt-rRNA genes.

1. Pre-rRNA by Conserved Gene Order

The conserved gene arrangement of mt-rRNA is expressed by the clustered $5' \rightarrow 3'$ gene order, where the 12 S mt-rRNA gene is located adjacent to the replication origin and 16 S mt-rRNA gene is separated by 10^2 bases. This order has been demonstrated in *Xenopus* (Wellauer and Dawid, 1973), *Drosophila* (Klukas and Dawid, 1976), and man (Robberson *et al.*, 1972; Wu *et al.*, 1972; Angerer *et al.*, 1976). The intervening transcribed spacer is usually short and may contain a tRNA gene. Theoretically, the pre-rRNA would be HMW RNA that is unstable and short-lived.

2. Pre-rRNA by Nonconserved Gene Order

The nonconserved gene arrangement is expressed by nonclustered $5' \rightarrow 3'$ gene order, where two rRNA genes are separated by a long (10^3–10^4 bases) intervening spacer, which may contain multiple genes for tRNA's and mRNA's. This nonclustered gene order has been shown for the rat (Saccone *et al.*, 1976), in which the 12 S rRNA gene is located adjacent to the origin of replication and the 16 S rRNA gene is far apart from 12 S rRNA gene. In yeast, physical mapping shows two intervening spacers: one is intergenic (between the 15 S rRNA and 21 S rRNA genes) and is half the genome (72,000 base pairs), and the other is an intragenic spacer within the 21 S rRNA gene (Nagley *et al.*, 1976; Bost *et al.*, 1978). Theoretically, the pre-rRNA would be a giant multigenic RNA that may be extremely unstable and short-lived.

D. Mitochondrial Ribosomes

1. Characteristics of Mitochondrial Ribosomes

In contrast to cytoplasmic 80 S ribosomes assembled from rRNA's and proteins coded by the nuclear genome, mitochondrial ribosomes are assembled from mt-rRNA's coded by the mitochondrial genome and by "imported" proteins coded by the nuclear genome. In addition, a number of peculiarities are observed (Leaver *et al.*, 1976; Mahler and Raff, 1975; Suyama, 1976a).

Mitochondrial ribosomes are smaller than extramitochondrial ribosomes. mt-Ribosomes of animals have a sedimentation value of 50–60 S and are made up of 25–35 S small subunits and 33–45 S large subunits. mt-Ribosomes of fungi and higher plants are larger (70–80 S) and consist of 35–40 S small subunits and 50–58 S large subunits. An exception is found in *Tetrahymena* 80 S ribosome, which is composed of two 55 S subunits, one containing 14 S mt-rRNA and the other containing 21 S mt-rRNA.

Mitochondrial rRNA's are smaller than the 18 S and 28 S extramitochondrial rRNA's. In addition, determinations by electron microscopy and by sedimentation analysis show that fungal mt-rRNA's (18 S mt-rRNA and 24 S mt-rRNA) are larger than bacterial rRNA's (16 S rRNA and 23 S rRNA), whereas animal mt-rRNA's (12–14 S mt-rRNA and 14–17 S mt-rRNA) are smaller. These interesting features suggest two opposing evolutionary trends in that the cytoplasmic rRNA's follow the evolutionary hypothesis of Loening (Loening, 1968), whereas the mitochondrial rRNA's follow a reverse trend.

Low molecular weight rRNA's such as 5 S rRNA and 5.8 S rRNA are not found in the large subunits of animal mitochondrial ribosomes. The 5 S rRNA is found only in mitochondria of higher plants (Leaver *et al.*, 1976) and "5 S"-like rRNA (3 S$_E$ RNA, less than 100 bases) is found in hamster cells (Dubin *et al.*, 1974). However, 5.8 S rRNA has not been observed in the latter.

2. Structure of mt-rRNA's

Although structural characterization of mt-rRNA's is not as advanced as that of cytoplasmic rRNA's, some important findings have been obtained.

The nucleotide composition of mt-rRNA's is characterized by a high AU content (Suyama, 1976a), which is different from that of prokaryotic and eukaryotic rRNA's. The evolutionary trend of increasing GC content (43–65%) of cytoplasmic rRNA's (Lava-Sandrez *et al.*, 1972) is, however, also seen with mitochondrial rRNA's (26–47%) (Mahler and Raff, 1975).

T1 RNase-derived oligonucleotide patterns show mt-rRNA structures distinctly different from cytoplasmic rRNA's (Cunningham *et al.*, 1976). However, the sequence at the 3'-end of 12 S mt-rRNA is similar to that of cytoplasmic 18 S rRNA (Dubin *et al.*, 1976).

Modified nucleotides of mt-rRNA's are mainly 2'-0-methylated nucleotides and show similarities to eukaryotic rRNAs (Cunningham and Gray, 1977; Dubin, 1974; Dubin *et al.*, 1978). m$_2^6$A, which is universally found in all 16–18 S rRNA's, is also found in 12 S mt-rRNA.

ACKNOWLEDGMENTS

The authors express their deep appreciation to Professor Harris Busch for his encouragement. The authors are indebted to Drs. N. R. Ballal, C. D. Liarokos, K. Randerath, and D. L. Robberson for their enlightening discussions.

REFERENCES

Adams, J. M., and Cory, S. (1975). Modified nucleosides and bizarre 5' termini in mouse myeloma mRNA. *Nature (London)*, **255**, 28–33.

Adams, B. L., Morgan, M., Muthukrishnan, S., Hecht, S. M., and Shatkin, A. J. (1978). The effect of "cap" analogs on Reovirus mRNA binding to wheat germ ribosomes. *J. Biol. Chem.* **253**, 2589–2595.

Adesnik, M., and Darnell, J. E. (1972). Biogenesis and characterization of histone messenger RNA in HeLa cells. *J. Mol. Biol.* **67**, 397–406.

Adesnik, M., Salditt, M., Thomas, W., and Darnell, J. F. (1972). Evidence that all messenger RNA molecules (except histone messenger RNA) contain poly(A) sequences and that the poly(A) has a nuclear function. *J. Mol. Biol.* **71**, 21–30.

Angerer, L., Davidson, N., Murphy, W., Lynch, D., and Attardi, G. (1976). An electron microscope study of the relative positions of the 4 S and ribisomal RNA genes in HeLa cell mitochondrial DNA. *Cell* **9**, 81–90.

Arnheim, N., and Southern, E. M. (1977). Heterogeneity of the ribosomal genes in mice and men. *Cell* **11**, 363–370.

Attardi, G., Parnas, H., Huang, M. L. H., and Attardi, B. (1966). Giant-size rapidly labeled nuclear ribonucleic acid and cytoplasmic messenger ribonucleic acid in immature duck erythrocytes. *J. Mol. Biol.* **20**, 145–182.

Attardi, G., Amalric, F., Ching, E., Constantino, P., Gelfand, R., and Lynch, D. (1977). Informational content and gene mapping of mitochondrial DNA from HeLa cells. *In* "The Genetic Function of Mitochondrial DNA" (C. Saccone and A. M. Kroon, eds.), pp. 37–46, Elsevier, Amsterdam.

Aujame, L., and Freeman, K. B. (1976). The synthesis of polyadenylic acid containing RNA by isolated mitochondria from Ehrlich ascites cells. *Biochem. J.* **156**, 499–506.

Baralle, F. E. (1977). Complete nucleotide sequence of the 5' noncoding region of rabbit β-globin mRNA. *Cell* **10**, 549–558.

Baralle, F. E., and Brownlee, G. G. (1978). AUG is the only recognizable signal sequence in the 5' noncoding regions of eukaryotic mRNA. *Nature (London)* **274**, 84–87.

Barta, A., Kuechler, E., Branlant, C., Sriwidada, J., Kvol, A., and Ebel, J. P. (1975). Photoaffinity labeling of 23S RNA at the donor-site of the *Escherichia coli* ribosome. *FEBS Lett.* **56**, 170–173.

Bastos, R. N., and Aviv, H. (1977). Globin RNA precursor molecules: Biosynthesis and processing of erythroid cells. *Cell* **11**, 641–650.

Battey, J., and Clayton, D. A. (1978). The transcription map of mouse mitochondrial DNA. *Cell* **14**, 143–156.

Bellemare, G., Vigne, R., and Jordan, B. R. (1973). Interaction between *Escherichia coli* ribosomal proteins and 5S RNA molecules: Recognition of prokaryotic 5S RNAs and rejection of eukaryotic 5S rRNAs. *Biochimie* **55**, 29–35.

Benecke, B., and Penman, S. (1977). A new class of small nuclear RNA molecules synthesized by a type I RNA polymerase in HeLa cells. *Cell* **12**, 1939–1946.

Berget, S. N., Moore, C., and Sharp, P. A. (1977). Spliced segments at the 5' terminus of adenovirus 2 late mRNA. *Proc. Natl. Acad. Sci. U.S.A.* **74**, 3171–3175.

Birnboim, H. C., Mitchel, R. E., and Strauss, N. A. (1973). Analysis of long pyrimidine polynucleotides in HeLa cell nuclear DNA: Absence of polydeoxythymidylate. *Proc. Natl. Acad. Sci. U.S.A.* **70**, 2189–2192.

Bispink, L., and Matthaei, H. (1973). Photoaffinity labeling of 23S rRNA in *Escherichia coli* ribosomes with poly(U)-coded ethyl 2-diazomalonyl-phe-tRNA. *FEBS Lett.* **37**, 291–294.

Bollen, A., Heimark, R. L., Cozzone, A., Traut, R. R., and Hershey, J. W. (1975). Cross-linking of initiation factor IF-2 to *Escherichia coli* 30S ribosomal proteins with dimethylsuberimidate. *J. Biol. Chem.* **250**, 4310–4314.

Borst, P. (1972). Mitochondrial nucleic acids. *Annu. Rev. Biochem.* **41**, 333–376.

Borun, T. W., Ajiro, K., Zweidler, A., Dolley, T. W., and Stephens, R. E. (1977). Studies of human histone messenger RNA. II. The resolution of fraction containing individual human histone messenger RNA species. *J. Biol. Chem.* **252**, 173–180.

Bost, J. L., Heyting, C., Borst, P., Arnberg, A. C., and Van Bruggen, E. F. J. (1978). An insert in the single gene for the large ribosomal RNA in yeast mitochondrial DNA. *Nature (London)* **275**, 336–337.

Bralant, C., Widada, J. S., Krol, A., Fellner, P., and Ebel, J. P. (1975). Nucleotide sequences of the T_1 and pancreatic ribonuclease digestion products from some large fragments of the 23S RNA of *Escherichia coli*. *Biochimie* **57**, 175–225.

Bralant, C., Widada, J. S., Krol, A., and Ebel, J. P. (1976). Extensions of the known sequences at the 3′ and 5′ ends of 23S ribosomal RNA from *Escherichia coli,* possible base pairing between these 23S RNA regions and 16S ribosomal RNA. *Nucleic Acids Res.* **3**, 1671–1687.

Brand, R. C., Klootwijk, J., van Steenbergen, T. J. M., de Kok, A. J., and Planta, R. J. (1977). Secondary methylation of yeast ribosomal precursor RNA. *Eur. J. Biochem.* **75**, 311–318.

Brawerman, G. (1974). Eukaryotic messenger RNA. *Annu. Rev. Biochem.* **43**, 621–642.

Breitmeyer, J. B., and Noller, H. F. (1976). Affinity labeling of specific regions of 23S RNA by reaction of *N*-bromoacetylphenylalanyl-transfer RNA with *Escherichia coli* ribosomes. *J. Mol. Biol.* **101**, 297–306.

Brenner, S., Jacob, F., and Meselson, M. (1961). An unstable intermediate carrying information from genes to ribosomes for protein synthesis. *Nature (London)* **190**, 576–581.

Brimacombe, R., Trupin, J., Nirenberg, M., Leder, P., Bernfield, M., and Jaouni, T. (1965). RNA code words and protein synthesis, VIII. Nucleotide sequence of synonym codons for arginine, valine, cysteine and alanine. *Proc. Natl. Acad. Sci. U.S.A.* **54**, 954–960.

Brownlee, G., Sanger, F., and Barrell, B. G. (1968). The sequence of 5S rRNA. *J. Mol. Biol.* **34**, 379–412.

Busch, H. (1976). The function of the 5′ cap of mRNA and nuclear RNA species. *Perspect. Biol. Med.* **19**, 549–567.

Busch, H., and Smetana, K. (1970). "The Nucleolus." Academic Press, New York.

Busch, H., Choi, Y. C., Nazar, R. N., and Ro-Choi, T. S. (1975). Nucleotide sequence of nuclear RNA species. *In* "Biochemistry of the Cell Nucleus: Mechanism and Regulation of Gene Expression" (E. J. Hidvegi, J. Sümergi, and P. Venetianer, eds.), pp. 125–138. Akademiai Kiado, Budapest.

Bubroff, L. M. (1977). Oligouridylate stretches in heterogeneous nuclear RNA. *Proc. Natl. Acad. Sci. U.S.A.* **24**, 2217–2221.

Caston, J. D., and Jones, P. H. (1972). Synthesis and processing of high molecular weight RNA by nuclei isolated from embryos of *Rana pipiens*. *J. Mol. Biol.* **69**, 19–38.

Catterall, J. F., O'Malley, B. W., Robertson, M. A., Staden, R., Tanaka, Y., and Brownlee, G. G. (1978). Nucleotide sequence homology at 12 intron-extron junctions in the chicken ovalbumin gene. *Nature (London)* **275**, 510–513.

Chia, L. L., Morris, H., Randerath, K., and Randerath, E. (1976). Base composition studies on mitochondrial 4S RNA from rat liver and Morris hepatoma 5123D and 7777. *Biochim. Biophys. Acta* **425**, 49–62, 1976.

Chiu, N., Chiu, A., and Suyama, Y. (1975). Native and import transfer RNA in mitochondria. *J. Mol. Biol.* **99**, 37–50.

Choi, Y. C., and Busch, H. (1970). Structural analysis of nucleolar precursors of ribosomal ribonucleic acid. *J. Biol. Chem.* **245,** 2954–2961.

Choi, Y. C., and Busch, H. (1978). Modified nucleotides in T$_1$ RNase oligonuleotides of 18S ribosomal RNA of the Novikoff hepatoma. *Biochemistry* **17,** 5551–5560.

Choi, Y. C., Ballal, N. R., Busch, R. K., and Busch, H. (1976a). Homochromatographic and immunological analysis of contents of nucleolar gene function. *Cancer Res.* **36,** 4301–4306.

Choi, Y. C., Lee, K. M., and Busch, H. (1976b). U-rich segments of ribosomal and pre-ribosomal RNAs of Novikoff hepatoma ascites cells. *Proc. Am. Assoc. Cancer Res.* **17,** 38.

Chow, L. T., Gelinas, R. E., Broker, T. R., and Roberts, R. J. (1977). An amazing sequence arrangement at the 5'-ends of adenovirus 2 messenger RNA. *Cell* **12,** 1–8.

Cory, S., and Adams, J. M. (1975). The modified 5'-termini in small nuclear RNAs of mouse myeloma cells. *Mol. Biol. Rep.* **2,** 287–294.

Cory, S., and Adams, J. M. (1977). A very large repeating unit of mouse DNA containing the 18S, 28S and 5.8S rRNA genes. *Cell* **11,** 795–805.

Cowan, N. J., Secker, D. S., and Milstein, C. (1976). Purification and sequence analysis of the mRNA coding for an immunoglobulin heavy chain. *Eur. J. Biochem.* **61,** 355–368.

Cox, R. A. (1970). A spectrophotometric study of the secondary structure of ribonucleic acid isolated from the smaller and larger ribosomal subparticles of rabbit reticulocytes. *Biochem. J.* **117,** 101–118.

Cox, R. A. (1977). Structure and function of prokaryotic and eukaryotic ribosomes. *Prog. Biophys. Mol. Biol.* **32,** 193–231.

Cox, R. A., Pratt, H., Huvos, P., Higginson, B., and Hirst, W. (1973). A study of the thermal stability of ribosomes and biologically active subribosomal particles. *Biochem. J.* **134,** 775–793.

Cox, R. A., Hirst, W., Godwin, E. A., and Kaiser, I. (1976a). The circular dichroism of ribosomal ribonucleic acids. *Biochem. J.* **155,** 279–295.

Cox, R. A., Greenwell, P., and Hirst, W. (1976b). Reactivation of the peptidyltransferase center of rabbit reticulocyte ribosomes after inactivation by exposure to low concentration of magnesium ion. *Biochem. J.* **160,** 521–531.

Cox, R. A., Godwin, E. A., and Hastings, J. R. B. (1976c). Spectroscopic evidence for the uneven distribution of adenine and uracil residues in ribosomal ribonucleic acid of *Drosophila melanogaster* and of *Plasmodium knowlesi* and its possible evolutionary significance. *Biochem. J.* **155,** 465–475.

Craig, R. K., Brown, P. A., Harrison, O. S., McIlreavy, P., and Campbell, P. N. (1976). Guinea-pig milk-protein synthesis. *Biochem. J.* **160,** 57–74.

Cunningham, R. S., and Gray, M. W. (1977). Isolation and characterization of ^{32}P-labeled mitochondrial and cytosol ribosomal RNA from germinating wheat embryo. *Biochem. Biophys. Acta* **475,** 476–491.

Cunningham, R. S., Bonen, L., Doolittle, W. F., and Gray, M. W. (1976). Unique species of 5S, 18S, and 26S ribosomal RNA in wheat mitochondria. *FEBS Lett.* **69,** 116–122.

Dahlberg, A. E. (1974). Two forms of the 30S ribosomal subunit of *Escherichia coli*. *J. Biol. Chem.* **249,** 7673–7678.

Dahlberg, A. E., and Dahlberg, J. (1975). Binding of ribosomal protein S1 of *Escherichia coli* to the 3'-end of 16S rRNA. *Proc. Natl. Acad. Sci. U.S.A.* **72,** 2940–2944.

Dalgarno, L., and Shine, J. (1973). Conserved terminal sequence in 18S rRNA may represent terminator anticodons. *Nature (London), New Biol.* **245,** 261–262.

Daneholt, B. (1975). Transcription in polytene chromosomes. *Cell* **4,** 1–9.

Darnell, J. E. (1968). Ribonucleic acids from animal cells. *Bacteriol. Rev.* **32,** 262–290.

Darnell, J. E. (1979). Transcription units for mRNA production in eukaryotic cells and their DNA viruses. *Prog. Nucleic Acid Res. Mol. Biol.* **22**, 327–353.

Darnell, J. E., Wall, R., and Tushinski, R. J. (1971). An adenylic acid-rich sequence in messenger RNA of HeLa cells and its possible relationship to reiterated sites in DNA. *Proc. Natl. Acad. Sci. U.S.A.* **68**, 1321–1325.

Dasgupta, R., Shih, D. S., Saris, C., and Kaesberg, P. (1975). Nucleotide sequence of a viral RNA fragment that binds to eukaryotic ribosomes. *Nature (London)* **256**, 624–628.

Davies, P. L., Ferrier, L. N., and Dixon, G. H. (1977). Sequence analysis of protamine mRNA from the rainbow trout. *J. Biol. Chem.* **252**, 1386–1393.

Dawid, I. B., and Wellauer, P. K. (1976). A reinvestigation of 5′–3′ polarity in 40S ribosomal RNA precursor of *Xenopus laevis*. *Cell* **8**, 443–448.

DeKloet, S. R., and Andrean, B. A. G. (1976). Methylated nucleoside in polyadenylated containing yeast messenger ribonucleic acid. *Biochim. Biophys. Acta* **425**, 401–408.

Deonier, R. C., Soll, L., Ohtsubo, E., Lee, H. J., and Davidson, N. (1973). Mapping ribosomal RNA and other genes on defined segments of the *E. coli* chromosome. *Fed. Proc., Fed. Am. Sco. Exp. Biol.* **32**, 663 (Abstr.).

Desrosier, R., Friderici, K., and Rottman, F. (1974). Identification of methylated nucleosides in messenger RNA from Novikoff hepatoma cells. *Proc. Natl. Acad. Sci. U.S.A.* **71**, 3971–3975.

Dube, S. K. (1973). Recognition of tRNA by the ribosome. A possible role of 5S RNA. *FEBS Lett.* **36**, 39–42.

Dubin, D. T. (1974). Methylated nucleotide content of mitochondrial ribosomal RNA from hamster cells. *J. Mol. Biol.* **84**, 257–273.

Dubin, D. T., Jones, T. H., and Cleaves, G. R. (1974). An unmethylated "3 S_E" RNA in hamster mitochondria: A 5S RNA-equivalent? *Biochem. Biophys. Res. Commun.* **56**, 401–406.

Dubin, D. T., and Shine, J. (1976). The 3′-terminal sequence of mitochondrial 13S rRNA. *Nucleic Acids Res.* **3**, 1225–1231.

Dubin, D. T., and Taylor, R. H. (1978). Modification of mitochondrial ribosomal RNA from hamster cells: The presence of GmG and late-methylated UmGmU is the large subunit (17S) RNA. *J. Mol. Biol.* **121**, 523–540.

Edmonds, M., and Winters, M. A. (1976). Polyadenylate polymerases. *Prog. Nucleic Acid Res. Mol. Biol.* **17**, 147–176.

Edmonds, M., Vaughn, M. H., and Nakazato, H. (1971). Polyadenylic acid sequences in the heterogeneous nuclear RNA and rapidly-labeled polyribosomal RNA of HeLa cells: Possible evidence for a precursor relationship. *Proc. Natl. Acad. Sci. U.S.A.* **68**, 1336–1340.

Efstratiadis, A., Kafatos, F. C., and Maniatis, T. (1977). The primary structure of rabbit β-globin mRNA as determined from cloned DNA. *Cell* **10**, 571–585.

Egawa, K., Choi, Y. C., and Busch, H. (1971). Studies on the role of 23S nucleolar RNA as an intermediate in the synthesis of 18S ribosomal RNA. *J. Mol. Biol.* **56**, 565–577.

Ehresmann, C., Stiegler, P., Fellner, P., and Ebel, J. P. (1975). The determination of the primary structure of the 16S ribosomal RNA of *Escherichia coli*. III. Further studies. *Biochimie* **57**, 711–748.

Eladari, M. E., and Galibert, F. (1976). Sequence determination of the 3′-terminal T_1 oligonucleotide of 18S ribosomal RNA. *Nucleic Acids Res.* **3**, 2749–2755.

El-Khatib, S. M., Ro-Choi, T. S., Choi, Y. C., and Busch, H. (1970). Studies on nuclear 4.5S ribonucleic acid III of Novikoff hepatoma cells. *J. Biol. Chem.* **245**, 3416–3421.

Enger, M. D., Walter, R. A., Hampel, A. E., and Campbell, E. W. (1974). Rich-ribosomal particles of cultured Chinese hamster cells. *Eur. J. Biochem.* **43**, 17–28.

Erdmann, V. A. (1976). Structure and function of 5S and 5.8S RNA. *Prog. Nucleic Acid. Res. Mol. Biol.* **18**, 45–90.

Erdmann, V. A. (1978). Collection of published 5S and 5.8S ribosomal RNA sequences. *Nucleic Acids Res.* **5**, r1–r13.

Erdmann, V. A., Sprinzl, M., and Pongs, O. (1973). The involvement of 5S RNA in the binding of tRNA to ribosomes. *Biochem. Biophys. Res. Commun.* **54**, 942–948.

Feldman, M. Ya. (1977). Minor components in transfer RNA: The location–function relationships. *Prog. Biophys. Mol. Biol.* **32**, 83–102.

Feldmann, H., and Kleinow, W. (1976). Base composition of mitochondrial RNA species and characterization of mitochondrial 4S RNA from *Locusta migratoria*. *FEBS Lett.* **69**, 300–304.

Fellner, P. (1974). Structure of the 16S and 23S ribosomal RNAs. *In* "Ribosomes" (M. Nomura, A. Tissieres and P. Lengyel, eds.), pp. 169–191. Cold Spring Harbor Laboratory, Cold Spring Harbor, New York.

Fiers, W., Contreras, R., Duerinck, F., Haegeman, G., Iserentant, D., Merregaert, J., Min Jou, W., Molemans, F., Raeymaekers, A., Van den Berghe, A., Volchaert, G., and Ysebaert, M. (1976). Complete nucleotide sequence of bacteriophage MS2 RNA: Primary and secondary structure of the replicase gene. *Nature (London)* **260**, 500–507.

Forget, B. G. (1978). Molecular lesions in thalassemia. *Trends Biochem. Sci.* **3**, 86–90.

Forget, B. G., and Weissman, S. M. (1967). Low molecular weight RNA components from K. B. cells. *Nature (London)* **213**, 878–882.

Forget, B. G., and Weissman, S. M. (1969). The nucleotide sequence of ribosomal 5S ribonucleic acid from KB cells. *J. Biol. Chem.* **244**, 3148–3165.

Frisby, P., Eaton, M., and Fellner, P. (1976). Absence of 5′-terminal capping in encephalomyocarditis virus RNA. *Nucleic Acids Res.* **3**, 2771–2787.

Fuke, M., Busch, H., and Rao, P. N. (1976). Evolutionary trends in 18S ribosomal RNA nucleotide sequences of rat, mouse, hamster and man. *Nucleic Acids Res.* **3**, 2939–2957.

Furuichi, Y., Morgan, M., Muthurkrishnan, S., and Shatkin, A. J. (1975). Reovirus messenger RNA contains a methylated blocked 5′-terminal structure $m^7G(5')ppp(5')GmpCp$-. *Proc. Natl. Acad. Sci. U.S.A.* **72**, 362–366.

Galibert, W. (1978). Why genes in pieces? *Nature (London)* **271**, 501.

Galibert, F., Larsen, C. J., Lelong, J. C., and Boiron, M. (1967). Position of 5S RNA among cellular ribonucleic acids. *Biochim. Biophys. Acta* **142**, 89–98.

Gelinas, R. E., and Roberts, R. J. (1977). One predominant 5′-undecanucleotide in adenovirus 2 late messenger RNAs. *Cell* **11**, 533–544.

Georgiev, G. P. (1967). The nature and biosynthesis of nuclear ribonucleic acids. *Prog. Nucleic Acid Res. Mol. Biol.* **6**, 259–351.

Gierer, A. (1963). Function of aggregated reticulocyte ribosomes in protein synthesis. *J. Mol. Biol.* **6**, 148–157.

Gierson, D., Rogers, M. E., Sartirana, M. L., and Loening, U. E. (1971). The synthesis of ribosomal RNA in different organisms: Structure and evolution of the rRNA precursor. *Cold Spring Harbor Symp. Quant. Biol.* **35**, 583–598.

Gilmore-Herbert, M., and Wall, R. (1978). Immunoglobulin light chain mRNA is processed from large nuclear RNA. *Proc. Natl. Acad. Sci. U.S.A.* **75**, 342–345.

Girshovich, A. S., Bochkareva, E. S., Kramorov, V. A., and Ovchinnikov, Y. A. (1974). *E. coli* 30S and 50S ribosomal subparticle components in the localization region of the tRNA acceptor terminus. *FEBS Lett.* **45**, 213.

Goddard, J. (1977). The structures and functions of transfer RNA. *Prog. Biophys. Mol. Biol.* **32**, 233–308.

Goldstein, L. (1975). Proteins and RNAs in nucleocytoplasmic interactions. *In* "Biochemistry of the Cell Nucleus Mechanism and Regulation of Gene Expression" (E. J. Hidvegi, J. Sümagi, and P. Venetianer, eds.), Vol. 33, pp. 189–204. Akademiai Kiado, Budapest.

Goldstein, L., and Ko, C. (1974). Electrophoretic characterization of shuttling and non-shuttling small nuclear RNAs. *Cell* 2, 259–269.

Goldstein, L., Wise, G. E., and Ko, C. (1977). Small nuclear RNA localization during mitosis. An electron microscope study. *J. Cell Biol.* 73, 322–331.

Greenberg, J. R. (1976). Isolation of L-cell messenger RNA which lacks polyadenylate. *Biochemistry* 15, 3516–3522.

Greenberg, J. R., and Perry, R. P. (1972). Relative occurrence of polyadenylic acid sequences in messenger and heterogeneous RNA of L cells as determined by poly(U)-hydroxyapatite chromatography. *J. Mol. Biol.* 72, 91–98.

Greenstein, G., and Schell, P. (1976). Isolation and sequence analysis of sea urchin (lytechninus pictus) histone H4 messenger RNA. *J. Mol. Biol.* 104, 323–349.

Greenwell, P., Harris, R. J., and Symons, R. H. (1974). Affinity labeling of 23-S ribosomal RNA in the active center of *Escherichia coli* peptidyltransferase. *Eur. J. Biochem.* 49, 539–554.

Gressel, J., Berman, T., and Cohen, N. (1975). Dinoflagellate ribosomal RNA: An evolutionary relic? *J. Mol. Evol.* 5, 307–313.

Hadjiiolov, A. A., and Nikolaev, N. (1976). Maturation of ribosomal ribonucleic acids and the biogenesis of ribosomes. *Prog. Biophys. Mol. Biol.* 31, 95–144.

Hall, B. D., and Doty, P. (1959). Robnucleic acid from microsomal particle. *J. Mol. Biol.* 1, 111–126.

Harwood, R., Connolly, D., Grant, M. E., and Jackson, D. S. (1974). Presumptive mRNA for procollagen: Occurrence in membrane bound ribosome of embryonic chick tendon fibroblast. *FEBS Lett.* 41, 85–88.

Hawley, D. A., Slobin, L. I., and Wahba, A. J. (1974). The mechanism of action of initiation factor 3 in protein synthesis. II. Association of the 30S ribosomal protein S12 with IF-3. *Biochem. Biophys. Res. Commun.* 61, 544–549.

Heckman, J. E., Hecker, L. I., Schwartzbach, S. D., Edgar, W., Baumstark, B., and RajBhandary, U. L. (1978). Structure and function of initiator methionine tRNA from the mitochondria of *Neurospora crassa. Cell* 13, 83–95.

Held, W. A. (1976). Role of 16S ribosomal ribonucleic acid and the 30S ribosomal protein S12 in the initiation of natural messenger ribonucleic acid translation. *Biochemistry* 13, 2115–2122.

Hellung-Larsen, P. (1977). Low molecular weight RNA components in eukaryotic cells. FADL's Forlag, Copenhagen.

Hendler, F. J., Padmanaban, G., Patzer, J., Ryan, R., and Rabinowitz, M. (1975). Yeast mitochondrial RNA contains a short polyadenylic acid segment. *Nature (London)* 258, 357–359.

Herman, R., Williams, J., Lenk, R., and Penman, S. (1976). Cellular skeleton and RNA messages. *Prog. Nucleic Acids Res. Mol. Biol.* 19, 379–401.

Hewlitt, M. J., Rose, J. K., and Baltimore, D. (1976). 5′-Terminal structure of poliovirus polysomal RNA is pUp. *Proc. Natl. Acad. Sci. U.S.A.* 73, 327–330.

Heywood, S. M., and Kennedy, D. S. (1976). Purification of myosin translational control RNA, and its interaction with myosin mRNA. *Biochemistry* 15, 3314–3319.

Hidvegi, E. J., Prestayko, A. W., and Busch, H. (1971). 65 S and 80S nucleolar RNA. *Physiol. Chem. Phys.* 3, 17–35.

Hirsch, M., Spradling, A., and Penman, S. (1974). The messenger-like poly(A)-containing RNA species from the mitochondria of mammals and insects. *Cell* **1**, 31–35.

Hochkeppel, M. K., Spicer, E., and Craven, G. R. (1976). A method of preparing *Escherichia coli* 16S RNA possessing previously unobserved 30S ribosomal protein binding sites. *J. Mol. Biol.* **101**, 155–170.

Holley, R. W., Apgar, J., Everett, G. A., Madison, J. T., Marquisse, M., Merril, S. H., Penswick, J. R., and Zamir, A. (1965). Structure of a ribonucleic acid. *Science* **147**, 1462–1465.

Holmes, D. S., and Bonner, J. (1974). Interspersion of repetitive and single-copy sequences in nuclear ribonucleic acid of higher molecular weight. *Proc. Natl. Acad. Sci. U.S.A.* **71**, 1108–1112.

Hori, H. (1975). Evolution of 5S RNA. *J. Mol. Evol.* **7**, 75–86.

Hunt, J. A. (1976). Terminal sequence studies of high-molecular-weight ribonucleic acid. The 3'-termini of rabbit reticulocyte ribosomal RNA. *Biochem. J.* **120**, 353–363.

Iatrou, K., and Dixon, G. H. (1977). The distribution of poly(A)+ and poly(A)− protamine messenger RNA sequence in the developing trout testes. *Cell* **10**, 433–441.

Inagaki, A., and Busch, H. (1972). Marker nucleotides for non-ribosomal spacer segments of preribosomal ribonucleic acid. *Biochem. Biophys. Res. Commun.* **49**, 1398–1406.

Ishikawa, H. (1976). The fragments from the 28S ribosomal RNA of *Galleria mellonella* with unesterified uridine at the 3'-termini. *Biochem. Biophys. Acta* **425**, 185–195.

Jacob, F., and Monod, J. (1961). Genetic regulatory mechanisms in the synthesis of proteins. *J. Mol. Biol.* **3**, 318–356.

Jelinek, W. (1977). Specific nucleotide sequences in HeLa cells inverted repeated DNA. *J. Mol. Biol.* **115**, 591–601.

Jelinek, W. J., and Darnell, J. E. (1972). Double-stranded regions in heterogeneous nuclear RNA from HeLa cells. *Proc. Natl. Acad. Sci. U.S.A.* **69**, 2537–2541.

Jordan, B. R., Jourdan, R., and Jacq, B. (1976). Late steps in the maturation of *Drosophila* 26S ribosomal RNA: Generation of 5.8S and 2S RNAs by cleavages in the cytoplasm. *J. Mol. Biol.* **101**, 85–105.

Kanamaru, R., Choi, Y. C., and Busch, H. (1972). Structural analysis of the hydrogen bonded B_3 fragment of 28S rRNA of Novikoff hepatoma ascites cells. *Physiol. Chem. Phys.* **4**, 103–124.

Kanamaru, R., Choi, Y. C., and Busch, H. (1974). Structural analysis of ribosomal 28S ribonucleic acid of Novikoff hepatoma cells. Primary sequence of the B_3-9 subcomponent. *J. Biol. Chem.* **249**, 2453–2463.

Karrer, K. M., and Gall, J. G. (1976). The macronuclear ribosomal DNA of *Tetrahymena pyriformis* in pallindrome. *J. Mol. Biol.* **104**, 421–453.

Kates, J. (1970). Transcription of the vaccinia virus genome and the occurrence of polyriboadenylic acid sequences in messenger RNA. *Cold Spring Harbor Symp. Quant. Biol.* **35**, 743–752.

Keith, G., Gangloff, T., Ebel, J. P., and Dirheimer, G. (1970). Establissement de la séquence des nucleotides de l'aspartet-tRNA de leinure de biere. *C. R. Hebd. Seances Acad. Sci.* **271**, 613–616.

Kenner, R. A. (1973). A protein-nucleic acid cross-link in 30S ribosomes. *Biochem. Biophys. Res. Commun.* **51**, 932–938.

Khan, M. S. N., and Maden, B. E. H. (1976a). Nucleotide sequences within the ribosomal ribonucleic acid of HeLa cells, *Xenopus laevis* and chick embryo fibroblasts. *J. Mol. Biol.* **101**, 235–254.

Khan, M. S. N., and Maden, B. E. H. (1976b). Conformation of mammalian 5.8 S ribosomal RNA: S₁ nuclease as a probe. *FEBS Lett.* **72**, 105–110.

Khan, M. S. N., Salim, M., and Maden, B. E. H. (1978). Extensive homologies between the methylated nucleotide sequences in several vertebrate ribosomal ribonucleic acids. *Biochem. J.* **169**, 531–542.

Kim, S. H. (1976). Three-dimensional structure of transfer RNA. *Prog. Nucleic Acid. Res. Mol. Biol.* **17**, 181–216.

Kim, C. H., and Sarma, R. H. (1978). Spatial configuration of the bizarre 5'-terminus of mammalian mRNA. *J. Am. Chem. Soc.* **100**, 1571–1590.

Kim, S. H., Suddath, F. L., Quigley, G. J., McPherson, A., Sussman, J. L., Wang, A. H. J., Seeman, N. C., and Rich, A. (1974a). Three-dimensional tertiary structure of yeast phenylalanine transfer RNA. *Science* **185**, 435–440.

Kim, S. H., Sussman, J. L., Suddath, F. L., Quigley, G. J., McPherson, A., and Rich, A. (1974b). The general structure of transfer RNA molecules. *Proc. Natl. Acad. Sci. U.S.A.* **71**, 4970–4974.

Klessig, D. F. (1977). Two adenovirus mRNAs have a common 5' terminal leader sequence encoded at least 10 kb upstream from their main coding regions. *Cell* **12**, 9–21.

Klukas, C. K., and Dawid, I. B. (1976). Characterization and mapping of mitochondrial ribosomal RNA and mitochondrial DNA in *Drosophila melanogaster*. *Cell* **9**, 615–625.

Kronenberg, L. H., and Humphreys, T. (1972). Double-stranded ribonucleic acid in sea urchin embryos. *Biochemistry* **11**, 2020–2026.

Kroon, A. M., and Saccone, C. (1976). Concluding remarks. *In* "The Genetic Function of Mitochondrial DNA" (C. Saccone and A. M. Kroon, eds.), pp. 343–347. Elsevier, Amsterdam.

Kuebbing, D., and Liarakos, C. D. (1978). Nucleotide sequence at the 5'-end of ovalbumin messenger RNA from chicken. *Nucleic Acids. Res.*, **5**, 2253–2266.

Kumar, A. J. (1967). Patterns of ribosomal RNA synthesis in *Tetrahymena*. *J. Cell Biol.* **35**, 74A.

Kumar, A. J. (1970). Ribosome synthesis in *Tetrahymena pyriformis*. *J. Cell Biol.* **45**, 623–634.

Kuriyama, Y., and Luck, D. J. L. (1974). Methylation and processing of mitochondrial ribosomal RNAs in poky and wild-type *Neurospora crassa*. *J. Mol. Biol.* **83**, 253–266.

Kurland, C. G. (1960). Molecular characterization of ribonucleic acid from *E. coli* ribosomes. I. Isolation and molecular weights. *J. Mol. Biol.* **2**, 83–91.

Lambowitz, A. M., and Luck, D. J. L. (1975). Methylation of mitochondrial RNA species in the wild-type and poky strains of *Neurospora crassa*. *J. Mol. Biol.* **96**, 207–214.

Larsen, C. J., Galibert, F., Hampe, A., and Boiron, M. (1969). Etudes des RNA nucléaires de fable poids moléculaire de la cellule KB. *Bull. Soc. Chim. Biol.* **51**, 649–668.

Lava-Sanchez, P. A., Amaldi, F., and La Posta, A. (1972). Base composition of ribosomal RNA and evolution. *J. Mol. Evol.* **2**, 44–55.

Leaver, C. J. (1976). Higher-plant mitochondrial ribosomes contain a 5S ribosomal ribonucleic acid component. *Biochem. J.* **157**, 275–277.

Lee, S. Y., Mendecki, J., and Brawerman, G. (1971). A polynucleotide segment rich in adenylic acid in the rapidly-labeled polyribosomal RNA component of mouse sarcoma 180 ascites cells. *Proc. Natl. Acad. Sci. U.S.A.* **68**, 1331–1335.

Levenson, R. G., and Marcu, K. B. (1976). On the existence of polyadenylated histone mRNA in *Xenopus laevis* oocytes. *Cell* **9**, 311–322.

Liau, M. C., and Perry, R. P. (1969). Ribosome precursor particles in nucleoli. *J. Cell Biol.* **42**, 272–283.

Lim, L., and Canellakis, E. S. (1970). Adenine-rich polymer associated with rabbit reticulocytes messenger RNA. *Nature (London)* 227, 710–712.

Littauer, U. Z., and Inouye, H. (1973). Regulation of tRNA. *Annu. Rev. Biochem.* 42, 439–470.

Lizardi, P. M. (1976). Biogenesis of silk fibroin mRNA: An example of very rapid processing? *Prog. Nucleic Acid Res. Mol. Biol.* 19, 301–312.

Lizardi, P. M., Williamson, R., and Brown, D. D. (1976). The size of fibroin messenger RNA and its polyadenylic acid content. *Cell* 4, 199–205.

Lockard, R. E., and Lingrel, J. B. (1969). The synthesis of mouse hemoglobin β-chains in a rabbit reticulocyte cell-free system programmed with mouse reticulocyte 9S RNA. *Biochem. Biophys. Res. Commun.* 37, 204–212.

Lockard, R. E., and RajBhandary, U. L. (1976). Nucleotide sequences at the 5'-termini of rabbit α- and β-globin. *Cell* 9, 747–760.

Loening, U. (1968). Molecular weights of ribosomal RNA in relation to evolution. *J. Mol. Biol.* 38, 355–365.

Loening, U. E., Jones, K. W., and Birnstiel, M. L. (1969). Properties of the ribosomal RNA precursor in *Xenopus laevis:* Comparison to the precursor in mammals and in plants. *J. Mol. Biol.* 45, 353–366.

Lund, E., and Dahlberg, J. E. (1977). Spacer transfer RNAs in ribosomal RNA transcript of *E. coli:* Processing of 30S ribosomal RNA *in vitro*. *Cell* 11, 247–262.

McClements, W., and Skalka, A. M. (1977). Analysis of chicken ribosomal RNA genes and construction of lambda hybrids containing gene fragments. *Science* 196, 195–197.

McCloskey, J. A., and Nishimura, A. (1977). Modified nucleosides in transfer RNA. *Acct. Chem. Res.* 10, 403–410.

McConkey, E. H., and Hopkins, J. W. (1969). Molecular weights of some HeLa ribosomal RNAs. *J. Mol. Biol.* 39, 545–550.

McReynolds, L., O'Malley, B. W., Nisbet, A. D., Fothergill, J. E., Givol, D., Fields, S., Robertson, M. and Brownlee, G. G. (1978). Sequence of chicken ovalbumin messenger RNA. *Nature (London)* 273, 723–728.

Maden, B. E. H., and Robertson, J. A. (1974). Demonstration of the "5.8S" ribosomal sequence in HeLa cell ribosomal precursor RNA. *J. Mol. Biol.* 87, 227–255.

Maden, B. E. H., and Salim, M. (1974). The methylated nucleotide sequences in HeLa cell ribosomal RNA and its precursors. *J. Mol. Biol.* 88, 133–164.

Maden, B. E. H., Forbes, J., de Jonge, P., and Klootwijk, J. K. (1975). Presence of a hypermodified nucleotide in HeLa cell 18S and *Saccharomyces carlsbergensis* 17S ribosomal RNAs. *FEBS Lett.* 59, 60–73.

Maden, B. E. H., Khan, M. S. N., Hughes, D. G., and Goddard, J. P. (1977). Inside 45S ribonucleic acid. *Biochem. Soc. Symp.* 42, 165–179.

Mahler, H. R., and Raff, R. A. (1975). The evolutionary origin of the mitochondrion: A nonsymbiotic model. *Int. Rev. Cytol.* 43, 1–124.

Mandel, J. L., Breathnach, R., Gerlinger, P., LeMeur, M., Gannon, F., and Chambon, P. (1978). Organization of coding and intervening sequences in the chicken ovalbumin split gene. *Cell,* 14, 641–653.

Marotta, C. A., Wilson, J. T., Forget, B. G., and Weissman, S. M. (1977). Human β-globin messenger RNA. *J. Biol. Chem.* 252, 5040–5033.

Martin, T. E., Bicknell, J. N., and Kumar, A. (1970). Hybrid 80S monomers formed from subunits of ribosomes from protozoa, fungi, plants and mammals. *Biochem. Genet.* 4, 603–615.

Melli, M., Spinelli, G., Wyssling, H., and Arnold, E. (1977). Presence of histone mRNA sequences in high molecular weight of HeLa cells. *Cell* 11, 651–661.

Milstein, C., Brownlee, G. G., Cartwright, E. M., Jarvis, J. M., and Proudfoot, N. J. (1974). Sequence analysis of immunoglobulin light chain messenger RNA. *Nature (London)* **252**, 354–359.

Molloy, G., and Darnell, J. E. (1973). Characterization of the poly(adenylic acid) regions and the adjacent nucleotides in heterogeneous nuclear ribonucleic acid and messenger ribonucleic acid from HeLa cells. *Biochemistry* **12**, 2324–2330.

Molloy, G., and Puckett, L. (1976). The metabolism of heterogeneous nuclear RNA and the formation of cytoplasmic messenger RNA in animal cells. *Prog. Biophys. Mol. Biol.* **31**, 1–38.

Molloy, G. R., Jelinek, W., Salditt, M., and Darnell, J. E. (1974). Arrangement of specific oligonucleotides within poly(A) terminated HnRNA molecules. *Cell* **1**, 43–53.

Monier, R. (1974). 5S RNA. *In* "Ribosomes" (N. Nomura, A. Tissiéres and P. Lengyel, eds.), pp. 141–168. Cold Spring Harbor Laboratory, Cold Spring Harbor, New York.

Moore, P. B., Engelman, D. M., and Schoenborn, B. P. (1975). A neutron scattering study of the distribution of protein and RNA in the 30S ribosomal subunit of *Escherichia coli*. *J. Mol. Biol.* **91**, 101–120.

Moss, B., Martin, S. A., Ensinger, M. J., Boone, R. F., and Wei Cha-Mer (1976). Modification of the 5'-terminals of mRNAs by viral and cellular enzymes. *Prog. Nucleic Acid Res. Mol. Biol.* **19**, 63–81.

Moss, B., Gershowitz, A., Weber, L. A., and Baglioni, C. (1977). Histone mRNAs contain blocked and methylated 5'-terminal sequences but lack methylated nucleotides at internal positions. *Cell* **10**, 113–120.

Muthukrishnan, S., Both, G. W., Furuichi, Y., and Shatkin, A. J. (1975). 5'-Terminal 7-methylguanosine in eukaryotic mRNA is required for translation. *Nature (London)* **255**, 33–37.

Nagley, P., Sriprakash, K. S., Rytka, J., Choo, K. B., Trembath, M. K., Lukins, H. B., and Linnane, A. W. (1976). Physical mapping of genetic markers in the yeast mitochondrial genome. *In* "The Genetic Function of Mitochondrial DNA" (C. Saccone and A. M. Kroon, eds.), pp. 231–242, Elsevier, Amsterdam.

Nakazato, H., Edmonds, M., and Kopp, D. W. (1974). Differential metabolism of large and small poly(A) sequences in the heterogeneous nuclear RNA of poly(A) HeLa cells. *Proc. Natl. Acad. Sci. U.S.A.* **71**, 200–204.

Nath, K., and Bollon, A. P. (1977). Organization of the yeast ribosomal RNA gene cluster via cloning and restriction analysis. *J. Biol. Chem.* **252**, 6562–6571.

Nazar, R. N., and Ray, K. L. (1976). The nucleotide sequence of turtle 5.8S rRNA. *FEBS Lett.* **72**, 111–116.

Nazar, R. N., Owens, T. W., Sitz, T. O., and Busch, H. (1975). Maturation pathway for Novikoff ascites hepatoma 5.8S ribosomal ribonucleic acid. Evidence for its presence in 32S nuclear ribonucleic acid. *J. Biol. Chem.* **250**, 2475–2481.

Nazar, R. N., Sitz, T. O., and Busch, H. (1976). Homologies in eukaryotic 5.8S ribosomal RNA. *Biochem. Biophys. Res. Commun.* **62**, 736–743.

Nikolaev, N., Schlessinger, D., and Wellauer, P. K. (1974). 30S pre-ribosomal RNA of *Escherichia coli* and products of cleavage by ribonuclease III: Length and molecular weight. *J. Mol. Biol.* **86**, 741–747.

Nikolaev, N., Birenbaum, M., and Schlessinger, D. (1975). 30S pre-ribosomal RNA of *Escherichia coli:* Primary and secondary processing. *Biochim. Biophys. Acta* **395**, 478–489.

Nirenberg, M. W., and Matthaei, J. H. (1961). The dependence of cell-free protein synthesis in *E. coli* upon naturally occurring or synthetic polyribonucleotides. *Proc. Natl. Acad. Sci. U.S.A.* **47**, 1588–1602.

Nishikawa, K., and Takemura, S. (1974). Nucleotide sequence of 5 S rRNA from *Torulopsis utilis*. *FEBS Lett.* **40**, 106–109.

Nomoto, A., Lee, Y. F., and Wimmer, E. (1976). The 5'-end of polio virus mRNA is not capped with m⁷G(5')ppp(5')Np. *Proc. Natl. Acad. Sci. U.S.A.* **73**, 375–380.

Nomura, M., and Held, W. A. (1974). Reconstitution of ribosomes: Studies of ribosom structure, function and assembly. *In* "Ribosomes" (M. Nomura, A. Tissiércs and P. Lengyel, eds.), pp. 193–223. Cold Spring Harbor Laboratory, Cold Spring Harbor, New York.

Ojala, D., and Attardi, G. (1974). Identification and partial characterization of multiple discrete polyadenylic acid-containing RNA components coded for by HeLa cell mitochondrial DNA. *J. Mol. Biol.* **88**, 205–219.

Okada, N., Shindo-Okada, N., and Nishimura, S. (1977). Isolation of mammalian tRNA^Asp and tRNA^Tyr by lectin-Sepharose affinity column. *Nucleic Acids Res.* **4**, 415–423.

Oliver, K. M., and Lane, B. G. (1970). The 3'-hydroxyl termini in yeast ribosomal RNA. *Can. J. Biochem.* **48**, 1113–1121.

Payne, P. I., and Dyer, T. A. (1976). Evidence for the nucleotide sequence of 5S rRNA from the flowering plant *Secale cereale* (rye). *Eur. J. Biochem.* **71**, 33–38.

Peacock, A. C., and Dingman, W. C. (1967). Resolution of multiple ribonucleic acid species by polyacrylamide gel electrophoresis. *Biochemistry* **6**, 1818–1827.

Pellegrini, M., Manning, J., and Davidson, N. (1977). Sequence arrangement of the rDNA of *Drosophila melanogaster*. *Cell* **10**, 213–224.

Pene, J. J., Knight, E., and Darnell, J. E. (1968). Characterization of a new low molecular weight RNA in HeLa cell ribosomes. *J. Mol. Biol.* **33**, 609–623.

Perry, R. P. (1967). The nucleolus and the synthesis of ribosomes. *Prog. Nucleic Acid Res. Mol. Biol.* **6**, 220–257.

Perry, R. P. (1976). Processing of RNA. *Annu. Rev. Biochem.* **45**, 605–629.

Perry, R. P., and Kelley, D. W. (1976). Kinetics of formation of 5'-terminal caps in mRNA. *Cell* **8**, 433–442.

Perry, R. P., Cheng, T. Y., Freed, J. J., Greenberg, J. R., Kelley, D. E., and Tartof, K. D. (1970). Evolution of the transcription unit of ribosomal RNA. *Proc. Natl. Acad. Sci. U.S.A.* **65**, 609–616.

Perry, R. P., Bard, E., Hames, B. D., Kelley, D. E., and Schibler, U. (1976). The relationship between hnRNA and mRNA. *Prog. Nucleic Acid. Res. Mol. Biol.* **19**, 275–292.

Pousada, C. R., Marcaud, L., Portier, M. M., and Hayes, D. H. (1975). Rapidly labelled RNA in *Tetrahymena pyriformis*. *Eur. J. Biochem.* **56**, 117–122.

Prescott, D. M., Bostock, C., Gamow, E., and Lauth, M. (1971). Characterization of rapidly labeled RNA in *Tetrahymena pyriformis*. *Exp. Cell Res.* **67**, 124–128.

Prestayko, A. W., Tonato, M., and Busch, H. (1970). Low molecular weight RNA associated with 28S nucleolar RNA. *J. Mol. Biol.* **47**, 505–513.

Prestayko, A. W., Tonato, M., Lewis, C., and Busch, H. (1971). Heterogeneity of nucleolar U-3 ribonucleic acid of the Novikoff hepatoma. *J. Biol. Chem.* **246**, 182–187.

Proudfoot, N. J. (1977). Complete 3' noncoding region sequences of rabbit and human β-globin mRNAs. *Cell* **10**, 559–570.

Proudfoot, N. J., and Brownlee, G. G. (1976). 3' Non-coding region sequences in eukaryotic messenger RNA. *Nature (London)* **263**, 211–214.

Proudfoot, N. J., and Longley, J. I. (1976). The 3'-terminal sequence of human α- and β-globin mRNAs: Comparison with rabbit globin messenger RNA. *Cell* **9**, 733–746.

Quagliarotti, G., Hidvegi, E., Wikman, J., and Busch, H. (1970). Structural analysis of nucleolar precursors of ribosomal ribonucleic acid. *J. Biol. Chem.* **8**, 1962–1969.

Rabinowitz, M., Jakovic, S., Martin, N., Hendler, F., Halbreich, A., Lewin, A., and

Morimoto, R. (1976). Transcription and organization of yeast mitochondrial DNA. *In* "The Genetic Functions of Mitochondrial DNA" (C. Saccone and A. M. Kroon, eds.), pp. 219–230, Elsevier, Amsterdam.

Raj, N. B. K., Ro-Choi, T. S., and Busch, H. (1975). Nuclear ribonucleoprotein complexes containing U-1 and U-2 RNA. *Biochemistry* **14**, 4380–4385.

Rao, M. S., Wu, B. C., Waxman, J., and Busch, H. (1975). Rigid structural requirement of the 5'-terminus of mRNA for translational activity. *Biochem. Biophys. Res. Commun.* **66**, 1186–1193.

Rao, M. S., Blackstone, M., and Busch, H. (1977). Effect of U-1 nuclear RNA on translation of messenger RNA. *Biochemistry* **6**, 2756–2762.

Reddy, R., Ro-Choi, T. S., Henning, D., Shibata, H., Choi, Y. C., and Busch, H. (1972). Modified nucleosides of nuclear and nucleolar low molecular weight ribonucleic acid. *J. Biol. Chem.* **247**, 7245–7250.

Reddy, R., Sitz, T. O., Ro-Choi, T. S., and Busch, H. (1974a). Two-dimensional polyacrylamide gel electrophoresis separation of low molecular weight nuclear RNA. *Biochem. Biophys. Res. Commun.* **56**, 1017–1022.

Reddy, R., Ro-Choi, T. S., Henning, D., and Busch, H. (1974b). Primary sequence of U-1 nuclear ribonucleic acid of Novikoff hepatoma ascites cells. *J. Biol. Chem.* **249**, 6486–6494.

Reeder, R. H., Higashinakagawa, T., and Miller, O. J. (1976). The 5' → 3' polarity of the *Xenopus* ribosomal RNA precursor molecule. *Cell* **8**, 449–454.

Reeder, R. H., Solliner-Webb, B., and Wahn, H. L. (1977). Sites of transcription initiation *in vivo* on *Xenopus laevis* ribosomal DNA. *Proc. Natl. Acad. Sci. U.S.A.* **74**, 5402–5406.

Rein, A. (1971). The small molecular weight monodisperse nuclear RNA's in mitotic cells. *Biochim. Biophys. Acta* **232**, 306–313.

Rich, A. (1977). Three-dimensional structure and biological function of transfer RNA. *Acct. Chem. Res.* **10**, 388–396.

Rich, A., and RajBhandary, U. L. (1976). Transfer RNA: Molecular structure, sequence and properties. *Annu. Rev. Biochem.* **45**, 806–860.

Richter, D. D., Erdmann, V. A., and Sprinzl, M. (1973). Specific recognition of GTψC loop (loop IV) of tRNA by 50S ribosomal subunit from *E. coli. Nature (London), New Biol.* **246**, 132–135.

Ritter, P. O., and Busch, H. (1971). The chromatographic comparison of cytoplasmic, nuclear and nucleolar valine and leucine tRNAs from Novikoff hepatoma cells and cytoplasmic tRNAs from rat liver cells. *Physiol. Chem. Phys.* **3**, 411–425.

Robberson, D. L., Clayton, D. A., and Morrow, J. F. (1974). Cleavage of replicating forms of mitochondrial DNA by *Eco*RI endonucleases. *Proc. Nat. Acad. Sci. U.S.A.* **71**, 4447–4451.

Roberts, R. B. (1958). Introduction *In* "Microsomal Particles and Protein Synthesis" (R. B. Roberts, ed.), pp. vii–viii. Pergamon, Oxford.

Robertus, J. D., Ladner, J. E., Finch, J. T., Rhodes, D., Brown, R. S., Clark, B. F. C., and Klug, A. (1974). Structure of yeast phenylalanine tRNA at 3 Å resolution. *Nature (London)* **250**, 546–551.

Ro-Choi, T. S., and Busch, H. (1974). Low-molecular-weight nuclear RNA's. *In* "The Cell Nucleus" (H. Busch, ed.), Vol. III, pp. 151–208. Academic Press, New York.

Ro-Choi, T. S., and Henning, D. (1977). Sequence of 5'-oligonucleotide of U1 RNA from Novikoff hepatoma cells. *J. Biol. Chem.* **252**, 3814–3810.

Ro-Choi, T. S., Moriyama, Y., Choi, Y. C., and Busch, H. (1970). Isolation and purification of a nuclear 4.5S ribonucleic acid in the Novikoff hepatoma. *J. Biol. Chem.* **245**, 1970–1977.

Ro-Choi, T. S., Reddy, R., Henning, D., and Busch, H. (1971). 5S RNA$_{III}$, a new nucleus-specific 5S RNA. *Biochem. Biophys. Res. Commun.* **44**, 963–972.

Ro-Choi, T. S., Reddy, R., Henning, D., Takano, T., Taylor, C. W., and Busch, H. (1972). Nucleotide sequence of 4.5S ribonucleic acid of hepatoma cell nuclei. *J. Biol. Chem.* **247**, 3205–3222.

Ro-Choi, T. S., Choi, Y. C., Savage, H. E., and Busch, H. (1973). Polyacrylamide gel electrophoresis of RNA. *In* "Methods in Cancer Research" (H. Busch, ed.), Vol. IX, pp. 71–153. Academic Press, New York.

Ro-Choi, T. S., Reddy, R., Choi, Y. C., Raj, N. B., and Henning, D. (1974). Primary sequence of U-1 nuclear RNA and unusual feature of 5'-end structure of LMW RNAs. *Fed. Proc., Fed. Am. Soc. Exp. Biol.* **33**, 1548.

Ro-Choi, T. S., Choi, Y. C., Henning, D., McCloskey, J., and Busch, H. (1975). Nucleotide sequence of U-2 ribonucleic acid. *J. Biol. Chem.* **250**, 3921–3928.

Rogers, M. E., Loening, U. E., and Fraser, R. S. S. (1970). Ribosomal RNA precursors in plants. *J. Mol. Biol.* **49**, 681–692.

Roop, D. R., Nordstrom, J. L., Tsai, S. Y., Tsai, M. J., and O'Malley, B. W. (1978). Transcription of structural and intervening sequences in the ovalbumin mRNA precursors. *Cell,* **25**, 671–685.

Rosen, J. M. (1976). Isolation and characterization of purified rat casein messenger ribonucleic acids. *Biochemistry* **15**, 5263–5271.

Rosen, P., and Eddman, M. (1976). Poly(A)-associated RNA from the mitochondrial fraction of the fungus Trichoderman. *Eur. J. Biochem.* **63**, 525–532.

Rosset, R., and Monier, R. (1963). A propos de la présence d'acid ribonucléique de faible poids moléculare dans les ribosomes d'*E. coli. Biochim. Biophys. Acta* **68**, 653–656.

Rottman, F., Shatkin, A. J., and Perry, R. P. (1974). Sequences containing methylated nucleotides at the 5'-termini of messenger RNAs: Possible implications for processing. *Cell* **3**, 197–199.

Rottman, F. M., Desrosiers, R. C., and Friderici, K. (1976). Nucleotide methylation patterns in eukaryotic mRNA. *Prog. Nucleic Acid Res. Mol. Biol.* **19**, 21–38.

Rubin, G. M. (1973). The nucleotide sequence of *Saccharomyces cerevisiae* 5.8S ribosomal ribonucleic acid. *J. Biol. Chem.* **248**, 3860–3875.

Rungger, D., and Crippa, M. (1977). The primary ribosomal DNA transcript in eukaryotes. *Prog. Biophys. Mol. Biol.* **32**, 247–269.

Ryskov, A. P., Saunders, G. A., Faraskyan, V., and Georgiev, G. D. (1973). Double-helical regions in nuclear precursor of mRNA (pre-mRNA). *Biochim. Biophys. Acta* **312**, 152–164.

Saccone, C., and Quagliariello, E. (1975). Biochemical studies of mitochondrial transcription and translation. *Int. Rev. Cytol.* **43**, 125–165.

Saccone, C., Pepe, G., Cantatore, P., Terpstra, P., and Kroon, A. M. (1976). Mapping of the transcription products of rat-liver mitochondria by hybridization. *In* "The Genetic Function of Mitochondrial DNA" (C. Saccone, and A. M. Kroon, eds.), pp. 27–36. Elsevier, Amsterdam.

Saini, M. S., and Lane, B. G. (1977). Wheat embryo ribonucleate VIII. The presence of 7-methyl guanosine "cap" structure in the RNA of imbibing wheat embryo. *Can. J. Biochem.* **55**, 819–824.

Sakuma, K., Kominami, R., Muramatsu, M., and Sugiura, M. (1976). Conservation of the 5'-terminal nucleotide sequences of ribosomal 18S RNA in eukaryotes. *Eur. J. Biochem.* **63**, 339–350.

Savage, H. H., Grinchishin, V., Fang, W. F., and Busch, H. (1974). Novikoff hepatoma nuclear 8–18S RNA. *Physiol. Chem. Phys.* **6**, 113–126.

Scherrer, K., and Darnell, J. E. (1962). Sedimentation characteristics of rapidly labeled RNA from HeLa cells. *Biochem. Biophys. Res. Commun.* **7**, 486–490.

Schibler, U., and Perry, R. (1976). Characterization of the 5'-termini of hnRNA in mouse L cells. *Cell* **9**, 121–130.

Schibler, U., Wyler, T., and Hagenbüchle, O. (1975). Changes in size and secondary structure of the ribosomal transcription unit during vertebrate evolution. *J. Mol. Biol.* **94**, 503–517.

Schneller, C., Schneller, J. M., and Stahl, A. J. C. (1976). Studies of odd bases in yeast mitochondrial tRNA: III. Characterization of the tRNA methylase associated with the mitochondria. *Biochem. Biophys. Res. Commun.* **70**, 1003–1008.

Sekeris, C. E., and Niessing, J. (1975). Evidence for the existence of ribonucleoprotein particles containing heterogeneous RNA. *Biochem. Biophys. Res. Commun.* **62**, 642–650.

Shatkin, A. J. (1976). Capping of eukaryotic mRNAs. *Cell* **9**, 645–653.

Shibata, H., Ro-Choi, T. S., Reddy, R., Choi, Y. C., Henning, D., and Busch, H. (1975). The primary nucleotide sequence of nuclear U-2 ribonucleic acid. The 5'-terminal portion of the molecule. *J. Biol. Chem.* **250**, 3909–3920.

Shine, J., and Dalgarno, L. (1974). Identical 3'-terminal octanucleotide sequence in 18S ribonucleic acid from different eukaryotes. *Biochem. J.* **141**, 609–615.

Shine, J., and Dalgarno, L. (1975). Determination of cistron specificity in bacterial ribosomes. *Nature (London)* **254**, 34–38.

Shine, J., Hunt, J. A., and Dalgarno, L. (1974). Studies on the 3'-terminal sequences of the large ribosomal ribonucleic acid of different eukaryotes and those associated with "hidden" breaks in heat-dissociable insect 26S ribonucleic acid. *Biochem. J.* **141**, 617–725.

Simsek, M., Petrissant, G., and RajBhandary, U. L. (1973). Replacement of the sequence G-T-ψ-C-G(A) by G-A-U-C-G- in initiator transfer RNA of rabbit-liver cytoplasm. *Proc. Natl. Acad. Sci. U.S.A.* **70**, 2600–2604.

Singer, C. E., Smith, G. R., Cortese, R., and Ames, B. N. (1972). Mutant tRNA[His] ineffective in repression and lacking two pseudouridine modifications. *Nature (London), New Biol.* **238**, 72–74.

Slack, J. M. W., and Loening, U. E. (1974). 5'-Ends of ribosomal and ribosomal precursor RNAs from *Xenopus laevis*. *Eur. J. Biochem.* **43**, 59–67.

Smith, M. J., Hough, B. R., Chamberlin, M. E., and Davidson, E. H. (1974). Repetitive and non-repetitive sequence in sea urchin heterogeneous nuclear RNA. *J. Mol. Biol.* **85**, 103–126.

Söll, D., Ohtsuka, E., Jones, D. S., Lohrmann, R., Nayatsu, H., Nishimura, S., and Khorana, H. G. (1965). Studies on polynucleotides, XLIX. Stimulation of the binding of amino-acryl sRNAs to ribosomes by ribonucleotides and a survey of codon assignments for 20 amino acids. *Proc. Natl. Acad. Sci. U.S.A.* **54**, 1378–1385.

Sonnenberg, N., Wilchek, M., and Zamir, A. (1976). Photoaffinity labeling of 23S rRNA by an analog of fMet-tRNA$_f^{Met}$. *Biochem. Biophys. Res. Commun.* **72**, 1534–1541.

Southern, E. (1975). Detection of specific sequences among DNA fragments separated by gel electrophoresis. *J. Mol. Biol.* **98**, 503–517.

Speirs, J., and Birnstiel, M. (1974). Arrangement of the 5.8S RNA cistrons in the genome of *Xenopus laevis*. *J. Mol. Biol.* **87**, 237–256.

Spradling, A., Pardue, M. L., and Penman, S. (1977). Messenger RNA in heat-shocked *Drosophila* cells. *J. Mol. Biol.* **109**, 559–587.

Srinivasan. P. R.. Ramanarayanan, M., and Rabbani, E. (1975). Presence of polyadenylate sequence in pulse-labeled RNA of *E. coli*. *Proc. Natl. Acad. Sci. U.S.A.* **72**, 2910–2914.

Sripati, C. E., Groner, Y., and Warner, J. R. (1976). Methylated, blocked 5' termini of yeast mRNA. *J. Biol. Chem.* **251**, 2838–2904.

Steele, W. J., and Busch, H. (1966). Increased content of high molecular weight RNA fractions in nuclei and nucleoli of livers of thioacetamide-treated rats. *Biochim. Biophys. Acta* **119**, 501–509.

Steitz, J. A., and Jakes, K. (1975). How ribosomes select initiator regions in mRNA: Base pair formation between the 3'-terminus of 16S rRNA and the mRNA during initiation of protein synthesis in *Escherichia coli. Proc. Natl. Acad. Sci. U.S.A.* **72**, 4734–4738.

Sturhmann, H. B., Haas, J., Ibel, K., DeWolf, B., Koch, M. H. J., Parfait, R., and Crichton, R. R. (1976). New low resolution model for 50S subunit of *Escherichia coli* ribosomes. *Proc. Natl. Acad. Sci. U.S.A.* **73**, 2379–2383.

Sugiura, M., and Takanami, M. (1967). Analysis of the 5'-terminal nucleotide sequence of ribonucleic acids, II. Comparison of the 5'-terminal nucleotide sequence of ribosomal RNA's from different organisms. *Proc. Natl. Acad. Sci. U.S.A.* **58**, 1595–1602.

Surry, S., and Nemer, M. (1976). Methylated blocked 5'-terminal sequence of sea urchin embryo messenger RNA classes containing and lacking poly(A). *Cell* **9**, 589–595.

Suyama, Y., (1976a). Properties of mitochondrial and cytoplasmic ribosomes and ribosomal RNA. *In* "Cell Biology" P. L. Altman, and D. D. Katz, eds.), Vol. I, pp. 224–228. Fed. Am. Soc. Exp. Biol., Bethesda, Maryland.

Suyama, Y. (1976b). Mitochondrial transfer-RNA species and cistrons. *In* "Cell Biology" (P. L. Altman, and D. D. Katz, eds.), Vol. I, pp. 228 2—230. Fed. Am. Soc. Exp. Biol., Bethesda, Maryland.

Suzuki, Y., and Brown, D. D. (1972). Isolation and identification of the messenger RNA for silk fibroin from *Bombyx mori. J. Mol. Biol.* **63**, 409–429.

Szer, W., and Leffler, S. (1974). Interaction of *Escherichia coli* 30S ribosomal subunits with MS2 phage RNA in the absence of initiation factors. *Proc. Natl. Acad. Sci. U.S.A.* **71**, 3611–3615.

Tilghman, S. M., Curtis, P. J., Tiemeir, D. C., Leder, P., and Weissman, C. (1978). The intervening sequence of a mouse β-globin gene is transcribed within 15S β-globin mRNA precursor. *Proc. Natl. Acad. Sci. U.S.A.* **75**, 1309–1313.

Tiollais, P., Galibert, F., and Boiron, M. (1971). Evidence for the existence of several molecular species in the "45S fraction" of mammalian ribosomal precursor RNA. *Proc. Natl. Acad. Sci. U.S.A.* **68**, 1117–1120.

Udem, S. A., and Warner, J. R. (1972). Ribosomal RNA synthesis in *Saccharomyces cerevisiae. J. Mol. Biol.* **65**, 227–242.

Valenzuela, P., Venegas, A., Weinberg, F., Bishop, R., and Rutter, W. (1978). Structure of yeast phenylalanine-tRNA genes: An intervening DNA segment within the region coding for the tRNA. *Proc. Natl. Acad. Sci. U.S.A.* **75**, 190–194.

Van Duin, J., and van Knippenberg, P. H. (1974). Functional heterogeneity of the 30S ribosomal subunit of *Escherichia coli. J. Mol. Biol.* **84**, 185–195.

van Holde, K., and Hill, W. E. (1974). General physical properties of ribosomes. *In* "Ribosomes" (M. Nomura, A. Tissiéres and P. Lengyel, eds.), pp. 53–91. Cold Spring Harbor Laboratory, Cold Spring Harbor, New York.

Vassart, G., Brocas, H., Lecocq, R., and Dumont, J. E. (1975). Thyroglobulin messenger RNA: Translation of a 33-SmRNA into a peptide immunologically related to thyroglobulin. *Eur. J. Biochem.* **55**, 15–22.

Volkin, E., and Atrachan, L. (1956). Phosphorus incorporation in *Escherichia coli* ribonucleic acid after infection with bacteriophage T2. *Virology* **2**, 149–161.

Wagner, R., Gassen, H. G., Ehresmann, C., Stiegler, P., and Ebel, J. P. (1976). Identification of a 16S RNA sequence located in the decoding site of 30S ribosomes. *FEBS Lett.* **67**, 312–315.

Warner, J. R. (1974). The assembly of ribosomes in eukaryotes. *In* "Ribosomes" (M. Nomura, A. Tissiéres, and P. Lengyel, eds.), pp. 461–488. Cold Spring Harbor Laboratory, Cold Spring Harbor, New York.

Warner, J. R., and Soeiro, R. (1967). Nascent ribosomes from HeLa cells. *Proc. Natl. Acad. Sci. U.S.A.* **58**, 1984–1990.

Warner, J. R., Knopf, P., and Rich, A. (1963). A multiple ribosomal structure in protein synthesis. *Proc. Natl. Acad. Sci. U.S.A.* **49**, 122–129.

Warner, J. R., Soeiro, R., Birnboim, H. C., Girard, M., and Darnell, J. E. (1966). Rapidly labeled HeLa cell nuclear RNA. *J. Mol. Biol.* **19**, 349–361.

Wei, C. M., and Moss, B. (1977). Nucleotide sequences at the N^6-methyladenosine sites of HeLa cell messenger RNA. *Biochemistry* **16**, 1672–1676.

Wei, C. M., Gerskowitz, A., and Moss, B. (1975). $N^6,O^{2'}$-dimethyladenosine, a novel methylated ribonucleotide next to 5'-terminal of animal cell and virus mRNAs. *Nature (London)* **257**, 251–253.

Weinberg, R. A. (1973). Nuclear RNA metabolism. *Annu. Rev. Biochem.* **42**, 329–354.

Weinberg, R. A., and Penman, S. (1968). Small molecular weight monodisperse nuclear RNA. *J. Mol. Biol.* **38**, 289–304.

Weinberg, R. A., and Penman, S. (1969). Metabolism of small molecular weight monodisperse nuclear RNA. *Biochim. Biophys. Acta* **190**, 10–29.

Weinberg, R. A., and Penman, S. (1970). Processing of 45S nucleolar RNA. *J. Mol. Biol.* **47**, 169–178.

Wellauer, P. K., and Dawid, I. B. (1973a). Secondary structure maps of RNA; processing of HeLa ribosomal RNA. *Proc. Natl. Acad. Sci. U.S.A.* **70**, 2827–2831.

Wellauer, P. K., and Dawid, I. B. (1973b). Measurement of mitochondrial RNA and RNA–DNA hybrids by electron microscopy. *Carnegie Inst. Washington, Yearb.* **72**, 45–46.

Wellauer, P. K., Dawid, I. B., Kelley, D. E., and Perry, R. P. (1974). Secondary structure maps of ribosomal RNA. II. Processing of mouse L-cell ribosomal RNA and variations in the processing pathway. *J. Mol. Biol.* **89**, 397–407.

Wellauer, P. K., Dawid, I. B., Brown, D. D., and Reeder, R. H. (1976). The molecular basis for length heterogeneity in ribosomal DNA from *Xenopus laevis. J. Mol. Biol.* **105**, 461–486.

White, B. N., Tener, G. M., Holden, J., and Suzuki, D. T. (1973). Activity of a transfer RNA modifying enzyme during the development of *Drosophila* and its relationship to the *su*(*s*) locus. *J. Mol. Biol.* **74**, 635–651.

Williamson, B. (1977). DNA insertions and gene structure. *Nature (London)* **270**, 295–297.

Woo, S. L. C., Rosen, J. M., Liarakos, C. D., Choi, Y. C., Busch, H., Means, A. R., O'Malley, B. W., and Robberson, D. L. (1975). Physical and chemical characterization of purification of ovalbumin messenger RNA, *J. Biol. Chem.* **250**, 7027–7039.

Wool, I. G., and Stöffler, G. (1974). Structure and function of eukaryotic ribosomes. *In* "Ribosomes" (M. Nomura, A. Tissiéres and P. Lengyel, eds.), pp. 417–460. Cold Spring Harbor Laboratory, Cold Spring Harbor, New York.

Wrede, P., and Erdmann, V. A. (1973). Activities of B. stearothermophilus 50S ribosomes reconstituted with prokaryotic and eukaryotic 5S RNA. *FEBS Lett.* **33**, 315–319.

Wu, M., Davidson, N., Attardi, G., and Aloni, Y. (1972). Expression of the mitochondrial genome in HeLa cells XIV. The relative positions of the 4S RNA genes and of the ribosomal RNA genes in mitochondrial DNA. *J. Mol. Biol.* **71**, 81–93.

Yamada, Y., and Ishikura, H. (1973). Nucleotide sequence of tRNA$_3^{Ser}$ from *E. coli. FEBS Lett.* **29**, 231–234.

Yukioka, M., Hatayama, T., and Morisawa, S. (1975). Affinity labeling of the ribonucleic acid component adjacent to the peptidyl recognition center of peptidyltransferase in *Escherichia coli* ribosomes. *Biochim. Biophys. Acta* **390,** 192–208.

Zapisek, W. F., Saponara, A. G., and Enger, M. D. (1969). Low molecular weight methylated ribonucleic acid species from Chinese hamster ovary cells. I. Isolation and characterization. *Biochemistry* **8,** 1170–1181.

Zimmermann, R. A. (1974). RNA–protein interactions in the ribosomes. *In* "Ribosomes" (M. Nomura, A. Tissiéres and P. Lengyel, eds.), pp. 225–269. Cold Spring Harbor Laboratory, Cold Spring Harbor, New York.

Zybler, E. A., and Penman, S. (1969). Mitochondrial-associated 4S RNA synthesis inhibition by ethidium bromide. *J. Mol. Biol.* **46,** 201–204.

Index